Cycle of Fire
Stephen J. Pyne

"Cycle of Fire" is a suite of books that collectively narrate the story of how fire and humanity have interacted to shape the earth. "Cycle" is an apt description of how fire functions in the natural world. Yet "cycle" also bears a mythic connotation: a set of sagas that tell the life of a culture hero. Here that role belongs to fire. Ranging across all continents and over thousands of years, the Cycle shows earth to be a fire planet in which carbon-based terrestrial life and an oxygen-rich atmosphere have combined to make combustion both elemental and inevitable. Equally, the Cycle reveals humans as fire creatures, alternately dependent upon and threatened by their monopoly over combustion. Fire's possession began humanity's great dialogue with the earth. "Cycle of Fire" tells, for the first time, that epic story.

"Cycle of Fire" is part of the Weyerhaeuser Environmental Books, published by the University of Washington Press under the general editorship of William Cronon. A complete list of Weyerhaeuser Environmental Books appears at the end of this book.

World Fire: The Culture of Fire on Earth

Vestal Fire: An Environmental History, Told through Fire, of Europe and Europe's Encounter with the World

Fire in America: A Cultural History of Wildland and Rural Fire

Burning Bush: A Fire History of Australia

The Ice: A Journey to Antarctica

Fire: A Brief History

FIRE IN AMERICA

A Cultural History of
Wildland and Rural Fire

Stephen J. Pyne

With a Foreword by William Cronon
and a New Preface by the Author

University of Washington Press

Seattle and London

Fire in America: A Cultural History of Wildland and Rural Fire
has been published with the assistance of a grant from the
Weyerhaeuser Environmental Books Endowment, established
by the Weyerhaeuser Company Foundation, members of the
Weyerhaeuser family, and Janet and Jack Creighton.

Copyright © 1982 by Stephen J. Pyne
Originally published in 1982 by Princeton University Press
Foreword and Preface to the University of Washington Press paperback edition
    copyright © 1997 by the University of Washington Press
Paperback edition published by the University of Washington Press in 1997
10 09 08 07 06 05 04   7 6 5 4 3

University of Washington Press
PO Box 50096
Seattle, WA 98145–5096, U.S.A.
www.washington.edu/uwpress

Library of Congress Cataloging-in-Publication Data
Pyne, Stephen J., 1949–
    Fire in America: a cultural history of wildland and rural fire /
    Stephen J. Pyne ; with a foreword by William Cronon ; and a new preface by the author
        p. cm. —(Cycle of fire)
    Originally published: Princeton, N.J.: Princeton University Press, c1982.
    Includes bibliographical references (p.) and index.
    ISBN 0-295-97592-X
    1. Wildfires—United States—History. 2. Fires—United States—History. I. Title.
II. Series: Pyne, Stephen J., 1949– Cycle of fire.
SD421.3.P96   1997   96–49191
304.2—dc21

"Where the Hayfields Were" and "The Danger in the Air," by Archibald Macleish, are reprinted
with the permission of Houghton Mifflin Company from Archibald Macleish, *New and Collected
Poems 1917–1976*, © 1976 by Archibald Macleish. "Fire and Ice" is reprinted by permission
of Holt, Rinehart and Winston, Publishers, from *The Poetry of Robert Frost*, edited by Edward
Connery Lathem, © 1923, © 1969 by Holt, Rinehart and Winston, © 1951 by Robert Frost. Two
lines from "Little Gidding, IV" are reprinted with the permission of Harcourt Brace Jovanovich,
Inc., from T. S. Eliot, *Four Quartets*, © 1943 by T. S. Eliot.

I started writing this book for Joe Alston and Richard "The Kid" Gonzales.

I ended up writing for Paul "Gummer" Engstrom, Peter "The Ape" Griffiths, Jack Guinan, Donnie "Greenback" Hackney, "Pinyon Foot" Kent Rethlake, Ralph "Rastus" Becker, Mike Gilbert, Tom Achterman, Tim McGann, Eric Brueck, Dana Tally, Randy "Go For It" Seeliger, "Uncle Jimmy" Owen, Ted Burke, Mark "Dash Riprock" Wright, Ken Castro, Jonathan "John-Boy" Lee, Jim "Hambone" Boone, Bruce "Pferd" Pferdeort, Lenny "The Dancing Pole" Dems, David "Ralph" Stiegelmeyer, and the "New Hampshire Hotshots" Rich Greer and Dan "The Man" Barter.

The other North Rim Longshots had a lot to do with it, too.

CONTENTS

LIST OF ILLUSTRATIONS

Unless otherwise noted, all illustrations are courtesy of the U.S. Forest Service.

FOREWORD
WILLIAM CRONON

When it was first published in 1982, *Fire in America* announced to the world one of the most remarkable scholarly projects ever undertaken by an American historian: a systematic rewriting of U.S. history that put fire at the center of the narrative. The book became an instant classic in the emerging field of environmental history, and it remains to this day the standard work on its subject.

Pyne weaves together three basic plot lines to construct his story of fire in America. The first is of nature and human agents bringing fire to the continent and reshaping the landscape through their use of this powerful tool. In contrast to "nature's fire"—lightning—which had been affecting North American ecosystems from time immemorial, we are given "the fire from Asia" brought by native peoples and the "fire from Europe" brought by European colonists and immigrants. Using fire, people hunted animals, cleared forests, and dramatically altered the world around them. Pyne's second plot structure cuts across and complicates the first by moving systematically across the different regions of the United States, exploring in fascinating detail the ways diverse ecosystems have adjusted to the presence of fire in their midst. Finally, the last of Pyne's plot lines traces the evolution of U.S. fire policy across the twentieth century, as managers, corporations, and government agencies embraced the notion that fire by definition "wasted" valuable resources and therefore needed to be controlled and suppressed. This last narrative takes up the bulk of the book and is a tale of great hubris, the full consequences of which we are only now coming to understand. Anyone wishing to understand the profound role that fire has played in shaping the American landscape, and the shifting ways that different people have used and interacted with fire across the centuries of the American past, cannot help but begin with this pathbreaking volume.

But *Fire in America* also launched a much larger and more ambitious project which even its author did not fully understand in 1982: a comprehensive history of the entire earth in which fire would join human beings as a crucial actor in shaping the terrestrial landscape and altering the course of human events. As Pyne explains in his Preface to this paperback edition, he wrote *Fire in America* with every expectation that it would be his last scholarly work. Unable to find an academic position during the terrible job market of the late 1970s and making ends meet by spending his summers as a seasonal firefighter on the North Rim of the Grand Canyon, he embarked on this book as a way to combine his passion for firefighting with his professional training as a historian. It was hardly a conventional topic, and by no means the safest way to scramble onto the tenure track--but it soon brought its author far more acclaim and recognition than a "safer" book would have done. In 1988, Pyne was awarded a MacArthur Fellowship, which gave him five years of funding in which to contemplate sequels to *Fire in America* on the grandest of scales. The result was "Cycle of Fire," the multi-volume series on world fire history which is finally being brought together and published in a uniform edition as part of the University of Washington Press's Weyerhaeuser Environmental Books series.

What Pyne demonstrates in *Fire in America* and in the "Cycle of Fire" series as a whole is that few natural forces have been more intimately a part of human history than fire. History as he narrates it is the story of the "fire planet" earth, where carbon-based organisms inhabit an oxygen-rich atmosphere to produce conditions that make combustion and conflagration a normal and recurring feature of life. Moreover, the peculiar role of humanity has been to employ fire as a tool for virtually every aspect of cultural existence: for heat and light, as a weapon and a source of power, for the humble rituals of the campfire and the heroic modification of entire landscapes. The story of fire goes so far back in the human past that one can plausibly view it—Prometheus-like—as the hearth from which all culture and civilization sprang. Such is the premise on which Stephen Pyne has based these extraordinary books.

Those accustomed to more conventional historical narratives may need to reorient themselves as they begin reading "Cycle of Fire." Fire is a protagonist and co-actor in these books to a much greater extent than would ordinarily be true even in an environmental history that pays close attention to the role of natural forces and ecosystems in the human past. Pyne has worked in each volume of the series to explore different rhetorical devices and techniques for telling the story of fire in different times and places. His intention throughout has been to make sure, as he says, that the "heroic role belongs to fire" as opposed to human beings. For people accustomed to placing humanity at the center of every story, this approach is at least disorienting, if not downright disturbing.

But the result is a truly unique perspective on the human past, almost as if one were viewing it through the disinterested eyes of a god or a visitor from outer space--or of fire itself. For fire as a protagonist lacks many of the qualities we ordinarily associate with heroes. Awesome and powerful it may be, but it lacks a consciousness with which we can easily identify. Although it has the power to act in impressive and sometimes devastating ways, we cannot easily see it as having moral agency, and we therefore have trouble assessing the rights and wrongs of the events it helps bring about. Perhaps most important, although its contexts and behaviors change and evolve over time, fire is not a hero that "learns" or "progresses" in the usual way we expect the human protagonists of our stories to do. It simply is, and the challenge of its history is learning to understand and accept the ineluctable, shape-shifting reality of its being. Any narrative that gives fire pride of place--at the center of the hearth, as it were—thus tends to challenge our most basic assumptions about agency, empathy, progress, and heroism—not just of fire, but of humanity itself.

From this beginning in North America, Stephen Pyne eventually wandered ever farther afield—to Australia, to Europe, even to the fireless continent of Antarctica—to construct his great "Cycle of Fire." With each new continent, he has experimented with new ways of telling fire's history, new ways of understanding human interactions with this most basic and terrifying of terrestrial phenomena. Because the hero of these stories is in so many places at once, shifting identities and roles like a trickster leaping across time and space, these books lack an easy linear flow to guide the reader effortlessly from beginning to middle to end in a straightforward and predictable way. Instead, they wander and loop back upon themselves, like fire itself, reveling in ironic anecdote and unexpected juxtapositions. Pyne is a brilliant but demanding writer, and readers who are willing to journey across the continents with him will be amply rewarded for their efforts. It is not too much say that the volumes in "Cycle of Fire" are among the handful of books that can forever change the way one sees and experiences the world.

PREFACE TO THE 1997 PAPERBACK EDITION

There is no spark like an idea, no tinder like personal experience, no fuel like long-fallow archives to combust into scholarship. Here—*Fire in America*—is where it happened for me. In the beginning, I didn't know where the idea would go; I certainly had no reason to believe that this study would lead to another. In fact, the book found ready kindling, and as the cautious flames flared upward they caught a tree-bending wind from environmental history and leaped over canyons and even continents. The book ran everywhere that could carry flame, and became the first of a suite, a kind of Leatherstocking Tales for which *Fire in America* serves as The Pioneers, the first volume published though set midway in the larger action.

When the idea of writing a fire history of America came, I was 27, a freshly minted PhD, academically unemployed, having completed my tenth season as a firefighter on the North Rim of Grand Canyon and working as a volunteer and winter seasonal at Desert View on the South Rim. My dissertation, a biography of the geologist Grove Karl Gilbert, was a study in Western American history, the history of science, and the larger cultural meaning of the place that mattered most to me, the Canyon. That winter, of 1976–77, gazing alternately at the Canyon and the San Francisco Peaks, I knew the time had come to bring my two worlds together.

The idea for a fire history became an enterprise when in 1977 a cooperative agreement with the U.S. Forest Service financed the travels I needed to visit sites and collect documents; but that contract specifically excluded salary or health insurance and my bride and I had to stay on the road to draw per diem. Our freedom had its price: for two years, I existed in almost complete intellectual isolation, and for the third, while the bulk of the writing was completed at the National Humanities Center, I had no real colleagues with whom to discuss the themes of what was an increasingly peculiar enterprise. Instead I remained with my firefighter companions, the North Rim Longshots, until the manuscript had passed through copyediting, and to them the book is dedicated.

In the beginning I had no clear conception what a fire history might actually be, or how it might reconcile a head trained at Stanford and the University of Texas with a heart committed to the Longshots. I sensed only that I had to apply the techniques of historical scholarship to the study of fire. Later I appreciated that I was trying to render fire into a topic for cultural analysis as

William Goetzmann, my supervising professor, had done for American exploration in *Exploration and Empire*. Instead of an exemplar, I had a sense of voice. In place of a thesis, I had an inchoate conviction that I should organize the book around fire, as in fact the life I knew best was organized around fire. There was no one to tell me otherwise.

My determination to use fire as an informing presence did not spring from philosophical conviction or historiographical insight. It simply reflected how I had learned to talk about fire. On the Rim we discussed fire endlessly: there was almost nothing else that mattered. We described our fires' quirks while hunched over ration coffee on late-night firelines, we compared our fires' ease and misery when we returned to the fire cache, we sang and cursed our fires at the saloon. They all, each one, had a personality. There were charmed fires and ugly fires, glorious fires and fires that were existentially wretched, fires rich with loose dirt and mean fires that burned amid nothing but roots and rocks. There were fires that hurt, fires that hummed, fires that inspired, fires that infuriated. The character of the fire determined our experience.

And that was how I would write about historic fires, about fire in toto. I made fire as central to American experience as it was to life on the North Rim. Later, armed with a richer understanding, I would grant fire a similar centrality to humanity at large, and I came to appreciate the power of fire to relocate history from an anthropocentric narrative. Fire could organize American geography, fire could date America's historical periods, fire could inform and animate American experience. I would redraw the geometry of historical narrative from its traditional single-centered circle into a double-centered ellipse.

The second fact is that, by the time the project was into its third year, I came to believe that this book would be my last as a scholar. Whatever I wanted to say about the history of fire, I had to say here. All the evidence argued that there would be no sequel. The manuscript's original title reflected this belief: "The Culture of Fire: A History of Wildland and Rural Fire in America." Many items that I might better have left out I felt compelled to put in, and many of those topics received allusions instead of the fuller discussion they merited. That, at least, is what the evidence in 1980 argued. Not for another year would I merit an interview for an academic appointment, and not until five years had passed after my doctorate would I get a university post. Only the equally dismal failure to land a permanent job in the fire community kept me in the academy.

Still, I resisted these conclusions and designed the manuscript with escape clauses. Primarily this meant adopting a format that substituted a kind of systems flowchart for genuine narrative. This way I might, from time to time, be able to revise the text piecemeal. The manuscript would resemble the loose-leaf training manuals and fire-danger rating systems I had known on the Rim, their dated sections constantly replaced by new insertions. Even better, a modular structure could enhance the stability of the text. It should be possible to quarrel over—even discredit, simply knock out—a good number of those modules before the entire edifice would collapse. As a side-benefit, the text would achieve a kind of narrative neutrality, my primary concession to prevailing historiographic criticism. The story could advance but without a narrative driver. The book's design (so I imagined) was a literary equivalent to the structural grid so fundamental to Modernist architecture. Still, there were literary stresses, and it is my guess that stringing those prose cables through the trusses accounts for 15 percent or more of the total text. I regard this as rather high.

So, the text grew, and in the absence of pruning colleagues and browsing critics, it ran wild. Besides, I reckoned that the fire community with whom I remained in contact, not academics with whom I had apparently lost connection, would be a primary audience, and not being book readers they would seize selectively on those sections that most interested them. The manuscript thus anthologized itself.

The book's blunders and omissions are perhaps obvious enough but a few deserve public explanation. To repeat the canard that some peoples did not have the capacity to make fire goes beyond embarrassment, even if it did draw on such authorities as Edward Tylor and A. R. Radcliffe-Brown. The introductory tutorial on fire behavior and ecology makes the point that a fire history should begin with fire, but the attention devoted to the section says more about an author immersed in the practice of fire management than to the imperatives of historical scholarship or the logic of narrative. Additionally, there are many, many new sources of hard data that deserve inclusion. Since the late 1970s, a tremendous research effort has produced fire histories founded on fire-scarred trees, charcoal sediments, comparative photographs, forest reconstructions based on age-structures, and the like. Over the past 15 years I have doubled the size of my research holdings. Any full-scale revision should include those studies. More seriously, European fire proved so interesting (and my first-draft discussion so defective) that I have had to write a separate book, *Vestal Fire*, to correct and elaborate that story.

A lapse sure to strike contemporary readers is the effective omission of the 1949 Mann Gulch fire in Montana. The explanation is simple. I passed through Missoula in the late fall of 1977 and found the documented record of Mann Gulch magnificently arranged in the Region 1 files because, I was informed, Norman Maclean had been through them that summer. I knew of Maclean as the author of *A River Runs Through It* and believed that he would be publishing his Mann Gulch study long before I completed my general history, so I photocopied the documents and left the story to him. In fact *Young Men and Fire* would postdate *Fire in America* by a decade. Besides, I had come to believe the Mann Gulch blowup, so vital to smokejumper lore, was a minor event compared to the fireball that followed two weeks later when the Soviet Union exploded its first atomic bomb; that, not a fiery hillside beyond the Gates of the Mountains, had mobilized the nation's cold war on fire. The Blackwater fire, the Rattlesnake fire, the Hauser Creek fire, the Inaja fire—all crew-consuming blowups—were of at least comparable significance. The whole smokejumper mystique was already approaching self-parody.

The postwar era was a dazzling period of equipment development, most spectacularly with aerial firefighting, most effectively with ground-based vehicles. A major omission in *Fire in America* is a description of the phenomenal orgy of road-building that carried those engines into the backcountry. Logging roads had become so prevalent that no comment seemed necessary, but such observations are exactly what a book like this should include. Roads and engines reworked the national forests as they did the rest of America, as the automobile did the geography of urban America. The significance of the postwar road boom did not impress me, however, until I began to study the reasons behind Sweden's silvicultural strategy and listened while Johann Goldammer described the density of forest roads in West Germany to Soviet fire colleages in Siberia. The still larger story is that with greater access came a decline in controlled burning as a tool of fire suppression. More and more, Americans elected to attack fire directly rather than backfire against it, so that even in fire control America substituted the pyrotechnologies of internal combustion for those of open flame. Like most of my reconsiderations, these derived from crosscultural comparison.

Perhaps the biggest surprise was the attention readers gave to American Indian fire practices. This was a minor feature of the book. I was far more interested in the transition from an agricultural to an industrial society than in pre-Columbian practices or the transition from American Indian to European colonist. When I completed the manuscript in 1980, I made fire the text's informing principle, not the principles of gender, ethnicity, and race that have come to dominate historical discourse in the intervening years. It was obvious that American Indians used fire widely. It was evident that the argument over whether or not they burned had entered into bitter controversies over appropriate industrial fire practices. Beyond that, I had little time to investigate.

It is possible now, however, to appreciate how extensively fire must have been applied, with what methods, and according to what system. There is little reason to doubt that pre-Columbian peoples used fire as pervasively as technologically comparable peoples elsewhere in the world. Rather, the burden of proof resides with those who would argue that American Indians, alone among the cultures of the planet, resisted the wholesale reconstruction of their landscapes by fire. The matter involves more than academic curiosity: it is increasingly evident that fire's exclusion is as ecologically powerful as fire's application, and to the administators of nature reserves it is becoming daily more apparent that the removal of anthropogenic burning has had dramatic consequences for historic ecosystems. The loss of Indian burning (without a surrogate fire) has seriously jeopardized many western biotas. These themes were not primary to *Fire in America*. They would be in any wholesale revision.

The most serious conceptual flaw, however, was the segregation of controlled burning and fire suppression into separate histories. That this was even imaginable speaks volumes for the state of thinking in the late 1970s: it demonstrates with brutal clarity how out of sync fire management had become. It is clear that the tragedy of American fire was not that wildfires were suppressed but that controlled fires were no longer set. That I could write a history in which these complementary practices could be disengaged speaks with more force and eloquence to what went wrong historically than any epigram I could muster now. It would be an act of useful, and necessary, scholarship to merge those distinctive stories into one. An essay in *World Fire* ("Initial Attack") attempts to begin this reconciliation.

Even as I finished, however, it was clear that I could learn more by comparative and crosscultural studies than by further plumbing American archives. In 1981 that prospect seemed wholly hypothetical. When I had the opportunity to write a new preface in 1988, I suggested comparative history as a direction for future work, and was in fact deeply immersed within the research for *Burning Bush: A Fire History of Australia*. (That second, temporarizing preface is not included with this edition since its topics have been incorporated into other works and because I committed a substantial portion of *World Fire* to update the themes left suspended in *Fire in America*.) The real sequel to *Fire in America* is the whole "Cycle of Fire," and the real marvel is that I found an outstanding publisher for the entire suite. Scholarship is only as good as its expression. The Cycle became a reality only with the generosity of the University of Washington Press and William Cronon, both of whom have my deepest gratitude.

The psychology behind the book's scholarship is thus easily explained: I simply made fire as central to humanity's history as it has been in my own. I hoped that the book might help me make the transition to an existence away from the North Rim. It did, and because of that I have continued a career of pyromanticism, although I now do my smokechasing as a scholar, and instead of tracking fires to Swamp Point, the Dragon, and the Walhalla Plateau, I've sought them in Umeå, Jonkershoek, Brasilia, Yakutsk, Chania, Freiburg, and Dehra Dun. Fires, I've learned, are everywhere. And searching them out is still the best of all lives.

Steve Pyne
Glendale, Arizona

PREFACE TO THE ORIGINAL EDITION: HISTORY WITH FIRE IN ITS EYE

Oh for a Muse of fire, that would ascend
The brightest heaven of imagination . . . —Shakespeare, *Henry V*

Mine has been a Promethean task. I have tried to bring fire to history.

Legend has it that Prometheus delivered fire to earth in a stalk of fennel. It was less obvious how a history of fire should be packaged. The book is designed for three audiences: fire managers, historians, and an interested public. It is organized, correspondingly, both topically and chronologically. Most historians are shrewd enough to write in one style or the other, but I adopted a mixed format because, first, I expected the book to serve fire managers as a reference volume, consolidating the many dimensions of the subject into various identifiable themes; hence, the value of a topical arrangement. But for historians and general readers, a chronological order and narrative thread were demanded. The result is a historical mosaic, one that, in defiance of historical perspective, extends into very recent times. "Fire giveth lyght to things farre off," wrote John Lyly in repeating the folk wisdom of Elizabethan times, "and burneth that which is next to it." The distant fires of history shine and warm; the fires of contemporary society can burn an interpreter by their proximity.

It may be useful to explain the logic behind the book's organization. Each chapter advances the general narrative line, from prehistoric to contemporary times. From the period covered, one or two topics will be drawn and discussed more or less intact. In the same chapter, a regional fire history, also based on the particular themes of the narrative line, will be told. The regional histories, moreover, will often be divided into two sections: the first will pick up the regional history at the time emphasized by the general narrative, commonly when the region assumed national prominence; the second portion will sketch the remaining history. Thus, Chapter 2 advances the general narrative summarizing the uses of fire by the American Indian ("Our Grandfather Fire"); relates the history of the region most affected by Indian fire, the grasslands ("These Conflagrated Prairies"); and discusses a topic of special significance, in this case the problems of transferring and adapting Indian fire practices by American settlers and foresters ("Paiute Forestry").

A word on terminology. I have tried to avoid both scientific jargon and occupational slang, but some terms common to fire management are particularly descriptive and conceptually useful. Though their meaning will become apparent as the book progresses, I offer some definitions to help get over the initial encounter. A *fire environment* consists of the fuels, topography, and weather within which a fire burns. When a fire environment combines with a consistent pattern of ignition, then a *fire regime* results, characterized by a particular vegetative ensemble and regular patterns of fire behavior. Such vegetative ensembles are often referred to as *cover types* or *fuel types,* and to transform vegetation deliberately from one cover type to another is known as *type conversion.* When, because of its fire pattern, a fire regime maintains a certain type of vegetative cover that, in the absence of fire, would give way to other cover types, then that biota is referred to as a *fire climax* and the particular vegetation as a *fire type.* When many fires burn in a region at one time, the result is a *fire complex,* and when a consistent pattern of reburning is established, the outcome is a *fire cycle.* Free-burning fire that is distributed over an area and responds to environmental influences is called *broadcast fire,* and it is a common form of *prescribed fire*—controlled fire that is introduced under predetermined conditions called a *prescription. Natural fire* is fire that starts from natural causes, notably lightning; it may be considered either as wildfire or prescribed fire. Under contemporary thinking, any fire that is not prescribed is considered as *wildfire,* and the suppression of large wildfires commonly relies on teams of specialists who assume supervisory roles—positions referred to as *overhead.*

Unlike Prometheus, I have had some generous support. A cooperative agreement (13–970) with the History Office of the U.S. Forest Service sustained my research travels, made accessible many sources of information that I might otherwise have missed, and made available many fire officers for review and criticism of early drafts of the manuscript. Since the early twentieth century, the Forest Service has exercised a unique position of leadership in the field of wildland fire protection. Its role is such that it has frequently supported projects that other organizations lacked the resources, vision, or courage to undertake. I believe this book belongs in that tradition. Fire protection, moreover, is fundamental to the history of the Service, and in turning over a history of it to someone who was both outside the Service and outside professional forestry, I believe it has demonstrated once again an admirable sense of public service. I was never pressured to write an official history. But I have tried to make the product worthy of the encouragement and confidence that the Forest Service has bestowed on me.

I have been aided, too, by a Fellowship to the National Humanities Center at Research Triangle Park, North Carolina. My residence at the Center for nine months not only measurably expedited the writing of the manuscript but also materially improved the quality of its contents. Much of the scholarly merit of the book must be attributed to the environment and services at the Center. That the Center would accept a young scholar with an unorthodox topic speaks well for it, and I am grateful. My tour at the Center will be long and warmly remembered.

The entire project, finally, would have been impossible without the forbearance and measurable assistance of my wife, Sonja. At nearly every stage of the project, she was an active participant. Except for her undue modesty, her name would deservedly belong on the title page with mine.

This book is dedicated to those who shared with me the summer smell of burning pine, wildfire and campfire both. Many a historian could have researched this subject, but probably only a firefighter would have thought of the project at all. My memories of former fire guards have given me plenty of reason to think the subject worthwhile and enough determination to see it through.

Among the many who contributed to the project in one form or another, I would like to thank especially David A. Clary, Dennis Roth, Frank Harmon, Carl Wilson, Beverly Ayers, Judson Moore, Ernie Balmforth, Al Bell, Art Brachebusch, Dave Butts, William Carver, Richard Chase, Raymond Clar, Dave Dahl, Henry DeBruin, Karen Eckles, Jim Fisher, Tom Price, Robert Gale, Nelson Grisamore, Robert Hall, Arnold Hartigan, Bill Hauser, Doug Helms, Brian Ingalls, Bill Johansen, Art Jukkala, Joe Kastelic, James Kerr, Bruce Kilgore, George Kitson, E. V. Komarek, Frank Lewis, Ronald McKibbin, Eugene McNamara, Ralph Winckworth, Rod Norum, Jack Puckett, Fred McBride, R. R. Robinson, H. E. Ruark, Mike Schori, Herbert Shields, Steven Such, Jerry Timmons, Jack Wilson, Lyle Adams, and the manuscript typists at the National Humanities Center supervised by Marie Long. A very special thanks goes to Craig Chandler, whose vision helped to launch the project and whose critical reviews of the manuscript helped to see it back to port, and to Gretchen Oberfranc, whose editorial skills increased measurably the coherence and literacy of the manuscript.

Thanks go to the following for reviewing portions of the manuscript at the request of the Forest Service: Larry Bancroft, Boone Richardson, John Dell, Edward Heilman, James Davis, Don Bauer, Charles Philpot, William Tikkala, Junius Baker, and Don Hansen.

ABBREVIATIONS

AFA   American Forestry Association
AFCS   Alaska Fire Control Service
ARPA   Advanced Research Projects Agency
BIA   Bureau of Indian Affairs
BIFC   Boise Interagency Fire Center
BLM   Bureau of Land Management
CCC   Civilian Conservation Corps
CFFP   Cooperative Forest Fire Prevention Program
CPS   Civilian Public Service
DASA   Defense Atomic Support Agency
DESCON Designated Control Burn System
DOD   Department of Defense
ECW   Emergency Conservation Work
EFF   emergency firefighter
FFFS   Forest Fire Fighters Service
FWS   Fish and Wildlife Service
GLO   General Land Office
ICS   incident command system
IR   interregional
MACS   multiagency command system
MAFFS   Modular Airborne Fire Fighting System
MEDC   Missoula Equipment Development Center
NAFC   North American Forestry Commission
NAS   National Academy of Science
NFDRS national fire danger rating system
NFPA   National Fire Protection Association
NIRA   National Industrial Recovery Act
NPS   National Park Service
NRDL   Naval Radiological Defense Laboratory
NSF   National Science Foundation
NWCG   National Wildfire Coordinating Group
NWS   National Weather Service
OCC   operational coordinating center
OCD   Office of Civil Defense
SAF   Society of American Foresters
SCS   Soil Conservation Service

SWFFF   Southwest Forest Fire Fighters
TTCP   The Technical Cooperation Program
USGS   U.S. Geological Survey
WFCA   Western Forestry and Conservation Association
WPA   Works Progress Administration

FIRE IN AMERICA

PROLOGUE: THE SMOKE OF TIME

Off to the north somewhere the woods were on fire.
The smoke tasted of time: the sad odor.
No one knew what country or how far.
—Archibald MacLeish, "The Danger in the Air"[1]

Wildfire is among the oldest of natural phenomena. As a product of lightning, wildland fires trace their ancestry to the early development of terrestrial vegetation and the evolution of the atmosphere. Fossil evidences of fires are buried within the coal beds of the Carboniferous period, and although there is no reason to suppose that these mark the origin of natural fire as an ecological force, there is certainly little doubt that such fires have continued unabated into contemporary times. Even today it is estimated that approximately 1,800 thunderstorms per hour are active across the globe, and the lightning strokes they generate are responsible for over 10 percent of the total number of fires per year in the United States alone. A phenomenon of such magnitude and longevity has unquestionably kindled profound evolutionary consequences. Lightning fires can be considered as a manifestation of climate, no less important than rain, sun, and frost.

The relationship is even more fundamental. Fire and life share a common chemistry of carbon and oxidation and a common geographic range, as well as a common origin in lightning. Hardly any plant community in the temperate zone has escaped fire's selective action, and, thanks to the radiation of *homo sapiens* throughout the world, fire has been introduced to nearly every landscape on earth. Many biotas have consequently so adapted themselves to fire that, as with biotas frequented by floods and hurricanes, adaption has become symbiosis. Such ecosystems do not merely tolerate fire but often encourage it and even require it. In many environments fire is the most effective form of decomposition, the dominant selective force for determining the relative distribution of certain species, and the means for effective nutrient recycling and even the recycling of whole communities. The knowledge of natural fire's ecological and evolutionary function grows almost daily; in wildlands set aside for management as primitive areas, wildfire is considered both inevitable and essential.

Equally, and perhaps even more profoundly, fire is a cultural phenomenon. It is among man's oldest tools, the first product of the natural world he learned to domesticate. Unlike floods, hurricanes, or windstorms, fire can be initiated by man; it can be combated hand to hand, dissipated, buried, or "herded" in ways unthinkable for floods or tornadoes. The natural regimes that might have resulted from lightning fire can be greatly exacerbated or ameliorated through the careless or systematic distribution of fire by human agents. Mankind is the primary source of ignition in the world, the chief vector for the propagation of fire, and the most significant modifier of the fire environment, notably its fuels.

It was fire as much as social organization and stone tools that enabled early big game hunters to encircle the globe and to begin the extermination of selected species. It was fire that assisted hunting and gathering societies to harvest insects, small game, and edible plants; that encouraged the spread of agriculture outside of flood plains by allowing for rapid landclearing, ready fertilization, the selection of food grains, the primitive herding of grazing animals that led to domestication, and the expansion of pasture and grasslands against climate gradients; and that, housed in machinery, powered the prime movers of the industrial revolution. Moreover, though lightning fire alone could rarely devastate a landscape, fire as an accompaniment to other human activities, such as farming, logging, land clearing, swamp draining, herding, or war, could be ruinous. The contemplation of fire has been a fundamental theme of folklore, ritual, philosophy, science, and even theology. It is difficult to imagine a component of human culture as elementary—and, at the same time, as ambivalent—as the relationship to fire. Fire is among the oldest of words in any language; its mathematical and physical study, among the most recent of modern sciences.

Fire came to North America in three great waves. Originally, of course, there was lightning, the most significant ignition source from the standpoint of natural history. Cultural history introduced two others: the fire brought across the Bering Strait from Asia by Pleistocene immigrants and the fire brought from Europe by immigrants in more recent centuries. The significance of fire as an environmental modifier of the first magnitude is difficult to dispute; but the relative impact of these general fire sources is almost impossible to determine. The vegetation that typifies the continent took shape with the warming trends at the end of the Pleistocene and with the retreat of the continental ice masses. This same period also witnessed the introduction of fire in probably widespread form by the new human immigrants. To discriminate between influences of climatic change, biotic migrations, natural fire, and aboriginal firing of the landscape is all but impossible. Even the impact of the more recent immigration of fire from Europe is difficult to determine. Not only was fire reapplied for new as well as for old purposes, but exotic flora and fauna were also introduced, resulting in profound cultural as well as environmental changes.

Cultures, too, were evolving. The great age of discovery coincided with the embryonic scientific revolution; the massive folk migration westward across the American continent was accompanied by an industrial revolution. The impact of fire on the environment and the perceived need for wildland fire protection were dictated by the form of this cultural evolution as much as by the shape of biological evolution or climatic history. Fire history cannot be understood apart from human history. The relationship between mankind and fire is reciprocal: fire has made possible most technological and agricultural developments and has provoked fundamental intellectual discourse; yet fire itself takes on many particular characteristics because of the cultural environment in which it occurs, just as it does in response to the natural environment of fuels, topography, and weather. The attempt to assimilate fire within the context of an industrial society has animated a one-hundred-year-old debate about proper fire practices. And it is the culture of fire—as distinct from its physics, chemistry, biology, and meteorology—that forms the subject of this study.

1 NATURE'S FIRE

> Gie me a spark o' Nature's fire,
> That's a' the learning I desire.
> —Robert Burns, 1786[1]

> And whereas it is generally conceived, that the woods grow so
> thick, that there is no more clear ground than is hewed out by
> labour of men; it is nothing so: in many places, divers acres being
> clear, so that one may ride a hunting in most places of the land, if
> he will venture himself for being lost: there is no underwood, sav-
> ing in swamps and low grounds that are wet . . . for it being the
> custom of the *Indians* to burn the woods in November. . . .
> —William Wood, *New England Prospects,* 1634[2]

It is the peculiar quality of fire that it is both natural and cultural. Nature
gave fire to man, presented an arena for its use that was to some extent
adapted to fire, and established limits, based on fire's behavior and effects,
to its potential exploitation by mankind. Man's ability to create and control
fire with relative ease makes his relationship to fire unique vis-à-vis those
other potentially destructive eruptions of energy, such as windstorms and
floods, which, like fire, cannot be separated from the landscape. An understanding of fire behavior and fire ecology is basic to any comprehension of
how fire has functioned historically and how policies for its management
must be shaped. Its origin in lightning, moreover, makes fire an essential
component of any environment that mankind purports to manage in a natural or wild state.

The fire history of the northeastern United States is a particularly apt
preamble to the fire history of the nation. The region presents a historical
kaleidoscope, previewing the problem fire types, the range of fire regimes,
and the responses typical of the American experience. In its coal fields are
the residue of ancient geologic fires. In its rich accounts of European settlement are excellent descriptions of Indian fire practices. It was the first
region to experience agricultural reclamation and logging on a grand scale
and the first to undergo industrial counterreclamation, that is, the abandonment and transformation of arable land. The Dark Days recorded in eigh-
teenth-century literature foreshadowed the spectacle of holocausts to come.
The 1825 Miramichi fire was the first conflagration to enter into the histor-

ical chronicle of great American fires. The 1947 Maine fires were among the first to suggest the volatile mixture of wild and urban areas created in post–World War II America. The New Jersey fires of April 1963, sweeping over more than a quarter of a million acres, testified to the endurance of one of the hardiest fire climax biotas in North America. The debate in the 1970s over wilderness fire management in Maine's Baxter State Park, shows the modern paradox that wilderness, like fire, is both natural and anthropogenic.

From the Northeast came many of the fundamental experiments in organized fire protection. New York, in particular, early enacted fire codes to regulate the agricultural uses of fire and established a rural fire warden system to protect against damaging escapes. The creation of the Adirondacks Reserve in 1885 set an important example for fire patrols not only in the Northeast but also in the Lake States and even the Far West. At Cornell and Yale the first schools of professional forestry were endowed. Fires in 1903 and 1908 had national repercussions in the campaign for organized fire control and conservation. The Weeks Act of 1911, by allowing the creation of national forests east of the Mississippi, helped to extend the example of the Adirondacks preserve throughout the Appalachians. Following the disastrous 1947 fires, the northeastern states set up the first interstate fire compact, a model for civil defense and rural fire protection throughout the nation. The fire history of the Northeast was, as often as not, a prototype for what was to come.

ELECTRICAL FIRE: A NATURAL HISTORY
OF LIGHTNING FIRE

The agent by which fire was first brought down to earth and made
available to mortal man was lightning. To this source every
hearth owes its flames.—Lucretius, *De Rerum Natura*[3]

This is the forest primeval.
—Henry Wadsworth Longfellow, *Evangeline*[4]

I

In the mid-1770s William Bartram undertook the travels that made him
famous as a naturalist and littérateur. Somewhere in the Carolinas he
"found friendly and secure shelter from a tremendous thunderstorm, which
came from the northwest and soon after my arrival began to discharge its
fury all around." It was "an awful scene; when instantly the lightning, as it
were, opening a fiery chasm in the black cloud, darted with inconceivable
rapidity on the trunk of a large pine-tree, that stood thirty or forty yards
from me, and set it in a blaze. The flame instantly ascented upwards of ten
or twelve feet, and continued flaming about fifteen minutes, when it was
gradually extinguished by the deluges of rain that fell upon it."[5] Electricity,
it has been observed, "came on the eighteenth century as unexpectedly as
nuclear power came on our century."[6] Bartram's naturalistic description of
lightning fire was complementary and nearly contemporary to Benjamin
Franklin's more physical analysis of "electrical fire." What each man shared
was a long-embedded cultural association of lightning with fire.

Electrical fire operated on the earth's surface as soon as an atmosphere
evolved and vegetation appeared. The relation between lightning and life
may be even more intimate. Work by Oparin, Urey, and Miller has suggested that lightning may have catalyzed the earliest organic compounds
out of a "primordial soup" of chemicals. Whatever the ultimate origin of
the relationship between life and lightning, the antiquity of this association
is impressive. E. V. Komarek has observed that "it has become apparent
that lightning fires are only one illustration of the ecological effects of lightning and that lightning itself is only one example of a basic natural and
ecologically important component of the universe: electricity."[7]

Lightning restores electrical equilibrium to the earth. Because air is a
poor nonconductor, some electricity constantly leaks to the atmosphere, creating an electrical potential. When the potential is great enough, electricity

moves back according to the gradient. During a thunderstorm, the gradient becomes very steep, and the electrical potential discharges as lightning. The discharge may move between any oppositely charged regions—from cloud to earth, from earth to cloud, or from cloud to cloud. It was calculated as early as 1887 that the earth would lose almost all its charge in less than an hour unless the supply were replenished; that is, on a global scale, lightning will discharge to the earth every hour a quantity of electricity equal to the earth's entire charge. Thunderstorms are thus an electromagnetic as well as a thermodynamic necessity. It has been reckoned that the earth experiences some 1,800 storms per hour, or 44,000 per day. Collectively, these storms produce 100 cloud-to-ground discharges per second, or better than 8 million per day globally. And these estimates are probably low. The total energy in lightning bolts varies greatly, but about 250 kilowatt hours of electricity are packed into each stroke—enough "to lift the *S.S. United States* six feet into the air."[8] Almost 75 percent of this total energy is lost to heat during discharge.

Two types of discharge patterns are commonly identified: the cold stroke, whose main return stroke is of intense current but of short duration, and the hot stroke, involving lesser currents of longer duration. Cold lightning, with its high voltage, generally has mechanical or explosive effects; hot lightning (also known as long-lasting current, or LLC), with higher amperages, is more apt to start fires. Studies in the Northern Rockies suggest that about 21 percent of all lightning moves from cloud to ground, of which about 20 percent is of the LLC variety. That is, in the Northern Rockies about one stroke in 25 has the electrical characteristics needed to start a fire. Whether it does or not depends strongly on the object it strikes, the fuel properties of the object, and the local weather. Ignition requires both heat and kindling. Lightning supplies the one with its current and occasionally finds the other among the fine fuels of rotton wood, needles, grass, or dustlike debris blown from a tree by the explosive shock of the bolt itself.[9]

The consequences of lightning are complex. Any natural force of this magnitude will influence the biological no less than the geophysical environment, and the secondary effects of lightning are significant to life. Lightning helps to fix atmospheric nitrogen into an organic form that rain can bring to earth. This may be important in regions of heavy lightning fire, because fires tend to volatilize organic nitrogen. In areas of heavy thunderstorm activity, lightning can function as a major predator on trees, either through direct injury or by physiological damage. In the ponderosa pine forests of Arizona, for example, one forester has estimated that lightning mortality runs between 0.7 and 1.0 percent per year. (The burned-area objective for fire control, by contrast, is 0.1 percent per year.) Other researchers have placed mortality as high as 25–33 percent. For southern pines, the figure may be even steeper. A study in Arkansas calculated that 70 percent of mortality, by volume, was due to lightning. These figures describe only direct injury, primarily the mechanical destruction of branches and bole; the other major causes of mortality—insects, wind, and mistletoe—are likely secondary effects brought about in trees weakened by lightning. All of these effects, in turn, may be camouflaged by fire induced by lightning.[10]

The process of "electrocution" is increasingly recognized. Lightning scorch areas of between 0.25 and 25 acres have been identified among southern pines in Georgia, Douglas fir in the Northwest, chestnuts in Virginia, saguaro cacti in Arizona, peat swamp forests and mangroves in Malaysia, and cabbage palms in Florida. Nor is the process limited to trees: it has been documented for grasses, tomatoes, potatoes, cabbages, tea, and other crops. Long attributed to inscrutable "die-offs" or to infestations by insects or diseases (often a secondary effect), such sites are now recognized worldwide as a product of physiological trauma caused by lightning. Interestingly, these effects are most readily visible in subtropical or tropical regions where lightning fire is rare and cannot mask the electrical effects of lightning alone. This may explain why, in areas where a lightning fire type, such as the southern pine in Florida, has been converted to crops or citrus groves, lightning kills are better recognized. Even where trees have been converted to barns, barn fires testify to the effectiveness of lightning predation.

The most spectacular product of lightning is fire. Except in tropical rain forests and on ice-mantled land masses like Antarctica, lightning fire has occurred in every terrestrial environment on the globe, contributing to a natural mosaic of vegetation types. Even in tropical landscapes lightning bombardment by itself may frequently be severe enough to produce a mosaic pattern similar to that resulting from lightning fire. Lightning fires have ignited desert grasslands in southern Arizona, tundra in Alaska, chaparral in California, swamplands in the South, marshes in the Lake States, grasslands of the Great Plains, and, of course, forests—especially conifer forests—throughout North America. Though the intensity and frequency of these fires vary by region, their existence is undeniable.

So is their persistence through geologic time. Fused inorganic tubes caused by lightning strokes to the ground, called fulgurites, are abundant in many portions of the earth. Ample evidence of fossil fires, called fusain, lies buried in the coal beds of all the coal-forming periods known to geology. For more recent geologic times, evidence of ancient fires can be found in peat. Lightning and fire scars have been identified on petrified trees. The geologic record even finds collaboration from the genetic record. Komarek has observed that "the antiquity of fire seems apparent in that the most ancient of tree families, such as the conifers, and the apparently oldest genera of grasses, such as *Aristida, Stipa, Andropogon,* etc., have the greatest concentration of those genes responsible for resistance and adjustment to a 'fire environment.'" Komarek has also suggested that intense lightning bombardment (and, indeed, intense fire) might act as a mutagenic agent, accelerating fire adaptability in zones of heavy lightning fire.[11]

The contemporary geography of lightning and lightning fire is equally impressive. Lightning behaves like other natural eruptions of energy. It exhibits a large number of discharges but a relatively small number of really intensive displays—a pattern that is repeated by lightning fire and by fire behavior. Komarek has tried to demonstrate some typical meteorological conditions that can distribute lightning to various regions of the United States. In one study he traced the passage of a cold front from Canada to Florida from April 30 to May 16, 1965. Using data only from national forests and grasslands (except for Florida, where full records were available), he identified 47 lightning fires—6 in South Dakota, 5 in Tennessee, 4 in Virginia, 3 in Nebraska, 2 in Georgia, 1 each in Michigan, West Virginia, and North Carolina, and a whopping 34 in Florida.[12] In another system, the summer monsoon typical of the Southwest, some 536 lightning fires were reported in Arizona and New Mexico between July 7 and July 16, 1965. Between 1940 and 1975 a total of 59,518 lightning fires occurred on the national forests of the Southwest, 79,131 on the forests of the Rocky Mountains, and 88,680 on the forests of California and the Pacific Northwest.[13] Though the effect of lightning fire may be masked when considering a continent or when amalgamating it with all fire starts, the local effect may be considerable. The evolutionary consequences are undeniable.

Lightning, however, is but one component of climate, and thunderstorms alone are inadequate to ignite fires. The heaviest lightning activity globally is in the tropics, where natural fire is rare; the lightest lightning loads are in the upper latitudes of the boreal forest, where natural fires, though infrequent, can reach conflagration size. In the summer of 1957, for example, some 5 million acres burned in the interior of Alaska, largely due to lightning. The most effective fire starters are "dry" lightning storms—thunderheads from which little precipitation reaches the ground and which commonly occur after droughts or dry seasons. The largest episode on record came during a 10-day period in June 1940, when 1,488 lightning fires broke out in the Northern Rockies. This is the heaviest known concentration by a factor of two; but from 1960 to 1971 the Northern Rockies and the Southwest regions of the Forest Service witnessed six separate 10-day outbreaks of 511 to 799 lightning fires each. Dry lightning storms occur in the Northern Rockies and Pacific Northwest several times each decade, whereas they are almost annual events in the Southwest (during the spring) and in Florida (during the winter). Major episodes are rare, of course, but as with floods, windstorms, and earthquakes, proportionately more change results from these larger eruptions than from the cumulative effects of minor events.

The appropriate lightning must interact, in turn, with other environmental conditions before fires can result. Natural fire regimes expand and recede, like the ebb and flow of glaciers, with fluctuations of climate and the effect of climate on fuels. Even with man's capacity to simulate lightning through ingenious ignition devices, there will be no fire unless weather and fuel are right. In areas somewhat resistant to fire it is often necessary to create conditions equivalent to drought before burning can begin. This can be done, for example, by draining swamps or marshlands, by killing or dessicating vegetation prior to ignition, or by altering microclimates through fuel type conversions.

Lightning fire brings a persistent and perhaps unalterable number of fires to the total fire load of a region. As a manifestation of climate, it is the basis for fire ecology, for fire behavior, and for the possession of fire by man. There are reports of fires starting from other ignition sources in nature—from branches rubbing together, stones striking against each other, volcanic discharges, and even spontaneous combustion (most evident in caves). But these sources cannot account for the widespread adaptations to fire by natural communities or for the universal capture of fire by man. The evolutionary reality of natural fire is an inescapable fact, and lightning fire is the philosopher's stone for nearly all contemporary thinking about the objectives of wildland fire protection.

The apparent rarity of lightning fire is a statistical phenomenon camouflaged by the ubiquity of anthropogenic fires. When it occurs, however, it can have tremendous impact. Many of the most stubborn and costly fires of recent years have been the result of lightning, often of multiple lightning fires in remote areas that burned together: the Alaska conflagrations, 1957; the Sleeping Child fire, Montana, 1961; the Elko fire, Nevada, 1964; the Glacier Wall, Trapper Peak, and Sundance fires, Montana, 1967; the Swanson River fire, Alaska, 1969; the Wenatchee fires, Washington, 1970; the Carrizo fire, Arizona, 1971; the Seney fire, Michigan, 1976; the Baxter Park fire, Maine, 1976; the Marble Cone and Hog fires, California, 1977; the fire complexes in southern Arizona and interior Idaho, 1979. Man-caused fires occur in areas accessible to people, which makes the fires equally accessible to control forces. Lightning fires are more randomly distributed, and remote fires, of whatever size, escalate suppression costs rapidly.

The relationship between lightning and life, like that between fire and life, has affected man no less than other elements of the biota. Lightning has been a persistent predator on man and his structures. More people die from lightning than from floods, and lightning damage to structures accounts for nearly 11 percent of annual fire loss—almost identical to the percentage of wildland fires ignited annually by lightning.[14] Man, too, has had to adapt to lightning, to exploit its most important consequence, fire, and to assimilate it intellectually. Lightning has long been a topic of folk-lore, mythology, and science. Almost invariably, only the pontifex maximus of the gods (or in the case of Thor, a trusted lieutenant) could wield the thunderbolts. It was with thunderbolts forged by Vulcan that Zeus estab-lished himself as supreme among the gods of Olympus and turned the siege of Troy to the favor of the Achaeans. (The association of lightning with volcanoes, evident in the personage of Vulcan, is also present in the South Seas in the form of the goddess Pele. Volcanic discharges can, in fact, result in massive lightning displays.) When Salomoneus pretended to be Zeus by hurling firebrands from his chariot, Zeus struck him down with lightning for the sacrilege. To the ancients, the place where lightning struck might be either blessed or cursed by the gods, depending on the augurs. A temple might be built on an unlikely site if Zeus or Jupiter had shown his favor there with a bolt. Rocks or trees might become sacred if struck. The Druidic worship of mistletoe, Sir James Frazer concluded, stemmed from the belief that mistletoe on oak was "a visible emanation of the celestial fires; so that in cutting the mistletoe with mystic rites they were securing for themselves all the magical properties of a thunderbolt."[15]

As the story of Salomoneus shows, the association of lightning with fire is long established in myth as well as in folklore. To possess either meant power, and both fire and lightning commonly resided in one personage. The ancient philosopher Lucretius devoted considerable time to the relationship. "As for *lightning*," he concluded, "it is caused when many seeds of fire have been squeezed out of clouds by their collision."[16] Folk customs generally related lightning to the "need fire"—the real and symbolic, initiating, and often sacred fire from which all other, profane fires would be drawn. Across the globe, custom frequently prescribed that need fires begin with the wood of a lightning-struck tree and, conversely, that when lightning began a fire on its own, all fires in the surrounding area would be extinguished and rekindled from the divinely sent source. Even in the eighteenth century, nat-ural philosophers like Benjamin Franklin described lightning as "electrical fire."

Many peoples sought protection from lightning in magic. The ceremonies were often identical to those invoked as security against the other chief pred-ators on mankind—notably wolves, witches, and fire. Only in very recent times, with the discovery and explanation of the electromagnetic fields of the earth, has lightning been perceived as an integral part of the natural equilibrium rather than as a paroxysm of a nature gone mad or as the lethal frenzy of a capricious god. But for early foresters, lightning fire remained a freak, a minor sport of nature amidst a landscape of anthropogenic fire.

Relics of lightning were believed to be prophylactics against further strikes. Lightning-struck trees and animals held magic powers; in the case of men, however, the bolts signified condemnation. It was recognized, too, that lightning seemed to strike certain species preferentially. The numerous rituals involving oaks reflect this fact. So does the proverb:

Beware the oak, it draws the stroke.
Avoid the ash, it courts the flash.
Crawl under the thorn, twill save from harm.[17]

The ancients held, as Pliny records, that "lightning never strikes the laurel." Thus laurel wreaths were awarded to the victors of the Olympic games and to Roman generals returning to claim a triumph: the laurel wreath would spare them the thunderbolts that might strike those guilty of hubris. The Emperor Tiberius, who had reason to worry, habitually wore a laurel wreath during thunderstorms.[18]

Similarly, woods solemnly charred in the various fire festivals held throughout Europe promised special protection. Here fire was explicitly associated with lightning. Perhaps the best known of these customs is that of burning an oak Yule log—oak, because that was the tree most often struck (and thereby favored) by Zeus; burned, so as to protect against lightning and fire; Yule, only after conversion to Christianity.

These shards of folklore amount to something more than an archaeology of relic beliefs and quaint customs. They represent an effort to understand and ultimately to control a powerful natural force; thus they complement the vastly more subtle and complicated intellectual assimilation of fire. Unlike fire, lightning has inspired awe, but rarely reverence. Lightning always served as a weapon of the gods or of conjurers or as an emblem of a secular power (the American eagle, for example, with lightning bolts in its talons). Never did it symbolize, as fire did, a principle of nature, a life force, or an inscrutable expression of the eternal.

It was not until the eighteenth century that modern science scrutinized these relationships. For his explanation of lightning in terms of Newtonian physics, Immanuel Kant exclaimed, Franklin was a new Prometheus who had stolen fire from heaven. Meanwhile, two contemporaries of Franklin, Joseph Priestley and Antoine Lavoisier, exploited the new technology of electricity and thereby discovered oxygen and made possible the advent of a scientific chemistry of fire. Thus fire was doubly challenged as a causative agent: it was removed from lightning in favor of electricity, and it was removed from combustion in the form of the imaginary element phlogiston. But fire remained in nature, and two centuries after Franklin's lightning rod appeared in *Poor Richard's Almanac,* the natural philosophy tradition of Franklin joined with the natural history tradition of Bartram under the auspices of the U.S. Forest Service to make explicit the relationship between lightning and fire. One of the first major projects of the Division of Forest Fire Research, called Project Skyfire, sought to discover possible mechanisms of lightning suppression as a means of fire prevention. Yet it was fire, not lightning, that motivated the research, and by the time the program concluded, changes had occurred in the perception of lightning fire. In a startling reversal of historical emphasis, lightning and its effects, notably fire, were sought out rather than avoided. No longer echoing the gloomy prophecy of Pubilius Syrus in the first century B.C., that there is no defense against lightning, land managers of the 1970s were often eager to incorporate lightning fires into plans for land management.[19]

II

Bartram's *Travels* was more than the travelogue of a professional naturalist. It became an important work of literature, and it helped to identify romantic nature with America. François Chateaubriand, Thomas Carlyle, and William Wordsworth, among others, looked to Bartram for inspiration on the state of nature wild. America encompassed not only the New World but also the oldest of worlds: it offered a site for futuristic societies, such as Bacon's New Atlantis, and it preserved the precivilized nature of primeval forests and noble savages. It projected a future and restored a past. A virgin, undisturbed, Edenic land—America was, in the language of a later day, a wilderness. It mattered not that such concepts were not only anthropogenic but also ethnocentric.[20]

It does matter, however, for a history of fire. Fire and man, wilderness and man, and fire and wilderness are not easily separated. Lightning-caused fire is the ultimate source of anthropogenic fire, and the fire adaptations it encouraged over geologic eons give the parameters within which anthropogenic fire must operate. Yet wilderness, like fire, is both a natural and a cultural phenomenon. The contradictions, paradoxes, and anomalies of the one are those of the other. Wilderness, like fire, may in theory exist apart from human society, but in fact, man is the chief source for both. As an idea and as a political fact, wilderness is a human artifact. The paradoxes of this alliance may be illustrated by considering two fires burning side by side in an official wilderness area. One fire has been set by lightning; the other, by man, either accidentally or intentionally. The lightning fire may be left to burn, while the anthropogenic fire is suppressed. To nature there is no distinction; to the culture there is. Wildfire, like wilderness, is a relative term, one defined in relation to human values and not according to some absolute natural standard.[21]

Natural fire has acquired considerable significance for fire management. Prior to the late 1960s fire in wildlands was managed according to whatever larger strategy of fire protection then prevailed. By the 1970s, however, wilderness fire had itself begun to dictate that strategy, though the question of fire protection on reserved lands has existed from the time wildlands were first set aside. Indeed, the modern era of fire protection begins with fire

suppression on newly reserved wildlands. Fire in the Adirondacks Reserve
brought New York state into fire control in 1885. Fire in Yellowstone
National Park involved the federal government in 1886, when Army troops
took over administration and protection duties. Many of the roads, trails,
and communication networks in remote areas were constructed for fire protection, and in many regions fire control—with its physical plant and organized crews—constituted the only real administrative presence of the land
agency. To extend fire protection to a remote region of little commercial
merit was to pay the supreme compliment to these lands: they were of equal,
perhaps greater, value than commercial lands. Commercial lands, after all,
returned the investment made in them through the revenue generated by
the sale of permits. The backcountry, by contrast, was protected by virtue
of a kind of noblesse oblige, patronage in support of esthetic and recreational
values. Virtually every advocate of wilderness values supported fire control.

Not until the late 1960s did this pattern change. Then it was deemed
more important to introduce fire into most wilderness areas than to withhold
it. So large had the issue of wilderness fire become that its concerns helped
to define the whole range of objectives and strategic interests for fire protection on all lands. Not only was natural, lightning fire considered essential
for wild lands, but, by analogy, fire of some sort was proclaimed useful for
lands of all sorts. There were practical reasons for accelerating programs in
prescribed burning, but there occurred an even more fundamental change
in philosophy: through prescribed fire the "goodness" of the wilderness and
its natural processes could be distributed to other landscapes.

The problem of adjusting fire practices to land use has existed for millennia. It is complicated by the fact that, because fire can be as influential by
being withheld as by being applied, there is no neutral position possible. The
question of wilderness fire, however, was an unprecedented issue because
the idea of wilderness was itself a recent American invention. Wilderness
has experienced two general meanings. The first belongs to an age of agricultural reclamation; the second, to the industrial counterreclamation. To
those intent on subduing the natural landscape for agricultural pursuits,
wilderness was something to be reclaimed. It was an antagonist. The
esthetic ideal of the reclamation was a pastoral harmony, a general uniformity or homogeneity of landscapes, not unlike the then perceived uniformity
of human nature across time and cultures. With the industrial revolution
came a counterreclamation. Its chief victim was the land of the American
reclamation, especially the marginal lands, and it generated a spectrum of
land uses that, by the standards of reclamation, could be collectively termed
wildlands. A special portion of these lands retained the name wilderness,
and the concept acquired complimentary rather than pejorative meaning.
Wilderness was something to be preserved, restored, even created; its
esthetic ideal was the sublime; it delighted in a stark juxtaposition of the

wholly synthetic and the wholly natural; it prefered a pluralistic mosaic of distinct types to a melting pot of landscapes.

Wilderness areas are assumed to represent natural templates against which man-made imbalances can be measured; genetic banks, where natural species and natural ecosystems retain molecularly coded wisdom acquired over eons of evolution; unspoiled sanctuaries, esthetically pure and uncontaminated by the impediments of civilization. America became the locus for these ideas because, unlike Europe, it possessed lands that were relatively unscathed by the Great Reclamation. Moreover, the intellectual heritage of Western civilization had long identified America with a precivilized, Edenic state of nature, and Americans had come to associate wilderness with a heroic pioneer past in need of preservation. In the words of the Leopold Report (1963) for the Secretary of the Interior, the national parks in particular should be maintained or recreated so as to represent "vignettes of Primitive America." The process of giving administrative definition to wilderness, as distinct from other wildlands (such as parks and forests) had begun in 1924 with the creation of the Gila Wilderness Preserve in the Gila National Forest, New Mexico. Not until a year after the Leopold Report, however, did the Wilderness Act give statutory definition and protection to lands destined for the National Wilderness Preservation System. An Eastern Wilderness Act (1975) extended the system into the eastern regions of the United States. By 1976 some 14.5 million acres had been incorporated into the system. The Forest Service managed 87 percent, the Park Service, 8 percent, and the Fish and Wildlife Service, 5 percent, though these proportions and the total acreage will change with a final determination of Alaska land classification.[22]

The question of fire management was not so easily solved. The paradoxes of wilderness fire are both ideological and operational. To begin with, "wilderness" is not a state of nature or a state of mind but the interaction between a continually changing state of nature and a perpetually evolving state of mind. A natural fire regime cannot be recreated simply by withdrawing fire control forces, for example. It was easy enough to restore at least partially a natural process like lightning fire, though the program did require a new definition of control. Earlier, fire was considered controlled only if it burned within the perimeter of firelines or fuelbreaks; now fire is considered controlled if it burns within the conditions established by a prescription. It was more difficult, however, to restore a natural landscape to pre-Colombian appearances. Because the concept of wilderness has been linked to the discovery and settlement of America by Europe, any enterprise that predates the reclamation, no matter how extensive, is considered wild. Thus Indian fire practices, which were enormously powerful as landscape modifiers, have been dismissed. The return of natural fire to wildlands is less likely to "restore" an ancient landscape than it is to fashion a landscape that has never before existed. The program is less a case of restoring a natural phenomenon so that it may interact with its natural environment than of managing one cultural and natural hybrid, fire, in its interaction with another hybrid, wilderness. The paradox is further compounded when, on the example of natural fire, prescribed fire is introduced to areas that have little fire history—the supposition being that if fire is good in wilderness sites, then it must be good elsewhere too.

Apart from such ideological paradoxes, profound practical questions remain about the management of fire in wilderness. Unwanted fire and smoke may enter a wilderness area from outside, and they may also leave a wilderness area. Some of the largest fires of recent years have begun as natural fires in wilderness sites, which then left their sanctuary, and some of the worst instances of air pollution both in wilderness parks and even in some urban centers (Phoenix, for example) have resulted from prescribed burning. Moreover, insofar as wilderness lands function as island sanctuaries for endangered species or rare habitats, they remain vulnerable to a large fire, whether it originates from lightning or from arson. Fire protection simply cannot be withdrawn altogether. It is no accident that, historically, the first administrative program for wild or remote lands was usually fire control and that fire management in some form will remain as the last presence to be reduced. Its forms will change, but it cannot be entirely eliminated if only because so many landscapes have become dependent on anthropogenic fire practices. Many tens of millennia ago mankind made a pact with fire. Its heritage of fire is one that mankind cannot repudiate, and this legacy imposes a responsibility that cannot be easily abdicated by appeal to a hypothetical state of nature. Fire is as effective by being applied as by being suppressed, and whether he applies or withholds fire, man cannot avoid responsibility for fire management. It was as keeper of the flame that man first became steward of the land.

Wilderness is distinct from wildland, and wilderness preservation and wilderness management developed uniquely in the United States. Wildland fire, in brief, carries a peculiar national and cultural significance. It has often been claimed that America's natural wonders, especially its great wilderness preserves such as Yellowstone and the Grand Canyon, are to American society what cathedrals, castles, or temples are to other cultures, that Americans look to nature for what other peoples find in the traditions of a common folk past or in the treasures of a bygone civilization. The preservation of wilderness sites, as distinct from the generation of wildlands, was often justified as the establishment of cultural memorials. "This is a plea," wrote Aldo Leopold, the great philosopher of wildlands, "for the preservation of some tag-ends of wilderness, as museum pieces, for the edification of those who may one day wish to see, feel, or study the origins of their cultural inheritance." For Americans, that inheritance is defined by a national creation myth in which its people evolved along a wilderness frontier. "It is only the scholar," Leopold concluded, "who understands why the raw wilderness gives definition and meaning to human enterprise."[23] He might have added that the reciprocal is equally true: it is the human enterprise, as preserved by the scholar, that gives definition and meaning to wilderness.

In ancient times many civilizations, Indo-European and Mesoamerican alike, maintained great national fires. The community fire originally grew out of practical necessity, a public utility from which people could replenish the fires of hearth, field, and industry. With time the sites of these communal fires acquired symbolic significance: they came to stand for the people themselves. Great temples were erected to contain the fires; priests and vestal virgins administered the proceedings. The fire became synonymous with the grandeur of the civilization—in the case of Rome, with the state itself—and was never allowed to become extinct. Political alliances were sealed by mingling the communal fires. Colonizers carried a brand from the ancient fire to their new lands. Greek and Roman armies never left their borders without carrying an altar with coals from the sacred national fire.

For Americans, wilderness sites have perhaps taken the place of those ancient temples, and the preservation of the eternal fires of nature, deposited by lightning, has created a great national fire. The ultimate function of these fires is perhaps to be cherished as a memorial to a creation myth—perpetuated as an especially visible emblem of wilderness and as an eternal flame to the European discovery of the New World. From the vestal fires on America's virgin lands people can rekindle symbolically the fires of civilization.

THE FIRE ENVIRONMENT: PRINCIPLES OF FIRE BEHAVIOR

It rejoiceth as a giant to run his course; and there is nothing to be hid from the heat thereof.—Book of Common Prayer

Fire and water have no mercy.—Old English Proverb

I

Fire takes two general forms in the landscape of the industrial revolution. It exists in a confined form, as with various fire appliances and engines, and its exists in a free-burning state, as with forest and field fires. For any fire, fuel, oxygen, and heat must come together in the right mixture; collectively, these make the fire environment. For confined fire, these elements can be closely regulated and the properties of the fire rigidly prescribed. Most scientific and engineering knowledge about fire deals with its behavior under such conditions.

Free-burning fire, however, is vastly more complex, its physics more statistical, and, in contrast to the fires of hearth and furnace, its engineering efficiency much reduced. The parameters of the fire environment cannot be finely tuned to sustain maximum efficiency or to modulate wild fluctuations of energy release and mass transfer. Instead of a single fuel element, wildfire responds to an ensemble of fuel complexes, grossly arranged; instead of a furnace designed to ensure maximum draft, it must follow broad topographic configurations; instead of a metered intake of oxygen, it must deal with traveling air masses superimposed over microclimates; and instead of careful engineering to ensure maximum heat transfer, wildland fire is propagated by a variety of mechanisms, often erratic and turbulent. By engineering standards, free-burning fire is inefficient. But by natural standards—measured against the magnitude of its task, by the amount of matter it transforms or transfers, and by the quantity of energy it releases—such fire is eminently satisfactory. Much as lightning discharges release in short order the stored electrical energy leaked to the atmosphere, so wildland fire releases the stored chemical energy gradually accumulated in the earth's vegetative cover. Fire is a violent form of decomposition, one that works to maintain a biochemical equilibrium rather as lightning does an electromagnetic equilibrium. It is precisely these properties, however crude by the standards of mechanical engineering, to which biotas have adapted and by which mankind has learned to manipulate fire for its own ends.[24]

To understand and predict wildland fire behavior, it is necessary to enlarge analogies drawn from confined fires and to create models for the components of the fire environment, such as fuels and weather, and for the mechanics of fire propagation. Wildland fire behavior multiplies probability with probability. Unlike astronomy, where it is possible to predict the position and velocity of individual objects with great precision, fire behavior deals with statistical ensembles—the limitless nuances of fuel complexes, the restless variety of topographic forms, and the maddening vagaries of weather, particularly on micro- and mesoscales. Wildland fire does not merely resemble a climatic or meteorological phenomenon; it results from them and is thus another order removed from simple determinism. Rather than presetting the key parameters, it is necessary to predict what they will be.

Wildland fire involves combustion on a grand scale. In the classic model for combustion the process advances by three stages: a preheating phase, a period of flaming combustion, and a state of glowing combustion. During preheating, or pyrolysis (literally, a breaking down by fire), the fuel is brought to its ignition point or kindling temperature. This transfer of heat can occur in several ways: by convection, which performs especially well on slopes, on aerial fuels, and on wind-driven flaming fronts; by radiation, particularly where energy release is high; and by conduction, the least effective, because wood is a poor conductor of heat. Pyrolysis vaporizes hydrocarbons, driving off free moisture and converting solids into flammable gases. With ignition of these gases, the period of flaming combustion begins, and this second stage propagates the heat flux needed to sustain pyrolysis in advance of the fire. It also completes the process of distillation from solid to gas. Combustion is a chain of chemical reactions; to sustain it, more heat must be generated than is absorbed.

Flaming combustion is an ephemeral state, varying with fuel properties and the local fire environment. It advances as a wave of flame, and its energy release—the greatest of any stage of the combustion process—can be described graphically as an asymmetrical curve or thermal pulse. The amplitude and shape of the pulse vary with fuel types and burning conditions. What limits the duration of the flaming state is primarily the slowness of conduction through wood. Wood surfaces may be heated by radiation and convection and hence are readily pyrolized, but the interior of the forest fuel can be exposed only by the relatively ineffective transfer of heat by conduction. By the time conduction exposes the unburned interior, the flaming front has already passed and with it the heat needed to sustain flaming combustion. The retarding effect of conduction accounts for the large quantities of charred debris left after a major fire and serves as a reminder that not all fuel is available for combustion and that not all fire energy contributes to fire spread.

Once the flaming front has passed, the remainder of combustion reverts to a glowing state. Oxidation occurs on the surface of the charcoal residue left from the flaming phase; the carbon burns, but as a solid rather than as a gas. In the case of heavy forest fuels, this phase may be protracted. Nevertheless, the combustion of forest fuels is rarely total. Even under the most intense conditions of incineration, such as mass fire, considerable residue remains. In many fires large quantities of dead material may be consumed, but equal or greater volumes of living matter may be killed, furnishing fuel for the reburns that are sure to follow.

Fire spread is more complex than this simple model of combustion suggests. Current mathematical models describe steady-state fires in which spread occurs uniformly under determinant conditions. Of more significance, however, is the problem of fire growth. Here the pattern of ignition, as well as that of combustion, is significant. Fires begin as point sources. The sources may be isolated or they may occur in multiples. In the latter case, the burning rate and proximity of starts will determine whether the resulting fires behave as separate events or whether they begin to interact synergistically with greatly magnified burning characteristics. In either case the first requisite is that ignition persist and that it result in a period of heat buildup during which an excess of heat is generated and made available to sustain fire spread.

As a fire spreads outward from its point of origin, it develops a flaming front along its perimeter and leaves behind a burned-out interior or one subject to glowing combustion. If the peak intensity of the fire is reached just prior to the interior burnout and then decreases in rate of spread and intensity as the flaming front advances, the growth of the fire will decelerate. But if in response to the conditions in the fire environment the fire can increase its rate of spread and intensity, it will accelerate. Of special importance in initial acceleration are the critical burnout time and the presence of surface winds or slopes. Both wind and slope have the effect of bending the flaming front and concentrating fire heat. That is, the heat must build up in excess of what is needed in order to sustain combustion, and it must be directed. If the the heat released by flaming combustion balances that required for pyrolysis, then the fire will stabilize and burn in a steady state with a constant rate of spread and constant rate of energy release. If conditions favor an excess of heat energy—and this is largely a function of fuel characteristics—and if the air mass is favorably structured to support a strong convective column, then the fire may further accelerate. In this event the discharge of smoke and convective heat no longer occurs as a diffuse plume but forms a tightly organized column—an atmospheric chimney that regulates, by its control over ventilation, the intensity of the fire below it.

In colloquial terms three general patterns of fire spread are recognized: the ground fire, which propagates largely by creeping and is sustained by glowing combustion; the surface fire, which spreads through ground fuels with a flaming front; and the crown fire, which is sustained by a surface fire but erupts into the canopies of forest fuels with often violent and discontinuous surges. It is tempting to equate these phases with stages of growth, but the transition from steady-state to accelerating fire involves much more than the term *crown fire* suggests. The transition requires, first, a tremendous accumulation of heat: the flaming front must be deep in relation to its length and large enough to act as an area source of heat rather than as a point or line source, and the rate of energy release must be of such a size that it can perturb the surface wind field. These conditions require a heavy fuel load, excluding, for example, such fuels as grass, tundra, and forest litter. Second, transition requires an air mass in which surface winds will not shear off a prospective convective column or contain an inversion that could prevent adequate vertical development. The convective column interacts with the surrounding air mass. To generate a large fire the energy release, or bouyancy, of the column must exceed that of the wind field. The fire then responds to the convective wind system generated by the fire itself rather than to its ambient wind field. Transition requires, finally, a triggering mechanism that affects burning intensity, such as spread into heavier fuels or breakup of an evening inversion layer.

Once initiated, transition can occur almost instantaneously. The fire is then said to have "blown up" or "begun a run." Between ignition and transition, however, the fire may persist in steady state for some time. Transition may occur within minutes after growth to a steady state or even precede (and preclude) steady-state burning. It may follow after days or weeks of steady-state combustion, either flaming or glowing. Or it may never come at all. A fire may make the transition once or a dozen times during its history, returning to a steady state after each transition.

Once a fire has advanced through the transitional stage, the process of combustion and fire propagation become seemingly discontinuous. Ignition may spread by saltation rather than by a steady entrainment of heat. That is, instead of continuous heat flux, new points of ignition deposited by firewhirls and showers of aerial firebrands create new heat sources, each of which scatters further brands. The sequence of combustion states can separate, with great clouds of volatile gases streaming through the atmosphere or exploding into flame like flak. Fire spread does not proceed continuously but in surges, sparing some areas entirely and devastating others with almost total burnout. Intensity accelerates locally, increasing by a factor of three or four, for example, with the development of firewhirls on the lee side of the convective column. Convective winds can acquire great velocity; turbulence and vorticity complicate convective drafting. A simple flaming front may break up into multiple heads. Fire propagation then depends no longer on quasi-deterministic processes of heat transfer by conduction, convection, and radiation but on other methods. Convection and radiation tend to dissociate, so that heat and winds may exist at some distance from the fire; the laminarlike flow of heat during a steady-state fire is broken into a turbulent ensemble. Depending on how the wind field and convective column interact, as mediated by topography, large fires exhibit an almost grotesque distortion of the combustion process that both magnifies fire behavior and fragments its continuity as a physical-chemical event. The transition is like that of a stream giving way to a waterfall. Many fires are large only in area, but large fires, properly understood, are great by virtue of their intensity.

What all large fires share is that they make the transition to a large fire from a point source. The transition results from the interaction of a single fire, its fuels, and the surrounding air mass. Another category of fire, however, begins with area ignition or the simultaneous ignition of many fires in a particular area. Under conditions that do not favor transition, these multiple fires burn as autonomous entities. But under conditions favorable to transition, they can interact with one another so that nearby fires become part of the fire environment. If fuels are light or burning conditions poor, the fires may slowly coalesce, at which point burning intensity reaches a maximum. But in heavy fuels and under favorable conditions, the coalescence, which begins with the upper convective columns, occurs rapidly. The resulting holocaust is a synergistic phenomenon of extreme burning characteristics known as *mass fire*. A mass fire may travel or it may be stationary. For the latter, the term *firestorm* may be applied; for the former, *conflagration*—though this expression might be better reserved to describe large, uncontrolled fires that continue to respond to the fire environment. A true mass fire tends to respond to its own internal energy dynamics. The properties of mass fire and its relation to area ignition were first recognized as a result of the firebombings of World War II and have been intensively researched since in the expectation that a thermonuclear war would result in a landscape of mass fire. But mass fire can be produced in nature, too, when aerial firebrands are thrown out in advance of a fire in great showers and saturate an area with simultaneous ignitions.

Most of the historic wildland fires of record were, by these definitions, large fires, and the classic accounts of their behavior dwell on these extreme properties. An especially detailed description of conditions in the boreal forest was given in 1889 by Robert Bell, assistant director of the Canadian Geological Survey. Bell observed that these fuels

> only await a spark of fire to give rise to one of the wildest scenes of destruction of which the world is capable. When the fire has once started, the pitchy trees burn rapidly; the flames rush through their tops and high above them with a roaring noise. Should the atmosphere be calm, the ascending heat soon causes the air to flow in, and after a

time the wind acquires great velocity. An irresistible front of flame is soon developed, and it sweeps forward, devouring the forest before it like the dry grass in a running prairie fire, which this resembles, but on a gigantic scale. The irregular line of fire has a height of a hundred feet or more above the trees, or two hundred from the ground. Great sheets of flame appear to disconnect themselves from the fiery torrent and leap upward and explode, or dart forward, bridging over open spaces, such as lakes and rivers, and starting the fire afresh in advance of the main column, as if impatient of the slower progress which it is making. These immense shooting flames are probably due to the large quantities of inflammable gas evolved from the heated tree tops just in advance of actual combustion, and they help to account for the almost incredible speed of some of the larger forest fires, one of which was known to run about 130 miles in twelve hours, or upwards of ten miles an hour.[25]

The rate of spread quoted may be excessive, but detailed case studies of major fires in recent times have yielded impressive figures. The Matilija fire (1932) in Southern California torched 20,000 acres within a few hours of ignition, and in one hour's time traveled a distance of 15 miles. The Harlow fire (1961) in the central Sierra foothills of California exploded over 30 square miles (13,200 acres) and two towns in the space of two hours. On the Nebraska National Forest the Plum fire (1965) traveled 13 miles through grass and pine plantations in a matter of hours. The Sundance fire (1967) in the heavy timber of the Idaho Panhandle crossed 16 miles in nine hours over a four-mile front; aerial firebrands further ignited spot fires 10 to 12 miles from their place of origin. The Little Sioux fire in northern Minnesota (1971) roared over 7 miles in six hours, crossing 9,000 acres of dense forest, roads, and muskeg. The Bombing Range fire on the coastal plain of North Carolina (1971) traveled over 2 mph during a four-hour run, with a maximum rate of almost 5 mph and a total run of 14 miles.[26]

Nevertheless, rate of spread is often a less important measure of a major fire than is intensity, which not only better describes the effects of fire and the difficulty of control but also furnishes a means of comparing fires. It can supply that discrimination among types of fires so lacking in common language and in the qualitative lexicons of fire management. Especially favored now are physical measurements—intensity, flame length, energy release rate—that describe the energy output of a flaming front. Byram's intensity (Btu/ft/s), for example, correlates directly with difficulty of control. An intensity reading of less than 50 Btu/ft/s describes the desirable range for most prescribed burns. A reading of 500 represents the limit for control by any means; beyond 1,000, large fire behavior can be expected, with spotting, firewhirls, crowning, and major runs with high rates of spread. The intensity of the Sundance fire varied with its pulsating runs, giving figures of 7,000,

11,000, and, ultimately, 22,500 Btu/ft/s. For the Bombing Range fire in North Carolina, the pulsing runs recorded intensities of 6,000, 8,000, and 18,000 Btu/ft/s. The theoretical limit for Byram's intensity has been calculated at about 30,000 Btu/ft/s. Fires with energy releases of this magnitude are stopped only by wholesale changes in the fire environment—new fuels, for example, or the onset of rain—or by their own exhaustion of fuels as a result of advance spotting. Mankind is capable of generating energy sources of this magnitude, but not energy sinks. A large fire may release the energy equivalent of a Hiroshima-type atomic bomb exploding every 5 to 15 minutes. But the only sink capable of absorbing such a quantity of heat is the atmosphere, which is one reason ventilation in the form of a convective column is so important in the establishment of a large fire.

These aspects of fire behavior are of fundamental significance for an understanding of fire history. Different regions and different historical periods have tended to show characteristic fire regimes. Within a particular regime, fires tend to be distributed in space and time with regularity. Thus, fires in southern temperate forests tend to be frequent but of low intensity; in northern forests, fires are episodic but of high intensity. Some regimes typically experience large fires, which tend to conform to a particular historic and geographic pattern. Geographically, fires rarely occur in isolation. Almost all historic fires were in reality fire complexes, and these complexes were a response to regional weather conditions. The October 1871 fire in Chicago was but a bubble in a vast cauldron of fire that stretched from the Ohio Valley to the Lake States and the High Plains. Farms burned in Indiana, villages were swept by flame in Wisconsin and Michigan, and towns in the Dakotas fell before fire. The celebrated 1910 complex in the Northern Rockies was but the most spectacular manifestation of a fire environment that saw fire break out across the northern temperate forests from Oregon to New Jersey. The great fire complexes were compounds of smaller fires, each of which made the transition at least once to the status of large fire and many of which actually merged according to the dynamics of mass fire behavior. Typically, dozens or even hundreds of small fires burned simultaneously until conditions favored the transition to large fire behavior, with a regional matrix of fire as the result. For forest fuels, runs on the order of 50,000 to 70,000 acres are probably close to maximum, though one fire may make several runs in the course of its history. According to historic record, at least 13 of these complexes in North America have each burned more than one million acres of forested land.[27]

Complexity also characterizes the distribution of fires through time. A fire is rarely an isolated historical event. More commonly a major fire initiates a cycle, both natural and cultural. It may create a more fire-prone fuel complex by replacing heavy fuel with fine fuel or by leaving massive amounts of charred heavy fuels to stoke a cycle of reburns. The 1910 fires, for example, set up a cycle of reburns that lasted into the 1930s. The Tillamook fire in Oregon (1933) began a fire cycle that did not end until the 1950s. Similarly, the stages by which a region was occupied by tribes or settlers introduced new fire practices that changed the fire regime characteristic of the area and established a fire cycle that might persist as long as the regime remained.

The complexity of fire growth ensures that only a small proportion of fires achieves large fire status. Fires show the same logarithmic distribution as other eruptions of natural energy, such as floods, windstorms, and earthquakes. Under both managed and natural conditions, the vast proportion of burned acreages and damages comes from a small number of fires. Nationally, more than 95 percent of annual burned area results from 2 to 3 percent of the total number of fires. In the Southwest, records show that 78 percent of lightning fires do not exceed one-quarter acre in size and that only 3 percent exceed 10 acres.[28] This logarithmic pattern is easily explained. The growth process for a fire is a random one; the number of individual probabilities that must be multiplied together is large, and their product, small. Even with man's intervention in the form of either fire suppression or fire ignition, the above relationship holds, though the value of the final product will either decrease or increase. The transition to large fire, moreover, may be almost instantaneous or long delayed. In the Radio fire (1977) in northern Arizona, an abandoned campfire made the transition within minutes. By contrast, the Forks fire (1951) on the Olympic Peninsula of Washington began during the summer as a 1,500-acre burn that was controlled but not mopped up. Patrol continued through the autumn, but so did drought. On a September afternoon the smoldering fire blew up and raced over 21 miles of mixed ownership lands in the course of a single afternoon. The resulting lawsuits went all the way to the Supreme Court.

Large fires present unique problems. They are responsible for the greatest damages, not only by virtue of their extent but also by virtue of their intensity. They account for the astronomical expenditures possible during fire suppression and are directly responsible for the establishment of a special emergency funding procedure. The need to forecast the probability of transition has underwritten much of the forest fire research effort and has sustained the drive to establish a national fire danger rating system. Invariably, large fires provide an occasion to reevaluate the effectiveness of past programs and stimulate major policy shifts. Following World War II, merging civilian and military concerns about mass fire made it the problem fire type of greatest interest to fire protection agencies.

The capacity of nearly all fires to make the transition to large fires has also dictated much of the threefold strategy of fire control—that is, to prevent ignition; to modify the fire environment, especially fuels, so that fires will burn with less intensity; and to suppress fires while they are small. The

classic strategy of fire suppression is perimeter control, to restrict the geographic spread of a fire by removing heat and fuel from along its spreading
front. But this strategy applies only to fires that have not yet made the transition to large fires. Even small fires can generate too much heat to be extinguished directly; they must be contained along the perimeter by interrupting
the transfer of heat or by breaking the continuity of fuels. A fire must be
controlled either before it can make the transition or after it has completed
a run and before it can begin another. The simple fact that fires begin as
point sources makes the former option the more attractive, and for most of
its history organized fire protection has sought to control the occasional
large fire by controlling all small fires. Even though the probability of a fire
making the transition to a large fire might be slight, the costs of controlling
it after it becomes large are such as to warrant the suppression of all small
fires. Nearly all of the major fire complexes of the twentieth century began
either as controlled fires or as wildfires that at some point in their history
were considered controlled. One of the primary reasons for the reduction in
burned acreages over the last century is simply that low-intensity, long-
burning fires are no longer tolerated.

Fire environments tend to exhibit certain regional and historic uniformi-
ties and to generate distinctive patterns of fire. Thus different regions at
different times show a characteristic mode of fire behavior. When this
behavior pattern is combined with a regular or systematic source of ignition,
the result is a fire regime. The analogy between confined (industrial) fire
and free-burning fire thus breaks down unless great simplifying assumptions
are made. It breaks down, too, because a fundamental component of the
free-burning fire environment and its associated fire regime is culture. Man
can deliberately alter wildland fuels to modify fire behavior; he can create
ignition patterns unlike those found in nature, which can magnify burning
intensities; and he can initiate fires under extreme weather conditions, when
natural ignition sources would be rare. Man can halt a fire that under natural circumstances would make the transition to mass fire, and he can promote mass fire when natural conditions might not have allowed for the
transition.

Even a full physical description offers only a partial understanding of a
fire. A fire whose flaming front burns with an intensity of 200 Btu/ft/s
might be a controlled burn in one context and a wildfire in another. Not all
damaging fires are large fires: a relatively light surface fire, for example,
can destroy young reproduction critical to a reforestation program. A fire
description must include a cultural referent as well as an inertial referent.

Similarly, there is a cultural as well as a physical form of fire control. The
proliferation of fire terms through the twentieth century has been without
precedent. Many have come from scientific research; others, from the new
sets of fire practices demanded by the industrial revolution. Each major pol-

icy shift by the U.S. Forest Service has spawned a new glossary of fire terms.[29] Though the physical properties of fires might be identical, the fires themselves are cultural phenomena; each is in a real sense different and must be identified as such. The decision over what to call a fire is often itself a statement of policy. "Light burning," "let burning," and "backfire," for example, have historical connotations that make them anathema, whereas "prescribed burning," "prescribed natural fire," and "counter fire"—all of which refer to the same physical phenomena—are encouraged. So far has the neologic process gone that naming has perhaps become a new form of fire control. All fires, even those of identical physical properties, are considered to be either wildfires or prescribed fires. For the former, the classic strategy of perimeter control still applies; the latter are considered to be controlled if they burn within a designated prescription describing their fire environment. At such times the cultural environment of fire creates the conditions by which the meaning of fire, like fire itself, can make the critical transition to a new, mass level.

II

According to a Naga tribe in east India, fire and water fought long, long ago. "The fire could not stand before the water, and fled and hid in bamboos and stones, where it lurks to this day," Sir James Frazer recorded. "But some day they will fight again, and fire will put forth all its strength, and the Great Fire *(Molomi),* which old men spoke of long before the missionaries came into the land, will sweep up from the banks of Brahmaputra and burn everything on earth."[30] Great fires and floods abound in mythologies across the globe, and commonly they are antagonists. Behind that mythological connection, however, lies a physical reality and a cultural significance that bears closer investigation.

As natural events, fire and flood share many similarities. Both show a similar distribution. Both respond to the same environmental variables— vegetative cover, topography, and weather.[31] A river regime shows a typical pattern of floods as a fuel complex does for fire. Both events are ultimately determined by weather and climate. Time and again, regardless of the density of cover type, flooding results from extreme precipitation, just as fires, regardless of the type and density of fuel loading, will burn under extreme atmospheric conditions. In short, fire and flood, as natural waves of energy, share important attributes. When these waves pass through human developments, they can produce similar disasters.

Wildland fires involve the transportation of heat, mostly by fluid convection. Energy is propagated during steady-state spread in a manner analogous to the entrainment and suspension of debris in a stream in a normal flow regime. During growth to a higher energy state, the mechanisms of fire and flood become more discontinuous and turbulent, with surges and trans-

portation of debris or firebrands by saltation. The profile of the thermal
pulse marking the passage of greatest intensity is analogous to the crest of
a flooding river. Fires and floods are similar, too, in their effects. Rivers
erode the potential mechanical energy stored in topographic relief; fires
decompose the chemical energy deposited on the forest floor. The energy
profile of each system is determined, in a sense, by the work required of it.
A river adjusts its channel geometry, pattern, and gradient so as to provide
just the right velocity needed for its waters to transport the debris delivered
to it. Fires behave in an analogous manner, displaying a shape and intensity
in accordance with the fuel available to them. The most stubborn fires burn
in fuel complexes most in need of decomposition; the fires most readily
extinguished are, in terms of the ecosystem's energy requirements, the least
important.

Wildland fires and wildland floods show the same logarithmic distribution
by size and frequency. One can speak of 10-, 20-, 50-, or 100-year fires as
one can of floods, and just as the actual slope of these flood curves varies by
region, so do the cognate curves for fire. There are fire regimes no less than
flood regimes. The combination of steep topography and heavy rain squalls
that causes periodic flooding in central Pennsylvania, for example, is exactly
analogous to the combination of rugged topography and high winds that
results in large periodic fires in Southern California. Disturbances of the
regime by human activity can either aggravate or mollify the typical energy
curve. Most flood damage is done by a relatively small number of large
floods, just as proportionately greater fire damage results from a small percentage of large fires.

Based on these similarities, the strategies for the management of fire and
flood show certain congruencies. The problem in each case is not one of
exclusion but of redistribution. Water must be drained from some areas and
stored for use on others. So with fire: the level of fire must be reduced in
some areas, augmented in others. Just as flooding serves—however chaoti-
cally from the perspective of mankind—to add nutrients to the soil, so does
fire. The management decision in each case is one of degree and technique.
Impounding rivers for the dual purposes of flood prevention and crop irrigation is analogous to managing fire in order to eliminate wildfire and to
introduce prescribed burning.

Efforts to limit the damages of fire and flood depend on similar considerations. Manipulation of vegetative cover to increase infiltration rates and
reduce sheet runoff for flood control has its counterpart in fuels management
programs that seek to reduce the type and amount of fuel available for combustion, thereby diminishing fire intensity. Dams, diversion canals, and lev-
ees have analogues in fuelbreaks and road networks. Efforts to lessen the
values at risk through zoning, while ideally reasonable, have met with indifferent success: high flood and fire regimes within and around cities continue

to attract developments, often expensive residential communities. Repeated devastations usually result in clamor for better engineering and suppression technology to export the calamity elsewhere or to delay it rather than in decisions to move the community from the traditional path of disaster. As suburbs penetrate further into less developed areas and as summer homes in wildlands increase in density, the hazard posed by fire and flood rises. When fire and flood do threaten, certain aggressive measures—evacuation, sandbagging, and fireline construction—can be taken to confine their spread.

Not least among the resemblances between fire and flood are the environmental problems that have been experienced in recent years as a result of protection efforts. A prominent function of rivers is to transport debris. A dam impounds the flow of sediment, and silting ultimately renders the dam useless. Similarly, a function of wildland fire is to decompose forest litter. But success in "damming" all wildland fires only impounds this litter with its nutrients into a large reservoir of fuels, slowing down the growth of many ecosystems and making possible future fires of catastrophic intensity. Both fire and flood protection seem to have reached a plateau. All of the obvious sites are protected, and the costs of noticeably improving the general system by further technological development appear exorbitant relative to potential benefits. Moreover, the possibilities for extending organized protection into more remote landscapes are economically and environmentally questionable.

Protection thus tends in some degree to be self-defeating. To a certain extent it introduces an element of instability into the system it proposes to regulate. Studies of rivers—for example, the Mississippi—have revealed a certain imbalance in their behavior during floods. Zealous attempts to eliminate the effects of all floods, including minor ones, has unwittingly encouraged conditions in which even moderate floods (as judged by volume of water) have exhibited effects associated with major floods. Apparently overdeveloped engineering structures intended to confine the river too closely have deprived the river regime of the flexibility it needs for self-regulating mechanisms to work.[32] Likewise, the refusal to tolerate small, low-intensity fires makes otherwise moderate fires likely to behave more erratically by increasing the amount and rate of energy released. But here fire and flood differ. Once rain ceases, the flood crest is determined by the characteristics of its channel. But a fire, however small, may make that critical transition to a large fire. Flood control devices must be in place prior to the flood; fire control, only soon enough to stop a fire before it has the opportunity to grow. Flood waters dissipate, flames propagate.

Fire and flood differ in several other important respects. Fires can be readily initiated by man, whereas floods cannot. Floods involve a transfer of mass as well as of energy. They begin as areas and concentrate into lines or

channels; fires begin as points and spread to lines or areas. The waters propelling a flood cannot be evaporated; they must, in some form, pass through
the river or its flood plain. A fire, however, can be extinguished. The thermal
pulse is a wave of energy, not of mass. As it passes, the pulse transfers and
transforms matter, but its control does not depend on the ability to store,
export, or in other ways direct an excess of matter. Because they begin as
point sources, fires are always, at some instant of their history, susceptible
to ready extinction. A flood, however small its crest, is large in area. These
facts make for a fundamental distinction in fire and flood protection. Flood
control relies on environmental engineering, on cover modification and
structures to confine or divert flood waters. Fire control, by contrast, relies
on firefighters. For most of its history fire management has meant the management of firefighters and the means by which to transport them quickly
to small fires. For inspiration and practical models fire control has turned
toward protection institutions, such as armies, that require the direction of
people rather than toward engineering organizations, such as the Bureau of
Reclamation, that deal with disasters by a system of fixed structural
improvements. Thus fire control tends to view its antagonists in more
anthropomorphic terms than those applied to floods or to natural disasters
like hurricanes, tornadoes, and earthquakes. The fact that most fires begin
by the hand of man has not discouraged this tendency.

It is this anthropomorphic quality of fire that is reflected in myth and that
helps to account for some of the cultural differences in the responses to fire
and flood. Fire and water were the first two public utilities, preserved in
special wells and temples. They are equally the two great traditional disasters: to have too much or too little of either is catastrophic. In nature, fire
and flood are often linked, with floods coming in the aftermath of mountain
conflagrations. In American history, wildland fire protection was first urged
as a means of assisting in flood control. Yet the perception of each phenomenon is different. *Water* is a neutral word, *fire* is not. Not every problem
with wildland hydrology is an emergency, but every fire is. Only a few floods
elicit the response given to every fire. Only a few areas can be engineered
for full flood control, but every fire can be extinguished by firefighters with
hand tools if they arrive quickly enough.

Water may destroy, but fire transforms. Water cleanses, but fire puri-
fies—even the etymology of the words in English have the same root. One
may be baptized by either fire or water, but one is tried and judged by fire.
Fire conveys a finality that water does not. Naga mythology, coming from
a people occupying a flood plain, held that water would ultimately triumph
over fire. But this was exceptional. For almost all peoples, from the Aztecs
to the Greeks, the great cycles of history—and the cycle of time itself—are
almost always held to end in fire. It may be this different perception, as

much as their different properties as natural events, that is responsible for the distinctive strategies that have evolved for fire and flood protection.

Christianity, too, has its version of the flood-fire sequence, the Mosaic Flood and the fiery Apocalypse. The Flood was but a premonition; it would be fire, the prophets foretold, that would usher in the final days, that would accompany the final Judgment. "And I will show wonders in the heavens and in the earth, blood, and fire, and pillars of smoke," cried Joel. "The sun shall be turned into darkness, and the moon into blood, before the great and the terrible day of the Lord come."[33] Accounts of historic fires commonly describe the weird, spectacular phenomena by allusion to the forthcoming Apocalypse. The parallels may be based on something more than panic and literary indulgence. The darkness and the roar, the streams of aerial fire and the scenes of devastation depicted by biblical prophets—all are properties of mass fires, and the literary descriptions probably derive from experiences with large fires in the brushfields of the Near East. Even the pillar of fire that guided the Hebrews through Sinai by night was probably based on firewhirls associated with mass fire.

Nor is modern mankind immune from the vision of holocaust. The development of thermonuclear weapons has, if anything, magnified the prospect of an apocalypse by fire and has perhaps helped to sustain an ancient perception of fire as destroyer. In fact, the problem of mass fire and thermonuclear armaments first led to the detailed investigation of the physics of fire behavior. There are, as E. V. Komarek notes, numerous words to describe the various states of water, but until the recent research into fire-behavior physics there was basically one word to describe fire.[34] Of the two great forces, fire and water, fire remains the more terrible, the least negotiable, and the most final.

And God gave Noah the rainbow sign.
No more water, the fire next time.[35]

FIRE AND LIFE: PRINCIPLES
OF FIRE ECOLOGY

The great majority of the forests of the world—excepting only
the perpetually wet rain forests . . . and the wettest belts of the
tropics—have been burned over at more or less frequent intervals
for many thousands of years.
—Stephen H. Spurr, Burton V. Barnes, *Forest Ecology*[36]

The earth, born in fire, baptized by lightning, since before life's
beginning has been and is, a fire planet.—E. V. Komarek[37]

I

Combustion requires only that fuels have certain physical-chemical properties. But an ecosystem is more than a fuel complex, and a fire does more
than simply rearrange the fuel elements and flammability of the system.
Fire is a profound biological event. The chemistry of combustion derives
from the chemistry of organisms, and the geography of fire is synonymous
with the range of organic matter. The combustibility of matter may depend
on the chemical properties of its fuel elements and on their physical arrangements, but that matter consists of biochemical assemblages shaped under
genetic direction. Fire may be measured by certain physical manifestations,
but its effects are described or reconstituted as organic entities or ensembles
of such entities. Fire and life, in brief, form a necessary kind of symbiosis.[38]

Fire is a mechanism of degradation. The heat it releases leads to thermal
degradation (pyrolysis), while the fire itself, as an oxidation reaction of great
rapidity, leads to chemical degradation. It breaks down the matter that
organisms have assembled from captured sunlight, releasing the stored
chemical energy. A fire thus discharges energy and delivers organic and
inorganic chemicals into the components of an ecosystem—its air, water,
and soil. An ecosystem must somehow cope with this flow of energy and
release of chemicals. The constant presence of lightning as an ignition
source makes this requirement an inevitability, but the adaptations that
have evolved over geologic time have often made it an opportunity. Some
organisms have adapted defensively to protect against the energy released
by fire, but others have adapted so as to seize on the nutrients released by
fire—and have consequently even encouraged properties promoting
combustion.

Wildland fire is thus part of a dynamic equilibrium between the production and decomposition of biomass. Its function is to recycle, both on the microscale of nutrients and on the macroscale of the community itself. The necessity for fire as an agent of decomposition varies with climate, however. A biota's nutrients are stored in various portions of its system—in the soil, in the standing vegetative cover, and in a mobile layer of dead organic matter called litter and duff. The geologic production of new chemicals proceeds too slowly to satisfy the needs of biological communities, and some of what new organisms require has a biochemical rather than a geochemical origin. They must come primarily from the mobile layer of litter, and it is this layer that is affected by fire. The necessity for decomposition on a grand scale is such that if fire did not exist, nature would have to invent it. In warm, humid climates growth is rapid, but so is decomposition by biological agents. In cold, dry climates both growth and decay are correspondingly slow. In both environments fire is relatively rare. In temperate zones, however, where growth rates generally exceed the rates of biochemical decomposition, fire is essential to liberate nutrients otherwise locked up in a reservoir of biologically inert matter. For landscapes like North America, as Spurr and Barnes affirm, *"fire is the dominant fact of forest history."*[39] Under natural conditions the intensity and frequency of fire varies according to the work required of it: the greater the litter, the more intense the fire; the more frequently litter is built up, the more frequent the fire. Only under extreme cases or in certain fire regimes that have adapted to and encourage this pattern does fire destroy the standing forest or the soil. Heat rises, and soil is rarely burned off or fried into impermeability unless fire has been accompanied by other anthropogenic practices. The association of fire with rituals of purification has its basis in fire's capacity to decompose and transform in nature. On a microscale, fire separates chemical compounds; on a macroscale, it drives off organic pests and unwanted flora and fauna.

The problem may be rephrased by noting the difficulty of inventing a fire retardant. Different retardants work on different phases of the combustion process: some retard pyrolysis or oxidation, some retard the production of flame, others retard the glowing state. To reduce the flammability of an object it is necessary to coat or impregnate the substance with special chemicals, to alter its chemical constituency, or to replace the substance with a synthetic counterpart built of artificial polymers. Natural organic materials, such as cellulose, are designed to burn, though some are more fire-prone than others. To halt the process it is necessary to intervene by adding special chemical inhibitors or by reconstituting the material into an artificial compound that incorporates such inhibitors or whose structure better resists degradation. It is not easy to do either without producing effects often more undesirable than those to be prevented.

What happens in the biochemical degradation of organic matter by fire occurs on a much broader scale in the effects of fire on an ecosystem. As a fire breaks down available fuel, it releases heat and nutrients. The heat may

kill many organisms, consume others, and reshape a microclimate by allowing more sunlight, wind, and so forth. Many organisms adapt against this
wave of heat by developing thick bark, storing food in tuberous roots, or
resprouting soon after a fire passes. Others, like certain insects with infrared
sensors, seek out the heat. Some plants seem to encourage properties that
promote a hot fire, thereby driving off less tolerant competitors; a number
of trees, for example, have serotinous cones that open to release seeds only
after violent heating. In a similar manner, the breakdown of biochemical
compounds causes some organisms to die out or to depart from a burn, while
others seize on the simpler compounds for the promotion of their own
growth. Sequoia seeds, for example, seem to germinate best in a sunny, ashy
soil. Many grasses sprout phoenixlike into luxuriant growth from the ashes
of their dead, decadent old stems. Legumes occupy burned sites readily,
enjoying a temporary advantage because of the volatilization of organic
nitrogen by a hot fire. Many berries thrive in the open sunny environment
of a burn, judiciously pruned of unnecessary growth by periodic fire, and
seize nutrients that in the postburn state are more accessible to them. Ash
released into water affects its chemistry, and ash discharged into the air can
often retard aerial-borne parasites, such as mistletoe. This whole process,
moreover, is self-reinforcing: the type of growth that occurs on a burn helps
to determine the nature of the fuel complex, and the resulting fuel complex
determines the intensity and frequency of fire and its future biological
effects.

On a broad scale, the effects of fire encourage the development of vegetative mosaics and the recycling not merely of chemicals but also of communities. Under ideal conditions, a kind of perpetual migration of successional stages and shifting geographic ensembles results. Fire and nomadism
are as much associated in the natural environment as in the human one. The
mosaic effect, moreover, functions at least in part as a means of regulating
fire spread and intensity: fire cannot be extensive because the fuels are broken. In reality the elements of the fire environment interact so as to create
a more or less definite fire regime, exhibiting a typical fire pattern and a
typical fuel complex. In many environments fire, anthropogenic or natural,
is the controlling agent of ecological dynamics, exerting an inordinant influence on the composition of flora and fauna, on their historical arrangements,
and on their contemporary energetics. Fire in natural, as in cultural, systems
is as effective an agent by being withheld as by being applied. The fertilizing
quality of ash is the basis for slash-and-burn (swidden) agriculture, and the
improved palatability and nutriousness of fired pasturage has been exploited
for millennia by hunting and herding societies.

As an agency for natural selection, fire has shaped and regulated biotas
for hundreds of millions of years. As evidence there is fusain—fossil charcoal—found in all varieties of coal and in all geologic eras. Commonly about

3 percent of coal consists of fusain, which gives coal its sooty, "dirty" texture. Fusain appears in both thick lenses and fine laminae.[40] A fire, however, produces more than charcoal, and the fire-related minerals that accompany burning may, as geochemical catalysts, be an integral part of coal formation. Charcoal itself is virtually indestructible except by fire and, like fire, it can purify. In nature the charred remains of a fire can become relatively impervious except to further reburns. Early man exploited this property by hardening wooden spears and tools over fire, and well into modern times charcoal has remained an important source of fuel and filters.

The genera most commonly associated with fusain are evolutionary relatives to many of the fire-tolerant forms abundant today. Among grasses and ferns, these include lycopodiales, lepidodendrils, cycadofilicales; among conifers, taxacae and pinaceae. Most coal deposits evidently developed in environments not much different from the coastal wetlands typical of the contemporary southeastern United States, a landscape of marshes, swamps, and pines. Even under present conditions and with contemporary suppression capabilities, fires are common in these landscapes. Fire can spread through marsh grasses above the waterline. Debris accumulates rapidly into thick organic mats, and when exposed by draining or drought, these reservoirs of fuel can burn with intensity and stubbornness. Peat continues to be a common source of fuel for man. Fossil remains in coal suggest that grassy rather than woody vegetation is typical of coal-forming sites, and such an association would also favor the fine lamination characteristic of fusain. If fire were not present, Komarek has observed, "most such plant communities such as marshes, etc. would be invaded by shrubs and trees." Ancient fire, in short, helped to perpetuate an environment favorable to coal formation. The mechanics of that process are still in evidence. When the drought of 1954–1956 exposed vast lakes of organic debris in the South, lightning ignited fires in the coastal pocosins of North Carolina, in the Okefenokee Swamp of Georgia, and in the Florida Everglades.

Nevertheless, adaptation can be an ambivalent concept. Very few terrestrial organisms can be utterly unadapted to fire and survive, if only because mankind has transported fire to almost every biota on earth. Fire can be likened to a plague, virtually obliterating any organism without defense against it. No one, for example, would seriously claim that DDT is essential to many species of mosquitoes because some have adapted to it. The same may be said of anthropogenic fire. Yet, because of its evolutionary antiquity, fire is a biological event of far greater complexity and significance: its effects are vastly more encompassing than those of pesticide or plague, and by transforming organic matter it liberates a chemical residue for other organisms to exploit. A surprising number of organisms, known as pyrophytes, have evolved to do just that; they are specialists at exploiting the opportunities made possible by a relatively constant natural process. It is this dualistic quality of fire that makes fire management so essential, even obligatory. Most of the biotas of greatest value to mankind are those that exist in fire regimes maintained by frequent, applied fire.

Fire has penetrated into virtually every environment of the United States, except the rain forests of the coastal Northwest and the mangrove swamps of the Everglades. Fire has swept wetlands like the cypress swamps of the South and the white cedar swamps of the Northeast, the marsh grass of Louisiana bayous and the glacial muskegs of Minnesota, the cane breaks of Alabama and the peat bogs of the Lake States. Fire has burned in arid environments as diverse as the arctic tundra and the Sonoran and Chihuahuan deserts. Fires have entered into and have helped to perpetuate such exotic communities as the palmettoes of the Carolinas, the sequoias of the Sierra Nevada, and the fanpalm oases of the Mohave Desert. Fire sustained and expanded grasslands from the Great Basin to the Great Plains. Fire presence typifies endemic brushfields like those in Southern California; its introduction can stimulate brush encroachment at the expense of forest, as in the Cascades; its suppression can encourage brush expansion at the expense of grassland, as in the Southwest. For deciduous forests, and particularly for coniferous forests, fire modifies composition and determines the mosaic of types and their successional levels. Especially in drained, upland areas, fire application may be routine and promote pyrophytic vegetation ensembles. Not only in the United States, but around the world, as Spurr and Barnes observe, "the dominance of pine and oak forests of virtually all species and in virtually all regions is due predominantly to fire."[41] Oak and pine are the two timber genera most useful to mankind, and the fires that perpetuate such forests are usually anthropogenic.

The biological response to a fire can vary widely. It will depend, first, on the physical properties of the fire—its intensity, size, frequency, and time of occurrence—all of which influence the chemical potential for combustion and determine the nature of the chemicals liberated by combustion. It will depend, too, on the genetic potential stored within a biota, which may also be released by a fire, and on the mechanisms or relationships for exploiting a fire that may exist within the biota. Even in a relatively simple case—the determination of forest species that succeed a fire—the outcome will depend on a series of probabilistic events. Fire application can often promote fire regimes in which fire will be more likely, but fire exclusion can sometimes create regimes in which fire can be reintroduced only after extensive preparation.

Natural fire is a climatic phenomenon. As the natural climate has fluctuated, fuel complexes and fire regimes have altered and migrated, and wildland fire has expanded and contracted its range. Modern vegetation and fire history begin with the waning of the Pleistocene, with a series of upheavals that transformed the earlier mosaics of natural fire regimes and witnessed the introduction of anthropogenic fire from Asia. Continental ice sheets physically displaced forests southward, and the global climatic changes that encouraged the development of glaciers interacted with local conditions to produce a kaleidoscopic pattern of vegetative types. With glacial retreat, forests reestablished themselves. But the ebb and flow of climate and vegetation, superimposed over local variations and microclimates, continued well beyond the abatement of the Wisconsin glacial some 10,000 gears ago. In the tenth and eleventh centuries, for example, coastal Greenland was sufficiently shorn of ice to support Norse colonies, and English chroniclers reported large fires burning across the British Isles. It was a time of land-clearing in Europe, and many new towns earned names like Brentwood, Burntheath, Brantridge, and Brindley. A cooling trend followed, culminating in the Little Ice Age. Alpine glaciers expanded, Norse settlers abandoned their colonies in Greenland, and fires in Britain receded into history. Postglacial climatic chronologies recognize five periods in New England, seven in Great Britain, and eleven to twelve in northern Europe. But it was not climate alone that shaped local fire regimes: fire, both natural and anthropogenic, can have the same effects. Applied fire can create fuels equivalent to the introduction of a warm, dry climate, and the suppression of fire can have the effect of initiating a cool, moist climate.

The effects of fire are magnified in marginal areas where types advance and retreat like glaciers under the oscillations of climate. The effects are also greatly exaggerated by the use of fire in conjunction with some other anthropogenic activities. Fire alone has rarely destroyed a landscape; evolutionary adaptations have seen to that. But fire and hoof, fire and ax, fire and plow, fire and sword—all magnify the effects by altering the timing of fire, its intensity, the fuels on which it feeds, or the biological potential for exploiting the aftermath of a burn. The greatest fire damages have occurred on marginal lands under multiple practices—on infertile farmlands carved out of forests, plowed, burned, and abandoned; on arid, overgrazed grasslands on which fire was either overapplied or too rigorously suppressed; on mountainous brushlands maintained by fire as browse for domestic livestock; on bogs burned during droughts or after drainage; on commercially forested areas where climate and economics encouraged a monoculture forest easily preyed on by fire; and on marginal urban sites used for residential developments, often on the fringe of prominent fire regimes.

The larger effects of fire on earth are thus really the effects of anthropogenic fire. They depend not merely on the genetic and ecological potential for exploiting a fire that is inherent in the natural system but also on the potential within the culture—on its domesticated flora and fauna, on its hunting and gathering preferences, on its perceived meaning of fire, on its understanding of fire behavior and its comprehension of fire's effects, on its ability both to apply and to withhold the fire of its own or of nature's making. The great fires, and the most important historic fire regimes, were the result of anthropogenic fire practices or of anthropogenic modifications of the fire regime.

The fire cycles initiated in the geologic past continue, and they are closely tied to human activity. Reburns persist. Under controlled conditions fossil fuels remain the primary energy source for industrial fire, though it is energy, not the recycling and reconstitution of matter, that is desired. In nature the material residue of fire is as important as its thermal pulse. But for human needs, the more thorough the conversion to energy, the more satisfactory is the combustion; ash and smoke are considered pollutants, not nutrients. Nor are all reburns controlled. In several coal regions mine fires have resisted all efforts to extinguish them. In the 1870s striking miners in southern Ohio pushed blazing ore carts loaded with wood down the mine shafts. The resulting fires continue to burn. During the 1930s CCC enrollees attempted to halt some of the subterranean fires by sealing off air passages, but without success. Some of the fires cause surface collapses as veins are burned out, and on occasion the resulting heat is adequate to ignite surface vegetation. It is an annual responsibility of the fire officer of the Wayne National Forest (Ohio) to map the progress of mine fires and to note the likelihood of forest fires from their spread.[42]

In ancient reburns, as in modern prescribed burning, there is a distinction between wild and controlled fire, but it is a discrimination made on cultural grounds. When mankind assumed responsibility for fire management, that obligation extended apparently to the management of ancient fires and fuels as well as to modern ones. If his use of more recent fire from the Pleistocene helped to differentiate *homo erectus* from other hominoid primates, ancient fires from the geologic past have helped to differentiate modern mankind from its ancestors in cultural evolution. It is fossil fuel, with its residuum of ancient fire, that powers the prime movers of the industrial revolution.

II

On January 22, 1974, the astronauts of Skylab Three reported unusual atmospheric developments in a broad belt north of equatorial Africa. "I've noticed it now for two days running," one observed, "and this is a pervasive and extensive cirrus development which extends off the coast of Somalia and the southern end of the Arabian peninsula out into the Indian ocean for 1500 to 2000 miles. As far as you can see. The particular character of the cirrus clouds is that they are dirty-looking, and appear much darker than other clouds in the same area." Clouds over West Africa "had the same appearance." The source of the clouds was identified several days later: fires set by slash-and-burn agriculturalists, by herders to improve pasturage, and by hunters to flush out game and sustain preferred habitats. Like the pillars of fire and dust that guided the Hebrews, the fires showed as flames by night

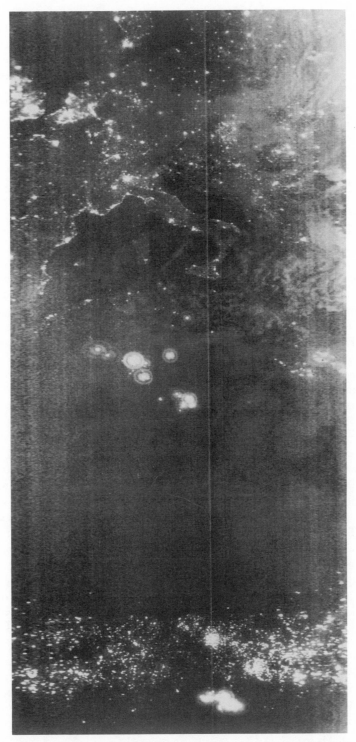

1. Evening satellite photo of Europe and the northern half of Africa, January 4, 1975. The pattern of lights across Africa is the product of agricultural burning for pasturage and shifting cultivation.

and as ash clouds by day—"Stone-Age images," one commentator called them, "flickering across a giant rounded map of Africa, like globes in movies that burst into flame to identify wars and native uprisings."[43]

In the late nineteenth century, with the introduction of forestry to the United States, the impact of lightning fire was slighted. The fire practices of frontier agriculture defined the problem for fire protection, and it was natural to emphasize the damages these caused for industrial forestry. The whole fire problem in America, Bernhard Fernow wrote in disgust, was one of bad habits and loose morals. By the late 1960s, however, the issue of wilderness fire had reversed that emphasis, and the prevalence and positive importance of lightning fire was accented. The problem of fire management seemed to be with fire suppression, an expression of bad ideas and loose money. Yet this contemporary reorientation should not obscure the very fundamental significance of mankind as a distributor of fire on a global scale and as a shaper of fire regimes on a grand scale. Combined with climatic oscillations and with intensive uses, such as the grazing of domestic stock, traditional fire practices could induce broad, perhaps irreversible changes in earth biota—which is exactly what Skylab reported in 1974 as the Sahara began a disastrous southerly encroachment. The prevalence of mankind owes much to its possession of fire, but it is equally true that the prevalence of fire owes much to its dependence on mankind.

Mankind is the primary source of fire in the world and is continually improving its capacity both to remove and to introduce fire in greater quantities than every before. Its proliferation of fire devices may in fact be taken as a measure of a culture's technological sophistication. The opportunities for accidental fire constantly increase. But mankind is also the primary vector for the purposeful distribution of fire. Anthropogenic fire represents not only a quantitative change from natural fire but also a qualitative one when it is introduced into areas rarely subjected to natural fire and timed in ways that may not coincide with natural fire cycles. As late as 1957 it was estimated that at least 25 percent of the earth is subjected to swidden agriculture, with perhaps one-seventh of that amount burned annually. Most of the grasslands of the nonindustrial nations are burned annually or biannually, usually in the spring and fall.[44] Other lands are burned for hunting and gathering activities. And much, of course, burns from carelessness or accident. In the United States by 1971, as much land annually came under controlled burns on industrial forests as was burned from all causes by wildfire. No terrestrial environment has failed to feel the effect, in some degree, of anthropogenic fire.

Mankind is the greatest modifier of the fire environment, especially its fuels. With or without fire, by design or by accident, a fire regime is shaped by human activities. The study of a "pure" fire regime, so dear to American ecologists, is a fantasy. No regime exists that has not to some extent been modified by human activity and has not to some degree adapted to anthropogenic fire practices. Some regimes rise and fall on changes in those practices. The history of regimes depends not merely on human migration or on the diffusion of exotics, such as domesticated livestock or cereals, but also on the internal evolution of civilizations. Fire is everywhere used with nomadic harvesting where land is extensive and population slight. Where conditions favor intensive management of natural resources, fire tends to be confined. Free-burning fire is useful in recycling nutrients, but where high yields are desired, nutrients must be added, not merely recycled.

No one has argued more eloquently than E. V. Komarek for the extent to which mankind is a grassland animal and a fire creature. "Our 'bread,'" he observes, "comes from cereals which are grasses, and present studies indicate that they were developed from fire-adapted grasses. The 'meat' in our diet comes largely from animals that eat grass, forbs, or shrubs and cannot in any manner be considered forest animals. Nearly all, if not all, of the major cereal food plants and our major domestic livestock apparently came from fire environments."[45] Even preferred forest products come from those biotas, like pine and oak forests, that are particularly adapted to fire. Yet, with the shift to more intensive cultivation, open fire has tended to give way to more confined forms of fire. Open burning is retained on landscapes given to more extensive management or to the cultivation of natural products like wildlife. Thus the pine torch was replaced by pots of burning pitch and pitch by distillates of similar chemistry. The use of smoke to stimulate the flowering of pineapples, for example, is replaced by the chemically active agent in the smoke, ethylene. Vitamin supplements and chemical fertilizers can fortify diets in ways not dependent on the recycling capability of fire. With sedentary agriculture and with concepts of fixed ownership of land, the requisite nomadism on which fire depends may vanish. Rather than burn field stubble, a farmer may mix the organic residue with nutrients to make cattle fodder. The money made from sale of the livestock can pay for the fertilizer, with a profit left. Cattle may be raised in feedlots rather than on the range, making broadcast burning of range grasslands less useful. Rather than harvest natural timber, special trees may be planted, and rather than burn slash or litter, forest debris may be chipped and harvested for pulp or wood products. The chemicals that fire would return to the natural system must, in short, be cycled as well through the pathways of a cultural system, and to power this activity the combustion chambers of industrial machinery may be more important than the free-burning fires of the field.

The human uses of the landscape often dictate competing sets of fire practices. In California, for example, stockmen wanted to burn off the brush to encourage the growth of grass, while foresters wanted to protect the brush from fire in order to encourage the eventual growth of timber. Earlier, the grasses that Spanish herders wanted to maintain as winter fodder were burned over by Indians interested in collecting seeds and driving rabbits. Agricultural draining of lowlands and swamps along the Atlantic seaboard opened up new fields for fire that had not existed to the Indians, while at the same time reclamation broke up the grassy "deserts" that Indian fire had maintained as wildlife pastures and allowed such sites often to revert to forests. The industrial revolution has, in its turn, greatly enlarged the amount of wild, extensively managed land at the expense of much marginal agricultural land. The fire history of any region is the history of such changes, as cultural mosaics and successive peoples replace the kaleidoscopic pattern of natural fire regimes with another pattern and substitute one set of fire practices for an earlier set. That fire is a natural process which may be used to further cultural goals is an important realization, but the value of that discovery for the fire history of a region is limited. It is known that plagues are a natural means for controlling populations, but mankind is reluctant to introduce prescribed plagues as a means of regulating its own growth.

It is not the peculiar relationship of fire and life that is, in the final analysis, so important; rather, it is the special relationship between fire and man. Mankind may properly be considered as among those species of pyrophytes that tend to promote fire as a means of ensuring their survival. The relationship is one of interdependence, with the existence of each one encouraging the expansion of the other beyond the natural bounds of space and time. Mankind continues to exploit the fires of the geologic past, and he will apparently expand the range of fire beyond its terrestrial home. The picture of Skylab perched over the smoke plumes of equatorial Africa dramatizes in one concise image the extent to which fire and man have been linked throughout their histories. Though the nature of fire and the purposes of its use will change, that ancient interdependence is destined to enlarge the realm of each. It was fire that brought mankind to the moon, but it was man who carried fire beyond its terrestrial limits.

THE BURNED-OVER DISTRICTS:
A FIRE HISTORY OF THE NORTHEAST

The aborigines of New England customarily fired the forests that
they might pursue their hunting with advantage. . . . The grounds
which were covered with oak, chestnut, etc., or with pitch pines
were selected for this purpose because they alone were in ordinary
years sufficiently dry. Such to a great extent were the lands in
New England, and they were probably burned for more than one
thousand years.
—Timothy Dwight, *Travels in New England and New York,*
1821[46]

New-England of the West shall be burnt over . . . as in some
parts of New-England it was done 80 years ago.
—Lyman Beecher, 1828[47]

I

Most of the fire regimes typical of American history are found in the
chronicles and fire practices of the Northeast. In its coal fields is the residue
of ancient fires, charcoal older than North America itself. The fire practices
of Indians, of agricultural reclamation, and of industrial counterreclamation
are well documented. Not only does the region contain the types of fire of
national significance, but many of these fires—and the responses to them—
have had important consequences for the evolution of a national fire policy
and of a national fire protection system. It is a region no less rich in fire
metaphor. The fire sermon became high art and popular excess. So often
were fire and brimstone called down upon the land that after the Second
Great Awakening Charles Finney declared western New York to be a
"Burned-Over District." Whether or not faith followed fire, fire certainly
followed the faithful. The burned-over district of New York, like areas of
New England many decades earlier, was burned over in fact as well as in
spirit.

The historic fire season of the Northeast arrives in the spring with the
passage of dry cold fronts and returns in the fall when stagnating high pressure leads to the famous Indian summer.[48] These annual cycles are aggravated by episodic phenomena that affect either the amount of available fuel
or the fuel moisture. Droughts, for example, preceded all the major regional
conflagrations, and when such climatic episodes are joined to massive accu-

mulations of fuel from disease or insect kill, from windstorms, from past fires, or from landclearing, the resulting fires may be invincible. The only practical method for disposal of fuel under quasi-natural conditions is fire. And it was just such realizations—backed by repeated historic demonstrations—that led to the enormous cleanup of debris left by the 1938 New England hurricane. Financed by federal monies and manned by CCC and WPA laborers, the Northeast Timber Salvage Administration undertook perhaps the most outstanding national enterprise in emergency hazard reduction. The feared conflagration never materialized, but residual debris that escaped immediate cleanup did contribute to the infamous 1947 fires in Maine, and the potential for fire was so alarming that many states in the region poured money into their fire protection organizations.[49]

Lightning fire is infrequent in the Northeast, ranging from 1 percent of all fires in Pennsylvania to more than 12 percent in Maine. The season for growing and decomposition is scant, however, and thus fires are typically episodic but of high intensity. Historically, the region has not had to rely on lightning fire. Since the Pleistocene, it had been occupied by Indians who not only shaped the vegetative mosaic by fire but also kept fuel loads in many areas down to a manageable level. In his biography of Champlain, Francis Parkman gave perhaps the classic account of the pre-Colombian forests: "The primeval woods," he called them, "ancient wilds, to whose ever verdant antiquity the pyramids are young and Nivenah a mushroom of yesterday."[50] So dense was the original forest, it was claimed, a squirrel might travel from the Atlantic to the Mississippi from tree limb to tree limb without ever touching the ground. Cleared of this nineteenth-century romanticism, the original accounts tell a different story. So open were the woods, one author advised with a touch of hyperbole, it was possible to drive a stagecoach from the eastern seaboard to St. Louis without benefit of a cleared road. The virgin forest seemed to many explorers not much different from the parks and champion fields they had known in Old England.[51] It was not the forests of the New World that startled them and strained their vocabulary, but the grasslands. The virgin forest was not encountered in the sixteenth and seventeenth centuries; it was invented in the late eighteenth and early nineteenth centuries. For this condition Indian fire practices were largely responsible.

The tribes of the Northeast cleared the woods for fuel and fields. So successful was their reduction of the forest that they inadvertently assisted the rapid deployment of European settlements. Nearly every colony occupied sites already cleared by Indians, and as immigration plunged deeper into the interior, European colonists continued to settle on sites often already open and clear or stocked with forests that had appeared between the time the Indians vacated a site and the time the Europeans occupied it. Some clearings represented the sites of villages. An entire site might occupy 100 to 150 acres, and the site might change every 10 or 20 years. Onondaga, capitol of the Five Nations, moved at least nine times between 1610 and 1780. Soil exhaustation, weed infestation, gradual reduction of game, steady increase of vermin, and, more than once, military objectives necessitated such moves.

Much of the clearing came from the Indians' prodigious demand for firewood. European commentators on the Iroquois and Delaware, for example, were shocked at the amount used. "So important was firewood in Indian economy," Gordon Day notes, "that the Narragansetts in Rhode Island thought the English had come to America because they lacked firewood at home."[52] The suggestion is not all that outrageous. England, after all, began to experience a timber famine in Tudor times: forests had been converted to heathlands, farms, and pastures; shipbuilding timber, notably oak and pine, was in critical supply (and a reason for interest in the American colonies); and firewood shortages forced the British Isles to turn to fossil fuels as a substitute. Settlers in the Massachusetts Bay Colony had to collect firewood from islands in the harbor because there was so little on the coastal clearings. Since the Indians practiced broadcast burning on areas around their villages, their wood supply must have come from forest clearings, and the observations on forest recession around villages would indicate that the wood came from standing timber. The supply would not have been that strenuous to provide: Indians were skilled with fire and stone axes, and experiments in Denmark using Neolithic axes have demonstrated that three men could clear about 200 square meters of forest in four hours.[53] Girdling, moreover, might have produced dead wood that later fell of its own accord.

Much of the incentive for clearing came with the expansion of maize agriculture, which required a slash-and-burn regime. The resulting ash was rich in nutrients for maize and created a soil chemistry that was alkaline, as favored by domestic grasses, rather than acidic, as favored by forests. When the extraction of nutrients by harvesting exhausted the fertility of the enriched soil, new sites were selected in a grand ensemble of shifting agriculture. Not all Northeast tribes—for example, the Algonkian—adopted maize agriculture in whole or in part, but by the advent of the Europeans much of the region was settled agriculturally. By 1535 the Iroquois had extended maize culture down the St. Lawrence Valley to the site of Quebec. Often European settlers merely occupied the sites of the old fields, and the earliest stages of colonial agriculture imitated the shifting, slash-and-burn agriculture of Indian predecessors.

Sometimes the fields came back to forests. In 1697 the Algonkian pointed out to de la Chesnaye the forest reproduction on former Iroquois villages. But often the fields were maintained, as they had been prior to maize agriculture, by systematic firing. Fire was widespread among hunting and gathering societies. It was used to encourage berries, to harvest natural grains and nuts, and to shape a habitat rich in game. Fire hunting was common in the fall, and the fires worked to sustain the herbaceous landscape frequented by elk, deer, buffalo, and turkey. The woods around inhabited areas like villages were periodically fired to eliminate underbrush and cover and to thin out the forest, thereby reducing the opportunity for ambush by marauding enemies. The price of these defense measures was a reduction in branchwood suitable for fuel.

There was little in the forest to attract Indians, and they succeeded wherever possible in replacing the forest with a mosaic of sites more to their liking. Where forests existed, they tended to be open, free of underbrush. Plymouth Colony, for example, found the woods "thin of Timber in many places, like our Parkes in England." The woods of Massachusetts were in many places so open that from a high place "cattle could be seen for a distance of three miles, and deer and turkeys a mile away." Near Narragansett Bay Verrazano observed "open plains twenty-five or thirty leagues in extent, entirely free from trees or other hinderances," and forests that "might all be traversed by an army ever so numerous." A resident of Salem, Massachusetts, exclaimed in 1630 that he was "told that about three miles from us a man may stand on a little hilly place and see divers thousands of acres of ground as good as need be, and not a tree in the same." More than once, military expeditions into the interior slowed to a crawl while extensive cornfields were fired. In 1656 Van der Donck reported large grasslands in New Netherland and remarked that "much more meadow ground" would be present were it not for the rapid reforestation that followed the removal of the natives.[54]

"The Savages are accustomed," explained Thomas Morton in his *New English Canaan* (1637),

> to set fire of the Country in all places where they come; and to burne it, twize a year, vixe at the Spring, and the fall of the leafe. The reason that mooves them to doe so, is because it would other wise be a coppice wood, and the people would not be able in any wise to passe through the Country out of a beaten path.... The burning of the grasse destroyes the underwoods, and so scorcheth the elder trees, that is shrinkes them, and hinders their growth very much: So that hee that will looke to finde large trees, and good tymber, must not depend upon the help, of a wooden prospect to finde them on the upland ground; but must seek for them, (as I and others have done) in the lower grounds where the grounds are wett when the Country is fired....

Morton noted that such practices forced the immigrants into a pattern of protective burning around their settlements "to prevent the Dammage that might happen by neglect thereof, if the fire should come neers those howses in our absence." Though he concluded that "the Salvages by this Custome of theirs, have spoiled all the rest [of the countryside]: for this Custome hath been continued from the beginninge," he also noted that "this custome of firing the Country is the meanes to make it passable, and by that meanes the trees growe here, and there as in our parkes: and makes the Country very beautifull, and commodious."[55]

In the higher mountains and the river bottoms and swamps, too moist for fire to penetrate routinely, forests grew relatively undisturbed, except during droughts. But in naturally drained territory, fire was applied widely, both by deliberation and by accident. Peter Kalm, for example, reported that in one place around Lake Champlain "the natives themselves are very careful" about the potential for escaped fires and that in another place "one of the chief reasons" for the decrease in conifer forests was "the numerous fires which happen every year in the woods, through the carelessness of the *Indians,* who frequently make great fires when they are hunting, which spread over the fir woods when every thing is dry."[56] Fire tended to be handled more carelessly by war parties in transit and by hunters far from their villages or on lands inhabited by an enemy; near to their own territories, a tribe used fire more circumspectly. Nonetheless, tribes could sustain the grasslands against reclamation only by means of fire: where fire or tribes were removed, forests sprang up. Where the period between removal and reoccupation was lengthy, "thick" forests appeared, which demanded laborious clearing.

In many cases the combination of clearing and fires stripped off the forests altogether. "Barrens," "clearings," and "deserts" were among the most common sights reported by early explorers. Undoubtedly, Indians maintained these deliberately as hunting grounds. Many of the clearings probably represented abandoned agricultural fields subsequently sustained as grasslands by annual broadcast burning. Whatever their origin, they were common at the time of discovery and were among the chief victims of settlement. Visiting some of the barrens remaining around 1800 in western New York, Timothy Dwight described how "the tract" around a wanderer "is seemingly bounded everywhere [yet] the boundary is everywhere obscure: being formed by trees, thinly dispersed, and retired beyond each other at such distances as that while in many places they actually limit the view, they appear rather to border dim, indistinct openings into other tracts of country. Thus he always feels the limit to be uncertain. . . ."[57]

With regard to the origin of the "peculiar appearance of these grounds," Dwight concluded that the "Indians annually, and sometimes oftener, burned such parts of the North American forests as they found sufficiently dry." The "object of these conflagrations was to produce fresh and sweet pasture for the purpose of alluring the deer to the spots on which they had been kindled." Although the Indian "destroyed both the forest and the soil, he converted them to the most profitable uses for himself. . . . Thus, in time, these plains were disforested to the degree in which we now see them, and were gradually converted into pasture grounds. It ought to be observed that they were in all probability burnt over for ages after they were disforested."[58]

To support his argument Dwight noted, first, that it was known that "the Indians customarily burned, every year, such parts of the forests as were sufficiently dry to admit of conflagration" and, second, that "wherever they have been for a considerable length of time free from fires, the young trees are now springing up in great numbers, and will soon change these open grounds into forests if left to the course of nature." He proceeded to show at some length that such conditions were common throughout New England at the time of settlement. His conclusion noted that the whole of southern New England, "except the mountains and swamps, was almost wholly covered with oak and pine forests"; that the proofs of early fires could still be discovered; and that "within my own rememberance there were in the township of Northampton spots desolated in a similar manner. These, although laid waste in an inferior degree, were yet so far destroyed as to be left in a great measure naked. Now they are completely covered with a thick forest."[59]

Early agriculturalists, especially those recently immigrated from Europe or affected by intellectual theories of fertility, often shunned the barrens initially in favor of sites with a heavier mantle of litter, which they subsequently burned, or in preference to areas featuring familiar hardwoods like the maple, which they had to fell and fire. Intellectuals educated in European agronomy, which tended to equate fertility with the depth of litter, condemned all such uses of fire. But farmers on the barrens soon showed that their cereals were generally superior, that is, richer in yield and more resistant to blight. Lacking an adequate term to describe such lands, however, settlers continued to refer to them as "barrens" or "deserts." When the Great Plains were encountered by the Long Expedition in 1819, its chronicler continued this bafflement by describing the grasslands as the "Great American Desert," distinguishing it from the numerous lesser varieties east of the Mississippi.

Not all of the Northeast was converted to "desert" or savannah. The upper mountains, the river bottoms, the swampy lowlands, and the denser boreal forests were more or less spared annual firings. But there is no reason to think that the process of converting forest to plain was not expanding. Lacking domestic livestock, Indians depended on wildlife for meat, and these anthropogenic fire plains were their pastures. Ironically, many of the forests that occupied the great pine and oak belt of southern New England and across the Appalachians were a byproduct of European settlement. In the short run, the pioneers adopted many of the Indian fire practices—fire hunting, slash-and-burn agriculture, broadcast fire for pasturage of wild and domestic stock, protective burning against other fires. But in the long run, suppression of Indian fire practices made possible the accidental and deliberate reforestation of the Northeast. Not only was there frequently no virgin forest to clear, but the forest that was cleared was often itself a product of the act of settlement. Though the larger cause for this transformation was the replacement of a hunting and gathering society by one dedicated to agricultural reclamation, the immediate cause was an exchange of fire practices.

II

In part, European reclamation only expanded the process of agricultural reclamation begun by the aboriginal tribes of the Northeast. More forested land was cleared and new villages were created; domestic grasses and managed pastures replaced the harvesting of natural foodstuffs and wildlife. One movement simply displaced and enlarged another. In greater part, however, the program of reclamation represented a dramatic change in land use. Areas where Indian agriculture and Indian fire had not penetrated, such as mountainous terrain and swamps, were exposed; exotic flora and fauna were introduced, from the relatively benign wheat to the more damaging livestock and the devastating gypsy moth and chestnut blight. With new markets for commodities and with economic ties across the Atlantic, new demands were made on forest products and on products dependent on reclamation of forested land to arable land. And new fire practices were developed. Not only was fire applied to lands previously inured against routine fire and for purposes, like the export of potash fertilizer, that previously had not been in evidence, but new sources of ignition were also introduced—for example, the friction match and the steam-driven locomotive—which greatly expanded the opportunities and range for accidental fire. The fuel complex and fire regimes of the Northeast were drastically revised—in some places expanding, in other contracting.

The fire practices of the initial settlers who moved into the interior were not far different from those of the Indian tribes they displaced. The reclamation of Europe had used fire in landclearing; field burning was common; and the maintenance of pasture and heathland by fire was well known, though in many portions of Europe it was stringently regulated. The forms of fire used in America for hunting, herding, farming, and logging, however, more closely resembled the practices of the New World natives than those of Old World ancestors, and they had a similar effect of encouraging a shifting or nomadic pattern for each use. In place of the semimigratory patterns of Indians, the colonists simply added an expanding frontier; rather than a recycled mosaic of land uses, a kind of historical or evolutionary mosaic was generated along a moving westward frontier.

Frontiersmen quickly adopted fire hunting, and it merged well into the practice of regularly burning forest pasturage for domestic stock. Dwight, for example, lamented that the early English inhabitants of the "barrens" in New England "frequently burnt" them and "foolishly followed this Indian custom in order to provide feed for their cattle in the spring."[60] To graze livestock in the common woods was an ancient European right, and it was one continued in the Americas. Whereas in northern Europe fire for pasturage was generally condemned, in America it was welcomed.

Among the earliest fire codes adopted in many states were regulations banning fire hunting, so efficient was this method for the extermination of valuable game. The practice had migrated west with the frontier. Thaddeus Harris, on a tour across the Allegheny Mountains in 1803, remarked "with regret and indignation, the wanton destruction of these noble forests. For more than fifty miles, to the west and north, the mountains were burning. This is done by the hunters, who set fire to dry leaves and decayed fallen timber in the vallies, in order to thin the undergrowth, that they may traverse the woods with more ease in pursuit of game."[61] Harris condemned the practice as destructive to wildlife and woods both and thus anticipated the century-long controversy over fire protection between those who used fire on the frontier and those who observed these practices from afar.

Early settlers practiced swidden agriculture. Dwight again provides the classic account of the succession of stages in frontier farming:

> A considerable part of all those who *begin* the cultivation of the wilderness may be dominated by *foresters* or *pioneers*. The business of these persons is no other than to cut down trees, build log houses, lay open the forested grounds to cultivation, and prepare the way for those who come after them. They accordingly cut down some trees and girdle others; they furnish themselves with an ill-built log house and a worse barn, and reduce a part of the forest into fields, half enclosed and half cultivated. The forests furnish browse, and their fields yield a stinted herbage. On this scanty provision they feed a few cattle, and with these and pernurious products of their labor, eked out by hunting and fishing, they keep their families alive.[62]

The process of landclearing itself was accomplished by either girdling or felling the trees:

> The latter has become almost the universal practice; and, wherever it can be adopted, it is undoubtedly to be preferred. The trees are cut down either in the autumn or as early as it can be done in the spring, that they may be so dry as to be easily burnt up in the ensuing summer. After they have lain a sufficient length of time he [the forester] sets fire to them, lying as they fell. If he is successful, the greater part of them are consumed in the conflagration. The remainder he cuts with

his ax into pieces of a convenient length, rolls them into piles, and sets fire to them again. In this manner they are all consumed; and the soil is left light, dry, and covered with ashes. These, so far as he can, he collects and conveys to a manufactory of potashes if there be any in the neighborhood; if not he leaves them to enrich the soil. In many instances the ashes thus gathered will defray the expense of clearing the land.[63]

Then the site would be broken by the plow and once broken, fired annually for pasturage.

In his "Short Description of Pennsylvania" (1692), Richard Frame put the process to poetry.

Then with the Ax, with Might and Strength,
The trees so thick and strong
Yet on each side such strokes at length,
We laid them all along.

So when the Trees, that grew so high
Were fallen to the ground,
Which we with Fire, most furiously
to ashes did Confound.[64]

It was subsistence agriculture at best during this stage, but the practice "had some important advantages," as Dwight conceded. The land "was eminently productive. Seldom were their crops innured by the blast or the mildew, and seldom were they devoured by insects. When the wheat was taken from the ground, a rich covering of grass was regularly spread over the surface, and furnished them with an ample supple of pasture and hay for their cattle." The forester, however, was "always ready to sell" to the less migratory yeoman and then to move on.[65] In this manner, and often by the hands of the same people at each new site, the early agricultural frontier advanced westward. Like the economy of its Indian predecessors, the colonial economy depended heavily on fire.

But it was a fire economy not always well managed. In his history of New Hampshire, Jeremy Belknap described how

in the spring, the trees which have been felled the preceding year, are burned in the new plantations. If the season be dry, the flames spread in the woods, and a large extent of the forest is sometimes on fire at once. Fences and buildings are often destroyed by these raging conflagrations. The only effectual way to prevent the spreading of such fire, is to kindle another at a distance, and to drive the flame along through the bushes, or dry grass, to meet the greater fire, that all the fuel may be consumed. In swamps, a fire has been known to penetrate several feet under the ground, and consume the roots of the trees.[66]

From such experiences the word *backfire* entered into the English language. Where settlers followed loggers, where the plow succeeded the ax, the uses of fire for landclearing could be especially tempting and particularly hazardous. The combination of fire with plow and ax was chiefly responsible for the great holocausts of both the Northeast and its protégé, the Lake States.

Settlers used fire, as did the Indians, to encourage the production of fruits and berries, especially blueberries, and escape fires from blueberry patches remained a prominent cause of conflagrations along the northern coastal plains well into the nineteenth century. Smoking fires to drive off bees and to flush out small game were common. Smudge fires set by hunters and fishermen, both professional and sportsmen, to drive off mosquitoes and blackflies also repeated Indian practices and led not infrequently to major fire complexes.

Fire was also employed as a weapon against predators, especially against wolves threatening livestock—a prophylactic measure promoted by eons of European folklore. It has been speculated that many of the barren hilltops in Maine have resulted from fire, though the soil profile was probably marginal at best, not being well developed after extensive Pleistocene glaciations. In the late 1740s, it is said, some villagers sought to destroy a troublesome wolf population by surrounding their mountain sanctuary with fire. It was a dry year, the fires burned well, and the mountain's soil as well as its wolves disappeared. The practice continued as late as 1800. To the eyes of one traveler it presented a "melancholy" scene with "remains of half-burnt trees which hung there on the sides of the immense steeps" and which "finished the picture of barrenness and death."[67]

Logging, too, like colonial agriculture and colonial urbanization, tended to become a migratory enterprise, not merely because of the exhaustion of resources but also because of the threat and fact of fire. Wood was cut in large quantities for shipbuilding, for furniture, for barrels, for urban construction, and for export. (*Lumber* was originally a slang term used to describe the masses of wood cluttering or "lumbering" up Boston wharves.) The demand for fuelwood was heavy. "The wood of this country," wrote one observer, "is its fuel. An Englishman, who sees the various fires of his own country sustained by peat and coal only, cannot easily form a conception of the quantity of wood, or, if you please, of forest, which is necessary for this purpose." Yet "all these forests renew themselves," and the forests of the New World settlements were thus young.[68] Much of the wood was converted into charcoal, a process that introduced not only vast strata of slash but also new sources of ignition. When fuelwood and charcoal gave way to coal as a fuel, the resulting mines consumed structural timber on a prodigious scale.

By the late nineteenth century, when such practices had combined in their worst excesses, the infamous "Pennsylvania desert" had been formed in the Alleghenies.[69] Even here, though, the benefits of fire were not ignored. Following the practice of the frontier forester, the ash from slash fires was often collected, stored in barrels, and exported as fertilizer.

Several causes contributed to the migratory nature of logging: tax laws discouraged long-term investments in forest regeneration, cheap uncut stumpage was available to the west, and forest insurance was nonexistent. Moreover, many believed that the plow would follow the ax, that the abandoned land would be converted into farms. Logging, it was felt, was in a sense a public service. Particularly in mountainous areas, where agricultural reclamation was repelled or where distance from markets made it uneconomical, logging companies commonly culled a forest several times, on each occasion searching for a different species. The initial cutting might be for white pine; later ones, for white birch (for pulp) or hemlock (for tanning). In each instance large quantities of slash were left in a thick veneer, and sooner or later fire would consume them. Such areas often burned several times, after each wave of logging. In some sites, as in Maine after the 1825 fires, the region might reseed to other tree types of commercial value (notably white birch) or convert to blueberry production (commercial canning in Maine began in 1866).[70] An area once logged and even farmed and then abandoned tended to reseed to white pine, and the logging industry around 1900 was able to exploit this regrowth. At least some of the pine logged in the nineteenth century was the result of earlier settlement, which had disrupted Indian firing practices. But after the later assault by fire hardwoods returned rather than pine, giving the region the particular growth typical of its twentieth-century appearance.

Fire, more than anything else perhaps, determined the migratory character of early logging. Fires in 1761–1762 temporarily destroyed the logging industry in southern Maine and led directly to settlement of northern coastal lands. Maine surrendered its timber supremacy between 1840 and 1860 to New York, and New York gave place in turn to Pennsylvania between 1860 and the 1870s. By the 1880s the Lake States replaced the Northeast as a national timber region.[71] As the source of timber migrated, so did fire. Industrial forestry—forestry based on sustained high yields from a fixed land base—was impossible until numerous conditions were satisfied, the foremost of which was fire protection. The enemy of industrial forestry was frontier fire practices, and, more often than not, the logging industry supported conservation measures to provide organized fire protection.

Early logging, however, was in a sense a component of the reclamation. With the eradication of Indian tribes by war, treaty, or disease, fire and ax and fire and plow exploded into a virtual vacuum, without many of the constraints under which they had formerly operated. Add to traditional fire uses those that had been adopted from the Indians, and it is not difficult to understand why the rapid expansion westward was accompanied by chronic fire problems and occasional holocausts. The great fires—wildland, rural, and urban—have occurred in areas experiencing sudden transitions. Wild-land fires have typically broken out in areas where forests were being logged intensively for the first time, where landclearing for reclamation was under-way on a broad scale, or where early industrial innovations (such as rail-roads) introduced new ignition sources. Not until the transition was com-plete did the new fire regime conclude, usually under a vastly different fuel complex.

The first premonitions of the large conflagrations to come were the famous Dark Days recorded in New England chronicles, when smoke palls blackened the skies. One occurred in 1716 and another, perhaps the most famous, on Black Friday, May 19, 1780. To the pious it often seemed that the Final Days had come. Black Friday was "a remarkably dark day," one witness recalled. "Candles were lighted in many houses; the birds were silent and disappeared; and the fowls retired to roost." The legislature of Con-necticut proposed to adjourn, but when the opinion of Colonel Abraham Davenport was asked, he answered, "I am against adjournment. The Day of Judgement is either approaching, or it is not. If it is not, there is no cause for an adjournment; if it is, I choose to be found doing my duty. I wish therefore that candles may be brought."[72] Those at the site of a conflagra-tion, however, found it less of an occasion for rhetorical bravura. To a sur-vivor of the Miramichi holocaust of 1825, "All it required to complete a picture of the General Judgement was the blast of a Trumpet, the Voice of the Archangel, and the Resurrection of the dead."[73]

The 1825 fires are the first historic holocaust of the reclamation. In the autumn of that year some 2 million acres burned along the Miramichi River in New Brunswick, and another 800,000 acres burned in Piscataquis County, Maine. Chroniclers indifferently blended the fires into a single com-plex, the Miramichi fire. The circumstances preceding the fire were typical: logging and landclearing had penetrated the region, leaving large amounts of slash; a drought had prolonged Indian summer conditions into October; and fires were abundant. Most were ignited by settlers eager to clear logged sites of slash. Such fall burning was commonplace. A witness to the events noted that until October 7, the day of the blowup, "no serious apprehension was entertained on account of the prevalence of conflagrations, which are so commonly put into action to commence clearances in the forest." ("Confla-gration" here meant, as it had from the Renaissance, simply a free-burning fire.) Hundreds of fires were burning on thousands of acres of man-disturbed fuels when conditions favored the transition to large, even mass fire. On October 7, as accounts repeated over and over, "the whole forest world appeared to be in flames." Survivors recalled terrifying roars and growls as the flames approached. A "dreadful hurricane" of wind and embers flung the fire forward. At least half a dozen villages vanished. Dozens of boats burned along the river fronts. Many inhabitants, possibly several hundred, lost their lives; an accurate count was never attempted. The smoke was so thick that ferrymen on the Penobscot River had to use compasses. It was said that more game perished in the fires than had been killed since the coming of the white man. When word of the calamity finally filtered down to Portland, relief efforts were promptly organized. Subscriptions were taken for the sufferers in both Maine and New Brunswick, and the response—as it would be for all the prominent fires in American history—was quick and generous. Settlement continued, the forests reseeded to species that acquired economic value in later decades, and a scenario was presented that would typify the large fires of the reclamation for a century.[74]

To the early settlers' sources of fire were added others, both tragic and comic. The Maine fire of 1837, which burned 150,000 acres, was started ironically enough by a state land inspector. Determined to halt timber depredations in the area of Sebois, the agent fired haystacks belonging to the poachers. The fire spread to nearby fields and forests, and the agent barely escaped with his life. The fires of the industrial revolution spread, too, into the landscape. Locomotives, notorious fire starters, were intimately associated with logging and landclearing and hence with disturbed fuel complexes. The use of wood fuel guaranteed that the train would belch an endless eruption of firebrands from its smokestack. The substitution of coal for wood and, under the pressure of early regulations and lawsuits, the maintenance of right-of-ways and the use of spark arresters brought some improvements. Nevertheless, a single faulty locomotive could ignite fires over miles of landscape. The territory on both sides of a track were proverbially burned-out wastelands. Even today locomotives remain a major source of fire in the Northeast and in the Lake States.

In 1880 Franklin Hough tabulated the major causes of fire. Leading the list, of course, was landclearing, followed by firing for pasturage and for hunting, and fire from transients and from locomotives. Fire was set by miners seeking to expose rock outcrops, by charcoal burners, by berry producers, by squatters who burned the brush for frontier economies and who "often select the driest and windiest weather to do so," and by men "with evil intent." Company lands were fired by arsonists and by men seeking revenge for real or imagined grievances. Fires were set to camouflage or destroy the scenes of illicit activities. Fire codes existed in most areas, but prosecution was rare. No one apparently wished to control escape or malicious fires so zealously as to prevent the many beneficial uses of fire required for the agricultural reclamation of the landscape. Most fires were controlled burns, and most of the burned area was not perceived as damaged, except by persons like Hough who wished to restore the forests. In 1880, for example, some 250,000 acres of Pennsylvania burned—an extraordinary amount for a state that today has a very low percentage of burning. Most of that fired land was retained as pasturage or as open woods, and though some of the fires might escape and become damaging, the common sentiment supported its use. As New England in particular suffered depopulation during the course of the nineteenth century, more and more of its landscape reverted to pasturage or to semiwild woodland.[75]

Prudence, of course, dictated that fire practices be regulated in accordance with common practices. Fire codes date back to 1631 in Massachusetts, and biblical scholars can find the principle of restitution for damages by fire in the Mosaic Law. Early codes sought primarily to restrict the use of landclearing and field fires to certain seasons, and to limit, wherever possible, the extensive practice of fire hunting. By the time of the American Revolution, only the southern states lacked fire control legislation. But fire control was a practical as well as a legal matter, and people familiar with techniques for using fire were also versed in techniques for controlling it; one was impossible without the other. In most areas custom required assistance on fires that threatened the community. In a few areas this practice acquired statutory force. In New York, for example, a statute of 1743 empowered anyone in certain counties to "require and command all or any of the neighboring and adjacent inhabitants to aid and assist him" in the suppression of a fire; anyone refusing could be fined. The law was extended to the entire colony in 1758. In 1760 New York inaugurated a fire warden system, and New Jersey followed suit in 1792.[76]

Whether dignified with legal language or enforced by community customs, some mechanism for fire control was imperative. Typically, though, a fire brigade would be activated only when a fire posed an imminent threat to the community and when, as often as not, it was already powerless against the large fires. Village celebrations in the 1760s, a chronicler records, were postponed because the menfolk were all fighting fires. James Fenimore Cooper concluded his 1823 novel of upstate New York, *The Pioneers,* with a savage forest fire roaring across the hillside. Leatherstocking calmed the distraught heroine by observing that "the men are use to fighting fires." When in the 1840s Thoreau let a campfire escape in the Concord woods, his neighbors rushed to extinguish it. In the fires of 1880, New Hampshire put some 200 men on a 600-acre fire, and New Jersey had the not unfamiliar experience for that region of having more volunteer firefighters on hand than it had organization and strategy. The relief efforts following occasional holocausts were likewise voluntary and spontaneous. In the case of the Miramichi fire contributions exceeded needs to such an extent that the surplus was used to construct a schoolhouse in the heart of the burn.[77]

Far from viewing fire as a dreaded foe, the agents of the agricultural reclamation saw it more often as an ally. Great outbursts like Miramichi were temporary phenomena, part and parcel of the often violent process by which America was overrun by a great folk migration, settled with breathtaking speed, and transformed from underdeveloped to industrial nation. Like other uncertainties and hazards of settlement, the holocausts would vanish when the land was at last reclaimed, when a landscape of wildland fuels suited for the hunting and gathering of wild products was reconstituted into a mosaic of domestic flora and fauna and into a more subdued fuel complex.

> Coming down the mountain in the twilight—
> April it was and quiet in the air—
> I saw an old man and his little daughter
> Burning the meadows where the hayfields were.
>
> Forkfuls of flame he scattered in the meadows.
> Sparkles of fire in the quiet air,
> Burned in their circles and the silver flowers
> Danced like candles where the hayfields were,—
>
> Danced as she did in enchanted circles,
> Curtseyed and danced along the quiet air:
> Silently she danced in the stillness, in the twilight,
> Dancing in the meadows where the hayfields were.[78]

III

By the latter part of the nineteenth century, agricultural reclamation was giving way to industrial counterreclamation. Its effects were generally first felt in the Northeast. The burned-over district, so important for American religious history, became no less significant for conservation history. The vision of sinners suspended precariously over hellfire was replaced by the vision of a landscape blasted by waste and fire, backsliding irrevocably into the nadir regions. If the counterreclamation had many causes—the abandonment of marginal farm land from the mid-nineteenth century on; national concern with conservation and with prospects of timber famine and of water shortages; alarm over the closing frontier; new industrial uses for wood products; new fuels to replace firewood, among others—it had one great consequence: the removal of land from the practices of reclamation. In particular, a veritable fever for forests swept the country. Nurseries were established and millions of acres reforested. In New York state parks alone, some 57 million trees were deliberately planted between 1902 and 1928.[79] The process continued through the 1930s under the New Deal impress of legislation that created such agencies as the CCC. The Tree Farm program after World War II extended industrial forestry and reforestation even to the farm woodlot. The establishment of forests had been as much a product

of agricultural settlement as landclearing, but such reforestation had come
inadvertently as a product of the suppression of Indian fire practices. The
counterreclamation proceeded deliberately, and it created yet another
mosaic of fuel complexes and demanded yet another set of fire practices.

New York led the way. In 1885 the state established a forest preserve in
the Adirondacks and followed with another in the Catskills. The primary
intention was to regulate the annual fluctuations of the Hudson River, but
the land was not to be further disturbed by logging or landclearing and
would be made available for recreational and esthetic pursuits. Above all,
the land would be protected from fire. The entire Northeast had suffered
from disastrous fires during 1880, and it was felt that removing settlers
would help to remove fire. Almost immediately, however, the state realized
that it was creating a fire protection problem, and perhaps a fire regime,
without precedent. By excluding settlement, it excluded the traditional
methods by which agricultural lands had regulated fire uses. A system of
"fire-ranging," fire patrols, and, eventually, fire wardens was inaugurated to
fill the vacuum. The result was an important model for the rest of the
nation.[80]

The first serious challenge to the new system came in April 1903, when
fires burned a large crescent from upstate New York to Maine. Total
acreage destroyed probably exceeded 1,000,000 acres, with some 600,000
acres burned in the Adirondacks. Ignition came primarily from railroads
and the smudge fires of fishermen, and the fires tended to cluster around
tracks and streams. Fallow field burning and outright incendiarism added
to the total, and even at this time there were men indicted for setting fires
in order to have themselves hired as firefighters. Control of the fires involved
logistical as well as tactical problems: laborers were hired or imported by
the hundreds from urban centers (400 Italians from New York City
manned one fire); railroads transported urban fire equipment for the protection of villages and developments; locomotives pulled special cars outfitted with pumps and hoses. Some $175,000 was spent attacking the fire—
which nonetheless was controlled only by heavy rains. The young U.S.
Bureau of Forestry, still without lands of its own to manage, sent a special
agent to investigate the fires and the success of the suppression forces
mounted against them. In 1908 the scenario was repeated, with much of the
same area reburning.[81]

The 1903 and 1908 fires did for the Northeast what the 1902 and 1910
fires did for the Northwest: they galvanized systematic fire protection, particularly on the state and regional levels. Already, as the 1903 fires demonstrated, impressive strides had been made in the financing, logistical support, and suppression technology of fire control on the newly designated
lands of the counterreclamation. If new industrial devices like the railroad
contributed to the ignition of the fires, they also assisted measurably in

suppression, allowing for the rapid mobilization of labor pools and heavy equipment. Locomotives even served as pumpers to contain fires near railroad right-of-ways. The widespread fires of 1870 and 1880 had brought little public outcry; the fires of 1903 and 1908 coincided with national concern over a probable timber famine, alarm about damaged watersheds, and excitement over the whole panoply of environmental issues, which crested when the Governors' Conference on Conservation convened at the White House in 1908. The disastrous Pennsylvania flood of 1907 gave added urgency to regional demands for adequate forest protection and fire control.

New York responded by establishing a model fire protection program for its reserved lands. Maine created another precedent in 1909 by organizing its famous "fire control district" over the unincorporated townships that composed most of its backcountry. As early as 1905 Maine and its logging industry had cooperated in the establishment of a lookout system complete with telephones, but now legal authority was brought to bear.[82] Vermont set up a similar lookout system in 1910, and state organizations in Maine, Vermont, Pennsylvania, Massachusetts, New Jersey, and Rhode Island were strengthened. Private timber companies formed protection associations in New Hampshire, Pennsylvania, and West Virginia.

In 1911 the federal government moved into the region with the passage of the Weeks Act. The act allowed, first, for the acquisition of land at the head of navigable streams as national forests and, second, for cooperative agreements and a system of matching funds between the U.S. Forest Service and state foresters to provide fire protection on similar lands, whether or not they belonged in federal hands. Most of the northeastern states promptly signed up; after all, the author of the act, John Weeks, was from Massachusetts. By 1920, all regional state foresters had joined. In 1924 the Clarke-McNary Act strengthened and expanded the program to include all watersheds, not just those sustaining navigable waterways. During the 1930s the CCC in the Northeast, as throughout the nation, tremendously boosted the muscle behind fire control and shaped a physical plant for fire control, in many areas almost overnight. Large fires seemed a dim shadow from the Dark Days of the past.

Then the shock came. The problem of fire control along the boundary of wild and urban lands, so peculiar to post–World War II America, was first dramatized in October 1947 as fires roared across some 220,000 acres in southern Maine. In four days' time more than 200 structures were destroyed, including a cancer research institute at Bar Harbor, and 16 lives were lost, though none to the fire directly. The fires began in drought conditions prolonged into late October under a stagnating high pressure cell; 108 days passed without rain. Fuel loads were frequently heavy: slash piles from the 1938 hurricane remained, debris from a heavy 1945 snowstorm added more tonnage locally, and accelerated logging during World War II and the postwar recovery period had left a heavy veneer of untreated fuels. By October 17, well past the time when fire control operations were normally shut down, protection organizations were being manned at full strength, and the governor proclaimed general woods closure to any fire. By October 20, however, 50 fires were reported, though all were small and easily controlled. The next day, October 21, strong winds broke several fires out of containment. The winds continued for four days, with those on October 23 reaching near hurricane velocity. More than 150 separate fires occurred during the week that followed, with about 9 making the transition to large conflagrations. The magnitude of the disaster simply overwhelmed local administrative capabilities. At Bar Harbor fires burned across seven and a half miles in 25 minutes. With the mainland causeway blocked by fire, "a second Dunkerque took place with fleeing residents rescued by Coast Guard patrol boats, the Navy and many private boats." The governor declared a state of emergency, President Truman concurred, and federal assistance was provided. In some places martial law went into effect, and the state found it necessary to set up an organization reminiscent of wartime civil defense.[83]

Under the circumstances, some large fires were probably inevitable. More than 40,000 acres burned elsewhere in the New England states during the occasion of the Maine debacle. But what became blatantly apparent was the difference between lands protected by the Maine state forestry department in what had become the Maine Forestry District and the lands protected by volunteer departments operated by organized towns. Of the nearly 250,000 acres burned, 200,000 were in areas under the jurisdiction of the towns. In the early days of the conflagrations the town departments simply failed to mop up and patrol the fires. When the winds whipped the smoldering ashes into mass fires, the towns—long bastions of local autonomy—could not coordinate efforts among themselves or with the state, nor could they assimilate the throngs of men and machinery that turned out to battle the fire. Not until the state forestry department assumed overall responsibility were control forces properly marshaled. From that point on, their efforts met with considerable success, though the fires were not contained until the winds abated, and they were not extinguished until the rains came on November 9.[84]

The consequences of the 1947 fires rippled out widely from coastal Maine. In 1949 the Maine legislature reorganized the state fire protection system. That same year the New England states and New York signed an interstate compact—the first in the nation—to provide for coordinated training and for material assistance in preventing and suppressing fires. In 1970 the compact became an international alliance with the admission of Quebec and New Brunswick.[85] Other regions copied the compact approach, particularly in the East, where the federal government was not a major landowner. In

1976 the northeastern states expanded their cooperative agreements by financing an equipment development center at Roscommon, Michigan.[86] The 1947 experience, moreover, became an object lesson for civil defense— not only for the coordination of services that a fire war might require but also for civilian disasters of all sorts. The lessons were most painfully transferred to Southern California beginning in the 1950s. Nor has the transfer been solely one of ideas. When large fires in California exhausted suppression resources throughout the West in the summer of 1977, dozens of organized crews developed for the national forests of the Northeast—many of them composed of volunteer firemen—were shipped to the firelines of the Ventana wilderness.

Perhaps a closer analogy to the Southern California fire regime can be found in southern New Jersey. Except for the coastal plains, the fire history of the Northeast most closely resembles that of the Lake States and the Northwest. But the plains themselves, a sandy wedge of white cedar and pitch pine from Cape Cod to New Jersey, can only be compared as a fire regime with the brushfields of Southern California. Smoke in the region was repoted by Henry Hudson, the result of "a great Fire" south of Sandy Hook. Throughout the seventeenth and eighteenth centuries smoke was common, especially during the spring and fall, when Indians customarily fired the plains to maintain the pine barrens as a hunting ground. In 1755 Ben Franklin reported 30 square miles of land on fire in the region. A New Jersey correspondent informed Hough in 1880 that "the whole country is overrun about every twenty years by fire." Large fires are recorded for 1820, 1829, 1832–1833, 1856–1859, 1865–1866, 1870–1872, 1875, 1880–1885, 1900, 1902, and 1908–1909. To the north along the coastal plain, wildfire on Martha's Vineyard in 1916 obliterated the last nesting grounds of the heath hen, and the extermination of the prized game bird soon followed. During the terrible drought of 1929–1930 much of the plains burned, swamp and pine barrens both. In 1936 a fire outside Chatsworth, New Jersey, burned over 58,000 acres (90 square miles) and took the lives of three CCC enrollees sent to control it. Twenty thousand acres of the same area reburned in 1954, and in 1957 fire swept some 64,000 acres and 50 structures in Plymouth City, Massachusetts.[87] Poor, sandy soil and high fire hazards repelled agricultural reclamation along the coast, leaving southern New Jersey, and in particular its famous Pine Barrens, as a sparsely inhabited enclave well into the twentieth century. Even as late as 1970, 46 percent of New Jersey remained forested, largely in the southern end of the state. "Whatever else they do," John McPhee observed in 1967, "men in the Pine Barrens are firefighters throughout their lives."[88]

It was during the prolonged drought of 1961–1965 that the fire potential of the coastal plains drew national attention. Three fires on Long Island— one in 1962 and two in 1963—burned a total of 7,000 acres and consumed

200 structures. In the spring of 1963 conditions were severe. Massachusetts reported 4,861 fires in one month, though none became major. Fires moved out of the woods and into suburbs around New York and Philadelphia and in a wide region encompassing Pennsylvania, Maryland, West Virginia, Virginia, and Kentucky. Conditions were the worst since 1930.

But it was in New Jersey that the real firestorm broke loose. On May 20, 1963, a series of fires began an 11-day rampage over more than 200,000 acres. Some 458 structures were destroyed, seven lives lost, and more than a thousand people left homeless. (By contrast, the largest single fire in California history, the result of two lightning fires that merged and then were supplemented by tens of miles of subsequent backfires, was 180,000 acres.) Most of the damage came on the twentieth, known ever after as Black Saturday. Fires spread, as researchers reported later, "rapidly across upland sites where there was relatively little fuel, as on areas where prescribed burning had been done one or two years earlier. On such sites a very light covering of pine needles was sufficient to maintain a fire. Oak leaves, where present, were blown across bare spots so that fires advanced rapidly even in scattered fuels." Near the Lebanon State Forest a fire advanced three and a half miles in two hours and nine miles in six hours. "High winds forced the abandonment of suppression attempts, even though the area had only a year's litter since the last prescribed burn."[89]

Many of the early state forestry operations began in the name of fire control, and except for New York and Maine, where forest reservations or forest districts were well established, this meant protection of the coastal plains as well. A report on the Pine Barrens at the invitation of the state of New Jersey was the first investigative assignment that Gifford Pinchot and Henry Graves conducted for the Bureau of Forestry, and their study impressed upon them powerfully the identification of fire with deforestation. In 1928 Cape Cod launched one of the earliest coordinated fire prevention programs and developed the "brush-buster" truck for controlling fires in the scrub oak and pine thickets crowding into residential and recreational areas. New Jersey, in particular, turned early to prescribed burning. Controlled fire was used in 1928 to create and maintain firebreaks, as it had previously been used to reduce fuel loads along railroad right-of-ways. The idea had been approved by Hough in his 1882 report, and by 1936 broadcast burning was employed for the treatment of areas and not merely lines. Twelve years later prescribed fire was approved for silvicultural treatment as well as for fire control. In recent decades the preferred cycle of firing, as it was in presettlement days, is on the order of one to three years.[90]

Another sort of prescribed fire has come into the region, too. In July 1977 lightning started a fire deep in Baxter State Park, a wilderness area in central Maine. The fire burned 3,500 acres before finally expiring along control lines. The probability of a fire had been forecast some years earlier, when a blowdown created natural slash heaps in the park. In accordance with a 1955 code allowing it to remove fire hazards, the state forestry department had contracted with logging companies for removal of the blowdown debris. A court injunction in 1976 halted the cleanup, however: the methods were deemed inappropriate, violating the purity of land designated for management as wilderness. Behind that rationale was another: fire was seen as appropriate in the natural world and essential to wildlands. The motives came to the Northeast from elsewhere, but the Baxter Park fire was the first regional dramatization of the innovation and an important contribution to the growing experience with the problem of greatest significance for national fire management, namely, wilderness fire.[91]

The original burned-over districts have faded into history—as a spiritual phenomenon, replaced by other regions and other enthusiasms; as natural features, remade into new vegetative mosaics, with their fires more a matter of historic record than of imminent resurrection. Yet the fire scenes on the coastal plains of New Jersey and in the interior of Maine testify to the unconquerability of fire. It may be transformed, but not eradicated. In its controlled states it can be productive; in its wild condition it is always unpredictable. The New England experience, moreover, was not forgotten. The settlers who spilled out of the Northeast carried fire practices as they did religious revivalism, until both became enduring features of the American frontier. The burned-over district was to be recreated many times over in the process of westward expansion.

2 THE FIRE FROM ASIA

Man has possessed fire so long that the inquiry as to whether it is a human characteristic has some point. —Walter Hough[1]

Man is himself a flame . . . —Loren Eiseley[2]

Early man did not invent fire, he discovered it. And in much the same way, fire did not invent culture, but it did discover and liberate many uniquely human potentials. Whether or not Hough was correct to assert that fire is a human trait, one can certainly agree with E. V. Komarek that man is a fire creature—not only because he uses fire as a technology but also because he tends to occupy environments for which fire is as basic as sunlight and water.[3] Fire is among the oldest of words, the most ancient of tools, and the most prominent of the means by which humanity projected itself onto the landscape. When European explorers encircled the globe, they discovered only three peoples who could not manufacture fire—the pygmies of the Congo, the tribes of Tasmania, and the Andaman Islanders—but even these peoples preserved and used fire with ease. In describing the ability of man to modify his environment instead of merely to conform to it, Jacob Bronowski observed that the leap of the gazelle never took it out of the savannah. Early man's use of fire enabled him not only to leave the savannah when he chose but also to recreate the savannah as he wished during his migrations.

Fire is a multiple phenomenon, and man soon adapted it for specialized domestic, agricultural, pastoral, and industrial uses. Domestic fire—the torch and the hearth—changed eating habits, defined the need for and the design of shelters, and revised social structures in accordance with the need to supply fuelwood and to preserve fire. A family is defined as those who share a fireside; the familial gods are the gods of the hearth; the last thing a president does before calling out the troops, one critic has observed, is to give a fireside chat. Agricultural fire allowed for the harvesting of cereal grains, berries, and nuts; with ash, man had a ready source for fertilizer, which allowed agriculture to penetrate into jungles, temperate forests, and northern coniferous forests (taiga). With fire, early man had a means of driving game, of baiting traps, and of creating a habitat favorable to those species he found most useful; fire hunting is among the almost universal purposes of broadcast burning. Pastoral fire probably helped the process by which wildlife was domesticated; wild herds could be moved from site to site as areas were burned and as palatable new growth appeared. The seasonal herding of livestock between two pastures (transhumanance) is almost universally accompanied by firing for the improvement of pasture. Industrial fire—the fires of the furnace—founded metallurgy, ceramics, and chemistry; the industrial revolution that so transformed the modern world had for its prime movers machines powered by fire. The list of the ways in which fire has influenced culture could go on indefinitely.

The discovery of fire involved more than simply the discovery of controlled heat and light. Wildland fire, as opposed to mankind's more domesticated variants, allowed early man to begin shaping the larger physical environment to fit his needs rather than merely to adapt by strict natural selection to the existing landscape. Broadcast fire made man almost instantaneously capable of wholesale modification of his environment. As a practical and intellectual problem, the discovery of fire by early man can only be compared in modern times with the discovery of the atom. With implements of stone and bone, man could improve his chances for survival; with fire, he had power.

Early man seems to have preferred an environment of grassland or savannah. Almost alone among the primates, man made the transition from forest to grassland, and once having "made the adjustment from jungle forest to grassland in Africa or Asia," E. V. Komarek observes, man "needed no further acclimation, for the grasslands gave him access, literally, to the entire world."[4] The bones of animals associated with early human sites belong to grassland browsers or grazers. Wild grains were a grassland product, and most of the livestock and cereals whose domestication made agriculture possible developed from grassland species. Many other foodstuffs available for annual gathering or domestication, such as acorns, berries, and beans, appeared along the burned edges of forests and grasslands. So did game birds, some of which, like the turkey, could be domesticated.

The importance of this relationship of early man to a grassland environment is that grasslands are fire regimes. Through the manipulation of fire man could adapt these regimes to his needs (for example, by transforming a natural pattern of summer burning into a spring-fall cycle); he could readily accommodate himself to and disperse himself across such regimes; and he could expand greatly the natural distribution of grasslands as a vegetative cover type, even creating savannahs to some extent against climatic gradients. It is difficult to fabricate swamps, forests, and tundras, but with only a torch one can create and maintain successful grasslands, savannahs, or forests so open as to feature many of the flora and fauna of true prairies.

As a hunter and gatherer, man used fire to hunt and harvest natural products, to ward off predators, and to maintain the habitat against the successional pressures that would convert the land into forest. Hearth fires rendered edible many plants and meats otherwise too tough or poisonous for consumption. As a herder, man used fire to create, sustain, and improve pasture; as a farmer, he used fire in swidden agriculture from the Amazon basin to the taiga of Finland; as a metallurgist, he resorted to fire mining in hardrock veins and to smelter fires for ore refinement. As *homo sapiens,* man found in fire a provocative source of myth, philosophy, religion, and science. Pyromancy was a primitive form of divination, and the fire sermon has persisted from the days of Zoroaster and the Buddha to modern Christian evangelicals and even to T. S. Eliot's *The Wasteland.* Ordeal by fire is a primitive form of law. The office of "fire keeper" was perhaps among the earliest in human society, and by ancient times it had evolved into a position synonymous with the state. When Athenians and Romans incorporated new tribes, for example, they sealed the treaties by mingling their sacred fires. Nearly all of the landscapes that primitive societies found most productive of food, fuel, and wildlife existed because of periodic fire. Without anthropogenic fire, such environments were commonly uninhabitable. Even industrial man remains dependent on fire, though in a different cycle and in more complex technological forms.

So simple and powerful are the consequences of wildland fire that it is hardly surprising that the migration patterns of early man followed grasslands. They presented glacier-free corridors during the Pleistocene. Moreover, they supported the game and foodstuffs most accessible to him and in an environment that he could easily manipulate to his own purposes. Man traveled to those areas where fire was possible and shunned those where it was difficult. It was such a migration across the Bering Strait, moreover, that brought a new source of fire to North America. Speculation about remote origins is always hazardous, but, however diverse his later cultural developments, early man came to this continent with a common fire heritage. For instance, although its object might vary, fire hunting was practiced ubiquitously throughout North America. Similarly, whether the crops consisted of wild or domesticated grains, berries or nuts, the techniques for harvesting by fire were remarkably uniform. Pleistocene extinctions were largely of grassland species, probably as a result of overhunting by a vigorous grassland predator, man, armed perhaps with a new weapon, fire. Today conversion of grasslands to forests and farms has meant that the most endangered fauna continue to be grassland species.

Loren Eiseley likened mankind to a flame because man, like fire, not only consumes but also transforms. There is a reality behind the metaphor: fire has been the primary tool in mankind's transformation of his environment and, hence, perhaps of himself. For millennia, a symbiosis has existed between man and fire. In Defoe's fable of Western man in the wilderness, Robinson Crusoe equates smoke with the presence of fellow men. The association was entirely natural: throughout history and prehistory, fire has been mankind's most powerful ally, and environments where fire could not be readily used were its most formidable enemy. Where man was, there was fire; but equally, where there were fire environments, there came man. Mankind belongs among the pyrophytes, species whose affinity for fire goes beyond passive adaption and whose selection of traits includes those that make fire more likely. Such species use fire to eliminate competitors as others use shade or toxic chemicals. But man the pyrophyte is unique in that he controls the source and timing of ignition. He can project fire rather than merely endure it. Most pyrophytes are phoenixlike species, dependent on their ability to recover quickly from self-immolation. It is more typical of man's relationship to fire that he invented the myth of the phoenix from his observations of fire.

It is little wonder, then, that in most mythologies mankind becomes something more than a primate brute only with the acquisition of fire and that fire is something that was bestowed by superior beings or had to be stolen from imperious potentates. Fire is power. It remains as a widespread symbol of culture: to pass the torch is to hand on civilization to a new generation. Fire imagery is among the most common of metaphorical associations. Fire is life. It is the fire of the sun that keeps alive the organic world. To have the inner fire go out is to cease to live. It was not uncommon among primitive societies to attempt to revive the recently dead with a fire, and even in modern cultures to be honored with an eternal flame is to be memorialized as having made a fundamental contribution to the life of civilization.

So enduring is the cultural legacy of fire and so compelling is the testimony of man's early use of fire in wildland environments that the effort to extricate fire from the landscape must be viewed as something of an aberration. Carl Sauer once remarked that wildland fire was a general feature of primitive societies; fire suppression, only of civilized ones.[5] But the use of fire has always been predicated on the ability to control it, and all societies have had to take measures to protect themselves from unwanted fire. The demand for organized wildland fire protection, however, is a very recent phenomenon, expressive not merely of "civilization" but also of industrial civilization. The apparent effort to exclude fire from the landscape is in reality only part of a general accommodation to the industrial revolution, the exchange of one set of fire practices for another. The immigration from Asia of man and fire superimposed a new and extensive fire regime over the existing natural one. Later immigrations exchanged, transformed, and superimposed other regimes. Industrialization worked an even more fundamental change, one without precedent. North American fire history is actually a palimpsest of fire practices and fire regimes, constantly in flux in accordance both with natural climatic change and with cultural changes due to migration or evolution. The thrust of Asian fire was, on the whole, to replace forests with grasslands; the impact of European fire, to replace grasslands with farms and forests.

In his short story "To Build a Fire," Jack London recreated in fictional form a Pleistocene landscape of ice, cold, and isolation, which became a fable of the survival of man as a newcomer to a natural world ruled under the harsh law of survival of the fittest. The success of this new creature, man, depended on his capacity to make a simple fire. The tenderfoot failed and perished. The semidomesticated dog that had accompanied the man and knew him only as a "fire-maker" wandered off. The Asian immigrants knew better how to handle fire, and by 9000 B.C. fire and man extended from the Bering Strait of Alaska to Tierra del Fuego, the Land of Fire, at Cape Horn.

OUR GRANDFATHER FIRE:
FIRE AND THE AMERICAN INDIAN

All burning was limited to certain Indians who were looked up to as leaders or who understood how it should be handled.
—Chief, Mewuk tribe, California[6]

Fire is, in all the Indian tribes that I have known, an emblem of happiness or of good fortune. It is kindled before all their deliberations.—Father Pierre DeSmet, 1857[7]

I

It is often assumed that the American Indian was incapable of greatly modifying his environment and that he would not have been much interested in doing so if he did have the capabilities. In fact, he possessed both the tool and the will to use it. That tool was fire. But even more than a wonderful instrument, without which most Indian economies would have collapsed, fire was a presence. To the tribes of the eastern United States it was known as Our Grandfather Fire.[8]

It is hard now to recapture the degree to which Indian economies were dependent on fire. In its domesticated forms, fire was used for cooking, light, and heat. It made possible ceramics and metallurgy. Its smoke was used for communication. It felled trees and shaped canoes. It was applied to the cultivation and harvest of natural grasses, such as the sunflower, of berries, such as the blueberry, and of nuts, such as the acorn and mesquite bean. Broadcast burning along the coastal gulf plains and in the Alaskan interior drove off mosquitoes and flies; the mosquito, it is often said, was responsible for the destruction of more Alaskan forest than any other cause. Fire was sometimes used to kill off broad expanses of forest, which might then be harvested for firewood. Broadcast fire could be used more circumspectly to produce delicacies like the carmelized confection that resulted from burning sugar pine cones on the ground. Of course, fire could also be handled carelessly: parties of warriors and hunters rarely extinguished campfires and signal fires, since they burned on someone else's land. It was common practice, for example, to set fire to a downed tree for a campfire and then to leave it, where it might continue to burn for days.[9]

Fire was used ceremonially as a part of actual or ritual cleansing or as sheer spectacle. Lewis and Clark reported an evening's entertainment by tribes in the Rockies that consisted of torching off fir trees, which then exploded in the night like Roman candles.[10] The Apaches were said to burn off miles of mountain landscape in the "delusion that the conflagration would bring rain." Josiah Gregg observed similar practices on the Great Plains, and these contributed in part to the popular theory by James Espy that smoke could induce precipitation. (Espy went so far as to recommend weekly burning of 40 acres of timber every 20 miles for several hundred miles from north to south during the summer to assist annual precipitation patterns.)[11] Fire was a practical purifier, too. Oregon Indians used smoke to harvest pandora moths, which infested pine forests; the moths would drop from the trees to the ground where the Indians could gather them up as food.[12] California Indians used smoke to drive off the mistletoe that invaded mesquite and oak.[13]

Broadcast fire sometimes served as a means of economic extortion, and it was commonly employed for military objectives, both as a tactical weapon and as a strategic scorched-earth policy. The Crees and Assiniboines attempted to drive off early Hudson's Bay Company posts on the plains by burning, but the traders took protective measures, and the tribes resorted to a more complicated scheme. By firing wide stretches of prairie in the fall, they hoped to drive off the buffalo from post hunters and then reap a handsome profit as middlemen when the company was forced to approach them for pemmican. But the fires ranged too far, the buffalo were driven deep to the south, some villages were overrun, and the Indian speculators found themselves indebted to the company for supplies to tide them over during the winter.[14] In the 1880s tribes in the Dakotas attempted a similar strategy against the incursion of cattle ranches, eventually burning out strips for hundreds of miles along the routes of cattle drives and forcing ranch hands into roles as firefighters.[15]

The Long Expedition of 1819–1820 reported an incident in which the Sioux forced some hapless Omahas out from heavy grass with fire. In Canada a "ragged Slavey hunter" turned the tide of battle against the Sioux and Crees by firing an area of heavy downfall forest and driving his "terror-stricken" enemies to destruction.[16] Such tactics were common on the plains. To camp in deep grass was to invite raiders to surround the site with fire in order to collect scalps or horses in the ensuing chaos. When Escalante's expedition of 1776 entered the Utah Valley, they found burned and burning plains. Knowing that the intruders were mounted and thinking them to be Comanches, the natives had fired the range so that, in Escalante's words, "lack of grass might force us to leave the plain more quickly."[17] Traveling through Texas in the 1850s, the young officer Albert Myer found it expedient to borrow a page from the Comanches who shadowed his patrol. When sentries reported skulking figures around the camp, the order went out to "Fire the prairie!" The fires proved so effective that they were routinely repeated at night whenever smoke signals from native patrols had been

2. Charles Russell, *Crow Burning the Blackfoot Range,* 1905.

sighted during the day.[18] Colonel Richard Dodge described how "setting fire to the grass in the vicinity of the camp at night was one of the Indian modes of annoying a party too strong for attack and too vigilant for a successful attempt at theft." He had "been followed for several days in succession by a party of Indians, who fired the grass to windward of my camp every night, forcing me to burn all around the camp every evening before posting sentinels. . . ."[19] Even more widespread was the use of broadcast fire to clear the surrounding woods of underbrush in order to prevent the unseen approach of a hostile force. The desire to open up fields of fire around villages, trails, and hunting sites was an almost universal reason given by Indians for broadcast fire.[20]

Fire was essential to those Indian agricultures not dependent on irrigation. In fact, the copious harvests returned under a hunting and gathering regime sustained by fire virtually precluded formal agriculture in many regions. H. J. Spinden observed that "the abundant harvest of wild acorns in California, of wokas in southern Oregon, of wappato along the Columbia, of camas and kous in the pleasant uplands of Idaho, and of wild rice in the lake regions of Minnesota and southern Canada were effectual barriers against the intervention or spread of agriculture among the tribes inhabiting these regions."[21] Where yield was light, the natural fertility could be maintained almost indefinitely by setting fires to recycle the unused debris. Where yield was high, as in maize culture, additional fertilizer had to be added or new lands cleared and burned. In either case, the absence of chemical fertilizer or manure from domestic livestock often made fire the only practical mechanism for replacing nutrient losses in the soil. What might appear to be a random firing and gathering of products, moreover, often represented the semidomestication of both plants and wildlife. The cycle of fire varied according to the crop and the intensity of yield: cereal grasses were fired annually; basket grasses and nuts, every 3 years; brush, perhaps every 7 to 10 years; large timber, for formal swidden agriculture, on a cycle of 15 to 30 years or even longer. Broadcast fire also served to rid the fields of vermin and disease.

Of all Indian uses for fire, the most widespread was probably the most ancient: fire for hunting. At night torches spotlighted deer and drew fish close enough to canoes to be speared. Smoke flushed bees from their hives, raccoons out of their dens, and bears out of their caves. Recognizing that new grass sprouting on a freshly burned site would attract grazers by its superior palatability, Indians placed snare traps on small burned plots—in effect, baiting the trap with fired grass. The Apaches used smoke to lure deer driven mad by flies and mosquitoes.[22] Fire hunting, properly speaking—that is, the strategy of surrounding or driving the principal grazers of a region by fire—was universal. In the East it was used for deer; in the Everglades, for alligators; on the prairies, for buffalo; along the tules of the Colorado River, for rabbits and wood rats; in Utah and the Cordillera, for deer and antelope; in the Great Basin, for grasshoppers; in California and the Southwest, for rabbits; in Alaska, for muskrats and moose. Captain John Smith reported the practice around Jamestown, where "at their hunting in the deserts" the Indians ("commonly two or three hundred together") used fire to drive deer within circles or off peninsulas where they could be easily slaughtered from canoes.[23] Toward the end of the seventeenth century John Lawson described the process in the Carolinas, where the Indians "commonly go out in great Number, and oftentimes a great many Days Journey from home, beginning at the coming in of the Winter," and again "they go and fire the Woods for many Miles, and drive the Deer and other Game into small Necks of Land and Isthmus's, where they kill and destroy what they please."[24] Cabeza de Vaca described similar practices in Texas in the early sixteenth century. "Those from further inland . . . go about with a firebrand, setting fire to the plains and timber so as to drive off the mosquitoes, and also to get lizards and similar things which they eat, to come out of the soil. In the same manner they kill deer, encircling them with fires, and they do it also to deprive the animals of pasture, compelling them to go for food where the Indians want."[25] Hennepin reported the Miamis burning the tall grass prairies of the Midwest for the hunting of buffalo; LaSalle and his company saw the practice used widely along the Gulf plains and Mississippi Valley. Indeed, as Carl Sauer reported after examining seventeenth-century literature, fires were set everywhere except in those areas that were too wet, and therefore inflammable, or too arid, and therefore too sparse of vegetation.[26] Lewis and Clark observed for the upper Missouri River country that the Indians set fires to improve pasturage for their horses and the buffalo, and they reported a clever variant of fire hunting, useful in the spring when the grass was dry and the ice still ran in rivers. "Every spring the plains are set on fire and the buffalo are tempted to cross the river in search of the fresh grass which immediately succeeds the burning." In the process they are often isolated on ice floes, float down the river, and are dispatched with ease by Indian hunters awaiting at convenient locales.[27] In the early nineteenth century Thomas Jefferson answered an inquiry from John Adams as to whether

> the usage of hunting in circles has every been known among any of our tribes of Indians? It has been practiced by them all; and is to this day, by those still remote from the settlements of whites. But their numbers not enabling them, like Genghis Khan's seven hundred thousand, to form themselves into circles of an hundred miles diameter, they make their circle by firing the leaves fallen on the ground, which gradually forcing animals to the center, they there slaughter them with arrows, darts, and other missiles. This is called fire hunting, and has been practiced within this State within my time, by the white inhabitants.

Jefferson shrewdly suggested that this practice was "the most probable cause of the origin and extension of the vast prairies in the western country."[28]

Such prairies were extensive and by no means limited to the western country. Apart from the general opening up of wooded areas that resulted from broadcast fire and clearings, special "deserts" or "barrens" (as early Europeans referred to them) were maintained for the harvesting of wildlife. The Indian had little use for closed forests: their main attraction was as a potential source of fertilizer for swidden agriculture. In the general absence of domesticated livestock, meat had to come from hunting, and through fire Indians maintained the reserves they required—the grassland or forest-grass ecotone (which maximized edge effects) that proved so productive of game. In this way fire hunting was not only a tactic of harvesting but also one that sustained and even expanded the habitat on which it depended.

It is likely that these fire practices and the environments they shaped were expanding at the time of European discovery The dominant vegetation type in America at that time may well have been grassland or open forest savannah. The role of fire in sustaining these landscapes is incontestable; when broadcast burning was suppressed as a result of European settlement, the land spontaneously reverted to forest. Rather than a climatic change, the general encroachment of grassland into forest may have reflected the penetration of nomadic hunting cultures enlarging the range of their prey. The buffalo, for example, crossed the Mississippi about A.D. 1000. By the sixteenth century the buffalo entered the South; by the seventeenth, its range extended into Pennsylvania and Massachusetts.[29] This expansion could have been accomplished only through a change in habitat, partly due to climate, but largely achieved through the application of anthropogenic fire. The prairie peninsulas wedging eastward from the central plains were the vanguards of nomadic cultures, which created a landscape favorable to the species they hunted. Even where the grasslands were not contiguous, the presence of smaller hunting grounds was ubiquitous—often, as with the Pine Barrens of New Jersey, in places where agriculture could not effectively enter, or, as with the Shenandoah Valley, where a natural thoroughfare existed. The sites were frequently some distance from established villages and were occasionally part of an annual cycle of seminomadic travels.[30]

In his report of the Long Expedition of 1819, botanist Edwin James commented perceptively that "in a country occupied by hunters, who are kindling their camp fires in every part of the forest, and who often like the Mongalls in the grassy deserts of Asia, set fire to the plains, in order to attract herbivorous animals, by the growth of tender and nutritious herbage which springs up soon after the burning, it is easy to see these annual conflagrations could not fail to happen."[31] Like the Mongols of Asia—the phrase is revealing. One of the characteristics that identified a Great American Desert was that the inhabitants were nomads. Time and again, Europe, and agricultural settlements in general, had been overrun by nomadic incursions, and wherever Europeans brought colonies, they sought to end nomadism among the natives. But, as James recognized, nomadism and fire have always been intimately associated: typically, to suppress one was to eliminate the other. In America, through broadcast fire, hunters were expanding the range of game just as, through slash firing, agriculturalists were expanding maize culture. New pastures could be added through successive firings, or, as remains the case in much of Africa and Asia, by burning abandoned fields to prevent forest regeneration. This pattern of wildland fire was surely a contributory cause for endemic nomadism, just as it was for migratory logging, urbanism, and agriculture in nineteenth-century America.

Initially, frontiersmen tended to adopt Indian fire practices—fire hunting, for example, and firing for pasturage, for reduction of brush and ease of travel, and for slash-and-burn agriculture. Indians understood the precepts of fire prevention, too. Debris was cleared from around villages; cooking fires were situated carefully; camp sites on long grass were avoided, or the grass was first burned; and, when necessary, fires were fought. When Wash-

3. Alfred Jacob Miller, *Prairie on Fire*, 1836.

4. A. F. Tait, *The Trapper's Defence, "Fire Fight Fire,"* 1845. This popular Currier & Ives print was one of several prairie fires painted by Tait.

ington Irving's party had a cooking fire escape and threaten their camp, panic ensued, and "no one thought of quelling the fire, or indeed knew how to quell it," Irving observed. "Beatte, however, and his comrades attacked in the Indian mode, beating down the edges of the fire with blankets and horseclothes, and endeavoring to prevent its spread among the grass; the rangers followed their example. . . ."[32] When prairie fires threatened a Hudson's Bay Company post in 1828, local Crees assisted suppression efforts by wielding a special pole and bag apparatus.[33] On the plains in the 1830s Alfred Jacob Miller painted a scene in which an Indian village was threatened by an approaching prairie fire; using blankets and brands, the tribe set backfires around its entire camp.[34] In *The Prairie* Leatherstocking discovers a Sioux who survived a tall grass fire by crawling under a heavy buffalo robe, in much the same manner as firefighters have done in recent years with aluminized fire shelters.[35] In accounts of the Peshtigo fire in 1871, there was, as Robert Wells notes, "a persistent note of grudging respect for the Indians' knowledge of how to survive a forest fire."[36] Most tribes lived in fire environments, often of their own making; ignorance of fire control methods or indifference to basic precautionary practices would have been disastrous. The evidence suggests that early European colonists were generally ignorant of wildland fire and of methods to control it, just as European foresters in the nineteenth century were ignorant of prescribed burning as a silvicultural technique. It was from the Indian—into whose fire environment they moved—that the European immgirants learned basic survival skills.

It would of course be misleading to generalize too broadly about the fire practices of particular tribes. "To simply note that all Indians used fire to modify their environments," as Henry Lewis cautions, "is no more an ecological generalization than to note that all farmers used plows."[37] Nor is it a meaningful ethnographic generalization. Fire practices were to some extent circumscribed by environmental constraints. Even more damaging to a simple picture of homogeneity is the recognition that tribes underwent migrations, experienced internal evolutions, and were forcibly dislocated. Their relationship to the land was thus a constantly changing one. Nor is there any evidence to suggest that the tribes lived in some perpetual ecological harmony with one another or with their environment, upset only by European intervention. The ruins of ancient civilizations—the mound builders in the Ohio Valley, the Anasazi of Mesa Verde and Chaco Canyon, and the Hohokam of the Southwest—testify to the existence of impressive cultures that rose and vanished long before the white man arrived.

The resulting mosaic of anthropogenic fire regimes is as complex as the historical geography of the cultures themselves. Maize culture and other agricultural practices radiated across the continent well before Columbus's arrivals. Plains tribes only began to acquire the horse in the late seventeenth century; throughout the eighteenth the horse was dispersed across the plains. The process was still underway when the collision with migrating Americans occurred in the mid-nineteenth. Most of the tribes along the eastern edge of the Great Plains were displaced from traditional lands in the Ohio Valley, the Southeast, or even the East Coast. The Five Civilized Tribes were forcibly moved from the southern Appalachians to the plains. The Seminoles fled to southern Florida. The Sioux completely abandoned the forests for the plains only in 1862, after perpetrating the New Ulm massacre in Minnesota. Other inhabitants of the "Indian Territories" carved out the Louisiana Purchase were fragments of tribes, such as the Delaware, shattered by European contact east of the Appalachians. In the Southwest, Apaches and Navajos gravitated around Spanish settlements, taking up careers as raiders and herders, respectively. Into the landscapes suddenly thrust upon the newly dispossessed, traditional fire practices might be of little use.

That they were of use, however, testifies to the fundamental role of fire for virtually all the tribes. Nor is this observation restricted to the tribes of North America. The Seminole practice of hunting alligators with fire is identical to the Indonesian method for hunting crocodiles; the Iroquois practiced swidden agriculture in much the same way as do modern peasants in northern Thailand. Pasturage fires for wildlife in America are indistinguishable from those set on the steppes of central Asia, on the veldt of East Africa, or on kangaroo ranges in southwest Australia. Not all tribes were equally prone to use broadcast fire, of course. The Sioux and Apaches, for example, apparently resorted to fire far more commonly and for wider objectives than other tribes.[38] But even if Indian practices had been uniform in intent, which they were not, or homogenous in technique, which was not possible, the fire seasons of the different regions of the United States do not coincide, and the fire regimes are intrinsically distinct. The prospect of a scorched earth across the continent as a result of Indian burning in the autumn, for example, was impossible. Fire was predominately local, though the multiplication of local effects could produce surprisingly extensive cumulative results. The modification of the American continent by fire at the hands of Asian immigrants was the result of repeated, controlled, surface burns on a cycle of one to three years, broken by occasional holocausts from escape fires and periodic conflagrations during times of drought. Even under ideal circumstances, accidents occurred: signal fires escaped and campfires spread, with the result that valuable range was untimely scorched, buffalo were driven away, and villages were threatened. Burned corpses on the prairie were far from rare.

So extensive were the cumulative effects of these modifications that it may be said that the general consequence of the Indian occupation of the New World was to replace forested land with grassland or savannah, or, where the forest persisted, to open it up and free it from underbrush. Most of the impenetrable woods encountered by explorers were in bogs or swamps from which fire was excluded; naturally drained landscape was nearly everywhere burned. Conversely, almost wherever the European went, forests followed. The Great American Forest may be more a product of settlement than a victim of it.

"They attribute to fire a sacred character," wrote Father Pierre DeSmet of the tribes he had known, "which is remarkable everywhere in their usages and customs, especially in their religious ceremonies." Before "consulting the manitous, or tutelary spirits, or before addressing the dead, they began by kindling the sacred fire. This fire must be struck from a flint, or reach them mysteriously by lightning, or in some other way. To light the sacred fire with common fire, would be considered among them as a grave and dangerous transgression."[39]

Though fascinated by the sacred uses of fire, it was the profane fire of the Indian that most interested the European immigrants who displaced them. Indian fire practices were well known and their effects widely recognized. In 1878 John Wesley Powell—destined to become the director of both the U.S. Geological Survey and the Bureau of Ethnology—wrote in his famous *Arid Lands* report that "the timber regions are only in part areas of standing timber" and that "this limitation is caused by fire." Fires annually destroyed "larger or smaller districts of timber, now here, now there, and this destruction is on a scale so vast that the amount taken from the lands for industrial purposes sinks by comparison into insignificance." Powell had personally witnessed two fires in Colorado and three in Utah, any one of which "destroyed more timber than that taken by the people of the territory since its occupation." Similar fires "have been witnessed by other members of the surveying corps."[40]

The "protection of the forests of the entire Arid Region of the United States," the former professor lectured, "is reduced to one single problem— Can these forests be saved from fire?" The chief condition for the fires, Powell noted, was the arid climate of the Far West, but the chief source of ignition was local tribes of Indians. Like other aborigines or pastoralists crowded out of their homeland by European expansion, the tribes distributed fire to their new settings in an effort to create a familiar and useful landscape. In this case most of the fires were set by hunters eager to accumulate furs that could be sold for cash. "The fires can, then, be very greatly curtailed by the removal of the Indians," Powell concluded, and "once protected from fires, the forests will increase in extent and value."[41]

The sequestering of Indian tribes into reservations helped to isolate a prolific source of fire. Whatever its social ills, the reservation system must be counted as one of the major mechanisms by which wildland fire was restricted and by which forests increased through European occupation. But although the reservations worked in one sense to restrict the range of traditional fire, they also worked to preserve it. The reservations were often a repository of traditional fire practices that, with the progress of industrial development, were lost to Western civilization. When prescribed burning was finally reintroduced to industrial forestry, its techniques came from two sources: the progeny of southern frontiersmen who had been isolated in the piney woods of the South and from Indians still practicing their ancient art on reservations in the Cordillera. Scientific forestry merely gave the old practices a new justification and shaped them into a new cycle of firing, one adapted to an industrialized society.

II

The conflict over fire practices has had its intellectual dimensions as well as its environmental ones. The evidence for aboriginal burning in nearly every landscape of North America is so conclusive, and the consequences of fire suppression so visible, that it seems fantastic that a debate about whether Indians used broadcast fire or not should ever have taken place. But so it did. For decades anthropology texts and ethnographic studies ignored fire, except as a tool of swidden agriculture. Enchanted by the powerful mirage of the virgin forest myth, historians also ignored aboriginal fire. The assumption of a "state of nature" at the time of the European discovery was essential to a judgment on Western civilization's presence in the New World—which, of course, was new only to Europeans. In America the story was further complicated by the persistence of romantic primitivism in the form of the noble savage and forest primeval myths and the enthusiasm for untainted wilderness. Only foresters bothered to examine the fire record.

Long before it was generally believed that fire invariably degraded the environment, it was widely recognized that Indians purposefully fired the landscape. Frontiersmen never doubted this fact. Even early foresters, bent on bringing scientific forestry to the United States from Europe, knew it too and disdained broadcast burning as mere "Paiute forestry." Only when the tenets of early industrial forestry triumphed and fire was considered an impediment to good management was it held that Indians did not intentionally burn. As the child of nature, the American Indian could not deliberately have damaged his environment, and by mid-century fire was generally considered an environmental evil. In more contemporary times, when prescribed fire has again been accepted as an appropriate tool in the management of natural systems, it has been discovered that, indeed, the Indian burned.

In each instance, the assumption was that primitive man was indelibly in tune with natural ways, that he was possessed of an almost mystical insight into the inner harmonies of the landscape, and that he therefore would do nothing to violate the delicate equilibrium of the ecosystem. When the "natural" way was determined to mean fire exclusion, it was evident—reports to the contrary—that Indians did not broadcast burn and that they were careful with fire. When the "natural" way included fire, it was determined that Indians in their ancient wisdom had burned. The corollary, of course, is that when the Europeans arrived, the landscape was in finely tuned balance, only to be disrupted by the corrupting habits of civilization. When fire was considered an environmental evil, the use of careless and broadcast fire by European and American settlers was excoriated; when it was considered an inevitable and even desirable part of the ecosystem, fire control practices were criticized. Controversies over whether to burn or under what conditions to do so are even yet clouded by appeals to the supposed fire practices of the Asian immigrants who had superimposed—with vast consequences—their cultural fire regime over whatever might have existed in nature.

For decades Western civilization seemingly forgot the values of broadcast fire known so well to hunting, gathering, and herding societies around the world and commonly recognized by European colonists living along frontiers. Foresters examining the question of whether or not American Indians burned concluded that, on the whole, they did not burn because there was no good reason for them to do so; nothing good could come of fire. Writing in 1935, Louis Barrett dismissed the argument that "light burning" was the "Indian way." Such fires were destructive, and "it seems very unlikely that our Indians had the time or the inclination to go into the forests and light-burn them as some claim they did before the whites came."[42] Publishing a massive study of forestry in California, Raymond Clar concluded in 1959 that "it would be difficult to find a reason why the Indians should care one way or another if the forest burned."[43]

Today, virtually the same historical record seems to show a widespread use of broadcast fire, because it is hard to imagine a society without fire. Fire exclusion could bring them no good. The evidence has not changed, but its cultural context has. Early foresters rarely had knowledge of tribes directly; their contact was with stockmen, miners, railroaders, and loggers, who often ingenuously insisted that they were only continuing Indian ways. Men trained in the eastern academies of forestry and prepared to look on such people as capitalist exploiters and robber barons were understandably skeptical and met appeals to remote Indian practices with cynicism. The continuance of Indian fire habits on the frontier, moreover, was inimical to industrial forestry. The history of modern fire protection is basically the story of how one fire regime, that of frontier economies, was replaced by another, that of an industrial state. Not until the 1940s did the Indian practices preserved in western reservations and the frontier habits perpetuated in the piney woods of the South work out a modus vivendi with modern forestry.

The debate has been doubly ironic. Through the use of fire, confined and broadcast both, Indian tribes created an environment favorable to their existence. It was an existence and a fire regime often well suited, at least for a transitional period, to the frontier that superseded them. The forests most commonly selected by Indian fire—pine and oak—were precisely those most valuable for logging. Grassland corridors like the Shenandoah Valley were maintained by fire and provided major thoroughfares for the dispersal of settlers. The lush wildlife that provided subsistence to early vanguards of frontiersmen depended on a habitat created and sustained by the Indian pattern of broadcast fire; as settlement progressed, converting "deserts" into farms and forests, the abundant game vanished. The habitat that supported so rich a natural population of grazers and browsers was ideally suited for the domestic stock introduced by Europeans and upon which their agrarian economy was so heavily dependent. In the same way that herders of domesticated stock occupied the openings previously maintained for the harvest of wild game, immigrant farmers moved into the former fields of the aborigines, fields cleared and fertilized with fire. The early agriculture of the American colonists resembled nothing so much as the shifting agriculture of the Indian, though at an accelerated rate. From Indian examples the colonists learned fire hunting and the techniques for controlling and surviving the fires so common in the high fire regimes they suddenly occupied. Just as European and American exploration of the continent was advanced by having native guides, so settlement advanced in good measure thanks to the preparation of Indian hunters, harvesters, and farmers.

The noble savage convention and the myths of the virgin forest and forest primeval are valuable literary and moral devices that have intruded into the debate over Indian fire practices. But it was in large measure owing to the Indian and his Grandfather Fire that the forest primeval had already been widely cleared, converted, and otherwise managed. That fact made the spread of Western civilization across the American continent much more rapid than it would have been had the myths been true.

THESE CONFLAGRATED PRAIRIES:
A FIRE HISTORY OF THE GRASSLANDS

"And if we can only get sight of a prairie on fire!" cried the
Count—"By Gar, I'll set one on fire myself!" cried the Little
Frenchman.
—Washington Irving, *A Tour on the Prairies,* 1836[44]

But that was the custom of the country, everybody dropped
everything that they were doing and went to fight fire.
—Cowboy, XIT Ranch, Texas, ca. 1900[45]

I

Shortly after reaching the Great Plains, the Lewis and Clark Expedition
"sent Mr. Durione the Sioux interpreter & three men to examine a fire
which threw up an emence Smoke from the Prairies on the N.E. Side of the
River and at no great distance from Camp."[46] Writing 40 years later, John
Charles Frémont described how "from a belt of wood which borders the
Kansas ... we suddenly emerged on the prairies, which received us at the
outset with some of their striking characteristics; for here and there rode an
Indian, and but a few miles distant heavy clouds of smoke were rolling
before the fire."[47] "These conflagrated prairies," as they appeared to the
chronicler of the Long Expedition, astonished by virtue of their immensity
and were almost always associated with fire.[48] Even in the 1830s, when
Washington Irving joined the early influx of tourists to the plains, a prairie
fire was as much a standard attraction as were native warriors and
buffaloes.

Fire and grass are genetically associated.[49] Taken in its broadest meanings to include plains, prairies, barrens, savannahs, and wetlands, grasslands
were probably the dominant cover type in North America at the time of
European discovery. Except for the High Plains, where the short grass
expanses were more or less determined by climate, nearly all these grasslands were created by man, the product of deliberate, routine firing. This
description includes even the tall grass prairies east of the 100th meridian,
which, combined with the High Plains, made the largest continuous grassland in the world. From this vast corridor—probably the means by which
early man dispersed throughout America—anthropogenic fire extended the
range of the grasslands outward. A series of prairie peninsulas brought
grasslands north into Wisconsin, Michigan, and Minnesota, east through

Illinois, Indiana, Ohio, and Kentucky, and south into central Alabama and Mississippi. As these manufactured grasslands radiated outward, their continuity broke down. Individual "barrens" appeared, like skirmishers, ahead of the main advance; these might be expanded in time and joined. Where heavier rainfall was encountered with the march eastward, bogs and bottoms prevented more or less complete conversions, leaving a mosaic of burned and unburned regions, of forest and grassland.

Continuous or not, grasslands followed the Indians nearly everywhere they took broadcast fire. Most of the coastal plain from Massachusetts to Florida to Texas was savannah. Wet grasslands typified many coastal marshes, glacial muskegs, and rivertine plains. Along the West Coast, grass flooded the Central Valley of California and the Willamette Valley of Oregon. In the Southwest grasslands spread from the pine savannahs of the mountains to the edge of the Sonoran and Chichuahuan deserts. In the intermontane region between the Sierra and the Rockies, grass was the basic cover type of the loess deposits of the Columbian Plateau and most of the northern Great Basin. And everywhere there were manufactured grasslands of great local importance: the Shenandoah Valley, a great grassy corridor at the foot of the Appalachians like a miniature High Plains; the celebrated Barrens of Kentucky; meadows in the Cordillera; and clearings carved out of even the dense forests of the Pacific Northwest. Where the land was drained or sandy, oak and pine forests typically formed savannahs that further enhanced the impression of grassland cover.

Knowledge of fire was a precondition to successful habitation on the plains and prairies; even nomadism was in part an adaptation to fire, both natural and anthropogenic. The adaptions by grass for fire and drought are much the same, and where those two phenomena could combine, the dreaded fire-flood cycle so damaging to soil in mountainous regions would be replaced on the prairies by a fire-wind cycle. Under the natural regimes, fires came with summer thunderstorms. There are numerous accounts of lightning igniting grass in the Great Plains, and lightning was responsible for the large range fires that hit Nevada in 1964 and for the Arizona fires in 1979–1980. After anthropogenic fire became dominant, the summer cycle was superseded by a pattern of spring and fall burning.

Plains tribes differentiated among types of fires and took precautions against those they feared. Near Fort Leavenworth, Kansas, in 1832 George Catlin paused to consider the differences. "The prairies burning form some of the most beautiful scenes that are witnessed in the country," he mused,

and also some of the most sublime. Every acre of these vast prairies (being covered for hundreds and hundreds of miles with a crop of grass, which dies and dries in the fall) burns over during the fall or early in the spring, leaving the ground a black and doleful color.

There are many modes by which the fire is communicated to them, both by white men and by Indians—par accident, and yet many more where it is done for the purpose of getting a fresh crop of grass, and for the grazing of their horses, and also for easier traveling during the next summer, when there will be no old grass to lie upon the prairies, entangling the feet of man and horse, as they are passing over them.

Over the elevated lands and prairie bluffs, where the grass is thin and short, the fire slowly creeps with a feeble flame, which one can easily step over; where the wild animals often rest in their lairs until the flames almost burn their noses, when they will reluctantly rise and leap over it, and trot off amongst the cinders, where the fire has past and left the ground as black as jet. These scenes at night become indescribably beautiful, when their flames are seen at many miles distance, creeping over the sides and tops of the bluffs, appearing to be sparkling and brilliant chains of liquid fire (the hills being lost to the view), hanging suspended in graceful festoons from the skies.

But there is yet another character of burning prairies that requires another Letter, and a different pen to describe—the war, or hell of fires! where the grass is seven or eight feet high, as is often the case for many miles together, on the Missouri bottoms; and the flames are driven forward by the hurricanes, which often sweep over the vast prairies of this denuded country. There are many of these meadows on the Missouri, the Platte, and the Arkansas, of many miles in breadth, which are perfectly level, with a waving grass, so high, that we are obliged to stand erect in our stirrups, in order to look over its waving tops, as we are riding through it. The fire in these, before such a wind, travels as fast as a horse at full speed, but that the high grass is filled with wild pea-vines and other impediments, which render it necessary for the rider to guide his horse in the zig-zag paths of the deers and the buffaloes, retarding his progress, until he is overtaken by the dense column of smoke that is swept before the fire—alarming the horse, which is wafted in the wind, falls about him, kindling up in a moment a thousand new fires, which are instantly wrapped in the swelling flood of smoke that is moving on like a black thunder-cloud, rolling on the earth, with its lightning's glare, and its thunder rumbling as it goes.[50]

When American settlers breached the Appalachians, they also began to encounter "deserts" of larger and larger extent, culminating in the Great American Desert itself. Where these lands were most extensive, a certain temporary breakdown occurred in traditional methods of settlement, in descriptive language, in concepts about their origin and their fertility, and in political assimilation. It was with the acquisition of the Louisiana Purchase that the political equilibrium between slave and free states began to collapse. A year before the Missouri Comprise of 1820, the Long Expedition began a scientific inventory of the new territory for the federal government, and within a year after the compromise, private enterprise traversed the plains in search of trade. The Ashley-Henry Expedition went up the Missouri in pursuit of furs, and the Santa Fe Trail was opened to the southwest. But many observers forecast that the land would prove too formidable for settled agriculture; it would remain the province of nomadic hunters and herders. In the sour words of one disgruntled Federalist, the plains would endure as "an omnium gatherum of savages and adventurers."

If it proved difficult to absorb the new lands into the political structure and economies of the young nation, there were equal problems in describing them at all. Almost on landfall in the New World, the English word *meadow* failed to convey the sense of what settlers saw. Folk terms like *barrens, openings,* and *deserts* were adopted; the more literate resorted to *champion fields,* after the French *champaigne* and by analogy to similar expanses in England. *Desert,* too, had a French origin, in the sense of abandoned land— emphasizing the perception that the land had been once cleared, then deserted. But when the prairie peninsulas were met, even these expressions proved inadequate. From the French Canadians came the term *prairie,* ultimately derived from the Latin *pratum,* meaning "meadow." The word came into vogue during the eighteenth century, first from the French in Acadia, then from the encounter of French explorers with the Great Plains. Though it means the same as the English term *meadow,* it was accepted, perhaps because it was, like the landscape, foreign. From the Spanish, meanwhile, came *savannah,* and this term was applied mostly to the grasslands of the southeastern coastal plains. The Spanish apparently picked it up from Caribs, and it was used in English by the seventeenth century. (The Spanish never adopted a single expression themselves, using *pampas* and *llanos* in South America and *parangs* in the Philippines.) But the settlers of the Ohio Valley and the explorers of the Great Plains came largely from the north and hence most often applied the term *prairie. Grassland,* an American contribution to English, did not appear until the early nineteenth century (Joel Barlow uses it, for example, in his epic poem "Columbiad"), and it did not come into wide currency until the mid-twentieth.[51]

The explanation for the origin of these lands likewise strained learned thought. For those familiar with actual conditions, however, there was never much doubt that they resulted from Indian firings. In 1819 R. W. Wells explained to the readers of the *American Journal of Science and the Arts* why, in contrast to speculation that prairies represented a flooding phenomenon, they resulted from fire. Many such lands, for example, "are ten to twelve miles in length, and three or four in width," and were found on the flanks of the Appalachian Mountains, especially the Alleghenies. They appeared only where the soil could support grass and where grass fire could gradually reduce the forest.

> The Indians, it is presumed, (and the writer, from a residence in their country and with them, is well acquainted with their customs) burn woods, not *ordinarily* for the purpose of taking or catching game . . . but for many other advantages attending that practice. If the woods be not burned as usual, the hunter finds it impossible to kill the game, which alarmed at the great deal of noise made in walking through the dry grass and leaves, flee in all directions at his approach. Also the Indians travel much during the winter, from one village to another, and to and from the quantity of briars, vines, grass, etc. To remedy these and many other inconveniences, even the woods were originally burned so as to cause prairies, and for the same and like reasons they continue to be burned towards the close of the Indian summer.
>
> Woodland is not commonly changed to prairie by one burning, but by several successive conflagrations; the first will kill the undergrowth, which causing a greater opening, and admitting the sun and air more freely, increases the quantity of grass the ensuing season: the conflagration consequently increases, and is not sufficiently powerful to destroy the smaller timber; and on the third year you behold an open prairie.
>
> Ordinarily, all the country, of a nature to become prairie, is already that state; yet the writer of this has seen, in the country between the Mississippi and Missouri, after unusual dry seasons, more than one hundred acres of woodland together converted into prairie. And again, where the grass has been prevented from burning by accidental causes, or the prairie has been depastured by large herds of domestic cattle, it will assume, in a few years, the appearance of a young forest.[52]

The scenario was repeated over and over throughout the United States. Timothy Dwight recounted it for the barrens of the Northeast. François André Michaux in 1802 described the process in the famous Barrens of Kentucky. "Every year, in the course of the months of March or April," he wrote,

> the inhabitants set fire to the grass. . . . The custom of burning the meadows was formerly practiced by the natives, who came in part to this part of the country to hunt; in fact, they do it now in the other parts of North America, where there are *savannahs* of an immense extent. Their aim in setting fire to it is to allure the stags, bisons, etc., into the parts which are burnt, where they can discern them at a greater distance. Unless a person has seen these dreadful conflagrations, it is impossible to form the least idea of them. The flames that

occupy generally an extent of several miles, are sometimes driven by the wind with such rapidity, that the inhabitants, even on horseback, have become prey to them.

And like Wells in the Ohio Valley, Michaux observed reforestation only with permanent settlement.

This example appears to demonstrate that the spacious meadows in Kentucky and Tennessee owe their birth to some great conflagration that has consumed the forests, and they are kept up as meadows by the custom that is still practiced of annually setting them on fire. In these conflagrations, when chance prevents any part from the ravages of the flame, for a certain number of years they are re-stocked with trees; but as it is extremely thick, the fire burns them completely down, and reduced them again to a sort of meadow. We may thence conclude, that in these parts of the country the meadows encroach continually upon the forests.[53]

More than anything else, the expedition down the Platte River led by Major Stephen H. Long influenced the national perception of a Great American Desert. Its chronicler, Edwin James, insisted that in fire "we have a satisfactory explanation for the cause of the present want of forest trees in extensive tracts of the Missouri, which appear, in every respect, adapted to the growth of timber." James repeated how "in the Autumn of 1819 the burnings, owing to the unusual drought, continued until very late in the season, so that the weeds in the low ground were consumed, to the manifest injury of the forest. Large bodies of timber are so frequently destroyed in this way, that the appearance has become familiar to hunters, and travellers, and has received the name of *deading*." It was James, moreover, who shrewdly realized that settlement would not automatically result in dense reforestations. "Whenever the dominion of man is sufficiently established in these vast plains, to prevent the annual ravages of fires, trees will spring up," he explained, "but we may expect that before forests, originating in this manner can arrive at maturity, the population along the banks of the Missouri will become so dense, as to require the greater part of the soil for the purposes of culture."[54]

The responses to grasslands, then, were mixed: in some locales Indian fire practices were continued by hunting and herding economies of the frontier; in others, farmers suppressed the fires and allowed for reforestation between fields. Everywhere, however, new techniques for fighting wildfire were developed, and new standards for the prevention of accidental fire were required. Josiah Gregg observed that "all those who have crossed the Prairies have had more or less experience as to the danger which occasionally threatens the caravans from these sweeping visitations," and he proceeded to narrate an episode in which "we were terribly alarmed at a sudden prairie conflagration." In one instance wagons loaded with gunpowder had to race wildly across the prairie, escaping into a spot of short grass "only when the lurid flames were actually rolling upon the heels of our teams."[55]

During the Mexican War, when Colonel Alexander Doniphan's column tried to outrace an escaped fire, it was necessary to run the artillery into a nearby lake, while cavalry units tried to trample down the grass ahead "by frequently riding over the same ground. They also rode their horses into the water, and then quickly turned them upon the place where the grass was trodden down, that they might moisten it, and thereby stop the progress of the fire, but still the flames passed over and heedlessly swept along." At last a captain of the horse guards ordered his men to dismount about two miles in advance of the fire, chop down the grass with their sabres, and backfire from the cropped swath; this time the strategy worked.[56]

Wildfire was a basic fact of life not only on the Great Plains but wherever savannahs, barrens, and prairies were encountered. To accept them was to become, like the Indian, seminomadic. But the newly acquired land was not long to be held in common, and nomadism was, for the majority of settlers, not considered to be a suitable condition for civilization. Fire had to be controlled, and the grassland thereby experienced a dramatic type conversion—in many cases even reverting back to forests. Using St. Louis as an example, Wells told of how the conversion frequently came as an inevitable byproduct of sedentary agriculture. "All the old French inhabitants will tell you," he reported, "that the prairies formerly came immediately up to those places. Now the surrounding country for several miles is covered with a growth of trees four or five inches in diameter, near the towns where the burning first ceased, and gradually diminishing in size as you recede, until you at length gain the open prairies." So it was, too, with "the barrens in Kentucky."[57] On the prairies, it was settlement that brought the forest, and, paradoxically, the more intense the development, the denser the woods that resulted.

II

The fire techniques adopted from the Plains tribes were suitable for transients—for trappers, hunters, explorers, military expeditions, sportsmen, and teamsters. But these were means for surviving in a permanent grassland fire regime, and, like those fire practices borrowed from tribes in the eastern woods, they were the tools of a transitional phase. The long-range settlement of the grasslands converted the landscape to farms, fields, and forests—replacing the wild grasses with domestic cereals, substituting domestic livestock for wildlife, and encouraging, both deliberately and inadvertently, the reclamation of the grasslands by woody species. The history of grassland fire is a history of habitat conversion. For this more profound occupation of the grasslands new fire practices had to be adopted or created.

The full agricultural reclamation of the Great Plains had to wait for the industrial revolution—for its barbed wire, windmills, and synthetic or fossil fuels; for its railroads, which tied interior farms to markets; for its fertilizers, which made possible the abandonment of forest farms, with their source of ash. The smaller grasslands were more readily converted, with settlers moving in from the perimeter, as occurred in the Barrens of Kentucky. The expansion of the prairies through fire had resulted in a virtual monoculture of grass. Settlement broke this expanse into a mosaic of fuels and a lattice of roads, stone walls, fields, and plowed firebreaks. Gradually, domestic grasses replaced wild grasses, and controlled burning of stubble replaced wildfires as the means for improving pasturage. A new cycle of burning was initiated, one integrated into the rhythm of agriculture rather than that of nomadic hunting. Heavy cropping of grasses by livestock or mowing by settlers prevented fires from approaching the locally grazed lands. Settlers organized fire brigades, and fire codes helped to regulate agricultural burning. Forests reoccupied the intervening land, that is, sites not yet put to the plow but still protected by heavy grazing or road fuelbreaks.

As the American reclamation moved west, so did this process of conversion. What had been perhaps the most extensive fire regime on the continent gradually became one of the least intense. The treacherous period was the transition, during which blocks of wild and domesticated grasses intermingled, along with the often antagonistic ignition patterns of farmers and hunters. Under these circumstances, it was easy for a farm to be burned out, and farmers quickly learned how to combat the fires that swelled out of the tall, dry grass. Laura Ingalls Wilder described some of the methods used when wildfire threatened *The Little House on the Prairie.* " 'Prairie fire!' shouted Pa. 'Get the tub full of water! Put sacks in it! Hurry.' " A furrow was quickly plowed around the house (several would be preferable, but there was not time: " 'That fire's coming faster than a horse can run' "), and backfiring was begun. The fire was lit on the outside of the furrow so that it could burn toward the main fire; fire escaping to the inside was quickly swatted out. The strategy was identical to that used by Plains Indians to protect their villages, with the plow replacing wet blankets. Had Pa Ingalls not been present, what would the family have done? " 'We would have gone to the creek with the birds and rabbits, of course,' Ma said."[58] With further settlement, however, extensive prairie fires became less common and eventually vanished—replaced by domestic crops and, where fire was retained, by a cycle of agricultural field burning. In the late nineteenth century, for example, the Texas Panhandle was the scene of stupendous conflagrations; today, thanks to artesian wells and irrigation, it grows vegetables, wheat, and melons.

Where the native grasses were to be retained rather than converted, as was the case with ranching, the problem of wildfires was more complex. Drought and fires were considered the major hazards of early plains herding. Fires caused a temporary loss of pasturage, especially acute when stock was held on the range over winter. They required roundups and transfer of cattle to unburned regions, and they compelled ranch hands to abandon other jobs and control the burns. These difficulties were less acute where open range laws were in effect, and well into the twentieth century fire and open range practices remain generally associated. Fenced enclosures, however, magnified the impact of fire: there might not be sufficient unburned grass to feed the herds, and livestock not infrequently burned to death because fences prevented their escape. Indian fire practices had been predicated on a migratory existence, where tribes and herds could roam among ranges alternately burned and greened. But that was no longer an option where land was fixed with individual owners and where the harvesting of livestock was more intensive than the hunting of natural wildlife.

Because wildfires constituted such a threat, ranchers on the Great Plains often kept records of them. In 1879 hundreds of thousands of square miles burned in west Texas. In 1888 and 1889 North and South Dakota towns like Leola, Jamestown, Sykeston, and Mt. Vernon were virtually incinerated by fires swelling out of the prairies. In 1894 the XIT Ranch in Texas suffered a four-day fire that burned clean over a million acres. An 1895 fire in the Panhandle devoured similar acreage, and in 1906 over 6 million acres burned in a single fire, and another million in a second, nearby burn. As late as 1947 two days of fire in South Dakota swept 380,000 acres. Though they lack the energy release of a brush or forest fire, grass fires can travel with incredible velocity. A 1910 fire in Nebraska was clocked at six miles in 40 minutes. An 1887 fire in Texas ripped through 16½ miles in two hours. Under such extreme conditions grass fires can leap even major streams. With settlement, other hazards were sometimes introduced: the exotic Russian thistle, which snaps from its taproot after the first frost, became a formidable fire propagator, and even cow chips, sailing like flaming frisbees before high winds, would distribute fire well ahead of the main blaze and were a major threat to isolated farms and towns.[59]

Ranching created a new fire regime that required new techniques of fire control. A good example of the measures taken comes from the XIT Ranch in west Texas. The rules of the ranch sternly enjoined that "in case of fire upon the ranch, or on the lands bordering on the same, it shall be the duty of every employee to go to it at once and use his best endeavors to extinguish it, and any neglect to do so, without reasonable excuse, will be considered sufficient cause for dismissal." After the ranch was devastated by massive fires in 1885, with an almost 25 percent loss among its herds, management instituted a system of plowed fuelbreaks, or "fire guards." In September teams of three men were sent out, two to plow and one to cook. Two or three strips were plowed about 100 feet apart. From mid-November to mid-

5. Grass fire suppression with sprinklers, blankets, and buckets, Oklahoma, 1908. A water wagon typically accompanied such an operation to replenish the sprinklers.

December the intervening grasses between the furrows were burned out. Three men with ropes soaked in coal oil spread fire, and several others followed with brooms or flappers to prevent the flames from leaping the furrow. Within a year, 1,000 miles of fire guards were emplaced.[60]

When fires did break out, a variety of methods was used to suppress them. If small, the fire would be attacked directly. If large or rapidly moving, it was worked along the flanks or halted by backfires from some fixed line. A common technique was the "beef drag," described by Evetts Haley. The nearest steer would be slaughtered, split open, and "then, tying to the lower fore and hind legs, two men took off, dragging the beef by the horns of their saddle. One took the grassy side, the other the side of the fire, and by straddling the path of the flame, they dragged the beef directly along it with the loose, wet hide flopping out behind and men on foot to finish the job. Often they rode at a trot while the men on the ground worked in relief: It was necessary, too, to change the horse on the burnt side at least every twenty or thirty minutes, on account of the heat on his feet." Where plows or fuel-breaks did not exist, cow paths were sometimes employed to begin a back-fire. Igniting a rope soaked in kerosene, a cowboy would drag it along the edge of the prepared line. Campaigns might last for several days, and chuck wagons would furnish logistical support.[61]

"Fires were of great extent and damage, grass being dense and nothing to break the sweep of the raging flames," recalled J. W. Armstrong. "Orders were to go to any fire within a radius of a hundred miles. Sacks, saddle blankets, brooms, or chaps were used to fight fires with till you got men and horses enough to kill a cow. . . ." For a beef drag, "the heavier the cow the better job, hence often a bull is slaughtered instead of a cow, this is if there are horses enough to drag the animals. You could put out more fire in this manner than fifty men could with sacks and blanket fighting. When the flesh of the cows were charred from the fire they were drug aside and another slaughtered to take their place." Behind the drag "men were placed in relays every hundred yards, to fight out the little spots of fire left." It took a good 10 men to fight a large fire.[62]

Strict rules against careless fire, such as smoking, went into effect during dangerous seasons. Special chain drags were kept at each camp, along with two large hides and a barrel of water—more efficient than the beef drag, and certainly easier on the herds. During extreme danger periods, a cowboy would frequently man a high windmill tower to spot for smokes. "I guess this [firefighting] and haying," another cowboy complained, "was the hard-est work we did, or it was what we hated worse anyway." There were stand-ing orders for all hands to respond to any smokes. And "neighbor ranch-men," C. W. Burrus recalled, "always came to help fight fires."[63] Nor were the lessons learned on the plains left there. A high percentage of early for-estry guards were hired out of the ranks of dispossessed cowboys, who brought with them techniques like the drag and flapper and terminology like "corralling" and "loose herding." In some of the open western forests logs replaced beef drags, and pine boughs substituted for wet blankets and chaps.

With more intensive management and the fuelbreaks of civilization, fire tended to disappear from ranching lands. Controlled burning rarely replaced it. Only in the forested lands where open range continued did firing for pasturage persist. The closing of the public domain; the emphasis on maximum production through importing special pasture grasses, mowing and storing grass for winter feed, and fattening stock at feedlots; the roads and other developments breaking up the continuity of grassland fuels into a new mosaic; and the heavy demand on grasses, which reduces fuel loads to the point where fire, however beneficial, may be difficult to propagate— all these factors mean that interest in the effects of fire on North American grasslands, as Richard Vogl observes, "is becoming more and more academic as it relates to range management." Vogl inventories other considerations:

> Factors discouraging the use of fire are the difficulties of executing controlled burns, the liabilities of escape fires, the need to remove livestock before and after burning, fear of fires, lack of burning experience, and economic pressures to run ranches to their fullest capacities. The poor results often obtained because of ineffective burns, burn-deteriorated ranges, adverse postburn climatic conditions, and the postburn presence of alien and often unpalatable plants as well as livestock.[64]

The chief uses for fire in the management of domestic stock currently are to eradicate noxious weeds, to convert brush to pasturage, and to retard encroachment by woods species—in brief, to enlarge potential pasturages. Once established, pasture lands, like farmlands, tend to experience a gradual reduction in the use of open burning in favor of mechanical or chemical treatments, or they are stocked with specially bred flora and fauna no longer well adapted to fire. The dominant objective for broadcast fire will remain, as it has been historically, to stimulate the production of wild game and to preserve grassland habitats within parks or refuges where high-yield extractions are not contemplated or where the land is to be "preserved" in a grassland state. So forceful are economic pressures that most cattle and hogs, for example, are fed high-protein mixtures and special grains at feedlots, not native grasses on the range. Such animals mature in a synthetic landscape, one that often is no longer even a domesticated version of the grassland.

The process of type conversion is actually much greater than that which is visible on the Great Plains. The traditional range land of early America, like that of Europe, was the woods. In America, on the Indian example, the woods range was perpetuated by fire in many places until the full brunt of either agricultural reclamation or industrial counterreclamation was felt. Then the woods typically were closed, the open range shut; and with their closing went the rationale for broadcast fire. Nor was that other technique of European and Indian herding, transhumanance, retained. Indian tribes often migrated between winter and summer ranges, firing the summer ranges either before or after they arrived to encourage wildlife. Mediterranean shepherds similarly migrated between summer (mountain) and winter (valley) pastures, likewise firing upon arrival and departure. This seminomadism largely vanished, as had open grazing, because of what appeared to be extravagant use of fire. Yet the conflict over closing the open range was among the most bitter in American fire history. Such patterns of life depended on abundant open land and on broadcast fire; to eliminate either was fatal. But in fact, both were shut down, the first often as a means to the second. In the interim, deprived of traditional ranges, herders and hunters tended to penetrate into new environments, spreading fire before and after them. In the Southeast, for example, the conflict settled down to one between white residents of the piney woods and foresters; in Utah, between Indian hunters and settlers eager to protect watersheds; in California, between Basque shepherds and government officials charged with guarding the parks and forest reserves.

If reclamation had the effect of pushing a grass and fire regime ahead of it, thereby increasing fire hazards, it also had the consequence of often increasing fire hazards to its rear. In the mid-nineteenth century most farms were located on forested lands or lands that naturally would support forests; a hundred years later, nearly all the farmland was situated on former grasslands. Thus the westward progress of agricultural settlement generally decreased fire hazards in the grasslands it occupied, but by inducing abandonment of woods farms and allowing them to revert to forests or wilds, it increased the fire hazard of the forests. Both problem areas were addressed before the end of the nineteenth century by the forces of the counter-reclamation.

III

The reclamation of vast portions of the grasslands resulted in the replacement of wild grasses with domestic cereals and of wildlife with domestic stock. In both cases, however, the land remained broadly identifiable as grassland. But on the margins of the interior plains—on the prairie peninsulas and barrens, on the multitudinous hunting grounds carved out by Indian fire—the consequence has been to replace grassland or savannah with forest, brush, or succulent desert plants. No region has escaped this process entirely.

Most of this change came inadvertently, simply as a result of settlement and grazing, and it was most effective along the marginal zones. Everywhere settlers reported trees growing where none had existed before.[65] In the Northeast the "barrens" gave way to "thick forests," often well enough developed by the late nineteenth century to be subjected to heavy logging. Oak groves appeared in quantity in the Ohio Valley. Pine forests invaded the clearings in the South until they gave way, under still more intensive protection, to hardwoods and "rough." In the Southwest pinyon and juniper spread over lands heavily grazed and rid of fire; forest stocking increased; and brush moved across the flanks of the mountains. Forestry officials in California applauded successful fire control. "In those districts in which the Indians were accustomed to set fires," the state board of forestry wrote approvingly, "the discontinuance of that custom has resulted most beneficially to the young forest growth."[66] In Oregon settlers remarked on the encroachment of woodlands on grassy plains subsequent to settlement. In the Great Basin the old grasslands succumbed to pinyon, juniper, and sage.

Writing of the Cross Timbers region of Texas, Josiah Gregg explained that it was "unquestionably the prairie conflagrations that kept down the woody growth upon most of the western upland" but that with settlement there were "parts of the southwest now thickly set with trees of good size, that, within the rememberance of the oldest inhabitants, were as naked as the prairie plains." Greg concluded that "we are now witnessing the encroachment of the timber upon the prairies, wherever the conflagrations have ceased their ravages."[67] Somewhat further west and later, W. L. Bray described "this struggle of the timberlands to capture the grasslands . . . an old warfare. For years the grass, unweakened by overgrazing of stock, and with the fire for an ally, held victorious possession. Now the timber has the advantage. It spreads like an infection."[68]

Aldo Leopold recounted how an almost identical process occurred along the northern boundary of the prairie. "If you doubt this story," he lectured, "go count the rings on any set of stumps on any 'ridge' woodlot in southwest Wisconsin. All the trees except the oldest veterans date back to the 1850s and 1860s, and this was when fires ceased on the prairie."[69] John Muir wrote similarly that "had there been no fires, these fine prairies, so marked a feature of the country, would have been covered by the heaviest forest."[70] Throughout most of the Midwest, farms were not carved out of forests so much as they created the forests.

Wherever sedentary agriculture went, forest reclamation was repeated to a greater or lesser degree. The process did not proceed uniformly, however. Where logging was intensive, such as in the Lake States and portions of the Northeast, cutting and the subsequent firing of slash often retarded reforestation temporarily and at times virtually obliterated the soil cover. Where grazing was prominent, particularly with open range policies, such as those operating in the South, grazing and fire retained or expanded grasslands; only the advent of organized fire protection, of industrial forestry, and of restrictions on open grazing allowed forests to reoccupy the land. Where Indian tribes vacated land well in advance of European settlement, as might happen by treaty, war, or plague, young forests sprouted up and required the landclearing traditionally associated with pioneer farming. Like the process that removed the woods, then, the process that restored them was predominately local but made large through its multiplication over immense regions. At times, and to an increasing extent, the process became deliberate rather than inadvertent. Orchards of fruit and nut trees were planted. Symbolized by the establishment of Arbor Day in 1872, tree planting was encouraged for decorative and esthetic reasons. It was officially promoted by Congress with the passage in 1873 of the Timber Culture Act. Later, watershed protection and timber famine were added to the list of fire protection concerns. On the plains themselves industrial forestry attempted reforestation in the Nebraska Sand Hills in the early 1900s, and in the

1930s government agencies advocated broadcast afforestation—for example, the Prairie States Forestry Project, also known as the Shelterbelt program—as a means of retarding soil erosion.[71]

Whether naturally or artificially; whether in advance of settlement or as an accompaniment to it; whether delayed by logging, grazing, or temporary occupation by farms; whether occurring as a natural consequence to settlement or as a deliberate policy of forest reservation or reforestation—the transformation of grasslands, prairies, and savannahs to forests is one of the most fundamental and widespread outcomes of European colonization. It replaced a fire regime adapted to short-cycle fires with one based on a much longer cycle—and thereby created problems and possibilities still unsolved. But in the early stages, control of existing fire practices was the guiding principle of conversion. In the case of the Sand Hills, for example, where foresters were determined to demonstrate how trees could reclaim the grasslands, fire was a serious obstacle. Twice—in 1905 and again in 1910—young pine plantations were obliterated by prairie fires entering the reserve from outside. Not until fire was contained through an elaborate network of fuelbreaks did a forest eventually survive.

Even where trees or woody species did not reclaim sites, the natural grasses that had supported fire were vanishing nonetheless. From the most remote meadow of the Sierra Nevada to the Great Plains, exotic grasses infiltrated, supplemented, and even displaced native species in areas that were fired, heavily grazed, or plowed. In the intermontane West, cheat grass, an exotic from the Russian steppes has spread like an epidemic, thriving on fire and seizing disturbed sites. On the plains, in particular, Russian thistle (tumbleweed) has become widespread, encouraging fire with its ability to become a wildblown firebrand. Not all exotics are pyrophytes, of course. Kentucky bluegrass, for example, and specially selected pasture grasses have been deliberately introduced at many sites to improve nutrition and to replace more pyrophytic native grasses. Heavy selective grazing has further reconstituted pre-Colombian grass complexes.[72]

Nor has the problem of grass fires entirely disappeared. For the Bureau of Land Management (BLM), which supervises the grazing lands established by the Taylor Grazing Act (1934); for the Fish and Wildlife Service (FWS), which manages a host of wet grasslands for natural wildlife; for the National Park Service (NPS), with its grasslands to be maintained in a natural state; for the National Grasslands managed by the Forest Service—for all such agencies and their lands some program of fire management is essential, a mixture of applied and withheld fire depending on whether a subsistence or an intensive yield is desired. For these agencies grassland fires have often had special historical significance apart from their role in habitat conversion. The conflict over herding and forestry focused on the pine savannahs of the South and the western Cordillera. The problem of sawgrass and pine in the Everglades led the NPS in the early 1950s to enter into fire research. The disastrous 1964 Elko fire in Nevada brought the BLM into big-league fire management. Its experiences with prescribed burning on wildlife refuges like St. Marks in Louisiana helped to shape the FWS's fire program, as did the marsh fire at the Seney refuge (Michigan) in 1976. Where suburbs crowd into grassy countryside, urban and rural fire protection has been challenged in regions as diverse as Idaho and California.

Writing in 1910, at the opening of the light-burning controversy, Hu Maxwell of the U.S. Forest Service surveyed with care the vast changes wrought in Virginia by Indian fire practices. Especially appalling to the professional forester were the extent of forest devastation and the advancing frontier of grasslands. To be sure, the migrations and displacements among the tribes themselves often gave the forests a chance to recover, even in advance of white settlement. But Virginia, Maxwell concluded with considerable hyperbole, "between its mountains and the sea, was passing through its fiery ordeal, and was approaching a crisis, at the time the colonists snatched the fagot from the Indian's hand. The tribes were burning everything that would burn, and it can be said with at least as much probability of Virginia as of the region west of the Alleghenies, that if the discovery of America had been postponed five hundred years, Virginia would have been pasture land or desert."[73]

The effect of European settlers was in general to reverse that frontier, first by occupying the land and then by bringing with them the forest, an environment that Indians found largely uninhabitable. Reforestation, primarily through direct or de facto fire control, has succeeded beyond the wildest dreams of the young aboriculture and forestry movements. But the expansion of forests has come with a price: the resulting habitat has not always been that which was desired or forecast; more woods and deeper forest litter have meant increased hazards of high-intensity fires; and some of the despised methods of "Paiute forestry"—of Indian and frontier fire practices—have had to be incorporated into fire management systems. Rather than a timber famine, many areas are choked with forest weedlots, and even prime timber sites have become impenetrable thickets. There is more to land than forests. And to save the land it has often become necessary once again to thin out the woods, to rediscover the virtues—so long known to early and primitive man—of grasslands and the fires that sustained them.

PAIUTE FORESTRY: A HISTORY OF
THE LIGHT-BURNING CONTROVERSY

The hand that kindles cannot quench the flame.
—Byron, "Lara"[74]

The hand that holds the brand will never be burned by the fire.
—Tuareg (North Africa) saying[75]

I

Wildland fire protection was an invention of the industrial revolution. Its successes rested not so much on its suppression of fires as on its suppression of fire practices. Wherever industrial forestry was introduced, the most immediate point of conflict in land use was usually fire. Nearly everywhere, the controversy had strong political overtones. In British and French colonies—for example, India, Burma, and Algeria—fire control was tied to colonial rule.[76] In the United States, where the conflict was complicated by democratic institutions, the debate related to the question of public control over timbered land. Most parks and forest reserves were carved out of the public domain and a good bit of the land was still in territorial status, administered from afar. The proposal to extend fire and forestry policies over even private holdings smacked of quasi-colonial economics. The decision over which fire practices to support and which to suppress had immediate implications for frontier economies, and fire control was often the first expression of administrative presence felt on what were formerly open lands—the national commons of the public domain. In England the forester had been a hated official since medieval times, an introduction of the Norman overlords and the means by which the folk were deprived of many traditional rights of access to the woods.[77] In the United States, too, though in a more benign form, he was again seen as the official of foreign powers and ideas and as the enemy of folk usage of forest and range. The issue of proper fire practices became more than a question of technique; it developed into a political protest against intrusion by professional forestry, a European import, and against the enclosure of common lands.

The conflict in America was a stubborn one, and at one point it enjoyed considerable public notoriety in the form of the famous "light burning" controversy in California. Light burning was in reality not a single technique or dogma but an expression that summarized generations of frontier fire practices. Local practitioners insisted that by broadcast underburning they were following "the Indian way" of wise forest management; foresters dismissed the proposal as mere "Paiute forestry." The controversy was at base a conflict between two sets of fire practices: one set learned largely from Indians and sustained by a frontier economy of hunting, herding, and shifting agriculture; the other set, better suited for industrial forestry. There was no a priori reason why American forestry should have rigorously excluded all forms of broadcast burning. Franklin Hough, for example, had advocated broadcast fire for the Pine Barrens of New Jersey, and the British in Asia learned after some decades of painful experience that native prescribed burning could be incorporated into industrial forest management.[78] American foresters eventually learned the same lesson, recognizing the potential value of broadcast burning under certain conditions. What they refused to admit, however, was that frontier laissez-faire fire practices could substitute for systematic fire protection under the direction of expert foresters. They saw light burning as a political threat, not as a management technique.

It was light burning, however, more than any other practice, that contributed to the establishment of systematic fire protection. The relationship exhibited a peculiar symbiosis. Not until government foresters began to control traditional fire practices did light burning as a program define its premises and refine its techniques, and only when they were confronted with light burning as a defined alternative did American foresters invent the program that became known as systematic fire protection. Both were American contributions to the history of fire. Gifford Pinchot, for example, learned nothing about the concept of light burning at the French forestry school at Nancy but first encountered it during a trip to Oregon.[79] At the same time, his formal training in France and Germany could not have led to a program of systematic fire protection. The fire scenes of western America and western Europe were irreconcilably different. Light burning, as distinct from pasture burning, appeared where there remained great belts of mature pine. It flared up in Oregon, Arizona, and South Dakota; it spread across the South; it was even debated for the lands around Washington, D.C.[80] The most celebrated incident, the one that focused national attention, occurred in the great pine forests around Mount Shasta and the northern Sierra Nevada in California.[81]

Advocates of light burning in California appeared in print during the 1880s. For the most part the spokesmen were settlers and timber owners, who saw in periodic surface fires a means by which to reduce fuels and the likelihood of conflagrations. The concept was also given wide currency through the pen of Joaquin Miller, a sentimentalist poet who saw in the technique a return to the "Indian way" of land management. Timber owners embraced it as a means to lessen fuel accumulations among their stands of mature timber, a practice long used by turpentine operators. Stockmen liked it for its encouragement of pasturage and its reduction of brush. Settlers saw in it an ancient tool for landclearing and for the management of fallow fields. "The people of the region regard forest fires with careless indifference," wrote an official with a boundary survey crew on the Plumas reserve in northern California.

Timber has been, until within a few years, of little value. Everyone has plenty for his own needs. To the casual observer, and even to shrewd men, who do not realize that the prosperity of a country may depend upon its capacity to grow timber, the fires seem to do little damage. The Indians were accustomed to burning the forest over long before the white man came, the object being to improve the hunting by keeping down the undergrowth, which would otherwise shelter the game. The white man has come to think that fire is a part of the forest, and a beneficial part at that. All classes share in this view, and all set fires, sheepmen and cattlemen on the open range, miners, lumbermen, ranchmen, sportsmen, and campers. Only when other property is likely to be endangered does the resident of or the visitor to the mountains become careful about fires, and seldom even then.[82]

In 1889 even John Wesley Powell—then at the height of his political powers—lectured Secretary of the Interior John Noble on the virtues of Indian fire practices. As Bernhard Fernow sourly recalled, "Major Powell launched into a long dissertation to show that the claim of the favorable influence of forest cover on water flow or climate was untenable, that the best thing to do for the Rocky Mountain forests was to burn them down, and he related with great gusto how he himself had started a fire that swept over a thousand square miles." Pinchot politely dismissed the latter episode as juvenile vandalism.[83] But it may well have been Powell's allusion to Paiute fire practices (he was an enthusiastic ethnographer of the Paiutes) that led to the subsequent denigration of light burning as Paiute forestry.

As forest reserves and parks were established and given stiffer guidelines for the protection of the land from fire, public protest became more vocal and specific. In 1902 the San Francisco *Call* published a letter from H. J. Ostrander attacking the "protectionist" policy of fire control as worse than ineffective because it allowed fuel hazards to accumulate. An editorial endorsed this position. Speaking for the opposite position, a special National Academy of Sciences committee on forest reserves urged a tough fire control policy; John Muir raged against promiscuous fires, particularly those set by herders in the Sierra; and Marsden Manson enunciated the protection policy of professional foresters and right-thinking conservationists in a 1902 essay published by *Water and Forest*.[84]

Two years later Ernest Sterling, an inspector with the Bureau of Forestry, reviewed the California fire situation and "the attitude of lumbermen toward forest fires." It was, he concluded, one of fatalism. Fires were con-

sidered inevitable, and they were usually ruinous. Without fire protection
there was little incentive for permanence and still less for sustained yields.
Sterling observed that in California, however, two experiments were underway to cope with the fire menace.[85] One approach, adopted by the Diamond
Match and the McCloud lumber companies on Sterling's advice, emphasized a fire suppression program. The other, exemplified by T. B. Walker,
a prominent timber owner around Shasta, involved light burning. The
"Indian burning system," as Sterling termed it in 1905, was considered
wholly impractical on a large scale; it could only maintain a status quo,
retarding the reproduction on which sustained yield depended. Worse, its
"moral" impact might be to sanction the laissez-faire fire practices so common throughout the countryside and so destructive in both fact and spirit to
modern conservation.

That same year, 1905, two other events supported Sterling's conclusion:
the Bureau of Forestry took over the national forest system, and the California legislature passed the Forest Protection Act, which included fire protection provisions suggested by Sterling. Among timber owners, the controversy festered. Proposals to incorporate a private timber protective
association in California on the model of those in Idaho, the Northwest, and
the Northeast floundered because of dissension over proper strategy for fire
protection. Even after an association appeared in 1910, it remained deeply
divided.

In 1909 T. B. Walker himself published an article for the National Conservation Commission in order to make public his experiences.[86] In the
August 1910 issue of *Sunset* magazine G. L. Hoxie, another timberman
from the Shasta area, went so far as to advocate that light burning should
be not merely accepted but made mandatory. Some agency of the government, he insisted, ought to broadcast burn all the forests of the state, with
the costs charged to each landowner.[87] Forest owners who did not practice
light burning were a menace to those around them. But this was too much;
even protectionists had not gone so far in proposing government intervention
on private lands. Unfortunately for Hoxie's cause, his extremist essay
appeared in the same month as the worst holocaust in twentieth-century
America. Perhaps 5 million acres burned on the national forests alone, some
3 million in Idaho and Montana. Even Walker had lost control of one of his
benign light burns; the Widow Valley fire exploded over 33,000 acres until
it was finally stopped at the boundary of a national forest.

The experience in the Rockies traumatized the young Forest Service,
compounding its humiliation at Pinchot's dismissal earlier in the year. When
the victor in the controversy that had resulted in Pinchot's ouster, Secretary
of the Interior Richard Ballinger, then lent his support to light burning as
a means of controlling the fires that the Forest Service was apparently
unable to contain otherwise, the Service could easily equate light burners

with its political enemies.[88] It knew, moreover, that the large conflagrations of the past half century, including the 1910 holocaust, had resulted from traditional woodsburning habits and that such practices were responsible, in the form of surface fires, for the deforested landscape that would lead the United States into a timber famine.

The Forest Service poured out bulletins written by its highest officials describing how to organize a fire protection system. The newly installed Chief Forester, Henry Graves, led the fight. "The first measure necessary for the successful practice of forestry is protection from fire," he wrote. "As long as there is any considerable risk from fire, forest owners have little incentive to make provisions for natural reproduction, to plant trees, to make improvement cuttings, or to do other work looking to continued forest production." Graves allowed for controlled burning of piled slash; for "broadcast burning . . . in making clear cuttings, provided the fire can be confined to small areas and fully controlled"; and for "deliberate burning of the litter as a protective measure," though "only under special conditions and only on selected areas," such as turpentine sites. But he made it clear that "annual burning for fire protection is never justified where it can not be systematically controlled." Where fire might be used profitably, it was directed only to the promotion of forests, not, as "in many parts of the South and West," for range improvement. "Merely setting fire to the woods without control is nothing less than forest destruction."[89]

Nevertheless, light burning, as practiced by Walker, was intended to serve fire protection goals, or so it was argued, and Graves even approached the timberman with the suggestion that Walker underwrite a chair of fire protection at the Yale School of Forestry. Walker agreed, contributing $100,000, and Graves authorized a broad range of field experiments to test the relative merits of light burning within parameters set by professional forestry.[90] Over the next decade, trials were conducted across the United States, especially in the pine belts of the West and South. The most celebrated were those in which the Forest Service met light burning on its own ground, and this confrontation was largely the work of Stuart Brevier Show. In the meantime, F. E. Olmstead, a ranger in California, replied specifically to the claims of Hoxie and Walker in a Forest Service circular titled *Light Burning in California Forests*. This essay fixed both the terms of the debate and the names of the antagonists; previously, "light burning" had represented an amalgamation of practices without a particular title or theory attached to it. In that same year Coert duBois, a ranger in California, began the studies that eventually culminated in his masterpiece, *Systematic Fire Protection in the California Forests* (1914).[91] Thus the two antagonists— light burning and fire protection—came into definition in tandem. Moreover, with the passage in 1911 of the Weeks Act, the Forest Service expanded into the piney woods of the South, and a southern version of the light-burning controversy immediately engulfed its management program for Florida.[92] Thus from the West Coast to the eastern seaboard, from the holocausts of the Northern Rockies to the roughage burning of the coastal plains, the Forest Service found itself undergoing a trial by fire.

First reports from the light-burning experiments were mixed. Show's initial reports attacked light burning on economic grounds (its costs were higher per acre). DuBois undertook a study on brush burning, but his results were equivocal. Believing no doubt that fuller research would demonstrate the folly of traditional burning and not wishing further to inflame local sentiment against official policy, the Forest Service suppressed duBois's results, though it allowed him to continue the program.[93] In the South, officials on the scene proclaimed their inability to manage the forests without some form of controlled burning. But it was in California that the issue became public, and there Show's continued studies provided the data for the Forest Service's official position. Show conducted some trials on lands of the Red River Lumber Company, owned by Walker, and initiated some other trials at Castle Rock, near Shasta. What his initial progress reports had made ambiguous, his subsequent reports made firm: beginning in 1915, he declared that light burning was fallacious as an economic policy, impractical as a method of forest management, and ineffective in comparison with fire control of the sort developed by duBois in his *Systematic Fire Protection in the California Forests*. Light burning sought an immediate reward, the preservation of mature forests, at the expense of long-range forest values, reproduction and soil protection. The Forest Service took the position that light burning was at best an expedient, to be used until formal protection could be established; at worst, it was a cynical sham promoted by timber barons in order to avoid their responsibility for the management of a public resource. In place of a poorly sytematized and somewhat emotional mélange of folk customs, the Service answered with the cool, powerful logic of duBois's program and Show's experiments.

But the issue would not subside. In 1916 light burning found a popular and articulate spokesman in the novelist and timber owner Stewart Edward White, and White was willing to take the controversy out of the Forest Service files and bring it to the public. White owned considerable timber land and became attracted to broadcast fire as much for its promised control of insect epidemics as for its fire protection through hazard removal. The controversy challenged the Forest Service on three levels: advocates of light burning attacked the Service's ability to make systematic fire suppression work, questioned the credibility of professional forestry as a repository of technical expertise, and, by implication, brought under scrutiny the whole apparatus of conservation by which important natural resources would be managed by government bureaus. More than a choice between various techniques of fire protection, the controversy—as perceived by the Forest Ser-

6. Coastal redwood forest converted to pasturage through successive burning, California, 1936.

vice—evolved into a choice between two diametrically opposed philosophies of forest management. As the lines of debate hardened, the possibility of compromise became less likely. Individual landowners could practice light burning, but systematic protection required organization, interlocking agreements, and research. The Forest Service had insisted that it should manage the forest reserves precisely because it offered something different from frontier practices. If it adopted the same methods, then the public would naturally ask why the Service was especially qualified to administer its lands.

In 1919 White was joined by Joseph Kitts, a former captain in the Army Corps of Engineers. A curious pattern of support for light burning was developing. With the exception of White, nearly all the public proponents of light burning from 1910 to 1930 were, like Hoxie and Kitts, civil engineers—men who were perhaps not especially impressed by the credentials of foresters but for whom the association of increasing amounts of fuel with increasing intensities of fire came naturally. Though his experience was limited to a handful of acres in Grass Valley, California, Kitts's articles promoting light burning as distinct from the "protectionist" policy of the Forest Service were viewed as a serious menace to "public confidence" in Service administration. The Service responded with a barrage of articles in both popular and professional journals and with new scientific experiments on fire behavior, fire effects, and controlled burning.[94] Show supervised the trials, which were as much demonstrations as experiments. Together with E. I. Kotok, he began an elaboration of duBois's *Systematic Fire Protection* with a series of bulletins that strengthened the hand of the Service. By now Show had little sympathy for light burning as either a technique or a philosophy of forest management.

By 1920 the controversy had reached the popular magazines. *Sunset* promised an exposé by White on the protectionist policy of the Forest Service and only agreed to run rebuttals by Chief Forester Graves in alternate issues after the Forest Service threatened lawsuits. The Service also promised legal action against the city of Seattle if its mayor allowed posters advertising White's position to continue to be displayed on public buses.[95] For *The Timberman,* meanwhile, Graves blasted light burning as mere "Paiute forestry," impractical for the management of industrial resources like timber and not worthy of a professional cadre of scientific experts. His successor as Chief Forester, William Greeley, followed two months later with an article on "Paiute Forestry or the Fallacy of Light Burning."[96] The first major crisis both men had faced was the 1910 fire, with Graves as Chief and Greeley as district forester. The last crisis in Graves's tenure as Chief, and the first in Greeley's, was the light-burning controversy. Though they often resorted to inflammatory rhetoric (Graves once titled a published reply "The Torch in the Timber"), both men argued on the apparent evidence: light burning damaged reproduction, thereby endangering future forests; it was more costly than systematic protection, requiring special treatment of mature trees to prevent against butt fires; and, in comparison with the rapidly improving organizations for systematic fire protection, it was less effective.

Nevertheless, the right choice was far from obvious. For the past decade, even within the Forest Service in California, there had been a wide range of experimentation to determine the best system. Only with the Mather Field Conference of 1921 did the duBois, Show, and Kotok school come out unequivocally on top within the Service. Outside of it there was considerable consternation. Some landowners began to doubt the wisdom of contracting with the Forest Service to furnish fire protection, and the enactment in 1919 of a highly unpopular state statute requiring mandatory fire patrol of some sort caused considerable resentment. After a bad fire season, the Service could count on a number of frantic owners to backslide into "the Indian way" of fire protection. The California controversy, moreover, was attracting national attention, and the Society of American Foresters, alarmed over

the tenor of the debate, offered its services in arbitration. It arranged for a
forum of three meetings, one private and two public, so that some amicable
solution could be worked out. In January 1920 the California Board of Forestry suspended its endorsement of either strategy until full investigations
were made.

"These meetings," as the *Journal of Forestry* reported in 1923, "accomplished little directly save expose the actual and potential danger of the
existing controversial situation, but it was largely as a result that the California Forestry Committee was organized."[97] The committee consisted of
members representing both sides, plus foresters and timbermen who were
nonaligned. One committee member, embittered over the result, sourly
recalled that its purpose was to "keep the agitation out of the newspapers
as much as possible."[98] In October 1920 the committee issued a verdict con-
demning light burning. The majority reported that, after inspection of sites
where burning was practiced, it had only learned about how and when *not*
to burn. Echoing Show's findings, it announced that it had not seen any
practical examples of successful burning.

One member dissented, however. A representative of the Southern Pacific
Railroad, which held extensive timber lands, he asked that light burning be
subjected to experimental tests. Southern Pacific that summer had suffered
a 10,000-acre fire along Moffat Creek in Siskiyou County, and it was proposed that portions of the burned acreage be used as a test site. The committee agreed. The state forester and Forest Service assigned top personnel,
included E. N. Munns and, of course, Show. While researchers waited
patiently, the entire 1921 season passed with unfavorable weather, preventing any attempt at light burning. In 1922 a small site in the Moffat Creek
area was eventually burned with "unsatisfactory" results. For the third year
more specific tests on fire effects and fire behavior were planned, again with
inconclusive results: conditions were either too damp or too dry in the spring,
and fall firing was "abruptly terminated by heavy rains."[99] Meanwhile, the
Forest Service had caused a sensation with its aerial fire patrols, conducted
with Army planes, and had ended a decade of uncertainty within its own
ranks with the Mather Field Conference of 1921. The conference was held
in California, but it overhauled and standardized fire protection for the
entire Service. Show and Kotok, too, consolidated a decade of research on
California fire with a formidable "analytical study" released in 1923.

The state committee reported its findings that same year through Donald
Bruce, professor of forestry at the University of California. Bruce carefully
listed the theses of both sides, and, in describing the committee's experiments, wrote perhaps the wisest observation to emerge in 20 years of ran-
corous controversy. "The issue was a practical one which involved not so
much the truth or fallacy of a theory as a practical and economic application
of whatever truth there might be therein." He continued: "The concrete

problem therefore, was not the correctness of certain theories but rather a determination of whether any modification in the existing system of fire protection could profitably be developed therefrom." But, of course, the issue had become much larger, and the controversy tended to reduce itself to a choice of one system or the other, each complete in itself. The protection system of the Forest Service, the committee found, was "definite, standardized, and well-understood." The light-burning theory, unfortunately for its advocates, "when translated into a concrete program of work was not a simple or single idea." Light-burning proponents "seemed to have in common only their opposition thereto [that is, to Forest Service policy] and their reliance on the . . . postulates already stated."[100] The "program" was simply a reworking of frontier fire practices, grounded in empirical folklore. The committee unanimously concluded that light burning was not superior to the protection policy of the Forest Service. The California Board of Forestry concurred later in 1924—a year that also witnessed another very bad fire season for California, saw the passage of the Clarke-McNary Act, and found the final consolidation of the scientific evidence for the protectionist policy in the publication of Show and Kotok's *Role of Fire in the California Pine Forests.*

Show and Kotok noted that "the role of fire in the pine forests is a long and complicated process, which has operated on an enormous scale for centuries, as the evidence in the remaining understocked forests and brush fields show." But they carefully detailed the insidious and cumulative effects of light burning. "Forest deforestation starts with the fire scar, of no great consequence itself, leads through the understocked but still merchantable virgin forest, the open forest with brush understory, the restocking brush field with scattered forest trees, the complete occupation of the soil by brush, finally with the aid of soil deterioration and soil removal to the chaparral type no longer capable of supporting forest trees." The two men distinguished between light burning, "an attempt to insure the safety of the merchantable timber," and promiscuous forest burning, "which disregards forest values, and aims only to improve grazing, facilitate prospecting, or render the forest more open." The latter was unspeakable; the former deserved at least a trial. What Show and Kotok did, however, was to demonstrate that the long-range effects of the two were often indistinguishable, that "existing difficulties in protecting and managing the forest area—difficulties due to past fires—will increase unless virtual fire exclusion can be put into effect." It was in the interest of the entire state, they concluded, "to recognize that fire is a destroyer and that protection to attain success must amount to fire exclusion particularly in brush fields and on cut-over lands."[101]

The California Forestry Committee took a less panoramic view but listed ten familiar reasons for deciding against light burning. In brief, it determined that the techniques for underburning were not systematic, being largely handled by folk practitioners, and that the costs of preparing and protecting a site could be high. Successful light burners, such as turpentine owners in the South, sometimes found it necessary to protect each individual tree. Graves had witnessed prescribed burning in the chir pine of India, but the British could rely on hundreds of native laborers at eight cents per day; American foresters could not. There was, too, the telling argument that light burning threatened reproduction, the future of American forests, and that it challenged institutional recognition for a national conservation program, as epitomized by fire protection, and thereby the future of American foresters.

To an intellectual elite trained in Ivy League schools and enamoured of the technology transferred from Europe, "Paiute forestry" could no more benefit an industrial civilization than would the Paiute diet of rabbits, grasshoppers, and seeds. Against folk wisdom that forests were better replaced by other land uses, against folk economies that relied heavily on habits of open fire, and against the folk fatalism that large fires were acts of God to which man could offer no resistance—the new intelligentsia argued that fire protection was practical and desirable. As formulated by White and others, light burning was something more than mere folk fire practices, but it was also something less than professional forestry. It was too closely allied with a heritage of frontier fire. As Graves put it, "light burning did not propose a system of controlled burning. It argued for a crude system of woods burning."[102] Foresters had proclaimed that fire protection was 90 percent of forestry. Any questioning of the tenets of fire control might jeopardize not only the administration of the national forest system but also the entire edifice of agreements between the Forest Service and state and private organizations, an edifice founded on fire and one that made national forestry programs possible.

The spirit behind both conservation and social thought during the Progressive Era, as later with the New Deal, was antithetical to laissez-faire practices of all kinds. It sought rather to organize, to protect, and to appeal to technical experts. To suggest that fire could be controlled by more fire would be like the new Bureau of Reclamation suggesting that some periodic mild flooding was beneficial, or the reconstituted Biological Survey recommending an increase in coyote and wolf populations, or the Food and Drug Administration urging limited dosages of toxic chemicals as a prophylactic against future massive doses. Recommending more fire in 1910 would be like encouraging more oil spills in the Santa Barbara channel today. After all, light oil seeps have contaminated the channel for eons; the ecosystem must have adjusted to such episodes by now and probably requires them to maintain an equilibrium. Instead of suppressing all spills, officials ought to distinguish between catastrophic spills and frequent low-intensity leaks. In fact, the capping of offshore wells has allowed pressure to build up in the reservoirs and may result in a blowout well. Better to release that pressure gradually through a schedule of calculated leaks. Maybe a program of prescribed light seeping would be in order. Such a program would be ridiculed, of course, and its proponents typed as cynical lackeys of oil magnates. The damage that oil spills do is obvious to all: they kill wildlife, destroy plankton (translate: soil), and blacken recreational areas into an unusable scenic blight. Oil is a precious industrial commodity, and it is probable that national resources will be shortly exhausted. What is needed is an aggressive program of spill cleanup and prevention. Even to suggest light seeping might compromise the credibility of the entire environmental movement.

As Donald Bruce observed, the issue in the light-burning controversy was not the ultimate truth of the proposition but its practical value. More properly, the issue was the larger cultural and historical context in which the controversy occurred. After 1924 light burning became an official heresy, though one that was never extirpated. It was possible eventually to employ nearly the same practice, but not to call it by its traditional name. In a 1927 letter to the supervisor of the Mendocino National Forest, for example, where local residents stubbornly refused to give up traditional range burning, Show referred to the "practice of administrative burning which has been done on the [forest] for many years," largely to appease locals. To Show it seemed "emphatically true that not only does the administrative burning fail to solve the fire problem to any great extent but its continuation is undesirable from the standpoint of National Forest policy," especially as it was understood by the public.[103] Whatever their private thoughts might be, or whatever local accommodations might have to be made, foresters presented a unified front to the public.

Like a deep ground fire, however, light burning continued to resurface periodically in California and to inflame public debate. In August 1928 Colonel John White, superintendent of Sequoia National Park, wrote an open letter to the Los Angeles *Times* congratulating the paper for not publicizing a fire that had burned into park lands. The state had intervened and after three weeks suppressed the fire. Not only did park officials withhold assistance; they complicated efforts by torching off a backfire under questionable circumstances. White insisted that the main fire was a fine example of light burning and was to be applauded. That same month, alarmed at the costs of suppressing large fires, Willis Walker of the Red River Lumber Company announced that his firm intended to return to a program of light burning. A few months later William H. Hall, a former state engineer, attempted to reopen the issue with appeals to the governor, the Commonwealth Club, and the California Board of Forestry.[104]

Nothing came of these efforts. Nationally, fire protection had been strengthened by the Clarke-McNary Act and the Forest Protection Board.

The National Park Service responded to Colonel White's challenge by appointing John Coffman, formerly supervisor of the Menodino National Forest, as a national fire control officer. The California Board of Forestry refused to reconsider the subject of light burning. The Forest Service created the Shasta Experimental Fire Forest as a model for the administration of systematic fire protection. Foresters looked on light burners with the smiling condescension mathematicians reserved for circle-squarers, and physicists for perpetual motion mechanics. When in 1950 someone again proposed a form of light burning to the California Board of Forestry, Kenneth Walker—grandson of T. B. Walker, nephew of Willis Walker, and a timber owner in his own right—promptly replied: "We don't want fire under any conditions in the woods."[105]

II

Of all the light-burning controversies, the one in California received the most publicity. But it was not the most stubborn or, in the long run, the most influential. As the Forest Service extended its administration onto new lands—like the cutover lands of the South and Lake States—it found it necessary to embrace new techniques, prescribed burning among them. This process was well underway by the late 1920s, and it was one reason that the light-burning controversy in California faded at that time from national significance. In a sense, the controversy moved to the South, though it assumed there a somewhat different form.

It had become an article of faith that southern forests had been cut over, grazed over, and burned over. The potential for southern forestry was stupendous, and logging had reached a peak by 1920. But, foresters insisted, until lumbering was regulated, hogs and cattle restrained, and fire eradicated, industrial forestry would never become a reality. Especially worrisome was the failure of the fabulous longleaf pine to regenerate. The case of the longleaf gave the South a specific focus, which in California had been lacking. Speculation about the longleaf's dependence on fire was outstanding; even Charles Lyell had mentioned the idea in his account of his visit to the United States in the 1840s. In 1907 Forest Service experts in silvics wondered out loud if fire might be useful in site preparation for longleaf, and H. H. Chapman, dean of the Yale School of Forestry, began his long study of fire and the southern pine, first in the Ozarks and then in Alabama. Many early timber owners, otherwise responsive to industrial forestry, advocated and often practiced some form of light burning as a means of fire protection by reduction of the famous southern rough.[106]

After the Weeks Act allowed the Forest Service to move into the South as both landowner and fire cooperator, the question of woodsburning almost immediately became a cause célèbre within the organization. Inman Eldredge of the Florida National Forest explained to the Society of American

Foresters in 1911 that the southern forest was unique within the national system, and that "the general methods of management in use throughout the Forests of the West can be applied to it but seldom and then only with modifications. This difference is perhaps most clearly shown in the matter of fire protection."[107] One of the modifications Eldredge proposed was a system of controlled burning, though it would be tied to the cycle of forest growth rather than to the annual cycle of grass. His article suffered from timing, however: others addressed the society on how systematic fire protection could prevent holocausts like the 1910 fires, and the publicity about the light-burning imbroglio in California shifted national attention to the pine belts around Shasta.

As elsewhere, the following decade was a period of experimentation, and many investigations into fire effects were conducted in Florida, Arkansas, Georgia, and Louisiana, both within the Forest Service and by men like Chapman. Sub rosa burning under the guise of "administrative experimentation" was apparently widespread, promoted by officers who felt the need for protective burning. It was during this efflorescence of research that burn plots were established at Urania, Louisiana, under the auspices of the Southern Research Station. One plot was fired annually, while the other was protected from both hogs and fire. The plots were intended, like some of Show's later studies, as much for demonstration as for experimentation: their purpose was to show the value of fire and grazing protection. Instead of periodic burning, as Chapman had come to favor, the burn plot was fired yearly. With this somewhat perfunctory effort, the Forest Service rested its case. Its evidence and arguments against light burning was being honed in California, where it held far more land, and it saw no distinction between light burning in the Far West and woodsburning in the South. It was determined to bring to the South the fire protection system that the California Forestry Committee had applauded. The great critique against light burning had been its damage to reproduction, and in the cutover lands of the South the need for regeneration seemed to provide prima facie evidence for the value of aggressive fire control.

Field men thought otherwise. In January 1919 the whole controversy surfaced amid a flurry of reports and memorandums. When the matter of controlled burning had appeared on the agenda of the national Service Committee a few months previously, the official opinion was that such a program "would establish a dangerous precedent."[108] But field officers in the South felt that fire control on the western model was inappropriate, expensive, and ineffective. The assistant and acting regional foresters appealed to the Washington office for a "somewhat radical change in policy," namely, that "we abandon our expensive and futile efforts at fire prevention and substitute for them a system of controlled burning." Having in mind, however, "the controlled burning agitation of the past," assistant regional forester

H. O. Stabler recommended that it was "desirable to make this radical change of plan in Florida in a quiet and unostentatious manner. . . . I am sure we can get by without stirring up trouble in the 'near east' or in far away Oregon and California. We are on the wrong track now and each succeeding year . . . adds to our risk of having a bad fire or fires in the wrong place at the wrong time." E. E. Carter, who forwarded the report to Chief Forester Graves, added that "this change need not be given wider publicity than is necessary and should, as far as possible, be considered as an experiment in applied forest management."[109]

The Washington office was reluctant to approve anything that might compromise its position in California, and most foresters considered protective burning to be an interim measure, acceptable only until a systematic fire protection program could be brought to bear. As a whole, the Forest Service thought of the regeneration of southern pines in terms of the reproduction of western pines, considered the southern rough as analogous to the California brushfields, and thought it saw behind the silvicultural argument for light burning the more insidious plans of herders to retain cutover lands for pasture, much as stockmen had supported protective burning in the Cordillera.

With respect to the southern pines, the torch, so to speak, passed to nongovernment foresters, most notably, Dean Chapman of Yale. In 1926 he published in the *Yale Forestry Bulletin* the results of his lengthy study of fire and the regeneration of the longleaf pine.[110] He credited periodic (not annual) surface burning as useful in site preparation, in fuel reduction, and in the control of brownspot disease. The Forest Service answered the challenge with a research program directed by E. N. Munns, who had supervised the burning experiments for the California Board of Forestry. Chief Forester Greeley, meanwhile, summarized official feelings in an August 1928 memorandum. "Light burning," he concluded, "is the most pressing forestry problem in the South today." The practice threatened "the effectiveness of organized fire protection," and "though there is an element of good in light burning," it created "conditions of uncertainty in timber growing" that could not be tolerated, particularly since broadcast fires rarely remained on the site at which they were initiated. The Forest Service, Greeley wrote, "is prepared to back up State agencies to the limit in their organized fire protection work." It recognizes that "it has a distinct responsibility of leadership—demonstration—in connection with this light burning problem, to be redeemed on the National Forests of the South as it was in California years ago when many people advocated light burning." And it "has the further responsibility to demonstrate that organized fire protection warrants the risks taken under this policy." Light burning on national forests would not be permitted "except as it may be necessary to comply with existing turpentine agreements, to carry on experiments, or to establish protective strips and fire breaks." Here, the Chief conceded, was a "big Public Relations program."[111]

Meanwhile, a 1927 ruling on the Clarke-McNary program withheld funds from those states that tolerated controlled burning. The research on fire and silviculture begun in 1928, moreover, coincided with the Southern Forestry Education Campaign sponsored by the AFA, which had as its goal the eradication of the southern woodsburning habit. But Chapman's example had inspired others. In 1929 S. W. Greene of the Bureau of Animal Husbandry published findings that, contrary to Forest Service propaganda, showed that burning did increase the palatability and nutriousness of southern pastures. Two years later he followed with a famous broadside, "The Forest That Fire Made," which argued for the desirability of periodic fire in the longleaf forests. Also in 1931, H. L. Stoddard, working with the Biological Survey, showed the value of controlled fire in habitat maintenance for wildlife management. A year later the Forest Service quietly modified its Clarke-McNary ruling, accepting controlled burning so long as it was done for specific objectives and by skilled practitioners.[112] By 1933–1934, moreover, its own Southern Research Station under Elwood Demmon had half a dozen articles ready with the same recommendation.

Unlike the light-burning controversy in California, the southern version divided foresters. It was debated in professional journals between experts, not in popular magazines between professionals and folk practitioners. Increasingly, too, a line was drawn within the organization between administrators and researchers. The manuscripts prepared in 1933 by the Southern Station, for example, were not published until 1939. Editorial pressure was strong within the Service for downplaying anything that might, however circumspectly, sanction the southern habit of promiscuous woodsburning or might play into the hands of light-burning enthusiasts in the West. Meanwhile, devastating fires ripped through the South during the drought years of 1932–1934. The Osceola National Forest in Florida lost 12,000 acres in an afternoon, and many cooperators and private companies hurried to introduce protective burning before they, too, were burned out.

When the controversy went public in 1935, it did so within the confines of the Society of American Foresters' annual convention, not within the covers of a regional magazine or within the context of public hearings. The Forest Service released a special paper that documented its research findings to date, and Chapman, as president of the SAF, succeeded in placing the topic of prescribed fire on the convention agenda. One participant noted damningly that "this is the first time that censorship on the subject has been removed and we have been told the facts." Among the speakers was Inman Eldredge, who observed trenchantly that "in silviculture, in utilization and in other phases of forest management, we foresters have fitted our practice always to the case in hand, but as regards fire, we have, for the main part, rallied behind the premise that the only way to manage any forest anywhere in the United States, from Alaska to Florida, is to cast out fire, root, stem and branch, now and forever. We have with closed ranks fiercely defended this sacred principle against all comers and under [all] circumstances and any forester who questioned its universal application was suspected of treason or at least was considered a dangerous eccentric."[113] Chapman carefully distinguished between the effects of fire in northern forests and those in southern pine.

The Forest Service replied in 1938 with the Loveridge-Fitzwater Report, which called for a comprehensive summary of current knowledge on fire and the longleaf pine. In 1939 the long-suppressed reports from the Southern Station found print at last. But already in 1935, even as Chapman and the SAF were airing their radical arguments for fire use, the Forest Service was announcing a no less radical strategy for fire control, the famous 10 A.M. Policy—a "continental experiment," as the Chief Forester called it, in aggressive firefighting and one that seemingly left little room for controlled burning.

But southerners considered their situation unique. Under the guise of administrative experimentation, a good bit of protective burning was already being conducted by worried forest officials. More bad fire years followed in 1941 and 1942, and the Civilian Conservation Corps, on which the Forest Service had counted to man its 10 A.M. Policy, vanished into the wartime draft. Chief Forester Lyle Watts inspected the southern forests the next year, and, convinced that protective burning was worth trying, he sanctioned its use in December 1943. A conference on prescribed fire immediately followed at Lake City, Florida, and roughage burning commenced soon afterward.[114] In 1946 the Forest Service published W. G. Wahlenberg's definitive *Longleaf Pine,* testimony to the beneficial role of controlled fire for silvicultural purposes. In the 1949 *Yearbook of Agriculture* Arthur Hartman announced on behalf of the Forest Service the value of periodic surface fires to southern forestry, some 45 years after Ernest Sterling had polled California timbermen for the same publication but with such different conclusions.[115]

The behavior of the Forest Service administration through the whole affair was often not merely hard-nosed but hardheaded. In his history of the controversy, Ashley Schiff noted that it was resolved finally as much by "administrators who, educated to the value of prescribed burning, commenced to persuade their colleagues, as it was by research personnel breaking down administrative barriers." He concluded somewhat rhetorically that "thus had evangelism subverted a scientific program, impaired professionalism, violated canons of bureaucratic responsibility, undermined the democratic faith, and threatened the piney woods with ultimate extinction."[116]

More was involved than mere evangelism, however, and the episode cannot be isolated from its larger historical context. The Forest Service's experiences with wildfire in the Northern Rockies and with the light-burning controversy in California had profoundly influenced the direction and rigor of its fire policy. To most officials the only differences between these regions and the South were "moral," that is, the woodsburning habit so deeply ingrained in southern society. But the situation on the West Coast was simply not equivalent to that along the Gulf Coast. In northern California, fire prevention meant that brush would succeed to timber. In the South, however, the absence of fire allowed timber to be overwhelmed with a choking cover of brush, the rough. The southern practice of woodsburning was actually a misnomer: it was the rough, not so much the timber, that burning was meant to control. To foresters, southern fire habits were a relic of those associated with the frontier, and it took a long time to sort out the practices that had little to do with forestry from similar practices that might sustain the objectives of forestry.

The process was unnecessarily slow. Schiff's analysis was intended as a study of public administration, particularly of the means by which science and policy interacted. Much more than scientific research was involved, of course. Among the federal agencies established during the Progressive Era to cope with industrialization, the Forest Service was unique. Most others, including the Food and Drug Administration and the Geological Survey, were mandated to research, to inform, and perhaps to regulate. As a result of the Transfer Act of 1905, however, the Forest Service was also required to administer land. The Geological Survey did not manage mines or rivers. The Food and Drug Administration did not manufacture and distribute food or drugs. But built into the Forest Service's administrative structure was a schizophrenia that gave it special strengths and also brought it special problems. It thought of research as a kind of engineering, a means by which larger policy could be made specific, not as a source of information from which a larger policy might flow.

With regard to controlled burning, the longleaf pine and the southern woods range were the points of ignition. Additional studies quickly extended the lessons of the longleaf to the other southern pines, and the western pines were not far behind. From the southern grasslands, too, there were lessons for the rest of the country, especially in the management of habitat for wildlife. The 1940s became a period of rediscovery. At least partly on the southern example, prescribed fire was extended to certain refuges administered by the Fish and Wildlife Service, to game habitats in Wisconsin under the control of the state department of conservation, to the ponderosa pine forests of Indian reservations in Oregon and Arizona, and even to California, where new legislation allowed brush burning for range improvement. Almost in parody of this recovery, the American military, so long disdainful of fire weaponry, rediscovered and widely deployed fire during the course of World War II. And so it went: across the country new experiments in controlled fire were undertaken by small clusters of forest researchers and academics, creating oases of refired lands. Symbolic, perhaps, of this process was the establishment of the Curtis prairie by the University of Wisconsin Arboretum. After considerable research its founders became convinced that fire was historically a part of prairie ecology, and a short period of attempted prairie restoration without fire confirmed that fire was essential to its management. Nor was the Forest Service immune. A 1945 report outlining plans for a separate Division of Forest Fire Research proposed the beneficial effects of burning as one of the topics for study.[117] Writing in 1949, H. T. Gisborne proudly read a roll call of leaders in prescribed fire for a commemorative volume published by the Society of American Foresters: H. H. Chapman, S. W. Greene, C. A. Bickford, and Arthur Hartman in the South; C. K. Lyman in the Northern Rockies; and Harold Weaver in the ponderosa pine of the Cordillera. "Prescribed burning," Gisborne concluded, "a distinctly American tool of forestry, was born, survived the usual attacks of infanthood, and has begun to walk alone in 40 years."[118]

It would nonetheless be wrong to generalize from the 1943 decision authorizing prescribed fire. During his visit to the southern forests Watts confessed that "controlled burning has me somewhat confused." He added: "Certainly I won't forget that in Florida there is an acute fire problem and that heavy equipment is one of the essential requirements for getting on top of that job."[119] It is no accident that the change in strategy announced in 1943 coincided with the development and dissemination of the tractor-plow—then, as now, the backbone of southern fire protection. Unlike light burning in California or its early versions in the South, prescribed fire was seen as a supplement to systematic fire protection, not as a substitute for it. The Forest Service was politically secure. Nor was prescribed fire advocated as a national program; as practiced in the South, it might have no more value in the western mountains than did the tractor-plow.

Yet it was largely from the South that the Service began to expand its approved range for prescribed fire. Writing to the regional forester in California, Assistant Chief Earl Loveridge described how "Hartman made quite an impression on us in Tallahassee with the truly scientific way in which he goes about the prescribed burning in R-8 [Southern region]." Loveridge proposed several ways by which Hartman's expertise might be extended to California. He concluded: "I hope the Forest Service is not dragging its heels on the use of fire in other sections of the country where it could and should be used to excellent advantage—as we dragged our heels in the South for so many years. Possibly we should plan to look in on that 'experimental burning' being done by the Indian Service in Arizona."[120] Hartman himself worried lest the swing of the pendulum might carry too far, that prescribed burning by enthusiastic but ill-informed practitioners might degenerate into thoughtless, faddish woodsburning.[121]

By 1956 A. A. Brown, director of Forest Service fire research, could describe to Hartman the progress made in California. Nine years earlier the University of California had hired Harold Biswell to teach range management courses. Based on some experiences in Georgia, Biswell immediately urged a program of controlled burning, and his arrival had coincided with a reversal in state policy that allowed the use of controlled fire on brushfields to improve range. Brown expressed "little sympathy with Biswell so far because he has made so little effort to be responsible or constructive." Biswell, he felt, "was very headstrong and very much an opportunist. He found a pretty strong group of extremists in California and got a lot of publicity for himself fronting for them. As you sense, there is a lot of old history. For many years the F.S. tried to hold the line against all burning with the exception of piling and burning logging slash. That helped to build up extremists for burning." Brown observed that "there are of course a few people in the F.S. who still have extreme aversion to using fire. But the old timers who fought the early battles against 'light burning' are all gone now. So it would not be correct to charge R–5 [California] with having a closed mind on the subject." There had been in fact a considerable change in attitude; burning was largely experimental, but "nearly every forest has a sample type project."[122]

III

The Forest Service—and through it, professional foresters at large—had never completely renounced the potential benefits of controlled fire, but they had always taken as their point of reference the influence of such fire on timber and watershed, not on land as a whole, and whatever type of controlled fire might be considered was always further compared with the reigning problem fire type that confronted the organization. One reason for the circumspection about extending prescribed fire from the South was that a new problem fire had emerged, one that commanded more attention from research and administration and gave new purposes and a new context to controlled burning. The light-burning controversy had occurred at a time when the Forest Service was struggling for its political survival and forestry for its scientific credibility. Light burning had national significance. The chief problem fire for the Service then was the mélange of frontier fire practices, and only those fire uses that did not resemble such practices were tolerated. The prescribed-burning controversy in the South came at a time when the Forest Service was expanding into cutover and abandoned lands and into the remote interiors of its western forests. Prescribed fire remained regional in character, a price of admitting the southern forests into the system. But in the same year that Watts approved controlled burning for the southern annexes and Weaver proclaimed the value of broadcast fire on the reservations of the West, Allied firebombing destroyed Hamburg, Germany.

A new problem fire, mass fire, came to dominate national attention. Even as Loveridge wrote, plans were underway in Southern California for Operation Firestop, a multiagency year-long investigation into mass fire behavior and control. Even as Brown explained to Hartman the absence of hostility toward prescribed fire, a new era of Forest Service research, centered on mass fire, was about to unfold. Fire protection left the timberlands of light burning and the cutover pineries of prescribed fire for the urban fringe; it expanded into regimes like the suburbs of Southern California and the interior of Alaska; it became a partner with the Office of Civil Defense; it investigated prescriptions for and measured the effects of mass fire against the analogy of thermonuclear warfare.

Among all the uses for prescribed fire, the Forest Service came to adopt two: the old one of hazard reduction, though in the form of fuelbreaks rather than of slash piles, a means of conflagration control; and a new one, the potential military use of controlled fire, a means of conflagration initiation. A Supreme Court ruling in 1957, moreover, held that the government could be liable for negligence not only when it acted in a "proprietary" capacity but also when it acted in a "uniquely governmental" capacity, such as in the role of a "public fireman." The case involved a fire that had been contained and then left to burn, only to explode into mass fire and leave the lands of the Olympic National Forest. The ruling, needless to say, had a dampening effect on the widespread use of controlled fire.[123]

For the era of conflagration control a new technique for fuel management was promulgated. In 1957 the Forest Service and the California Division of Forestry launched the Fuelbreak Research and Demonstration Program, which had at its core a new concept in fuelbreak design. Technically, earlier projects of similar intent had constructed firebreaks, a fireline built in anticipation of a fire. The 650-mile-long Ponderosa Way and Truck Trail, proposed by Show in 1929 and built with CCC labor, was of this genre, a Maginot Line along the Sierra Nevada separating brush from timber. The new program, however, emphasized fuelbreaks, a strip of reduced flammability between 100 and 300 feet wide and generally the result of type conversion to grass. Fuelbreaks were more esthetic, less susceptible to erosion, and easier to maintain and they often used fire for construction and maintenance. The 1957 program sought to improve conflagration control by conversion to and expansion of a system of fuelbreaks. Originally, the program was intended for Southern California, but by 1962 it had been enlarged to include the lands of the northern Sierra. The presence of inmate labor made available through California's conservation camp program substituted for the cheap manpower earlier provided by the CCC. By 1972 it was estimated that some 1,950 miles of fuelbreaks wider than 100 feet existed in California. Most of the Ponderosa Way, which deteriorated after the demise of the CCC, had been restored, and the famous International Fuelbreak had been established along 41 miles of the Mexican border.[124]

Prescriptions for mass fire also took forestry to the fringe as the Forest Service explored the possibilities for controlled burning for objectives quite beyond forestry, such as military and civil defense considerations. The doctrine of conflagration control, as epitomized by the fuelbreak program, was a policy of containment, part of a cold war on fire. The program sought to segregate red zones, or high fire regimes, from safe zones, to prevent fire from breaking out of one into the other. Collectively, the larger programs formed a kind of demilitarized zone between incompatible ecologies. Civil defense interest took fire agencies into another kind of boundary, the suburban fringe that crowded into wildlands, and military interest brought mass fire to the geopolitical boundaries of Southeast Asia. In such a context the old quarrels over light burning seemed merely quaint.

By the end of the fuelbreak program a new problem fire had appeared, wilderness fire, and it promised to take prescribed burning from the perimeter of land management to its interior. Agencies like the National Park Service and the Forest Service redesigned policies to accommodate natural fire in primitive areas. They retitled fire control to fire management and sought to extend the perceived benefits of such fire into other regions. The Francis Marion National Forest in South Carolina, for example, adopted a Designated Control Burn System (DESCON). So long as its prescription was met, DESCON accepted any fire, regardless of source, as a means of advancing its management goals.[125] In 1972 the American Forestry Association, as it had after 1910, devoted an issue of *American Forests* to the subject of fire, and a special task force appointed by the SAF in 1975 to report on recommendations for fire management in the Northern Rockies urged greater use of prescribed fire.[126] Prior to the 1970s the range of uses for prescribed fire had always been circumscribed. National strategy focused on certain types of problem fires, and only within the context of those fires was prescribed burning conducted. By the 1970s, however, prescribed fire had become itself the problem fire of defining interest for fire management, and fire management agencies were actively extending its scope—even to areas, like the Alaskan Panhandle, where there was little natural precedent.

In the early years the consensus among foresters was that forestry would be impossible if surface fires were tolerated. By the 1970s it was asserted with equal conviction that forestry and land management would be impossible if prescribed surface fires were excluded. So complete was the conversion that when threats to prescribed burning appeared, foresters—with the Forest Service in the lead—rushed to its defense. The arguments advanced in favor of prescribed fire were in many cases indistinguishable from those put forward by the proponents of light burning. The techniques and concepts had not changed, but their context had. The concept of light burning had assumed a new form, one wholly accepted within the context of the counterreclamation.

The problems with prescribed fire are now two. Escape fires have become larger and more common, reversing a long period of attrition, and the volume of wildland smoke has increased after nearly half a century of decline and in the face of an aggressive national program to control air pollution. Smoke has become a prominent effluent of industrial forestry. In some places, like the Central Valley of California and the Willamette Valley of Oregon, air contamination from prescribed burning has led to heavy regulation and a practical ban. Smoke has obscured views for weeks on end at Yosemite and Grand Teton National Parks. Air pollution alerts have struck a number of cities; Phoenix, for example, was seriously afflicted in 1975 and again in 1979 by broadcast burning in the watershed of the Salt River. Foresters have rushed in with smoke management plans to rescue their prescribed fire projects, programs that they now consider essential to silviculture and even—in ironic contrast to the position taken against Stewart Edward White—to the control of forest parasites like mistletoe and brownspot disease.[127]

The problem of escape fires, too, has been met with similar resolve. Among the largest fires of recent years have been those that, like the great fires of history, began as controlled burns and then, as weather changed, became uncontrollable. The biggest burn of 1979 began as a prescribed natural fire in Idaho. In Michigan a 1980 spring burn to improve the habitat of the Kirtland warbler raged over nearly 50,000 acres, took one life, and destroyed numerous houses. A prescribed burn outside Rocky Mountain National Park escaped and forced the evacuation of a nearby community. But whereas the prospect of escape fires had alarmed early foresters, they are now considered, like smoke, merely a cost of doing business, a small price to pay for a reduction of fuel hazards and for desired ecological engineering. Where escape fires were previously made public to show the hazards of light burning, they are now quietly shelved to retain support for the promises of prescribed fire. Like local juries who were reluctant to prosecute violators of rural fire codes lest in restricting the misuse of fire its valid uses would also be limited, committees of foresters tolerated the errors of using fire in order to prevent what they perceived to be the larger error of not using it.

3 THE FIRE FROM EUROPE

Horridae quondam solitudines ferarum, nunc amoenissima
diversoria hominum [What were formerly frightful wastelands fit
only for beasts are now pleasing habitats for men].
—Carolingian scribe, ca. 800[1]

It may be said, That in a Sort, *they began the world a New.*
—Jared Eliot, on American husbandry, ca. 1750[2]

When the Portuguese discovered the island of Madeira in 1420, they named
it in honor of the dense conifers that clothed it. To improve access to the
interior for agricultural clearing, early settlers touched off fires. The fires
soon roared out of control, and settlers fled for their lives, many dashing into
the ocean. Accounts vary as to how long the fires persisted; some say for
seven years, others only six months. But if the seven-year figure does not
apply to a single fire, it does probably describe the continued firings associated with landclearing. The island eventually became a prosperous Portuguese colony, valued for grapes and sugar cane. Fires decreased in intensity and frequency. By the time the smoke vanished, the vegetative cover of
Madeira was permanently altered.[3]

The violent settlement of Madeira encapsulates, after a fashion, the
European settlement of that much larger island, the New World. It had
been customary for colonists in the ancient world to carry fire from the
mother city or state with which to ignite their new communal fires. The
world of the Renaissance did the same, though its fires were those maintained over millennia of landclearing. The torch was not merely carried; it
was also applied to the landscape of the colonies.

The continuity of Old and New World history is aptly symbolized in how
two historical processes, the reconquest of Spain and the reclamation of
Europe, were transferred to the New World. In the same year that Columbus discovered the West Indies, the last Moorish stronghold in Spain, Gran-
ada, fell to Spanish arms. The reconquest had created a class of warrior
gentry, the conquistadors, and "for seven centuries," as Bernard DeVoto
summarized, "the way to lands, competence, and distinction had been to go
out and conquer them from the Moors."[4] With the discovery of the Americas, the conquistadors exchanged the Moors for the Mexicans. The military
conquest of the New World was achieved by peoples trained in centuries of

border warfare in the Old. The year 1492 also witnessed a new charter of privileges from Ferdinand and Isabella to the Mesta, the awesome sheep monopoly of Spain. In the name of pasturage and transhumance, the Mesta doomed Spain's forests to destruction for centuries.[5] This fate, too, was available for export: fire and hoof would do to North America's vegetation what fire and sword would do to its peoples. When the Spanish took possession of the Valley of Mexico and its forests, for example, they soon converted the landscape into something like the plains of central Spain.

Similarly, the expansion of Europe that had launched its crusades coincided with the Great Reclamation, a heroic assault on the woodlands of interior Europe in which forest, swamp, and estuary were converted to arable land. By 1300 the drive, spearheaded by Germanic peoples, had exhausted itself. Wars and plagues temporarily stymied further advance. By the sixteenth century, however, the reclamation of lands began again, and when wood shortages appeared, the overseas colonies offered an opportunity to export the reclamation as they had the reconquest. Fire, ax, and plow would join fire and hoof in altering the modern history of the New World as they had the medieval history of the Old. The transfer would gain tremendous acceleration, however. What occurred on Madeira was so sudden and violent as to become almost a caricature of the process. But it had its truth. What had taken Europe two millennia to accomplish would occur in the United States in two centuries. Not until the maturation of the industrial revolution would there appear a systematic counterreclamation to restore arable land to forest, brush, swamp, and wilderness. The fire regime of the New World would never be quite the same again.[6]

New World colonists, particularly those from northern Europe, often entered fire regimes unlike any they had known in Europe. In New England the difference between Old and New was not great, but in the grasslands, and especially in the South, colonists encountered a fire environment for which they had little preparation. In order to survive, they frequently borrowed the techniques of their Indian predecessors. The frontier thus became a blend of European and Indian fire practices, the one for farms and the other for fire hunting and range. More than in any other region of the United States, this amalgamation persisted in the South, giving it an incendiary tradition without parallel elsewhere in America.

The cultural compound of old and new fire practices was not always stable. Some of the techniques transferred from Europe had disastrous consequences when directly applied to the American scene, especially when fire was combined with intensive use of grazing, farming, and logging. The frontier adoption of Indian habits like fire hunting could also be problematic, leading to overuse of fire and overexploitation of wildlife. The earliest fire codes in the colonies reflected this mixed fire heritage, seeking to regulate both old and new. In this respect they epitomize the problem of fire prevention at large. The simplest way to eliminate fire damages is of course to prevent the fire. The process is complicated, however, because new ignition sources are created at ever accelerating rates and because the capacity to put them into contact with fuel steadily increases. Though most anthropogenic fires even yet come from traditional causes—debris burning, campfires, incendiarism, and so on—all of these represent the persistence of a frontier and agrarian heritage within an industrialized society. These practices are, almost by definition, predictable: all begin as controlled anthropogenic fires that subsequently become wildfires. But many of the most hazardous fires now come from the products of industrial technology and from the fuel complexes created by the counterreclamation. These fires are truly accidental and only indirectly anthropogenic, coming instead through the medium of machinery. The strategy for their prevention is akin to that employed against lightning.

A bizarre episode that occurred in Southern California illustrates this point. As an investigator from the Angeles National Forest reconstructed the scenario, a snake seized a gopher and was in the process of swallowing it when a hawk spied the scene, swooped down on its sluggish victim, clutched the writhing snake with its talons, and prepared to take to the sky. The snake, still struggling to swallow the gopher, upset the hawk's flight as it veered near some high-voltage power lines. Either the snake wriggled free, or, in order to protect itself, the hawk released its complicated victim. The snake fell across the power lines, the lines arced, and sparks fell on the tinder-dry brush below. When suppression crews arrived, they found a charred snake on the wires with a gopher, still very much alive, in its fried jaws.[7]

In its own way that incident summarizes the latest fire history of America. The fire from Asia was eventually extinguished or isolated into reservations; the fire from Europe receded with the advance of industrialism and the counterreclamation. But the fires of the industrial revolution no longer come simply from the hand of man. Machinery and the landscape can interact on their own with a stormy independence like that of the electrical fire of lightning. The "arts of man," founded, as Prometheus claimed, on his theft of fire, are in a peculiar way returning that theft.

PROMETHEUS BOUND: FIRE AND EUROPE

"That is my record. You have it in a word: Prometheus founded
all the arts of men."
—Prometheus, in Aeschylus, *Prometheus Bound*

If fire breaks out, and catch in thorns, so that the stack of corn,
or the standing corn, or the field, be consumed therewith; he that
kindled the fire shall surely make restitution. —Exodus 22:6

I

The coming of the Great Reclamation to the New World did not proceed
from a vacuum or enter one. Europe had experienced an ebb and flow of
wild and reclaimed land since at least Neolithic times, and the American
Indian had done considerable landscaping prior to European discovery. The
reclamation of American lands from Indian tribes recapitulated in dramatically compressed fashion, however, the most recent reclamation of European lands from the hunting, gathering, and herding tribes that had spilled
into Europe from central Asia during the breakup of the Roman Empire.
The history of this process may be considered in terms of three regions: the
Mediterranean, western and central Europe, and eastern Europe. For the
most part, eastern European immigrants to America arrived late in the settlement process and were crowded into urban centers. The Mediterranean
and Spanish pattern of transhumance never really became established,
except in post–Civil War days in the great cattle drives on the plains and in
the widespread, if temporary, invasion of the Cordillera by shepherds, many
of them Basque. Both patterns required vast stretches of public domain, and
both soon succumbed to enclosure by barbed wire and forest reserve. The
evils of fire and hoof were in fact a source of greater concern to many foresters and conservationists than were those of fire and ax. The American
pattern of open grazing developed more out of the tradition of the English
commons than out of the transhumance prominent in the Mediterranean.

Of all the European regions, the Mediterranean has undergone the most
stunning changes, and the role of fire in those changes is well documented.[8]
Mediterranean agriculture was limited by climate and topography, but
herding could and did spread throughout the basin, especially with the settlement of the eastern portions by nomadic tribes from the steppes. Even in
classical times, the forests of Greece were vanishing before ax, hoof, and
fire. The demands of agricultural landclearing, the insatiable appetite of the
shipbuilding industry, and the need for fuelwood put Mediterranean forests

into a recession from which they never recovered. They might have regenerated but for the introduction of livestock, notably the goat. Burning at the hands of herders became widespread, and fire and overgrazing left Greece impoverished, not only of forests but also of soil. Plato wrote despairingly that Attica, compared with former days, was "like the skeleton of a sick man."[9] In the Levant, where similar grazing pressures operated, the forests also vanished, and with them the noble cedars of Lebanon. "O Lord, to thee will I cry," exclaimed the prophet Joel (1:19), "for the fire hath devoured the pastures of the wilderness, and the flame hath burned all the trees of the field." In the *Aeneid* Virgil described fires ignited by pastoralists "when the wind is right."[10] Seneca reflected that "in a moment the ashes are made, but a forest is a long time growing."

The explosion of Bedouins across North Africa and into Spain in the name of Islam further spread pastoral burning practices. During the late Middle Ages transhumance and seasonal burning were all but universal. The deliberate conversion of forest to flammable browse occurred through the Apennines and the Balkans. The Mesta, the great grazing monopoly on the Iberian Peninsula, slowly eroded Spanish forests, transforming them into a vast treeless plain.[11] Fire in the hands of herders effectively—and in many cases irretrievably—deforested the eastern Mediterranean, much of the Adriatic, North Africa, and Spain. Once destroyed, a forest only rarely recovered. Brush was in many places endemic, and its range expanded throughout the Mediterranean at the expense of forest. The "red belt" across southern France and Corsica, for example, harbors a fire regime similar to that common in other Mediterranean climate regions, including California and southwestern Australia, but one complicated by severe overbrowsing and broadcast fire. When reforestation or even outright afforestation of pine plantations has been introduced, the plantings have suffered grievous fire losses—much as forest regeneration projects in the United States where open grazing was allowed. Where overgrazing persisted, all vegetation vanished and the soil with it.[12]

There were, of course, other sources of fire. In his *Georgics* Virgil debated the value of field burning for fertilizer and for pest control. Silius Italicus described the "multitude of fires that the shepherd sees from his seat on Monta Gargano [Apulia] when the grazing lands of Calabona are burned and blackened to improve the pasture."[13] Lucretius dwelt on the peculiar "baleful roar of crackling flames" that results when "a flame is swept over laurel-crowned hills by a squall of wind."[14] St. Matthew used well-known practices with field fires as a metaphor for the final harvest of souls: "As therefore the tares are gathered and burned in the fire; so shall it be in the end of this world."[15] In surmising the origin of metallurgy, Lucretius summarized the range of fire uses. Metals, he suggested, were discovered "when fire among the high hills had consumed huge forests in its blaze."

The blaze may have been started by a stroke of lightning, or by men who had employed fire to scare their enemies in some woodland war, or were tempted by the fertility of the country to enlarge their rich ploughlands and turn the wilds into pasturage. Or they may have wished to kill the forest beasts and profit by their spoils; for hunting by means of pit fall and fire developed earlier than fencing round a glade with nets and driving the game with dogs. Let us take it, then, that for one reason or another, no matter what, a fierce conflagration roaring balefully, has devoured a forest down to the roots and roasted the earth with penetrative fire. Out of the melted veins there would flow into hollows on the earth's surface a convergent stream of silver and gold, copper and lead.[16]

In advancing this origin for metallurgy, Lucretius may very well have had in mind the spectacle of Spanish miners who torched off most of the Pyrenees in their search for minerals. The practice continued to be a common one—soon to be as visible in the American West as in the South American highlands of pre-Columbian times.[17]

In western and central Europe landclearing was a slower process, conducted equally in the name of pasturage and cultivation. In Neolithic times farmers practiced *Brandwirtschaft,* or slash-and-burn agriculture. Even today, swidden agriculture continues in remote regions of Scandinavia: it is estimated that perhaps 80 percent of Finno-Scandinavia has been swaled, or slashed and burned. When done two or three times on the same site, the product is an almost pure stand of even-aged pine—the basis of the region's great timber resources.[18] With decreased mobility, however, slash-and-burn was replaced by forest-field rotation and finally by scientific crop rotation. In the forest-field pattern, portions of the fields would be allowed to revert to woods, and after a period of years the woods would be felled and burned. The cycle was shorter than that demanded by shifting cultivation, and it required less migration. In many areas farmers entered the woods in search of branches that could be placed on the fields and burned to supplement the ashy reservoir of nutrients left by fired stubble, thus allowing for more intensive harvesting. The procedure was regulated and required licenses. Later, potash was imported from North America to fertilize European fields without firing European forests.

From classical times onward, the expansion and contraction of woodlands paralleled the ebb and flow of the political and economic systems occupying the lands. The conversion to sedentary agriculture of the seminomadic tribes that filled Europe after the collapse of Rome set the stage for the Great Reclamation. A decree from Charlemagne, for example, urged his liegemen "whenever they found capable men, to give them woods to clear," and elsewhere it demanded "that our *silva* and *foresta* be well-guarded: and where there is a place suitable for clearing have it done, not allowing the woods to increase in the fields."[19] The establishment of a feudal order tied vassals to their fiefs and serfs to the land; often the only mobility allowed was the conversion of wasteland to farm and pasture. Viking marauders and Magyar invaders upset this evolution for a time. But the subsequent expansion of Europe in the eleventh century supported, and was in turn based on, the Great Reclamation. For 200 years villages expanded arable fields. New communities were promoted—in effect, leading to an interior colonization of Europe. Monasteries—first the Benedictines, then the Cistercians—set about to reclaim the wilderness in an agricultural crusade. A description of Cistercian monks in northern Germany in the twelfth century relates how the abbot took ritual possession of the untamed land in the name of Christ, then loosed three groups of laborers to transform wasteland to arable field. The first group *(incisores)* felled timber and cleared brush, the second *(extirpatores)* removed the trunks, and the third *(incensores)* fired the slash.[20]

Collectively, the whole process of agricultural reclamation took a millennium, and though its results could be extraordinary, even heroic periods of acceleration like the Great Reclamation could give way to epochs of stagnation or reoccupation of tillage by forest. Much of the forest belonged to the nobility, church, or crown. During periods of state weakness, peasants cut into the forests on their own, and during periods of depopulation, the forests returned. The vicious depopulation during the Hundred Years War led to the saying that "the forests came back to France with the English." The same happened to stretches of Germany devastated by the Thirty Years War. For the Great Reclamation proper, a recession of activity set in about 1300. The plagues and Mongols, like the Vikings and Magyars before them, further stalled progress, but these were overcome by 1500. Renewed activity during the sixteenth century reached such a pace as to cause alarm over the reduction of the forests. By the eighteenth century reclamation had penetrated into remote mountain slopes, encroached on sandy barrens along European shorelines, and crossed oceans to distant colonies.

Each period of accelerated reclamation also brought concern that deforestation was unwise, and efforts were undertaken to preserve some sites and to regulate the use of others. Forests had provided a multitude of products— tree bark for tanning and cork; firewood and charcoal for fuel; timber for ships and naval stores; timber for mine shafts and smelting; and mast as feed for livestock, notably swine. The nobility and crown set aside preserves, though these were primarily intended to ensure a supply of wildlife for the hunt. Even the monastic orders had second thoughts and sought to leave some portions of the landscape aside.[21]

It was from such efforts that the word and concept of "forest" derived— from the Latin *foresta*, which means not "woods" *(silva)* but "reservation."

By being reserved from reclamation and traditional fire practices, a site came to be stocked with trees, and thus the expression took on its modern usage. The English word *park* had a similar etymology. Originally it meant "enclosure," usually for hunting. Since most hunting was done with hawks or dogs or on horseback, the landscape was open, and the expression "park" or "parklike" came to mean "open and grassy." Foresters were thus guards, game wardens who watched against theft of fuelwood, poaching of wildlife, or modification of forest cover—crimes against "vert and venison."

Naturally there was considerable resentment among the peasantry about the establishment of forests to which they were denied access (such as the traditional *jus ad pascendum*) or in which they were stringently regulated, and the forester—especially in England—became a hated figure.[22] Despite the hostility, efforts were made to halt wanton destruction in the name of national economic interest or military security rather than merely for the king's pleasure, and both dimensions of the forest reservation system were transferred to colonial America. There again, the real antagonists of the forester were the folk farmers and herders of the reclamation.[23] The pattern of encroachment during periods of unrest was as true for the New World as for the Old, with the important difference that the administrative apparatus was even more remote and ineffectual, and in the United States a state of civil unrest became a more or less permanent feature of national history.

However impressive such preservation efforts were as an administrative invention, they were limited in extent. By the late Renaissance fuelwood was becoming locally scarce, and most of Europe began to import naval timber from the Baltic states. Substitute materials were found—coal for fuelwood, for example—and the New World colonies picked up much of the slack on naval stores and mast timber. At times, government legislation, such as the Code Colbert in France, restricted forest use. But enforcement was usually difficult, and peasants gnawed into forest reserves for fuelwood, pasturage, and arable land. The penetration of reclamation projects into the Alps, moreover, led to watershed deterioration by the end of the eighteenth century. A study of the subsequent flooding led French and Italian engineers to the articulation of modern hydraulics and hydrology. Impressive preservation schemes had been launched meanwhile in southwestern France, along the Baltic coast of Germany, and in the lowlands of Belgium and Britain, but their primary purpose was to retard the encroachment of shifting beach sand and to increase traditional forest uses in the service of agriculture. The project looked back toward the Great Reclamation, with its conversion of marsh and wasteland, rather than forward to the counterreclamation, with its promotion of industrial forestry. The latter did not appear until the late nineteenth century, when forest protection became a national objective in many countries. The agricultural revolution had proceeded so far, in fact, that in the latter half of the eighteenth century it inspired a distinctive group of philosophers, the Physiocrats. At the same time, sustained by the coming of industrialization, the forester changed from a guard to a manager of forested or reserved land. But as European states attempted to reserve land in their colonies, the forester once again became a figure of emnity.

Throughout all the centuries of landclearing, fire was apparently employed wherever possible to dispose of slash, to fertilize fields, to improve pasturage, and to remove stubble. Certainly escape fires occurred with changes in weather and with prolonged droughts, and many European place names testify to the presence of wildfire in the past. Legal codes prescribed stringent penalties for incendiarism and for escape fires from agricultural burning, as they had done since at least the codification of the Mosaic Law. Charters for the new towns created by the Great Reclamation usually had explicit prescriptions for protection from fires set in field and forest and for coping with escaped fires. Peasants attacked wildfire with brooms, backfires, and firelines.[24]

Deforestation did not always replace woods with farms. Vast cutover areas were left overgrown with broom and heath. Even in the thirteenth century, thoughtful observers discriminated between wood and brush; they knew that brush, fire, and livestock would become an endless cycle but that with protection the forest would return. Especially in northern Europe, there were often harsh penalties for fires set by pastoralists. As a means of discouraging fire for range improvement, it was not uncommon to prohibit grazing for 10 to 15 years on a forested area that had been burned. Goats were rigorously excluded—a fact for which northern Europe and North America may be eternally grateful. The herding economy of the north relied on swine, a creature who feeds on mast and roots—forest products—rather than on grass or browse. (An exception was Scotland, where sheep replaced cattle and heath burning persisted on a 10-year cycle.) The penalties for pastoral burning helped to ensure that in northern Europe, unlike the Mediterranean, the farm rather than the flock would be the primary economic unit of agricultural society. The persistence of forest grazing was another legacy of a hunting and herding economy only partially converted by the feudal order and only partially modified by the Great Reclamation. With further reclamation and enclosure, livestock left the woods almost entirely for cultivated pastures, mowed grasses, and prepared troughs. Animal husbandry replaced herding, and in leaving the open commons, herdsmen left fire.[25]

The pattern of private ownership and enclosure that characterized intensive agriculture also worked against the preservation of ancient fire practices. Hunting was restricted to certain reservations, like the heathlands and moors of England and Scotland, that were perpetuated by fire. Broadcast fire characterized seminomadic hunting and herding economies, and by the late eighteenth century both were curiosities from distant lands rather than basic units of production within European agricultural economy. In fact, one may question whether the typical livestock husbandry practiced in western Europe created its fire practices or whether insistence on certain fire practices effectively dictated the nature of husbandry. In either case, pasturage was gradually segregated from broadcast burning. A similar decision would be needed in America, where sedentary reclamation would be superimposed on a hunting society and mixed with a herding economy. Where the open range was sustained, as in the South, woodsburning became almost endemic.

English law became the fountainhead for American law, but the heritage of British fire offered little precedent for American foresters. When Caesar marched on Britain, he reported it to be largely forested. Roman settlement reduced some of the land to cultivation, but, except for domesticated livestock, the cultural state of the indigenous peoples was not much different from that of the peoples whom European colonists found in the New World. After withdrawal of Caesar's legions, agricultural settlement by Scandinavian invaders gradually reduced the forest further. Forests were left for the production of wild game and as "pasture," providing mast for domesticated swine, and some woods received special royal protection as game preserves. The Norman invasions beginning in 1066 coincided with the general expansion of Europe and with the Great Reclamation. A system of royal forests was established, so extensive at one point that they embraced nearly a quarter of England. Special courts administered forest laws, which emanated directly from the king and were thus outside common law. Violations—by fire, theft, or poaching—were considered forms of trespass, and in the early centuries transgressors were punished harshly. Villages proliferated amid the new landclearings, and their charters, like that of Ipswich in 1200, provided for fire protection measures: tubs of water were to be strategically placed; citizens could expect a call for duty as fire watchers and for service in firefighting. A Forest Charter wrested from Henry III in 1217 restricted the boundaries of the royal forests, limited the prerogative of the foresters, and guaranteed traditional rights of access to forest lands, much as Magna Carta had guaranteed traditional political rights and restricted the operatives of the king.[26]

Land not actually farmed was held in common, usually for fuelwood or pasturage. Not until 1483 were people who lived in the forest allowed to enclose their holdings and to market wood. Enclosures of forests and fields intensified during Tudor times, though the purpose of enclosure was primarily to protect pasturage for sheep. Access to woods was restricted, and many "wastelands" of the lord's demesne were given over to hunting preserves. Heath and moor developed from grazing (or hunting) and fire in Britain as they did elsewhere around the North and Baltic Seas, and efforts were made to regulate them. Fire, moreover, had never been a device of

English sport hunting, and from early times the English turned to dogs or native beaters rather than flame to drive game, though Shakespeare has King Lear wish for a brand to use against his enemies in order to "drive them hence like foxes."

As alarm spread over shortages of local fuelwood and of national stocks of ship timber, there were further prohibitions. The Puritan Revolution swept aside many of the ancient laws that had governed the use of forested lands, though some of the proscriptions were reinstated by the Stuarts. In 1664, for example, John Evelyn reported in his *Silva* that Parliament had legislated against moor and heath burning between April and September in many counties; too many fires had escaped into fields and pastures, smoke had reached obnoxious proportions, and wildlife had been harmed. The program was not unlike many provisions of the Code Colbert adopted by France at nearly the same time.[27] But changing circumstances, political and economic, caused such edicts to fall into disregard. Further enclosures followed during the late eighteenth and early nineteenth centuries, leaving remaining forests almost exclusively in private hands. By then, scientific crop rotation schedules had eliminated much of the need for broadcast fire for range improvements. At the start of the First World War only 5 percent of Britain was forested, and the country was in desperate straits for much of its war material. Estate forestry dominated the management of the remaining forests, which were, like the other semiwastelands of Britain, managed primarily for sport game. Arboriculture flourished, decorating Britain with exotic ornamentals but not adding much to its forest reserves. When the growth of the British Empire had demanded large-scale management of forests, Britain had been forced to turn to the Continent for professional advice and had directed most of its attention to its distant possessions in India, Burma, and Canada. Americans seeking precedents for their policies looked to British India, which was then under the direction of the German-trained Sir Dietrich Brandis.

Immigrants to the New World consequently brought different heritages of fire with them. Finns and Scandinavians who settled in the north woods often tried to apply swidden agriculture, and their landclearing burns led to some of the worst holocausts on record. Settlers from pastoral economies like Spain's were familiar with broadcast fire for range improvement, and the rapid deforestation of Mexico was more a repetition of Spanish experience than a cruel innovation. The reconquest had been succeeded in Spain by the extension of grazing cooperatives like the Mesta. The conquest of Mexico followed suit, and the pattern was extended throughout the Spanish borderlands, from Florida to California. Most of the conquistadors, in fact, hailed from Estremadura, in the heart of the lands dominated by the Mesta. The Great Reclamation, by contrast, bequeathed a northern European heritage of fire in the service of landclearing and sedentary agriculture. Swine were allowed to roam the woods in search of mast, but other grazing animals were generally tied to individual farms or village commons; mowing gradually replaced burning as a form of pasture management. By the time of large-scale colonization, European agriculture was in the process of enclosure and intensification, not of expansion. Nomadism and expansion have always favored fire; enclosure and high-yield farming have always restricted it. The New World offered virtually unlimited lands for reclamation and thereby for the recapitulation of the fire history of western and northern Europe. Although wildland fire was for many as much a New World curiosity and danger as were its wild beasts and wild peoples, the functions of controlled fire and the mechanisms of protection against it had long been a fundamental part of European heritage.

For many, however, this was a heritage lost. By the nineteenth century the Great Reclamation had long since transformed the forests of France and Germany into farms and fields, and the hardwood forests left as hunting preserves and woodlots offered little opportunity for wildfire. In England large fires had been known from the ninth century to Tudor times and were reflected in town names like Brentwood and Burnham. But an end to extensive landclearing and the advent of a cooler climate made such fires, like the great fires of the Continent, only a memory from the Middle Ages. Plenty of burning remained on the fringes of western Europe, but such practices were considered aberrations: the swidden fires of Finland, heath burning in northern Scotland, the pastoral brush burning along the Mediterranean, the grassland fires set for range improvement on the Spanish plateau and the Russian steppes. They were employed by people for whom sedentary agriculture was an exception. For an industrializing western Europe, Prometheus was again bound.

Nineteenth-century observers considered the American forests to be analogous to those on other frontiers of Western civilization—the chir pine, teak, and tropical hardwood forests of India and Burma; the cork forests of Algeria; the taiga of the Canadian shield. Most Americans saw little value to forest or wildland fire protection: it was technically impossible, environmentally undesirable, and surely indefensible on economic grounds. A special committee of the National Academy of Sciences thought that British India perhaps furnished the best model for the United States to follow.[28] In fact, the revolution in forest reserve policy that the British brought to India came in 1885, making it exactly contemporaneous with developments in Canada, where Ontario adopted a program of "fire-ranging," and in the United States, where the Army undertook fire control on the national parks and New York modified its fire warden system to accommodate the Adirondacks preserve. Like their counterparts throughout the world, too, American foresters considered the great challenge of fire protection to be the elimination of traditional fire practices that had the effect of replacing forests with wild or domesticated grasses for pasturage or crops. They pointed accusingly to the havoc wrought by traditional firing and found in promiscuous fire the cause for the downfall of ancient civilization: slash-and-burn firing preceded the Mayan collapse; forest felling and firing brought the soil erosion that slid China into irreversible decay; brush burning caused the endemic impoverishment of Greek and Mediterranean cultures.[29] When timber owners and stockmen proposed that foresters adopt the fire practices of the American Indian, the suggestion was met with ridicule and incredulity.

The American situation offered some novelties, however. The national experience had loosened traditional restraints on forest exploitation, leaving land strictly as a commodity; a relatively weak state and a more or less permanent condition of unrest encouraged laissez-faire logging and land-clearing; and the advent of industrialization helped to compress into two centuries a process that had taken two millennia in Europe. In America industrial logging companies replaced monastic orders as the great land clearers. With new land continually opening up, a pattern of shifting, slash-and-burn cultivation became normal; new forests were felled and fired as fertility declined on old fields. Even George Washington, who tried so diligently to introduce the latest scientific concepts of crop rotation onto his estate, found that, although the system was appropriate for 300 acres, it broke down when applied to 3,000. The very size of the country meant that there was plenty of opportunity for shifting agriculture, and agrarian pioneers moved from site to site—clearing, burning, plowing, and then moving on again. Much of the reclamation of Europe had been predicated on a feudal order with restricted mobility; in America there was little trace of feudalism, and mobility was high—prime conditions for fire. Fires of natural, accidental, or incendiary origin were part and parcel of a complex of frontier violence and waste. Not only was the geographical scale of reclamation in America much vaster than that in Europe, but its historical scale was also much shorter. Agriculture barely had time to establish itself before the industrial revolution encouraged a counterreclamation. Some areas went from forest to farm to forest within a hundred years, and some from wilderness to forest reserve to wilderness with hardly a blemish from hoof, ax, or plow. After the Appalachians were breached in the last half of the eighteenth century, American settlers compressed centuries of European land-clearing into decades. By the 1830s Indian resistance to settlement became negligible, and the folk migration spilling across the Appalachians exploded into a virtual vacuum.

What made this movement so different from the reclamation of Europe was that it coincided with the incipient industrial revolution. The machinery of industry, like the steamboat and the railroad, allowed agricultural settlement to proceed rapidly, and by providing distant markets it encouraged agricultural settlement of lands that might otherwise have remained as wasteland. The industrial revolution also assured that most agricultural land would be occupied only temporarily, until conversion to an urban and industrial society was completed. Lands in a transitional state of occupation are always more prone to fire than lands that are settled and stable for long periods, and the mixture of migration, immigration, and industrialization meant that few American lands enjoyed much stability. The American landscape typically experienced a series of rapid conversions, not simply a single period of reclamation. At the same time, the very rapidity of the enterprise left some portions of the landscape virtually untouched in a wilderness state and other portions preserved either by law or social custom in a frontierlike condition. A hastily assembled mélange of fuels and ignitions covered the country.

The speed and scale of this folk migration propelled settlement faster than government supervision could follow. Surveying invariably lagged behind actual occupation. As with other controls over frontier violence, fire protection passed into local hands. Volunteer associations modeled on landowner or cattlemen's associations were eventually formed by timber owners to protect their investments; volunteer fire brigades furnished rural protection; individual landowners took precautions to see that firebreaks were plowed and arranged to coordinate burning schedules with their neighbors. In many states sheriffs or fire wardens were empowered to impress citizens for fire duty during emergencies, a kind of *posse comitatus*. Not until the late nineteenth century was federal and state fire protection apparatus developed, and this was limited to reserved parks and forests.

Land ownership became mixed. Government land reservations for forest and park were a relatively late phenomenon and not always successful, as repeated violations of Indian reservations by the populace demonstrated. In Europe the land had belonged to the crown, nobility, state, or church, and access to forest or wasteland was regulated from the beginning. The Cistercians, for example, took possession of a site with an elaborate ceremony and jealously protected their claim against trespass. In the United States, possession of title was likely to come through land speculators. Logging companies functioned as the national *incisores,* individual farmers as *extipatores* and *incensores*. The government saw its duty in the disposal of land, not in its reservation. Land, moreover, was a commodity, property to be bought and sold. In Europe inherited customs as well as legal prescriptions had guided traditional rights and responsibilities toward forest and wasteland. The only American equivalent to the German *mark,* for example, was the concept of wilderness, and that was, relatively speaking, a long time coming.

The counterreclamation has resulted in an ironic reversal not only of land use but also of technology transfer. Especially after World War II, Europe undertook massive reforestation programs on its devastated forest lands, on marginal agricultural lands, and on reclaimed marshes, sand dunes, and brushfields. The de facto depopulation of portions of Europe by virtue of urban immigration and the creation of flammable cover types by means of reforestation (and in some cases by afforestation) has created a fire regime not unlike that known in Europe's former colonies. The industrialization of the Old World is, in many places, creating conditions not unlike those in the preagricultural New World. The long-term effect of the Great Reclamation was to replace forests and grasslands with farms; the long-range consequence of the counterreclamation has been to replace farms with forests, recreational suburbs, and wildlands. To cope with its new fire regimes Europe has solicited help from those former colonies that have become major fire powers, especially the United States, Canada, and Australia. The original accommodations that European-educated foresters had to make with traditional fire practices in the New World are being incorporated into the management of lands in the interior of Europe—loosening a little the chains that professional foresters had wrapped around the long-suffering Prometheus.[30]

II

When European colonists carried the torch to the New World, they were recapitulating, after a fashion, the most famous of their myths about the origin of fire. Prometheus had "honored man" and, at great personal sacrifice, had stolen for an enfeebled humanity "his great resource, his teacher in the arts, a spark of fire"—so, at least, Aeschylus had him proclaim. The legend was not common to all of Europe, and among the ancient Greeks there was also disagreement. The Argives insisted that their ancient King Phoroneus had discovered fire, and well into the Pax Romana they continued to honor his memory with a sacred fire at the great temple of Lykios Apollo. Even the story of Prometheus varies according to the license of chroniclers like Hesiod, Aeschylus, and Plato. The European colonists to the New World brought with them more than a technology of fire: they also transmitted a special culture of fire lore, both in the rituals of folk knowledge and in the literature of high culture. This corpus of legend and ceremony conditioned a perception of fire and, in the absence of formal science, constituted an explanation for mankind's unique relationship to fire. Acquired and modified over millennia of Old World experience, this knowledge, no less than the technology that lay beneath it, underwent serious trials in the New World and became an object of controversy.[31]

According to Hesiod, a jealous Zeus hid fire from mortal man. Prometheus, however, was sympathetic to humanity, so he pilfered some of Zeus's heavenly fire and carried it to earth in a stalk of fennel—a common, bamboolike reed that held flame like an enclosed candle, frequently used as a slow match in ancient times. Prometheus was the son of the Titan Iapetus and thus an enemy of the Olympians, led by Zeus, who had overthrown the Titans to achieve supremacy of the universe. For his rash act Zeus punished the hero by chaining him to a peak in the Caucasus Range. Each day without fail an eagle would appear before the helpless Promethus and devour his liver or heart. Each night the organ would grow whole again. This torture continued for thirty or forty thousand years, until Hercules finally liberated the yet defiant Titan. It was this version that Aeschylus used for his famous tragedy, and, not surprisingly, it was this vision of the rebellious hero that later attracted the romantics. *Promethean* entered the English language as an expression for the larger-than-life deeds and aspirations of the romantic hero, an appropriate symbol for a civilization seemingly bent on shattering the ancient bonds of geography, history, and knowledge. The liberation of Prometheus' gift, fire, also reached new frontiers.

Plato offered a more philosophical version. In the *Protagorus* he described how the gods fashioned mortal creatures from compounds of earth and fire, two of the world's four elements. Creation took place underground at the direction of Hephaestus, god of fire, and Athena, goddess of the arts. Once the creatures had been rudely fashioned, the gods assigned Prometheus and his brother Epimetheus the duty of refining and delivering them to the surface. As the etymology of their names suggests, Prometheus could think ahead; Epimetheus, only after. When the time came to equip men and beasts with their requisite powers and functions, Epimetheus convinced his brother that he could handle it. Foolishly, Epimetheus began distributing the valuable but limited skills to the animals as they appeared. By the time man arrived, there was nothing left. The day fast approached when the finished creatures should be disgorged onto the earth; there was no time to rectify their errors. But Prometheus was friendly to man, and he reasoned that if man had fire and mechanical skills, he could survive. The Olympian fire was too closely guarded by the savage warders of Zeus, so Prometheus stole into the workshop of Hephaestus and stealthily removed fire from his forge. (Hephaestus himself and his fire had come from the heavens after Zeus hurled him into banishment.) Thus Prometheus could proudly claim that he had founded all the arts of man.

Among myths relating to the origin of fire, the Greek myths are distinguished by their anthropomorphism. In other respects, their motifs are familiar. Fire was almost always granted to a grateful humanity by a beneficent potentate, or, more often, it was something that humanity had to steal. Fire was power, and it was not easily acquired. Frazer distinguished three periods to which most myths of fire refer: a Fireless Age, when humanity was little better off than the beasts of the field; an Age of Fire Used, during which man had the use of fire but not the secret to its creation; and an Age of Fire Kindled, when man acquired the ability to manufacture fire. But even more than presenting a technological problem, fire brought intellectual, political, and moral problems. Indo-Europeans had a fire cult centered on the figure of Atar, who in India became Agni, at one time chief of the Hindu pantheon. In Persia this cult was reworked into Zoroastrianism, the first of the great monotheisms. In Greece, through Heraclitus and Empedocles, fire became one of the four basic elements, and in Celtic and Teutonic societies it became the need-fire, of which the Yule log is a modern relic.[32] Fire burned in hearths, the symbol of family; in national temples, the symbol of the state; in eternal flames, the symbol of civilization. As a physical event, fire inspired profound speculations in natural philosophy; as an emanation of divine presence, it raised fundamental religious questions and was incorporated into religious ceremonies; as a symbol of power, it sealed political treaties in the ancient world (the fires of various tribes would be mingled), and a portion of the sacred fire from the homeland would be carried by departing armies and colonists. The discovery of fire may aptly be compared as a cultural problem with the discovery of the atom in the twentieth century.

Fire became the focus of some of the most ancient of European folk ceremonies. These fire rituals were of two basic kinds: the need-fire, which reenacted the discovery and distribution of fire; and various purification rites, in which fire and smoke were applied to crops, orchards, herds, and people. Throughout Europe the need-fire ritual appeared during times of distress or misfortune, and especially when herds suffered epidemics. It was also the source for all the more elaborate fire ceremonies. Among Teutonic peoples, the need-fire was sometimes referred to as the "wild fire"—the ancestor to the more domesticated species used by men. It was most commonly kindled in the open. Wooden sticks were required; flint and steel or more modern ignition devices were prohibited. Sometimes special kinds of wood and kindling were prescribed, and rules governed the choice of persons to ignite the new fire—the participants were typically young and unmarried and were often required to be naked. From the need-fire a giant bonfire was established, and for a certain distance around the need-fire, say, a parish or village, all fires would be extinguished. When the ceremonial blaze subsided, sick animals were passed through the smoke or coals. Ashes would be smeared on the participants; a procession would march to the village; and the hearths would be rekindled with embers carried from the need-fire.[33]

A host of other fire festivals evolved from this one. Certain Celtic rituals on Halloween and May Day (Beltane), for example, also involved ceremonial fires. All such rituals had as their purpose to protect and to purify, though portions of the fire might be carried home and preserved as a talisman against lightning and hail as well as against fire. The themes of the rituals testify to their origin among primitive groups of herders, hunters, and gatherers who flourished well before the Great Reclamation. But just as the "wild fire," like wild beasts, had been domesticated, so the fire of hunters and herders was tamed for the services of sedentary agriculture, its fire rituals absorbed into a solar calendar and agricultural cycle, and its value as a purifier expanded from livestock and woods to field and orchard. The greatest of the ancient fire festivals came at midwinter and midsummer. Christianity at first condemned, then partly assimilated these rites into its own liturgical calendar, with fire festivals prominent during the first Sunday of Lent and Easter. (Something similar happened with the Gospel story of Christmas: the Magi—Zoroastrian priests—come to investigate the new "fire" in the heavens, and thus the birth of Christ coincides with the fire symbolism of Ahura Mazda.) That such ceremonies antedated the Church is attested by the fact that remarkably similar practices existed in North Africa, persisting, like their counterparts in Christian Europe, even after the Moslem conquest and conversion to Islam.

What is perhaps most interesting about the ceremonies is how they have been interpreted. The great compendium of European folklore was Sir James Frazer's *The Golden Bough,* first published in 1890. Frazer catalogued the known fire festivals in considerable detail. He concluded that

the custom of kindling great bonfires, leaping over them, and driving cattle through or around them would seem to have been practically universal throughout Europe, and the same can be said of the procession or races with blazing torches round fields, orchards, pastures, or cattle-stalls. . . . As the ceremonies themselves resemble each other, so do the benefits which the people expect to reap from them. Whether applied in the form of bonfires blazing at fixed points, or of torches carried about from place to place, or of embers and ashes taken from the smouldering heap of fuel, the fire is believed to promote the growth of the crops and the welfare of man and beast, either positively by stimulating them, or negatively by averting the dangers and calamities which threaten them from such causes as thunder and lightning, conflagration, blight, mildew, vermin, sterility, disease, and not least of all witchcraft.

In Frazer's day there were two theories by which to interpret the fire festivals. The solar theory held that the ceremonial fires, on the principle of imitative magic, sought to mimic or encourage the sun. The purificatory theory relied on the explanations the peasants themselves preferred: the fires drove off harmful influences. Frazer sided with the latter. Wolves and witches, he noted, were the great fears of herdsmen, and in its dim Indo-European past the population of Europe had evolved from hunting and herding societies ranging out of the central Asian grasslands. Fire offered practical protection against wolves and talismanic security from witches. Dread of witchcraft, Frazer thought, originated the fire festival, and it was accepted that witches would be executed (if only in effigy) by fire, which could purify as well as kill, rather than by other means.[34]

What is most intriguing about Frazer's analysis is that no actual value is attributed by him to fire and smoke as purificatory agents. They existed only symbolically as brazenly superstitious rituals, as magical weapons to be directed against supernatural powers, and as symbolic purges partially sublimated by using effigies in place of condemned witches. *The Golden Bough* appeared in the same year as William James's *Principles of Psychology,* and it was the temper of the age to perceive such events as symbolic and psychic. Yet there can be little doubt that what was enacted ritualistically in the fire festivals was, in the distant past, actually practiced, not merely as a tool of exorcism but also as an instrument of primitive land management and animal husbandry. Early European peoples undoubtedly broadcast burned fields, pastures, and woods in an attempt to fertilize, improve, and purify them. By the Middle Ages, when Church attempts to eradicate persistent pagan rites prompted the first records of fire festivals, the broadcast fires of antiquity had been reduced to token torches, rows of religious candles, and symbolic bonfires. Wildfire had been tamed, enlisted in the service of the Reclamation and integrated into agricultural cycles, much as fire ceremonies had been absorbed into the liturgical calendar of the Church. But there is no reason to doubt that among seminomadic hunting and herding peoples or practitioners of *Brandwirtschaft* such fires were actual practices rather than merely semireligious ceremonies. Nevertheless, although fire might cure the malady, it could not always prevent it; hence, the exorcism of witches was incorporated into the ceremonies as a supernatural prophylactic. By the time fire customs were recorded and folklorists like Frazer had the opportunity to interpret them, only shards and relics of the original practices remained, like the Yule log and the bonfire rally. Interpreters could no longer imagine any empirical rationale for these ceremonies.

A year after the *Golden Bough* was published, the United States began to set aside forest reserves. Following recommendations from a National Academy of Sciences investigating committee, the Forest Management Act of 1897 set forth some early guidelines for the conduct of forestry. As much as in the field of folklore, the possible value of open broadcast burning was ignored. Woodsburning belonged with medieval superstitions like witch burning. When light-burning advocate Stewart Edward White recommended smoking forest trees as a precautionary measure against insect infestations, his arguments must have sounded like a page from the *Golden Bough.* Likewise, southern farmers who argued that woods fires retarded chiggers, ticks, and boll weevils actually had the force of ancient custom behind them, though they faced equally the resistance of modern opinion. Open burning in agriculture had faded in purpose and practice over the centuries.

It is a testimony to the youth of forestry that it has had to rediscover the techniques and objectives of free-burning fire and that from the beginning it has insisted on purely scientific data. It often ignored and ridiculed evidence of Indian and folk practices that suggested a hidden value to prescribed fire, not only in America but throughout the lands that Europe colonized. Such fires are no longer considered a quaint vestige: they have become a vanguard in the modern management of extensive forest lands. Rather than serving the ritualistic prescriptions of folk practitioners, fire now submits to the often equally ritualistic prescriptions of forest engineers.

OUR PAPPIES BURNED THE WOODS:
A FIRE HISTORY OF THE SOUTH

Woods burnin' 's right. We allus done it. Our pappies burned th'
woods an' their pappies afore 'em. It war right fer them an' it's
right fer us. —Southern Appalachian resident, 1939[35]

You've got the money, but we've got the time.
You cut the hardwoods, and we'll burn the pine.
—Poster found on a southern national forest, 1930[36]

I

During the last stages of the prescribed-burning controversy in the South,
the U.S. Forest Service assigned a psychologist, John Shea, to investigate
the causes for persistent woodsburning in the region. Shea visited several
national forests, conducted hundreds of interviews, and concluded that
woodsburning continued because it was a folk custom. Inveterate burners
listed scores of reasons for firing. "Fires do a heap of good," one insisted.
"Kill th' boll weevil, snakes, ticks, an' bean beetles. Greens up the grass.
Keeps us healthy by killin' fever germs." Asked whether forests might not
be more productive without fire, the informant replied: "Might hol' the
floods a mite and make a few more squirrels, but it ud make living harder
and we'd see more rattlesnakes."[37]

Like Frazer writing half a century before him, Shea gave no credence to
local explanations; woodsburning, he realized, was "a survival of an old culture." But whereas Frazer found hidden motives for fire ceremonies in
superstition, Shea identified the source of incendiary fire in psychological
necessity: "With the closing in of the agrarian environment, it has become
predominantly a recreational and emotional impulse." Shea concluded that
"the light and sound and odor of burning woods provide excitement for a
people who dwell in an environment of low stimulation and who naturally
crave excitement. . . . Their explanations that woods fires kill off snakes, boll
weevil and serve other economic ends are something more than mere ignorance. They are the defensive beliefs of a disadvantaged culture group."[38]
Despite its somewhat higher degree of secularization, Shea's conviction was
identical to Frazer's: woodsburning was a ritualistic practice, a mere superstition perpetuated from generation to generation.

The South has long dominated national fire statistics, leading in both frequency and acreage burned. Some of the earliest reports on the New

World—for example, those given by Verrazano and Drake—speak of large fires and thick smokes along the coastal plains from Virginia to Florida. Drake observed "one special great fire, which are very ordinary all alongst this coast." Searchers for Raleigh's lost colonists in 1590 sailed from smoke to smoke, hoping in vain that one would be a signal fire instead of "grass and sundry rotten trees burning." English adventurers on the Chesapeake in 1607 reported "great smokes" from burning woods on all sides.[39] Indian burners created vast pine and oak savannahs along the plains and piedmont, used the peninsulas of the tidewater region for fire hunting, and maintained abandoned agricultural clearings as open prairies. Bartram spoke nonchalantly of the "annual firing of the deserts."[40]

Perhaps nowhere else in the country were Indian burning practices more thoroughly adopted and maintained than in the piney woods, in the remote hills, and on the sandy soils where rice or cotton plantations failed to penetrate. For many of the Scotch-Irish immigrants who settled these regions, the socioeconomic environment was not unlike that in Scotland, which had helped to perpetuate a herding and hunting economy that routinely used broadcast fire. Here pioneer livelihoods—predominantly hunting and herding—persisted. And here woodsburning endured. Even in the early twentieth century, it was reported that 105 percent of Florida burned in one year—the improbable figure resulting from combined spring and fall firing for range improvement. Inman Eldredge, supervisor of the Florida National Forest in the early twentieth century, observed with an almost Faulknerian blend of amazement and outrage that "the people right down on the ground, the settlers, the people who lived in the woods, the turpentine operators, and so forth, were completely uninformed and were the greatest, ablest, and most energetic set of woods burners that any forester had to contend with."[41] The Appalachian residents interviewed by Shea were the impoverished progeny of this tradition.

What the psychologist Shea failed to realize, like the folklorist Frazer before him, was that firing the woods was not merely a socioeconomic phenomenon mindlessly and violently perpetuated through the ages. It had environmental and historical logic. Socioeconomic changes had made the old pattern of woodsburning obsolete, even detrimental; but environmental considerations—notably the silviculture of the southern pine, the sprouting of the "rough," and the needs of wildlife—have made prescribed fire all but mandatory. The woodsburning tradition is intricate—at once violent, shrewd, and subtle. It had been practiced more for hunting, grazing improvement, and pest eradication than in the interests of forestry. But once adaptions were made through work like Chapman's, prescribed burning made the transition from an agrarian to an industrial tool. Wildland fire has been progressively removed from the hands of folk practitioners and placed in the grasp of professional foresters. *Woodsburning* has consequently remained a pejorative term—the vicious weapon of the arsonist or the relic practice of an antiquated agrarianism.⁴² Forestry in the South would have been impossible if promiscuous and malicious woodsburning had continued. But equally, it would have been impossible if conducted on a dogmatic policy of fire exclusion. Woodsburning was one of the few frontier skills that had real meaning in industrial society: Europe had lost the art, and Indians either were considered too primitive or were confined to reservations. Woodsburning, however, was both widely known and widely applied. What the Indian was to the light-burning debate in California, the woodsburner was in the prescribed-burning controversy in the South. Like the maligned and coveted piney woods themselves, woodsburning was a resource that had to be managed. As H. H. Chapman himself observed, "between proper use of fire and promiscuous burning there is all the difference between success and failure."⁴³

II

The fire history of the South is in good part a history of its fuels. Geographically, the South divides into three general provinces: the coastal plains, the piedmont, and the Appalachian Mountains. The coastal plain is dominated by pines and pocosins; the Appalachians, by hardwoods; and the piedmont, by a mosaic of pines and hardwoods concentrated in the moister zones, such as river bottoms. But whereas fire problems in most regions result from heavy fuel accumulations in the wake of human activities, in the South fuel accumulated rapidly in its natural state. In the North the natural fuel load rose slowly, punctuated by occasional blowdowns or insect kills. Fire protection meant preventing fire from entering logging and landclearing slash or from leaving it. But in the South, fire was essential simply to keep the annual growth in check. To be sure, fuel loads could reach fantastic tonnages from logging, hurricane blowdowns, ice storms, pocosin drainage, and insect invasions, but it is the forest understory—the rough, with its tall grass, hardwood saplings, reproduction, and vines (and on the coastal plains its gallberry and palmetto)—that is the typical fire hazard. Within a handful of years the rough can present an impenetrable jungle of vegetation. If allowed to mature, it presents an increasing fire hazard, chokes wildlife, and eventually allows hardwoods to replace the economically important pine altogether. It was the rough, not the forest, that traditional woodsburning sought to destroy. In northern California and the Northwest, where light burning was debated, the brush understory was preserved from fire in the expectation that timber reproduction would eventually succeed it. In the Lake States the unwanted brush deserved protection because it was in actuality often forest reproduction. But the southern rough, through fire or succession, ultimately destroyed the pine forest that was most valued, the pasturage most palatable to livestock, and the habitats most productive to game birds. Most large southern fire complexes thus coincide with drought rather than with landclearing. The regular firing of the woods prevented the fuel buildups that encouraged episodic fires elsewhere, and the fire history of the South is remarkable for the absence of conflagrations until the advent of industrial forestry in the 1930s.

Early settlers on the coastal plains learned broadcast burning from local tribes. As they moved inland, crossing some of the premier fire regimes of North America, pioneers carried their fire habits with them. The northern woods might be cleared and settled without fire, but not the southern rough. Skill in broadcast fire was essential to southern frontier survival: nearly all dimensions of southern agrarian economy relied on it—for landclearing, for hunting and habitat maintenance, and for range improvement. It was employed for fuel reduction in naval stores operations, the antecedent to industrial logging, and it was used by homesteaders to protect themselves from the fires that others were sure to light. Fire protection was even built into the architecture of frontier cabins: the cleared yards around wooden structures acted as firebreaks and as points for igniting protective back-fires—doing double duty, as fish ponds did for rural houses in New England. Fire practices were incorporated into the fabric of frontier existence. What made the South special, however, was the confluence of economic, social, and historical events that worked to sustain this pattern of frontier economy long after it disappeared elsewhere in the United States, a pattern that created a socioeconomic environment for the continuance of woodsburning.

Indian agriculture had reached considerable sophistication in the South, and European and American settlers were able to occupy old fields and even broad prairie belts.[44] On the better sites, shifting agriculture developed to farm cash crops like rice, tobacco, and, of course, cotton, and eventually these small subsistence farms gave way to plantations aimed at crop monocultures. This pattern differed from that common to the North, where marginal farmlands were often carved out of forests and then abandoned in the rush for better land to the west. In the South the marginal lands were never really cleared, and their inhabitants never really left. The use of fire for pasturage and game continued. What other regions had known at most as an ephemeral system became in the South more or less permanent. The usual statutes against fire hunting appeared early in the South, as they did everywhere, but general injunctions against broadcast burning were very slow to develop. Agricultural fires were too extensive; they were used to prepare sites, to manage fallow fields, to purge sites of pests, and to dispose of debris like cane stalks and field stubble. Mark Twain described how the whole region between Baton Rouge and New Orleans would become "an impenetrable gloom" each fall with the firing of "bagasse" (cane stalks), which "burn slowly, and smoke like Satan's own kitchen."[45]

Such a regimen quickly exhausted the soil. Three years of tobacco and not many more of clean-furrowed crops like cotton were enough to wear it out. A pattern of shifting agriculture became mandatory, with new soil resources to the west constantly exploited. After the War of 1812 Indian lands were ruthlessly seized, and a "Great Migration" swept across Georgia, Alabama, and Mississippi. Without new lands, the cotton culture would have collapsed. In 1817 an "Alabama boom" brought thousands across the piedmont and red hills. Flush times had come to the South. As Ray Billington notes, the entrenched cotton plantation culture "doomed the deep South to an agrarian economy that ran contrary to the national trend."[46] Plantation owners and aspiring yeomen seized the best agricultural lands. Small farmers found themselves constantly crowded out, and they resettled further west. The remaining hills—the piney woods, sandy barrens, and worn-out cotton fields—became home for the "poor whites," "crackers," "hillbillies," and "sandhillers." Without annual firing, these lands would be reclaimed by the southern pine. The devastation of the Civil War and Reconstruction helped to industrialize the North, but it left the South committed more than ever to agrarianism. A pseudo frontier economy persisted in the woods, just as a pseudo plantation farming evolved with sharecropping.

That earlier economy had been devoted to hunting, herding, and fire. Typically, the agricultural frontier of the South was spearheaded by itinerant herdsmen of horses, cattle, and hogs, organized loosely into a "woods ranch." A seventeenth-century traveler colorfully recounted that "they go in gangs ... which move (like unto the ancient patriarchs or the modern Bedowins in Arabia) from forest to forest in a measure as the grass wears out or the planters approach them."[47] Swine and cattle had been introduced early from Europe. DeSoto brought 13 hogs with him, and after marching them across the southeast and even over the Mississippi River, he ended up with some 700. Whether the "piney woods rooter" descended from this herd or from later English importations is undetermined, but hogs became a staple of southern herders. The English practice of leaving swine to forage the woods for themselves continued on the open ranges of the South on a vastly larger scale. By the 1820s the Asheville Basin of North Carolina had developed a substantial corn-livestock complex. Cattle, hogs, and poultry were driven from Kentucky and Tennessee through Asheville en route to the eastern seaboard. By the 1830s, Merle Prunty records, "several hundred thousand animals, including 140,000 hogs, were fed in-transit annually in the Basin."[48]

This sort of pastoral economy invariably encouraged broadcast fire. Cattle herders followed old Indian clearings and both maintained and extended them through fire. The range, of course, was open, and the herder with 40 acres and 400 head of cattle became a common figure in southern folklore, exploiting the open pine savannahs as his ancestors had the oak openings of England and the heathlands of Scotland. The pattern of livestock grazing thus contributed powerfully to the spread of woodsburning in the South, leaving both a fire regime and an economy unique in the United States. Many of the southerners who migrated westward in such large numbers after the Civil War took their herding and fire habits with them.[49]

Wildlife biologist Herbert Stoddard recalled the pattern of fire and grazing as it existed in the South during the late 1890s. At that time Florida was a major cattle producer.

The cattlemen believed, with reason, that the woods had best be kept open and *ground cover short* so that they could drive their cattle to and from the cowpens, from one prairie or savannah to another, and most important, so that much of the upland terrain would produce a maximum of the most palatable grasses, legumes, and other herbaceous vegetation for their grazing livestock.

Neither would they tolerate brushy pinelands. Like the Indians before them, they ranged the woods barefoot or rode horseback either day or night. They wanted the woods as nearly free of ticks and chiggers as possible, and observations and common sense told them that frequent burning kept these pests in greatly reduced numbers, though it did not exterminate them. They wanted the ground cover, mainly of Saw Palmetto and wiregrass, short and open so that they and their livestock could see and avoid the dangerous Diamondback Rattlesnakes and Cottonmouth Moccasins. . . .

They enjoyed eating the huckleberries and blueberries and had observed that they fruited most abundantly when occasionally pruned back by fire. In addition the berry patches localized the bears for the fruiting period each year. As bears were enemies of their semi-wild hogs, this assisted in finding them, so they could be more easily killed. The people well knew how much stronger the cooling breezes blew during the heat of summer in the open woods, as compared to brushy jungle. . . .

The frequently burned-over woods were easily traversed on such expeditions, and they well knew that the burning increased the growth of Partridge Pea and the flowering of the Saw Palmetto and other honey-producing plants.

These children of nature also knew that large areas of dense jungle harbored "varmits" such as bears, wolves, pumas, bobcats, snakes, etc., and they wanted no such hiding place near their homes, where barefoot children played the year round, or near their free-ranging poultry. So *nearby* woods were burned over *annually*. Naturally there was little or no pine reproduction; none was expected or desired in the immediate vicinity of their homes. The distant cattle ranges were not burned nearly so closely, as the cattlemen wanted plenty of well distributed "rough" as well as fresh green grass on recently burned ground. Hence

Longleaf Pine came in strongly following large "mast" crops, and
replaced natural mortality from over-age. . . . Where the few seed trees
[after logging around 1900] included Slash, it took over most of the
Longleaf sites. This was the most tragic angle of the fire exclusion, and
was largely responsible for the terrific damage done by the wildfires
during the great droughts of the 1930s and the 1950s. . . .

The "wimmin folks" and the children of the pioneers frequently
assisted in burning around headquarters, and it was a hard, dirty,
though interesting, job; great care had to be taken to avoid damaging
the rail fences that encircled each corn, cane, sweet potato, and cowpea
patch, and the ever-present cowpens. . . .

The forefathers of these cattlemen had come in from the North Carolina mountains long before the nineties. . . . As they rode the ranges,
they set fires at intervals when conditions were right for light burning,
from early fall to late spring. They *knew* from the way cattle gravitated
to the fresh burns that the tender grass would make them grow and
fatten. It put them in shape to market, or survive the hardships of the
coming "dry season," and occasional severe cold.[50]

Change, Stoddard recalled, came with "stunning swiftness," the result of
a growing timber industry based on the southern pines. The flush times of
cotton were to be repeated for pine. In the 1880s industrial logging began
moving to the South from the Lake States and mid-Atlantic region. Earlier,
the southern woods had been given over to other uses. The pineries of the
coastal plain, in particular, had been exploited since the seventeenth century
for naval stores—tar, pitch, and turpentine—and fire was used in nearly all
stages of the operation. The woods, if choked with rough, would be burned
off to make passage easier and to lessen the fear of snakes. The site would
be developed and then periodically burned to reduce fuel. With their slashings dripping in pitch, the pines could easily be ignited by wildfire and the
site ruined. Controlled burning for fire protection thus became standard procedure. Needles and debris would first be raked from the base of the tapped
trees, then the area ignited. The procedure would be repeated annually.
Since the slashings were often crude, the trees would die before long and
the whole operation would move. In 1834 the copper still was introduced,
with an effect on naval stores not unlike the impact of the cotton gin on
plantation farming. Nearly half of all the naval stores at the time were
exported from North Carolina, but by 1844 the turpentiners were vigorously
spreading throughout the South in a migratory exploitive pattern not unlike
shifting agriculture. With the advent of industrial logging, the economic
importance of naval stores as a forest product diminished. But it contributed
one important legacy to southern forestry: the concept and practice of systematic protective burning.[51]

More timber was logged out of the South than from any other region.

7. Louisiana pinery largely converted to pasturage through some logging and much fire, ca. 1920. Although the woods remain, the effect is more that of a savannah.

Cutting reached a peak in 1909. By 1920 the virgin pineries were virtually gone, and little regeneration had appeared to restore them. The South, momentarily prosperous, had collapsed again. It was gutted with more than 92 million acres of unproductive cutover land, and the ghost towns of old mills littered the landscape like the mining camps of Nevada. The tax structure of the time encouraged the abandonment of cutover lands, which simply passed into tax delinquency. Fires to stimulate pasturage for cattle and heavy browsing by hogs prevented regeneration. Even farmland decayed rapidly, hit by the boll weevil. The cycle of turpentining, logging, and fire thus temporarily expanded the rangelands of the South. As professional foresters moved in to encourage another crop of timber, they came into violent conflict with the southern herding economy. The most visible point of the confrontation centered on proper fire practices.[52]

When Inman Eldredge addressed the Society of American Foresters in 1911 on the fire problems of the South, he enumerated the local forest population in Florida: "homesteaders, cattlemen, turpentine operators, and negro turpentine hands. The homesteaders and cattlemen are nearly all native Floridians of the 'cracker' type." He then described clearly how these different groups responded to fire.

> The popular sentiment of the residents within the Forests, in common with nearly all of the South, is unqualifiedly in favor of the annual burning over of the pineries. The homesteader and the cattleman burn the woods to keep down the blackjack undergrowth and to better the cattle range. The turpentine operator burns over his woods annually, after raking around his boxed trees, and at a time when the burning will do the least harm, in order to protect his timber from the later burnings that are sure to occur. He burns also with the idea of keeping the turpentine orchards clear of undergrowth and free from snakes, in order that his negro laborers may gather the gum with ease and safety. The camp hunters, of whom there is a large number during the fall and winter months, set out fires in order to drive out game from the thickets. All of these different classes of people have for a great number of years been accustomed to burning the woods freely and without hindrance of any kind, and it is done without the knowledge or the feeling that they are breaking the laws or in any way doing damage. On the contrary, they all have the most positive belief that burning is necessary and best in the long run.
>
> The turpentine operator burns his woods and all other neighboring woods during the winter months, generally in December, January, or February. The cattlemen set fire during March, April, and May to such areas as the turpentine operator has left unburned. During the summer there are almost daily severe thunderstorms, and many forest fires are set by lightning. In the dry fall months hunters set fire to such "rough" places as may harbor game. It is only by chance that any area of unenclosed land escapes burning at least once in two years.[53]

The beleaguered supervisor proposed a sensible compromise as a working fire plan: logged areas would be protected by organized patrols until reproduction could establish itself; the remainder of the forest would be burned annually in January; unauthorized or uncontrolled burning would be prohibited. The cost for the system would be meager. And, Eldredge noted approvingly, "it would not be a difficult matter to obtain the hearty cooperation of most of the settlers within the Forest in carrying out such a plan as this."[54] Similar problems plagued the entire South. To remove fire was to unravel the fabric of the rural South. Wildland fire protection was in a sense a vanguard of industrialism.

Even timber owners and foresters argued bitterly among themselves over the proper strategy for fire protection. Without wildfire protection, industrial forestry would have been doomed, and much of the argument was actually a debate about socioeconomic changes, of which conflicting fire practices were only a highly visible symbol. H. H. Chapman, who did so much to advance prescribed burning, saw clearly the necessity to prevent wildfires if industrial forestry was to succeed. Reporting in 1912 on some Arkansas and Louisiana lands owned by Crossett Lumber Company, he adamantly recommended that "fires should be absolutely kept out of recently cut-over areas after slash is burned, for a period of at least five years, by employing rangers and providing an organization to fight and put them out."[55] Some 32 percent of company lands had already been destroyed by uncontrolled fires. In 1923 Chapman reported that establishment of a fire control organization outdistanced all other forestry concerns combined. Although the company purchased new lands, fires on lands it already owned were causing losses of "probably three times as great a value" by eliminating reproduction.[56] The debates over the torch in the timber took place on several levels: between those who opposed industrial forestry over other land uses, and between those who were committed to eliminating wildfire in industrial forests but who favored different means for breaking the cycle of southern woodsburning. Forestry created a necessity for fire protection that had not previously existed. It demanded a different cycle of fire and brought forestry organizations into violent conflict with local fire practices.

Abandoned land in the South, both from logging and from cotton fields devastated by the boll weevil, became a national scandal in the 1920s. By the 1930s chemists had developed methods for converting the fast-growing southern pine into pulp. Government and industry began purchasing lands for forests, and they began to exclude fire. Throughout the South broadcast fire and an open range had been mutually interdependent. With the acquisition of lands for forests came enclosure, either directly through fences or de facto through modifications in the vegetative cover. Fire protection allowed the rough to thrive, and the rough gradually smothered the range. Stockmen complained bitterly that they had been dispossessed. In a sense, they had: the industrial conquest of the South was not unlike the agrarian conquest that European and American settlers had imposed on Indians only a century earlier. The Resettlement Act, the closure of open hunting and range lands, and the acquisition of tax-delinquent lands bore an uncanny resemblance to the means by which whites had displaced Indians. Pine had simply replaced cotton. A legacy of resentment and confusion was often the consequence. It is not surprising that the natives who failed to make the transition, who failed to acculturate to the new economy, should strike back or simply refuse to change their old habits. With fire they could, at one instant, destroy the new and restore the old.

The question of fire practices extended as well to that other half of the traditional southern economy of the frontier, hunting. Like herding, southern hunting may be dated back to DeSoto, who introduced greyhounds and bloodhounds as well as hogs. Both hunting and herding were practiced on the open range and were in a sense complementary. In the wake of the Civil War, southern agriculture came apart socially and environmentally. A social consequence was the agricultural system of sharecropping, which exploited soil and people about equally. No less profound were the environmental changes that resulted as the woods crowded back onto former arable land, leaving a mosaic of fields, forests, and brush—ideal conditions for the production of game like fox, racoon, and quail. As happened with the heathlands of ancestral Scotland, the fields of the South converted from herding to hunting. Hunting plantations sprang up on the British model, and in the 1880s railroads began encouraging northerners and well-heeled foreigners to enjoy the South as a winter resort.[57] The idea of establishing plantations to exploit a cash crop came easily to the South, and sites appeared from the Carolinas to Florida and Arkansas. The crop was quail and turkey. The old patchwork of fallow fields, woods, grass, and brush encouraged game bird populations, and this vegetative pattern was sustained by a tradition of land use and fire.

By the early 1870s, hunting camps for red grouse in Britain began to experience a dramatic decrease in the population of game birds, largely attributable to the suppression of fire by wardens on formerly grazed heathlands. The South began to take up the slack. But with the advent of fire protection in the South, game birds decreased much as pasturage had and as grouse populations had in Britain. The vegetative ensemble that sustained maximum populations gave way to roughage and woods. By 1923 hunting plantations in southern Georgia and northern Florida were in decline. Desperate owners agreed to pool resources for scientific research into the question. The acreage concerned was not large, but the issue, as it turned out, had national significance. Out of this concern came the Cooperative Quail Study Investigation. Direction of the project was given to the U.S. Biological Survey, which handed it over to Herbert Stoddard in 1924. Stoddard completed his field work four years later, and in 1931 he published a classic in wildlife management: *The Bobwhite Quail, Its Habitats, Preservation, and Increase*. His report to the Biological Survey was reviewed by the U.S. Forest Service, and his comments on fire proved so challenging that Forest Service pressure required the chapter to pass through five editorial drafts, each successively watering down the conclusion. Stoddard determined that quail populations in the South depended on land management practices and that in this complex process "fire may well be the most important single factor in determining what animal or vegetable life will thrive in many areas."[58] Stoddard's quail study thus concluded at the same time as S. W. Greene's investigations on fire and pasturage and his *Bobwhite Quail* saw

publication in the same year as Greene's "The Forest That Fire Made."
Thus, southern fire practices were vindicated for hunting and herding at
nearly the same time.

What made this work especially significant was that it gave scientific
credibility to the subject of wildlife management through fire. Stoddard
removed the topic from the realm of "cracker" folklore, much as Chapman
had done for timber management and Greene for livestock management.
Woodsburning had to cease because of socioeconomic changes, manifested
by enclosure, not because it was environmentally degrading to the habitat
for game, timber, or stock. Prescribed burning for wildlife management in
refuges became respectable, just as it did for silviculture and range management. Where the natural products of the ecosystem were being harvested, fire once again demonstrated its utility, even its necessity. Wildlife
managers have generally remained among the most active advocates of prescribed fire, and wildlife refuges in the South were among the first federal
reservations to implement prescribed burning. Moreover, it was through the
Cooperative Quail Study that Stoddard, and later the Komareks, organized
the Cooperative Quail Association, which served as a consultant for hunting
plantation managers throughout the South. In turn, it was succeeded by the
Tall Timbers Research Station, which served in the 1960s as a critical
forum for research into the ecology, techniques, and philosophy of prescribed fire.[59]

The quail plantation was the product of settlement patterns, of historical
events, of fire practices learned and maintained, and of environmental
potentials. In the broadest sense, the quail population was a cultural creation, and so it has been with the southern fire regime at large. For reasons
of historical accident, cultural tradition, and environmental potential the
South pioneered in the rediscovery of controlled wildland fire. In complicated ways the woodsburning birthright helped to bridge low-productivity
agrarianism with high-yield industrial forestry and land management.

Woodsburning, like moonshining, became obsolete, quaint, and dangerous. But it persisted. When the psychologist Shea conducted his interviews
in the late 1930s, industrial forestry was only 10 to 15 years old in the South,
and most of the southern national forests represented New Deal acquisitions
(the oldest dated no earlier than to the Weeks Act). Change had come with
terrific abruptness, not unlike the changes thrust on the Five Civilized
Tribes a century earlier. It would have been surprising if the fire practices
that were built up over generations of frontier experience had disappeared
within a few decades. Nor were northern foresters much more responsive.
The sensible compromise proposed by Eldredge was not implemented for at
least a good half century. Instead, the habitat was simply converted: the
human inhabitants suffered no less than its trees, wildlife, and livestock. The
Indian whose hunting grounds had vanished into farms had a counterpart

in the agrarian southerner whose hunting and herding lands were swallowed by industrial forests. Especially with the advent of prescribed burning and strategies like DESCON, and with the successful accommodation to industrialization—not only by forestry but also by society—the rationale for illicit woodsburning is disappearing. The changing circumstances that have made prescribed burning acceptable have made traditional woodsburning an anachronism. It threatens to become a mere manifestation of malingering frontier violence—progressively antiquated, increasingly less useful, and sadly more malicious.

III

Until the advent of industrial forestry in the 1930s, there is in southern history a curious nonchalance about large fires. The most notorious illustration may be the 1898 fire in North Carolina, which, according to the state geologist, passed over 3 million acres and barely made it into the back pages of the Raleigh newspapers.[60] Evidently the fire began in traditional ways, but as it swelled out of control, hundreds of protective fires were set. Although the acreage was large, the damages were not. When fires swept over portions of South Carolina during the winter of 1949–1950, a similar response was apparent. In terms of acreages it was the worst fire season on record; as drought worsened, more than 2,000 fires broke out over 72,000 acres in February alone. Almost 250,000 acres burned in all, and about half the fires were incendiary. The alarmed state forestry commission issued press releases on the fires' progress, and UPI phoned the mayor of one town threatened by a mile-long fire front for his comments. His reply was that "we don't know anything about any forest fire here." All he saw was some "ordinary little old woods fires."[61]

Organized fire protection began at the turn of the century. A few lumber companies experimented with fire control, and two—the Crossett Lumber Company in Arkansas and the Urania Lumber Company in Louisiana—were the sites for the Yale Forestry School research under Chapman. Cooperative associations developed in West Virginia and North Carolina. A few states created forestry agencies, but most state organizations were little more than figureheads until passage of the Clarke-McNary Act in 1924, which greatly expanded the provisions of the Weeks Act. The U.S. Forestry Service entered the South as landholder and cooperator, and eventually state forestry departments acquired jurisdiction over most state forested lands, rural and wild both. The extent of these charges could be large: Georgia, for example, protects more forested lands than does California. The state programs, moreover, were integrated into rural fire departments from the beginning. Rather than the sharp segration of urban and wildlands typical of the West, the South more typically offered a mosaic of rural landscapes.

The southern fire years correspond not only to periods of drought but equally to socioeconomic changes. The first major fires of record appeared in 1930–1931, a time when drought ravaged the entire eastern seaboard and even allowed the Dismal Swamp to burn uncontrollably. But that was also the year in which Charles Herty and others discovered the means of converting the southern pines to pulp. With the arrival of the pulp industry, southern forestry assumed its colossal value, and the specter of abandoned land could no longer be tolerated. That same year Stoddard and Greene published their celebrated reports on the value of prescribed fire. Although their findings were accepted warily, they at last provided a meaningful discrimination between wildfire, prescribed fire, and woodsburning, something that did not exist before the writings of Chapman.

The 1930–1931 drought sustained fires primarily in the states bordering the South proper. Kentucky and Virginia were hardest hit: the Virginia Forest Service went into virtual bankruptcy and had to suspend control efforts on more remote fires. In 1932, on the other hand, Georgia and Florida suffered heavy losses. The Forest Service amended its guidelines on the allocation of Clarke-McNary money to allow state cooperators to engage in protective burning. But as acreages of abandoned, tax-delinquent lands increased, state and federal governments acquired larger holdings and converted them into forests. The arrival of the New Deal and programs like TVA and the Resettlement Administration assisted in the transition, and the Civilian Conservation Corps began to create a physical plant for fire control—something that had never before existed—and to provide hundreds of fireline workers, an equally rare resource. Through the Clarke-McNary program, the CCC could be put to work on the lands of state and even private cooperators. The CCC did not end the debate over prescribed fire, but it did fashion a fire protection organization that could accommodate controlled burning. Prescribed fire could be conceived of as a supplement to, rather than an alternate for, a fire control program. When the CCC left, the South quickly seized on mechanical replacements—the tractor-plow and ground tanker, for example—and later adopted air tankers, helicopters, and, for a time in Virginia and Arkansas, smokejumpers.

Drought brought serious fires to Kentucky and Arkansas in 1938. Arkansas's forestry agency, like Virginia's during the drought of 1930–1931, nearly folded under the pressure, and in Kentucky hundreds of families and their livestock were evacuated from fire areas. But it was the drought years of 1941–1943, after the CCC had left, that proved decisive. Missouri in 1941 and Arkansas in 1943 struggled to contain widespread outbreaks of fire, while in 1943 Florida broke all previous regional fire records. After inspecting the fire scene, Forest Service Chief Lyle Watts reversed earlier prohibitions against prescribed burning on the southern national forests.[62]

Further droughts in the mid-1950s induced other changes. Fires in east

Texas in 1947 burned 55,000 acres but were overshadowed by the great oil fires at Texas City and by the more devastating Maine fires. South Carolina experienced widespread fires over the winter of 1949–1950. But it was a run of fires that swept the South between 1952 and 1957 that produced the most memorable record. The Indian summer of 1952 brought drought to Arkansas, Kentucky, and West Virginia. As rainless day followed rainless day, hundreds of fires broke out in Arkansas. Over the course of several months more than 150,000 acres burned. The epidemic strained local firefighting resources to the limit. The state forester declared flatly that "the forest fire situation is out of control" and informed the public that "your lives are in danger." Further east, especially in Kentucky, organized fire protection— which in the face of local tradition had never been strong—simply collapsed. About 2 million acres blazed in eastern Kentucky and West Virginia. The fires had multiple sources, and in late October as many as 80 to 100 new starts a day were reported. Squirrel-hunting season was in progress and many burns began from abandoned campfires or game-smoking fires. Regulations against debris and brush burning during daytime hours were widely ignored. Arsonists fired timber and coal lands owned by large industrial concerns. And a considerable portion of all fires simply represented protective backfires ignited by worried rural dwellers in conformity with time-honored frontier tradition. Volunteers and local fire brigades attacked fires threatening farms and villages. Coal miners, union and nonunion alike, reported to the firelines. Ohio activated its National Guard, supplemented by honor camp inmates, to fight fires in the southeastern part of the state. When the fires spread onto Fort Knox, regular troops beat them back. A sheriff described his Kentucky county as one "ball of flame." In wooded counties like Wayne and Taylor, where there was no formal protection, fires simply ran free. Where national forests or environs were threatened, organized crews were sent to the line. In West Virginia forest protection organizations, many of them dating back 40 years or so, contributed manpower and equipment. Numerous backfires set by woods dwellers more than once virtually surrounded weary firefighters, forcing them to back off hastily from their prepared lines. In the third week of October the governor of Kentucky declared a state of emergency. Hunting season was cut short, picnicking and camping in state parks were banned, and attempts were made to enforce brush-burning laws.[63]

The total burned acreage in 1952 is the result of perhaps several thousand fires consuming a few hundred or thousand acres each. Much of the fuel was scrub or hardwood, not explosive conifers. Thus the complex seemed to follow a typically southern pattern. In the major fire complexes of the West and North a handful of rampaging mass fires usually accounts for most of the damage. In the 1952 outbreak, fire was widespread, but few individual fires really dominated. The emergency was sustained more by continued

protective backfiring and persistent brush and debris burning than by
blowup fires. The situation was, as public officials reiterated almost daily,
"critical." But losses were not as great as acreage figures alone suggest, and
the episode gives some insight into how large fire complexes must have
behaved in the South during frontier times and, in areas depopulated by the
Civil War, during the late nineteenth century. Explosive mass fires were
limited to remote areas with heavy fuel, such as reclaimed pocosins and
swamps drained by drought.

The drought moved south and east. In 1953, faced with increased fire
responsibilities, Everglades National Park began its controlled burning
experiments. In the winter of 1954–1955 fires broke out in Georgia. When
a lightning storm ignited five fires near Okefenokee Swamp, nearly 500,000
acres burned. The rains that in the West would have threatened soil erosion,
here merely refilled the swamp, allowing the fire scars to heal without untold
damage. In 1955 it was North Carolina's turn. Some 600,000 acres burned;
within 10 days two fires on the coastal plains roared through 115,000 and
290,000 acres, respectively. In the wake of this disaster the state's entire fire
organization was overhauled and a new master fire plan approved.[64]

Individual state solutions were not enough, however. In February 1956 a
controlled burn near the Osceola National Forest in Florida escaped. It was
driven into a swamp, where it continued to burn on an island. In March the
Buckhead fire escaped from its sanctuary and under the influence of a dry
cold front raged over 110,000 acres. Georgia suppression units halted the
fire along the state border.[65] There had been plenty of reasons before to
support an interstate compact on the New England model; if doubts
remained, episodes like the Buckhead fire swept them aside. The western
states could establish de facto contracts through mutual agreements with
large federal landholders like the Forest Service. In the East and South this
was not practical; in any case, thanks to their responsibilities for rural as
well as for woodland protection, the state organizations were far more powerful than their local federal counterparts. Industry, too, wanted better protection. Before 1956 ended, the Southeastern Interstate Forest Fire Protection Compact was signed.[66] The South thus became the second area to opt
for regional strategy. In the last year of the drought, 1957, Georgia and
North Carolina were again hit hard. Early in this cycle of fire a study by
George Byram offered one of the first physical explanations for mass fire
behavior, and two years after the cycle passed, the nation's first forest fire
laboratory was established by the Forest Service at Macon, Georgia.[67]

There have been other large fires, of course, though none seems to have
had the historical unity or political impact of the 1952–1957 burns. In 1963
heavy fire losses were reported in Arkansas, West Virginia, Virginia, Maryland, Kentucky, and North Carolina. But national attention focused on the
coastal plains of New Jersey, New York, and Massachusetts. In 1966 and

again in 1967 conflagrations raged on the coastal plains of both North and South Carolina, but bigger acreages and larger numbers of fires struck the Northern Rockies.[68] More large fires burned on the coastal plains in 1971. In the autumn of 1978 the traditional southern fire complex again asserted itself. A stubborn drought settled over the Deep South, and fires began to appear in Georgia, Mississippi, and Alabama. Most were incendiary. In Alabama the drought was the worst since 1904. Despite bans on outdoor burning, as many as 200 fires a day were reported. The National Guard was mobilized, and Birmingham was placed under an air pollution alert as a high pressure cell stagnated over the region. Rains eventually ended the emergency in November.[69] Even during this fire-intensive period in the South, however, Southern California offered competition as thousands of acres burned in Malibu from a fire set by an arsonist. One hundred sixty homes were destroyed, with perhaps a hundred more damaged. In terms of property damage it was the worst disaster since the infamous 1961 Bel Air conflagration, and there were dire predictions of mudslides to follow.

The difference between the two fire complexes, California and Alabama, is instructive. It involves more than media coverage and publicity. As two of the dominant fire regimes in the nation, each region had adopted dramatically different strategies. The contrast between the light-burning and prescribed-burning controversies is but a point of departure. In both cases, "brush" was at the core of the issue. In Southern California especially, the public demanded that the brush be protected at all costs, notably from destruction by fire; in the South, the demand was for the rough to be removed by any means, preferably by fire. There is the strategic difference, too, between the pattern of a few large fires typical of the West and that of an epidemic of smaller ones common to the South. California's rugged setting added to the illusion that this was a sublime natural disaster; the Alabama fires looked more like a riot. They occurred in a rural rather than a semiwild landscape and were resolved with determined endurance rather than heroic drama. The southern fires responded to socioeconomic conditions and historical prejudice, not simply to the challenges of constructing expensive homes in a natural fire regime.

Its peculiar fire heritage helped the South to train the rest of the nation in the art of prescribed burning. From the Florida-based Tall Timbers Research Station, beginning in 1962, came an influential series of annual fire ecology conferences, and four years later the Southern Fire Lab of the Forest Service sponsored a program of prescribed-burning seminars. The nation's leading fire region also became a leader in fire protection—mechanized, sophisticated, and no longer locked into parochial issues. North Carolina, a state in which a 3 million acre fire in the nineteenth century could go virtually unreported, offered a splendid example. In 1970 the state responded to pleas for help from the Forest Service, then overwhelmed by fires on the West Coast. In 1976 the Interior Department made the request, and North Carolina shipped some of its special high flotation plows to the Seney fire in the upper peninsula of Michigan. It marked only the second occasion that heavy equipment was airlifted between states for fire suppression. A year later, North Carolina sent nearly 100 men to California for fire duty, and at national level training courses sponsored by the National Wildfire Coordinating Group, the fire overhead team from North Carolina swept all honors in the field of "fire generalship."

FIRE WOLVES AND SMOKEY BEARS:
A HISTORY OF FIRE PREVENTION

They are loosed from their hiding
And the red wolves are riding—
There is blood and blast and fury in their eyes—
And their packs go a-crashing
There's a crackle and lashing
Breathing smoke and sparks and splinters to the skies.
—Anthony Euwer, "Red Wolves"[70]

Remember, Only You Can Prevent Forest Fires. —Smokey Bear

I

As the southern scene so clearly reveals, and as an emphasis on lightning fires in the backcountry of the Far West tends to mask, the prevalence of fire worldwide is due to the prevalence of people. The vast majority of fires are anthropogenic, and though many such fires exceed their intended scope, they represent practices of controlled burning that reach back through millennia. Most contemporary fires, big and little, have their origin in traditional fire habits—campfires, hunting, fires in warfare, fire in pasture improvement, debris burning, arson—though the scale of open burning has been reduced in many instances, and such fires persist in often diminished forms. Debris burning, for example, replaces landclearing, smoking fires replace fire hunting, and smudges replace open bonfires. A prevention program must address not only certain techniques for handling fire but also the cultural environment in which fires occur. In the early years of systematic fire protection, prevention programs were directed almost entirely against traditional fire practices and tended to lack a separate identity. Certain problems with industrial machinery—the "donkey" steam engine and the railroad locomotive—were added, but the commanding issue was to educate the public to the meaning of the counterreclamation and its demand that traditional fire practices cease. Where education failed, local, state, and federal authorities enacted and enforced tough fire codes.

The simplest fire uses were often the most intractable. Escaped campfires, for example, were commonplace. Hunters, travelers, teamsters, miners, transients of every sort—all are credited with starting fires in all parts of the country. Copying Indian practices, frontiersmen and tourists alike commonly torched a convenient log and rarely extinguished it. Such a brand might burn for days, and as the weather changed, the fire often spread. The literature on escaped camp and cooking fires is legion, and such fires not infrequently returned to threaten their careless originators. One of the most famous literary accounts is in Mark Twain's *Roughing It*. A summer's idyll on the shores of Lake Tahoe ended as Twain looked up and "saw that my fire was galloping all over the premises."[71] Eventually the fire swept over the distant vantage points and drove Twain to a boat for safety. Similar episodes were especially prevalent on the grasslands, and, as described by Colonel Richard Dodge, travelers exercised routine precautions.

> Under such circumstances the camp fires should as far as possible be made on the leeward of the camp and grazing ground. If this cannot be done without too much inconvenience, holes should be dug in the ground, large enough to build the fires in, the long grass near should be cut (with a spade in default of a better implement) and carried off, and the earth taken from the holes spread over to leeward. It is a common custom to burn off a space sufficiently large for the fires, but this is very dangerous if the grass be long and the wind high. Even though men stand around with blankets to whip the fire out when necessary, it sometimes gets beyond the control of the best directed efforts.

Dodge then recounted how one escaped fire "in five minutes from first alarm . . . reduced [us] from a well-armed, well mounted aggressive force, to an apparently half-armed, half-mounted, singed, and dilapidated party."[72]

Such incidents receded as settlers replaced transients and as farmers abandoned the open fire for the confined hearth. But for the lands deliberately reserved from settlement, such fires did not disappear. The problem of wandering tourists reached such proportions at Yellowstone that in 1889 and 1890 the Army established regular campgrounds and forbade camping outside these zones as a means of concentrating the fire hazard.[73] The concept was adopted for the national forests, too, with the additional requirement that visitors carry shovel, ax, and bucket to extinguish their campfires—or the escaped fires that might result. Failure to comply could bring prosecution under trespass laws, and guilty parties could even be made to pay for the costs of suppression.

The problem could be particularly acute where frontier and forest were not legally or practically disentangled. The sequestering of sightseers into fixed campgrounds was followed on a much vaster scale by the development in the West of reservations for Indian tribes and in the South by enclosure movements that tried to segregate frontier forest users from industrial users. Where the two remained mixed, fire was not only a visible measure of their differences but also a frequent weapon in their struggles, to which the nightly guerrilla warfare on the Plains was only a prelude. Pointing to fire lookouts on the Clark National Forest in the Ozarks, a resident suggested darkly that "I think these towers are for war, so the government can conquer the people." As late as the 1930s, when another resident was asked whether the forest officers were friendly, he replied, "Yes, they are. They wouldn't ever dare to get tough or we would burn them out." Seeing no harm in their woods fires, locals could delight in the frantic endeavors of northern foresters to suppress them. Where CCC enrollees from outside the region furnished organized crews, fires might be started just to "keep those Yankee boys busy." Another informant observed that "the farmers didn't burn at all in here in 1935 and '36, but we got quite a bit burned off this year and will do better next year. You go along in a car and throw out matches, or take a burning rubber inner tube and ride through the woods horseback and let the pieces drop off and start a fire wherever they drop. Yes, the grasshoppers and ticks are a lot worse. We use to burn every year."[74] However attractive such fires were to farmers and herders accustomed to open grazing in the woods, they appalled the foresters charged with overseeing the national forests.

Nor were retaliatory fires absent from other portions of the country. The process of separating frontier economies from the woods came slower to the South than to the West, but the mixture had been common and explosive throughout the Cordillera, where transhumant shepherds practiced their ancient arts. Members of the Industrial Workers of the World set fires on timber company lands in the Northwest during the labor unrest of World War I. Coal miners in Ohio propelled burning ore cars down mine shafts to ignite underground coal seams. In Pennsylvania coal company timber lands were fired as well. Stockmen in the Coast Ranges of northern California, outraged at state and federal burning restrictions, ignited so many fires that local suppression forces were left exhausted and near bankruptcy. During 1927 in Lincoln County, New Mexico, the scene of bitter frontier range wars in the nineteenth century, incendiary fires were constantly being set around a certain ranch. Angry over fire restrictions, the settler vowed to keep firing and threatened to shoot anyone who tried to stop him. When firefighters were indeed met with rifle shots, the sheriff and local forest supervisor set out after the unrepentant incendiarist. In the ensuing shootout an innocent Forest Service clerk, commandeered as a driver, was killed along with the rancher.[75]

Frontier fire practices also included fires for spectacle, again on the Indian example. In nineteenth-century Oregon it was something of a sport to see who could start the largest fire in the Cascades, and in the 1890s it was reported that tourists around Mount Rainier in Washington "frequently set fire to the resinous fir trees for the pleasure of seeing their lives go out in sudden flashes of flame. This wanton custom . . . has greatly injured the beauty of the foreground of one of the noblest and most impressive pictures in the United States."[76]

There was considerable local resistance toward efforts to establish restrictive fire codes and vigorous enforcement programs. Fire was so fundamental to frontier economies that residents and local juries tended to tolerate abuses for fear that tougher prosecution would deprive them of the fire they needed. This is not to say that mores and statutes were not in effect to regulate fire practices, but only that such codes were extended to the lands of the counterreclamation by foresters, not by pioneers. Massachusetts Bay Colony, for example, enacted a law in 1631 that forbade any burning prior to March 1. Plymouth Colony had a similar statute in 1633, and 50 years later legislation on burning practices extended from Maine to Pennsylvania. By the time of the Revolution, only Maryland, Virginia, South Carolina, and Georgia lacked protective legislation—all colonies, it should be noted, in the South.[77] A few practices were outlawed, such as fire hunting, but most were merely regulated. There were penalties for escaped fires, including fines, imprisonment, and, of course, restitution for damages; and there were provisions for mandatory firefighting service in case of an escape. But prosecution was typically lax. Until lands were removed from settlement, there was little interest in extending such provisions to remote sites—and considerable resistance against it where it interfered with traditional woods use, such as grazing.

Unwanted fires that threatened resources valuable to the community would be met with frontier justice, however. Following an outbreak of forest fires in Idaho, local townspeople threatened to lynch the offender if found. During the struggle over enclosure of the Plains, fires were often directed at those fencing formerly open range. In 1884 Texas was compelled to make deliberate range burning a felony. When the XIT Ranch in the Texas Panhandle suffered from cattle rustling, it hired Ira Aten, a former Texas Ranger, to patrol its eastern boundary. Word was leaked to him that he would be burned out if he persisted with his strong protective measures. "I told them I could not help it if they did," Aten recalled. "But if I caught one doing it, I was going to kill him if it was the last thing I did." When Aten later rode off to confront one particular arsonist, gun on hip, the word went out. And that, Aten concluded with satisfaction, "was the last time my range was set a-fire maliciously."[78]

Such direct methods were not available to the Forest Service. But one reason for its persistent requests in the early years for Army troops to be stationed on the forests during the summer was for the "moral effect" that their presence would have. Nor could the Service count on certain expressions of conscience such as struck William Pixley of Marin County, California, or the brothers Louis and Philip Audiget of Los Angeles County. Pixley started a brush fire on Mount Tamalpias in 1881 that quickly escaped control and swept over 65,000 acres. It was reported that when Pixley realized that the fire was out of control, he dropped dead on the spot. Even more macabre was the fate of the Audigets in 1890. When a brush fire approached, they started a backfire that soon got beyond their control and did considerable damage to their neighbors. Apparently believing that California conformed to the Code Napoléon, the two Frenchmen thought their deed was punishable by death and elected for suicide instead. Lying under a tree, they shot themselves in the head—Philip once, and Louis three times. Two days later they were discovered, still alive, and were rushed to the hospital.[79]

In truth, American fire codes, even on paper, were fairly lenient. In Japan incendiarism was a capital offense. In Scandinavia and Germany it was practically unknown. Spain adopted strict measures against broadcast forest fire in the late nineteenth century in a desperate attempt to intervene in the irrevocable erosion of Mediterranean soils. In Greece promiscuous burning did so much damage that in 1900 the Greek Orthodox Church finally threatened anyone convicted of incendiarism with excommunication.[80] But the enforcement of even existing codes in the United States was indifferent at best, a situation that outraged professional foresters. In 1890 Bernhard Fernow wrote in disgust that "the whole fire question in the United States is one of bad habits and loose morals. There is no other reason or necessity for these frequent and recurring conflagrations."[81] Pinchot, too, castigated the "moral effects" of incendiarism. "There is no doubt," he sermonized, "that forest fires encourage a spirit of lawlessness and a disregard for property rights."[82] That many of the fires on the western lands were set by transients and frontier riffraff did nothing to lessen the impression that fires were irresponsible as well as destructive.

Gradually, however, the separation of lands resulted in a separation of fire codes and an exchange of fire practices. But in eliminating many traditional causes for deliberate and accidental fire, fire protection itself added some others. Prescribed burning, of course, is one instance. Another is "job hunting," the setting of fires so that one may be employed to fight them. In its way an analogue of arson for profit, the job-hunting fire relies on the nature of wildland fires, which may last for days or even weeks, and on the availability of emergency accounts to finance suppression. The practice is apparently as old as paid firefighting. There were charges that some of the 1903 Adirondacks fires had been set to gain employment, and the conviction was apparently widespread that the same was true for the 1910 fires in the Northern Rockies. In his famous lectures at the Biltmore School in North Carolina, the German forester Carl Schenk listed job hunting among a dozen prominent causes for fire. When the Depression deepened in 1930, fires became epidemic in certain areas, particularly in logging regions like the Northwest. Job-hunting fires became in effect a form of emergency government relief. The sudden infusion of dollars from a major fire could transform a stagnant hamlet into a boom town for a week or two. The situation became so acute in Idaho in 1931 that portions of the state were placed under martial law. In Washington a state fire warden not infrequently arrived at a fire to discover a whole crew of local townsmen leaning on their shovels. The fire, obviously set deliberately, had not been touched, while the prospective crew waited to be hired by the warden. Throughout the West the pattern was repeated: job-hunting fires may have reached at least 30 percent of the total fire starts in the Northwest and were intimately associated with the giant Tillamook fire of 1933.[83] What contained the outbreak were New Deal programs that provided some economic relief and the arrival of CCC camps, which replaced local towns as a source of labor.

That lesson was not lost on fire officials, and the problem of job hunting has influenced the character of the organized fire crews that succeeded the CCC. Wages are kept low, for example, and crews are brought from some distance to major fires in order to reduce the incentive among locals and crews both to start more fires. A smokejumper contingent stationed at McCall, Idaho, ingeniously bypassed this prescription by releasing burning briquets out of their plane while en route to fires, thus ensuring uninterrupted employment. Nor have job-hunting fires lacked diabolical qualities, too. The Rattlesnake fire in California (1953) which took the lives of 15 firefighters, was set by a social misfit who hoped to be hired as a camp cook.[84]

Fire burns in a cultural as well as in a natural environment, and wildfire—even when set by lightning—is a relative concept. Fire and mankind enjoy a symbiosis: most fires are set, directly or indirectly, by man, and even natural fires are tolerated according to human criteria. Anthropogenic fires ebb and flow with human migrations and socioeconomic changes. As a form of technological pollutant, industrial fires continue to grow in variety. The number of heat and ignition sources in a culture may be taken as a measure of its technological sophistication, and fires, like pollutants, are an inevitable byproduct of industrialization. Yet most fires take on a human face—often misguided, like promiscuous woodsburning; occasionally tragic, like jobhunting fires; and sometimes comical, like the fire set by a tourist harassed by hornets, a fire that ran 700 acres from the troublesome nest. Many fires stem from inept practices, and a few from illicit ones, like moonshining. Fires have escaped from moonshiners' cooking fires; stills have exploded and spread other fires; and fires have been set to divert the attention of snooping revenuers.[86] Fire may be an object in and of itself, a true accident, or an occasion to other ends. Consider, for example, the pattern of fires that began outside an Oregon town in the 1930s. After a CCC camp had been established nearby, fires mysteriously broke out like clockwork on Friday and Saturday afternoons. The entire camp was routinely dispatched. Officials puzzled for weeks over the incidents until the cause at last became apparent. The CCC enrollees, it seems, had given the local boys a good deal of competition for the attentions of the village maidens. By igniting fires on the weekend, the local boys cleverly disposed of their romantic rivals.[87]

II

Until wildfires could be controlled, there was little point in trying to prevent them, and until one could discriminate between good fires and bad fires, there was little foundation on which to base an argument for prevention. The counterreclamation created just such a discrimination by contrasting fire practices suitable for the reserved and forested lands with those typical of frontier economies. In the early years, prevention meant posting notices that advised forest users of their obligations, or warned them of forest closure, or lectured about the evils of careless fire and light burning. All of early forestry was in a sense a campaign at fire prevention through public education—that is, by training people in proper fire habits, advertising fire codes, giving people a sense of personal proprietorship in the public lands under protection, and admonishing locals to abandon traditional uses of controlled fire for a program of fire exclusion. The task had national goals but local focus; posters in California, for example, had to be printed in eight different languages in 1914.

Among the first systematic programs was that developed by the Western Forestry and Conservation Association (WFCA), a consortium of private timber protective groups in the Northwest and California. At the core of its conservation program was fire protection, and under E. T. Allen it immediately undertook a "campaign of publicity and education." Allen selected fire among all forestry issues because he believed that it was the "strongest game . . . [and] because it is easiest understood as well as desirable." He considered "the organization of a general news bureau to furnish copy to all available press outlets as a matter of prime importance." Pamphlets, posters, essays, and educational material were distributed to trade magazines, popular periodicals, newspapers, schools, granges, and clubs. No outlet or medium was overlooked. Allen himself, perhaps with the aid of his novelist wife, Maryland, designed "sequence posters" that delivered fire prevention messages to roadway travelers. The 1910 conflagration focused favorable attention on the association and helped to legitimize its credo that fire protection was the foundation of forest conservation. In 1912 the public education program flooded the Northwest with 35,000 "Keep Them Green" pamphlets. Periodic fire bulletins were sent to members and newspapers. What the WFCA did for the organization of fireline suppression by sponsoring W. B. Osborne's *Western Fire Fighter's Manual,* it did for prevention programs through the precedents set by its media campaign.[88]

State organizations could be equally aggressive and inventive. Michigan, in conjunction with the Northern Forest Protection Association, sponsored a special lecturer in 1911; his 402 lectures on fire prevention were translated into Finnish, Italian, Romanian, Serbian, and Hungarian. Later in the year, at a suggestion of the governor, the Michigan Forest Scouts were organized among boys aged 8–18. By the next year the scouts had "extinguished or reported 509 fires, 55 without adult aid, at least 3 of which would have taxed the ingenuity and intelligence of the best men fire fighters." Posters had been widespread since 1903, most either advising the populace of existing fire codes or reminding them of the legacy of horrible Lake States fires. A common slogan ran: "One Tree Can Make A Million Matches—One Match Can Destroy a Million Trees." Around 1928, "Keep Michigan Green" came into vogue, though the fact that a gubernatorial election was in progress and that the incumbent's name was Green may have influenced its selection. By 1921 the immediacy of motion pictures with names like *The Red Poacher* and *Nature's Gangsters* was being tried. During the mid-1930s a specially outfitted railroad car, "The Wolverine," toured the state to promote a fire prevention message.[89]

National Fire Prevention Day first appeared in 1914, but although it commemorated equally the Peshtigo and Chicago fires of 1871, its emphasis remained with urban fire. Forest fire prevention came into its own nationally in the 1920s. Inspired by the success of professional advertising, the Forest Service experimented with new techniques to sell the idea. Many of the messages were aimed at children. The Service published a model "Forest Fire Prevention Handbook for Children" in 1926, and similar works were prepared for schools in Oregon and Washington. Other messages were directed at rural audiences still uneducated to the perils of light burning. Here the Service turned to films, which it distributed widely among its cooperators. One state ranger in Missouri alone wore out five prints of *Trees of Righteousness*. He took the film around his district in a Model T that hauled a trailer outfitted with a projector and generator. The use of films and traveling shows became commonplace throughout the South in particular, a style not unlike that of itinerant revivalists. Building on this experience, Missouri subsequently developed the long-running "Showboat" program, which used vans, films, and lecturers to bring the word to the Ozark woods. Electricity was a novelty as great as films, and Showboat performances could draw an audience from miles around.[90] Rural electrification and television eventually outmoded such programs, but their successes were undeniable. It was as if the campaign against the vestiges of frontier fire habits required a combination of salesmanship and entertainment reminiscent of the patent medicine drummers, peripatetic revivalists, and traveling circuses of the old frontier.

Other would-be educators appreciated the value of simple visual messages. By 1926, for example, copies of Denishoff-Uralsky's famous painting of a fire in the Siberian taiga, *The Untamed Element,* were being circulated with a fire prevention message. Copies could be purchased from advertisers in *American Forests,* where it was promoted as a "Sermon in Color." The prints were widely distributed, and a good many primitive painters (Grandma Moses among them) reproduced and sold versions of it.[91] Equally common were references to "Demon Fire," in which flames would be shaped into diabolical forms, and to the "Fire Wolf," thus appealing to the ancient association of wolves and fire that saturated European folklore. The fire wolf became a stock figure in an endless parade of poems and doggerel. By the late 1920s, there were so many fire poems that John Guthrie of the Forest Service collected them into a book.[92]

The Service also employed lecturers. Pinchot, of course, provided a fine model, and commercial advertising might well have borrowed a few pages from the Forest Service. The agency was frequently accused of being more intent on lobbying Congress and public opinion than in actually managing forests, and few items were more amenable to publicity than forest devastation by fire and the heroic efforts made to combat it. From 1924, H. N. Wheeler headed the Service's lecture circuit. His "Billy Sunday emotionalism," as one official described it, was embarrassing to many in the Forest Service, and when in a rage of pyrophobic fundamentalism he published a tract entitled "Forest Devastation in the Bible," many foresters felt obliged to protest openly. But it was the era of *Elmer Gantry* and Amy Semple McPherson, and Wheeler's methods seemed to bring results, especially in the intractable South. His long service record, moreover, was unimpeachable. His itinerary was so packed that he lived for years on the ragged edge of nervous and physical exhaustion. More than once a complete breakdown was averted only by recourse to all-night vigils with the Bible. After World War II and Wheeler's retirement, his peculiar pamphlet was succeeded by an even more popular version, "Forest and Flame in the Bible," which was distributed by the millions.[93]

A concerted national fire prevention program did not follow, however, until two experiments in 1928: the Cape Cod Forest Fire Prevention Experiment and the Southern Forestry Education Campaign. The Cape Cod program involved 110,000 acres of scrub oak and pitch pine, and it was intended to compare the costs of prevention and presuppression with those of suppression. It addressed almost all techniques short of actual suppression—lectures, films, pamphlets, posters, fuelbreaks, and patrol rangers who made many public contacts. The experiment lasted for three years under the sponsorship of the Massachusetts Forestry Association, the Massachusetts Division of Forestry, and the U.S. Forest Service, the latter having just modified its policy to support further presuppression expenditures as a prophylactic against oversized suppression bills. The study discovered that an increase in prevention funds by 20 percent led to an 80 percent reduction in total expenditures. The site thus became a model for the internal allocation of protection funds, much as the Shasta Experimental Fire Forest, also set up in 1928, did for various methods of administration and strategy.[94]

The outcome of the Southern Forestry Education Campaign was much less decisive. To begin with, its subject was not the internal distribution of agency funds but the promotion of fire protection as a concept. Nor was it concerned with the question of transient visitors; it addressed instead the stubborn question of southern woodsburning. The idea was initially proposed by Ovid Butler, executive secretary of the American Forestry Association in 1925. Since at least 1923 the association had experimented with various mediums for communicating a fire prevention message in addition to its periodical, *American Forests,* and in 1928 Butler's idea was approved. Outfitted with the techniques of the Showboats and the peripatetic Wheeler, the "Dixie Crusaders" hit the road. Their strategy was to concentrate on the states of Georgia, Florida, and Mississippi; to promote their message effectively without relying primarily on printed material; and to demonstrate to the unregenerate southerner that his future belonged with forests rather than with a traditional subsistence economy of agriculture, herding, and hunting. The program ran for three years on a budget of $150,000, some of which was contributed by the states involved.[95]

With W. C. McCormick directing field operations, the Crusaders would roll into town with a caravan of vehicles whose sides were emblazoned with the crusade's slogan: "Stop Woods Fires—Growing Children Need Growing Trees." They distributed souvenir memorabilia printed with fire prevention messages; put up posters throughout the countryside; sought out local cooperators; sponsored essay contests; supplied newspapers with a steady stream of copy; and toured some areas in a specially equipped Pullman made available by the Georgia and Florida Railroad. A pictorial leaflet titled "Woods Fires, Everyman's Enemy" was handed out to participants at local rallies; pledges were obtained from adults, as though at a prohibition convention; and, of course, there were films, often the first ever seen by some rural residents. Initially the Crusaders relied on existing Forest Service films, but these proved so humorless and didactic that they were utterly ineffective. The AFA found it necessary to produce replacements, such as *Pardners* and *Danny Boone,* which relied on plot, humor, and romance, or *Burnin' Bill,* which offered melodrama.

McCormick estimated that the campaign reached some 3 million people, traveled over 300,000 miles, and disseminated some 2 million pieces of literature. He assured the readers of *American Forests* that "the evidence is clear in word and action however, that the woodsburning habit inherent with southern ruralists has been staggered by the assault of the Dixie Crusaders. . . ." Other commentators confirmed that "a great tide of indignation was sweeping out over the Piney Woods, mobilizing sentiment against the woodsburner."[96] Pledges notwithstanding, there was no doubt a good deal of backsliding.

The issue was clouded further, for just as the Crusaders were adding up their new converts, the scientific evidence in favor of regulated controlled burning began to mount. Fred Morrell wrote agonizingly of Greene's works on the longleaf pine: "I fear they may nullify the Forest Service efforts to stop widespread woodsburning"[97] The Service's chief of the Division of Fire Control, Roy Headley—certainly no fundamentalist, but baffled by the persistence of southern incendiarism—wrote that "H. H. Chapman cannot evade a share of responsibility for the bad results from this agitation. Perhaps he would cooperate in stemming the tide as a means of protecting his ultimate standing as a forester." Headley, for one, wanted firmer distinctions for controlled and laissez-faire burning but realized that propaganda does not thrive on close distinctions.[98] The fear that the public could not (or would not) differentiate between promiscuous and prescribed fire, the feeling (strong since the days of Pinchot) that "good psychology would dictate making an issue of something simple to preach," and the orientation of most programs to illiterate ruralists and children—all encouraged forestry and the Forest Service to pursue a simple, tough message against fire.

The coming of the New Deal led to an explosion of money and manpower for fire protection schemes. With the CCC and the creation of an emergency presuppression account, the emphasis of the period was on labor-intensive fuel modification projects—fuelbreaks, like the grandiose Ponderosa Way; right-of-way cleanups along roads and tracks; wholesale slash removal, epitomized by the emergency salvage operations after the 1938 New England hurricane. But with the reforms in the mid-1930s that reconstituted Forest Service fire protection, there also came a new emphasis on fire prevention. During the 1936 Spokane Fire Conference, Headley and David Godwin both argued eloquently for a national campaign that would rely on acknowledged successes of applied psychology and would explore the "possibilities of using soundly conceived commercial advertising practice, the effectiveness of which is not questionable in principle." They suggested that "a high class 'opinion manager' of the Ivy Lee type (the man whose pen powers canonized John D. Rockefeller) . . . be found and hired." "PR in Washington ought to be strengthened," they concluded, "and its efforts unified with the field."[99]

It was time, they felt, for something more than superficial contacts and didactic education. Those who would inculcate fire prevention ethics were instructed to discover the "cause behind the cause." What this meant was explained in a "Fire Planning Letter" that was circulated to the field offices and reprinted in *Fire Control Notes*. *"Active cooperation with psychology departments of Universities,"* it urged, "should be planned, looking to better understanding of human behavior and focusing particularly on possibilities for changing habits of carelessness in the use of fire. Forest plans should specify the dominant aspects of prevention which need study from the psychological angle."[100] Within a year the Chief Forester himself addressed a letter to the American Association for the Advancement of Science soliciting help from its social scientists. The "deepening concern with the social implications and responsibilities of forestry" prompted his request that the

AAAS join with the Forest Service in establishing several councils to discuss topics of mutual interest. An organization committee was appointed in time for a 1938 meeting. The Forest Service liaison was John Shea, and among the anthropologists selected by the AAAS were Bronislaw Malinowski, Frederick Osborn, and Clark Wissler. Under the guidance of the AAAS–USFS Advisory Council on Human Relations, an experimental education program was developed at Pratt Institute on the subject of the "prevention of the destruction of our forests by fire."[101]

Meanwhile, under Shea's direction the Forest Service initiated field investigations in 1938 on the Cumberland (Kentucky), Clark (Missouri), and Deerlodge (Montana) National Forests, and in 1939 the "psycho-social" method was extended to Alabama. Shea also made contact with the psychology departments of a dozen universities. The University of Kentucky, in turn, supplied James Curtis for the study on the Cumberland; the University of Missouri sent Harold Kaufman to the Clark; Shea himself went to Alabama; and, perhaps as a kind of control, the Service sent a seasonal employee with a master's degree in educational psychology, Homer Anderson, to the Deerlodge National Forest. The methodology was no more sophisticated than the participant-observer approach then in vogue in anthropology. The researchers circulated among the local populace under various pretexts and interviewed unsuspecting natives in the course of general conversations. Only Shea made any pretense of much more. But at a time when Headley despaired that fire protection had lost its economic bearings, the psychological approach to prevention offered to restore some sanity to Service policy and practice, especially in its backcountry and cutover lands.[102]

"Your Psychologist," as Shea styled himself, popularized his own findings in his famous potboiler for *American Forests,* "Our Pappies Burned the Woods." In 1942 and 1943 another researcher, George Weltner, extended the methodology to the national forests of Mississippi. Shea himself planned a five-year study on the Clark National Forest to test the relative values of education and law enforcement in challenging the woodsburning habit, but wartime pressures shut down the program in 1943. In the late 1950s, the southern strategy was revived as a cooperative venture between the Forest Service and the Social Science Research Center at Mississippi State University, and it continues to this day. By 1966, however, the program had at last gone beyond the question of woodsburning to include fire starts from causes other than incendiarism and to integrate engineering, fuels management, and other physical means of prevention into a total fire management program.[103]

But there was more to the psychological approach of the late 1930s than Shea's celebrated experiment. Determined to rally popular sentiment nationwide, the Forest Service listened to its field men as well as to its new

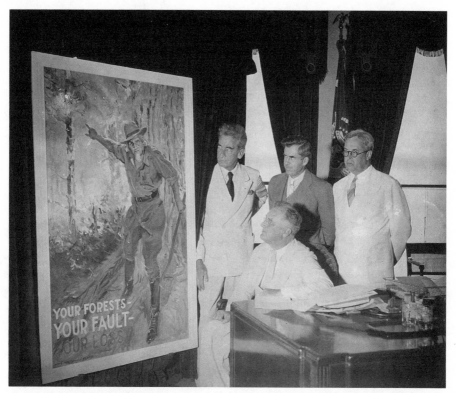

8. A James Montgomery Flagg poster for the national fire prevention campaign begun in the late 1930s. Admiring the poster are Flagg (standing left); Earle Clapp (standing right); and Franklin D. Roosevelt.

psychologist advisors: simple, emotional pictures were far more effective than complicated written messages. In 1937 it contacted James Montgomery Flagg, one of the outstanding commercial artists of the day and an old hand at patriotic posters. It was Flagg who had drawn the most successful recruitment posters of World War I and who had invented the face of Uncle Sam (modeled on his own physiognomy). Flagg executed a series of posters that won the approval of both FDR and the Forest Service. In 1939, under commission to the American Forestry Association, Flagg produced another new poster, and on the eve of World War II he sketched posters equating forest protection with national defense.[104]

The AFA moved to attack the national fire problem as it had the southern question a decade earlier. It dedicated an entire issue of *American Forests* (April 1939) to fire, and the text opened with a letter from FDR himself. Every agency contributed articles. The newly commissioned Flagg poster was used for a cover, and the AFA promised to distribute a million colored prints throughout the nation. Ovid Butler editorialized, apparently without irony, that this issue was to be a "torch of truth and knowledge." He explained that "forest fires once started are fought like wars—on the firing line. And as with wars, forest fires are prevented by a vibrant public will to prevent them." The totalitarianism that threatened America from without was matched by an insidious enemy within. "There are still millions of people," Butler wrote in a confused metaphor, "who, informed and made fact-conscious that the cry of 'fire-wolf' means red wolves in never ending packs—as run the fire wolves today—are ready and eager to become Paul Reveres on the public opinion highways of our country." To make the analogy even more explicit the magazine announced the creation of the American Forest Fire Medal for Heroism.[105]

It was the real war, however, that finally nationalized fire prevention programs. Forest products acquired value as war material. Incendiarism was tantamount to sabotage. Suspicious woods arson was even investigated by the FBI. Along coastal areas evening fires could illuminate shipping for enemy submarines and furnish expedient smoke screens for hostile landings. Fire could shut down gunnery and aircraft training; it diverted manpower and equipment from essential war industries. It is no coincidence that the war created three national fire prevention programs, two of which continue to the present day: the Keep Green program, grounded in private industry; the Cooperate Forest Fire Prevention Program (CFFP), a federal-state undertaking assisted by the Advertising Council; and a short-lived campaign in 1945 and 1946 by the Red Cross, which was absorbed into the CFFP.[106]

The Keep Green movement was a counterpart of the embryonic Tree Farm program. The greater portion of American forests are in small woodlots, not government forests. The Tree Farm program—enthusiastically endorsed by William Greeley—sought to reach this group of owners, and it was in this same spirit that Keep America Green began. The idea for a publicity program was first broached in 1939 as a Pacific Northwest contribution to the AFA campaign. The American Legion responded by inaugurating a Junior Forest Warden program, directed largely at teaching children forest fire prevention. In 1940 the governor of Washington convened a meeting of interested parties to explore how tree farms could be made to work, especially given the fire hazard. The American Legion program, it was suggested, could be broadened to include adults. It was during the course of this conference, according to Greeley, that the slogan and concept of Keep Green were invented. "Someone referred to Washington's pride in being the 'Evergreen State' and said it was the job of all of us to keep it that way." During ensuing discussions, "Keep Washington Green" was organized. The phrase "Keep Green" itself represents a case of independent discovery, or rediscovery: it had been used in the Northwest since at least 1912, when the WFCA had distributed pamphlets of that title; in Pennsylvania, it emblazoned state forestry letterheads ("Keep Penn's Woods Green"); and in Michigan, the phrase had been used in 1928. Stewart Holbrook, author and ex-lumberjack, agreed to write copy for the program, and out of his work came the first—and for decades the only—history of wildland fires in America. Blood-and-thunder journalism though it might be, *Burning an Empire* (1943) popularized the fire problem to a public much broader than that touched by forestry journals. On Washington's example, Oregon and Minnesota followed with Keep Green programs in 1941 and 1944, respectively. Ultimately, a national Keep America Green program emerged in 1944 under the auspices of the American Forest Institute.[107]

Time and again fire prevention programs aimed at children had enjoyed success, and in 1942 the freshly minted Keep Oregon Green committee created the "Oregon Green Guards." The Guard movement was aimed at youths, ages 8–18, and was designed to supplement existing organizations like the Boy Scouts and 4-H Clubs. No adult leadership was extended, and none proved necessary. As an inducement to join, a Green Guard kit was assembled, stocked with a manual, membership card, arm emblem, posters, and so forth. Interest spread immediately. The response to a single news release statewide threatened to overwhelm the secretarial and financial organization of Keep Oregon Green, and frantic pleas went out not to release more publicity. Within a few years, simply by word of mouth, 30,000 Green Guards were registered. Guards reported dozens of fires, and a "Service Under Fire" certificate was awarded to those youths who actually assisted in fire suppression. (During the war years, it was not unusual for firelines to be manned by crews of high school students.) The sponsors of the program proudly pointed to the story of 11-year-old Patricia Barnes as a symbol of the movement. A fire on her grandparents' farm inspired the child on her own initiative to hitch, start, and drive a tractor (something she had never done before) across a broken field to deliver water and gunny sacks. Perhaps the greater lesson here was the potential for a child-centered ad campaign of fire prevention, such as the CFFP.[108]

The Cooperative Forest Fire Prevention Campaign began in 1942, the same year as the Green Guards and not too long after a Japanese submarine lobbed a few shells at an oil depot near Goleta, California. Fire agencies had recognized the vulnerability of the West Coast to an incendiary attack, and forest defense councils had already been established. A journalist on the staff of the Angeles National Forest, Arnold Larson, conceived the notion of soliciting ideas, assistance, and posters for a prevention campaign from prominent advertising agencies. The Young and Rubicam agency in New York, in turn, suggested that the idea be submitted to the newly constituted Wartime Advertising Council, and the agency took the liberty of forwarding Larson's letter to the council's managing director. The council alerted

Don Belding of the Lord and Thomas agency in Los Angeles, who met with William Mendenhall, supervisor of the Angeles, and together they agreed on a general program. In May 1942 Secretary of Agriculture Claude R. Wickard approved the plan and announced it to the nation during a radio broadcast in July. "In wartime," he lectured, "the control and prevention of forest fires is a first line defense job on the home front. . . . We cannot forget . . . that the British Royal Air Force found it worthwhile to start great fires in the forests of Germany. Every fire in our fields or forests this year is an enemy fire." That was the theme of the early years: "Careless Matches Aid the Axis" and "Your Match, Their Secret Weapon" read the captions beneath the leering, flame-illuminated faces of Hirohito and Hitler.[109]

For the first time, under the pressures of the war, professional advertising talent was mobilized behind fire prevention. The Foote, Cone, and Belding agency of Los Angeles assumed charge of the program at the end of 1942, and, through the Ad Council, has continued to furnish free services ever since. The early wartime posters were naturally intense and uncompromising. After the success of Bambi, and perhaps recalling the familiar symbol of the predatory fire wolf, the Wartime Council secured permission to use Bambi and friends for its 1944 campaign. The animals proved popular, and the message became one of general fire prevention, not simply one of wartime preparedness. But there were problems with the licensing of Bambi, and it was decided to create an independent symbol unique to fire prevention. Squirrels, monkeys, and other creatures were tried. Opinion finally settled on a bear. Its shape was not easily arrived at. "The bear," Mal Hardy recalled:

> was to be a characterization with a short nose (Panda style), brown or black fur, with an appealing expression, a knowledgeable but quizzical look, "perhaps wearing a campaign hat that typifies the outdoors and the woods." The artist was warned to avoid simulating bears drawn by Cliff Berryman of the Washington Star (Teddy Bears); used in Boy Scout publications; used by Piper Cub airplanes; the bear that symbolized Russia; or the bears on a Forest Service bookmark then in use.[110]

But a bear it was to be. Albert Staehle did the original artwork. A uniform was added, and a name followed—"Smokey," reportedly inspired by "Smokey Joe" Martin, a New York City fireman. The first Smokey poster appeared in 1945. When the war ended, the Ad Council continued—and so did the popular fire prevention campaign, sustained by the growing alarm over mass fire, much as the question of mass fire was to prolong the tough 10 A.M. Policy that guided Forest Service suppression. In 1947 a new slogan was suggested: "Remember, Only You Can Prevent Forest Fires," and a Washington disc jockey, Jackson Weaver, created Smokey's radio voice by speaking into an empty wastebasket. The modern form of the bear appeared in 1948, when Rudy Wendelin, a career artist with the Forest Service, took over the drawings.

The real stroke of genius came in 1950. It had been suggested by the Ad Council that a live Smokey would be an asset to the program. That summer, following a large fire on the Lincoln National Forest in New Mexico, an orphaned bear cub was discovered on the burn. A state game officer, Ray Bell, bandaged and nursed the bear at his home until Forest Service officials realized that here, at last, was "Little Smokey." Amid great fanfare he was whisked by plane to a home in the Washington National Zoo and ultimately to his own zip code, a Saturday morning cartoon show, and a recognition level so great that a national advertising research report in 1968 stated that Smokey was "the most popular symbol" in the United States. Smokey Bear was better known than the President. As early as 1952 it became necessary to protect Smokey's name: Congress passed the Smokey Bear Act regulating the commercial exploitation of Smokey. Twenty years later revenue from licensing under the act reached its first million dollars. So readily was Smokey identified with the cause of fire protection that the Golden Smokey Award was created in 1957 for outstanding work in fire prevention, and in 1970 the first National Smokey Bear Workshop was held. A mate, Goldie—like himself an orphan from New Mexico—was furnished Little Smokey in 1961. No offspring resulted, so when another orphan cub from the Lincoln National Forest was discovered, he was adopted in 1971. Suffering from old age, the original Smokey was officially retired in 1975 and died two years later.

The appeal of Smokey Bear became legendary, especially among children, and with time the program became more and more directed at preadolescent youths. That generations of American children had been reared with Teddy Bears made Smokey easy to identify with and made the selection of a bear as the spokesman for fire prevention an especially happy choice. Thanks to Smokey, the CFFP became one of the great success stories in America. Urban fire protection, for example, has nothing like it. And although the program is jointly sponsored by the Service, the Interior Department, and the National Association of State Foresters, with generous assistance from the Ad Council, Smokey Bear is peculiarily (and rightly) identified by the public with the Forest Service. In advertising fire prevention, Smokey advertised the Service. When *Newsweek,* for example, ran a cover story on fire protection in 1952, it equated the "fabulous bear" with the "famous service." Once again, however, its fire mission put the Forest Service in an awkward public relations position: it was glad to be in the public eye but sorry it was so consistently identified only as a firefighting agency.

The success of the wartime ad campaigns suggested other postwar programs. In 1947 President Truman convened a national conference on fire prevention. The alarming fire regime being fashioned in Southern California, meanwhile, led the Forest Service to designate a special fire prevention research unit for its Pacific Southwest Station in 1955, thereby complementing its traditional program in the South. In 1956 the Southern Forest Fire Prevention Conference brought the southern fire scene again to the vanguard of interest for professional forestry. For the first time since the AFA had sent the Dixie Crusaders south, the whole gamut of fire and southern forestry was reviewed. The success of Smokey Bear, in particular, suggested that professional advertising might be mobilized in a like manner against the chronic incendiarism in the South. Smokey was directed largely at children; the southern program would aim at adult arsonists and careless debris burners. The idea simmered until 1958, when the Southern Cooperative Forest Fire Prevention (SCFFP) program resulted. Advertising services were donated by Liller, Neal, Battle, and Lindsey of Atlanta. The Forest Service and southern state foresters (assisted with Clarke-McNary money) contributed to funding and staffing. The program and the character of Mr. Burnit were intended to complement Smokey Bear. Unlike the Smokey program, however, the SCFFP had to discriminate between wildfire and prescribed burning. Until 1972 the SCFFP program was directed out of the CFFP office.[111]

The postwar period saw experiments as well with forms of prevention centered on the modification of the fire environment. Fuel management projects came largely under the purview of presuppression, by now an autonomous category of fire protection. In addition, wartime developments and spectacular advances in meteorology suggested the possibility of weather modification. In 1947, the same year as the presidential conference on fire prevention, the WFCA passed a resolution urging that weather modificaton— specifically, rainmaking and lightning suppression—be investigated as an aid to forest management. A cooperative agreement resulted between Vincent Schaefer of General Electric Laboratories and the Forest Service. In 1949 Schaefer concluded that super-cooled clouds were common in the Northern Rockies during the summer and that cloud seeding as a means of modification was feasible. The Forest Service consequently promoted a broad cooperative research program dubbed Project Skyfire. The primary participants were the Forest Service and the Multinalp Foundation, a private research organization specializing in meteorology; others in the program included the Weather Bureau, the National Park Service, and the General Electric Research Program. The program flourished throughout the 1950s, and in the 1960s it received some National Science Foundation funds. Trials were conducted in the Northern Rockies and in Alaska with mixed results.[112]

But weather modification was alarming to many and was considered to be unnecessary by others. By the early 1970s fire protection was undergoing a broad transformation in goals and policies. A concern with wilderness fire suggested, for example, that the target of the Skyfire program—suppression of lightning fires in remote sites—might be undesirable in the larger perspective of wilderness management. Prescribed fire was being widely disseminated in the name of silviculture and fuel reduction. This led to some embarrassment about the legacy of the "enemy fire" propaganda that had blanketed the country with enormous success in the postwar years and suggested that the simple message being communicated by such programs as Smokey Bear might need revision. Wilderness fire, not mass fire, was the informing fire problem. Fire protection seemed to have lost its identification with national defense, and it was deemed more essential to expand the realm of prescribed fire than to circumscribe further the range of wildfire. The emergence of rival fire agencies with different missions made the simple prevention campaigns less widely useful and seemed to recommend that here, as elsewhere, Forest Service hegemony would have to be broken. And finally there was the question of skyrocketing suppression and presuppression costs. It was hoped that a vigorous prevention program might reduce some of these extravagant expenditures. Prevention, in short, was ready for a new purpose. In 1973 the Chief Forester asked the CFFP program to assign a task force to reevaluate wildfire prevention programs.

The groundwork had been developed over several years. A cooperative agreement with the University of Southern California had produced a new study design for assessing prevention programs, and shortly before the Chief Forester's formal request two studies on fire-flood prevention programs in Southern California strongly recommended that prevention be strengthened vis-à-vis presuppression and suppression. It was the first quantitative study since the Cape Cod experiment.[113] The fires of 1970 in the Northwest and Southern California provided the final stimulant, as they did for so many aspects of fire prevention in the decade to follow. Among the early responses was a prevention research task force under Forest Service direction.

The prevention reanalysis was quickly absorbed into the general drive for interagency cooperation on all fronts and into the redefinition of fire as either wild or prescribed. A National Wildfire Prevention Analysis Task Force vigorously supported the creation of the National Wildfire Coordinating Group (NWCG), and the task force published its conclusions in 1976, the same year in which the NWCG was chartered. Among the standing committees of the NWCG is one on fire prevention, and an advanced fire prevention course is among those taught at the National Advanced Resource Technology Center. Thanks to Forest Service liaisons, fire prevention has taken on international dimensions. Smokey Bear was early exported to Canada; in Mexico, outfitted with a sombrero, he became Simon El Oro; in Turkey, a country conscious of the Russian bear to the north, Smokey was metamorphosed into a deer. Missionaries to the Belgian Congo even introduced Smokey to the jungle, where children "were intensely interested in the bear that wore a hat and wondered if all the animals in America wore hats."[114]

The fire prevention message was in danger of being a bit scrambled in the United States as well. One way to look at the prescribed-burning controversy is as political history, the transfer of the torch from one group (frontier agriculturalists) to another (foresters). The techniques and even the purposes of the two groups were often identical, but their political contexts were not. Industrial forestry, advancing against great resistance, had achieved real success against the popular belief that fire was useful and that fire prevention was unnecessary or misguided. Then foresters found that they had to reeducate the public to something like their former beliefs. But this time it was a particular policy that was at stake, not the political existence of professional forestry. Having convinced the public that it should keep America green, the Forest Service now found it necessary to inform the same public that black was beautiful. It had to explain to a skeptical populace the nature of its newly found detente with fire and why this was essential to survival. By the early 1970s more acres were being burned for prescribed fire than were burned from wildfire. The DESCON program tolerated all ignitions regardless of origin. Even Smokey Bear slipped in his popularity ratings, finding rivals for children's attention among such television celebrities as the Jolly Green Giant and Big Bird.

Fire prevention entered into a program of reeducation as great as that which brought systematic fire protection in the first place. Like the new fire policies, with their pluralism of responses, fire prevention programs had to address local or regional problems rather than attempt a single national message. They had to cope with the relativity of wildfire, allowing for "professional" exploitation without giving license to a potentially careless public. They had to account for the inevitability of natural fire in natural systems and the desirability of prescribed fire in managed systems without increasing the likelihood of wildfire in either. What prevention had to address, in short, was the nature of the ultimate source of anthropogenic ignition, the inevitable friction between human beings and their natural environment.

4 THE GREAT BARBECUE

There are three things that are never satisfied, yea, four ... The
grave; and the barren womb; the earth that is not filled with
water; and the fire that saith not, It is enough.
—Proverbs 30:15–16

If the State can stop the wanton destruction of forests by fire, it
will do more for the cause of forestry than it ever has done.
—Report of California Board of Forestry, 1890[1]

"**Congress** has rich gifts to bestow—in lands, tariffs, subsidies, favors of all
sorts," V. L. Parrington caustically wrote of the post–Civil War era, "and
when influential citizens made their wishes known to the reigning statesmen,
the sympathetic politicians were quick to turn the government into the fairy
godmother the voters wanted it to be. A huge barbecue was spread to which
all were presumably invited." To a frontier people, "what was more democratic than a barbecue, and to a paternalistic age what was more fitting than
that the state should provide the beeves for the roasting. Let all come and
help themselves."[2]

The conclusion to the Civil War left the South with the prospect of being
reconstructed and the West with the likelihood of being more or less plun-
dered. Both smoldered with resentment at what they considered neo-colonial
status to eastern capital and politics. The Gilded Age witnessed the full
onslaught of the industrial revolution. Its tycoons have been likened to robber barons and the splendiferous merchant princes of the Italian Renaissance. No other period of American history saw quite so wholesale and
tumultuous an exploitation of natural resources. When the era began,
America was overwhelmingly agricultural; when it ended, the Census
Bureau declared that no frontier line of settlement was discernible and that
the country was on the verge of finding the bulk of its population in cities.
The transformation had been always extravagant, often violent, usually
accompanied by great social and economic dislocation, and alarmingly
wasteful of natural resources like wood, soil, water, and wildlife. The United
States was no longer a developing nation, but a young industrial giant.

To thoughtful observers, the extravagance of the Great Barbecue brought
to the fore the question of how the national wealth in natural resources
should be exploited and the problem of who should control this process. The
second was a political question; the first, seemingly a problem in engineer-

ing. Out of this debate came a philosophy of conservation.[3] If industrialization was wasteful of social and natural resources, it also brought with it the necessity and means for a more efficient utilization of both. However paradoxical it might seem on the surface, the transformation from frontier to industrial state included a new sensibility toward landscape and a scientific interpretation of "nature's economy" that demonstrated how both social and natural landscapes could be efficiently engineered. Thus the era exhibited some of the worst speculative excesses in the national history, a nadir in its political consciousness, and an orgy of resource abuse—all epitomized by the enormous land grants lavished on the railroads—and, at the same time, some of the nation's finest acts in the stewardship of its landed heritage—the creation of government scientific bureaus, the reservation of forest lands from the public domain, and the establishment of the national park system.

It was in this enthusiasm for the scientific management of natural resources that professional forestry made its appearance. It came as a species of technology transfer from Europe, from industrialized states to a developing nation. The U.S. Geological Survey had already broken trail for conservation reforms; forestry followed its example in nearly all points. Forestry, like geology, promised to reduce waste through the application of scientific engineering. The U.S. Forest Service strongly relied on the Survey's institutional examples and adopted some of its most promising intellectual arguments, notably the need to manage water resources. Later, foresters developed a theme more specifically their own: the concept of a "timber famine." But what really put forestry in a special category was the transfer of the vast forest reserves to the Bureau of Forestry in 1905. At one stroke the Transfer Act made an obscure government agency into a landholder with an estate larger than that of many nations. Likewise, the new responsibility compelled a somewhat arcane and foreign subject, forestry, to expand its interests and vision, placing it center stage for a while in the great national drama of the conservation movement. And it subjected both forestry and the Forest Service to a literal trial by fire.

The ambivalence of the era is thus well synopsized by its fire history. The Great Barbecue had its charcoal supplied by half a century of devastating conflagrations, mostly in the pineries of the Lake States. Industrial logging was a source of almost unprecedented holocausts that seemed to epitomize the most shameful of laissez-faire excesses. But these same fires helped at last to raise public indignation to its kindling temperature. The process of working out an accommodation to the industrial revolution is well encapsulated in the struggle to replace a fire regime based on agriculture and frontier economies with one appropriate for an industrialized society. Fire was a visible medium for expressing these tensions.

It was recognized at the time that, although frontier settlement for ranching or farming temporarily increased fire hazards, the further progress of settlement would ultimately banish large fires. The trick to surviving fire on the frontier was to make it through the transitional period, especially where industrial logging was practiced. The Lake States were perhaps the first to be visited with the full onslaught of the industial juggernaut, a scene illiminated by the flames of the worst conflagrations in North American history. But within a few decades the very process of logging created economic and ideological conditions that would no longer tolerate waste on such a scale. When logging became permanently fastened to the land, it could no longer ignore routine holocausts. Even in the Lake States, it was the mixture of agriculture and logging, of frontier and industry, that was most deadly. Where the two were finally disentangled, when the conversion to industry was complete, the big fires disappeared.

The doctrine of conservation was a complicated concept, but everyone could understand that it attacked what Gifford Pinchot labeled the "Dragon Devastation."[4] Fire was as much an organic part of this dragon as it was of mythical ones. Fire control was a leading edge of industrial forestry in regions like the South where forestry came into conflict with other land uses. It was a serious test of the government's ability to regulate, but it was, for the most part, politically neutral. Agreement on the need for fire control often made it the one political topic on which virtually all participants on the side of industrialization could agree. The real conflict was between frontier and agrarian fire practices on one hand and industrial fire practices on the other. Conservation was most opposed by groups and in regions where the industrial exploitation of resources had not worked its transformations. It was frontiersmen, farmers, and pastoralists who protested most violently against fire control, not the socioeconomic groups and interests, such as logging, created by industrialization. A controlled fire by frontier standards was often a wildfire by the canons of professional forestry. Consequently, the conservation program that took shape during the early decades of the twentieth century found fire control among its most popular, least politically controversial tenets. It was often the consensus fashioned by fire protection, moreover, that led to subsequent reforms in forest protection and regulation.

During this time, too, wildland fire control, as distinct from rural fire protection, received its first practical tests. These came on wildlands reserved from settlement by states like New York and by the federal government on its national parks and forest reserves. Fire control became in turn an *experimentum crucis* on the feasibility of forest conservation and on the concept of management through government agencies staffed with technical experts. If the government and forestry failed to protect a reserved area from fire, they were self-evidently incapable of managing it in more sophisticated ways. It seemed to many of the principal participants that if professional forestry failed, so would conservation. Wildfire was an especially visible and public measure of success or failure. The most direct way to end the Great Barbecue was to extinguish its fire.

THE DRAGON DEVASTATION:
FIRE AND THE COUNTERRECLAMATION

And there was too much fire. Which last nobody could deny.
—Gifford Pinchot, *Breaking New Ground*[5]

For the Forest Service to be without several copies of this book
[*Breaking New Ground*] is equivalent to a church without Bibles.
—History Office, U.S. Forest Service, 1971[6]

I

In *The Fight for Conservation* Gifford Pinchot remarked that "as a forester I am proud to think that conservation began with forestry."[7] He was quite wrong. The conservation philosophy came largely from geologists; the issue that organized conservation concern was not forests, but water; and it was the U.S. Geological Survey, not the Forest Service, that pioneered in the politics of resource management. Even as the West was being exploited, its resources were also being inventoried by geographical and geological surveys, which were eventually consolidated in the U.S. Geological Survey. As a philosophy of land management and as an institutional presence, conservation developed from the experiences and personalities of the Survey, "the mother of bureaus."[8]

This is not to say that forests were an inconsiderable element in conservation reform; rather, they were one aspect among many, and not always a primary one. Agitation for reform appeared in the 1870s with memorials from the American Association for the Advancement of Science warning of resource destruction and forest wastage; with books and lectures by thoughtful observers and alarmists alike; with nearly annual warnings from the Secretary of the Interior about imminent timber famines if present trends of disposal continued; with the establishment of Yellowstone National Park in 1872 and the American Forestry Assocation in 1875; with geological surveys to the West, culminating as a political force in John Wesley Powell's celebrated *Arid Lands* report in 1878; and, in that bumper crop year of 1879, with the Public Land Commission review and the creation of the U.S. Geological Survey. Congress enacted several land laws in the 1870s, which, despite the outrageous frauds perpetrated under them, recognized the special value of forested land and granted timber a quasi-legal status not really present before. In 1876 Franklin Hough was hired as a "forestry agent" by the Department of Agriculture, and in 1881 a Division of Forestry was created under his direction. Four years later California and Colorado established state forestry commissions, and New York created the Adirondacks Forest Reserve. But the primary function of the Adirondacks experiment was to stabilize the watershed of the Hudson; forestry practices were in fact forbidden. Perhaps the event of greatest long-range significance for conservation in 1885 was the brilliant performance by John Wesley Powell in defending government scientific bureaus before the Allison Commission of Congress.[9]

Geology was an intellectual precedent for forestry, just as the Geological Survey was an institutional exemplar for the Forest Service. Until the early conservation movement nearly expired, forestry remained an ancillary field. It was geologists who lobbied for the creation of Yellowstone National Park. Geologists and hydraulic engineers staffed the Mississippi River Commission and the Sacramento Debris Commission. Powell's *Arid Lands* report was the first conservation document to prescribe institutional and legislative remedies for resource management. It contributed directly to the creation of the Public Lands Commission, of the Geological Survey, and, after Powell took over as director, of the famous Irrigation Survey. It was the *Arid Lands* report, and the decades of western surveys it subsumed, that became the model for the scientific inventory of natural resources, not merely of minerals or water but of land as well. Moreover, geology pioneered in technical education (forestry schools came nearly four decades later), and geologists and mining engineers furnished the model for a professional caste of experts on natural resources, an example that foresters tended to emulate.[10]

The practice of reserving forests grew out of concern for water resources, especially for irrigation in the West and navigable rivers in the East. The Geological Survey supplied data for both. The Forest Reserve Act of 1891, which allowed for public lands to be withdrawn from settlement by presidential order, had as its first purpose the protection of watersheds and was only secondarily concerned with the maintenance of a permanent supply of timber. A similar concern with watersheds motivated the establishment of the Adirondacks Reserve in New York and the purchase of lands under the Weeks Act of 1911. No foresters were needed: timber was primarily important as vegetative cover, not as an industrial material. When the American Forestry Assocation retitled its publication *Irrigation and Forestry,* it made the ties explicit. A Forestry Committee established in 1896 by the National Academy of Sciences to inspect the forest reserves and recommend principles by which to administer them was dominated by men trained in geology. It was a geologist and protégé of Powell, WJ McGee, whom Gifford Pinchot credited with being "the scientific brains of the conservation movement."[11] Charles Walcott, successor to Powell as director of the Survey, put Pinchot on the NAS committee and, according even to Pinchot, rescued the 1897 Forest Reserve Act from legislative extinction. When the forest reserves were surveyed between 1898 and 1900, the Geological Survey conducted the inventory.

The conservation showcases of Theodore Roosevelt's presidency were the Inland Waterways Commission and the Governors' Conference, both in 1908. The first was organized by McGee with help from the Survey and the Bureau of Reclamation (newly independent from the Survey), and the second led to a national resource inventory conducted largely by geologists assembled by Charles Van Hise, professor of geology at the University of Wisconsin. Even the experiences of the Survey in regulating water monopolies, reducing flood damages, and overseeing the productive storage and distribution of irrigation water anticipated to an uncanny degree the later trials of the Forest Service in reducing the destruction and monopolization of timber. Pinchot's dogmatic confusion of forest conservation, a technical question, with nationalization of forest industries, a political issue, nearly wrecked organized conservation as had Powell's analogous treatment of water resources with the Irrigation Survey many years before. By the turn of the century, compared with the scope of Survey activities in the field of resource management, the Bureau of Forestry seemed parochial and forestry itself rather esoteric. Yet almost overnight, as a result of the Transfer Act, which gave the Bureau responsibility for the management of the forest reserves, forestry rushed to the vanguard of organized conservation.

II

Forests had never been far removed from the larger program of conservation, but only in the late nineteenth century did foresters successfully create a separate identity for professional forestry. They achieved this by bringing together three strands of the conservation movement: a popular enthusiasm for reforestation, which emphasized the value of "forest influences" for agriculture and which found spiritual solace in the woods; the industrialization of timber production, which demanded a more stable resource base and economy and which brought fears of a crippling timber famine; and the counterreclamation, which began the process of actually withdrawing land from agricultural use. All three appeared at nearly the same time. In 1872, for example, J. Sterling Morton inaugurated the first Arbor Day, industrial logging had penetrated the north woods of the Lake States to such an extent that widespread fires had resulted the year before, and Congress established Yellowstone National Park. But not until nearly the turn of the century, and after the establishment of large forest reserves from the public domain, did these movements become compelling. By then it was necessary for some institution to manage them and, from among many candidates, forestry assumed that charge.

Reclamation had not been oblivious to trees. In many regions the effect of settlement may well have been a new increase in woods as a consequence of suppressing Indian fire practices. Elsewhere reclamation had replaced wild woods with orchards for the production of fruits and nuts. Certain trees had been planted for sentimental or esthetic reasons—the trees often, like the settlers themselves, being émigrés from Europe. To these traditional reasons for reforestation were added others by the mid-nineteenth century, under the concept of "forest influences." It was held, for example, that an increase in forests would moderate climate and increase rainfall and that sound forests could best protect watersheds important for the irrigation that was essential for the reclamation of the arid West. Enthusiasm for arboriculture led to programs for the protection of existing forests, especially those guarding watersheds, for the replanting of acres lost to agriculture or logging, and for the afforestation of semiarid landscapes like western Nebraska.

Not all of the planting involved native trees, or came deliberately, or had benign consequences. Much of the popular arboriculture involved exotics, many of which contributed directly or indirectly to fire hazards. Import nurseries inadvertently brought Dutch elm disease, white pine blister rust, the gypsy moth, the tussock moth, and other plagues from Europe; from Asia came the chestnut blight. Where the die-offs were heavy, fire hazards increased. Australian pine augmented fire problems in south Florida. Tamarisk, an ornamental from the Near East, was introduced into the Southwest in the 1880s, spread along the water courses, and created a fire hazard without historical precedent in the Inner Gorge of the Grand Canyon. When George "Lord" Bennett helped to found the town of Bandon on the Oregon coast in 1873, he introduced more than the name of his Irish home: he brought gorse, a woody ornamental used as a hedge. The gorse spread throughout the town of 2,000 and its surburbs. When on September 26, 1936, slash fires from loggers and campfires from hunters burned in the woods surrounding Bandon and the relative humidity plummeted to 8 percent, fires entered the gorse, and within minutes a mass fire was in the making. Coast Guard cutters began evacuations by the only exit left open to the townspeople, the sea, but 11 perished nonetheless, and within four and one-half hours Bandon was a smoking ruin.[12]

Nor was the planting of exotics restricted to sentimental immigrants. In the twentieth-century South, government officials with the Soil Conservation Service were responsible for the wholesale introduction of kudzu, an ivylike vine first imported as an ornamental from Japan in 1876. During the New Deal kudzu was widely planted as an erosion control measure, usually along the banks of railroad and highway right-of-ways. Unfortunately, the first killing frost turns the otherwise lush kudzu into a flash fuel, a fuse to nearby fields and timber. Eradication necessitates burning, plowing, and "salting" the soil with herbicides.

In California private initiative was enough to oversee the wholesale importation of another pyrophyte. Eucalyptus first appeared in 1856 as an ornamental curiosity. Between 1890 and 1910, however, amid national alarms about a timber famine and awareness of regional needs for a good hardwood, spectators promoted eucalyptus plantations as an opportunity to sell timber and land both. The resulting bubble was perhaps rivaled only by the tulip mania that swept seventeenth-century Holland. In Southern California eucalypts were planted as a windbreak for citrus orchards; in central California, as a source of fuel, hardwood furniture, and timber. Railroads invested as a future source of ties, and the bark was even advertised as a prophylactic against malaria. The University of California and the state Board of Forestry issued handbooks.[13]

The bubble began to burst in 1910, when it was finally learned that eucalyptus is a generic name applied to over 700 species, that the trees transplanted to California were commercially worthless, and that they constituted a serious fire hazard. By 1920 the eucalyptus craze had evaporated into that nirvana of wornout California enthusiasms, but it left behind millions of closely stocked eucalypts. More than once that legacy in Alameda County has threatened to burn Berkeley to the ground. In 1923, following a freak snowstorm that brought the brittle branches to the ground, Berkeley suffered one of the worst urban conflagrations in the twentieth century, and nearly a quarter of the campus burned. A freak frost in 1972 killed over a million eucalypts from the root collar up, and only good fortune prevented another holocaust.[14]

The perceived value of forest influences had a spiritual component, too. Woods reserved from reclamation were endowed with a special power to uplift the sentiment and inspire the communicant. For this, the romantic revolution was responsible. Only with romanticism did forest gloom become a desirable commodity, did impenetrability, remoteness, and awful isolation become precious. The concept of a forest primeval became acceptable and its presumed loss, a source of melancholy. For many preservationists the greatest horror of fire was its destruction of a noble scene. Though writing in 1927, Robert Marshall, founder of the Wilderness Society, conveyed this perception well. "There were some scenes of desolation," he wrote of the Northern Rockies, "that pretty nearly drive an imaginative person crazy. ... A pessimist would conclude that one summer's fires destroyed more beauty than all the inhabitants of the earth could create in many years, while an optimist would go singing through that blackened, misshapened world rejoicing because the forest will look just as beautiful as before—in two or three centuries. Take your choice."[15] It was a sentiment toward fire unimaginable before the romantic revolution.

The greatest material changes in forests were induced by the industrial revolution. With the opening of the plains by the railroads, agricultural commodity production tended to leave the forests for the grasslands, and the depopulation of many eastern lands gave way to spontaneous regeneration of forests. Industrialization also brought a gradual shift away from wood fuel, which also helped to promote increased stocking of forested lands. In 1850, for example, over 64 percent of the total energy consumed in the United States came from fuelwood. Coal began to make heavy inroads only in the 1870s; by 1910, just 10 percent of American energy came from wood; and by 1970 fuelwood contribution was negligible. A low estimate of per capita consumption of fuelwood in 1850 is 0.8 cords per annum. With a population of 23 million, national consumption per year amounted to some 18.4 million cords, or almost 2.4 billion cubic feet of wood. Steamboat and railroad consumption must have been considerable. During the Alaska Gold Rush steamboats on the Yukon burned over 300,000 cords per year. Large reserves of wood went into the manufacture of charcoal. The conversion to fossil fuel by the end of the century removed a heavy burden for wood production, liberating land or forests for other uses. Industry created a great demand for wood as a raw material, particularly after pulp became important. Agricultural reclamation had ascribed certain economic values to forested land, such as for fuelwood and fencing, but industrialization transformed those values and added many others of greater significance. Forests were proclaimed to be essential for industrial survival.[16]

Despite Pinchot's insistence that the public forests would be managed as a public utility, and despite his idiosyncratic view that the timber industry ought to be nationalized, professional forestry found that its great ally was industrial logging and its great enemy the frontier. So long as logging was considered as a form of frontier landclearing, there was little need for foresters. Only when it became linked to an industrial economy dependent on raw forest materials—a condition that brought the need for continued or sustained yield production and for certain standards of efficiency—did logging cease to migrate. It became instead part of the counterreclamation. For industrial logging the great impediments were an unsuitable tax structure and fire. "To understand how and why industrial forestry developed in the United States," Henry Clepper wrote in 1971, "one must understand the role of fire which up to a quarter-century ago was the greatest obstacle to forestry."[17] Forest fire, William Greeley concluded in 1943, "has often played a leading part in the economic progress of the forested states beyond the Alleghenies." It had much to do "in shaping the character of forest-using industries, in accentuating their migratory habits, in creating the lumberman's tradition of 'cut out and get out.' It was a risky business whose raw material might be swept out any autumn day. . . . It is not surprising that timber came to be regarded as an asset for quick liquidation—rather than a resource for permanent industry."[18] For Henry Graves, writing in 1910, fire prevention was 90 percent of forestry; by Greeley's time it was only three-fourths of forestry. "For the great coniferous forests of the Lake

States, the South, and the West this is literally true," Greeley affirmed. "There could be no forestry in any regional sense until woods burning was brought under some measure of control." The habits and negligence "of pioneer days still carry over. There is the great enemy of American forestry."[19]

It was control over the forest reserve system that made forestry a national force in conservation. Under a slogan of "timber famine" it could claim responsibility for the regulation of timber industries. Under the doctrine of watershed protection, it absorbed to some extent the program of water conservation. Under the concept of forest influences, it came to assume de facto charge for the large amount of popular reforestation and arboriculture. But above all, and alone among the conservation bureaus, it had a land base: it controlled, as no other agency did, the raw material of the counterreclamation; it could integrate, as no other single agency could, the various concerns of resource conservation. Especially after the Weeks Act gave it the power to enlarge its domain beyond the public lands of the West, the Forest Service put itself visibly before the public as the spokesman for conservation at large. It shifted attention from frontier economy to industrial economy, especially industrial logging. For both economies, however, fire policy was fundamental: nothing else so immediately and powerfully influenced access to forest resources as the regulation of frontier fire practices, and on the value of fire control—and often on very little else—industrial logging, professional forestry, and the Forest Service could agree. For the frontier, then, fire control presented the terms of conflict; for industry, the conditions of compromise.

III

That forestry today should be the main repository of knowledge about fire, and that forestry agencies should furnish the main institutional framework for wildland and rural fire protection in the United States is something of a historical accident. It is directly attributable to the Transfer Act of 1905, which removed responsibility for the protection of the forest reserves from the General Land Office (GLO) in the Interior Department and gave it to the Bureau of Forestry (soon renamed the Forest Service) in the Department of Agriculture. For this circumstance Gifford Pinchot was almost wholly responsible and is rightly honored as a prophet of professional forestry. Forestry at the time was a minor field of study. Only two colleges gave forestry degrees, and the first class from Yale (a school endowed by the Pinchot family) had graduated but a year before. The reserves, moreover, existed for watershed protection and other purposes in addition to timber production—purposes about which foresters were not especially well informed. Forester-historians have pointed with incredulity to the paradox of a Bureau of Foresty with no land to manage and a GLO with no foresters.

Yet the Bureau of Forestry was not unlike a dozen small scientific agencies that appeared about the same time, none of which could do more than advise and research. There was no more reason to give the reserves over to forestry control than to give them to the Biological Survey, which could scientifically manage game, or to the Bureau of Reclamation, which could better administer the watersheds, or, considering the volume of grazing done on the reserves, the Bureau of Animal Husbandry, which could manage the ranges. At best, forestry was prepared to give technical assistance on only one aspect of the reserves' management. The National Academy of Sciences committee on forestry saw the problem of the reserves as one of law enforcement and fire protection and therefore recommended that the reserves be turned over to the Army at least temporarily. In retrospect it is surprising that the reserves were not given to the Geological Survey: no other scientific bureau at the time was so well equipped to assess resources and to offer policy decisions.

But the determination of Gifford Pinchot, then chief of the Bureau of Forestry, and the rise of Theodore Roosevelt (with whom Pinchot had a close personal and working relationship) to the presidency won the reserves for the Bureau. The Transfer Act is rightly considered to be the day of Genesis for the Forest Service, and Pinchot became a national figure in conservation. As Inman Eldredge recalled, "I think without exception the young men who went into the Service at that time believed that Gifford Pinchot was a prophet, a god, that he was unselfish and far seeing."[20] These were young men. The decision to staff the administrative posts with only professional foresters created an immediate market for the fledgling graduates of the new schools and inspired a dynamism among the newly initiated. "I often think what a wonderful thing it was," Elers Koch reminisced, "to have a government bureau with nothing but young men in it."[21] Youth—not only of staff but of forestry itself—made for great energy and great conviction. But the Bureau was less willing to have its credibility or that of professional forestry questioned. It could initiate controversy but was not responsive to criticism; it was too homogeneous by training and temperament, and too self-conscious about its newly won political and intellectual stature. Nowhere else were the strengths and flaws of this young organization more apparent than in its fire policy. The institutionalization of professional forestry was probably the greatest determining cause for the direction that fire protection took in the Forest Service and in the United States at large.

When Pinchot railed at the Dragon Devastation, the image was well conceived: it evoked the heroic figure, the righteous knight errant who deals summarily with a beast that breathes fire. Agitation for better fire protection had been a staple of conservation reform, and whatever else forest users might argue over, on fire protection they were usually united. Virtually every memorial urging forest protection began with a program of fire control. Even by 1881 Bernhard Fernow, the Chief of the Division of Forestry, felt that the fire question had been "discussed to satiety."[22] John Muir voiced the opinion of the NAS forestry committee when he wrote that "bad as is the destruction from logging, it is less well known that 10 times as much destruction occurs annually from running forest fires that only the federal government can stop."[23] Powell had witnessed two large fires in Colorado in 1867 and 1868 that in his opinion had destroyed more timber than logging had removed from the time of the earliest white settlement.[24] More than once his surveying parties in the West had shut down operations because of heavy smoke. His *Arid Lands* report included a lengthy section on Indian burning practices and the threats they posed to forests and watersheds. His map of Utah for the report included as burned acreages an extent of land that exceeded the forested land of the territory. His resource classification of Utah set a pattern of land classification and burned acreage maps that the Geological Survey followed during its inventory of the forest reserves. By 1890 he had published half a dozen broadsides in the major national magazines arguing for an integrated federal land policy to govern grazing, land distribution, forests, irrigable waters, minerals, and so forth. Fire control in the service of watershed protection was prominent among the eight points he advocated.[25] But, unlike professional foresters, he was not instinctively hostile to all forms of controlled burning: his long studies of the American Indian for the Bureau of Ethnology apparently convinced him that the ancient practices had some value.

In 1896 the Secretary of the Interior directed three questions to the NAS forestry committee. First, was it "desirable and practicable to preserve from fire and to maintain permanently as forested lands those portions of the public domain now bearing wood growth, for the supply of timber." The second asked about forest influences, and the third solicited recommendations for remedial legislation. Obviously, the latter questions were meaningless if the answer to the first was negative. "Fire and pasturage," the committee solemnly concluded, "chiefly threaten the reserved forest lands of the public domain." The first was commonly a product of the second, and "in comparison with these the damage which is inflicted on them by illegal timber cutting is insignificant." For the western states and territories, the committee noted, no statistics existed to show the fires' annual burned acreage, "but nearly every summer their smoke obscures for months the sight of the sun over hundreds of square miles, and last summer your committee, traveling for six weeks through Montana, Idaho, Washington, and through western Washington and Oregon, were almost constantly enveloped in the smoke of forest fires." The fires were intensive as well as extensive, and the committee soberly reported that "no human agency can stop a Western forest fire when it has once obtained a real headway."[26]

The strategy of fire control, however, had been shown "conclusively" to

be workable. Since 1886 the Army in Yellowstone and since 1885 the province of Ontario had adopted fire patrols with great success, and standing
behind both experiments was the protection of the Adirondacks Reserve in
New York. Quebec subsequently initiated "satisfactory" patrols, and the
Canadian government recommended a compulsory extension of the system
throughout its dominion. Among other exemplars, the committee examined
the "difficult sylvicultural conditions" and the "great pecuniary success" of
fire protection in British India. Although fires there were "at least as difficult to control as the United States," the cost of protection was less than
half a cent per acre per year. But it was the rousing success of the Army in
Yellowstone that most impressed the committee. It urged that the Army
extend its jurisdiction over the forest reserves as well as the national parks,
at least during fire season; that the principles of forestry and land management be taught at West Point by Academy instructors trained in Europe;
and that this system should continue until a civilian organization of sufficient depth could be created to supplant it. "Many of the duties" required
in managing the reserves, the committee noted, "are essentially military in
character, and should be regulated for the present on military principles."
Foremost among these was fire control.[27]

The outbreak of the Spanish-American War and the endowment of a forestry school at Cornell University the following year intervened to prevent
the Army from assuming control, and management of the reserves passed
to the General Land Office. The Geological Survey's reports on the forest
reserves, meanwhile, both in word and by map, showed the prevalence of
fire in the West. For the poor condition of the Black Hills, an investigator
concluded that "forest fires are directly responsible, and the present aspect
of the forest is the result of long abuse and the struggle of the forest to
reestablish itself. The broken condition of the forest, the large proportion of
defective trees, the many wind breaks, the prairies, parks, and bald ridges,
are due to the destructive forest fires which have swept the hills periodically
for years and probably for centuries."[28] It was reported that 70,000 acres
burned in the Bighorn Mountains of Wyoming that summer alone, "a number several times as great as has been taken from these mountains by every
other means since the white man struck the first blow of his ax. . . ." The
immediate cause was a line of abandoned campfires from the Shoshone
Indians, and the entire deforestation of these mountains "would be a short
process if fires continue at this rate."[29] In the Northern Rockies, "one meets
with burnt areas everywhere—in the old growth, and where the seedlings
that are beginning to cover the deforested areas have just commenced to
obtain a fair hold. The burnt tracts are in large blocks, thousands of acres
in extent, and in small patches of 15 to 150 acres which extend in all directions through the forest, which at a distance is apparently green; sometimes
they are in broad swaths, sometimes in narrow, tortuous windings just suf-

ficient to open a land for the destructive high winds to tear the living forest down."[30] Where the climate was semiarid, the burns came back to brush or grass; on the higher, more humid elevations where the reserves where largest, they reseeded to lodgepole pine and aspen, which were of little commercial value.

The Survey's agents discovered that where settlement or commercial interests had stabilized and looked to long-term use of the reserves, there was some interest in controlling fire. More typical, however, was the sentiment of a "less responsible cowpuncher": "Well, I guess Uncle Sam can stand the racket, if the whole shootin' match burns up." In areas where frontier economies were threatened with stiff control, the agents commonly heard sentiments such as "If the Government intends to guard and preserve the timber from fires and prevent unlimited cutting, we will try to burn up what is left as soon as possible," or "Since the reserve has been set aside every prospector carries an extra box of matches along to start forest fires with." "Tenderfoot tourists" were blamed for other wildfires. But even where ignorance or retaliation were not at fault, fire was simply too much a part of frontier economy for its traditional uses to be controlled by the creation of reserves.[31]

Fresh from his participation on the committee and recently installed as chief of the Bureau of Forestry, Gifford Pinchot promptly directed his staff to study fire damages. "We began an intensive study of forest fires and their history," he recalled, "with the sound idea of finding out how much they were costing the Nation." After reaching a tentative figure of $20 million, Pinchot shelved the work in favor of more pressing obligations, but the choice of initial topics was revealing.[32] In his 1899 *Primer on Forestry,* with which he hoped to educate the nation, Pinchot wrote that "all of the foes which attack the woodlands of North America, no other is so terrible as fire."[33] Meanwhile, assisted by Henry Graves, Pinchot investigated the Pine Barrens of New Jersey, which he found to be an appalling testimony to the stunting power of recurrent fire. Agents of the Bureau were dispatched to make firsthand reports of major conflagrations, including the 1902 fires in the Northwest and the 1903 fires in the Northeast. Forest management plans were compiled for private companies who requested them, and nearly all were founded on fire control. The Bureau also assisted the GLO with forest management plans, one result being the *Forest Reserve Manual* approved in 1902, a direct antecedent to the *Use Book,* which spelled out Forest Service policies after 1905. Both manuals admonished forest officers on their "three chief duties: To protect the reserves against fire, to assist the people in their use, and to see that they are properly used."[34] The order of charges is significant.

So, too, was the apparent contradiction between the charge to eliminate fire and that to assist local use of reserve resources, nearly all of which depended on fire. Writing of his experiences in the West, Pinchot expressed the common opinion of professional foresters. He recalled "very well indeed how, in the early days of forest fires, they were considered simply and solely as acts of God, against which any opposition was hopeless and any attempt to control them not merely hopeless but childish. It was assumed that they came in the natural order of things, as inevitably as the seasons or the rising and setting of the sun. Today we understand that forest fires are wholly within the control of man. . . ."[35] Pinchot wrote that in 1910, the year that brought the great fires to Idaho and Montana and the light-burning controversy to California. It was not simply, as Pinchot assumed, that people did not believe that fires could be controlled; rather, they did not want all fire practices eliminated. It was the holocaust they wanted suppressed, not traditional fire uses.

But foresters were not prepared to discriminate among such matters. The Transfer Act was more than a political statute: it symbolized a whole process of technology transfer by which European forestry was brought to the United States. The attitude toward fire that professional foresters embraced was one that Pinchot well conveyed, that the forestry school his family endowed at Yale helped to promote, and that the Forest Service offered ample opportunities for converting into practice. Unlike geology, which by 1900 had established an internationally recognized "American school," forestry necessarily looked to Europe for precedents. What William James observed for philosophy in 1902 held many times over for forestry: "It seems the natural thing for us to listen whilst the Europeans talk. The contrary habit, of talking whilst the Europeans listen, we have not yet acquired."[36] It was to northern Europe that foresters looked for models and knowledge. "General principles of forest management," as John Curry and Wallace Fons observed in 1940, "were established over 200 years ago in Central Europe and the task of the American foresters has been largely to adapt these practices to the management of the vast, rough, and inaccessible natural forests of this country." Yet there was one fundamental difference: "Where forestry originated, forest fires were relatively rare and unimportant, while in the New World, fire protection is the most serious of all problems facing American foresters."[37] And apart from technical guidelines, "for which there is little precedent," the annual spectacle of conflagrations was a profound embarrassment to American foresters, both as patriots and as nascent professionals. For a modern industrialized country to have its landscape blackened by a scourge of fire was as demeaning as to have it depopulated by plagues.

By 1900 Europe's fire menace had long vanished into history. Forestry had become the study of silviculture and economics. The real tyranny of technology transfer in this instance was that a self-proclaimed science demanded the repudiation of frontier folkways. Centuries of practical experience painfully and empirically acquired by American settlers in a range of fire regimes was abruptly sacrificed, only to be rediscovered later. American conditions, of course, were recognized as unique, and in some matters Americans tended to overstate their independence. Writing in 1950, H. H. Chapman even considered as a "handicap ... the feeling that European experience in silviculture is inapplicable to American species and economic conditions. No such prejudice exists against the laws of physics, chemistry, soil science, or agriculture."[38] Perhaps this was the case for silviculture and economics, but on questions of land management and of fire protection, America posed unprecedented difficulties—as Chapman discovered for himself in the South. In any case, fire protection was hardly a science on the order of physics and chemistry. What European forestry could and did provide for American forestry was the vision of a relatively fireproofed forest. The desire of American foresters to demonstrate their professionalism did the rest.

The situation may be illustrated by the story of *Bambi*. The original novel by Felix Salten was translated into English in 1929; *Bambi's Children* followed 10 years later. The woodland setting was Austrian, and the chief threat to Bambi and his companions came from poachers. In the American version animated by Walt Disney Studios, however, the climactic moment features a great forest fire. The scene was so powerful that Bambi was for a time used as a fire prevention poster, but the wildfire had to be fabricated—and, equally, it was expected. It became necessary for American foresters, like the American animators, to introduce fire into the forest, with or without European precedent.

"Throughout the West, as in other timber regions of the United States," Henry Clepper concluded in his history of professional forestry, "forestry began with fire control. Indeed, there could be no forestry without fire control."[39] The observation applies as much to forestry as an institutional creation as to the practical realities of silviculture. Time and again it was the need for fire control that created the demand for organized forest management. Even timber barons like F. E. Weyerhaeuser could agree. "To save the forests," he informed a congressional committee in 1908, "the main thing is to make laws to prevent fires."[40] Fire control was politically neutral in ways that other means of combating the Dragon Devastation were not. When congressional hearings in 1923 bogged down on questions of political control over logging practices, William Greeley was asked to summarize the main problem with the forests. "Stop the fires," he replied.[41] On that, everyone could agree. The Clarke-McNary Act, a means of extending the influence of federal forestry to the states, was the result.

But fire control was a two-edged sword. It helped to bring political power to the Forest Service, but it also made the successful control of fires into a highly visible public test of whether forestry, however attractive in theory, could really work. With respect to fire policy, there could be little room for uncertainty. When challenged, as it was in 1910, the Forest Service tended to adopt a single policy line. By then, thanks to Pinchot, forestry and the Forest Service were in the process of appropriating leadership over the conservation movement. To attack fire policy or the possibility of fire control was to question the credibility of professional forestry and the aptitude of the Forest Service. To doubt their abilities was to call the entire edifice of organized conservation into question. Thus fire protection acquired a significance and a character quite unlike that which developed in other countries where European forestry was introduced.

In the United States, professional forestry and the Forest Service reinforced each other, and the practical necessity of managing a vast national estate left little opportunity for intellectual refinements of borrowed technology. That both appeared almost overnight by administrative fiat meant that there had been little chance to work out in advance a modus vivendi between frontier fire traditions and the precepts of forestry. Professional forestry was transferred as a more or less complete corpus of thought and techniques, from which fire was generally excluded and within which fire could be accommodated only by painful modifications.

Its dazzling emergence in 1905 established another tradition for the Forest Service. Almost immediately it had to solve problems outside the purview of traditional foresters. Concepts like multiple use, for example, which seem pedestrian and obvious to contemporary land managers, came to American foresters as a growing revelation. The need to expand its domain has paradoxically given forestry both great depth and a stubborn provincialism. Whenever a need developed—for research, for equipment, for policy—forestry has sought a solution from within its own ranks. Even its histories tend to be written by foresters. Thus while forestry enlarged its own domain, it also retained unusual cohesiveness and a resistance to outside pressures. It would expand from the inside out. Fire policy changes would result from debates within, not by arguments advanced by those outside the profession.

Forestry and the Forest Service clung tenaciously to those rationales and cherished those experiences that had thrust them into their new sphere of influence. They continued propaganda on the value of forests for watersheds long after the Geological Survey, the Bureau of Reclamation, the Corps of Engineers, and even the Weather Bureau had modified or abandoned the traditional stance on "forest influences." They continued the cry of timber famine long after it had lost political and economic meaning. They perpetuated the old quarrel over the nationalization of the timber industry well after the issue had lost significance for conservation as a whole. And they persisted in a single attitude toward fire in spite of mounting evidence that varieties of controlled burning could be useful and were in some instances mandatory. Some form of watershed protection and some degree of regulation over frontier fire practices was essential; they had become precepts of conservation. But what form that protection should take was not obvious. Because of certain events that culminated in 1905, these decisions were given to foresters. Because of events in 1910, the Forest Service strongly circumscribed the potential range of choices, both in objectives and techniques. Had the Geological Survey been given the reserves or had they remained with the GLO, a different policy and a different attitude toward fire would have been the likely result.

But amidst the extravagence of the Great Barbecue, against a projected horizon of resource depletion, and against a nineteenth-century backdrop illuminated by the flames of unspeakable holocausts, doubts were not meaningful. The fire program of the early Forest Service and conservation movement must be measured relative to what it reacted against, not what it led to. "Every member of the Service," Pinchot recalled of those halcyon days in 1905, "realized that it was engaged in a great and necessary undertaking in which the whole future of their country was at stake. The Service had a clear understanding of where it was going, it was determined to get there, and it was never afraid to fight for what was right."[42] Whatever uncertainties its policies might face in the future, there were none at the creation.

SKY OF BRASS, EARTH OF ASH:
A FIRE HISTORY OF THE LAKE STATES

Thus sped the days—fearful days—but they brought no relief.
The sky was brass. The earth was ashes.—Frank Tilton, 1871[43]

I

The Treaty of Ghent, which concluded the War of 1812, opened up the plains bordering the Great Lakes as well as the plains along the Gulf Coast. But important regional differences retarded northern development: the northern prairies lacked the dynamic force of the plantation system; the severe winters made pioneering more a seasonal occupation; and hostile Indian tribes stalled progress still further, in many places until after the Black Hawk War. What catalyzed settlement was the opening of the Erie Canal in 1825. Michigan and Wisconsin were populated by immigrants from New England, the mid-Atlantic states, and northern Europe. By 1850 the prairies of the Old Northwest Territory were filled. Running across the middle of the Lake States, however, was an oak savannah that separated the prairies of the south from the pineries of the north. The agricultural penetration to the north did not come until after the Civil War, when industrial logging could clear the land wholesale and railroads could provide transportation and markets. It was assumed that the plow would follow the ax, that New England farmers would follow New England loggers, both abandoning the exhausted lands of the Northeast for new opportunities. Instead, fires followed each. The first railroad entered northern Wisconsin in 1870. It was succeeded a year later by the worst forest fire disaster in American history.

For about 60 years this pattern was repeated throughout the north woods. Fires of unprecedented size and intensity rampaged over small villages and towns of moderate size and thereby earned names as historic events. Half a dozen holocausts achieved special notoriety because of damages to property and loss of life. Any one of them could qualify as perhaps the worst fire disaster in American experience. The fires were the product of a particular set of conditions: wholesale logging, which made the Lake States from 1880 to 1900 the chief source of timber and an unrivaled tinderbox of abandoned slash; farmers looking for cheap, easily cleared land and not adverse to using fire for landclearing; and railroads, whose transportation potential made both logging and farming economically feasible and whose brakes and smokestacks were a frequent source of ignition. Most of the communities

overrun by the fires were railroad and mill towns. This great era of holocausts began about 1870 as all three elements first came together; its last
deadly outburst came in 1918; and it concluded only in the 1930s with the
exhaustion of virgin timber and the abandonment of agricultural settlement.
Logging had supplied the fuel; agricultural landclearing had furnished the
ignition; and the railroad had been the catalyst for their interaction. No
other region dramatizes quite so well the transition from agriculture to
industry as it affected fire, and no other era has its fire problem quite so
precisely circumscribed by historic events.[44]

This relative homogeneity is borne out by the uncanny similarity among
the great fires. For 50 years the fires were virtually interchangeable: the
names, dates, and locations varied, but otherwise the account of one fire
could substitute for that of another. There will be little violence done to
history to construct from the various accounts a single composite narrative.
It can apply with equal verity to the Wisconsin (Peshtigo) and Michigan
fires of 1871; the Michigan fires of 1881; the Minnesota (Hinckley), Wisconsin, and Michigan fires of 1894; the Minnesota, Wisconsin, and Michigan (Metz) fires of 1908; the Minnesota (Baudette) fire of 1910; the Michigan (Ausable) fires of 1911; and the 1918 fires in Michigan and Minnesota
(Cloquet). Nor were the fires restricted to the vicinity of the towns they
overran. The Black Year, as the *Chicago Tribune* referred to 1871, saw fires
burning almost simultaneously in Illinois, Indiana, Wisconsin, Michigan,
North Dakota (where several small towns were wiped out), and, of course,
Chicago. The reported acreages are surely underestimations.[45]

Fires were prominent earlier, of course. Studies of fire scars and stands
of even-aged pines at Itasca State Park, Minnesota (legendary source of the
Mississippi), show conflagrations in 1714, 1803, 1811, 1820, 1865, and
1886, before protection arrived in 1894,[46] and surveys in Carlton County,
Minnesota, give dates of 1819, 1864, 1874, 1885, and 1894.[47] Chronicles of
similar periodicity could be constructed for all of the north woods. Indian
fire practices were as prevalent here as elsewhere. The conflagrations came
in the fall, following a summer drought or a rainless Indian summer. An
observer in 1881 wrote that "these fires which so devastate and utterly ruin
so many thousands of acres of large pine forests are said to be set by Indians,
purposely, and assisted to spread, to kill the timber, and so give better feeding ground for the moose and deer which abound in the area."[48] Like the
white farmers who came after them, the natives had little use for unbroken
expanses of forest.

The traditional chronicle of large fires ends around 1918 because whole
settlements were no longer being swallowed up by the flames, but hundreds
of thousands of acres continued to burn. In 1925 fires swept some 1.4 million
acres in Wisconsin, Minnesota, and Michigan. Wisconsin mobilized its
national guard, but too late; all control forces watched helplessly until the

rains finally arrived. More fires broke out in the terrible drought years of the early 1930s. But by 1936 the era of the big fires was over. Zoning laws and state and federal forests carved out of the remaining public domain and from tax-delinquent lands effectively reversed 60 years of attempted agricultural settlement; industrial forestry settled down to harvest second-growth crops; and fire control forces existed in strength.

The circumstances preceding the great fires were remarkably uniform. Extensive logging created mountains of slash, and natural windfalls added more. Fierce burns left a tangle of downed heavy fuels and thick underbrush. In some areas the debris rose 12 to 15 feet high. Even after the 1871 fires in Michigan, it was reported that a man from Port Huron walked for a mile on the trunks of fallen trees without touching the ground.[49] Not surprisingly, this area was visited by even more terrible fires a decade later. Still, the land was attractive. In the midst of chronicling the devastating 1881 fires, the Detroit *Post* paused to observe without irony that here was a real "chance for new settlers . . . where the fires have raged the forests have been killed, the underbrush burned and the ground pretty effectively cleared. There are square miles and whole townships where the earth is bare of everything except a light covering of ashes; and other square miles where all that is needed to complete the clearing is to gather up a few scattered chunks per acre and finish burning them."[50] No professional promotor could have boosted the devastated lands more enthusiastically.

Similar broadsides issued from land speculators and railroads. Far from being depicted as treacherous capitalists turning a fast buck at the sometimes fatal expense of honest settlers, the logging companies were applauded. By their collective axes they were clearing land that would break the backs of individual pioneers. They spearheaded the critical railroad lines, which tied farms to markets. For their part, the lumber companies ingenuously claimed to be victims of the agriculturalists. The *Lumberman's Gazette,* the Detroit papers recorded in 1881,

> disputes the theory that our pine forests are being cut off too rapidly; and argues that the pine must be cut speedily to save it from being destroyed by forest fires. It is a question whether this valuable timber shall be saved to be used for the convenience of human beings, or be wasted by destructive forest fires. If it is to be saved, it must be cut as fast as possible. It cannot be husbanded and preserved for the future. As the country becomes settled, fires started in clearing lands are more frequent, the underbrush becomes thicker, swamps and water courses dry up, and the fires are not only more numerous but more destructive.

The fire pattern in the north woods, in short, was like a self-reinforcing dynamo: the more forest that was cut, the greater the influx of landclearing farmers and the greater the fire hazard; the greater the hazard, the more rapidly and wastefully logging had to proceed. "At least, so argue many who have studied the subject with some care."[51]

There is little doubt that the slashings supplied the initial fuel for the great holocausts and that landclearing fires were the primary ignition source for the conflagrations. What had not already burned could be fired deliberately in the spring before the summer rains and in the fall before the winter snows. The railroads added other ignition sources, but the vast extent of the fires and their ability to encircle communities so readily can only be explained by the prevalence of dozens of landclearing and protection fires set on all sides. This complex of events also helps to account for one of the historical puzzles surrounding the fires: the seeming indifference of the townspeople to the fire threat. Until the blowup actually arrived, residents evidenced only moderate concern. Few heeded even the warnings passed down the tracks by trains piled with panicked refugees from nearby towns. More than once residents would be hurriedly rescued from a threatened community and deposited a few miles distant, only to be reevacuated as the holocaust sprawled across the countryside. Nearly all the towns consumed by the conflagrations had ample warnings.

This seeming indifference appears inexplicable only because the historian knows what followed. In fact, landclearing fires were a common phenomenon, as predictable as autumn itself. They were a welcome sign of progress. During severe conditions, precautions were taken by the prudent: farmers plowed firebreaks around their houses, barns, and fields; railroad gangs patrolled tracks to suppress fires threatening wooden trestles and ties; volunteer fire departments fought fires that crept to the outskirts of villages; mill workers suppressed forest fires that threatened the lumberyards. Nor were landclearing fires totally irresponsible. Most were set by immigrants from Finland, Germany, and Poland. Scandinavians, in particular, were familiar with the principles of swidden agriculture and debris burning. But about the volume of fuel made available by logging, about the meteorological conditions peculiar to the region, and about the tremendous potential for conflagrations, they were woefully naive. After the holocausts of 1871, the specter of '71 burned bright in the memories of local Cassandras who later issued dire warnings. We will have "rain or ruin," they would say portentiously and shake their heads.

Fires were common, holocausts rare. "There have always been more or less forest fires in Minnesota," one realist proclaimed. The very volume of landclearing around many towns gave residents an exaggerated sense of security. The large fuels had already been burned, the nearby timber long ago removed, the towns surrounded by protective firebreaks of low-density fuels. A Michigan observer wrote in 1880 that "no important fires have happened in this county since the great fires of 1871, and as the county is now well settled up, nothing like the forest fires of any general character will ever again be likely to occur within our limits."[52] The next year hundreds of thousands of acres that had burned in '71 were reburned with, if anything, increased intensity. Nor were the townspeople inexperienced in coping with forest fires. When Escanaba was threatened in 1871, the menfolk turned out 1,000 strong, cut a swath 500 yards wide, and backfired. Some fires might burn for months; only when they threatened property were they attacked, usually with backfires or by herding into swamps. The strategy was entirely defensive; fortifications were thrown up and counterattacks launched only when fire posed an imminent threat.

Writing in 1871, one commentator summarized the fire situation in Wisconsin that year:

> The surrounding woods were interspersed with innumerable open glades of crisp brown herbage and dried furze, which had for weeks glowed with the autumn fires that infest these regions. Little heed was paid them, for the first rain would inevitably quench the flames. But the rain never came, and finally valiant battle was waged far and near against the slowly increasing fires. In this, as in other towns, the danger was thought well warded off by the general precautions. The fire had raged up to the very outskirts of the town weeks before that fatal Sunday, and the fires were set outward to fight the enemy. Everything inflammable had apparently been taken out of harm's way on that memorable Sunday. One careful citizen traversed the western outskirt, and assured his people that no danger could come from that quarter.

The situation seemed so tame that citizens of several towns refused to do more because of the Sabbath.[53]

For weeks in advance of the 1871 blowup on October 8 and 9, fires swept the "prairies and openings of all that part of Wisconsin lying northward of Lake Horicon, or Winnebago marsh, which was itself on fire."[54] An observer commissioned to report on the conflagration noted that "farmers, saw-mill owners, railroad men, indeed all interested in exposed property, were called upon for constant and exhausting labor, day and night, in contending against the advancing fires." Mills were protected by earthworks and ditches. "In this labor of fighting fire, the millmen, farmers, and others were engaged through October, the exhausting work going with good cheer, in the constant hope that either the welcome rain would come, or that finally the ground would be wholly burned over and leave nothing for the flames to feed upon." The Chicago and Northwestern Railroad kept its tracks open through the burning region by employing a "large force of men stationed along the line." In expectation of disaster, "many devices were resorted to for the protection of life." Excavations were dug and outfitted with earth-covered roofs, in which persons sought refuge. Many fled to wells that had gone dry from the long drought. And much property was taken from houses and placed in the open fields for safety—though it was not uncommon for such caches to burn while the houses themselves escaped. Towns beefed up fire brigades, and Green Bay hired extra police forces. The "fire-beleaguered people . . . for weeks past had in mind penciled places of safety in case of defeat in their hazardous flight." As flames approached, ditches and firebreaks were constructed; water was hauled by wagon and delivered to the fire with buckets; and backfires were set. But as time drew on and the ground was burned over, the long harassed people began to take breath, "believing that the worst had passed."[55]

For most towns and against most fires, such measures would have been adequate. But these were mass fires, and the townspeople had taken precautions only against traditional surface fires. Survivors recognized that the holocausts that swept their towns and farms were something rare. They described them in terms of great storms and invented wild meteorological explanations to account for the unearthly conflagrations. In fact, the great fires were extreme cases of mass fires, so violent and intense that they became discontinuous in behavior: long strips of forest bordered gutted landscapes; towns and fields were left untouched next to devastated areas. The mosaic of light and heavy fuels was all but meaningless and ineffective in regulating the conflagrations' fury. Spot fires ignited in advance of the main fires, flaring and throwing more spots. Heavy slash piles and large structures might be unscatched, light fields and small buildings totally consumed. Areas "fireproofed" by earlier surface fires reburned with devastating crown fires. Even combustion itself became discontinuous: gases, flames, heat, and convective winds became separated and could be individually felt. Most of the victims died from asphyxiation or the inhalation of toxic gases as they sought refuge in root cellars, wells, and other ad hoc bunkers; few died from actual burns. In modern times the only comparison is to the firebombings of Dresden, Hamburg, and Tokyo. Indeed, the origin of the term *firestorm* goes back to accounts of the 1871 fire; its use in 1943 Germany was a case of independent discovery when confronted with analogous horrors.[56]

The first sign of impending doom for most communities was a preternatural darkness. To those from New England, the scene recalled the famous Dark Days. To survivors of '71, it called up memories of the awful scenes to follow. Descriptions suggest that the darkness came from clouds not unlike the towering swirls of dust that swept out of the Great Plains in the early 1930s, except that these were charged with soot and firebrands instead of earth. The blackness was "Egyptian"; noon became midnight; a man could not see his hand before his face. The darkness contributed measurably to the panic and confusion that followed. Families became separated. Disoriented people dashed in wrong directions.

When the blackness was rent by violent firewhirls brightening the sky like flashes of lightning, it seemed that the apocalypse was at hand, that the Day of Judgment had come. For thousands, it had. At Peshtigo, Wisconsin, one survivor recalled that many who believed that the judgment had come

> fell upon the ground and abandoned themselves to its terrors. Indeed this apprehension, that the last day was at hand, pervaded even the strongest and most mature minds. All the conditions of the prophecies seemed to be fulfilled. The hot atmosphere, filled with smoke, supplied the signs in the sun, and in the moon, and in the stars; the sound of the whirlwind was as the sea and the waves roaring; and everywhere there were men's hearts failing them for fear, and for looking after those things which are coming on the earth; for the powers of heaven shall be shaken.[57]

On the opposite side of Green Bay throngs crowded into a "conventual school" in the belief that the world was being consumed, and "falling upon their faces crawled round and round it with long-continued prayers." But for many the ways of judgment were strange. A grandmother survived the fire on a large stone while sparks and the flaming corpse of a woman fell beside her and 50 others perished close by in wells and cellars. "I cannot pretend to understand the providence of God," she pondered later, "which spared me, an old woman, with my days fulfilled, and took my sons and daughters."[58] Like the fires, even providence was magnified in its capriciousness.

Out of the darkness came "currents of air on fire," a "sirocco," a withering blast of heat. In areas otherwise untouched by the fire itself, heat and heated gasses claimed lives. So fierce was the heat that it alone drove hundreds to shelter in root cellars, wells, and stone buildings. The vast majority of those who fled to cellars perished from asphyxiation; those in wells, from asphyxiation and incineration; and those in buildings, from suffocation and flame. Some lived but it is difficult to understand why even panicked masses would seek refuge from fire in a combustible building. Heat was the apparent cause: the landscape for miles ahead of the flames was violently preheated and dessicated.

The fire itself was preceded by a barrage of firebrands and a thunderous roar. A "storm of fire brands, cinders, and ashes" showered the landscape with "fire flakes." Other people spoke of "balls of fire" and "fire balloons," probably bubbles of heated gasses distended from the main burn itself and ignited by radiant heat. Many of the "balloons" exploded above ground, blasting fire like shrapnel. An island half a mile out in Lake Michigan erupted into fire. One terrified eyewitness described the fires as a "veritable cyclone of flames. There came, as it seemed to me, great balls of fire from the sky, and when they were within 20 feet of the ground, they burst, sending down a heavy rain of flashing sparks, like a mighty sky rocket exploding with a brilliant display of flashing light."[59]

Overpowering everything was the roar. It sounded like thunder, the pummeling of a dozen cataracts, the pounding of heavy freight trains, and "all the hounds of hell." It was unleashed like a "heavy discharge of artillery," and one Civil War veteran thought for an instant that he was returned to Chancellorsville to face again the Confederate batteries. The noise so resembled the sound of a cyclone that some fled to holes, as though they needed protection only from a windstorm.

When the fire arrived, it came not as a wave or a surge of flame but as though suddenly dropped from the sky. The landscape was instantly enveloped in a "tornado of flame," a "hurricane of fire." Firewhirls traveled 6–10 mph, "the pine tree tops were twisted off and set on fire, and the burning *debris* on the ground was caught up and whirled through the air in a literal column of fire." One witness exclaimed, "it was a waterspout of fire."[60] Its winds, like its heat, sometimes preceded, sometimes coincided with the advent of the main fire and reached staggering velocities. Surface winds were rarely excessive, frequently 15–40 mph, but the turbulence from the violent convection was awesome. Winds of 60–80 mph uprooted trees like match sticks; a 1,000-pound wagon was tossed like a tumbleweed. Papers were lofted by the winds from Michigan across Lake Huron to Canada. The peculiar physics of mass fire had multiplied its fury into a maelstrom of energy equivalent to the chain reaction of a thermonuclear bomb. There was no defense for the populace but flight.

And flee they did. A stream of people poured into rivers, ponds, gravel pits, and even the Great Lakes. But the droughts had drained the marshes and lowered the streams to a trickle. Near the town of Hinckley, Minnesota, in 1894, 126 people took refuge in a marsh, only to have the fire sweep over them and convert the sedge into a vast crematorium. Hundreds fled to a stream at Peshtigo, Wisconsin, in 1871, but so did terrified cattle that trampled many women and children. And when a building upstream collapsed in flame, its fiery debris floated down the river into the screaming victims. The exact body count will never be known. Perhaps 1,500 perished in the Peshtigo fire and another 750 in the Humboldt fire, both in Wisconsin, 1871; at least an additional 10 died in Michigan at the same time. For the 1881 fires in Michigan, figures differ wildly, from 138 to several hundreds. In the 1894 fire at Hinckley, Minnesota, the official count was 418, with another 13 contributed by the nearby Phillips fire across the border in Wisconsin. In Minnesota in 1910, the Baudette fire swept away 42 lives; the Cloquet fire, 551. These numbers do not include isolated farm families trapped by the fire or remote villages of Indians.

The survivors in almost all cases—and this was true for nearly 50 years— were saved by locomotives. The stories of reckless gallantry and of unselfish bravery in the face of fire that the twentieth century has bestowed on members of its organized fire crews, the nineteenth century gave to its railroad engineers and conductors. From 1871 to 1918 the heroes of the great fires

9. The exodus from Peshtigo, as reported by *Harper's Weekly*, 1871.

were operators who defied smoke and flame to evacuate hundreds of panicked refugees in daring rescues over burning ties and smoldering trestles, past flaming woods whose heat blistered the paint off the sides of cars, and through fire and smoke that brought intense personal suffering to the exposed engineers. No fire was without its narrative of relief trains flying courageously into the inferno to evacuate trapped settlers.

One of the most famous of these rescues occurred during the Hinckley fire of 1894. The train for the St. Paul and Duluth consisted of one combination car, one coach, two chair cars, and an engine. As the crew and passengers approached Hinckley, unaware of the extent of the fire, the smoke thickened, and it became necessary to use the headlight and to slow to a crawl. In this alone there was nothing abnormal. Engineers had become as accustomed to smoke as the local farmers and townspeople. Breathing became difficult, and as the train glided into the far outskirts of Hinckley, it met the flames and roar of the fire. Passengers became excited, and as one survivor recalled, "the quieting words of the trainmen fell like seeds upon rather stony ground." Panic increased as fleeing refugees from the town flagged down the engine a little over a mile from Hinckley; perhaps 150 boarded the train. The engineer, Jim Root, paused to consider his options. At first he thought to run through the fire under full steam. Then he recalled a small marsh pond known as Skunk Lake adjacent to the tracks and about four miles to the rear. He put the throttle in reverse and began the ride that would propel him and his crowded throng of passengers into the lore of American fire.

The flames at times outraced the engine. Heat, smoke, and even fire enveloped the tiny train from front to rear. Flames came through the ventilators at the top of the passenger cars and through cracks along the windows. When the rear coach took fire, its occupants fled into the next car. One by one the heat blistered the exterior paint of the other cars, then burst them into flame. The stifling heat caused even the interior paints to blister, then run. The interiors were swallowed in darkness, broken only by screams and shouts. The window glass cracked. At last the insufferable heat and congestion drove one man mad. Shrieking, he threw himself out the window and vanished instantly in a cauldron of flame. Another followed his example, then another still. The remaining passengers all but gave themselves up to complete panic. Only the unflinching example of the Negro porter, John Blair, calmly working a fire extinguisher to suppress sparks on women's dresses prevented what might have become a general and fatal exodus.

Conditions were even worse in the cab. Root and his fireman, Jack McGowan, suffered horribly. At one point McGowan plunged himself into the manhole of the watertank to quench the fire on his clothes. He and Root thrust their swollen, inflamed hands in a pail of water. Periodically, McGowan doused Root with water to revive him. At last heat burst the glass plate at the cab window and a piece of flying glass slashed Root near the jugular vein. He bled profusely, and to heat and smoke and a crushing sense of responsibility for his passengers was added simple blood loss. Root slumped to the floor, his hand still on the throttle; pressure dropped to an alarming 95 pounds. McGowan, himself all but overcome, revived the fading Root. Again they collapsed. The front of the cab was burning: fire seized the wooden handles of the steam connections, scorched the seats, and melted the cab lamp. But still the train moved, and again McGowan and Root revived.

The flames pulsed in great waves, and when Skunk Lake was at last reached, there was a providential pause. McGowan carried the disabled Root to the pond, a sump a mere 18 inches deep at center. Passengers

10. The aftermath at Hinckley, 1894.

poured out of the blazing cars, many utterly disoriented; the heat was crippling, but there was enough time for everyone to drop into the muck before another surge of flame passed over. The train erupted into fire, the coal tender with it. During another pause, the faithful McGowan separated the engine from the tender, ran it to the middle of the tracks across the marsh, and saved the engine from annihilation. For four hours nearly 300 people huddled in the water and mud, swathed in an Egyptian darkness torn only by exploding ribbons of flame. But all lived. Only the three who had dived out of the train and two Chinese who stoically remained in their seats perished. When the fire at last subsided, the conductor, Thomas Sullivan, struggled back to the nearest station to warn a freight of the dangers and track debris ahead. After sending the message, he collapsed. The passengers from the Hinckley fire were fortunate. A relief train sent to Metz, Michigan, in 1908 tried to make a similar run through flames, only to have the fire reach such intensity as to burn the ties and warp the rails. Derailment resulted, and 17 people in a steel gondola were roasted to a horrible death. In the Hinckley episode, Root, McGowan, and Blair were accorded well-deserved hero's honors.[61]

The railroads also spearheaded early relief efforts after the fires. As soon as it learned of the disaster at Hinckley, the St. Paul and Duluth, for example, readied a train stocked with food, clothing, and supplies. Progress toward the stricken town was slow. Rails, culverts, and even ties had to be repaired or replaced. The fire had passed in early afternoon; by 3 A.M. the next day, the relief train arrived. The Eastern Minnesota Railroad, which had also successfully evacuated Hinckley residents in the face of the fire, hurried a relief train as well. Indeed, the entire relief effort was not the least of the remarkable similarities shared by fires of this epoch. Relief committees were established both locally and at large urban centers like St. Paul and Detroit. Not infrequently, competition ensued among various committees, and some larger governing body, usually appointed by the state, was assigned overall responsibility for coordination. The principles of such a body has been well established since at least the days of the celebrated civilian Sanitary Commission founded during the Civil War.[62]

Contributions were solicited; supplies of food, seed grain, clothing, medical goods, tents, and lumber for housing were distributed to thousands of homeless and destitute settlers. As the holocausts invariably came in the fall, desperate victims required aid throughout the long, brutal winter. Railroads shipped relief supplies free of charge for weeks, even months. The amount of money collected could be impressive, especially in the early fires, when nearly all came from voluntary pledges, not government assistance. Nearly $500,000 was raised for the Wisconsin fires of 1871, over a million for the 1881 Michigan disaster, and $185,000 for Hinckley. For the devastating Cloquet fire of 1918—the last of the holocausts—a state-appointed committee eventually distributed over $3.1 million to the victims: $1 million from voluntary contributions; $300,000 from a state calamity fund; $75,000 from the Red Cross; and the remainder from the sale of Fire Relief Certificates of Indebtedness. The state assumed the burden of rebuilding roads and bridges; the Army contributed blankets, clothing, and tents as an expedient. The Cloquet fire is particularly interesting, because the victims successfully sued the federal government for an additional $12 million. At the time of the fire the government was running the railroads under its war powers authority. By showing that at least some of the fires originated from locomotives, the victims demonstrated liability on the part of the government. Appeals kept the suit in court until the 1930s, but in an age antedating federal disaster relief programs, the residents of Cloquet had discovered an ingenious substitute.[63]

Relief did not end the cycle of fire. One of the arguments for relief, in fact, was the fear that farmers would shun the north woods if their plight was ignored, that the tide of progress, capital, and settlement would move elsewhere. Land speculation was big business; boosterism, a reflex arc. The generous outpouring of voluntary contributions was overwhelmingly humanitarian, but its ultimate goal was to keep the farmer where he was, not to move him to a less fire-prone environment. "These lands offer the best inducements for new settlers," the Detroit *Post* informed its readers.

These lands are now in such a condition that they are all ready for seeding to wheat, merely requiring the harrow to be used upon them, in case there is not time to plow. The rich salts of the former vegetation are preserved

in their ashes, which are an excellent top-dressing of manure, soon to be
washed into the earth by rains, and absorbed under cover of the snows of
winter. The trees, the underbrush and all the impediments to agriculture,
it usually costs so much in toil for the pioneer to remove, have been swept
away, and the rich land lies open and ready cleared for the settler. . . . All
that the new settler who buys these lands now has to do is to build his
cabin and go right to work putting in his fall wheat. He can put fences
up at his leisure. Next spring, he will find his land cleared and ready for
spring plowing and spring crops.

There are other great advantages, also. The insects and forest pests of
the farmer are nearly all extinct. There will be no potato bugs, no weevils,
or army worms, no curculio, very few birds or squirrels for several years
to come on these lands. Fences will be little needed, because the cattle,
sheep and hogs have been largely destroyed. Settlers need not to be
deterred by fear of future fires. There will be no danger of any more great
fires, unless these lands are permitted to lie idle for at least ten years, and
grow up to underbrush again. There can be no more fires, because there
are no more brush or swamps to burn; and destructive fires can not visit
the region again till these are once more grown. If the lands are soon
settled, these will never grow again; and it will be the safest region in the
state from forest fires.

Other advantages for new settlers are the facts that there are roads
already made throughout the district; and so much has to be rebuilt that
there will be work and wages in plenty for several years to come. . . .[64]

Scavengers even collected potash from the burns for export elsewhere.
Once an area had been overrun by a mass fire, residents felt confident that
the region was fireproofed. Towns were rebuilt, again entirely of wood—
corduroy roads; wooden sidewalks, roofs, and structures; sawdust on the
streets. No slash heap was more combustible. And, of course, there were
souvenirs. A dozen local authors churned out memorials and eyewitness
accounts, local societies of survivors were formed, and a museum was dedicated at Peshtigo. As with Chicago after 1871, a small trade developed in
memorabilia and trinkets salvaged from the awful wreckage.

II

In their fire experiences, the Lake States were not unique. The holocausts
were an intensified expression of the settlement process, accelerated and
magnified to the point of lethal parody. Far from being hostile outbursts of
an inscrutable nature, the conflagrations that swept the Lake States from
1870 to 1920 were an integral part of the attempted farming of the north
woods, and they concluded only when the land was withdrawn from agricultural settlement. The impact of the fires, moreover, did not end with their

flames. For the Lake States this tide of conflagrations resulted in massive environmental changes and socioeconomic reforms; for the United States as a whole, they coincided with the growing agitation for some rational form of accommodation to the industrialization of the landscape, for conservation, and for organized fire protection. In their search for a suitable model, the Lake States, like much of the nation, looked to New York: the Empire State's environmental setting resembled theirs; logging had migrated to the north woods from upstate New York; the Erie Canal had long linked the economic interests of the two regions and had supplied many settlers; and as early as 1885 New York had taken steps to administer large blocks of forest reserves and thus furnished an exemplary organization for fire protection.

Yet the most interesting comparison is perhaps with the South. The fire history of the Great Lakes Plains nicely complements that of the Gulf Plains. The vegetative mosaic was similar for each, with bands of prairies, pine woods, and muskeg or swamp. The flammable pine occupied sandy, shallow soils poorly suited to farming. The prairie and savannah farmlands were seized first; woods were opened only with railroads and heavily capitalized logging. The economic future of both the north woods and the piney woods depended on satisfactory treatment of lands cut over and burned over. Both regions experienced stubborn incendiary problems and considerable hostility toward wildland fire suppression. What stimulated both to establish vigorous protection programs was the reclamation of cutover lands for industrial logging, especially for pulp. Yet it is a mark of their differences that the history of large fires in the Lake States concluded at nearly the same time that large fires in the South began.

The reasons are both environmental and socioeconomic. The plantation system of the South pushed less successful settlers into the woods, perpetuating a frontier economy of hunting and herding. The logging of the north woods had the reverse effect, concentrating farmers onto clearings. But both groups, in a sense, were socioeconomic islands amid the rising tide of industrialism. Both consequently persisted in frontier fire practices—the one for habitat maintenance and herding; the other for landclearing and fertilization. So long as the woods were occupied under these conditions, fire would remain a more or less chronic phenomenon. The differences in the intensity of the fires, however, was environmental: the South could reduce its fuels buildup through light winter burning; the North had to burn under riskier spring and fall conditions. Natural decomposition was slower in the North and fuel loads heavier. The result was a strongly episodic pattern of holocausts coinciding with prominent droughts. In the South, mass fires appeared as threats only with the advent of fire protection; in the Lake States, fire protection tended to eliminate the mass fire. The tough winters and denser woods of the North replaced open-range grazing with farm-fed livestock or dairy cattle. The association of open ranges, game hunting, hogs, and cattle so integral to southern fire practices did not exist in the North; fires were limited primarily to agricultural settlement, with logging and railroads a secondary source. The trapping of fur-bearing animals was a winter and spring occupation and one not readily tied to burning, as was southern hunting.

It long remained the hope of the Lake States that their cutover lands would be converted into farms, that successful settlement in itself would eliminate the fire hazard. Amid the spring tide of reform during the early decade of the twentieth century, state forestry bureaus were established. Wisconsin created a system of town fire wardens, and, prodded by the Lake States Forestry Congress, private timber owners from Michigan and Wisconsin organized the Northern Forest Protective Organization in 1910. General Christopher Columbus Andrews, a former United States ambassador to Sweden, railed against his fellow Minnesotans with the familiar refrain that "the careless use of fire that has been habitual in the forest regions of our northwestern states would not be permitted in a country like Germany."[65] But the states rarely funded the agencies they created. After a disastrous fire, interest would be rekindled, only to disappear within a few years; rangers would be hired, then dismissed. Nonetheless, a Lake States Forest Fire Conference was held at St. Paul in 1910, probably the first national meeting devoted exclusively to the fire question. The governors of Wisconsin and Minnesota attended, and nearly every organization involved in fire protection sent representatives. The fire protection systems of the West were discussed, but New York and Maine offered more suitable models, despite their commitment to wildlands qua wildlands rather than as a prelude to farms.[66]

The federal government entered the picture in 1908 with presidential proclamations that created national forests from the yet unsettled public domain. Minnesota, Wisconsin, and upper and lower Michigan all had lands transferred to the U.S. Forest Service. Cooperative agreements were established with state and private organizations. By 1914 all three states received federal fire funds under the Weeks Act. The states, too, began to acquire land. Minnesota, which had asked a stiff price for its lands, had not sold a great deal; on both state and federal lands the timber had been cut or sold but not the land. With the exhaustion of the pineries around 1900 and the exodus of many lumbering operations to the South and West, tax-delinquent cutover lands accumulated rapidly. In the South such sites largely passed into private hands; in the North, the states acquired or retained title. Each of the Lake States held millions of acres and gradually administration— that is, fire protection—was extended over them.

After the catastrophic Minnesota fires of 1918, major periodicals editorialized that "the prevention of all forest fire in such a country as northern

Minnesota is probably an impossibility; the prevention of a great conflagration is not altogether possible, but the occurrence of such a fire is criminal."[67] Cloquet, after all, had not been a minor, obscure hamlet; it was a famous sawmill town on the outskirts of Duluth. Other states had demonstrated the feasibility of controlling railroad fires. And at long last the mirage of the plow following the ax through the north woods began to dim. Cloquet was rebuilt in a year.[68] But the terrible influenza epidemic of 1918 hit fire victims hard. Hail ruined crops the next year, and the potato harvest failed the year after that. Even more damning, the pattern of fire and plow had devastated the topsoil. Especially on sandy lands, on drained peat bogs, and on lands heavily veneered with logging slash, agricultural firing for site preparation not only occasionally exploded into mass fires but also annihilated the organic soil. In some places many feet of organic matter were stripped off, leaving only barren stretches of sand or rock. Such fires did not benignly recycle nutrients: they irreversibly vaporized them.[69]

The other events surrounding the Cloquet fire pointed to the future. Many of the structures consumed by the fire were summer homes maintained by the affluent of Duluth and Superior, and many of those killed were in transit with their automobiles as the fire struck over the weekend. For the first time, in fact, much of the evacuation of threatened areas was handled by motorists volunteering their services, reminiscent of (and perhaps inspired by) the transportation of the French army to the Marne in taxicabs a few years prior. The moral was clear: the automobile made summer homes and recreation more valuable than woods farms, and this complex created an imperative for fire control that had not previously been in force.

The second development emerged in the debate over whether to rebuild Cloquet or to let it vanish into the obscurity of mining and logging ghost towns that already littered the nation. When lumbermen elected to rebuild, they were committing themselves equally to the reconstruction of the regional timber industry. After a burn as severe as that of 1918, the pine forests would be replaced by "weed" species—aspen, balsam, and jack pine. Already vast portions of the Lake States were flooded with their reproduction, and they would dominate the second-growth forests. The jack pine in particular was a pyrophyte; its serotinous cones and volatile flammability created a fire environment highly favorable to its perpetuation and distinctly hostile to competing, less phoenixlike species. Harry Hornby and Rudolph Weyerhaeuser hired two of the best wood chemists in America—Dr. A. W. Schorger and Howard Weise—to investigate the fiber properties of aspen, balsam, and jack pine with the ideal of finding commercial applications.[70] By 1920, in addition to pulp and paper, an insulation fiber and synthetic lumber made of fibers and particles were developed. Two years later Weyerhaeuser constructed an impressive wood conversion plant on the ashes of its sawmills. The state soon joined in with an experimental forest to explore the silviculture of these trees. As in the South a decade later, the new technology transformed worthless reproduction into raw materials of real economic value. In the South, wornout sharecropper fields gave way to pine plantations; in the North, subsistence farms turned to tree farms. In both cases, fire protection became a necessity and a reality.

Yet this first wave of enthusiasm was in some ways only another false dawn. Formal fire control organization reached a sufficient level so that in 1921 the Northern Forest Protection Assocation was disbanded. But heavy fire years plagued the Lake States through the 1920s and early 1930s. Not until the fires of 1925 were the states roused to action on the scale demanded. Relief of the serious drought of 1930 was financial as much as meteorological: rains and the CCC came at nearly the same time. In those portions of Wisconsin with severe soil damage due to drainage and fire, many bogs (or what was left of old marshes) were deliberately flooded— and, in a sense, that was exactly what the inundation of federal money and state involvement did for the burned-over districts. Almost as suddenly as it began, the great epoch of holocausts terminated.

Underwriting the transformation was a fundamental reassessment of land policy. Wisconsin enacted county zoning laws that closed 5 million acres to agricultural settlement and authorized purchase of lands owned by isolated settlers. The Resettlement Administration of the federal government, eager to take marginal lands out of production, assisted in the relocations. Collectively, these and similar events in Michigan and Minnesota amounted to a remarkable reversal of the frontier philosophy that had led the states to see their future largely in agriculture and to promote logging and landclearing as a prelude to woods farms. The natural division between forest and prairie was reinstated, with farms on the former grasslands and logging and recreational use in the second-growth north woods. In nature, the boundary between the two provinces had ebbed and flowed with changes of climate and fire frequency. Settlement had mixed the two into an unstable amalgamation, a volatile compound ready to explode into episodic holocausts. In its reclaimed state the former boundary was restored somewhat, and the old biota of the north woods assumed at least a shadowy resemblance to its former grandeur. State forestry bureaus acquired jurisdiction over vast acreages of tax-delinquent lands; with the exception of New York, no other state had so much land under its immediate control. In the West the lands remained under the aegis of the federal government; in the South, the cutover lands were purchased by industrial logging concerns. Responsibility for these lands gave state forestry bureaus unusual strength. With the agriculturalist either banished from the woods or converted to tree farming, the Lake States avoided the incendiarism that endured in the South, where continuing residence perpetuated a way of life and fire habits generally inimical to industrial forestry. In the North debris burning by summer recreationists replaced landclearing by farmers.

When fire suppression at last took hold, it did so with great tenacity and ingenuity. The Lake States mechanized early. Over a decade in advance of the South, Wisconsin and Michigan experimented with plows drawn by crawler tractors. In 1929 Michigan established its famous Forest Fire Experiment Station.[71] Initially, the station readied a far-flung program of research into fire effects, fire behavior, and equipment development. It soon specialized in heavy equipment, set the pattern for the equipment centers that would come after World War II, and even today is unrivaled as a source for state and rural equipment development, serving both Michigan and a consortium of northeastern states. This early commitment to mechanization, together with their responsibilities for fire protection over vast acreages of state land, has given the Lake States unusually strong suppression organizations. Although all belong to the National Wildfire Coordinating Group, there has been no effort within the region to form an interstate compact for mutual assistance. Between the NWCG and cooperative agreements with the U.S. Forest Service, perhaps no other arrangements are really needed.

The cultural impact from a heritage of holocausts was no less significant than the environmental changes wrought by the fires. The fires coincided exactly with the great debate over conservation: they began in the 1870s and concluded with the end of the Progressive Era. The entire controversy over forest conservation and wildland fire protection took place against the violent backdrop of the worst fire disasters in the national experience. The national consciousness could expect no more brilliant and sickening demonstration of the hazards of wildland fire than what occurred in the north woods. The damage and casualty figures made the notorious Johnstown flood look like a minor incident. It is no accident that the first national conference devoted to fire was organized in the Lake States. The social costs in transition from agriculture to industry, so often described in terms of urbanization, had a no less vivid environmental illustration in the unhappy mixture of agriculture and industrial logging.

Smoke often closed the Great Lakes to navigation. Pleas for relief contributions touched all the major metropolitan areas of the East. Numerous accounts of the great fires were published, though most had local circulation. Ironically, the Chicago fire, which has since tended to overshadow the other 1871 fires, at the time actually helped to publicize its rural counterparts. The stories of the Chicago disaster appended descriptions of the Wisconsin and Michigan fires and gave them wider circulation than they might otherwise have achieved. During the 1881 Michigan fires, two other national liaisons were established: stimulated by the disaster, the American Red Cross was finally incorporated and made its first contributions for relief; and the newly constituted Army Signal Service, the beginning of a national weather service, sent an observer to the burn. Sergeant William Bailey's report was the first excursion by the Signal Service into disaster monitoring, and it marks the beginning of fire weather analysis.[72]

The environmental consequences of this epoch of conflagrations were enormous. Ax and fire effectively stripped the north woods of its great pineries; ax and plow often vaporized its soil cover. In former bogs as much as a dozen feet of organic mat had been burned away, leaving bare soil and rock as a residue. In many former pine sites only sand remained, a landscape not unlike the glacial outwash left by the retreating ice sheets. The fires prevented reproduction of commercial species, and, with a few courageous exceptions, the logging industry moved on. As in the South, chemistry eventually showed how the weed species could be used as raw materials for manufacturing. But the greatest asset of the land was to be its fish, game, and recreational potential, not its timber. The land is now managed largely for habitat—both wild and human. For its summer homes and campgrounds, this means fire protection; for wildlife, it increasingly means prescribed burning.

The larger and most vexing fires of recent years have resulted from just such prescribed burns. In 1976, near the end of a severe drought, a lightning fire began on the Seney National Wildlife Refuge in the upper pennisula of Michigan. Local Officials "managed" the fire by treating it as if it were a natural prescribed fire, while they continued to wrestle with a one-acre prescribed burn, ignited a short time before, that stubbornly resisted control. The lightning fire eventually blew up over more than 70,000 acres, thus becoming (along with some fires in Minnesota) the largest fire complex in the Lake States in 50 years, and it created a political maelstrom when it left the refuge for private and state lands. The experience led to a national reorganization of fire management within the Fish and Wildlife Service, but not until a staggering $9 million had been spent in extinguishing the burn.[73]

In 1980 a prescribed burn set by the Forest Service in the spring to improve habitat for the Kirtland warbler escaped control, raged over some 45,000 acres and several structures, and took one life. The warbler had promised to do for prescribed burning in the north woods what the bobwhite quail had done for the piney woods of the South. An endangered specie, it depended for its habitat on stands of jack pine that were at a certain stage of development. The invasion of jack pine following the era of holocausts had greatly expanded the range of the warbler, and now that the stands had matured, the range of warbler habitat was shrinking. Prescribed fire was urged as a means of restoring the earlier state in which jack pine had replaced white pine.[74]

The reintroduction of fire to the north woods promised to prove as difficult for twentieth-century foresters as it had been for nineteenth-century settlers. The warbler fire suggested that there were limits to the extension of prescribed burning under the doctrine that fire was good and natural, and the Seney fire recommended that organized fire protection had also somehow lost its bearings, that the management of fire control resources themselves might become a greater problem than the fire. Both debacles, moreover, were the responsibility of federal agencies, not of state organizations, and both underscored the fact that the malaise they epitomized was national, not merely local. That both occurred in the Lake States showed that with the restoration of a "wild" environment—a habitat suitable for desired wildlife and thus required by recreational interests—the potential for large fires was also being restored. In short order the Lake States had once again presented national challenges to modern fire practices—to suppression and prescribed fire both. But there were unbridgeable chasms between these fires and those of the past; the cultural circumstances surrounding the fires had changed no less than their environmental context. The warbler fire was less an important fire than it was a tragic fire, and for all its size the Seney fire was less great than it was merely costly. The sky, it seemed, was filled with slurry, the earth with greenbacks.

FROM FIRE AND AX: PRIVATE AND
EARLY GOVERNMENT FIRE PROTECTION

The most important duty of the superintendent and assistants in
the park is to protect the forests from fire and ax. . . .
—House of Representatives, Report on Yellowstone National
Park, 1886[75]

The county is thickly settled, and when a fire breaks out the
whole neighborhood turns out to fight it.
—W. W. Tully, New Albany, Indiana, 1880[76]

I

Wildland and rural fire protection did not begin with the introduction of
professional forestry. What foresters did, on the contrary, was to adapt rural
fire techniques to their particular ambitions, organize the skills, and give
them a new purpose, even a sense of mission. The use of fire in reclamation
had been predicated on the ability to control fire; a knowledge of fire was as
much a requirement of frontier and agricultural settlement as it was of
urban development. Most settlers took prudent steps to protect themselves
and their property from damage. Farmers, ranchers, loggers, townspeople,
and even railroads plowed and burned firebreaks, organized fire brigades,
developed equipment, and, through basic landclearing as well as through
broadcast burning, reduced fuel loads and transformed dangerous fire environments. In 1880 Franklin Hough conducted a county-by-county survey of
forest and fire conditions through correspondents. The extent of burning,
both controlled and wild, was impressive. But so were the measures taken
to cope with escape fires.

Throughout the country a fairly consistent pattern emerged. Fire codes
and social mores regulated the seasons for controlled burning. In the South,
for example, where woodsburning for fuel reduction was commonplace,
farmers would inform their neighbors, and all would burn around the same
time. A Georgia correspondent assured Hough that "we seldom hear of even
a panel of fence being burnt." In North Carolina, after firelines had been
raked, the pine woods were burned, and "the farmers turn out with all hands
and see to the careful management of the operation." Mississippi settlers
took advantage of roads and streams and plowed lines where necessary.
"Ordinary good sense and caution are all that is required for preventing
damages from these fires." In grasslands like the Dakota prairie, where fires

were considered the "scourge of the country," plow lines and fuelbreaks were routine precautions. The state of Minnesota appointed an official to supervise the burning of prairies in those counties "invaded by grasshoppers." New Jersey enforced statutes that prevented anyone from hindering the owners of meadows and salt or fresh marshes from burning them over.[77]

A careful distinction was made between controlled fires and wildfires. Arkansas farmers often advocated woodsburning, but hunters' fires "set by unprincipled men" caused "the farmers a good deal of trouble and considerable expense in having their fences burned and otherwise."[78] The "otherwise" included the destruction of numerous farms where fuelbreaks were not satisfactorily emplaced. Where structures, fields, or villages were threatened, settlers would readily turn out for fire call whether or not state law gave county fire wardens the power of legal impressmen. In Texas, stockmen organized their ranch hands into fire brigades, and nearby ranchers rushed to common fires. Some 200 men reported to large fires in New Hampshire. When fire threatened turpentine sites in North Carolina, all the naval stores workers responded. In Jefferson County, Missouri, "residents turned out en mass and fight the fire day night until it is completely subdued." Even in hotbeads of incendiarism like Kentucky, prior to heavy settlement "whole neighborhoods" of Shelby County "would be rallied to stay the ravages of the flames, to prevent the destruction of decayed timber and fencing." In Indiana, communities would respond to fire calls to plow firelines, throw dirt, and backfire. Michigan settlers delivered water on wagons and tossed it with buckets. In San Mateo County, Californians responded with wet sacks, corridors of felled trees, and horse-drawn plows. Two watchmen patrolled for fires in Ohio's Lawrence County, and Clinton County witnessed "large gangs of men" control a 1,000-acre fire. A New York newspaper reported that the eastern slope of the Catskills was aflame and that "all the surrounding citizens of Woodstock village have been ordered out to fight the fire."[79]

In 1935 two Forest Service officials in California published a fire history of the state based on old records, especially newspapers. The purpose of the exercise was to put light-burning proponents permanently to rest, but the result was a fascinating cross-section of fire practices in the late nineteenth and early twentieth centuries. Settlers had considerable experience with controlled fire, and when fires escaped, the community could generally be rallied to make an attempt at control. Time and again settlers near sequoia groves rushed to save the Big Trees from fire. Without such response, the State Board of Forestry concluded in 1890, the "scorching . . . would have destroyed half the trees." In Yosemite National Park proprietors of a hotel at Wawona fought fire for as long as two weeks to clean the air of smoke that obscured mountain scenery and made breathing difficult The "line of fire" in Riverside County, a stagecoach line reported in 1874, was "nearly a mile long with a strong east wind." The "settlers," it noted, "were all out fighting fire." In 1877 an immense fire raged in the Santa Monica Mountains and "burned over 1,000 acres destroying a number of dwelling houses. Between four and six miles are now on fire and the residents are exhausted and discouraged." In Placer County a timber fire started from gun wadding; the San Francisco *Examiner* reported that "about 200 miners and ranchmen are now fighting the flames." In San Mateo County (1879) "citizens are already taking measures to start backfires as the burn is one of the largest ever known in the county." The fire burned in chaparral, and a similar burn in Amador County elicited "great efforts. . . . The fire is very near Ione and the citizens expect to be obliged to fight hard tonight to save the town." In 1880 in Mono County "a forest fire broke out near Towle and the Central Pacific sent a force of 1,000 men to subdue it." Near San Gabriel " 'Lucky' Baldwin had 200 men fighting a large fire" that threatened his ranch. When fire raged in Marin County (1881) along a line 12 miles long, the *Chronicle* reported that "the people are gathering from all the adjacent country to fight the flames." The fire continued under a strong offshore wind, "although every available man that can be found has been pressed into the service to subdue it." When large fires broke out in Santa Cruz County the next year "many men" were said to be "fighting the flames. 140 have gone out from here [Santa Cruz] to aid in an endeavor to arrest the progress of the flames. The Santa Cruz Water Company has 500 men fighting here." The 1887 fires were widespread. Near Santa Rosa, flames traveled at better than a mile an hour and claimed the life of one firefighter; in Alameda County "many succeeded in saving their dwellings by plowing around them and keeping watch night and day on the flames"; in Sonoma County "the flames have traveled ten miles in the past twenty-four hours destroying everything in their path. Families are homeless and miles and miles of timber and vineyards are destroyed. Hundreds of men are fighting the fire. The loss will be immense." Elsewhere in the county, a fire "was fought by the mountain tenants whose homes were threatened. It swept through the canyon thick with underbrush and pine trees. Several men barely escaped the fire." In 1888 another fire in Santa Cruz saw "armies of fire fighters . . . organized to give battle—There were over 300 woodsmen and ranchers at work before dark Friday night." In Sierra County (1889) "the citizens of Red Dog saved their little town by backfires," while in Sonoma County "the farmers . . . sent to town for help and a body of men went out to fight the fires." Even intellectuals were not exempt: when a fire threatened Mt. Hamilton in 1891, the "observatory people" were obliged to fight fire for several days. And so it went. Convictions for carelessly handling debris fires—the main cause of fire in settled areas—were almost impossible to obtain. Evidence was hard to come by, and local sentiment was almost entirely on the side of the burners. In a field as in a factory, smoke meant progress. But in the bad years, when fires did escape, settlers and townspeople were ready and able to protect themselves.[80]

It would be grossly inaccurate to assume that because conflagrations burned regularly in the backcountry no provisions existed for rural fire protection. The mere quantity of controlled burning presupposed the ability to regulate fire use and to control escaped fires, and it frequently constituted in itself a fire protection practice. But clearly this ad hoc voluntarism could break down, and it was incapable of coping with the holocausts responsible for most damages. Voluntarism did not generally deal with the irresponsible fire starters—the tramps, migrant hunters, teamsters, and other frontier riffraff—and it could not exert reliable pressure, strategically applied on the fireline. Often, unless danger was imminent, no response was taken at all. As a correspondent in Idaho succinctly put it, "No efforts are made to check them [the fires] except when they threaten individual property, and more than this would be useless." Hough concluded that "as long as fire is half a mile from a farmer's field, he will rest quietly at home; and when it approaches too closely, he may, perhaps, go with a few hands to run it around upon the tract of another. . . . The great trouble is that those owning these lands, when they see a fire started at some distance from their lands, expect that someone else whose lands are nearer will go to extinguish it. Thus the fire is left alone, except by those whose property is in immediate danger, and their efforts are too often not equal to the task." In San Fernando Valley (1878), for example, brush fires threatened grain fields until a hard fight by 70 men "turned the flames in another direction." The brush was left to burn north of Tujunga Pass. In northern New York, "responsible men, who would not think of endangering a neighbor's house by a bonfire, in their garden, think nothing of letting loose their fallow fires into adjoining timber." A California correspondent observed that most fires began on public land or common land, not on land owned by individuals.[81] "Wherever land is occupied and fenced, forest fires are feared and fought. . . . But the unoccupied public domain is devastated and blistered with impunity. . . ."[82] Even when a fire posse was assembled, competing interests and philosophies often retarded a concerted effort. "The means chiefly employed for stopping the progress of the fire," a New Jersey correspondent wrote in disgust,

> was by backfiring on the line of the roads; those nearest the fire being used first, and if that failed, the next. In most cases, however, this was a failure, chiefly because all present and assisting were interested in the saving of their own land, without regard to the interests of neighbors or the interest of the whole. Some would insist that one road be taken, and others would as strongly insist upon another. In the meantime, before the matter could be settled, the fires had already crossed the road, where a defense could be made, and an opportunity that might have proved valuable was lost.[83]

Local fire suppression arrangements were neither binding nor formalized. They did for fire protection what sheriffs' posses did for law enforcement. Nothing more was considered necessary: extreme fire hazards were regarded as only temporary conditions, an ephemeral byproduct of frontier settlement. Hough noted that "it is observed everywhere, that in the first beginnings of a settlement, whether it is a forest region or upon the prairies and the plains, the dangers from running fire is the greatest, and that this gradually diminishes as the region becomes thickly settled and well-cultivated." The notorious fire starters were from that "class of population" in the vanguard of opening up new land, "this unstable and transient class, the first beginners ... generally persons of slender resources ... [who] have little to lose." "Habitually careless and improvident," Hough lectured, "they do not hesitate, where there is a motive and opportunity, to apply fires to lands not their own, for the purpose of improving and extending the range for their cattle, or to clear lands for cultivation, and sometimes to destroy the evidences of their own trespass and depredations." Moreover, during the early stages of their existence, "whole communities regard these fires with satisfaction, providing that they escape personal damages. . . ."[84]

Wildfire would go the way of wildlife and wild tribes as settlers converted wilderness to civilization. Where conditions remained unsettled, as in the Lake States, large fires persisted; where socioeconomic isolation perpetuated a frontier existence, as in the piney woods of the South, frontier fire practices continued; where Indians were isolated in reservations, an older pattern of fire was also preserved; and where land was withdrawn from settlement altogether or given only to regulated uses, as in the national forests and parks of the Far West, wildfires likewise thrived. In the unstable equilibrium of early settlement and on permanent reservations, formal fire protection organizations became a *desideratum*. Their effect was either to accompany socioeconomic reforms by spearheading industrial forestry, as in the South, or to perpetuate essentially primitive environmental conditions, as in the Far West. The moral was that, until fire practices and environmental circumstances reached an accommodation, damaging wildfires would continue. Settlers, however, expected that the normal evolution from frontier to agricultural community would achieve a stable equilibrium. Fire protection was best advanced by continued settlement aimed at complete conversion of cover type: incomplete conversion created sanctuaries favorable to fire; hence, it was not surprising that many of the Lake States holocausts flared out of bogs and marshes where they had burned for weeks.[85]

Until conversion had been completed, conflicting practices could exist side by side, as dictated by migrations and socioeconomic change. In Tennessee, for example, fires were routinely set for improvement of pasturage and hunting, mostly in the mountains. In Washington County such burns were controlled when necessary by a series of backfires. In Loudon and Rutherford counties, however, "public opinion is against firing the woods"; law enforce-

ment was stringent; and "at the first outbreak of such a fire the whole neighborhood goes to work, and makes a quick job of it." Where a people long
accustomed to mild burning—for example, the Poles and Finns—were
introduced into heavy slash concentrations accumulating in the Lake States,
the mix resulting could make a deadly fire regime. Conversely, when a company of West Virginians migrated to the Northern Rockies in the early
1900s, the introduction of southern fire practices was not received favorably,
either by the environment or by prior settlers. Through various legal maneuvers, the intruding mountaineers were successfully removed.[86]

With advanced settlement, fire protection could be expected. As hamlets
became towns, fire brigades were organized, fire wardens appointed, and
fire codes enforced. As more and more range lands were enclosed, stockmen
took precautions against fire. In Southern California "private enterprise"
extinguished a fire in 1896 that had burned 40 square miles of brush in the
San Gabriel Mountains.[87] Railroads likewise took steps to protect their
physical structures, such as trestles, tunnels, and ties, and to conform to
statutes requiring certain fire prevention measures. As early as 1833 Maryland deemed railroads legally liable for damages caused by locomotive fires.
Companies responded with spark arresters, patrols by section gangs during
high fire danger periods, and right-of-way maintenance. But the railroad
magnate shared with the settler the belief that fire protection was only a
temporary expedient at best. New equipment, fuels less spark-prone than
wood or coal, and settlement along the tracks would soon eliminate the irk-
some need for protection. For many areas this scenario was vindicated. But
where the trains passed through lands inimical to settlement, lands logged
but not settled, or lands reserved from settlement, large fires continued. The
Southern Pacific in 1878, for example, established a lookout at Red Mountain in what is now the Tahoe National Forest for the exclusive purpose of
reporting fires threatening its expensive snow sheds.[88] Engineers developed
special tank cars with hoses and pumps to cope with dangerous fires through
forest reserves like the Adirondacks and through logging areas with heavy
slash accumulations. Section gangs were trained in firefighting and were
available to national forests as part of the use permits issued to the railroads.

It would be tempting to deduce from this roll call of early endeavors that
wildland fire protection began under private sponsorship and was later
adopted by state and federal governments. In fact, however, that evolution
was often reversed. Local private measures ultimately depended on the laissez-faire proposition that the progress of civilization itself would eradicate
the scourge of fire as it would disease, locusts, wolves, and poverty. Certainly
there was little to recommend the extension of frontier fire control into
remote, unsettled, wilderness areas. Extraordinary measures were simply
not needed, and forest fire protection on a large scale in unreclaimed forest
areas was impractical in technical terms, unjustifiable in its economic costs,

and undesirable in its consequences for the environment. Many frontiersmen lived in a high fire regime and wanted to keep it; the rest knew that the inexorable evolution from settlement to civilization would obviate the need for anything more than what tradition prescribed.

II

The examples of New York State in its Adirondacks and Catskills reserves and of the Army in Yellowstone National Park first demonstrated that fires remote from settlements could be attacked and destroyed, that the safest strategy was to extinguish all fires, however removed, and that fire protection could come at reasonable costs. The Province of Ontario, at exactly the same time, proved the same propositions across the international border.[89] The concept of fire patrols—or "fire ranging," as it was commonly termed—was not new; the AFA had proposed it in 1882, for example.[90] But until some agency was made responsible for the active administration of these largely vacant public lands and not merely for the supervision of their disposal, fire protection was neither desirable nor feasible. The New York and Yellowstone experiences, in turn, demonstrated the technical possibility of fire control and established an exemplar for the management of the national forest reserves. At the same time, professional forestry emerged to infuse forest fire protection with its intellectual, moral and political muscle. The old justification for fire suppression was the self-evident hazard to fields and structures; the newly reserved land would by definition be void of settlement and demanded other rationales. Forestry legitimated the concept of fire suppression even in unsettled lands as a means of preserving beneficial forest influences.

New York's commitment to fire protection went back at least to 1743, when it adopted a statute to organize the counties of Albany, Dutchess, and Suffolk and the manor of Livingston. Anyone who discovered a wildfire could "require and command" all or any of the neighboring and adjacent inhabitants "to aid and assist him" in suppressing the fire. Refusal, neglect, or tardiness could result in a fine. Fifteen years later, the provisions of the act were expanded to cover the entire colony. In 1760 a fire warden system was inaugurated; the towns, manors, and precincts of Albany and Ulster counties could elect "fire men" to supervise fire control efforts, analogous to the election of a sheriff to replace local vigilance committees. The system was extended gradually to other counties as settlement pressures dictated. In the interim local and state fire codes were enacted to regulate fire practices for hunting, pasture improvement, and field burniing.[91]

When New York created the Adirondacks Reserve in 1885, followed by one in the Catskills, it created a new fire protection quandry, one that paralleled the federal dilemma in the public lands of the West.[92] Ordered to preserve the lands from the ravages of ax and fire, the state expanded its existing warden system to cover the lands absorbed into the preserves. The state would pay half the costs incurred during fire suppression, with the affected town or county contributing the rest. In fact, the 1885 enabling act extended the fire protection responsibility of the Forest, Fish, and Game Commission to forested lands throughout the state. A chief fire warden was appointed to superintend the organization. But there were serious weaknesses in this arrangement. For one thing, the warden system was only activated after a fire began; it could do nothing to prevent fires or enforce codes. For another, the system was really effective only around villages and towns where manpower could be assembled. It was thus based on a contradiction: to work, it required settlements; but the goal of the preserves was to maintain a wilderness environment. In effect, the system was only a more formal and better financed version of frontier protection practices. Its deficiencies became painfully evident after the introduction of railroads through the preserves in 1888. Locomotives belching fire transported hunters and fishermen whose camp and smudge fires further aggravated the regional fire loads without adding to suppression capabilities. Logging on private lands enclosed within the reserves added unhealthy tonnages of untreated slash.

The results were predictable enough: New York experienced fires not unlike those raging across the Lake States under similar circumstances. In 1903 and again in 1908 the fires reached catastrophic proportions. The report of H. M. Suter, a U.S. Bureau of Forestry agent sent to observe the 1903 conflagration, offers a striking contrast to a similar analysis of the terrible 1902 fires in the Pacific Northwest. In Oregon and Washington, aside from the hit-and-run efforts of settlers to save their farms and the work of rangers hired for fire patrol by the GLO, little organized control was attempted, even though many of the fires burned on the Cascades within sight of Portland. In upstate New York, on the other hand, efforts were undertaken to control a major holocaust that burned more than 600,000 acres. As Suter noted, "had there been no such organization the losses of this year would have been much more severe and extensive." Firefighting continued for nearly six weeks without cessation. It was "only the timely appearance of heavy rains, beginning June 7, that brought the fires under control," Suter observed with a mixture of dismay and admiration. "Hundreds of men dropped their tools that day and slept the sleep of utter exhaustion. Another week of strain would have beaten down all defense."[93]

At the fires' height some 6,000 men were allied against them. Hundreds of Italian laborers were hired and freighted in from New York City. Railroads whisked city fire engines to towns and lodges threatened by the flames. After the major fires broke out along their tracks, the railroads brought their performance up to code: spark arresters were installed, ash pans and dampers inspected, freight trains reduced to half their former tonnage to discourage the expulsion of heavy sparks, and patrols along the track scheduled. Special tank cars outfitted with pumps, hoses, and water supplies roamed endangered sections of the line and assisted local control units. On the ground wardens adjusted their attacks to morning and evening, when the fires burned less vigorously. Firelines were scratched or dug down to mineral soil. Where water was plentiful, the line would be wetted down, then defended with shovels of dirt or sand. Grass fires were swatted out with branches of spruce or balsam. Lines were constructed by plows where soil permitted. And when direct methods failed, backfires were set. In their "gallant fight" men worked 15 hours a day for many consecutive days, then collapsed from complete exhaustion or were prostrated by sickness. Individual crews dug dozens of miles of trench. Some $175,000 was spent in the control effort. All in all, however compromised the end result, it was a surprisingly modern endeavor. It showed that the basic techniques of fire suppression were well understood and that all the proposition required was money, purpose, and a system more aggressive in the enforcement of codes and in the patrol for fires not close to settlements.

The whole fire spectacle was repeated in 1908, but this time all parties clamored for better protection, and the state fire organization was thoroughly overhauled. The forest reserve was divided into four fire districts, each with a supervisor, observation towers, and paid patrolmen. The land itself, not merely the towns within it, was made the basis of fire control. Reforms were also promoted in the field of fire prevention: locomotives were required to burn diesel fuel, state inspectors ensured railroad compliance with fire codes, the state enacted tougher statutes against careless campfires, and the legislature empowered the governor to close the reserves during fire emergencies. At this time, too, Maine fashioned its famous Maine Forestry District, in many ways analogous to the New York organization; private timber protective associations were organized in the Far West, Pennsylvania, and New Hampshire; and the U.S. Forest Service was on the brink of major reforms. If no longer unique, the New York plan had nevertheless achieved its purpose: it had shown how, in broad terms, large forests under the administration of state governments could be guarded from fire. It was no accident that both Maine and the Lake States looked to New York for expertise and inspiration.

Even as New York first wrestled with the implications of its new arrangements, the modern style of wildland fire protection was made operational in the Far West. In 1883 Congress authorized the Secretary of the Interior to request assistance from the War Department in the administration of Yellowstone National Park, and in early 1886 the Army was asked to assume control. Depredations and fires only highlighted the impotence of a civilian administration hamstrung by the absence of legal powers, funds, and manpower. "To protect the forests from fire and ax" had been a prominent charge to park officials; it was the one the Army inherited.[94] For the first time in the United States fire control was practiced on wildlands. One agency was responsible for fire protection, and it was not dependent on crews manned by local volunteers. Precautions were taken to prevent fires from occurring, rules and codes were enforced, and patrols were dispatched to seek out and extinguish fires throughout the park. However inadvertently, the Army not only launched federal fire protection but also demonstrated conclusively the techniques by which all wild and forested lands could be managed. The difference between fire protection in the eastern and western United States is vividly contrasted in the experiences of the Army at Yellowstone and of the New York fire warden system in the Adirondacks. It was, in a sense, a lesson that federal fire agencies have never forsaken.

The Army assumed control in August 1886. A tour of the park quickly followed, with detachments of troops stationed at six stations. Then Captain Moses Harris, the new superintendent, with the remainder of M Troop, First U.S. Cavalry, began to extinguish fires. The civilian superintendent just prior to the arrival of the military had frantically wired the Secretary of the Interior: "Three large fires raging in the Park beyond my control."[95] The civilians had tried before to control fires, mostly around developed and scenic areas where tourists left campfires. There are records of tour groups voluntarily suppressing fires in the course of their visit. But the Army's position was different. Many of the fires were the result of "unscrupulous hunters" who torched park lands in an effort to drive game outside the legal boundary of the park. "The Park," Captain Harris soon learned, was "surrounded by a class of old frontiersmen, hunters and trappers, and squawmen" who had only disdain for park rules.[96] Not a few of the fires confronting the Army were deliberately set simply to taunt and embarrass the civilian administration. Moreover, without roads and developed campgrounds, tourists roamed at large, leaving abandoned campfires to flare up as weather conditions changed. The suppression of fires was thus intimately connected with the establishment of Army administration: fire suppression was a visible, material, and symbolic expression of Army determination to rid the park of destruction and vandalism of all sorts, to regulate tourism, and to confront and remove the lawless class of poachers. The regulation of people and the control of fire were complementary duties. It is ironic that the establishment of wilderness meant the expulsion of frontier elements. But so it did, and fire was one manifestation of this necessity: Yellowstone was but another illustration of the maxim that wherever the frontier lingered, fires persisted. Like almost everyone else at that time, the Army made no distinction between man-caused fires and lightning fires: their orders were to protect the park from fire.

Over 100 fires broke out in 1888, though burned acreage was kept to an astonishingly small 5 acres. Some 61 fires were suppressed the next year, and Captain F. A. Boutelle, the new superintendent, spirited off a series of complaints to the Secretary of the Interior about inadequate firefighting equipment. When his telegraphs requesting funds for the purchase of 20 axes and 20 rubber buckets went unanswered, the disgusted captain—who had been "personally fighting fires for some days and nights"—explained his predicament to quizzical visitors. A "Mr. Leavis of Pennsylvania" donated the money for buckets out of his own pocket. In 1889, 70 fires burned, some over large acreages. Boutelle described the persistence with which a fire above Madison Canyon was fought. "It took me just an hour to walk around it in making an examination with a view to determining whether it was possible to do anything with it. Concluding that it was worth the trial, I called up all the available men—29 in number—and by night a clearing was made entirely around the fire." The fire was "surrounded and controlled.... High winds prevailed almost every afternoon while this fire was burning, and at times the flames would jump the cut-off and get beyond control, but as soon as the winds subsided another cut was made, and at the end of the three weeks the fire was out."[97]

The fire problem became so all-absorbing that Boutelle turned to means of prevention. Organized campgrounds in safe locations were one result; another was increased patrols along frequented routes; still another was rigorous enforcement of a regulation that required expulsion of any visitor who abandoned an unextinguished campfire. A few expulsions a year were found to be adequate as a deterrent. By the mid-1890s these aggressive measures had, in the words of the current superintendent, "brought about the particularly good results of which we can boast."[98] As the Army assumed administration over other national parks, it brought its fire experiences with it; in the Sierra Nevada, for example, the problem of fire and poachers was replaced by fire and herders. The successes were sufficient to impress the National Academy of Sciences, to send soldiers for temporary duty in several national forests at the request of the Secretary of the Interior (prior to 1902), and to have troops dispatched to fires as needed. In 1905, for example, some 5,000 soldiers poured out of the Presidio in San Francisco to fight a fire on Mount Tamalpais.

But whereas the national parks that the Army administered until 1916 were opened only to tourists as pleasuring grounds, the forest reserves that were given to the GLO were to be utilized in accordance with strict regulations after 1897. For its first year of operation, until fiscal year 1899, the reserves had no appropriation at all for protection and administration. Yet almost immediately, the GLO campaigned as best it could against the fire menace. Special agents and forest supervisors were admonished that fire protection was among their foremost assignments and that they should take emergency action to control reported fires. The commissioner underscored this message in his annual report by observing that "depredators denude the

public domain of much of its timber wealth, but fire is the greatest enemy."
Posters and instructions were promptly distributed. In 1898 Commissioner
Bingham Herman wrote to his field men that "a forest officer upon receiving
information that a fire is in progress which needs his attention should report
promptly to the locality and use all reasonable means to extinguish the
same. . . . You will direct all forest officers under your charge to endeavor
to enlist public sentiment in favor of the Government's efforts to protect the
public forest from fire, and secure their cordial cooperation in bringing
offenders to justice." In 1900 the commissioner urged enactment of "immediate legislation to place all unreserved forest lands under the watchful care
of a disciplined ranger and fire force." But the hapless GLO encountered
only congressional apathy without and stubborn bureaucratic inertia
within.[99]

In 1901 Secretary of the Interior Hitchcock issued an outline of the principles and practices that should govern the supervision of the reserves.
Among them was the unblinking charge that "the first duty of forest officers
is to protect the forest against fires. Rangers should be ordered, as they are
now forbidden, to leave their own beats when necessary to assist in extinguishing fires on adjoining beats. . . ."[100] Hitchcock also created a Forestry
Division within the GLO to advise and coordinate the administration of the
reserves. Filibert Roth, E. T. Allen, and several advisors transferred from
the Bureau of Forestry, and the results—despite the red tape, centralization,
excessive legalism, and politicized nature of the GLO—were such that
Allen enthusiastically remarked that "the improvement was greater than
would be thought possible in so short a time."[101]

In 1902 a *Forest Reserve Manual* was published to consolidate the
administration of the reserves and in September 1902 a circular described
how patrols should be conducted. The fire situation improved measurably.
The next year the acting commissioner of the GLO circulated a general
letter to all rangers congratulating them on their vigilance and success in
the previous year. Supervisors were admonished that they "cannot too ear-
nestly impress upon the rangers the necessity for a strict watch for forest
fires during the danger season. You must not confine your ranger force too
closely to cutting trails or attending to other business to the subordination
of the fire question." Rangers were exhorted to post fire warnings liberally,
to remind parties granted permits on the reserve that they were obliged to
aid in fire suppression, and in general "to give this matter your best thought
and attention." In the absence of a "satisfactory explanation supervisors and
rangers will be held personally responsible for any fire that is allowed to
escape." And in a passage that illustrates perfectly both the GLO's determination and its naiveté toward the problem it confronted, the acting commissioner pondered: "It is not understood why forest fires would get away
from the rangers, or rather why they do not find them and extinguish them

more promptly. It seems reasonable that a ranger provided with a saddle horse and constantly on the move, as in his duty, should discover a fire before it makes much headway. This statement is made knowing that some of the ranger's districts are extremely large."[102]

From 1902 onward, the GLO made strenuous efforts to control fires. Traditional lore has dismissed the whole history of the reserves prior to the Transfer Act as one of bungled opportunities, technical ineptitude, and outright corruption. But within the limits of its appropriations, the GLO had its successes. To William Cox, who investigated the 1902 fires in the Northwest for the Bureau of Forestry, it was obvious that "commendable work has been done by the government rangers in the forest reserves, and the absence of serious fires in them should serve as an object lesson to the states of Oregon and Washington."[103] Though nearly a million acres eventually burned in those fires, only a portion of the reserves was damaged. Ranger patrols and organized crews of settlers and permittees kept fire losses to a respectable level. On the Pecos Reserve in the Southwest, a young ranger just assigned to a district in 1903 sighted two smokes. Ranchers had already seen one smoke and were gathering tools when he arrived to collect them. They had the first fire controlled by the next morning. Immediately, without any sleep, the ranger rode to the village of Agua Negra, rousted out the alcalde, and asked for help. The alcalde refused, but the ranger persisted, eventually changed the alcalde's mind, collected 15 to 20 men, and attacked the second fire. It was controlled overnight.[104]

By the time of the Transfer Act, the GLO, as S. T. Dana acidly commented, "ended its administration of the forest reserves, as it started that administration, without a single professional forester in its employ."[105] True enough. But it was hardly true that foresters were the obvious, much less the only, candidates to administer the reserves. Insofar as the chief duties in the reserves were law enforcement and firefighting, the military was probably better equipped to handle the federal lands. Many of the actual rangers hired on the districts after the Transfer Act were the same sort—and in many cases were the same people—employed by the GLO. Insofar as the reserves were established for watershed, the U.S. Geological Survey and its offspring, the Bureau of Reclamation, were better candidates for administering them. Nor was professional forestry especially equipped to cope with prospects of controlled burning; it was as intellectually retarded in its acceptance of prescribed fire as a silvicultural tool as the GLO was administratively shackled as a land management bureau. Moreover, forestry's vaunted technical skills amounted to little in the actual practice of fire control on the reserves. E. T. Allen blasted the GLO because "general instructions to subordinates were few and brief, and no one was able to advise his subordinates. . . ."[106] Yet, following the transfer a fire guard on the Gila National Forest remarked in 1909 that "all the instruction I had was to go up there and look out for fires—and put them out."[107] He even had to bring his own tools and gear. Things were no better at Coeur d'Alene, Idaho. A guard recalled how he was assigned to patrol the railroad tracks for fire. "I had no previous training whatever in fighting forest fires and only instructions from Ranger Haun to put out any fires that might occur in my district. . . . About the only directive given me by Haun was that results were all that he was after, and that the guard the previous year had not gotten them so he had had to fire him."[108] So much for forestry as a storehouse of technical expertise.

Forestry as an intellectual discipline could only lend theoretical support to the generally accepted proposition that wildfires ought to be controlled. The practical skills and strategies for controlling fires were adapted from the fire practices of the reclamation. In 1900 the only real models of efficient wildland fire protection in the United States were the fire warden system in New York and the military patrol system in the national parks. Forestry was essential to neither.

Admittedly, however, the GLO had acute problems. Some stemmed from Congress, which offered only stifling, token appropriations with which to administer the reserves. Until 1905 rangers had no power of arrest without a court warrant, a situation that made a mockery of attempts to enforce regulations. The land laws of the 1870s all but invited fraud and left the GLO both vulnerable and seemingly corrupt. Other problems plagued the GLO from within. Giving such a bureau control over land management was like giving the customs office in the Treasury Department control over Alaska—which was also done at this time. Against its own inertia, and in defiance of 100 years of service in disposing of land, the GLO suddenly attempted to manage it as an economic system. From 1902 onward, partly out of desperation, partly because of Pinchot's prodding, it turned to the Bureau of Forestry for advice on timber and grazing matters. The Morris Act of 1902 further committed the Bureau to practical investigations for the Interior Department, this time on the Chippewa Indian Reservation in Minnesota. The Department of the Interior turned to Pinchot's staff for working forestry plans for a number of its reserves. Meanwhile, land frauds scandalized the GLO, and in 1903 a Public Lands Commission (the first since 1879) was appointed to review the impossible legal morass of land legislation. All of these circumstances seriously compromised the ability of the GLO to administer the vast reserves. But the greatest weakness of the GLO was not its legal confinements but its sheer lack of vision. It saw itself as a mere caretaker of the land, not as an ideal, a movement, a public utility, and a political power. The greatest difference between the GLO and the Bureau of Forestry was not professional stature or technical training but this vision—one that the zealous Gifford Pinchot had in abundance. So it was to him in 1905 that the reserves were given.

The drive for organized fire protection soon extended to private lands as well. The successes on large public reserves demonstrated the economic plausibility of fire suppression. Companies that before had undertaken limited defensive measures on their own lands became more aggressive and, more significantly, banded together or cooperated with state and federal organizations. In 1902, for example, the Lehigh Coal and Navigation Company of Pennsylvania actively assisted the states in prevention, patrol costs, and the establishment of lookout towers. The fires of 1903 in Maine led to ad hoc patrols by worried timber owners. In 1904 the Stockman's Protective Association of California organized for mutual fire suppression in Alameda and San Joaquin counties; the Linn County Fire Patrol Assocation in Oregon combined timber owners and agriculturalists; the Pocono Fire Protective Assocation in Pennsylvania was established; and four protective associations were launched in various parts of Idaho. Across the country the logic of private timber protective associations dedicated to cooperative assistance became more compelling. But the real breakthrough came in Idaho. The 1902 fires had rightly alarmed timbermen. Several years of debate resulted in the formation of the Clearwater Timber Protective Association in 1906. The idea escalated. After the Idaho Forest Law of 1907, four associations existed in the north and one in the south of Idaho. The "Idaho idea," as it became known, spread throughout the logging centers of the country. Within 10 years 40 formal organizations existed in California, Oregon, Washington, Idaho, Montana, Michigan, Wisconsin, New Hampshire, Maine, Pennsylvania, West Virginia, and Kentucky. Dozens more were set up in the South during the 1920s; by 1933, Georgia alone boasted of 87. In part, the movement manifested the simple logic of consolidation in the face of common danger; in part, it was a response to legislation—to compulsory patrol statutes passed in Oregon and California, to the absence of adequate state organizations, and to the Weeks Act. The private associations, through the state forester as a liaison, could qualify for federal assistance. At a later date, in states with feeble fire organizations, the private protective associations qualified the land for inclusion in the CCC program.[109]

Most of the associations faded with time, and state and federal organizations assumed their burdens. But some have endured, and in their day— especially in the Northwest—they were both an important forum and a critical catalyst for later developments. The most famous is the still flourishing Western Forestry and Conservation Assocation. The association grew out of a 1909 fire conference attended by representatives of four states and the Forest Service to review the experiences of the four forest fire protective associations of north Idaho. The meeting was chaired by George S. Long, manager of the Weyerhaeuser Timber Company and president of the Washington Forest Fire Association. Out of the discussions came the Pacific

Northwest Forest Protection and Conservation Association, a consortium of the private organizations in the region. When the California Forest Protection Association joined, the name was changed to the Western Forestry and Conservation Association. By 1912 some 16 associations were included, and British Columbia also became a partner.

The association held its first meeting in 1910. E. T. Allen, a district forester with the U.S. Forest Service, was elected as permanent secretary, and Long, speaking for the Board of Trustees, instructed Allen "to find out what is the right thing to do and then go ahead and do it regardless of whose interest it may effect."[110] Allen and the association followed that courageous advice with a campaign of public education, legislative lobbying, and mutual fire assistance. The association's annual meetings became a national sounding board for all aspects of fire control. The association was responsible for persuading President Taft to order federal troops into the 1910 fires. It was prominent in publicizing the 1910 holocausts, in promoting early fire prevention campaigns, in advocating reform of forest and fire codes, and, above all, in extending cooperative fire protection. It was through the Western Forestry and Conservation Association that such basic equipment as the Osborne fire finder and the pulaski tool became widely known and that W. B. Osborne's *Western Fire Fighter's Manual* achieved publication and wide circulation. And the association successfully prodded the Weather Service to begin fire weather special forecasts.

The "right thing" in 1910 meant fire control, but the organization soon expanded its interests. In 1924 a research department was created to cope especially with the woefully confused forest tax structure. In the early 1930s an equipment research committee encouraged numerous improvements in design and distribution. Like forestry itself, achievements with fire sustained the association in its endeavors to expand into other conservation and forestry fields. Yet its historical ties to fire have remained. A permanent committee exists to review fire topics annually, and the association serves as an umbrella organization for a spate of "fire councils"—cooperative interstate institutions dedicated to coordinating and disseminating fire research and policies. All are located in the West and are in a sense analogues of the interstate fire compacts typical of the East.

Not all the fire protective organizations devolved from forestry. Particularly in California, associations arose to protect range lands, watersheds, and settlements. In 1914, for example, Forest and Home Protective Associations were incorporated for Santa Cruz and Placer counties. The Tamalpais Fire Association in Marin County was established after the 1913 fire to protect some 40,000 acres; the association relied on volunteer crews of up to 500 men and donations from property owners of 10 cents an acre. In 1917 the California legislature created the Tamalpais Forest Fire District, the first such district in California, which allowed the county to levy taxes for its support. In 1922 the Contra Costa Hills Fire Protective Association was organized in Berkeley to promote better fire control around the town. Foresters at the University of California were prominent in the effort, but their warnings went unheeded until the conflagration of the next year.[111]

Nor were the foresters always in agreement over the objective of such organizations. Most were created by large timber owners or consortiums who wished to protect their substantial investments in mature timber. Until tax reforms in the 1920s created an incentive to nuture a second growth, there was little interest in fire protection for reproduction. The federal forest agencies, however, took the opposite stance, arguing that the protection of future crops was more important than merely the current harvest. Hence, almost immediately the value of various strategies of fire control was debated. The growth of timber protective associations advanced stride for stride with agitation for light burning. Private owners in California and Oregon liked the concept; it seemed cheaper; it appeared to be as effective as fire patrols; the smoke itself was deemed valuable; and after all, it was the "Indian way." Foresters were horrified at the prospect of fire protection organizations—for which they had campaigned so long—subverted to principles with which they could not agree. The internecine warfare that resulted was especially bitter. Having mobilized public opinion in favor of fire control, foresters were aghast lest the public apply it in the wrong forms or to the wrong goals. Pinchot himself warned that "to achieve this or any other great result, straight thinking and strong action are necessary, and the straight thinking comes first."[112] Forestry in the early twentieth century may have had little to contribute to the technique of fire control, but it was confident of its straight thinking as to the ultimate objectives.

Soon after it advised the struggling GLO on the administration of the reserves, the Bureau of Forestry began a cooperative program with private timber companies in the South and in California. The Bureau provided working plans that would bring the companies into conformity with the best practical wisdom of the day. Naturally, fire protection was built into the design of these plans. Two in particular achieved notoriety. The Diamond Match Company study and the McCloud Lumber Company study, both in northern California, became centerpieces in the smoldering debate over light burning. The plans were intended as examplars that could motivate and furnish models for surrounding lands, and they contributed to the growing conviction that wildland and forest fire protection was both possible and economical.[113]

By 1905, however, the Transfer Act had given the Bureau a much vaster stage on which to perform. While continuing its liaisons to private associations, the Forest Service directed its unquenchable zeal to the protection of the national forests. The *Use Book,* a revision of the *Forest Reserves Manual,* stated unequivocally that "officers of the Forest Service, especially Forest Rangers, have no more important duty than protecting the Reserves from forest fire."[114] Nonetheless, until 1910 the state of fire protection on the reserves was not vastly different from that under the GLO. For one thing, manpower was hopelessly inadequate. In 1905 George Vanderbilt employed more people on his forestry estate at Biltmore in North Carolina than the Forest Service had for the entire country. Of the 734 personnel available to the Service, more than a third were ensconced in the Washington office. An average reserve had a million acres and a staff of 8, including clerks. For some time Forest Service fire guards and rangers were as dependent on permittees, settlers, and railroad section gangs for manpower as the GLO had been before them. "The assignment to fire duty was one of trials and terrors of a forester's life," one veteran of the northern region recalled. In words that the troops at Yellowstone and Yosemite could readily echo, he concluded that "most forest work was a pleasure, but fire fighting was a nightmare."[115]

Given the circumstances, one could hardly disagree. A fire guard in Montana during 1908 patrolled between Velcour and Libby "on foot, work trains, freight trains, hand cars, and sometimes on self-propelled speeders." His equipment consisted of a canteen, canvas water bucket, shovel, and carrying case in which he kept some food.[116] But he was fortunate. A fire off the tracks or road might take days to reach. In Southern California one fire burned for three days while the forest supervisor and a fire guard struggled to reach it. If he discovered a small fire, it was easier for a ranger to extinguish it himself than to try to find help.[117] Where permittees or settlers were organized, they could be enlisted for larger fires, but the absence of radios and telephones made it sometimes difficult to round them up.

Rangers were instructed to extirpate fire, and they made a gallant attempt. A diary from the northern region described the rather fantastic efforts made by one desperate ranger who found himself on a fire without any tools. He badly skinned his knees reaching over a cliff, burned and scabbed his hands on the fireline, and endured pain and frustration for two days before at last suppressing the fire with nothing but his fingers, feet, and rocks. "Damned if I'll ever fight fire with my bare hands again," he jotted in his field diary.[118] But he *would* fight fire again. And it was this determination, perhaps more than the additional manpower, equipment, trails, and plans that the Forest Service committed to fire control, that actually distinguished its early management from that of the GLO or its private cooperators.

By 1908, conditions were improving. Systematic fire plans were prepared, appropriations increased available manpower, and an act of May 23 of that year allowed for deficit spending to cover the cost of fire emergencies. But the Armageddon of fire control loomed ahead with the great fires of 1910. In retrospect it is hard to imagine a conflagration occurring at a more critical instant. Forest conservation seemed at a crucial divide. Gifford Pinchot had been dismissed by Taft earlier in the year. Charles Van Hise published *The Conservation of the Resources of the United States,* and the National Conservation Commission (established by the 1908 Governors' Conference) released its report, which included a lengthy special study of forest fire. Congress scheduled hearings to investigate the entire spectrum of national conservation policies. Forest Service arguments about forest influences on flooding were under sharp attack by the Corps of Engineers and the Weather Bureau. With the Ballinger-Pinchot brouhaha malingering viciously, the conservation movement threatened to splinter into quarreling factions. "The United States," Pinchot thundered after his firing, "has already crossed the verge of a timber famine so severe that its blighting effects will be felt in every household in the land."[119] For fire control the summer of 1910 was no less critical. The light-burning controversy finally flamed into public forums. Frederick Clements published his *Fire in the Lodgepole Pine Forests,* arguably the beginning of fire ecology in the United States and a milestone in the theoretical scientific justification for fire protection. The fires of 1910 themselves left fire control organizations in shambles and made a mockery of proclamations that only tough government foresters could lance the Dragon Devastation.

And perhaps not the least in this confluence of events, the American philosopher William James published his celebrated essay "On the Moral Equivalent of War." It was printed twice, first in August 1910 at the time of the fires and again afterward. Writing within the memory of the Civil War and in the shadow of Social Darwinism, the pacifist James admitted that the martial spirit was inexpungable, that "war is the romance of history," and that "militarism is the great preserver of our ideals of hardihood, and human life with no use for hardihood would be contemptible." The strenuous life, so popular a theme in the era of Theodore Roosevelt, was worth preserving in social as well as in natural communities. "The martial values," James conceded, "although originally gained by a race through war, are absolute and permanent human goods." The preservation movement was one side of the equation; it sought to maintain the primitive struggles of nature. The other side was social. "If now—and this is my idea—," James wrote inspiredly, "there were, instead of military conscription a conscription of the whole youthful population to form for a certain number of years a part of the army enlisted against *Nature* ... numerous other goods to the commonwealth would follow." The participants "would have paid their blood-tax, done their own part in the immemorial human warfare against nature; they would tread the earth more proudly, the women would value them more highly, they would be better fathers and teachers of the following generation." Not coincidentally, James directed his paean of the strenuous life to the youth of the affluent and educated—precisely the type attracted to the settlement houses of Jane Addams, to the Rough Riders and the Boone and Crockett Club of Teddy Roosevelt, and to the Yale Forestry School of Gifford Pinchot.[120]

James said nothing about firefighting, and it is more than doubtful that many a patrol ranger or fire guard read his essay—or even knew of him. But both existed in the same climate of opinion, and nothing could better depict the cultural impact of the 1910 fires than their simultaneous appearance with "On the Moral Equivalent of War"—for such was forest firefighting to become. The technique of organized patrol and fire prevention worked out by the Army in the national parks was not the only point of comparison between the troopers and their civilian counterparts in the national forests. In an intangible but utterly real way, firefighting accrued to itself more than an analogy to military policing action, or a militarylike espirit de corps, or a military organization: it became a moral equivalent of war itself. Like national defense, fire protection became more than a mere question of economics and evidence. The Forest Service felt itself challenged from the outside by local and national politics and from the inside by fire. And like a country in a time of perceived national threats, the Forest Service—almost a nation itself with its land management mission—turned to its own paramilitary branch, fire control. An arduous, dirty job became improbably noble and even glamorous. The conservation crusade that the charismatic Pinchot helped to inspire found in forest fire a worthy foe, and in fire control the young crusaders who staffed the Forest Service discovered a suitable arena for the conduct of the strenuous life.

5 THE HEROIC AGE

I preach to you, then, my countrymen, that our country calls not
for the life of ease but for the life of strenuous endeavor. . . .
Above all, let us shrink from no strife, moral or physical, within
or without the nation, provided we are certain that the strife is
justified, for it is only through strife, through hard and dangerous
endeavor, that we shall ultimately win the goal of true national
greatness. —Theodore Roosevelt, "The Strenuous Life," 1912[1]

The martial type of character can be bred without war. Strenuous
honor and disinterestedness abound elsewhere. . . . The only thing
needed henceforward is to inflame the civic temper as past history
has inflamed the military temper.
—William James, "On the Moral Equivalent of War," 1910[2]

The Northern Rockies fires of 1910 left a burned swath across the memory
of a generation of foresters, not unlike the effects of the Great War on the
intellectual class of Western civilization. In the summer of 1910, 5 million
acres burned on the national forests, 3 million in Idaho and Montana alone.
The timing of the fires was exquisite: no event could be better programmed
than the conflagrations that occurred between Pinchot's stormy dismissal
and the passage of the Weeks Act. The holocausts gave fire protection an
overbearing role within the U.S. Forest Service and brought the Service
unexcelled power in the field of national fire policy. As Elers Koch blandly
understated, "The 1910 forest fires in the Northern Rocky Mountain
Region is an episode which has had much to do with the shaping of fire
policy not only in that region but the whole United States."[3]

By any standard, the fires were impressive. But their historical impact far
exceeded the values at risk. Neither industrial logging nor agricultural settlement had yet reached the Northern Rockies. Most communities and
many homesteaders were the residue of past mining days. Railroads crossed
the mountains, but with nowhere near the density typical of the Lake States,
Northeast, or South. The salient trait of the landscape was its remoteness:
it was vast, rugged, virtually inaccessible, and much of it had been withdrawn from the public domain before real settlement could arrive. There
was little state apparatus to cope with fires, and, except in portions of Idaho,

little more private machinery. The bulk of the timberlands lay in forest reserves, and what fire suppression machinery existed there rested in the hands of the U.S. Forest Service. For the first time the Forest Service could not blame ruthless loggers, misplaced agriculturalists, enfeebled state organizations, or an inept GLO. The responsibility was entirely its own.

And for the first time the Forest Service had to confront the reality of remoteness and inaccessibility. Settlement had always been considered the bulwark of fire protection; the large fires of history were only typical of the transitional stage between wilderness and civilization. Yet the very success of the Forest Service in encouraging reserves meant that it had inadvertently created a fire problem with few eastern precedents. Its lands often passed directly from wilderness to national forest. The Service could not rely on communication and transportation systems developed in the natural evolution of settlement. Networks would have to be created specifically for fire control, and in many cases invented. Rather than depending on local settlers to man fire brigades, special fire crews would have to be formed. Rather than integrating fire protection closely with logging, grazing, and settlement patterns, fire suppression would necessarily become a preliminary to future forest management practices. Fire control opened up the lands. It was fire that dictated early timber management, which was often restricted to fire salvage sales and replanting on burned sites. The mountains that contributed to the inaccessibility of the region also introduced new variables into fire behavior, fuels distribution, and wind fields. For the first time lightning fires dominated the scene: man-caused fires stayed close to settlements, roads, and railways; lightning pounded the vast backcountry, igniting sweeping holocausts through fuels unbroken by landclearing.

The 1910 fires brought home the magnitude of the fire problem as nothing else could. All operations of the Service's Northern Region were subordinated to its solution, and only California contributed so heavily to the early national debate about the demands and limitations of fire policy. Of the five Forest Service chiefs after Pinchot, four wrote long articles about the need to improve fire control after the 1910 debacle, and two, William Greeley and Ferdinand Silcox, were, respectively, the regional and assistant regional rangers for the area at the time of the fires. When light burning was proposed in California, it was countered by the spectacle of fire in the Northern Rockies. When prescribed fire was advocated in the South during the 1930s, it was pitted against the 250,000-acre Selway fires from the North. The Forest Service's internal debate over the role of fire control in wilderness began with Koch's lament on the passing of the Lolo Trail; its experiment in permitting natural fires to burn came in the Selway-Bitterroot Mountains. When the SAF assigned a task force to evaluate fire policy in 1970, the group focused on the Northern Rockies. The acceptance of the total mobility concept and the promise of imminent changes in fire policy were announced by the Forest Service in 1974 at a national conference held, appropriately enough, in Missoula, Montana.

Yet all this attention given to the Northern Rockies is something of an anomaly. Other regions have higher fire loads. Most have higher values at risk. For many, fire protection is demanded by the public. Other regions have been seared by holocausts with greater losses of life and property. California and the Pacific Northwest were more likely candidates for a commanding role in establishing national fire policy. But a combination of environmental circumstances and historical accidents isolated the Northern Rockies as an arena for demonstrating the technical practicality of fire control.

That test began in 1910. For the first time, in a concerted way, fire suppression went on the offensive. The inhabitants chose to fight, not flee. For the first time firefighters dominated the casualty lists: of an official death toll of 85, nearly all were firefighters, and 100 others were hospitalized. The anniversary reunions of the great burn were attended by surviving firefighters who had steadfastly pursued the fires into remote canyons and rugged mountain tops, not settlers who had survived by hiding in root cellars or bogs. Confronted with an enormous deficit (over one million 1910 dollars), a sad and frightening casualty list, an appalling acreage loss, hostile critics, and a traumatic change of leadership, the Forest Service might have been terminally demoralized. Instead it stiffened its resolve. The 1910 fires were its Valley Forge, its Long March. Henceforth, it would throw all it had into the fire problem.

And it had a lot. The prospect of an imminent timber famine enveloped the fight to save forests in an aura of patriotism. Conservation had become a byword, a popular national movement, even a crusade. The doctrine of the strenuous life found wide currency—in the enthusiasm for saving wilds, preserving the passing frontier, and fostering habitats for big game hunting; in the nascent Boy Scout movement; in the novels of Jack London and Edgar Rice Burroughs; in the personality of President Theodore Roosevelt and that of his sidekick in rough sport, Gifford Pinchot; and in the high culture of William James's essay on the moral equivalent of war. A generation of artists, writers, naturalists, and, yes, foresters went west to reinvigorate the Eastern Establishment with the hard endeavors of the frontier.[4] Not least among its legacies, the 1910 fires participated in this national drama by bequeathing a lore of high adventure amid a Lewis and Clark backdrop of wilderness splendor and by contributing to the tradition of courageous action in the face of national perils. The wistful proposal of William James could not have found a more concrete vindication than in the tales of daring that poured out of the smoking mountains. With these exploits and with the political events that surrounded them, firefighting entered its heroic age.

1910: A FIRE HISTORY
OF THE NORTHERN ROCKIES

A forester in the Northwest dates the events of his life by fire years. The 1910, 1917, 1919, 1926, and 1931 fire seasons each have a character of their own, and in each year are individual fire campaigns which the forester remembers as the soldier remembers the separate engagements of the war.
—Elers Koch, Supervisor, Lolo National Forest[5]

Suitable headstones with bronze tablets were created over as many of these "heroes of peace" as could be traced, for they died as truly in the service of their country as did those on Flanders' poppy-covered fields. —*American Forestry,* 1923[6]

I

The great fires came with little warning. Fire loads had been small from 1905 to 1909, and rangers were preoccupied with boundary surveys and reorganization. Firefighting expertise was not great, certainly not with large fires. Nor was there the traditional reservoir of skilled manpower. The settler class skilled in controlled burning and fire suppression was excluded from the reserves themselves. When fires accompanied railroad construction in the Lolo and Coeur d'Alene national forests in 1908, forest officials succeeded in organizing 12-man fire crews among the workers. But railroad gangs and miners were often an unstable mixture of adventurers, drifters, and recent immigrants, few of whom spoke English and many of whom warily passed under assumed identities. Forest guards received little training. No detection system existed except that of patrol. Few trails penetrated the backcountry. There were no equipment depots. The only communication was by messenger. Agreements, however, were entered into with the railroads and with whatever private timber protection associations were on the scene. In 1907 Pinchot declared a little prematurely that "the measures adopted for detecting and extinguishing fires on the national forests are efficient." After a tough fire year in 1908 and a reorganization of the Service, Clyde Leavitt wrote for the National Conservation Commission a year later that, although "the force is still much too small,"'the Forest Service "may justly pride itself" on its fire record and its demonstration "that destruction of forest property by fire may be almost entirely prevented by the adoption of a suitable system of patrol."[7]

Following normal snows during the winter of 1909–1910, a drought set in at the beginning of April. The mountains browned. Crop failures throughout the region reached such an extent that by July 10 the Northern Pacific Railroad was laying off men by the thousands. Southwest winds passing over the arid Columbia Plain dessicated the vegetation further. And fires broke out. Fires appeared in June, increased in frequency through July and August, exploded into conflagrations on August 20—the celebrated Big Blowup—and only subsided in September with winter rain and snow. The greatest fire source was the railroads: probably 56 percent of the total fires occurred within their right-of-ways, many set by the locomotives and by hoboes who followed the tracks. But these fires caused little damage, because access to them was simple, the routes were routinely patrolled, and section gangs were trained in the art of fire suppression. Other fires resulted from landclearing and campfires. Official reports mention little incendiarism, but unofficial estimates were high. In Glacier National Park it was rumored that as many as 75 percent of the fires were set by "hoboes and others who want work." Especially in view of the massive layoffs by the railroads and the prospect of ready employment by the Forest Service, it seems probable that many fires were deliberately set. The striking fact, however, was that the major damage resulted from a scant 15 percent of the fires—all of them in remote locations and virtually all ignited by lightning. The large number of accessible fires tied up control forces, and without lookout systems, communication networks, and trails, it was not uncommon for backcountry fires to burn for days before any effort could be made to contain them. A severe lightning storm on July 23 deposited fires in distant locations and strained resources to the breaking point.[8]

Even by July 15 over 3,000 men were employed as firefighters. The labor reserves at Missoula, Butte, and Spokane were drained. After the storm of the twenty-third an additional 1,000 men were somehow collected, and for the next 45 days an average of 4,000 men were on the fireline. Before the campaigns concluded, some 10,000 men served as firefighters. The Western Forestry and Conservation Association viewed the scene with consternation: the fire complex threatened to overrun its members' lands as well as those of the U.S. Forest Service. As in 1871, the fires' effects were regional: heavy losses were recorded in Washington and Oregon (1 million acres), in California, South Dakota (Black Hills), Nebraska (Sand Hills), and Minnesota (Baudette fire), as well as in Idaho and Montana (3 million acres). The impact on Forest Service administration was no less broad. The call for overhead to take charge of the thousands of laborers on the firelines swelled out from the Rockies into even remote forests. Sharlot Hall reported the distant response from the Arizona Strip.

> He was out on the Kaibab when the word came, but he rode in to Fredonia as fast as his horse could take him, got food, outfit and

fresh horses, said goodbye to wife and children and parents, and
was on the road again, riding through the night without sleep or
rest, still pushing on by day and night the long distance to the
railroad, then still on to the post of duty far in advance of what
had been thought the quickest time he could make—and his service there befitted the [pioneer] blood of which he came. . . .[9]

On August 8, President Taft bowed to mounting pressures and authorized
the regular Army to support fire suppression efforts. Troops were used in
the California Sierra, in Washington and Oregon, and, of course, in the
Northern Rockies. These were in addition to the hundreds of troops already
gathered to suppress fires burning in Yellowstone and other national parks.
Some 60,000 acres burned in Glacier National Park alone. Fires breaking
into the Colville and Flathead Indian reservations brought pleas for still
more troops. Support meant pack trains, medical supplies, and Army surgeons as well as line troops. Ten companies from Fort Missoula, Fort Hel-
ena, and Fort George Wright were dispatched to the Coeur d'Alene, Lolo,
and Flathead national forests. Idaho mobilized its militia, but to little effect:
they were recalled almost as soon as they assembled (just prior to the
blowup) so that they could vote in a special election for the governor who
had thoughtfully called them out. Montana also ineffectually mobilized its
militia.

The first troops arrived before the Big Blowup, and the Army committed
others after the conflagration had run its course. The medical supplies and
surgeons proved invaluable. Army pack trains replaced the dozens lost in
the holocaust. Soldiers assisted in evacuations and introduced reforms on
the fireline; they practiced constant patrols, experimented with explosives
for line construction and mop-up, employed a system of messengers, and
demonstrated (if any additional proof was really necessary) the value of the
telephone. The telephone was a major technological innovation in fire control, and the 1910 holocaust was its first conclusive test. The successful evacuation of many towns relied on it. The soldiers' performance elicited praise
on all sides. Ranger Thaddeus Roe summed up the admiration of many
when he confessed that after his experiences with a platoon at Wallace his
"attitude toward the Black race has undergone a wonderful change. . . ."[10]
The soldiers were not in the hills when the blowup occurred, but their presence staved off complete disintegration until the civilians could be reorganized. In California, Oregon, and Washington they were often in the front
lines. Not surprisingly, after labor problems plagued the Rockies, the use of
soldiers was not infrequently urged. A year after the blowup, Greeley proposed the establishment of military outposts on the national forests during
fire season for the express purpose of supplementing fire manpower.

Just before the troops arrived, the situation brightened. On August 9, both the regional forester and the local supervisors informed the press that the fires seemed under control. The next day the weather made a mockery of their predictions. High winds and low humidity conspired to break fires out of their control lines from the Bitterroot Mountains to the Lolo Pass. Firefighters rallied, however, and as the weather improved, the fires were again corralled, though the whole administrative edifice was grossly over-extended. But there was little choice in the matter. From the beginning the Forest Service had determined to make a stand. Secretary of Agriculture James Wilson backed it to the hilt. He voiced a collective opinion when he wrote afterward: "I was confronted with the problem of either putting out the fires or being directly responsible for what would have been one of the worst fire disasters in the history of the country. Without hesitation I called upon the forest officers to stop the fires and to make such expenditures as seemed absolutely necessary to accomplish this result. Every source of help was called in."[11]

And so they were. In Glacier National Park, Lieutenant Mapes, for example, lamented the worthlessness of a band of Greeks furnished by the Great Northern Railroad.

They could not be trusted alone—and on one occasion they even sat down along side the trail or trench and loafed all the afternoon and when my soldiers came back to camp at 6 P.M. found them bathing their feet in the river. The soldiers saw that the fires had jumped the trenches in a number of places so drove the Greeks back to work and then found that the fire had burnt up the tools the Greeks had been using. We had dug four lines of trenches already so I formed my entire command and passed buckets a greater portion of the night until the fires were completely out. The remaining three days at that camp I sandwiched the Greeks between the soldiers and got very good results. They undid our work three times by trusting them to finish a little work alone and then abandoning it as soon as out of our sight. We left them at Essex digging trenches to save the railroad.[12]

The brunt of the tough assignments fell to Forest Service regulars, who quickly became veterans in fire suppression, and to troops tested in the national parks.

Again the fires seemed under control. Thousands of men were in position in lines surrounding virtually all the burns. By August 19 weary firefighters and exhausted forest officers looked forward expectantly to a subsidence of fire before the advent of an early winter. But on August 20 and 21 gale winds roared out of the southwest. Mass fires stormed across the landscape, disintegrating prepared firelines like tissue paper tossed into a furnace. Fires swept 30–50 miles across mountains and rivers, oblivious to natural barriers and terrain. The special properties of very large fires, so painfully learned in the Lake States, were retaught to the Northern Rockies. Winds felled trees as if they were blades of grass; darkness covered the land; firewhirls danced across the the blackened skies like an aurora borealis from hell; the air was electric with tension, as if the earth itself were ready to explode into flame. And everywhere people heard the roar, like a thousand trains crossing a thousand steel trestles.

It is a measure of the fire's significance that the literature it spawned—in every way as extensive as that from the holocausts of the Lake States—is dominated by the response of firefighters to the fire, not by the fate of fire refugees. In the Lake States, responses mattered little, being dictated largely by the capricious behavior of the fire. It was as if the people were visited by a plague or cursed by a swarm of locusts before which they were helpless. They scrutinized the fires to discover clues by which to forecast future catastrophes. In the 1910 fire, the focus shifted from the fire to the men who sought to contain it; the behavior of the firefighters and of fire organization was of more interest than that of the fire. But it is certain that the fire followed a typical pattern of smoldering and blowups, repeated time and again as small, dry cold fronts passed over the region. If the fires burned like the intensively studied Sundance fire of 1967, and they almost certainly did, they could make runs of more than 16 miles in a single burning period, engulf over 50,000 acres at a shot, throw firebrands 10 miles in advance of the flaming front, induce turbulence of up to 80 mph, and burn—where stalled by natural crucibles—with a staggering intensity of 22,500 Btu/ft/s, the equivalent of a Hiroshima-type (20 kiloton) bomb exploding every two minutes.[13]

From all accounts the 1910 complex behaved in just this way, except that dozens, perhaps hundreds, of fires burned simultaneously. The worst fires were ignited by lightning in remote locales and escaped detection for several days. When attacked, they were held to respectable acreages. The passage of a trough encouraged flare-ups; the fires made runs before firefighters rallied and contained them again. The large sizes of the fires prevented mop-up. The fires were patrolled, supply lines strengthened, and trails cut. Then came the Big Blowup. What happened on the Sundance fire could be multiplied by the score. All resistance crumpled; crews fled from the hills; camps disintegrated into ash; pack trains vanished. A light rain and snow quieted the flames on the twenty-third, marking the passage of the front. Then burning conditions returned. For the third time the fires boiled over, though with somewhat less intensity. Despite earlier tragedy, district forester Greeley grimly reorganized. New trails were cut, new control lines dug, and new crews assembled. Then at long last the rains came. Beginning on the night of August 31, the rains brought a grateful end to the fires.

The essence of the 1910 fire season is distilled into the two days of the

Big Blowup. When the fires erupted, thousands of men were in the field, hundreds were near the forest, and scores were directly in the fire paths, as were a number of towns. The villagers showed the peculiar fatalism of the Lake States residents. The town of Wallace, Idaho, was typical. Preparations were made, of course: the fire brigade was strengthened, retooled, and reorganized in view of the fact that most of the menfolk were fighting fire in the hills. Large fires burned to the south and west, and as early as August 13 firebrands ignited awnings in the town from six miles away. The next day the town's insurance brokers did a land office business writing fire policies for one and all. The flurry of policies continued until the blowup was imminent. The railroads, too, kept a wary eye on events.

Fortunately, the fire swept out of the hills at night, perhaps under less severe burning conditions. It entered the town about 9 P.M. The eastern third of the city melted before the flames. But the town's fire department—in a rare burst of bravura—halted the fire at this point with the loss of 100 buildings and two lives. At 10 P.M. the Northern Pacific relief train pulled out of Wallace loaded with women and children and guarded by regular troops. More refugees were picked up to the east, and the train arrived in Missoula the next morning. Meanwhile, looting broke out in Wallace, and martial law was declared at midnight.

Other trains had more harrowing flights, with soldiers stoically enduring heat to protect the wooden cars and quell panic. Bridges and rails burned out. At the last possible instant, a work train swept up 1,000 people from the town of Avery, steamed over burning trestles, and raced into a tunnel, where it weathered the firestorm that consumed everything outside. Avery was miraculously spared when a ranger and eight businessmen lit a successful backfire from within the town itself. The neighboring towns of Taft, Deborgia, Haugan, and Tuscor crumbled before the advancing flames.

But the traditional story of the 1910 holocaust is the story of how the men in the hills confronted the fire. To the crews in its path, the fire came in different forms. To some it announced itself as a great roar and a towering column of smoke. To others it appeared as a "falling star" that ignited fires nearby. To others it was a whirlwind of flame. To still others it came as words tumbled from the mouths of panicked messengers and scouts. To the Christians it must have seemed that the Apocalypse had arrived. To Hindus—if any there were in that polyglot babel of humanity—it must have recalled the words of Krishna in the *Bhagavad-Gita:* "I am become death, the shatterer of worlds. . . ." Whatever its avatar, all knew they faced a harrowing trial by fire. Those who had time buried their camp. Some fled to streams, some found caves; some sought shelter in root cellars; some cooly backfired from hastily prepared lines. Some went insane, babbling deliriously, and some became suicidal, preferring a bullet to burns. As a rule, those who separated and fled perished; those who stayed by their foreman survived. Their individual stories have filled volumes. But one in particular has succeeded in moving from history into legend, and it has come to epitomize all that 1910 meant to fire protection.

Ranger Edward Pulaski was a direct descendant of Casimir Pulaski, the Polish hero of the American Revolution. At the time of the fire he had been mining in the region for nearly 25 of his 40 years. In the words of his supervisor, he was a "man of most excellent judgment, conservative, thoroughly acquainted with the Region." Pulaski had charge of about 150 men between the Coeur d'Alene and St. Joe rivers. On August 20 a wind "so strong it almost lifted men out of their saddles," the roar "of a thousand freight trains," and smoke and heat "so intense that it became difficult to breathe" amply warned Pulaski that it was time to evacuate. He rounded up the 45 men of his crew who were immediately threatened by fire and started off to Wallace.[14]

They found their path blocked by more fire. It may have been a spot ignited in advance of the main burn behind them or the backfire impetuously set from Wallace. It hardly mattered. The crew found itself in the eye of a firestorm. Trees crashed to all sides, and blackness swallowed the landscape. The crew of pickup laborers was frantic. Shouting over the roar, Pulaski calmed and bullied them equally. He knew of two prospecting tunnels nearby, he informed them, and promised to lead them to safety. With a wet gunny sack over his head he dashed through the woods, located the tunnels, determined they were safe, and returned. He led his crew to the larger of the two shafts, the War Eagle mine. One man lagged behind and perished beneath a crashing tree. The rest staggered into the mine shaft. At Pulaski's direction they retreated to the rear of the mine where a small seep leaked water. They dampened their clothes and placed wet clothes over their mouths and noses. Pulaski stripped the blankets off his horses, wetted them, and stood his ground at the entrance, determined to keep the fire out of mine timbers whose collapse would seal them all to eternity. The crew was seized by fear. The roar was insufferable; smoke and gases filled the small cave. The fire lashed at the mine with the fury of a tidal wave. Some men prayed, some wept, and some panicked. One man tried to dash outside. Pulaski drew his revolver and promised to shoot the first man who tried to leave. Swept first by fear, then cowed by Pulaski's stand, the crew retreated to the seep, where all eventually succumbed to the heat, smoke, and fumes. Everyone passed out. Pulaski slumped into unconsciousness at his post by the mine shaft entrance.

One man revived late that evening, crawled out of the shaft, and straggled into Wallace about 3 A.M. with the news. A rescue party was quickly organized. Meanwhile, all but five of the others in the mine regained consciousness. They worked their way to the entrance about 5 A.M. There they encountered the body of Pulaski crumpled in a heap. "Come outside boys,"

11. The War Eagle mine, where Ranger Edward Pulaski held his crew during the Big Blowup, 1910.

one weary crewman said, "the boss is dead." "Like hell he is," came the voice of the indomitable ranger. Their ordeal had utterly drained them. They crawled to a creek but found to their dismay that the water was hot and alkaline with ash. The fresher air heartened them, however, and they struggled on toward Wallace, soon to meet the rescue party. Pulaski was all but prostrate with respiratory problems and temporary blindness. For two months he convalesced in a hospital; for a decade his eyesight threatened to leave permanently. In later years he became reticent about his feat, declining further interviews. But for the nation he became a celebrity, a symbol of the strenuous life bravely battling the reckless waste of natural resources, a backwoods Leonidas. For firefighters he became a folk hero.

The impact of the fire did not end with the coming of September rains. It was estimated that seven to eight billion feet of marketable timber had been destroyed. Sheet erosion became severe in places, and some rivers, like the St. Regis, showed unstable behavior for many years afterward. The scars of the burn are still visible. As in the Lake States, natural reseeding replaced the commercial species with a pyrophyte; white pine, larch, and fir gave way to expanses of almost pure lodgepole pine. The burned lands, moreover, smoldered with the possible outbreak of insects or disease. Bark-beetles invaded the burns and in 1914 crossed into green timber, soon to be followed by blister rust. Confronted with fire, rust, and beetles, the western white pine slouched into permanent decline. As the infected stands died off, fire hazards rose proportionately. Combined with the heavy downed timber and snag patches left by the holocaust, these vast sites became a breeding ground for reburns for nearly two decades. The reburns, in turn, spread to green timber and created still more reburn potential in a vicious cycle. The fires thus had a multiplier effect on the regional fire load. Koch sadly esti-mated that "it is not at all impossible that the burned area since 1910 has been twice as great as would have happened if the 1910 had not occurred." Looking back from 1954, regional forester P. D. Hanson observed that as a result of this "milestone" burn, "the economy of three states was disrupted almost overnight and the ecological balance of an area two-thirds the size of the State of New Jersey was so upset that the effects will be evident for years to come."[15]

Despite redoubled protection efforts, fire became a dominant fact of life in the administration of Northern Rocky forests. Timber management con-sisted of fire salvage sales, replanting on burned sites, control of insects and disease outbreaks spreading from burns, and snag felling as a fuel modifi-cation measure. The fires had destroyed many witness trees, and boundary surveys had to be rerun. The imperatives of fire control opened up the remote backcountry. Fire control built roads and trails, strung telephone lines, constructed observation towers, influenced the direction of forest research and equipment development, and made the region a national authority on the fire question.

In the same way, the Forest Service, which had been something of a late-comer in the field of practical fire protection became a national leader. The 1910 fires did little to damage the Service's reputation for forest protection and contributed to the passage a year later of the Weeks Act, which granted the Forest Service the right to purchase land for national forests and to engage in cooperative funding with states for fire protection on watersheds feeding navigable rivers. The effect was to nationalize fire protection under the aegis of the Service. Even before passage of the Weeks Act, Henry Graves, William Greeley, Ferdinand Silcox, and Earle Clapp—all future Chief Foresters—wrote articles on the theory and practice of fire control, taking the 1910 fires as a starting point. And so they were, nationally speak-ing. *American Forests* devoted its entire November 1910 issue to detailed accounts of the fire. The meteorology and physics of the fire were not well understood, but all the survivors agreed that a prominent cause was the abundance of uncontrolled or merely contained fires scattered over the landscape. When California timbermen took their polemics on light burning to the public in August 1910, the Forest Service could only meet their arguments with incredulity. No one who had lived through the holocausts in the Northern Rockies could seriously accept the proposition that fire should be liberally distributed throughout the woods. It was not until Lyle Watts succeeded Clapp—the last of those who had weathered the fires—as Chief that the Forest Service reversed its stand on prescribed burning.

If the government was magnanimous in offering new responsibilities to its bureaus, it was niggardly in providing for its servants. The Lake States fires had burned on private lands, and private capital was generous in addressing the plight of victims. The 1910 fires burned on government lands, and the victims were largely at the mercy of the government for aid. Little was forthcoming. No employment compensation act was in effect. The Red Cross made a $1,000 contribution toward hospitalization expenses, and a collection circulated among meagerly paid Forest Service employees brought more. Townspeople were aided by other towns, and the railroads, as usual, furnished free services. The Army contributed medical supplies. But the dead firefighters were buried in potter's fields. Not until 1912 did the government at last awaken to its duty. Special legislation provided compensation for victims and covered costs of interment. A deadline was set for claims, and the Forest Service began a complicated correspondence intended to identify victims and locate relatives. This became a global undertaking, for the work pool of miners and railroad gangs included Italians, Austrians, Persians, Bulgarians, Greeks, Irish—virtually every nationality on earth.[16] The total cost of claims came to $23,192.90. Inexplicably, many heroes of 1910 never benefited. Pulaski's eyes troubled him for years, and when an operation was deemed necessary, he found that the claim date had expired. He was lamely referred to the Carnegie Foundation for an award out of its hero program. None was forthcoming. In 1923 he finally gained some compensation for his trials by winning an essay contest sponsored by *American Forests* on "My Most Exciting Experience as a Forest Ranger." Meanwhile, Congress had neglected to allocate money for maintenance of the graves it had allowed the dead firefighters. The omnipresent Pulaski then took it upon himself to keep up the appearance of the sites. Eventually, in 1923, the Forest Service was granted maintenance funds. A plaque was erected. Ten years later the Forest Service officially closed its files on the 1910 fire, and a central burial plot for firefighters killed in the line of duty was established at St. Maries, Idaho.[17]

The 1910 complex easily lends itself to superlatives. Its record, however, is best chronicled in the words of those who were individually touched by it. To Mrs. Swaine of Wallace, Idaho, "it was a terrible ordeal, but I wouldn't have missed it for anything."[18] Pulaski himself was more humble: "My experience left me with poor eyes, weak lungs, and throat; but thank God, I am not now blind."[19] For William Greeley, "a young forester, thrown by chance into a critically responsible spot on a hot front, that summer of 1910 brought home the hard realities of the job. . . . I had to count the cost in hardship and sweat, in danger and human lives. . . . The conviction was burned into me that fire prevention is the No. 1 job cut out for us. . . ."[20] To Joe Halm, a youthful ranger and former football star whose exploits rivaled those of Pulaski, the aftermath brought the voice of determination: "First we'll dig out our tent, salvage the grub, then look the fire over. We'll order more men and equipment and hit the fire again."[21] To the troops sent into the mountains to locate the bodies of dead firefighters, the scene was perhaps more familiar: the landscape gutted, smoke hanging pungently among the trees, a ghastly battlefield in the war on nature. Over a mass grave the soldiers fired a volley of rifles while a trooper blew taps.[22]

II

Fires swept the Northern Rockies long before 1910 and have continued long past it. Studies of fire scars and even-aged stands of old timber show a consistent pattern of fire frequency from at least 1600 to 1900. Heavy bombardment by lightning has been a climatic feature since the end of the Pleistocene, and it has remained the great problem incendiarist of the region. Indians burned widely for traditional reasons, particularly east of the Continental Divide. Explorers and surveyors spoke often of "parks" in the Black Hills, Big Horns, and Central and Northern Rockies that tribes had created and sustained by fire. Especially where grassland tribes moved easily into the mountains for summer game or lodging or passed through on trails like the Lolo or Nez Percé, fires were common. Fire hunting by Indians was prominent in the fall, and in 1842 unusually large conflagrations resulted when the fires got out of hand. The landscape around major Indian trails resembled the wasteland around the right-of-ways of railroads.[23]

Everywhere explorers met with enormous tracts of forests fired by nature or man and then felled by high winds into impenetrable tangles. On its 1871 survey of Yellowstone, the Hayden Expedition found passage through the burned expanses of lodgepole pine as difficult as forcing a deliberately constructed barricade.[24] Writing for the Geological Survey in 1898, F. E. Town explained that "there is abundant evidence from these appearances that every acre of these mountains [Big Horns] has been burned over at some time, and probably many times successively in the past." The chief source of that summer's fires was "a band of Shoshone Indians."[25] John Leiberg noted that on the Priest River Reserve "one meets with burnt areas everywhere—in the old growth, in the second growth, in the young growth, and where the seedlings that are beginning to cover the deforested areas have just commenced to obtain a fair hold." Fires on the Bitterroot Reserve, Lei-

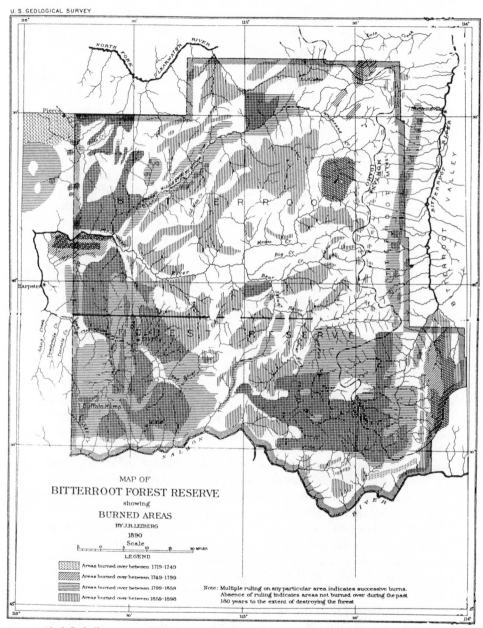

MAP OF
BITTERROOT FOREST RESERVE
showing
BURNED AREAS
BY J.B.LEIBERG
1890
Scale

LEGEND

Areas burned over between 1719-1749
Areas burned over between 1749-1799
Areas burned over between 1799-1859
Areas burned over between 1859-1898

Note: Multiple ruling on any particular area indicates successive burns.
Absence of ruling indicates areas not burned over during the past
180 years to the extent of destroying the forest.

12. J. B. Leiberg, map of Bitterroot Forest Reserve, 1898. Most maps of the reserves showed
areas burned and reburned at the time of investigation.

berg observed, "have been as extensive as elsewhere in the West, but have done less damage to the merchantable timber" owing to the properties of the ponderosa pine. He concluded that the aftereffects of fires in the Selway Basin "are various, but are always evil, without a single redeeming feature." Fire had determined "the age of the timber stands in the reserve, and in some degree the species of trees that compose them. Most of these fires burned centuries ago." The more recent fires of the white man "have influenced the complexion of the growing forest in a minor degree only." The effect of centuries of fire was to create "large open hillsides covered with grass, sedges, or bear grass." Where forest regeneration occurred, it tended to come back as the pyrophytic, commercially worthless lodgepole pine.[26] On the Flathead Reserve, H. B. Ayres judged sadly, "actual prairie making has been accomplished."[27] Further east, in the Black Hills, a mountain island surrounded by an ocean of prairie, Henry Graves described in detail the "history of a park" made by successive firings and remarked that "aside from a few restricted areas, practically the whole forest shows traces of forest fires, and usually some actual injury caused by them. There have been periods, separated by about half a century, when the whole or a large part of the hills has been burned over."[28]

The white man came first for furs and then for precious metals. Placer strikes were made in the 1860s; hardrock mining began in earnest during the 1880s. Railroads were constructed to support the mines. Logging supplied shaft timber and fuelwood. Agriculture came slowly; most foodstuffs were transported into the region. Prospectors contributed fire inadvertently from abandoned campsites and deliberately in an effort to strip off vegetation and expose outcrops—a practice, as Lucretius reminds us, of ancient lineage and global scope. In their search for legendary lost mines in the Lochsa Basin, prospectors repeatedly fired whole mountainsides, sustaining, and in some cases enlarging, the fire regime inherited from the Indian. Slash from limited logging activities and railroad construction furnished fuel for other burns. Large fires were reported in Wyoming during the 1860s and 1870s. Further north, heavy fires swept the region in 1889.[29]

The infusion of Americans into the region had a mixed result. Where settled existence appeared, so did fire protection. For example, when fires threatened one of its camps in 1897, the Homestake Mining Company hired a large body of men, who succeeded in extinguishing the blaze. The Anaconda Copper Mining Company launched a major fire prevention campaign, and the larger ranchers on the White River Reserve were "often active in the prevention of forest fires." But where transients persisted, so did broadcast fire: prospectors, cow punchers, tourists, sheep herders (Basques, in particular, entered the Rockies in the 1880s), teamsters, aging frontier hunters, patrols of soldiers, bands of traveling Indians, and so forth were responsible for widespread fires. The "open defiance" of government regulations by these classess—or, from their vantage point, the protection of ancient folk rights—was commonly cited by observers as the greatest threat to the preservation of the reserves.[30] Though fires in ponderosa pine were typically surface burns, elsewhere, even under natural circumstances, fires were large, cramming into a few brief weeks their vast job of decomposition and tending to fashion forests into enormous blocks of even-aged timber.

In the Northern Rockies, as in the Lake States, there is one fundamental problem fire, of which the 1910 complex is an archetype. As a natural phenomenon, that complex was remarkable only in degree, not in kind: it multiplied the number of large fires at one instant but did not alter their typical behavior. As a historical event, however, its impact was quite the obverse: its timing rather than its size or damages distinguished it. The fire load of the Northern Rockies was reduced through the steady application of improved fire control techniques, research, and suppression hardware, not through a dramatic conversion of land use patterns, as in the Lake States, or by socioeconomic changes, as in the South. Timber protective associations were operating by 1910, the Army had responsibility for the national parks, and both Montana and Idaho were participating in the Weeks Act program as early as 1913. But the story of fire control in the Northern Rockies has to an unusual degree been the story of the U.S. Forest Service. And, in turn, the region has exerted an inordinate influence on the course of national policy. It was no accident that, after the approval of the 10 A.M. control policy in 1935, the national conference called to discuss its implementation was held in Spokane, or that, when a new policy was imminent 40 years later, it was announced in Missoula. Still brooding over the region and its chief federal administrator into the 1970s, however improbably, was the specter of the 1910 holocaust and its reburns.

The administrative history of the Northern Rockies is dominated by the paraphernalia of fire control. This was something new not only to the Forest Service and to forestry but also to the history of the American West. Fire control became in a sense a frontier institution. New lands had been settled by missions, ranchers, miners, farmers, speculators, traders, and Army posts, but never by forest firefighters. Yet this was an incontestable and unexpected result of the forest reservation system. The lands were put to no immediate use, or to limited use by permit; they remained as wild or forested land, suspended indefinitely for some future generation. Nevertheless an administrative presence was required: the public demanded signs of "management," which it perceived from the public words of foresters to mean fire protection; and the canons of professional forestry insisted that fires had to be tamed before the land could be used for the expressed objectives of scientific stewardship, that is, of conservation. There seemed no point in preserving the lands if they were not protected. It was largely out of the necessities of fire control that the country was opened up, settled, and administered.[31]

Later fires have differed from the 1910 complex only in size. The pattern of fire behavior and starts has remained virtually unchanged. In 1914 and 1917 over 100,000 acres burned collectively. In 1919 almost 2,300 fires burned 1.3 million acres, much of it reburns. The season was marred by labor agitation and the dark suspicion that fire control was being sabotaged by incendiarists affiliated with the IWW. Half a million acres burned in 1926 and 1929 combined, more of it reburns. The Forest Service sent its toughest fire officer, Major Evan Kelley, to the region with orders to stop the big fire. Drought came with the early 1930s. In 1931, indendiary fires forced the governor of Idaho to declare martial law in several counties and to mobilize the national guard. In South Dakota federal troops were called out to the Black Hills.[32]

The coup de grâce came in 1934. The scenario was eerily similar to that of 1910. Lightning fires broke out on July 24 and left crews and overhead exhausted. A man-caused fire appeared on August 8, and 19 lightning fires followed on the eleventh. The 20 fires totaled 252,000 acres. Two of them—the Pete King and McLendon Butte fires on the Selway National Forest—accounted for 95 percent of the total damages, reaching 161,000 and 81,000 acres, respectively. Both represented the eventual merger of several closely packed lightning fires. The Pete King fire blew up almost immediately. By evening the Forest Service had amassed 1,059 men from the National Industrial Recovery Act rolls (used in blister rust control), from CCC camps, and from local employment agencies; ultimately, 5,000 firefighters were on the line. Bulldozers were used where appropriate. The fire was contained until August 17, when it again blew up and made a major run; another followed on the eighteenth. A dry cold front converted the long flank of the fire into its new head. Crews fled, camps were lost, and narrow escapes were the order of the day. Once more surrounded, the fire blew up again on the twenty-second. The pattern was repeated on the thirtieth and thirty-first of August. Each time control lines vanished in flame. Heavy rains on September 20 mercifully brought the nightmare to an end. The Selway debacle enjoyed as prominent a position in the policy debates of the 1930s as the 1910 fire had in the policy debates of its day. The 10 A.M. Policy was designed to address just such fires.[33]

Slowly, against terrific adversity, the tide turned. The region's fire officers showed both resolve and ingenuity, for example, when they created the celebrated Remount Depot to supply the large quantities of pack mules demanded by fire control. More dramatically, they adopted aircraft—building landing fields in remote mountain valleys, experimenting with paracargo drops, and adopting the young smokejumper program. By 1945 aircraft had so transformed fire protection that the Continental Unit was established in the Rockies as an experiment in which all detection, suppression, and support would be by air. Trials were conducted in the aerial bombing of fires with chemical agents, and Project Skyfire, an exercise in the aerial suppression of lightning itself, began its long tenure. Even the region's failures had national consequences. The Blackwater fire (1937) led to reform of CCC fire control and helped to redirect the aerial fire control experiment from California to the Northwest and from bombing to smokejumping. The Mann Gulch fire (1949) resulted in reform of the entire smokejumper program and the creation of the Aerial Fire Depot at Missoula. Missoula, in fact, became something of a Forest Service company town. It housed the regional offices, the offices of the Lolo National Forest, the Northern Forest Fire Lab, the Missoula Equipment Development Center, and a regional fire cache, as well as the Aerial Fire Depot. Only in Southern California is there a similar concentration of fire power.[34]

Despite flare-ups, fatalities and burned acreages fell almost annually. The massive lightning barrage of 1940 brought 3,109 fires, some 1,400 in one week, but none became large. In 1960 to a small degree, and in 1961 to a wider extent, the old pattern reappeared. But as trails were constructed, roads blasted into the wilds, radios and telephones distributed, lookouts manned, hand and power tools acquired and amply stored, pack trains and motorized vehicles and specially equipped aircraft held in readiness, and as fire behavior expertise grew—the abundance of mass fires steadily receded. As bad as the 1934 season seemed, its burned acreages were one-tenth of those of 1910; in 1961, they were only one-thirtieth.

Then came the real test. The 1967 season showed classic form. Rains ceased by mid-July, and soon afterward fire danger registered "extreme" for 59 consecutive days. The condition was regionwide, extending over the Pacific Northwest as well as the Inland Empire. In all, more than 5,000 fires were reported; 30 exceeded 1,000 acres. For the Northern Rockies the fires came in traditional busts. On July 12, 131 fires were reported, 118 from lightning. The storm was accompanied by rain, however, and burned losses amounted to only 600 acres. On the twelfth of August a dry lightning storm set 167 new fires. Others followed. Ninety-seven began on September 6. On September 21 another 167 fires resulted. Before the season concluded, lightning had ignited more than 1,400 fires, the majority crowded into concentrations that threatened to overload protection systems. The statistics and timing were much like those of 1910. In 1910, 80 percent of the reported fires were held to a small size; in 1967, 95 percent of all fires were held to less than 10 acres. Only 9 exceeded 1,000 acres, and 3, including the mammoth Sundance fire, accounted for 84 percent of the total burned acreage. Major runs were induced by the passage of dry upper-level troughs.[35]

The mobilization statistics are impressive. Nearly 13,000 firefighters were employed; at peak, some 7,000 were on the line at one time. Some 520 organized crews were assembled—Forest Service hot shot crews, Indians from the Rockies and the Southwest, Hispanics from the Snake River Valley, Eskimos from Alaska, job corpsmen and recruits from everywhere. It was another testimonial to the ability of fire pay to attract polyglot adventurers. Smokejumpers set new records with 1,691 individual jumps. The regular military and state national guard furnished support services with 700 personnel and stockpiles of specialized equipment and stood by to evacuate endangered towns. An aerial armada of more than 200 aircraft delivered record quantities of chemical retardants and ferried men by the thousands and cargo by the millions of tons. Suppression costs totaled $22 million, and emergency rehabilitation funds added to the final bill.

Perhaps the greatest difference between these figures and those of 1910 lies in the quality of the suppression effort. Instead of 3 million acres, the fires burned only 90,000—56,000 of that from the Sundance fire. Instead of 78 fatalities, there were 3—1 due to heart failure, and 2 to burns when a tractor operator and spotter were overtaken during the Sundance run. Take away the Sundance fire, and the season would have brilliantly vindicated the patient, aggressive triumph of fire protection over its historic challenges in the region. Fires that might have burned for days before discovery in 1910 were detected and attacked within hours, sometimes within minutes. Oblivious to smoke, aerial infrared scanners mapped fire perimeters. A team from the Northern Forest Fire Laboratory offered fire behavior predictions. Mobile units from the National Weather Service hurried to project sized fires. Radios, telephones, and teletypes furnished almost instantaneous communications, and the National Fire Coordination Center at Arlington, Virginia, provided interregional and interagency liaisons. The region was declared a national disaster area, and the Office of Emergency Preparedness joined the coordination effort. The declaration had the beneficent effect of obliterating legalistic distinctions of jurisdiction and made possible the transfer of responsibility for the Sundance fire from the Priest Lake Timber Protection Association—in whose jurisdiction the fire began—to the Forest Service, which finally stopped it. To prevent man-caused fires and to simplify the movement of men and material, an unprecedented full closure of national forests was ordered, analogous to a declaration of martial law. So impressive was the speed and scope of the emergency response that the Office of Civil Defense contracted to have the fire season studied for possible lessons for wartime fire defense. The report's benchmark for comparison was 1910.

The 1967 season was the last conflagration in the Northern Rockies to be handled under the more or less exclusive direction of the Forest Service. It brought to conclusion a cycle of fire protection initiated by the 1910 holocaust. A few years afterward the Boise Interagency Fire Center went into operation, and following the 1970 fire complexes in Washington and California, the Forest Service, Bureau of Land Management, and National Weather Service became full partners. The concept of total mobility came into vogue. Nevertheless, the imposition of further administrative superstructures was approaching a limit of effectiveness. The Civil Defense report hinted at the same paradox, concluding that in the event of radioactive fallout it would be unlikely that the elaborate mobilization so impressively executed in 1967 could be matched. Instead, fires would necessarily be attacked by relatively autonomous crews, much as they had been in 1910.

The drive for better integration of support facilities, overhead, equipment, and even jurisdiction was only part of the transformation after 1970. The Northern Rockies also became a proving ground for methods of prescribed burning. In this it confronted several peculiar dilemmas. Whereas in the South the landscape had been subjected to routine anthropogenic surface fires, a crown fire was the norm in the Northern Rockies. The fine fuels that carried fire in the South were ground fuels; in the forests of the Northern Rockies they were commonly aerial fuels. In the South prescribed fires could be lit virtually year round; for the Inland Empire, these were possible only during late summer or early autumn—prime conditions for an undesired blowup. Fire protection in the South meant the management of people; for the Northern Rockies the problem fire was the lightning strike in a remote hinterland. Prescribed burning had long been practiced locally on insect-infested and disease-killed sites and for disposal of logging slash, but, except for ponderosa pine belts, the strategy of broadcast underburning had little environmental or historical rationale in the Northern Rockies. The primary achievement of fire control had been to reduce the cycle of heavy reburns.

In 1970, at the same time that control forces were moving to greater coordination, the SAF appointed a special task force to investigate fire in the Northern Rockies. Its report observed that fire could not be excluded from these forests, that fire could advance conditions useful to the general imperative of land management, and that prescribed fire should be increased.[36] At the same time, the Forest Service revised its 1935 fire policy to tolerate fuller use of natural fire. In the Whitecap drainage of the Bitterroot-Selway Mountains plans were advanced to permit natural fires to burn in a broad wilderness area. Nature willingly provided a test. Fires outside the zone were attacked vigorously; within the zone, they were allowed to burn unrestricted unless or until they threatened to leave their sanctuary. Inaccessibility—long the curse of fire control—now became an article of salvation for fire management. At least on a limited basis, the very remoteness of the landscape allowed for natural prescribed burns. In anthropogenic fire regimes, broadcast fire had to be disseminated by humans; the fire most useful to the Inland Empire could be distributed by lightning, as it had been for millennia.

THE FORESTER'S POLICY:
A HISTORY OF FIRE POLICY IN
THE U.S. FOREST SERVICE

A little fire is quickly trodden out, which, being suffered, rivers cannot quench.—Shakespeare, *Henry IV*

When immediate control is not thus attained, the policy then calls for prompt calculating of the problems of the existing situation and probabilities of spread, and organizing to control every such fire within the first work period. Failing in this effort *each* succeeding day will be planned and executed with an aim, without reservation, of obtaining control before ten o'clock of the next morning.—10 A.M. Policy, *National Forest Manual, 1935–1978*

The problem of fire oppressed and preoccupied the Forest Service from its beginnings. In 1898, when the Service was still a bureau with only advisory powers, Pinchot warned that "like the question of slavery, the question of forest fires may be shelved for some time, at enormous cost in the end, but sooner or later it must be met."[37] After 1910 there was little danger that the problem would be shelved. Chief Forester Henry Graves declared in 1913 that "the necessity of preventing losses from forest fires requires no discussion. It is the fundamental obligation of the Forest Service and takes precedence over all other duties and activities."[38] This belief became so pronounced that Pinchot himself feared that the fire question would divert attention from other issues. Fire loomed so large in the normal experience of forest rangers, became so much a part of organizational folklore, and so readily identified the Forest Service mission to the public mind that it was never far out of administrative sight. What a forester at a 1941 conference stated may well be projected over the course of Forest Service history. "We have been so industrious in our crusade against fire," he lamented, "that the public generally recognizes us as a fire organization rather than a forest organization."[39]

The history of fire policy and that of the Forest Service go readily together. Both may be understood in terms of four problem fire types that at different times have challenged the fire protection mandate of the Service and have evoked different responses. All aspects of fire protection tended to cluster around the particular concerns of the reigning problem fire, and each fire type brought with it both a strategic concept for its control and a certain tactical emphasis, a means by which the larger strategy might best be achieved. These problem fires might be characterized as the frontier fire (1910–1929), the backcountry fire (1930–1949), the mass fire, (1950–1970), and the wilderness fire (1970–present). Moreover, the Forest Service sought to extend its policies and solutions to its cooperators. In fact, it may be said that fire was usually the first, and often the only, basis for the vast cooperative programs in forestry that have evolved nationwide.

Wildland Fire Protection: The U.S. Forest Service Experience

Date	Problem Fire	Policy	Fire Control		Research
			Strategic Concept	Tactical Emphasis	
1910–1929	Frontier fire	Economic theory	Systematic fire protection	Administration	Fire as forestry Economics, planning, statistics of fire
1930–1949	Backcountry fire	10 A.M. Policy	Hour control	Manpower	
1950–1970	Mass fire	10 A.M. Policy	Conflagration control	Mechanization	Fire as physics Laboratory Field experimentation
1971–present	Wilderness fire	Fire by prescription	Fuel modification	Prescribed (broadcast) fire	Fire as biology Natural laboratories Simulation experiments

Each of these periods may also be characterized in another way: as a response to certain types of abundances that became suddenly available to fire protection. For the era of the frontier fire, the surplus was one of the land—the result of the Transfer and Weeks acts—and of money—a consequence of the Act of 1908. For the era of the backcountry fire, there was an overabundance of money and manpower as a result of New Deal conservation programs, especially the Civilian Conservation Corps. This confluence was replaced during the era of mass fire by a military and civil defense liaison that brought large quantities of war surplus equipment to fire control. By the 1970s an almost abnormal, or at least unassimilable, amount of information had been produced on the subject of fire, particularly on fire ecology, and policy reforms can be imagined as a means of coping with that "surplus." The exploitation of these abundances did not involve simply the opportunistic acquisition of new means: the means at hand were often so powerful as to dictate to some degree the ends to which they might be applied. They upset an evolving internal equilibrium between policy and programs, problems and opportunities.

Each of these fire types suggested, too, a potential range for controlled burning and proscribed, as well as prescribed, certain fire uses. During the frontier fire period, slash burning in piles was acceptable, but broadcast burning was not—it too closely resembled the light burning of the frontier and the careless disposal of logging slash by landclearing fires. As systematic fire protection extended its limits, so too were limits set on traditional fire uses; and as systematic protection was successfully achieved, as in the South, so it accommodated certain uses of free-burning fire. During the era of mass fire, a prescription for mass fire was a prominent objective of fire research, and the application of mass fires for military purposes was tested. The era of wilderness fire, however, divided all the fires into either wild or prescribed fires, and it sought to expand greatly the range of prescribed burning. The issue, that is, was never one of simple fire exclusion but of fire use only within the confines and context of the dominant problem fire type. For the periods of frontier, backcountry, and mass fire the ruling concern was control over these particular types of fire, and it had the effect of limiting many potential uses of fire. With wilderness fire the overriding concern became the dissemination of a particular set of fire applications, and it had the effect of limiting certain forms of fire suppression.

I

The *Use Book* that guided Forest Service management from 1905 to 1911 envisioned small fires handled by a solitary ranger or guard, or at most a forest officer in charge of a *posse comitatus* of ranchers, farmers, miners, and loggers. An early ranger was instructed to "take horses and ride as far as the Almighty will let you and get control of the forest fire situation on as much of the mountain country as possible. And as to what you should do first, well, just get up there as soon as possible and put them out."[40] The ranger examination included questions on fire control, but answers to questions on how to handle a "top fire" (crown fire) were likely to be met with replies like the one given in the Southwest: "How'd you fight a top fire? There's only one way; I'd run like hell and pray for rain!"[41]

The frontier wisdom embodied in the early programs broke down completely in the face of widespread holocausts. The 1910 fires came as a shattering revelation to the Forest Service. But although they were the *casus belli,* they were not themselves the enemy. The more basic threat came from frontier fire practices—light burning in California, and, after the Weeks Act, folk woodsburning in Florida. The effect of both sets of fire types was to demonstrate the inadequacy of Forest Service fire plans: the one challenged its technical ability to control big fires, and the other its understanding of the meaning of fire for forestry. The two could be blended as well. After reviewing the 1910 scene, Secretary of the Interior Ballinger publicly recommended that "we may find it necessary to revert to the old Indian method of burning over the forests annually at a seasonable period."[42] But most foresters sided with the German observer, Professor Deckert, who noted that "devastating conflagrations of an extent elsewhere unheard of have always been the order of the day in the United States." He observed with satisfaction that its success with fire prevention formed "a brilliant vindication of the forestry system of middle Europe," and he concluded that "measures must be taken in the future" to staff the reserves with better trained rangers and ultimately to convert the wild forests into European-like woodlots.[43]

It was not obvious how this should be done. Europe had achieved its successes after centuries of reclamation, a time frame that America could not afford and a process that American foresters had decided to exclude. American foresters recognized that they would have to invent the methodology and techniques of fire control for themselves. From California Coert duBois lectured that "American foresters have found that they have a unique fire problem, and they can get little help in solving it from European foresters whose work has been popularly understood and approved for 300 years." Instead, duBois urged, "We must work it out for ourselves. We must gather our own facts, arrange them according to our own classifications, and draw our own conclusions."[44] The outcome of these labors became one of the most significant contributions by Americans to international forestry.

In addition to a complex of unique historical events, including the 1910 fires, American foresters faced a unique political and economic matrix as well. As a result of the Transfer Act (1905) the Forest Service had inherited responsibilities for a national estate larger than many European countries, and as a consequence of the Weeks Act (1911), which allowed it to purchase lands from outside the public domain and to enter into cooperative arrangements with states for fire protection, it found its cares further enlarged. This almost instantaneous acquisition of lands created special problems for administration, particularly for fire control, which were both eased and aggravated by a 1908 act that permitted deficit spending in the event of forest fire emergencies. After the Forest Service submitted vouchers to the General Accounting Office verifying that the money had been spent for fire suppression, Congress would enact supplemental appropriations to cover the budget deficit. Otherwise a large fire might bankrupt a forest, or even the whole Forest Service. The first test of the legislation came, naturally enough, during the fires of 1910, when a staggering $1.1 million deficit accumulated. Secretary of Agriculture James Wilson approved the expenditures, and Congress sustained the Act of 1908. The resulting precedent created a double system of accounting: one set of economic criteria applied to normal fire seasons, subject to budgetary constraints; another set applied to catastrophic seasons, subject only to the perceived needs at the scene.[45]

The act produced a profound ambivalence among professional foresters.

It was recognized as necessary, but regrettably so. Silviculture and economics were the fundamentals of forestry, and Pinchot, demanding that the national forests be conducted as a public utility, told the nation that the forests would, like any other business, become a paying proposition. Large fires were thus not only a publicly visible failure of field operations but also a violation of sound business principles. "After all," as Roy Headley never tired of repeating, "fire control is a form of economic behavior."[46] But the national forests were a political institution, not an economic one, and fire control successfully resisted efforts to apply strictly economic criteria for its conduct. As the Act of 1908 illustrated, economic considerations were in a fundamental sense secondary to the protection of the resources for the public and future good. The calculation of resource values under protection, moreover, proved exasperatingly intractable to any economic calculus known to forestry; nor were there any insurance companies that could, as they did for urban fire services, provide a standard of values. The potential for financial irresponsibility was great, and not a little of the debate about proper fire control objectives centered on the need to restrain the capacity for almost unlimited spending.

From 1911 to around 1930 the dominant theme of fire protection was the determination to organize, to search for models, and to isolate the governing concepts by which to transform folk practices into a formal bureaucracy capable of bringing uniformity and expertise to the protection of national natural resources. The decade following the 1910 fires, in particular, was a period of experimentation in methods and objectives; by the early 1920s there was a toughening of policy to show that professional forestry could manage fire control and, indeed, could not be conducted without it. The upper echelons of the Service—Chief Forester Graves and his lieutenants—immediately proposed means to cope with fire problems like those posed in the Northern Rockies. But the more fundamental problem, the question of transforming frontier fire practices into something that could accommodate industrial forestry, fell to field men whose studies went well beyond operational questions like fire patrols and telephone lines and who confronted head on the larger planning needs and the clarification of objectives with which fire control could meet its most stubborn critics. In Florida in 1911 Inman Eldredge proposed a system of fire management that could be integrated into the traditional cycle of southern rough burning.[47] In Arkansas in 1912 Daniel Adams proposed a general scheme of fire planning and crammed his study with dozens of ingenious devices, from fire shields to gas-powered locomotive pumpers.[48] In 1912 the *Use Book* was overhauled into the *National Forest Manual,* a compendium of Service policy, and it included a lengthy section of fire. All of these programs represented pragmatic reforms in fire control operations, an acceleration of trends, and a magnification of techniques already understood. Almost certainly this evolutionary process would have continued had not an alternative program developed in northern California. "Systematic fire protection," as the California concept became known, answered exactly those needs that American foresters as a whole clamored for: it outlined simultaneously the tenets of theoretical inquiry, a methodology of practical research, and a program of administrative reorganization. Within a decade of its invention, the concept was triumphant in California; within two decades, it would underwrite national fire planning efforts.

Why California? National forests were extensive there, and so was fire. With the aid of fire, brushfields had advanced over broad expanses of forest. Equally prominent were vast acres, privately owned, of mature pine. Nowhere else was light burning so loudly drummed as a solution to fire protection, and nowhere else was the Forest Service—and professional foresters in general—so ringingly denounced for "protectionist" policies. Out of this controversy came the concept of systematic fire protection: light burning was the forge and fire protection the hammer. All the program needed was someone to wield the proper instrument. California had that, too, in the person of its regional forester, Coert duBois.

An elder duBois much later described himself at the time of the "crusade" as a "young, impressionable, and enthusiastic idealist" who, "metaphorically crying 'Gifford Pinchot le vult!'" went into forestry "head down and tail up." Studying the brushfields of northern California, he realized that "there wasn't money enough in the U.S. Treasury to reclaim these brush fields by tree planting" and that fire control alone could reforest them. With scarcely a moment's rest after his honeymoon, duBois entered the 1910 fires in California. His experience on the Stony Creek fire near Tahoe haunted him for years. In 1911 he produced a fire plan for the Stanislaus National Forest, "which was to serve as a model for similar Plans for the other Forests. It was like a war plan which might have been called 'Operation Hell.'" Indeed, duBois looked to military operations as a possible exemplar. He initiated research on the economics and effects of light burning, a study he turned over to Stuart Brevier Show, his heir apparent. And there were dozens of special studies on all dimensions of the fire problem. From 1911 to 1913 the California forests were given virtually free rein to experiment with different systems of protection. In late 1913, when "all the chips were in and the rainy season on," duBois recalled, "I collected all the reams of data and special studies; Headley's on finances; Godwin's on discovery; Coffman's on causes; Redington's, Ayres' and Charlton's on educational work; Kelley's and Ryder's on organization of forest forces and Benedict's on tools and equipment, and retired to my cell, like John Bunyan, to write THE book on Fire Protection." In the spring of 1914 he emerged with the classic *Systematic Fire Protection in the California Forests*. "I consider that it took the hardest mental work I ever did," he declared, "and was the most important contribution to the public I ever made." It was the most important contribution to emerge from the entire era.[49]

DuBois proceeded by analogy to the efficiency studies then popular among industrial engineers, a methodology that broke down factory operations into minute components and then reassembled them according to predetermined intervals of time in order to maximize efficiency of operation. Time was thus the measure of efficiency, and efficiency the standard of value for factory output. In his search for "mathematical certainty" and a standard criteria by which to evaluate the success of protection organizations, duBois accepted just such a time standard, gradually transforming objectives based on burned area into a set based on time of control. The approach harmonized well with the accepted precept that fires were best controlled when small. In duBois's hands the method touched every conceivable dimension of fire protection. At its most sophisticated, it was used to give quantitative definition to rates of spread in different fuels; at its silliest, it precisely identified six different gaits for horses, with instructions for applying the proper gait to particular situations. So precise and comprehensive was the document that it simply overwhelmed skeptics, both inside and outside the Forest Service. It made California critics of organized protection, with their allusion to the "Indian way" of laissez-faire management, look like woolgathering folk philosophers. The book quickly became required reading.[50]

It was a peculiar document. Fire research remained, as it had begun, a field of forestry, grounded in a tradition of silviculture and economics. Yet duBois revolutionized its procedures by introducing other models, such as those of industrial engineering; by establishing the individual fire report as the fundamental unit of information; and by demonstrating how, through careful statistical milling of such information, a fire program could attain rigor. With its passion for Cartesian logic, its quest for "mathematical certainty," and its insistence on irrefutable atoms of scientific fact, *Systematic Fire Protection* did not exemplify American pragmatism at work in American forests or symbolize the transfer of German forestry into the hands of its former students; it meant the triumph of French positivism. Within a year after its publication the Forest Service undertook a major reorganization, established a separate Division of Fire Control within its Operations unit, and created an independent Branch of Research—all of which had the effect of crystallizing the need for a fundamental approach to fire protection systems. Into this urgent, somewhat amorphous ferment came the Cartesian rationalism of duBois's treatise. Though not apparent at first—while it remained as one among several approaches—the completeness and rigor of *Systematic Fire Protection* would prove as challenging to internal critics of the Forest Service's fire program as to external ones.

The issues went to the very fundamentals. In 1914 Graves released the first of a series of memorandums on the calculation of fire damages. The idea was that the amount expended for fire protection ought to bear some relationship to the value of the resources under protection and the cost of protecting them.[51] In 1916 Graves, Ferdinand Silcox, and Roy Headley circulated proposals for economic objectives in fire control. Graves revised his formulas.[52] Silcox proposed that as a standard for effective control annual burned acreages should not exceed 0.1 percent of the area under protection—a standard based, of course, on conditions in the Northern Rockies. Using California conditions, duBois had proposed 10 acres per fire as a standard for timber lands and 100 acres for brush, though he then converted these area objectives into time standards. Headley proposed an "economic theory," which became known as the "least cost plus loss" concept and which insisted that the combined sum of damages and suppression costs should be held at a minimum. The theory was nothing more elaborate than a restatement of profit maximization criteria, but it helped to verify forestry's insistence that forest management was a form of economic behavior, and it promised to furnish guidelines for the use of emergency suppression funds. If duBois's work gave fire protection its fundamentals for operations, Headley's study gave it an economic rationale. Both endeavors, however, underwent an endless number of permutations and reincarnations as each era reshaped their tenets to meet its own particular concerns.[53]

Nor were the two in complete harmony. DuBois implicitly recognized that the national forests were a political institution for which fire control was a necessity, though one that could be made more efficient; Headley insisted that the forests were economic institutions for which fire control might be too costly. Though nearly all foresters applauded the economic theory, it proved almost worthless in practice: no accurate assessment of damages could be made, especially where values included something more than timber, and the theory made assumptions about fire behavior that were not realistic. It broke down because fire seasons came episodically, fires were distributed logarithmically, and fires grew allometrically. The theory assumed the existence of an "average bad season" upon which budgets could be based, and it posited the existence of an average fire whose size represented a ratio between its driving force to expand and the resistance offered to it by control forces. But fires do not grow steadily, they blow up; and the vast percentage of damage and suppression costs result from the small fraction of fires that make this transition. A fire does not behave as if it were in rational economic competition with the Forest Service. A fire is often caught small or not at all, and for this reality duBois's efficiency criteria based on time of control were the only practical guide.

The economic theory was first published in the "Fire Suppression Manual" issued to all California forests in 1916. For the next three years it was accepted as a working objective and led, with Headley's approval, to a program of "let burning" in more remote regions. Although the costs of suppression dropped (it was really the only economic variable that could be manipulated), fire damages increased, and detailed studies by Show led to the restoration of a standardized, intensive control program in 1919. Show concluded that "these facts clearly point out the danger of any policy of protection that emphasizes low costs as the main objective, for this policy may not merely lead to a larger proportion of class C fires [10 acres plus], and therefore of damage, but may defeat its own purposes by greatly increasing the least controllable element of cost—that of suppression."[54] So great could suppression costs become that larger expenditures for initial attack were more than warranted. Not incidentally, Show was also engaged in continuing studies of light burning at the time, and it was difficult for the Forest Service to argue against light burning as both damaging and uneconomical while at the same time it tolerated natural light burns on its own lands. Meanwhile, California enacted a compulsory fire patrol statute in 1919, and, thanks to duBois, the Forest Service joined with the Army Air Service for a program of aerial fire reconnaissance in California. Fire suppression had a glamorous new tool to answer what increasingly looked like tired old arguments. Shortly afterward, duBois resigned to join the Consular Service and was succeeded by the man who had become the most vocal advocate of "systematic fire protection" and the most important innovator of its methods, Stuart Brevier Show.

The decade of experimentation came to an end with the Mather Field Conference near Sacramento in 1921.[55] The first national conference held by the Forest Service on any topic, it showed Service resolve to attack light burning on its own grounds, and it displayed the organizational genius of the new Chief Forester, William Greeley. Prostrated twice by overwork, Graves had resigned in early 1920, and Greeley, who had been regional forester for the Northern Rockies in 1910, assumed command. Greeley's autobiography begins revealingly enough with an account of his experiences in the 1910 fires: "the conviction burned into me that fire prevention is the No. 1 job of American foresters—both to keep trees on the land and to put faith in the men who own it." He openly professed that he considered "smoke in the woods" as the yardstick of progress in American forestry.[56] During his tenure as Chief (1920–1928) no aspect of fire protection went unregarded. The Mather Field Conference, which he arranged, was not only the first all-Service meeting on fire but also the first on any subject held by the Forest Service.

The convocation sought to establish standards for a subject that had for a decade been groping toward an intellectual foundation equal to its growing practical skills. The conference consolidated, standardized, and disseminated the accumulated expertise of the Service. Some 68 topics were proposed, assigned to committees, reported on, and approved, with personal commentary, by the Chief Forester himself. Discussion included such rudi-

13. Mather Field Conference, 1921. The conferees included: *rear row:* W. B. Osborne (2nd from left), Lyle Watts (6th from left), H. N. Wheeler (7th from left), Stuart B. Show (11th from left); *middle row:* Fred Morrell (6th from left,) William B. Greeley (7th from left); *front row:* Aldo Leopold (3rd from left), E. I. Kotok (6th from left), Evan Kelley (9th from left), Roy Headley (10th from left), Howard Flint (11th from left).

mentary topics as terminology, fire classifications, report forms, and statistics. Engaging many of the best minds in the Service, it brought to an end the period of uncertainty and experimentation in fire control. Another such conference was not needed for 15 years.

Of particular interest were the topics of financing and objectives. At issue was not only the total level of protection but also the means of distributing funding among different forests and, with the Weeks Act still adding cooperators, allocating the available federal funds among the various states. Again, the tendency was to separate the efficiency of fire control from the values protected. Greeley commented at length on the interpretation of a central concept discussed at the conference: elapsed time—another version of duBois's speed of control. "The time which elapses between start of a fire and attack is without question the ruling element which controls the success of fire control, personal equation of fire fighting force eliminated," the Chief confirmed. "Therefore, I believe that the study of elapsed time figures is [the] most useful known method of investigation to determine for various types of forest cover, and of topography, the proper distribution of fire control forces, which means, in other words, the allotment of funds for protection. For such studies, the elapsed time record is of primary value. Secondarily, the use . . . as a means of checking the performance of individuals, is an unquestionably sound reason for keeping accurate elapsed time records."[57]

Elapsed time standardized basic fire statistics, ensured uniform performance, and, perhaps unintentionally, tended to give undue weight to those areas most inaccessible to control and of less economic value, that is, to the remote backcountry where elapsed times would be lengthy. Fire protection would in a sense settle these areas. As early as 1912, for example, heavier grazing of backcountry lands was advocated within the Forest Service with the twofold purpose of reducing grassy fuels and of placing firefighters (the permittees) into these remote locations. There was little discord among participants at the conference over the various proposed objectives in fire protection; elapsed time was simply a means to stay within what Silcox had set as an acceptable annual burned acreage loss and what Headley had theorized as an economic goal. All three proposals were in fact endorsed by the conference. The intellectual and practical success of the conference marked the beginning of a national extension of systematic fire protection methods and the beginning of the end for light burning.

The Mather Field Conference only inaugurated Greeley's many real achievements in fire protection. More than merely directing the national forest system, Greeley advocated a national vision of cooperative forestry, founded on cooperative fire protection. In 1924 he helped to mastermind what became the crown jewel of his administration, the Clarke-McNary Act, which strengthened and expanded the provisions of the Weeks Act. With his assistance the Weather Bureau established a fire weather forecasting service in 1926, and in 1927 the Forest Protection Board was created, with Greeley as director, to coordinate activities among the federal land agencies. Through the board, Forest Service policies and procedures were disseminated among federal fire agencies as they were distributed among the states through the Clarke-McNary program. The McSweeney-McNary Act of 1928 gave the Forest Service responsibility for federal forestry research, including fire research.[58]

The problems of the national forests were never ignored. The 1924 fire season struck California especially hard, and suppression costs reached new highs. During hearings in 1926, Greeley admitted that "from a purely business standpoint it is obvious that we should do anything that can be done within reason to cut down these large emergency expenditures which are necessarily wasteful because they are made under emergency conditions that involve great haste and stress, and which, after all, simply represent the stopping of great destruction. They are not constructive expenditures." Greeley promptly appointed a committee to revise fire damage appraisal estimates and arranged to hold a financial conference to set better guidelines for the next year.[59]

Other experiments, both from field and study, were developing to address the same questions. From California Show and E. I. Kotok produced a steady flow of dense statistical studies showing in effect the ultimate logic of suppressing all fires while small. Evan Kelley of the Eastern Region, meanwhile, established a simple fire code based on time principles, which directed that each fire should be controlled before the next day's burning period, that is, by 10 A.M.[60] Within a year the Southwestern Region undertook an experiment in all-out control. "Let burning," "looseherding" in low-value backcountry, and ad hoc "take a chance" decisions about whether or not to attack certain fires in rugged hinterlands had plagued the local organization from the start.[61] The results of both regional experiments were resoundingly conclusive: both costs and burned acreages dropped with aggressive action on all fires, regardless of cause or location. What was happening was that the methods developed at Mather Field for primarily front-country lands of high-value timber were being extended into further regions. It made little sense for the Forest Service to ignore fires on its own low-value lands while campaigning, for example, for fire prevention on cutover lands that could not sustain mature timber for several decades. After all, the clinching argument against light burning—that it sacrificed future forests for present gain—could be directed equally against the Forest Service if it tolerated let-burn practices in its own backcountry or on its newly acquired forests in the South.

After the 1926 season Greeley personally inspected the fire damages suffered in the Northern Rockies. All of the national forests had been hard hit, except, with ironic perversity, the Selway. There the forest supervisor, in defiance of existing guidelines, had doubled manning and directed that all fires be attacked quickly. Losses were dramatically less than those of surrounding forests. The veteran of 1910 was impressed. He quickly authorized increased presuppression expenditures and arranged for yet another study of damage appraisals prior to the finance conference. When he left office in 1928, there were new guidelines for fire financing, and the *National Forest Manual* and the economic theory had been revised to read that the "objective of fire control is to reduce to a minimum the sum of the cost of fire prevention, presuppression, fire suppression, and the damages caused by fire."[62] Greeley's personal energies and the momentum that had been built up in California since 1910 allowed his successor to establish a special advisory committee on fire control, which would report to the Chief, and to create the Shasta Experimental Fire Forest as a showcase for the administration of systematic fire protection. Both the committee and the experimental forest were determined by members of the Show cadre.

At his departure Greeley left a story that deeply impressed his listeners—the men who would have to guide the extension of his program into more remote lands, cutover lands, and lands prized for their intangible values. When he returned from France with the 10th Engineers, Greeley noticed that every time the anchor was dropped and withdrawn, without exception, each link was tested. Even when only a few hours separated the tests, they were infallibly made. The captain explained to the curious forester that experience over the centuries had demonstrated to shipmasters that *no* chances with the anchor chain, on which depended the safety of the ship, could ever be taken. What seemed like an indiscriminate obsession was actually fundamental to the success of the voyage. To Greeley it seemed that fire control was the anchor of forest management and that by attacking all fires, regardless of season or location, officials were merely testing the links of the chain.[63] It was a parable whose significance would be more widely recognized when the coming of the New Deal multiplied the number of links and enormously magnified the means for testing them.

II

The turning point in fire control policy may well have come during the regional foresters conference at Washington, D.C., in 1930. Dozens of topics were assigned to committees for reports, but the Fire Control Committee, chaired by E. I. Kotok, was especially influential. Its ruling concern was expressed in the opening topic proposed to it: "What are we after in fire control and what expression of our objective or objectives will be most useful in making allotments for fire control by Forests and Regions?"[64] But that question was only a front for another: How far could systematic fire protection be extended—geographically, technically, administratively, and financially?

To guide its deliberations the committee continued a precedent set by the Mather Field Conference and classified the national forests into three groups according to their fire problems: critical, marginal, and acceptable. It urged that programs be allocated so as to upgrade critical forests into the marginal category, and marginal forests into the acceptable. Before this could be done, however, it was necessary, first, to determine if such a transformation was technically feasible and, second, to decide on what grounds the forests would be ranked—that is, to set new objectives for fire protection. Again the decisions made for the national forest lands could be (and were) equally applied to lands of cooperators under state jurisdiction and to lands managed by other federal agencies.

To answer the first question, Show and Kotok were preparing their *Determination of Hour Control for Adequate Fire Protection in the Major Cover Types of the California Pine Region*. Despite its precise title, the study was actually a brilliant extension of duBois's methodology, and it presented a general model for designing fire protection systems based on the concept of hour control. It was followed by other Show and Kotok studies that demonstrated how to rationalize all dimensions of fire protection, from smoke detection to fire effects. The hour control study coordinated well with the influential *Transportation Planning* by T. W. Norcross, chief engineer of the Service, and the new Chief Forester, Robert Stuart, gave his blessing to Show's and Kotok's synthesis. "The transportation studies," he agreed, "should be tied in closely with a determination of hour control needs by cover type."[65] The program was granted high priority along with other needs of fire protection. In this the report of the Improvements Committee, chaired by Evan Kelley, concurred: "In the opinion of this Committee, development and utilization projects should necessarily take a subordinate position in a priority rating until fire control needs are met."[66] Road and improvement projects for fire protection were approved first, and thus the studies collectively furnished a blueprint for the future development of the backcountry of the national forest system. When the future unexpectedly brought a floodtide of federal money and labor with the New Deal, the Forest Service was in an excellent position to transform plans into practice.

The Show and Kotok hour control program demonstrated the technical grounds for extending fire control, but the issue remained whether or not it ought to be extended and, if so, on what terms. The Fire Control Committee agreed that the "objectives in fire control" should be based on consideration of the variations in fire damage among various forest types. The committee then consolidated the assorted existing proposals into an index or scale. The more valuable forest species, pine and spruce, were rated at an index value of one, and since "complete fire exclusion within the next decade" was "obviously impractical," the committee set their permissible annual burned acreage goal at 0.1 percent.[67] Other forest types followed in relative proportion, with the acceptable burned acreages adjusted accordingly. But this goal, suitable for valuable timber on frontcountry lands, did little for deciding the real level of protection for backcountry lands or for lands with values other than marketable timber at stake. Nor could it guide execution of the hour control program, which, even at a primitive stage, would likely involve greater expenditures than those justified by the committee's index of values.

The conference hovered on the border of two problem fires. It upheld the values established in the fight against frontier fire while it cautiously explored the possibility of extending similar protection to more remote sites as means permitted. Its handling of backcountry fires—a topic it chose to address directly—was thus somewhat contradictory. Analyses of the question were presented by Earl Loveridge and Evan Kelley. As principles of management, "let burning" and "loose herding" were soundly denounced; the consensus was for "prompt and thorough protection or no protection."[68] The careful index of relative values prepared by the conference was impressive but, like Headley's economic theory, often practically meaningless. The calculus of time standards necessarily tended to replace the calculus of values as an operational goal. Thus a double standard persisted: one set of criteria was based on time, which could be measured uniformly and accurately, and another set was based on damages and value appraisals, which could not. The two were meant to be complementary, and a natural equilibrium of available means and ends tended to prevent one from absorbing the other.

The Fire Control Committee so impressed the Chief Forester that he retained it in an advisory capacity for several years. He also granted it the experimental forest it desired for testing different methods of fire protection. Thus was born the Shasta Fire Control Project. Directed by an all-Service committee, the project brought researchers and administrators together, advertised the hour control strategy, and launched the careers of many notable fire officers and researchers both. The committee met several times at various locations. At its third rendezvous, in Spokane, where the Forest Service maintained a large warehouse of fire equipment, it debated the question of equipment standardization and a possible equipment laboratory. The committee, in brief, was a kind of privy council on fire for the Chief Forester, one that perpetuated the concepts of the 1930 meeting.[69]

The Forest Service was thus poised to enter the various backcountry lands for which it had responsibility. But as with its occupation of frontcountry territory earlier, it hesitated over how much protection it could justify. Headley was now chief of the Division of Fire Control, and he proposed a national test of the let-burning program he had experimented with in California. With Kelley's support, he proposed in 1932 that minimum protection be extended to the high fire regimes in the remote mountains of the Northern Rockies—the very scene of the 1910 holocaust. A special board of inquiry convened. Its intentions somewhat paralleled those of the California light-burning board, but its format was dramatically different. For two weeks many of the top minds of the Forest Service undertook an elaborate pack trip through the controversial backcountry of the Northern Rockies, each night stopping for prolonged, quasi-formal discussions, all carefully recorded. The conversations read like the imitation Platonic dialogues common during the Renaissance. But instead of Castiglione and Galileo, the authors had names like Headley, Show, Kelley, and Loveridge. During this curious expedition, no aspect of the troublesome backcountry management question went unexamined, and in the end the gathering ruled against Headley. At the next regional foresters conference in 1935 the 10 A.M. Policy was adopted.[70]

Headley was horrified. But to Show, who had argued against this let-burn scheme as he had against its California predecessor, the episode was yet another illustration that "the whole field of fire control economics was barely touched by serious study, and that dangerous pseudo-economic thinking had not been laid to rest."[71] Show and Company had a better alternative, they believed, simmering on the backburners of Forest Service planning; the New Deal set it aboil.

The conservation programs of the New Deal were both a boon and a burden to Forest Service fire control. They brought an expansion in the amount and type of lands under protection, an enlargement of incentives, and a magnification of means. They pushed fire protection physically into the backcountry—not only of the Forest Service but of the nation—by resettling marginal farmland and by acquring cutover forests, tax-delinquent lands, and deteriorated grasslands for the national forest system. Despite their lack of immediate value, such lands were to be protected as fully as frontcountry lands. For political reasons, that is, the economic theory was rendered worthless as a guide; it could not apply satisfactorily to lands that did not have an immediate value, nor could it incorporate the "social" goals that were important to the new legislation. The new programs pushed protection into the backcountry backhandedly, too, by requiring that the abundance of money and manpower made available under CCC and WPA programs be put to use. By dumping money and manpower in larger quantities than could often be used, the New Deal swept away economic objections to an expansive program of fire control. The emergency suppression account was even enlarged to encompass presuppression activities, and for the first time thousands of civilians, already organized into crews, were available for active fire duty. Pleas for the military to station troops on the forests over the summer were finally dropped: the Army-run CCC sufficed. The blueprints prepared by the hour control program were filled out. Roads, trails, telephone lines, lookouts, fuelbreaks, hazard reductions, and guard stations—all appeared almost overnight. Thanks to cooperative fire programs, these benefits were extended to the states. It was amid this unprecedented outpouring of men and material that the Forest Service launched a complete modernization of its fire program. In the three years between 1935 and 1937 the Service sought to complete in full the program with which it had struggled piecemeal for three decades.[72]

The 10 A.M. Policy contained a simple directive, but its creation was not a simple event. There were, first of all, unusual environmental conditions. For over a decade a dry weather cycle had extended throughout the West, deepening disastrously in the early 1930s. Between 1905 and 1918 only one bad fire year occurred, 1910. Between 1919 and 1934, however, there had been five—1919, 1924, 1926, 1931, and 1934. In each of the three years immediately preceding the 1935 regional foresters conference, a conflagration of major proportions had raged in the backcountry: the Matilija fire of 1932, the Tillamook fire of 1933, and the Selway fires of 1934. The Tillamook fire was to forestry at large what the Dust Bowl was to farming. In the face of such extraordinary challenges, unprecedented control measures were called for. The response in fire control was not so different from those programs frantically rushed into being to cope with the blighting drought on farmlands in the East, with the Dust Bowl and soil erosion in the Great Plains, and with flooding in the Mississippi Basin.

With unprecedented challenges there came unprecedented opportunities. The Tillamook fire engaged the CCC on a fire assignment for the first time in large numbers, and it led to official sanction for their use on the fireline. The tidal wave of federal Emergency Conservation Work (ECW), of which they were a prominent part, brought opportunities for organized fire crews, for the implementation of the hour control program, and for reforestation on an unprecedented scale.[73] But as the Service became more dependent on various emergency funds, it saw its regular program dissolve and its appropriations slashed. Volunteer fire crews withered away, CCC enrollees replaced regular fire guards. When the CCC program ended, the fire organization was decimated. Many labor-intensive projects, including roads and fuelbreaks, permanently decayed into disrepair. Fire control was thus both a prominent beneficiary of New Deal conservation and, unwittingly, a victim of it. The flood tide of federal dollars carried all before it. The Forest Protection Board, for example, was dissolved. In its place ECW funds acted as a coordinating medium. With the Forest Service controlling the lion's share of CCC camps, it assumed a commanding position in the national fire picture; interagency cooperation was achieved largely through cooperative agreements with the Forest Service. The Service was left, in fact, as virtually the sole authority on the subject of wildland fire.

The sheer presence of CCC labor and ECW monies demanded, moreover, that they be used. It was easy enough to divert them into fire projects already on the boards, especially when these were granted high priority. But such schemes were soon exhausted, and the demand for more projects persisted, particularly large-scale, labor-intensive projects. Without the CCC, the debate about opening the backcountry was academic: there was simply not enough money and labor to construct roads, trails, and the rest. With the advent of the CCC, however, such projects became all but mandatory. Fire improvement schemes like the Ponderosa Way and the blueprints of the hour control program were a major sink for the abnormal quantities of manpower suddenly made available. Under the guidance of Show, California led the way in directing the legions of Roosevelt's Tree Army toward fire protection measures, providing model for the rest of the country and a demonstration of the hour control program. The 1935–1937 reforms in fire policy, organization, and physical plant were more than the opportunistic reactions of a bureaucracy gleefully expanding; they were, in subtle ways, forced on the organization by outside political, social, and economic conditions over which it had little control.

Overtly, the Forest Service was delighted. With respect to trained crews and physical facilities, ECW programs created in years what might have taken decades in a normal evolution of events. In the long run, less tangible aspects of the fire program, such as policy, were swept along as well. Instead of a slow evolution of policy and objectives, of thought and action mutually reacting in complex equilibrium, the construction of facilities and the appearance of organized crews far outstripped the debate about values and goals. With regard to policy, there occurred a fatal separation of means and ends, the balance between the feasible and the desirable. New goals had to be designated by administrative fiat, goals commensurate with the grandeur of the means at hand. Like the wave of precious metals that flooded out of the New World to the coffers of Spain, the emergency money momentarily enriched the economy and eventually impoverished the spirit of the Forest Service. New goals were created almost overnight, positing objectives that would not otherwise have been reached for decades and that in the course of ongoing debates, might have been revised or discarded. It seemed possible by one bold, revolutionary stroke to attain objectives that might otherwise be unreachable. Like Spain, blinded by its new wealth to the dangers of undertaking foreign, quasi-religious wars, the Forest Service expanded its fire program into its remote domains. By 1935, the objectives that had been regarded by the 1930 regional foresters conference as "obviously impracticable" seemed reasonably plausible. As the Chief Forester pointed out, here was an opportunity for experimentation on a "continental scale."[74]

The Selway fires deep in the Northern Rockies provided the immediate stimulus for a new round of debates in 1934. The fires were notable on several counts: they were the largest since 1910 and they saw the use of a fire danger rating meter, the first to forecast severe conditions with reasonable accuracy. The general failure of suppression forces to halt the fires raised uneasy questions about the technical feasibility of fire control in the backcountry, questions that in turn raised doubts about the proper objectives of fire control in these lands and that resulted in an important board of review

The fire danger rating meter was the invention of H. T. Gisborne. It was intended to restore better financial control over fire expenditures by providing a standard for comparing fires and performances and by furnishing a guide to presuppression funds, which, since Greeley, had become an accepted part of the theory behind fire economics. But the spirit and opportunities of the age were otherwise. Instead of restraining the emergency suppression account, the meter led to the creation of yet another emergency account, this time for presuppression. The account was activated as part of the 10 A.M. Policy. Thus the economic objections against extending system fire protection into the backcountry melted away. "As long the money is plentiful," observed a participant in the policy debates, "it is not necessary to worry about values; if money becomes scarce, highest protection to greatest values naturally follows."[75] There was plenty of money.

The Selway board of review removed another objection, that the low-value lands were simply not worth protecting regardless of the means at hand. The arguments here were both more subtle and more damaging, and they came, somewhat surprisingly, from Elers Koch. After the 1910 holocaust, Graves had singled out Koch as an exemplary administrator, a man who had confined fire losses to acceptable limits by an ambitious system of trails. But it was another trail that now concerned Koch, and although it had much to say of values, it had nothing to do with the economics of fire suppression. "The Passing of the Lolo Trail" was his passionate lament over the vanishing wilderness. In the name of civilization the ancient Lolo Trail, over which Lewis and Clark had crossed the Continental Divide, had disappeared before bulldozers and automobiles. That was hardly news, but civilization in this case had come in the form of fire control. "So it is everywhere," Koch signed. "The Forest Service sounded the note of progress. It opened up the wilderness with roads and telephone lines, and airplane landing fields. It capped the mountain peaks with white-painted lookout houses, laced the ridges and streams with a network of trails and telephone lines, and poured in thousands of firefighters year after year in a vain attempt to control forest fires." In a series of damning rhetorical questions Koch asked—as others of this era were asking of the Dust Bowl—whether "all this effort and expenditure of millions of dollars added anything to human good? Is it possible that it was all a ghastly mistake like plowing up the good buffalo grass sod of the dry prairies? Has the country as it stands now as much human value as it had in the nineties when Major Fenn's forest rangers first rode into it?"[76]

Dismayed over the Selway fires of 1934, Koch—a man with an unimpeachable reputation as a firefighter and as the first historian of the 1910 fires—sadly answered in the negative. "I am not criticizing the efforts of others," he explained. "I have personally taken a considerable part in four major fire campaigns on the Lochsa River . . . but so far as results are concerned there is little difference between 1919, when crews of thirty or forty men, in a vain but courageous gesture, were trailing the leeward end of each of five or six gigantic fires, and 1934, when firefighters were counted by the thousands and the fires swept 180,000 acres." Against a major fire, Koch declared, "the whole United States Army if it was on the ground, could do nothing but keep out of the way. After years of experience I have come to the considered conclusion that control of fire in the backcountry of the Selway and Lochsa drainage is a practical impossibility. I firmly believe that if the Forest Service had never expended a dollar in this country since 1900 there would have been no appreciable difference in the area burned over." Fire control "has resulted in greatly modifying and to a large extent destroying the special values of a unique and distinctive wilderness area," Koch concluded, while "the results of fire control have been negligible."[77]

Koch wrote the essay in white heat after the board of review presented its findings and after discussions with Headley and Kelley. He mailed it to the *Journal of Forestry,* which gave it to *American Forests.* Then Koch withdrew it, feeling that "the article is indiscreet and might be quoted to the disadvantage of the Forest Service if it appeared in a popular magazine"— a good illustration of how the fraternity of professional foresters could work to censor itself without official pressures. Instead, he mailed it to the Chief in November 1934 "merely as a memorandum of a viewpoint which I believe is held at least in part by a good many other foresters."[78] Apparently, however, the only others to voice such apprehensions were Aldo Leopold and Robert Marshall. Both mourned the breakup of wilderness in the name of "protection roads" for fire and predator control, which had become especially pronounced with the advent of the CCC. Despite Koch's self-censorship, Headley urged publication of the essay, and it was run shortly before the 1935 conference.

Koch's lament was an eloquent plea for the preservation of wilderness values. But it was seriously isolated: the idea of wilderness is an evolving concept, and even the most outspoken proponents of wilderness and national parks were in their day strong advocates of fire control. Wilderness had no place in the index of values computed by the economic theory, and by claiming that fire protection was a "practical impossibility," Koch fatally flawed his argument that, in effect, cultural values outweighed environmental values.

The 10 A.M. Policy and the hour control program were designed explicitly to overcome just such "practical" obstacles. In a detailed reply Loveridge systematically broke down the latter argument. Failures were attributed to "*a lack of a clear cut, readily understandable fire-control policy for the 'backcountry' under discussion.* ... Imagine the effect on military officers and troops under a similar handicap. Not being sure of their objective, a fatal hesitancy, a lowering of alertness, would be bound to result. This has been true, at times, with the fire control forces in this section."[79] Let-burners belonged in the same category as light-burners, people who refused to grapple with the challenges at hand, however seemingly intractable, and who preferred inaction to aggressive management. The strenuous life still beckoned; Koch and Headley belonged to an old guard not yet accustomed to the possibilities inherent in the massive amounts of money and manpower unleashed by the New Deal. To Koch's "radical" proposal, Loveridge countered with one equally radical: the 10 A.M. Policy.

Headley, too, left the Selway board of review with mixed feelings, and he expressed them in an eleven-page confidential memorandum to Silcox. "Any survey of the present status of fire control in Idaho is disconcerting," he glumly confessed. "We were never so well equipped and in certain ways, never so helpless as in 1934." During the review, he reported, three fire protection policies were examined: first, that low-value country be withdrawn from protection; second, that the 1930 conference's burned area objectives, however inadequate, be met, but with only small increases in financing; and third, that the Service adopt a "keep every acre green policy." No one really agreed with either Koch or Loveridge, at the two extremes. "Pride," as Headley reported it, was the reason Koch gave as to why fires were fought in some areas at all. Headley conceded, moreover, that Koch was unquestionably right "that the country is in worse shape now than when we took charge of it." But the old staff men from the Washington office also recognized that Koch's proposal could never command official support. Too much pride was indeed at stake, and too many appropriations. "Bureaus and professions do not voluntarily withdraw under such circumstances. It simply is not done."[80]

Support for the other two alternatives was split nearly evenly, but in practice there was likely to be little difference between them. "Even Kelley admits that physical things and human tendencies being what they are," Headley noted with dismay, "the policy that Region 1 proposes comes to about the same thing in the end as the 'keep every acre green' policy. . . . The distinction between 'present policy' and the unlimited protection policy is therefore an unreal one." But it was one that Headley was reluctant to abandon. "At this point I want to record my lack of respect for the 'keep every acre green' policy or any similar financial policy," he informed the Chief. "If I could see any practical point in opposing the trend, I would want to do so with all the power at my command. But when I find Kelley and Stockdale advocating policies and plans which call for 100 percent or more increase in the cost of fire control, it cuts the ground from under me. . . . It seems to me more intelligent to go along with the trend and await the possible imposition of a more social point of view by forces outside the organization." Headley placed faith in his "favorite hope" that the techniques of aerial firebombing would be developed as "the only escape from the foregoing financial and social policy." The Chief declined to sign approval for the memo.[81]

The only objection left to 100 percent suppression was a refurbished version of the light-burning controversy. The special requirements of southern silviculture and fire control demanded some sort of broadcast burning. The land base was different there, too—most of it was in private and state hands—and the objective was the regeneration of cutover or abandoned lands rather than the protection of existing, if remote, stands. After 1923 the Forest Service had relegated light burning to the trash heap of outmoded fads and heresies. But as the Service and its cooperators assumed responsibility for more cutover, tax-delinquent, and resettled lands in the South, the issue rekindled and posed a challenge as serious as that advanced by Koch. As systematic fire protection spread over more southern lands, it had to incorporate some burning programs. In 1932, while let burning heated up the Northern Rockies, the Forest Service allowed its southern cooperators to engage in protective burning and still qualify for Clarke-McNary funds, provided the burning was done according to an approved plan.

In 1935, while a regional foresters conference debated the 10 A.M. Policy, the Society of American Foresters sponsored a session on controlled burning chaired by H. H. Chapman. Coming from professional foresters, the arguments for controlled burning might have been successfully disengaged from the opprobrium of light burning and developed into an alternative to systematic fire control, one organized around broadcast fire. Or at least, as with Koch's plea, they might have prevented the establishment of a single national program. But it was never the intention of these foresters to overthrow systematic fire control, only to supplement it with controlled burning. They tended to emphasize the uniqueness of the southern scene and the need to liberalize the fire control and fire research methods being introduced into the South, not to promote a national program of burning based on the southern experience. Their arguments were regional, based in fact on a careful contrast between northern and southern fire: southern forestry, they implicitly argued, was as unique as the southern political, economic, and social milieu. By its nature the protest did not offer an alternative for a national program of fire protection.[82]

Yet a national program was exactly what the Forest Service demanded and what the new means at hand suggested was possible. Like the light-burners and let-burners, the southern advocates of broadcast burning never fashioned coherent systems on the order proposed by Show, Kotok, and duBois. Instead, they offered pragmatic advice on the management of such seeming exotics as the longleaf pine. Like Koch's eloquent paen to wilderness, the challenge was turned aside, not by virtue of flaws intrinsic in its logic, but because of its regional specificity and its political context. Yet as fire control expanded its southern domain, it had to assimilate more environments than those manifest in the Cordillera, and forms of prescribed burning were part of that accommodation. The protest was not really an objection to the natural policies of expansion but a statement of the terms by which expansion had come to this region. The proposal was advanced as a means of improving fire control capabilities, not, as Koch's seemed to be, as a means of removing it as technically incompetent. Just as special fire prevention campaigns had been designed for the South, so the new fire protection program would also incorporate elements of folk idiom. For a national model, however, the Forest Service looked to the Northern Rockies as earlier it had focused on California.

It was within these contexts that Ferdinand Silcox convened his regional foresters in April 1935, and Silcox himself was hardly inclined to veto the results at the last minute. The Chief Forester was a firefighter of repute and a man deeply committed to the spirit of the New Deal. That the Forest Service directed so many CCC camps is at least partly attributable to Silcox's public enthusiasm for the Roosevelt administration. Silcox had originally proposed 0.1 percent as an annual burned acreage objective, but he responded warmly to the new possibilities opened by the New Deal. Like other reforms of the era, the 10 A.M. Policy was promulgated as an experiment—pragmatic, humanitarian, and dependent on massive planning and funding by the federal government. It sought to protect the nation's lands and citizens from an unprecedented plague of social, economic, and environmental disasters, though, like many other New Deal programs, it was destined to persist long after its period of experimentation, still reliant on emergency deficit spending.

At the conference Earl Loveridge proposed that, at least as an experiment, a new fire policy be adopted aiming at control by 10 A.M. the day following the report of a fire. If the fire escaped control, plans would be made to control it by 10 A.M. the next day; if it escaped again, control would be planned for 10 A.M. the following day; and so on. The idea, of course, was not new, but the prospect of enlarging it as a blanket prescription over the nation was. A dozen arguments were advanced in support of the proposal: "accessibility" was only a matter of degree; low-value land today might have higher value in the future; land not worth protecting from fire should not be in the national forest system; "intangible" forest values increased the worth of so-called low-value areas; fires often broke out of remote lands into high-value lands; large backcountry fires crippled fire detection systems by flooding forests with smoke; temporary and seasonal laborers hired as fire guards were not adequately trained to decide which fires to attack aggressively and which to let burn; forest rangers, too, needed—and wanted—explicit guidelines; ambiguity led to confusion and tardiness; the ultimate cost of suppressing a large fire was many times the cost of extinguishing a good many small ones; the major conflagrations of recent years were attributable to a lack of aggressiveness; and "the sum total of costs plus losses, based on a full consideration of values, will be less under an all-out suppression policy than under the policy previously in effect."[83] With the vigorous advocacy of Show, the resolution passed unanimously.

Silcox distributed the new policy, soon known as the "Forester's Policy" or the 10 A.M. Policy, through two letters to all rangers, a short one on May 7 and a long one on May 25. In concluding the short letter, Silcox wrote: "No fixed rule can be given to meet every situation; the spirit implied in the policy itself will determine the action to be taken in doubtful situations." The Chief further instructed that "subject to the action required to meet the above quoted policy, expenditures for preparedness and suppression will be held to the absolute minimum, and will vary with the total of the tangible and intangible values," though no method existed for determining just what values were endangered or how to calculate them.

In elaborating the spirit communicated in the short letter, the long letter introduced confusion. Its emphasis, first of all, was to confirm the established policy of keeping fires small. Perhaps buoyed by the grandiloquence of the idea, the Chief then indulged in an unfortunate bit of rhetoric. "Beyond that point the policy is new and embodies important changes. It includes all National Forests in its scope. It emphasizes rapid suppression of all fires. In these respects it treats all areas on an equal basis. It eliminates our current appraisals of tangible forest values, of themselves, as the basis for different intent and driving force in fire control. It simplifies both the making of key decisions on the ground, and the inspection and review of suppression and pre-suppression action." Many interpreted this statement to mean that values were thrown out of consideration. In fact, it only dismissed tangible values, thus granting backcountry lands equal stature with frontcountry lands. According to both letters, responses to reported fires were to be determined through a "calculation of probabilities." But much to the dismay of even avid supporters, including Loveridge, the long letter became the foundation for the policy.[84]

The policy was intended to turn fire protection around in the face of drought and holocaust, to rally the program as other New Deal legislation would revive the slumping economy. Neither achieved its explicit goal. The big fires continued, and the economy stagnated. It was wartime production that brought the economy out of the Depression, and it was postwar technology that gave fire suppression the mechanized tools it needed to reduce burned area figures and to meet hour control objectives. The interim results were all too often merely inflationary—an inflation of price and bureaucracy on the one hand and of fuels and total program costs on the other. Even the spirit of the continental experiment was eroded. The long letter of May 25 perplexed many field men and led to mere gargantuanism and unjustified extremes. But as the years passed, the policy entered folklore and acquired a mystique that made discussion of it difficult. It took on the aura more of a papal encyclical than of a circular letter from the Chief Forester. It seemed above reproach, and whatever its difficulties in particulars, its liberal spirit had to be sustained.

When it was issued in 1935, the policy was an administrative tour de force. It standardized firefighting for the first time. It was intended to improve fire suppression, especially in backcountry areas, but by implication it came to stand for a Service-wide attitude toward fire and the fire protection mission. Its adoption immediately catalyzed the Division of Fire Control: Headley made an inspection tour of the western forests; the Forest Service declared 1935 as fire prevention year; and major conferences were scheduled. Three national meetings were held in 1936 alone, two in Washington, D.C., and one in Spokane. The Spokane Fire Control Meeting was the first comprehensive conclave since Mather Field, and, like its predecessor, it left no aspect of fire protection unexamined. Fundamental reforms emerged: the Foresters' Advisory Committee on Fire Control was reactivated; *Fire Control Notes* was founded as a quarterly periodical to be published by the Forest Service; a committee began compiling the first fire equipment handbook (issued in 1938); a firebreak conference was scheduled for Los Angeles; preparation of an all-Service training handbook was begun; experiments were undertaken with chemical retardants and aerial bombing, which led eventually to the smokejumper program; and James Montgomery Flagg was commissioned to paint his celebrated fire prevention posters. Inspired by Forest Service enthusiasm, the American Forestry Association promoted further programs in 1939. In 1937 a special conference for fire officers was held at Portland, Oregon, and a conference in Washington launched a three-year national replanning program for fire control—with the Forester's Policy as its goal and the hour control program, as revamped by Lloyd Hornby, as its methodology. Out of the ferment came the psychosocial research of John Shea, agitation for an equipment development center, and a contagious enthusiasm that the fire question was at last approaching a manageable state. Follow-up conferences were held in 1940—a Service-wide one at Ogden and another for the Southeast—and a fire research conference was held in 1936, with a follow-up at Priest River in 1941.[85]

At the 1936 Spokane conference Kelley and Loveridge were the prime defenders of the new policy, noting that its accent was on adequate presuppression measures, that the index of efficiency was based on time of control, and that values would be satisfied automatically by keeping all fires small. The policy was well received by most field men; explicit standards relieved them of uncertainty and made for fairer inspections. The Forest Service is a unique blend of autocracy and autonomy: it establishes standard, centralized policies, but decentralizes the power to implement its programs. To most men on the districts the very rigor of the 10 A.M. Policy thus became an asset.

Not everyone was pleased with the direction of this new energy, however. For the most part they were men, like Headley, still committed to the economic theory. To most foresters, however, the economic theory, as originally formulated, was already on the ropes. John Curry declared the position of most when he observed that "the minimum cost theory has largely failed as a guide to fire protection needs," and he revised it to include presuppression (or "preparedness") costs.[86] Total costs were to be minimized by concentrating on those variables that could be measured, notably expenditures and burned acreages, rather than on nebulous values at risk, which could not. An emphasis on presuppression could in theory reduce total expenditures, and speed of control could minimize the burned area. Thus Curry gave economic justification for precisely those points—presuppression and quick initial attack—that were the soul of the 10 A.M. Policy. The rethinking bothered some nevertheless, a perplexed Roy Headley among them. The 10 A.M. Policy "undoubtedly had an effect" on fire control, he conceded in 1943. "Whether this effect was desirable or not is hard to prove."[87] Whether or not the policy could be justified, that is, it seemed to be working.

A singular cause for alarm appeared in the memorandums that announced the policy. Whereas research and past field experimentation had made an increased emphasis on presuppression only seem prudent, in May 1935 the Chief explicitly made emergency suppression funds available for presuppression activities. The use of emergency funds—an extension of the Act of 1908—seemed only to aggravate the old question of accountability. Throughout the 1930s federal fire control became increasingly dominated by emergency funding programs existing outside regular, budgeted appropriations and by CCC, ECW, WPA, and NIRA funds as well. In some important respects firefighting never recovered from the experience. Before the New Deal, the use of the emergency fund had been something of an embarrassment, an admission of extraordinary events, something that better fire protection measures and accounting procedures would eventually eliminate. The states, for example, which were precluded by law from deficit spending, financed fire emergencies through other, budgeted means. After the New Deal, emergency funding tended to replace regular appropriations matter-of-factly. The 10 A.M. Policy was then both a product of emergency monies and a prod to use them further.

Massive emergency spending, like massive manpower reserves, proved to be both a blessing and a curse. The 10 A.M. Policy would have been unthinkable without virtually unlimited drafts from the U.S. Treasury behind it. But the potential for reckless spending on a prodigious scale made mandatory an explicitly worded policy and rigid standards for control. This the new policy provided. Ultimately, and if rigorously administered, it would result in a reduction in total costs. Most fire officers saw no conflict between the instructions of the 10 A.M. Policy and the revised version of Headley's economic theory. Most thoughtful critics would probably have agreed with Kenneth Davis that "studies—Hornby's for instance—certainly have shown that a strong initial attack is generally the cheapest, and first period control [by 10 A.M.] does have a certain psychological stimulus."[88]

Such reasoning did not impress H. T. Gisborne, who was appalled to find that his invention was used to expand funds rather than to control them. In 1941 he exclaimed that the policy "undoubtedly rates either a milepost or a tombstone on our 35 year road of progress. If and when that policy becomes clearly recognized as a temporary expedient, I believe that it will rate a milepost. If, however, it has become or ever does become the death knell of all previous objectives based on damage, then it rates a tombstone executed in the blackest of black granite. It has already cost us six years of attention to variable damage as an objective," he fumed. But he admitted, too, that "it seemed to have achieved something else which may have been, at the time, worth more than the little thought which might have been given to damage." He even accused the policy of bureaucratic hypocrisy. It "is not fully enforced," he insisted, and "the framers themselves never intended that it should be. It therefore says one thing but means another." In a calmer moment Gisborne observed that the 10 A.M. Policy "actually had the same objective as the Show and Kotok minimum damage theory of 1923; to wit: stronger prevention and presuppression action so as to catch small fires rather than stronger suppression action. . . ."[89] And only the year before, Gisborne had lent support to the policy with a comparative study of intensive versus limited suppression based on the great lightning storms of 1940. In the United States, all fires were promptly attacked, and burned acreages were minimal. In Canada, however, the fires were fought selectively, with heavier losses and ultimately with higher suppression costs.[90]

Most of the policy's critics left the scene during or shortly after World War II. For the succeeding generation the 10 A.M. Policy was not an experiment whose spirit would dictate appropriate decisions but an entrenched part of its bureaucratic reality and inheritance. Postwar events tended to prolong the value of the policy. Not until 1967 would it be discussed in depth again. When that time came, the arguments for and against the policy had changed little, but their context had altered dramatically.

That context is fundamental for understanding how the policy first came about. The promulgation of the 10 A.M. Policy was a unique event, compounded from other unique events. The policy was created, in part, because of a dry weather cycle that resulted in an abnormal series of heavy fire years. In this it resembled other panic legislation and large-scale programs, like the Shelterbelt, which also responded to the terrible drought of the early 1930s. In part, the policy addressed a special environment of fuels—vast cutover pineries in the South and the Lake States; heavily burned areas in the Northwest and Northern Rockies; infestations by exotic fungi and insects that left massive tracts of dead timber; and everywhere great volumes of untreated logging slash. The policy addressed the values of lands then remote from commercial exploitation and without practical recreational or wilderness uses. It appeared during a unique moment in national history, too, a mixture of politics and economics that made the New Deal and the CCC possible. And then, of course, there were personalities: a President keenly interested in forest protection; a Chief Forester who began his career in the smoke of 1910; an Old Guard who still remembered Gifford

Pinchot and the crusading origins of the Service; Yale men who, in search of the strenuous life, sought out the great fires of the West and who made their way up through the ranks by their ability to handle crises, of which fire was a prime example.

In almost dialectical fashion, the 10 A.M. Policy helped to create an environment in which it was no longer needed. When the policy appeared in 1935, fire protection was still expanding as a philosophy and as a system. When the policy was replaced in 1978, that traditional expansion was completed. The transformation from agricultural fire practices to industrial fire practices was itself a historical process without precedent, and it may be said in the final analysis that the 10 A.M. Policy was, as its originators intended it to be, an experiment on a continental scale in administering that transition.

III

World War II took away most of the means, but not the goals, of the 10 A.M. Policy. The war caused a severe drain of manpower and money, and when it ended, so did the New Deal. But the continental experiment embodied in the 10 A.M. Policy endured, even thrived. Rather than replace the policy, the Forest Service was given new incentives and discovered new means to sustain it—initially, through the wartime emergency; then, through the imperatives of the Cold War. It expanded the policy's geographic domain once again, this time to include rural fire protection, the territories, and international contacts. It acquired new incentives from civil defense and from the creation of new fire regimes, such as the one in Southern California. Both had the effect of introducing still more values, including national defense, that could not be absorbed by the simple calculus of the economic theory, and both gave fire control a further sense of urgency, even a new moral energy. And through its wartime liaisons with civil defense and the military the Service discovered new means to replace those lost with the waning of the New Deal, contacts that would grow stronger with the deepening of the Cold War. Office of Civil Defense (OCD) and Department of Defense (DOD) money helped to establish a fire research division, and war surplus equipment—after conversion experiments done in cooperation with the military—became the basis for the strategy of conflagration control. The emergency accounts were maintained, and organized crews were developed to replace the CCC. The upshot was that fire control became nearly a paramilitary service of national defense, wildfire tended to be typed as enemy fire, and the Forest Service more than ever became the centerpiece in the national system of fire protection. The 10 A.M. Policy was never questioned. The genius of the coming era was to discover and elaborate the techniques by which the old policy could be sustained.

Soon after Pearl Harbor the Forest Service became increasingly involved with civil defense. Fire detection systems along the West Coast were put on alert for enemy aircraft, and special fire protection districts were created in sensitive areas along the eastern seaboard. Fear of incendiary attacks and sabotage became locally important. Civil defense officials organized volunteer crews into the Forest Fire Fighters Service, and agreements between the military and the Forest Service provided for mutual assistance in fire control and sought to limit the number of fire starts resulting from military training maneuvers. Hundreds of marines were mobilized, for example, to combat fires near San Diego in 1943 and 1944. When the Japanese launched long-range incendiary attacks by balloon, Operation Fire Fly detailed regular troops to the Forest Service for fire control.

In 1943, partly under the pressure of reduced manpower, the Forest Service allowed broadcast burning for fire control on the national forests and entered into a cooperative agreement with the Department of the Interior for mutual assistance on fires. Despite its reduced resources, the Service was also assigned responsibility for coordinating rural fire defense activities.[91] A year later Chief Forester Lyle Watts circulated two proposals for postwar consideration: one concerned the prospects of an equipment development center under Forest Service direction, and the other, the establishment within the Service of an autonomous Division of Forest Fire Research.[92] That same year the Forest Service was authorized to sell and distribute supplies, equipment, and materials to other government agencies and to state and private cooperators in an effort to bolster local fire forces.[93] The Service thereby became a conduit for equipment as it was for policy, planning, and federal matching funds. When the war ended, John Coffman of the National Park Service suggested that the old Forest Protection Board be reinstated, but no one else considered it necessary: the Forest Service had achieved de facto integration of the national fire establishment.

By 1950 the wartime liaison connecting the Forest Service fire programs with civil defense and the military was strengthened. All shared a common problem: mass fire. World War II had been a fire war, and officials grimly forecast that the next war could also be a fire war of probably even greater devastation. Mass fire thus intensified old Forest Service fire problems and introduced new reasons for expanding its response: mass fire moved the question of suppression techniques out of the backcountry and into the urban fringe, and it added new values at risk quite beyond the traditional realm of forestry. As the Cold War worsened, the alliance between fire control and national defense improved. Conflagration control replaced hour control as a strategic concept. Particularly with its use of fuelbreaks, the doctrine of conflagration control became a policy of containment, not unlike that popular in contemporary geopolitics. Its new contacts never reached the extent that the Service had known under ECW programs, but they were ample—and for fire research and equipment development, they may well have been decisive. With the help of these new liaisons, and amid growing national affluence, the Service was able to develop new means commensurate with the grandeur of its established ends.

The accent of the new era became evident even as the war ended. In 1945 an Equipment Development Meeting in Washington, D.C., explored ways of converting newly developed and soon to be released military surplus into hardware for fire control. Diminished manpower among forest industries during the war had led to successful mechanization; the chain saw and tractor-plow, for example, became locally important. Between 1947 and 1948 another burst of administrative energy set the tone for the era of mechanized fire control, much as the 1935–1937 years had for the labor-intensive era. The Clarke-McNary program was reexamined and overhauled with national meetings, and its financing greatly expanded. The war gave many men a new awareness of organization, and fireline procedures were accordingly revamped with the publication of the first national fireline notebook. A joint meeting of fire control officers and fire researchers was staged in Washington; an equipment development meeting promised an updated fire equipment handbook; and region-by-region reports were analyzed for "trends in fire control." Cooperative tests were scheduled with the military to ascertain the value of helicopters for fireline operations and the suitability of air tankers for bombing wildland fires with chemical retardants. A follow-up conference was held in Ogden in 1950.

In 1954, the same year that it acquired national responsibility for rural fire defense, the Forest Service also convened two fire conferences—one at Ogden and one at Washington—and persuaded half a dozen agencies, including OCD and DOD, to participate in a one-year crash program in fire behavior research and technology transfer. Operation Firestop was conceived, in a sense, as the Manhattan Project of fire suppression.[94] The allusion is not facetious: the specter of thermonuclear war gave an urgency to fire protection that could not have evolved out of the Service's own resources. It forced the Service to consider fire in a context other than forestry and to evaluate its successes in terms other than those proposed by the economic theory. When a new study of values at risk was conducted in 1956, it was contracted to an organization outside the Service, and it operated implicitly on the assumption that a fire would be a mass fire and probably an enemy fire.

Especially after the Cuban missile crisis, fire research blossomed with military and civil defense funds. Three fire laboratories were established, and the conceptual description of fire was completely revised. For equipment development, too, it was a golden era, and the Forest Service funded two centers for equipment research. The Federal Excess Equipment Act extended the wartime authority of the Forest Service to furnish hardware to cooperators, which eventually meant not only military surplus but also purchases through the General Services Administration.

Thus the Forest Service came to dominate the national fire establishment by virtue of its mandated responsibilities, its disbursement of Clarke-McNary funds, its supervision of the Cooperative Fire Prevention Program (Smokey Bear), its control over the production of essential information and research, and its responsibility for fire equipment development and distribution. It was responsible, too, for the organized crews that emerged around 1950—for their development and for their deployment during fire emergencies. Because of the Service's near monopoly of responsibilities and resources, wildland and rural fire protection achieved an impressive degree of integration and, to a lesser extent, of standardization.

But power had come to the Forest Service both by deliberation and by default, and as that power grew, the Service found itself subtly corrupted in spirit and imagination. The progress of events since the New Deal had isolated the Forest Service as something of a benevolent dictator in the field of fire management. This position proved doubly problematic because the Service had historically demonstrated itself to be responsive to ideas from within its ranks and distrustful of those from outside—a sentiment reinforced by its insistence that managerial positions would go only to professional foresters, a fraternity with closely shared values. As a bureaucracy, the Service showed an interesting mixture of autocracy and autonomy, of tradition and innovation—tradition in the realm of policy and innovation in the ways to implement it. Its new liaisons did not challenge its inherited fire policy; rather, in strengthening that program, they tended to narrow it. New funds took fire research into the physics of fire, but not into its ecology; into military uses of fire, but at a time when the new environmental movement argued for uses of fire in wildland and forest management; into the context of urban fire and disaster preparedness for civil defense, but not into the area of environmental pollutants and land ethics. It had long been an article of faith among American foresters that fire control was the beginning of forestry. But it seemed to many during the era of the 10 A.M. Policy that it had become the end of forestry as well. The nation's Chief Forester looked more and more like the nation's fire chief.[95]

As it became more entrenched as a firepower, the Forest Service became more estranged from public interest in its public domain, and when challenges to policy appeared, they came from outside the Service—from new legislation, from federal competitors in fire control, from nonforesters who questioned the logic of Service attitudes, and from a new problem fire, wilderness fire. Not until 1967 was there a serious policy review; not until 1971 were there bona fide amendments; not until 1978 was the venerable 10 A.M. Policy superseded by a wholly new policy.

IV

By the early 1970s the hegemony of position and policy that the Forest Service had enjoyed was seriously questioned and, over the course of the decade, replaced by a pluralism of means and ends. Wilderness fire succeeded mass fire as the topic of informing interest. With the Cold War in twilight, the liaisons and shared concerns that had linked the Service with OCD and DOD lapsed into insignificance, and with them went the moral energy that had helped to sustain the rigor of the old policy. Instead of a continued geographic expansion of areas under its protection, the Forest Service saw its fire control machinery removed from certain wildland sites, replaced by prescribed burning in others, and no longer needed by other federal agencies, which had developed independent suppression organizations of their own. The abundant means that had underwritten the 10 A.M. Policy either dissolved or came at the cost of a loosening monopoly of resources. The emergency presuppression account was disbanded; the investment in manpower and equipment reached a point of diminishing returns; and wholesale modification of the fire environment (that is, of fuels) became the tool of preference. The Boise Interagency Fire Center (BIFC) provided a collegiality of shared suppression resources, while the National Wildfire Coordinating Group furnished a common market for the exchange of knowledge about all dimensions of fire management. Though it fully supported both organizations, neither was an institutional invention of the Service. New sources of funding like the National Science Foundation (NSF) replaced the one directed by the National Academy of Sciences Committee on Fire Research, which had been set up by the Service and the OCD. New legislation removed a degree of administrative discretion and autonomy from management of national forest lands and subjected Service fire policies to review by agencies and professions outside the Service. In brief, new ends as well as new means came into being. Fire policy had to contend with new areas of interest, like wilderness; with new means, like prescribed fire; with a new research emphasis, largely fire ecology; with new standards set by environmental legislation and reviewed by outside agencies, like the Environmental Protection Agency; and with a new problem fire, wilderness fire. The cumulative effect was to replace the 10 A.M. Policy.

The agitation for policy reform came most visibly from revived interest in the backcountry as a result of the Wilderness Act (1964) and from a series of fire ecology conferences (1962–1975) organized by the Tall Timbers Research Station, a private Florida laboratory. The first focused on the proper goals of fire protection in official wilderness areas, particularly on the beneficent role of natural fire; the second, on the beneficial aspects of consciously applied prescribed fire in a variety of environments. The Wilderness Act slowly relieved fire officers from the practical question of fire control in remote areas and from the bureaucratic problem of appearing not to manage the lands should the Service's only real presence, fire control, be withdrawn. But if the act removed the imperatives of hour control and conflagration control, it also voided the logic of the economic theory: the economic values of land, intangible or otherwise, were secondary to cultural considerations. Moreover, as scientific evidence mounted for the positive benefits be derived from natural fire in wilderness sites, prescribed natural fire helped to legitimate prescribed fire of all sorts. Traditional fire control techniques were put on the defensive. Instead of being applauded for opening up the backcountry, fire control, especially in its mechanized forms, was attacked for despoiling it. A number of regional seminars on prescribed fire were organized by the Southern Forest Fire Lab of the Forest Service, but outside the South controlled burning still meant, for the most part, slash disposal and site preparation—topics within forestry, not biology.

In February 1967 a Fire Policy and Procedure Review Committee meeting in Washington sustained the 10 A.M. Policy for normal fire seasons but permitted leeway for pre- and postseason fires.[96] The committee reached its decision five months before the 1967 fire crisis in the Northern Rockies. With the even larger busts in Washington and California in 1970, the time for reform had come. Between 1971 and 1978 fire protection was once more overhauled. At a Fire Policy Meeting in Denver in 1971 some 31 recommendations were reported on and assigned to committees. Center stage was the 10 A.M. Policy. The opening recommendation reaffirmed the policy but authorized preplanned exceptions, if approved by the Chief, for wilderness areas and for periods of low fire danger. Though loosened in this respect, the policy was tightened in another. The meeting was ambivalent, split between two problem fires: mass fire, for which the 10 A.M. Policy could still be considered suitable, and wilderness fire, which demanded other guidelines. Thus at the same time that it promoted natural fire in wilderness areas, the meeting also approved a 10 Acre Policy to guide presuppression planning and expenditures, much as the 10 A.M. Policy governed suppression. The objective of presuppression, it announced, was to contain all fires within 10 acres. This goal would direct the national fire planning effort conducted between 1972 and 1975.[97]

Like other efforts to control spending by spending more, the 10 Acre Policy proved a costly failure. Real as well as inflationary costs for fire protection had spiraled upward. Since World War II fire costs had increased at an 8 percent annual growth rate, or four times the rate of population growth. Within 100 years protection costs would equal half the national debt, within 150 years, they would equal the entire gross national product. Fire protection was hardly alone in such vigorous growth, but neither could such expansion be maintained. In trying to reduce emergency suppression expenditures by substituting more emergency presuppression funds, the 10 Acre Policy resulted in a wild surge of presuppression costs, a process accelerated by the passage of the Forest and Rangelands Renewable Resources Protection Act (1974). To meet the goals of the act, the Forest Service sought to reduce by 2 percent the number of fires exceeding 10 acres, but the effort required a 90 percent increase in presuppression expenditures. Presuppression costs had swelled from $6 million in 1965 to $11 million in 1970, then from $25 million in 1973 to $85 million in 1976. Suppression costs increased almost as rapidly. Virtually all of Forest Service fire protection, that is, was financed through a system of deficit spending, though the same was also true for the fire programs of other federal agencies. Traditional fire suppression and presuppression strategies had clearly reached a point of diminishing returns. The use of fire for fuel modifications and the legitimation of prescribed natural fire looked attractive in economic terms.[98]

A host of new programs and conferences immediately followed the Denver meeting. In 1974, at a joint meeting of the Tall Timber Fire Ecology Conference, the Intermountain Fire Research Council, and the Forest Service Land Use Symposium, the Service publicly proclaimed its dedication to fire *management,* as opposed to fire control; to the total mobility concept, thereby joining the Boise Interagency Fire Center rather than further developing its own National Fire Coordinating Center; to collegiality, as manifest in the National Wildfire Coordinating Group; to land management as a basis for fire planning, in accordance with the slough of environmental legislation being produced through the decade; and to a new appreciation for prescribed fire in both wilderness and nonwilderness areas, as exemplified in its fire plans for the Bitterroot-Selway and Gila wilderness areas, for DESCON in the South, and for fuel management through the use of prescribed fire.[99] Within a year new standards were established for all Forest Service fire positions. To meet its new goals, the Service established a Fire Research, Development, and Application program at the Missoula laboratory and instigated a new review process for field compliance. Meanwhile, its paramilitary connection with civil defense was reduced by the advent of the Rural Community Fire Protection Program; its ties to military research were broken by congressional decree in 1972; its exploration of urban fire was limited by the creation of the U.S. Fire Administration; and its research programs through the National Academy of Science Committee on Fire Research were terminated by the dissolution of the committee. The larger goals of forest administration were dictated by legislation like the Forest Resources Management Act (1976).[100] Cooperative fire protection experienced additional shocks from legislation like the Cooperative Forest Management Act (1978), which affected the Clarke-McNary program.

It became necessary to revamp the theory and practice of fire financing, too. The Chief Forester and the Office of Management and Budget (OMB) requested an analysis of fire planning methods and a study of presuppression effectiveness. The resulting staff report, issued in late 1977, recast the economic theory once again. The report insisted that "loss" should mean "net loss," in recognition that not all fire effects are detrimental and that many prescribed burns are beneficial, though it proved every bit as difficult to quantify perceived benefits as it had been to calculate perceived losses.[101] In fact, although the report failed to mention it, the economic theory was proving to be a more accurate description of controlled burning than of wildfire; wildfires did not spread in the linear fashion required by the theory, but prescribed fires could. Nonetheless, as in the past, the fundamental decisions about fire protection were made on the basis of what were, for the age, self-evident propositions, intuitively obvious statements. Every decade had revised Headley's economic theory to express new values, until it resembled a Hindu avatar. This is not to say that such studies were a bureaucratic sham whose outcome was administratively decreed in advance; rather cost accounting studies were but one among many attempts to reify and articulate what was fundamentally a philosophical position.[102]

New fire policy alternatives were selected at a regional foresters meeting in July 1977. In August, a scant two years after completion of the last planning project, another National Fire Planning Meeting was scheduled for Denver. The next year the *National Forest Manual* replaced both the 10 A.M. and the 10 Acre policies. The new version encouraged a pluralistic approach to fire, a policy of fire by prescription. Even for suppression, once initial attack failed, alternatives were to be considered by the fire boss, possibilities that might or might not mean further efforts at suppression. At the insistence of OMB the emergency presuppression account was replaced by a more budget-accountable fire management fund. A review process was established to oversee program effectiveness.[103]

Behind this innovation was an attempted restoration. Beginning with the 1971 Denver meeting, determination had grown to reduce the relative autonomy of fire control, to subordinate it to general land use objectives under the direction of professional foresters, not fire managers. Large fires had always been an embarrassment to the Forest Service. They presented the most dramatic difference between American and European forestry and the greatest impediment to traditional forestry practices. Yet its fire mission had immeasurably strengthened the Service. Like so many human enterprises, forest management began with fire management. In the United States it was fire that had helped to create professional forestry and had directed the paths of cooperative forestry. Fire control advertised and dramatized forest conservation—and the Forest Service role—as no other public message could have done. Forestry had not brought fire protection; on the contrary, it was the need for fire control on the lands of the counterreclamation that often created the need for a technical bureau to oversee the protection of these lands, and it was professional forestry's good fortune to take charge of that bureau. That is, professional forestry in the United States developed in large measure not in spite of fire but because of it. The Forest Service survived in its early years because it learned to control fire; it has endured because, with time, it has learned to accommodate fire. The Service's greatest nemesis has, in many respects, been its greatest benefactor.

A MORAL AND LEGAL CHARGE:
A HISTORY OF FIRE POLICY IN
THE DEPARTMENT OF THE INTERIOR
AND INTERAGENCY ORGANIZATIONS

All forests under federal ownership of jurisdiction should be protected from destruction by fire. . . . As the owner of property, the Federal Government is morally if not legally charged with the duty not to maintain a nuisance.
—Forest Protection Board, 1929[104]

I

Federal fire protection began in the national parks, and the invention of the public campground by the Army is symbolic of what in national park policy has most distinguished it from the administration of the national forests. From the origin of the park system, management of the parks has meant the management of people, not the management of natural resources, and from the start, important ideas and techniques in fire protection have been imported from outside the parks' civilian administration. Initially, fires were attacked as a manifestation of human impact, and after 1968 fire control was gradually withdrawn for the same reason. The primary intrusion onto park lands came in the form of visitors, not, as on the forest reserves, in the form of fire, insects, and storm. If the Forest Service determined to "settle" remote areas through fire control, the Park Service resolved to do it with tourists. For both agencies, to ignore the inaccessible backcountry was to invite charges of administrative malfeasance. Forest Service fears that failure to maintain a good fire posture could cost it its lands was matched by Park Service apprehensions that failure to make remote areas accessible might cause public rejection of the park idea. The Park Service thus saw its backcountry problem as primarily one of recreational access; the Forest Service, of resource protection, notably fire control.

From the beginning of Army administration in 1886, the parks enjoyed fire protection. The failure of civilian superintendents to control fire was in fact one of the arguments for Army intervention.[105] The National Academy of Sciences committee thought highly enough of the experiment to recommend in 1897 that the Army take over the forest reserves as well, and in 1909 the National Conservation Commission noted that "on the National parks protection from fire is as complete as on the military reservations. . . .

Where danger from fire exists a regular patrol is maintained, and a sufficient body of troops is available to promptly extinguish any fire that may occur."[106] When the 1910 fire complex trespassed onto Glacier National Park, the Army was ready for it.[107]

Following the creation of the National Park Service in 1916, fire control expertise in the parks came from forestry and the Forest Service. The historic relationship between the two agencies has been one of wariness, rivalry, and cooperation. The bitter fight over the dam at Hetch Hetchy highlighted their contrasting philosophies of preservation and conservation, and the distinction between people management and resource management was reflected in the personalities of their two formative directors. Pinchot prided himself on being a professional forester, whereas Steve Mather made his reputation in business advertising and promotion. Pinchot sought to run the national forests as a public utility, managed under a program of scientific engineering. Mather used his powers of persuasion to "sell" the parks to the public, administering them as public campgrounds and pleasuring grounds. The Park Service sought to monopolize the recreational resources of the public domain, much as the Forest Service wanted to nationalize its forest resources. The Forest Service often regulated access to its lands as a means of controlling fire; the Park Service tended to promote access, even at the risk of fire.[108]

The Park Service relied heavily on the Forest Service for fire protection. The parks never developed a professional class analogous to the graduate forester, and after the military occupation, many park rangers came out of forestry schools or transferred from the Forest Service. They brought with them beliefs about fire and techniques for fire control. Nor has the Park Service created theorists from its own ranks on anything like the order achieved by the early Forest Service. The great statements about wilderness values and parks were always brought in from outside the organization, and this was no less true for fire management. Most of the early parks, moreover, were carved out of national forests, and national forests continued to surround them. Many parks inherited at least the rudiments of the fire control organization that the Forest Service had previously created—lookout towers, roads, trails, and telephones. Mutual assistance pacts helped to guard borders, and cooperative agreements eventually included mutual training, exchange of overhead on campaign fires, and shared support of aircraft, lookouts, and even a smokejumper base at West Yellowstone. In 1976 the two agencies signed an agreement to tolerate prescribed natural fires that might cross their contiguous wilderness areas in the Tetons.

It was foresters, acting through the Forest Protection Board, who rebuilt the NPS fire organization in the late 1920s and who laid the foundation for CCC fire development programs in the parks. It was a small corps of professional foresters who for decades directed fire protection within the parks.

That is, the Park Service owed most of its fire protection apparatus to the Forest Service, either directly by contribution or indirectly by example. But the distinctiveness of its mission and its locus in the Interior Department also allowed the NPS greater flexibility: it administered its own fire program, however meager, never contracting with the Forest Service to provide fire services. This, plus its emphasis on people rather than resources, made it easier for the Park Service to revise its fire policies. In fact, thanks to widespread enthusiasm for wilderness, it led federal land management agencies in the philosophy (though not the technique) of prescribed natural fire.

The NPS had long identified fire as a threat to the scenic and recreational values of the parks. In 1921, for example, its acting director admonished the superintendent of Glacier that "forest fires are our greatest danger to this park, and this act of Christiansen's [brush burning] should be punished with the greatest severity, not only as a punishment for this particular offense, if a case can be made against him, but to serve as a salutary warning to others."[109] Later in the 1920s officials of Glacier complained to the Forest Service that fires burning to the south were discouraging tourist travel.[110] But it was not until 1922 that the Park Service received a special appropriation for fire control—in effect an emergency account to be used only in the event of fire. In 1926 this appropriation was combined with others into a general fund to cope with disasters of all sorts and to rehabilitate physical improvements damaged by them. Previously fire costs had been covered from other budgeted accounts, and all purposes suffered from a lack of money until Congress appropriated supplemental funds. But deficit spending in anticipation of supplemental appropriations—the Forest Service practice—was not allowed; nor was the emergency fund to be used for any presuppression activities. Meanwhile, the Park Service hired Ansel Hall, a forestry graduate from the University of California, as chief of its Division of Education and Forestry. The emphasis of the position was on the interpretation of forest resources to visitors, but with the coming of the Forest Protection Board in 1927, Hall's duties were expanded to include fire planning.[111]

The pivotal year was 1928. During that summer a large fire in the vicinity of Sequoia National Park extended over lands of mixed jurisdiction. The fire took three weeks to control and burned about 1,000 acres on the park, but the Park Service contribution had been to retire to the tree line separating the brush from the forest and backfire toward the Central Valley. To state and Forest Service firefighters, the act was irresponsible at best: it intensified the fire, prolonged the campaign, and actually endangered crews working below. Feelings worsened when the park superintendent, Colonel John White, an expatriate Briton and former soldier of fortune, denounced control efforts in an open letter to the Los Angeles *Times* and expressed a preference for light burning. As a result of the fire the first interagency board of review convened and issued promises to establish joint general headquarters in the event of future fires in areas of joint responsibility.[112] That same year Horace Albright took over as director of the Park Service, and John Coffman, supervisor of the Mendocino National Forest, was hired by the NPS as a national fire officer. The agency was then completing its second year as a member of the Forest Protection Board, and with Coffman as a sympathetic liaison, the Park Service, along with the other participants on the board, prepared a comprehensive fire prevention plan that detailed the requirements for adequate facilities and patrols for every park in the system. Congress made the first national appropriation ($10,000) for fire protection in the parks; previously the money had been assigned to individual parks. The need for better presuppression capabilities was also stressed, and at last the Park Service was allowed to spend fire funds for presuppression work; fire lookouts were not authorized until 1931, however. And finally, Sequoia, whose superintendent had brazenly advocated light burning, began an expensive campaign, financed by special congressional appropriation, to remove by hand flammable debris from around its groves of Big Trees.[113]

The Forest Protection Board itself could not put its new plans for the Park Service into effect, but the CCC could. In 1933 Coffman became chief forester of a new Branch of Forestry and assumed charge of the Emergency Conservation Work program.[114] With the national fire plan as a blueprint, New Deal conservation programs gave the parks their first overhaul since the days of Army and Forest Service occupation. The liberation of money and manpower intoxicated the NPS as it had the Forest Service: it inflated the organization terrifically. In 1929 the permanent fire organization of the Park Service had consisted of one national fire officer, a special fire organization at Glacier, and a fire guard at Sequoia. Fire duties otherwise remained ancillary to visitor-related responsibilities and general maintenance. Ten years later the fire facilities of the parks were more or less on a par with those of the national forests. Some 650 camps of 100 men each were stationed on park lands. Logistical support and administrative services required the Park Service to hire more than 7,000 employees—a figure not exceeded even in the early 1970s.[115]

In preparing its plan for the national park system, the Forest Protection Board reasoned that the parks were "an economic service in the form of national education and recreation of a value probably already even greater than an equivalent area of the choicest commercial forest." The duty of the Park Service "as custodian is to preserve, unmarred by fire or other agency of destruction, the picture dedicated to public use and placed in its charges." The Board recommended as "not unreasonable" that the "standard established for forest fire control within the National Parks should be at least equal to the standard established for areas of comparable hazard and highest economic value within the National Forests."[116] That standard was the one spelled out in the 1928 *National Forest Manual,* and this interpretation guided NPS thinking in directing its CCC camps.

In a way, the board's statement was an effort to demonstrate the importance of seemingly low-value wildlands, which presented as serious a problem for the national parks as for the national forests. By granting them equal stature with high-value commercial lands, the board was doing with the national parks what, after a fashion, the regional foresters were then doing with the national forests, that is, obliterating artificial distinctions between sites based on intangible values and bestowing equal protection to all lands. Insisting that such lands deserved fire protection was a means of officially recognizing their worth. Arno Cammerer, NPS director during the New Deal, declared that "fire and its consequences are the major causes of loss of forests usable for recreation." Those who caused fires, Cammerer insisted, were guilty of "murder." Nonetheless, he recognized that "the conflict of objectives between preservation of natural wilderness areas free from roads, trails, structures and facilities, and the need for prompt action on all fires, creates a problem of relative values." By allowing recreationists into the backcountry, the Park Service was augmenting the fire hazard; "accordingly some means must be provided of counteracting these human encroachments and of preventing abnormal damage to these factors of mature forests that help make wilderness attractive."[117] That is, fire control was a means of regulating the impact of visitors, and in the days before aircraft, fire control demanded certain encroachments of its own.

The problem of value was especially acute for the national parks. Foresters had always insisted that trees were crops and would grow back, even from fire; the parks, however, were repositories of unique treasures, irreplaceable artifacts, and esthetic scenery as priceless as great works of art. Once lost, their value was irrecoverable. An article published in *American Forests* in 1929 opened with such a sentiment: "Fire is today, without a doubt, the greatest threat against the perpetual scenic wealth of our largest national parks, which, bereft of their trees and foliage, would become haunts only of those interested in the study of desolation."[118] Gisborne, for example, endorsed the Park Service's need for the most efficient fire control organization in the United States. "The inspirational and scenic values of park lands cannot be measured in dollars and cents," he concluded, thus putting the parks outside the economic theory.[119] Not until the concept of preservation changed its emphasis from the products of nature to the processes of nature was the imperative for fire protection diminished. Until the 1960s virtually every advocate of wilderness and every director of the Park Service demanded a strong fire program.[120] Prior to 1968 the standing directive for the national parks was that firefighting took precedence over all other activities except the safeguarding of human life.

All of these values were incorporated into the strategy of CCC employment. The North Rim of Grand Canyon National Park is a good illustration of what resulted. During the summer months two or three camps operated on the North Rim and were relocated to the South Rim or the Inner Canyon during the winter. The residence houses, warehouses, maintenance shops, and fire cache (for both structural and forest fires) were all constructed by CCC labor. Main roads were improved and primitive roads punched to remote sites. One permanent metal lookout was moved to a better location and a second constructed. An auxiliary network of tree towers, complete with metal scaling ladders, was erected. Telephone wires joined camps, cache, and command post. Water holes were created, both for wildlife and for fire control. Roadside fire hazards were cleaned up. And, of course, CCC crews were dispatched to fires. The facilities for fire management on the North Rim today are essentially those bequeathed by the CCC. A summer crew lives in the same houses, operates out of the same cache, and uses the same roads and towers.[121]

The war years hit the national parks hard. The CCC was gone, and, unlike the national forests, the parks did not house essential raw material of strategic military value. Even its staple product, recreation, was devastated by wartime rationing. Some emergency fire appropriations helped to carry the parks through World War II, and a 1943 memorandum of understanding between the Secretaries of Agriculture and the Interior made Forest Service resources available to the parks.[122] In terms of facilities and services, however, the parks ebbed into decline. In retrospect this period of restricted access coincided with the evolution of new park values based on limited access (or carrying capacity), the correlation of park lands with wilderness, and the concept of wilderness as something more than an esthetic and recreational resource. But these concepts percolated imperceptibly through the organization from the outside, and their consequence for fire policy was at first ambivalent.

For the Park Service the great postwar event was a 10-year Marshall plan for rehabilitating park facilities. The ambitious scheme was launched in 1956 and dubbed Mission 66 to signify the intention to conclude on the fiftieth anniversary of the NPS. Meanwhile, without the funds and manpower liberated by the CCC, fire control was again absorbed into ranger activities, with forestry made a division of resource management. By 1956 nearly all park operations were subsumed under Mission 66. Whereas Forest Service plans in those years led to institutional developments for fire equipment and fire research, Park Service officials emphasized expansion of such facilities as visitor centers for people management.

For resource management in general, and for fire in particular, reforms began in 1962, when the Secretary of the Interior appointed an Advisory Board on Wildlife Management in the National Parks, chaired by A. Starker Leopold, professor of wildlife management at the University of Califor-

nia and son of Aldo Leopold. The committee's report of March 1963 amplified its specific charge into an eloquent summary of park goals and a concept
of park management. The board exploited the idea, made popular by Aldo
Leopold, that wildlife management meant the management of habitat, and
it proposed as a goal that "the biotic associations within each park be maintained, or where necessary recreated, as nearly as possible in the condition
that prevailed when the area was first visited by the white man." In its celebrated phraseology, the board recommended that "a national park should
represent a vignette of Primitive America."[123] These suggestions were
endorsed by the Secretary and became the foundation for a complete overhaul of NPS administrative policies. That same year Congress created the
Bureau of Outdoor Recreation, which broke up the Park Service's longstanding monopoly of federal recreation management and prepared it, following the Wilderness Act, to assume a new role as chief spokesman for
wilderness management.

The implication for fire management were enormous. Fire control in itself
was now considered inadequate—indeed, ruinous—as a program of
resource management. No matter that the determination to use the landscape prior to the advent of European man as a standard for "naturalness"
was just as arbitrary as a fire policy that sought to control all fires within
10 acres. The new policy was an open invitation to use prescribed fire. The
advisory board's report observed that "of the various methods of manipu-
lating vegetation, the controlled use of fire is the most 'natural' and much
the cheapest and easiest to apply."[124] As an example, it cited prescribed
burning experiments conducted in the Everglades. It is worth noting that
the committee had been charged with evaluating wildlife management, not
forest management; that there were no foresters among its members; and
that its attitude toward fire was entirely one of fire effects, of fire as a biological phenomenon.

The report additionally recommended that the Park Service inaugurate
a research program to support its management objectives. Traditionally,
NPS research had supported interpretive functions, a process of visitor
education. Traditionally, too, that research had been predominately geological; the parks, after all, were usually centered around such geologic spectacles as mountains, canyons, and volcanoes. The advisory board proposed
instead that research be conducted as a guide for resource management and
"that every place of management itself be under the full jurisdiction of
biologically trained personnel of the Park Service." Why biologists were
especially equipped to manage parks was an injunction as curious as the one
maintained by the Forest Service that only professional foresters could manage range, recreation, and watershed. Perhaps a feeling lingered that the
landscape had to be "managed"—which almost always means manipulated—and it was certainly easier to modify the biota of a landscape than

its geology. Within a few months of the Leopold Report, a National Academy of Sciences-National Research Council committee published a similar recommendation that the Park Service support a research branch. The combined effect was sufficient to create an office of chief scientist, which Starker Leopold initially occupied as a consultant. More important, Forest Service hegemony over research into questions of resource management was eroding.[125]

Thus, while the Forest Service explored the physical equations of fire behavior, the Park Service undertook research almost exclusively on the biology of fire. Not surprisingly, most of the early research came from students of Leopold, nearly all of them wildlife biologists, and like wildlife researchers throughout the century, most were enthusiastic about prescribed fire. Furthermore, they had the example of Harold Biswell, a professor of range management at the University of California and since the early 1950s a strong advocate of prescribed fire. Park Service researchers had at least one example from within their own ranks, too. Between 1951 and 1952 Everglades National Park had hired William Robertson as a fire control aid, but with the understanding that he would do research on fire effects. Robertson, a biologist, completed his report in 1953. He recognized that the peculiar biology of the Everglades represented an equilibrium between fire and water. The problems of drainage and fire damages (and of fire control damages) had to be solved concurrently; heavy drainage north of the park was creating droughtlike conditions that demanded fire control at least temporarily "to maintain the ecological *status quo* of south Florida." With success on the drainage and diversion problem, however, experiments in prescribed burning were begun in 1958—15 years after the Forest Service, based on its own Florida experiences, officially approved prescribed burning.[126]

Two studies conducted soon after the Leopold Report focused on fire and the giant sequoia. One, investigating the relationship of fire to sequoia regeneration, was headed by R. J. Hartesveldt of the University of California. The other, a survey of fuel hazards around sequoia groves, was directed by Biswell. Both led to recommendations for prescribed burning, and reports of both were published in the proceedings of the Tall Timbers Conferences, which became a major publication outlet for Park Service experiments with fire.[127] The sequoia, in fact, had long posed a conundrum for fire policy. The desire to protect the Big Trees from fire was instinctive among thoughtful conservationists. Pinchot described how in 1891 Kaweah colonists proudly informed him that they had saved the grove from burning up 29 times in the past five years, before the groves became part of the national park system. To this Pinchot wryly inquired, "Who had saved them during the remaining three or four thousand years of their age?"[128] Indians and miners had burned the area without undue damage to the giants, and loggers, recognizing the asbestoslike quality of their bark, had simply burned slash *in situ* around the felled redwoods, expecting the downed trees to weather the resulting firestorms. The practice horrified conservationists, and even a died-in-the-wool light-burner like Colonel White opted for hand removal of debris accumulated beneath the trees. By the late 1960s, however, prescribed burning had become acceptable.

All of this ferment was incorporated into the new policy books released by the Park Service in 1968. The informing principles of the administration of the natural areas were those enunciated in the Leopold Report. The policy recognized fire as a natural phenomenon, encouraged the practice of letting natural fires in predetermined areas run their course, allowed the use of prescribed fire as a substitute for natural fire, and expected control of any fire not advancing management goals. The new policy handbooks appeared soon after the Forest Service policy meeting and anticipated by 10 years Forest Service adoption of a similar policy of fire by prescription. But where the Forest Service sought to subordinate fire managers to foresters and fire management to land management, the Park Service sought to subordinate its fire organization to biologists and people managers. It hired more resource managers but not more fire managers. It retained its emergency presuppression account and remained heavily dependent on it. The description of a natural prescribed burn at Yellowstone showed how this new commitment readily meshed with its old one. A researcher and a fire manager described a prescribed natural fire four acres in size. "The herd of elk remained in the lush, green meadow below the fire throughout the afternoon and evening," the writers observed, "giving park visitors a once-in-a-lifetime opportunity to easily view and photograph elk against a natural fire backdrop." Interpretive contacts were expanded around the fire. "A conservative estimate is that over 3,000 visitor contacts were made at the fire scene up to this time." Wildfire had become a tourist attraction, a thrill for visitors much like the region's bears and geysers.[129]

In the aftermath of the new policy manual there began a period of experimentation with fire both by research and administration. This laissez-faire approach had the advantage of introducing variety and emphasis on local peculiarities, but it had the disadvantage of being fragmented and sometimes ill-informed. On the national level the NPS joined BIFC and NWCG. It welcomed the emergence of a strong BLM fire organization, which bolstered the collective hand of the Interior agencies. It saw in interagency cooperation a means to promote park values and park fire philosophy. For the first time since Army days, national interest in wilderness put the Park Service into the vanguard of a national debate about fire policy. Nor was the move toward a more sensitive fire program damaged by the spectacle over the next decade of bulldozers on the mountains of Glacier, of heavy tracked vehicles in the Everglades, and of mechanized line equipment amid the ruins of Mesa Verde and Bandelier—all unleashed in the name of fire control.

Slowly, experimentation gave way to a consistent national program. A smoldering natural burn in the Tetons smoked in Jackson Hole, incensed local residents, and obscured the peaks. Always sensitive to public opinion, the Park Service issued a set of interim guidelines to give more specific standards for the conduct of its fire program. In late 1976 a task force was appointed to reevaluate the various plans in effect. Two years later a new handbook advised on all aspects of resource management in the parks, including fire.[130] Thus the Park Service and the Forest Service achieved their full conversion of policy at about the same time. Guided by the dazzling philosophy of the Leopold Report, the Park Service had advanced a policy too far ahead of its knowledge and technical skills; the Forest Service, with expertise and information in abundance, had lagged in policy.

The interim fire management program announced that "the incorporation of natural ignitions from lightning into management programs has been led by this Service. Particularly in this latter program there is no precedent technology, we must develop it to meet our needs." The Park Service thus ignored the great strides made since 1970 by the Forest Service in this same field, though elsewhere the interim program urged use of "existing research and technology of other agencies."[131] If a little smug, the remark nonetheless showed the special spirit with which the Park Service conducted fire management, its belief—magnified by the enormous value ascribed to parks and wilderness in the 1960s and 1970s—that the parks were unique places and that the Park Service was uniquely qualified to administer them.

The management of resources remained subordinate to the management of people. The Teton smoke controversy had followed a more significant confrontation for the agency between rangers and a crowd in Yosemite Valley. In an ensuing riot the rangers were ingloriously routed from the field. The Park Service responded with a vigorous training program in law enforcement and "visitor protection," even giving training in SWAT tactics and hostage negotiation. Whereas the Forest Service assumed responsibility for resource management on its land but left people management largely to local officals like sheriffs and to institutions like hospitals, the Park Service took over those functions for itself, including search and rescue, law enforcement, and emergency medicine. Whereas the Forest Service strove to integrate fire and resource management through national programs and data banks like FIREBASE, FIRESCOPE, and FOCUS, the Park Service turned to the National Crime Information Center, eager to join national law enforcement efforts. The trend was reinforced by the Washington office's promotion of urban parks.

The rising tide of law enforcement officers in the Park Service followed an ebb tide of foresters. In its heyday fire control in the parks had been supervised by professional foresters in park employ. As resource managers, these men were replaced for the most part by people trained in law enforcement, in recreation management, and in wildlife biology. The Park Service's moment of national preeminence in fire management was destined to recede both from forces within and from pressures without, ironically assisted by the very interagency liaisons it had encouraged. With few fire managers coming through the lower ranks, interagency programs would soon swamp its involvement, and the Park Service would return to its traditional status. It had found itself thrust into public limelight by values coming from outside the organization, but it would retire by the force of the traditional values that yet remained in it. In revising its policies the Park Service, like the Forest Service, reached back to its origins—and determined to be people managers first, and fire managers second.

II

If the Forest Service had emphasized resource management and the Park Service people management, the Bureau of Land Management and its antecedents have specialized in the management of legal title. For most of the Bureau's history, "protection" has meant the protection of patent; "trespass," the infringement of title, not encroachment by fire or vandalism. Unlike the Bureau of Indian Affairs (BIA), the BLM did not manage lands in perpetual trust; unlike the NPS, it was not mandated to preserve lands unimpaired for future generations; unlike the FWS and the Forest Service, it was not established to manage natural resources. For the better part of its bureaucratic existence, it was merely a holding company for federal lands. Its domain would eventually be disbursed to the public or distributed among other federal land management agencies. Its concern was thus with land and its title, not with its resources. The very process of disposal virtually ensured that the BLM would retain those lands least economically valuable, that it would rarely oversee lands in large continuous blocks, and that it would normally retain lands for a limited period of time. Where the Forest Service assumed the lead in federal resource management and the Park Service in recreation, the BLM was responsible for cadastral surveys, land patents, and mineral leasing rights. Charges of ineptitude in forest management and of legal clerkishness were in reality a compliment: only slowly was the BLM given responsibility for actively managing the resources on the unappropriated federal lands.

It is somewhat surprising that the BLM should administer the major fire protection organization of the Department of the Interior. The evolution of that apparatus, like the history of the BLM itself, is one of slowly aggrandized responsibilities. In 1812 Congress established the General Land Office to oversee the disbursement of federal lands. With the creation of the Interior Department in 1849, the GLO became the main administrator of the public domain.[132] Where the Forest Service would be embarrassed by large fires and the NPS by visitor scandals, the GLO was haunted by land fraud. It encountered a public reluctant to prosecute title trespass, as the Forest Service found with fire trespass, and it confronted a Congress that discouraged adequate protection of title, as the early Forest Service found with inadequate protection of resources.

With the creation of forest reserves, the GLO was charged with their legal protection against unauthorized preemption. No scheme existed to manage the reserves otherwise. The lands were simply withdrawn from settlement. The Forest Management Act of 1897 established guidelines for the administration of the reserves and directed the GLO—as the largest federal landholder—to administer them. The GLO, in turn, looked to the Geological Survey and the Bureau of Forestry for advice in technical matters. The USGS surveys included maps of burned areas, and the Bureau of Forestry advisors created the rudiments of a fire protection system. But the reserves were anomalies within the GLO, and its failure to satisfy Forest Service standards of resource protection, its supervision by a malleable commissioner, and the politics of the Pinchot-Roosevelt friendship saw the reserves transferred to the Forest Service in 1905.[133] The historic role of the GLO as a vehicle for the disposal rather than for the management of lands was thus reenacted. With a terrific expansion in the national forest and national park systems and with the creation of wildlife refuges under the Biological Survey (later, the Fish and Wildlife Service), the cream of federal lands passed out of GLO hands, and with them passed an early commitment to fire control.

The GLO never completely abandoned its involvement with fire protection, however. Lands in Oregon that had been granted to the Oregon and California (O&C) Railroad were returned to federal government in 1916. Their prime timber sites compelled the GLO to administer these lands as commercial forests, and the agency was granted limited powers to classify its holdings according to their resources. Congress appropriated some money for fire protection, but the checkerboard pattern of land ownership discouraged the GLO from creating a major fire establishement. Instead it contracted with states or with local national forests to provide protection. The Protection Act of 1922, which gave the NPS its first fire budget, also allocated money to the GLO, and the completion of the Alaska Railroad from Seward to Fairbanks that same year involved the GLO with fire problems in the Alaska interior. From 1921 to 1933 the Field Division of the GLO hired two fire guards to patrol along the tracks of the Alaska Railroad and the embyronic road network that appeared with it. The question of fire protection in the interior remained "a matter of grave concern" to the Forest Service for many years, and in 1927 it proposed a fire protection plan for Alaska to the Forest Protection Board. As with the national parks, fire protection again became a prominent issue in interbureau rivalry as well as in interagency cooperation.[134]

The GLO had been a charter member of the Forest Protection Board, but its position was recognized as peculiar. GLO lands were scattered and in imminent danger of being removed from GLO administration through patent or proclamation. The board concluded that "the best use of the General Land Office funds for the protection of forested public lands can be secured through cooperation with such other agencies as may be functioning for fire control purposes." Accordingly, agreements were made with the states of California, Oregon, Washington, Idaho, and Montana. The GLO assisted with the cost of patrol and lookouts, and the states absorbed the cost of suppression. In other states, aid was extended for fire suppression without formal agreements, "it being sometimes necessary to procure a deficiency appropriation when expensive fires occur." In 1931 a total of $60,000 was appropriated for fire control, and some 70 fire guards were supported in five states—almost all in Idaho, Oregon, and Washington. Of particular concern was the presence of large coal fires in Colorado, Wyoming, and Montana. The GLO had considerable expertise in leasing mineral rights, but only abortive experience in the management of biotic, surface resources, such as wildlife, range, and timber. A coal fire was something within its ken.[135]

The GLO recognized that, for the most part, its role was restricted to that of custodian of legal title. But is also recognized that it was obliged to furnish its timber "reasonable protection from destruction." To the board the GLO argued that this responsibility stemmed from "three different sources, all equally compelling": that there existed "Government trusteeship over the public domain"; that "as the owner of property the Federal Government is morally, if not legally charged with the duty not to maintain a nuisance"; and that "the Federal Government should be a reasonably good neighbor and show a reasonable spirit of cooperation in protecting a region jointly occupied by the Government and the holdings of its citizens." Whereas fire protection on both the national forests and national parks was part of an almost ideological crusade on behalf of larger political and social ideas on land management, the GLO offered a more humble rationale: the simple responsibilities incumbent on any landowner by virtue of legal title. With most of its forested lands protected by other agencies, the GLO admitted that "this, in turn, practically forces the acceptance of the standards already established by the cooperating agency"—in effect, the fire policy of the Forest Service.[136]

The New Deal had far less impact on fire protection by the GLO than on that by the Forest Service and the Park Service. In 1934, however, Congress passed the Taylor Grazing Act, which closed the public domain, allowed some land classification by the GLO, and established a federal Grazing Service, administered through local grazing boards composed of ranchers to oversee livestock use on unappropriated public lands. The advent of the Grazing Service committed the Interior Department to attempt at least some form of resource management, and with the closing of the public domain in the continental United States, the GLO lost some of its historic role as merely the disposer of public lands. It was compelled to manage, or at least protect, its holdings. Nor did the CCC do much to advance fire protection on lands managed by the GLO or the Grazing Service. Most camps were monopolized by the Forest Service and the Soil Conservation Service, both in the Agriculture Department; in the Interior Department, the NPS took the lion's share of CCC manpower, and much of that went toward its newly acquired historic sites. Most GLO holdings lay in the arid West—the low-value backcountry of the public domain that lacked even the scenic magnificence that could qualify it for national park status. The GLO's main use of the CCC for fire protection was in the attempted suppression of coal fires.

During the 1930s the Interior Department under Harold Ickes lost a major battle to recover the national forests from Agriculture. Perhaps in response, Interior set out to manage its lands as the Forest Service did the national forests; in fact, the administration of the new programs fell largely to renegade foresters from the Forest Service. The Grazing Service was one such program. So was the 1937 O&C Act, which provided for sustained yield forestry on O&C lands. And so was the Alaska Fire Control Service (AFCS) established in 1939. With the introduction of professional foresters, aggressive fire control was certain to follow. It was the shared values of the forestry profession, as much as wartime pressures, that led to the 1943 memorandum of understanding between the two traditional rivals, Agriculture and Interior, and provided for cooperation in fire protection.[137]

The logical source for the development of an autonomous GLO program was Alaska, where, except for some national forests and parks south of the Alaska Range, it was the sole custodian of territorial lands. But it is indicative of the status of fire protection within Interior that a separate agency had to be created to control fires on the public domain in an area without formal cooperators or permittees. Between 1940 and 1942 the fledgling AFCS even supervised the CCC program in Alaska. Though logistics and appropriations strongly circumscribed it, the AFCS had ambitions of great scope. Alaska, after all, contained over half of the lands still in the public domain.[138]

In 1946 an executive order consolidated the Grazing Service, the General Land Office, the O&C lands, and the Alaska Fire Control Service into the Bureau of Land Management. The AFCS was soon absorbed into a new Division of Forestry. The reorganization streamlined executive problems in administration but did not provide a real charter for the BLM. Nor was the executive alliance always an easy one, either for bureaucratic structure or for policy. The Grazing Service had been strongly decentralized, and the GLO quite the opposite; the Grazing Service sought to regulate resource use, the GLO largely title. The O&C lands were managed more or less as a national forest. The AFCS was strictly designed for fire protection, without a particular land or resource base. Rather than organically integrating these various functions, the reorganization order that created the BLM for the most part simply stirred them together.[139]

For fire protection the consequences were indeed mixed. Neither the GLO nor the Grazing Service had really desired or been designed for fire protection. With their patchwork quilt of land ownership patterns, they were content to contract for protection from the states and to cooperate with neighboring national forests. Neither expedient was possible in Alaska. The territorial lands in Alaska were not much different from those withdrawn from settlement in the late ninteenth century, and the BLM struggled to protect Alaska lands as the GLO had the original forest reserves. The immediate impact on the AFCS was to suppress its influence and ambitions no less than its autonomy; only with the approval of a new five-year plan in 1949 did the BLM in Alaska receive special firefighting appropriations. But in having absorbed the AFCS, the BLM controlled its own fire suppression machinery, no matter how limited, and inadvertently began the process of nationalizing fire protection throughout its lands. The ultimate result was to bring fire control from the interior of Alaska to the public domain of the Far West. Faced with big fires in 1937 and a 300,000-acre fire in the Snake River Valley in 1943, the GLO and the Grazing Service could only seek to strengthen their cooperative agreements with the agencies that actually worked the fireline for it. Confronted with large fires during the early 1960s in the Great Basin, the BLM could transplant a vigorous fire control organization nurtured in Alaska. The dominance of federal fire protection by the Forest Service as a result of New Deal programs was thus challenged in the mid-1960s on two fronts: in policy by the NPS, and in fire control technology by the BLM. Out of this rivalry came a new era of interagency cooperation.

The BLM made few advances in the decade after its consolidation. The key to its Alaska operations was aircraft: helicopters, air tankers, paracargo transports, and reconnaissance planes would do for Alaska what the CCC did for the Lower 48. Not until 1955 did the BLM really take on a major fire in the interior, and not until fires during the 1957 season burned nearly 5 million acres did it enter big-time fire control. With Alaska on the verge of statehood, the fires became a national scandal. The BLM abandoned the least-cost-plus-loss theory underwriting its fire policy in favor of a 10 A.M.

Policy, and in 1958 it was allowed to spend emergency funds for presuppression activities. It was determined that all fires in the interior could be suppressed.

What followed was a spectacular case of technology transfer. In 1959 the BLM established a smokejumper base at Fairbanks, developed an air tanker program, created a small air force of detection and transport planes, and organized Eskimos and Alaska Indians into fire crews. The Forest Service began studies in fire behavior, fire weather, statistics, and a fire danger rating system suitable for Alaska. With the coming of Alaska statehood, moreover, the BLM had a sounder reason for protecting surface resources: it would protect the land on behalf of native corporations, other federal agencies, and the state until the final disposition of the lands could be decided within the twenty-year period stipulated by Congress. Within a decade the BLM moved into the major leagues of fire suppression. The 70,000-acre Swanson River fire on the Kenai Moose Refuge, which it fought for the Fish and Wildlife Service, introduced the BLM to the modern definition of "big fire"—not big by virtue of size and environmental havoc so much as by virtue of cost ($11 million), complexity of internal organization, and damages inflicted by suppression technology.

Fires had also begun to embarrass BLM operations in the Lower 48. The Seebring Ranch fire in Montana (1959), the Rattlesnake fire in Wyoming (1960), and the Wheeler Pass fire in Nevada (1961) all showed a growing threat from range fires. The range itself was changing. Heavy overgrazing by horses, cattle, and sheep and changing fire patterns had replaced native grasses with tough, fire-hardy exotics like cheatgrass and tumbleweed, had allowed tremendous expansion in the range and density of sagebrush (and of fuel loads), and had encouraged the spread of juniper throughout the western range. The result was a mixture of fuels that were both flashier and more heavily loaded. Ranch and even surburban developments also crowded into former rangelands, increasing the values at risk. Yet the BLM retained the local volunteer fire organizations it inherited from the Grazing Service. It is said that when the BLM mobilized its firefighting resources to meet the 1960 Rattlesnake fire, the total hardware emptied out of its fire cache consisted of 14 pulaskis, 14 shovels, and 1 pickup truck.[140] After the 1961 fire season the BLM resolved to recruit a fire organization for the Lower 48. Most of its experienced fire officers came from the Forest Service. Its cooperative agreements were revised and strengthened, including (in 1963) the agreement among Interior, Agriculture, and the Forest Service that had been in effect since 1943.

The first test of the new organization came with the Nevada fires of 1964. On August 16 lightning ignited some 40 fires near Elko. Two days later, after the suppression of some fires and the merger of others, 6 major fires were burning over nearly 300,000 acres. The district officer activated cooperative agreements with the state and the Forest Service, and control forces quickly poured in from 10 western states—2,500 men, 64 aircraft, 280 trucks and tractors. The governor declared a state of emergency. But fires broke out elsewhere in the West, too. The Forest Service in Utah discovered that it had overcommitted its resources to the BLM and requested the return of some crews in order to control fires on its own lands. The BLM refused to release the crews. It was obvious to all that a central dispatch service was imperative to prevent collapse of fire suppression efforts throughout the West. The Nevada fires were draining men and equipment into the Great Basin as fast as a telephone could place fire orders. On August 18 the national director of the BLM established a Western Fire Coordinating Center at Salt Lake City for the duration of the fire. The Forest Service maintained a liaison. The center allowed for a wider perspective on fire problems and resources throughout the western United States.[141]

The 1960s were a time of tremendous expansion in government institutions and of frequently costly adventurism in nearly all aspects of American life—economics, social reform, education, environmental legislation, technology, science, and the military. The BLM and fire protection were no exceptions. As a land management institution, the BLM was strengthened in purpose by the passage of the Clarification and Multiple Use Act of 1964, by the creation of the Public Land Commission (1964–1970), and by the establishment of special service centers. With assurance of more stable land tenure, the BLM moved into active resource management on its lands and sought, as did other land agencies, to integrate fire protection into its total program objectives.

As a fire control organization, the BLM worked to build up its internal capacity to suppress fire without recourse to cooperative agreements and at the same time to improve interagency operations when mutual control was necessary. In accordance with its new national guidelines, the BLM reexamined its fire program in depth and elected to adopt a system not unlike that used on the national forests. A report released providentially in the same month as the Nevada fires recommended that in order to satisfy its management goals, the BLM would do better if it provided fire protection from within its own "force accounts" rather than through contracts with other agencies.[142] The BLM position on fire control was thus identical to that reached by the NPS on fire research at the same time: each preferred to construct its own independent program. Following the Forest Service example throughout the West—and building upon its Alaskan experience— the BLM developed organized crews of firefighters, notably out of migrant labor pools in the Snake River Valley and nearby Indian tribes. These were first used in 1964. Among the federal service centers created nationwide was the Great Basin Fire Center, a more permanent version of the Western Fire Coordinating Center thrown together during the duress of the Nevada fires.

In 1965 the Great Basin Fire Center was transferred to Boise, Idaho, renamed the Boise Interagency Fire Center, and placed under the responsibility of R. R. Robinson, chief architect of the BLM's Alaska operations. When range fires threatened the suburbs of Boise itself the next year, the BLM, the Forest Service, and state of Idaho joined forces to control them.[143]

The original scope envisioned for the BIFC was, at best, regional. It would provide a fire service center for BLM fire operations in the Great Basin and, by maintaining liaisons with the Boise National Forest and Forest Service regional offices (Region 4), it would avoid the dispatching confusions that had marred the Nevada fires. But in transporting experience from Alaska, the BLM also transferred a different vision of its range and purpose. The 1964 fires did for the Great Basin what the 1957 fires did for interior Alaska. The BLM promised to create for the one an organization almost identical to that fashioned for the other. It committed BIFC to aircraft, even stationing its physical plant adjacent to the runway of the Idaho Air National Guard as BLM Fairbanks offices rested next to the runways of Fort Wainwright; it established a smokejumper training base and explored the idea of using smokejumpers in Utah. It experimented with all-terrain vehicles suitable for the Great Basin. And it urged a larger interagency involvement. Except for some national forests in the south and southeast, the BLM had exclusive fire responsibilities in Alaska; the state and other federal agencies contracted with it for fire services. Robinson sought to bring a similar level of integration not only to the Great Basin but also to the United States as a whole. The BIFC would serve as a national dispatcher, coordinator, and general support center for the entire United States. Even as it officially opened in 1969, BIFC coordinated support for the Swanson River fire in Alaska.[144]

The Forest Service balked at the proposal. From its beginning it had entered into cooperative agreements for mutal support, and it had chaired the defunct Forest Protection Board. To commit its own elaborate national resources to a *nouveau riche* agency that a decade earlier would have been hard pressed to equip a solitary crew with shovels was a hard decision. Most BLM fire operations imitated Forest Service practices and policies and depended on the transfer of Forest Service personnel and technology. Suspicions were not lessened by insistence that the director of BIFC come from the BLM. Aircraft were expensive, and air tankers little more than a decade old. The warehouse at BIFC was no larger than any one of several of the Forest Service's own regional fire caches, such as those at Missoula. Moreover, the National Fire Coordination Center at Alexandria, activated during the 1967 emergencies, promised to do what BIFC claimed it alone could do. The old rivalry between Agriculture and Interior surfaced and had to be squelched at the Secretary level. Policies, procedures, and regulations differed. And, of course, there was the matter of simple pride. The Forest Service was perhaps jealous at seeing another agency match the size of its smokejumper complex and indulge in the elaborate air attack operations in which the Forest Service had pioneered. For the better part of the century the Forest Service had been the focus of a national effort at wildland fire protection; it had itself virtually achieved a national integration of purposes, broken trail in equipment development, and dominated fire research. Now, from several sources, it found that leadership challenged. The nightmarish fires of 1970 in California and Washington, however, decided the issue. For reasons of economics if nothing else, the old pattern of mutual assistance pacts proved inadequate. The Forest Service joined the National Weather Service and the BLM as a full partner at BIFC.[145]

As the BLM undertook its first bureauwide fire planning effort, it suffered more large fires in 1973. If the 1970 fires had compelled the Forest Service to join BIFC, these new burns stimulated the formation of the National Wildfire Coordinating Group. BIFC existed solely for logistical support on going fires; NWCG worked to integrate all aspects of fire management, including multiagency fire planning and training. A year later the Interior Department released a new fire policy for its agencies. The 10 A.M. Policy adopted by the BLM was replaced by policies not unlike those worked out by the Forest Service at this same time. In 1976 the BLM at last received a congressional charter with the Federal Land Policy and Management Act, and for the first time it moved east of the Mississippi River to the Seney fire on the upper peninsula of Michigan. Despite its more flexible policy, the BLM's handling of that fire was uncannily like its management of the Swanson River Fire on the Kenai Moose Refuge. BIFC even enlisted many BLM Alaska veterans for overhead. Aircraft were heavily employed, and costs swelled to $9 million. The Fish and Wildlife Service was once again left wondering if fire suppression had not done as much environmental damage on the refuge as the fire itself.

Even as it submitted environmental impact statements in the late 1970s for its bureauwide fire program, the BLM, like the Forest Service and the Park Service, was returning to its roots. The heartland of its fire operations had been Alaska, but as the 1970s came to a close, the BLM was rapidly divesting itself of its Alaska lands. What lands the BLM retained in Alaska, as elsewhere, would be remote, low-value sites. Even in the Lower 48 it was converting the better scenic areas into wilderness and primitive areas. It was now expected to surrender its Alaska empire, and it was ironically compelled to conform to the interagency pattern that it, as a newcomer, had advocated for the Lower 48.

The BLM and its antecedents had historically held land only temporarily, assuring protection of title and, to a lesser extent, protection of resources. The story in Alaska recapitulated that pattern. That the BLM should become so heavily involved in fire control was something of an anomaly. Its adoption of the 10 A.M. Policy was attributable to the affluence and technological potential of the 1960s, as Forest Service adoption of the policy had been dependent on federal emergency monies and manpower in the 1930s. With the inevitable transfer of its lands, with astronomical increases in suppression costs, and with conversion from fire control to fire management (the BLM had little experience with prescribed fire), justification waned for the sort of wholesale fire program that the BLM had so suddenly manufactured in Alaska and transplanted to the Great Basin.

Alaska had been special, a bit of ninteenth-century public domain preserved into the twentieth. Without it the BLM would never have created a fire organization on the scale it did, and as Alaska slipped out its hands, so did its brilliant experiment in fire control. In the Lower 48 the BLM could, where necessary, contract out its fire problems and, where desirable, could subject many lands to the intensive land management directives embodied in its 1976 charter. But in Alaska the BLM was destined to return to its traditional concerns—land title, cadastral surveys, and mineral leasing— much as the GLO had done following its brief, if influential, experience in administering the forest reserves.

III

The interagency institutions that became such a common feature of the fire scene in the 1970s had a long history. So long as independent agencies tended to draft firefighters out of the same pools of unemployed laborers and to adopt a similar policy of control, cooperative agreements for mutual assistance were adequate. Not until individual agencies developed their own suppression resources and articulated distinct philosophies of land management did the need for formal integration arise. The interagency story is consequently the product of two competing tendencies: one toward autonomy and fragmentation, and the other toward cooperation and synthesis. As agencies acquired unique objectives, they felt the need for control and research organizations solely responsible to their particular policies. But the similarity of most fire problems and the sheer costs incurred in suppressing large fires and in maintaining adequate presuppression facilities and manpower repeatedly pushed federal bureaus into interagency cooperation— with the states, with each other, and with the National Weather Service.

Fire is not simply a surface event. It is also a meteorological phenomenon. Just as contiguous territory forced land management agencies into mutual assistance, so a shared atmosphere joined the National Weather Service to wildland fire protection. As early as 1881 the Weather Service (then the Signal Service under Army administration) sent an observer to the Michigan fires. Sergeant William Bailey's report was the first in a long series of published *Signal Service Notes*. A decade later the Signal Service was ceded to civilians as the Weather Bureau, which in 1913 began to issue special fire warning forecasts for high winds east of the Columbia River gorge—another consequence of the 1910 fires, when gale winds resulted in the Big Blowup. A year later the forecasts were extended to Washington, Idaho, Oregon, and California, and by 1916 all the western forecast centers were authorized to issue fire weather warnings. The service was conducted through a cooperative agreement with the Forest Service and was largely dependent on weather observations taken on national forests.

The fire weather warning program coincided with a general expansion of special services by the Weather Bureau, including programs for the prediction of hurricanes, floods, frosts, and severe storms. The fire weather service, however, was among the earliest. Its expansion was facilitated by the fact that at this time the Weather Bureau was part of the Department of Agriculture, which made exchanges with the Forest Service much simpler. By 1924 two employees, one in Seattle and one in Portland, were issuing special forecasts during the fire season. As a result of the Berkeley conflagration, the importance of relative humidity in fire behavior was recognized, and forecasts were expanded to include such parameters in addition to the traditional alerts for high winds.[146]

The 1920s was a great era for cooperative fire protection. The same forces that pushed through the Clarke-McNary Act and created the Forest Protection Board also assembled in the interests of a better alliance between forestry and the Weather Bureau. A conference sponsored by the Western Forestry and Conservation Association in 1926 petitioned Congress for more assistance in forecasting. The resulting plan for fire weather services relied on cooperative agreements with the Forest Service and was initially confined to the fire-prone areas of the Far West and Lake States. Several states contributed money to assist the project. The next year the Weather Bureau joined the Forest Protection Board and issued its forecasts through seven district offices. In 1929, with funds provided by the Forest Service and the California Division of Forestry, a truck was converted into a mobile fire weather unit. Staffed by meteorologist L. N. Gray and a radio operator, the unit was dispatched to large fires, took weather readings at the fire site, and tailored area forecasts to the specifics of the fire at hand. The devastating fires in the drought-ridden landscape of the 1930s motivated the fire weather service to expand into the Appalachians. Funds were allocated in 1937 to add four additional mobile weather units. For many years fire weather forecasters and fire officers in the West met annually for conferences.[147]

Over the next two decades, however, involvement stagnated. The Weather Bureau had no lands to protect, and there was little it could do to manage the atmosphere, though lightning suppression and rain-making experiments were conducted with the Forest Service after World War II. Instead, outfitted with new instruments to measure the structure of the atmosphere and motivated by the demands of a military progressively committed to aircraft, the Weather Bureau reached into aviation forecasts. It saw itself as simply delivering a service for fire and forestry, not unlike other agricultural and severe storm forecasts. Special research, particularly into fire danger rating systems and into localized as opposed to regional forecasts, would have to come from the fire agencies themselves, which meant the Forest Service. In late 1948 the Service organized a Division of Forest Fire Research; within 20 years the division, like the Division of Fire Control, had taken to the air to become the Division of Forest Fire and Atmospheric Science Research.

The investigation into fire behavior principles made such progress that by 1958 the Forest Service was ready to sponsor formal training sessions. The first was scheduled to be held at Missoula, and the Service requested help from its old cooperator, the Weather Bureau, in designing and conducting the course. Both parties were interested in reviving the dormant fire weather service, and over the next two years a National Fire Weather Plan was assembled.[148] It received funding soon afterward, thus coinciding with the development of powerful new weather observation tools, such as satellites, and with the construction of three fire behavior laboratories. The Weather Bureau detailed several of its meteorologists to the labs for a range of investigations. The Bureau never challenged Forest Service research into the meteorology of fire as the NPS and Tall Timbers Station did with regard to the biology of fire, and without lands it had no reason to manufacture a firefighting organization as the BLM did. The common complaint was not that the Weather Bureau upset the status quo but that it failed to do so, that its involvement was marginal.

For several years the plan developed on schedule, and the western fire weather meteorologists even revived their conferences in 1964. In 1967 the Weather Bureau consolidated amendments into a Federal Plan for a National Fire-Weather Service. Beginning a year later the American Meteorological Society and the Society of American Foresters began a series of biennial symposiums on fire and forest meteorology. But the increased demands of the BLM, especially in Alaska, drained the system and stalled implementation of the national plan. Eventually the BLM contracted with a private company to provide its forecasts in Alaska. In 1970 the alliance came to a climax of sorts: the Weather Bureau was reorganized into the National Weather Service (NWS) and amalgamated into the National Oceanographic and Atmospheric Administration; its entire forecasting program was overhauled; it joined BIFC as a full partner and centered its national fire weather headquarters there; an airborne weather kit was created to supplement the old ground mobile units; and Mark Schroeder of the Weather Service and Charles Buck of the Forest Service, assisted by OCD, published one of the milestones in fire research, *Fire Weather*.[149]

Fire Weather was more than a fundamental handbook for fire managers and a training manual for fire behavior courses. It was a brilliant compendium of almost 90 years of research, beginning with Sergeant Bailey's observations of the Michigan fires for the Signal Service and culminating in a powerful partnership for research and management between NWS and the Forest Service. It was a symbol of institutional no less than intellectual hybridization. At the Fourth Conference on Fire and Forest Meteorology, the chief of the Meteorological Division of the Weather Service declared: "We don't really see any major changes in the National Weather Service policy for Forest Weather Service. Simply stated, 'The NWS will do what it does best to help Forestry interests do what they do best.'"[150] Competition over the atmosphere never reached the stage that competition over land did. In 1975 the Forest Service and NWS jointly established the National Fire Weather Data Library, which became indispensable to a host of sophisticated computer programs that guided decisions on prescribed fire, fire suppression, and fire planning.

Nor was the exchange one-sided. Its liaison with the Forest Service expanded both the institutional domain of the NWS and the intellectual realm of meteorology. Meteorology is a statistical science; it deals on a macro scale with ensembles of atmospheric events. Yet fire management demands highly particularized forecasts—the microclimate on different sides of a mountain, the exact paths of lightning-laden storms, the moment of wind shifts, the pattern of smoke dispersal, the climatic nuances of various fuel types. Fire demands consideration of unique phenomena as well—local wind patterns, meteorological events peculiar to a particular site. And, of course, there is fire itself—a meteorological phenomenon of at least as much local importance as storms, both a product and a creator of weather. Fire protection opened new fields of study and new opportunities for service to the NWS. Like lightning, its ultimate source, fire helped to bring meteorology out of the clouds and back to the earth.

The 1920s was the first great era of interagency experimentation on a national scale; the fire weather service was but one example among many. The Forest Protection Board (1927–1933) did for federal land bureaus what the National Association of State Foresters (1920) did for the states, what the Clarke-McNary Act (1924) did for federal-state cooperation, and what the National Forestry Program Committee did for cooperation with private interests. William Greeley was a prominent member of the committee, and he infused it with the same cooperative spirit that he brought to the Forest Service during his tenure as Chief. In 1926 the committee porposed that the interests of government efficiency demanded better coordination among the federal land agencies. The Office of Chief Coordinator agreed, and in January 1927 the Forest Protection Board was created.[151]

The board was purely advisory, its purpose to ensure a "coordination of effort through a central agency" by which "to facilitate cooperation among these [federal] agencies as well as with state and private protection services."[152] The national board assembled three agencies from Interior—the National Park Service, the General Land Office, and the Indian Service—and five from Agriculture—the Forest Service, the Weather Bureau, the Biological Survey, the Bureau of Entomology, and the Bureau of Plant Industry. Seven regional boards composed of representatives from these same agencies were established in the West and Lake States in an effort to extend the cooperative endeavor to local issues. For some time the national board convened weekly. It was chaired by Greeley, then by Robert Stuart, Greeley's successor as Chief Forester. Though it purported to examine all aspects of forest destruction, its overwhelming concern was fire.

The board existed for a little over five years before it was swept away by the coming of the New Deal. In a concluding summary of its history, the board listed eleven accomplishments. It had, for example, eliminated much senseless duplication of services and facilities. It had prepared comprehensive forest protection plans that helped to secure congressional appropriations for its members and to pool resources, plans that during the New Deal would largely serve as a blueprint for expansion. It created a mechanism for shared logistics, based on the central purchasing of equipment and on central warehousing. In the Northern Rockies, for example, nearly all the agencies purchased supplies through the Forest Service warehouse at Missoula. The member agencies strengthened mutual aid agreements, extended support during large fires (including overhead), organized interagency boards of review, and emplaced facilities (notably lookouts) so as to serve the needs of more than one agency. Among its early assignments, the board even constructed financial plans through 1929 for all of its members and projected budgets through 1932. By 1933 the federal land agencies were well advanced on a program of fire cooperation; practically all of the concepts embodied in the 1970s by BIFC and NWCG were present in the programs of the Forest Protection Board.

Early in its career the board concluded that although it existed to promote consistency of policy and efficiency of operation, "it is agreed that coordination of forest protection policies means consistency rather than uniformity. The major objectives of the responsible agencies differ and such differences will be reflected in differences of policy."[153] Out of sheer sensibleness rather than coercion, the agencies tended to adopt as a standard the existing policy of the Forest Service, as embodied in Greeley's 1927 memorandum a "General Plan for Protection of the National Forests from Fire" and in subsequent amendments to the *National Forest Manual*. Forest Service fire plans became an exemplar for other federal agencies. In fact, virtually all the cooperative fire programs of the 1920s were promoted in some fundamental way by the Forest Service.

Yet the board and its programs sank without a trace, eliminated in 1933 by executive order. The concepts it embodied had to be reinvented decades later. ECW and CCC became, in effect, the new media for interagency cooperation. The Forest Service controlled the lion's share of CCC camps, and with them it was able to develop an autonomous program. It did not need the resources of other agencies; in fact, it was in a position to extend to others the new wealth it enjoyed. But by intensifying its own organization, the Forest Service became somewhat isolated from the others. The 10 A.M. Policy segregated the Service from other federal agencies in policy, and the emergency presuppression account eliminated the economic rationale for cooperation. Early interagency cooperation was based on the premise that no one agency could cope with fire singlehandedly, that it was necessary to pool resources. After New Deal reforms had run their course, each agency—but the Forest Service more than most—had a separate organization and in time separate emergency accounts.

By the 1960s the general blossoming of affluence in American society and the willingness of the federal government to direct this wealth allowed ample funds for other agencies. As these created separate research units, as they upgraded their control forces (through emergency presuppression accounts of their own), as suppression costs escalated, and as distinctive agency policies were redefined, often by new legislation—Forest Service hegemony was effectively challenged, and the growing need for formal institutions to direct interagency cooperation was recognized. For the federal agencies the two most prominent manifestations of this need were the Boise Interagency Fire Center and the National Wildfire Coordinating Group.

BIFC developed after the Elko fire debacle. The NWCG, however, traces its immediate origin to the aftermath of the 1970 fire season. The Secretaries of Agriculture and the Interior requested an option paper to explore better ways to coordinate fire-related interests. In 1973 a meeting of Forest Service and Interior agency representatives discussed the possible shape and scope of a coordinating group. A name was selected and ten working teams established. The composition of the resulting group reflected the realities of federal fire protection: of its nine members, two came from BLM, one each from NPS, FWS, and BIA, and four from the Forest Service (its BIFC liaison, and one member from each of its three fire interests—fire management, fire research, and cooperative fire). A follow-up meeting was held in 1974, during which, at Forest Service insistence, a state representative was added (to be selected by the National Association of State Foresters). The meeting drafted a charter that was approved by the Secretaries of the Interior and Agriculture in March 1976. In the meantime, the U.S. Fire Administration also joined, though as a nonvoting member.[154]

In words that unconsciously paraphrased the credo of the Forest Protection Board, the NWCG announced that it would accomplish its "goal by coordinating the programs of the participating agencies so as to provide a means of constructively working together."[155] Central to the objectives of the Group was the concept of total mobility—the complete interchangeability among agencies of crews, equipment, and overhead on the fireline. This led on the one hand to an Interagency Fire Qualifications Rating system based on the old Forest Service "red card" and on the other hand to the conversion of a Forest Service training facility at Marana, Arizona, into the National Advanced Resources Technology Center for interagency fire training. Like the regional boards erected under the auspices of the Forest Protection Board, local coordinating programs spontaneously appeared as working miniatures of the national organization.[156]

The pressures for integrated fire protection resulted in an intensification of many local and state operations as well as coordination on the national level. Among the most spectacular examples is FIRESCOPE which also grew out of the 1970 fires. Seven hundred fire starts were reported in Southern California within 13 days; 17 major fires burned half a million acres and 700 structures. Efforts at centralized planning and communications almost broke down. A state task force was immediately organized after the debacle to investigate better procedures during major emergencies, and Congress passed a special appropriation to augment fire command and control systems research, which was to be centered in Southern California at the Riverside laboratory. The next year the Forest Service, the California Division of Forestry, the California Office of Emergency Services, and the fire departments of Los Angeles city, Santa Barbara County, and Ventura County agreed to "participate in a Research and Development Program specifically aimed at increasing the effectiveness of the southern California protection agencies in future multiple fire or similar emergency situations."[157]

The result was FIRESCOPE. The Riverside laboratory of the Forest Service assumed responsibility, and the program was projected as a five-year enterprise. More than a question of centralizing logistics or of coordinating political control, FIRESCOPE for the first time applied integrated information technology to the fireline and blurred the distinction between urban and wildland fire. Fire, in fact, was only an exemplary emergency: the systems analysis of organizational structure, of terminology, and of information flow could apply to any "incident," individual or multiple, single agency or interagency.

Two major fireline systems resulted: an incident command system (ICS) and a multiagency command system (MACS). In actual practice an operational coordinating center (OCC) would come into play at the regional level. The gist of the systems approach was to improve the collection, processing, and dissemination of information—intelligence about the fires, predictions about fire behavior, assessment of anticipated damages—and to develop a communications system capable of relaying such information. To further its environmental intelligence, FIRESCOPE planned to install a network of automatic weather stations and developed a telemetry linkage that allowed the transmission of airborne infrared scanning to ground units immediately after a fire was mapped. The ultimate design contemplated a package of sophisticated computer models to predict fire spread and to evaluate alternative suppression and tactics. Fundamental portions of FIRE-SCOPE became operational during the large fires of 1977.

NWCG and BIFC were independent rediscoveries of ideas submerged during the 1930s. FIRESCOPE, on the other hand, was part and parcel of an unbroken evolution of institutional cooperation between the states and the Forest Service. What all three shared—and what every national program of cooperation in the twentieth century has required—was the participation of the U.S. Forest Service. Everywhere it provided a common denominator. Whether or not it was the initiator of a particular program, and it usually was, whether or not it served as the medium of communication, and it commonly did, no program could succeed without its support, and no program of note failed for lack of that support. The successful integration of fire resources, and often of programs and policies, on all levels of involvement is overwhelmingly the result of Forest Service participation or of the transfer of ideas or personnel from the Service.

That integration had not come without great resourcefulness. As American foresters quickly realized, the scale and remoteness of the lands created by the couterreclamation made European precedents for the management of these lands often impossible. Scores of unique American innovations resulted, and several of these inventions have had lasting effects—the national park idea and systematic fire protection among them. In this creative process the Forest Service might well have acquiesced, as the other federal agencies often did. But it saw its mission in larger script, as a national no less than a bureaucratic role and as a moral no less than a legal charge. To those other ideas for the management of American wildlands may be added another, for which the Service is almost wholly responsible, a hybrid of the technical issues of systematic fire protection and the political realities of American federalism: namely, the concept of cooperative fire protection. It represented an innovation in the history of ideas no less than in the history of government. Ironically, it was through its willingness to join with others that the Forest Service itself achieved such hegemony.

6 A CONTINENTAL EXPERIMENT

I propose to create a civilian conservation corps. . . . This
enterprise is an established part of our national policy. It will
conserve our natural resources. It will pay dividends to the
present and future generations. It will make improvements in
national and state domains which have been largely forgotten in
the past few years of industrial development.

More important, however, than the material gains will be the
moral and spiritual value of such work.
—Franklin Delano Roosevelt, March 21, 1933[1]

The Great Crash of '29 seemed to announce a series of environmental disasters, spearheaded by drought, that was destined to devastate the natural
environment as fully as the Depression did the social and economic ones.
The New Deal sought to redress both catastrophes. As William Leuchtenburg has observed, the "word which appears most frequently in the writings
of the New Deal theorists is 'balance.'"[2] Capital and industry had precipitated a panoply of imbalances—political, economic, social, and environmental—for which a new equilibrium had to be created. "Men and Nature,"
Roosevelt argued, "must work hand in hand. The throwing out of balance
of the resources of Nature throws out of balance also the lives of men."[3]

The arena where these imbalances were most magnified was the rural
landscape. Roosevelt himself, as Rexford Tugwell observed, "always did,
and always would, think people better off in the country and would regard
the cities as rather hopeless. . . . "[4] The physical and spiritual erosion of
rural America was echoed in the major art of the period—in books like John
Steinbeck's *The Grapes of Wrath,* James Agee's *Let Us Now Praise Famous
Men,* and Erskine Caldwell's *Tobacco Road;* in the hard realism of photographs by Margaret Bourke-White and Walker Evans; in the revived landscape genre of painters like Thomas Hart Benton; and in the propaganda
films of conservation, such as Pare Lorentz's *The River* and *The Plow That
Broke the Plains.*

The intended reclamation of rural America underwrote many of the New
Deal's more important, and most durable, programs. As social reform, this
reclamation comprised the Agricultural Adjustment Act, the Resettlement
Act, the Rural Electrification Administration, the Farm Tenancy Act

(which created the Farm Security Administration), and the establishment of rural and suburban model communities—from Greenbelt City, Maryland, to the Matanuska Valley in Alaska. As environmental reform, the reclamation of rural lands rather than of rural life motivated a host of conservation legislation—the Omnibus Flood Control Act, the Shelterbelt program, the Soil Conservation Service, the Taylor Grazing Act, and the creation of the Fish and Wildlife Service. Vast acreages of mostly tax-delinquent land passed into federal and state hands. By 1936 the federal government had acquired more than twice the acreage in forest lands as had been purchased in the previous history of the national forests. Above all, as the epitome of its ambitions for the rural landscape, there was the Tennessee Valley Authority, in which the reform of social and environmental systems would proceed simultaneously and in symbiosis. Society would regenerate the rural landscape, and the rural landscape would in turn help to regenerate society. All these efforts were perhaps best symbolized in the great tool created almost at the start of the New Deal, a tool that was as much as anything an expression of its resolve to reclaim the rural landscape and, in turn, to use the rural countryside as a mechanism for curing social ills: the Civilian Conservation Corps.

Roosevelt was himself an amateur forester, an honorary member of the American Forestry Association, and a man profoundly influenced in his youth by a Pinchot lecture on deforestation and soil erosion in China. That presentation "started me on the conservation road," he informed the Yale School of Forestry.[5] Firefighting, in fact, had been among his last acts as a fully healthy man. On August 10, 1921, while sailing off the shores of the Bay of Fundy, Roosevelt spotted a forest fire, left the boat to beat it out with pine boughs, and, to clean off the grime, dived first into the warm waters of Lake Glen Severn and then into the icy waters of the bay itself. That evening, exhausted and complaining of chills and aches, he retired early to bed. The next morning paralysis began. In the future Roosevelt, though on an enormous scale, would fight fire by proxy.[6]

Ferdinand Silcox, the Chief Forester, was an equally ardent exponent of the New Deal. When Roosevelt assigned a majority of the CCC camps to the Forest Service, Silcox was hardly prepared to protest. And when the President declared that "one of our great difficulties all over the country is with fire," the veteran of 1910 and the architect of the 10 A.M. Policy would hardly disagree.[7] A week before Roosevelt announced his intention to create the CCC, the Forest Service had released *A National Plan for American Forestry*, popularly known as the Copeland Report. Based on the debates inaugurated by a congressional request in 1928 and on the 1930 regional foresters conference, the Copeland Report reiterated the fundamental, perhaps controlling, role of fire in American forestry. It was the largest survey on forestry questions since the 1920 Capper Report. The section devoted to fire, authored by E. I. Kotok, Evan Kelly, and E. F. Evans, argued that "the prevention and control of forest fires is a basic requirement of forestry, whether the purpose of management is timber production, watershed protection, or game and recreational development." The thoughtless, greedy practices of the past had ravaged the forested landscape as fully as they had the economic and social ones, and the forest required similar restitution. "One of the major problems in American forestry," the report observed, "is to rebuild depreciated forest lands that have already suffered severely from overcutting and burning, and success in recapturing such forest values must be predicated on keeping fires entirely out or within reasonable check."[8]

The restoration motif so common in New Deal social and environmental programs thus figures prominently in forestry thinking, too. The methods of restoration would be reforestation and protection, notably fire control. The Forest Service responded with its "continental experiment." Previously reclamation had meant the conversion of wildland to agriculture; now it meant the restoration of marginal cropland to forest and marsh. But conservation still meant, as it had in John Wesley Powell's day, the control of natural forces, essentially those upset by an industrial economy, and its focus was still rural America. The New Deal programs were themselves as much a reclamation of old programs as was the landscape they hoped to produce. For fire protection this meant that the 10 A.M. Policy was a more logical outcome than was a reassessment of new, fundamental principles.

Floods in the Ohio and Mississippi valleys brought distress to thousands. But it was the droughts of the early 1930s that were most widespread and ruinous. The Dust Bowl in the Great Plains was flanked by fires along the Atlantic and West coasts. Incendiary and swamp fires nearly ruined state forestry departments along the eastern seaboard. The mammoth Matilija fire was for decades the yardstick of destruction in Southern California. And towering over all was the great Tillamook Burn of Oregon. Fire was already identified with the wastage of natural resources: it caused the floods, soil erosion, and rural impoverishment that added so much urgency to New Deal conservation programs. As a symbol, the Tillamook fire was to forestry what the Dust Bowl was to farming. Even more than the devastation wrought by the fire itself, the remarkable emergency salvage program that followed ties the event most firmly to the spirit of the New Deal—a spirit of recovery, restoration, and reclamation.

Though federal CCC camps were dispatched to the fireline, the attack on the Tillamook Burn of August 1933 was a state operation. Not since the great fires of the early 1900s in the Lake States and New York had a state organization been compelled to cope with a fire of such magnitude. The success of the Oregon Department of Forestry—and the contrast to state fire control 20 to 30 years earlier—marked the maturation of state forestry programs. The early conflagrations came before basic federal assistance was

14. The Tillamook fire in eruption, 1933. During the main blowup, the smoke column punched through a layered atmosphere to heights of 37–40,000 feet.

available in the form of the Weeks Act (1911); the Tillamook came nearly a decade after the Clarke-McNary program and on the eve of the tremendous improvements conducted by the CCC and made available to state cooperators by the Forest Service. Many states had enacted tough fire statutes, and the Northwest had compulsory patrol laws. Indeed, industrial logging had shifted its center of gravity to the Pacific Northwest, and the cooperative fire program advanced with it stride for stride. The reclamation of cutover and burned-over lands that had lapsed into tax delinquency coincided with a new wave of abandonment of marginal farmlands. Together these lands formed the basis for state forestry departments. This circumstance, in turn, has allowed rural fire protection, for the most part, to move in from the forests rather than out from the cities, reinforcing the significance of state forestry services for fire protection and binding the states still further to the national network of fire services coordinated by the Forest Service. Not only in the Northwest, but also in the Lake States and the

South, the 1930s marked the coming of age of state forestry and state fire organizations.

At the Tillamook burn, moreover, the CCC received its baptism by fire. Not the least of the CCC's achievements was that it revolutionized the structure of manpower in fire control. This accomplishment demanded a de facto revolution in fire protection itself—symbolized by the adoption of the 10 A.M. Policy by the Forest Service. Practically all of the organized crews so essential to modern fire control have evolved in one way or another out of the CCC experience. It was the CCC, moreover, that dramatically created a physical plant for fire control almost overnight.

Yet for all the innovations they stimulated, the fires of the Tillamook cycle were fought more in the spirit of the old conservation era than of the new; they were more a continuation of old fire problems, though quantitatively magnified, than a confrontation with problems that were qualitatively new. They represented the extension of systematic fire protection into the nation's backcountry—to lands remote in space (like those in the Northern Rockies) and to those remote in time (like the cutover pineries of the South). The Roosevelt administration looked back with longing to an imagined rural past. Its conservation reforms were, in spirit, a revival of those proposed by Powell and Pinchot, and the period was less the beginning of a new deal in fire protection than the conclusion of an old hand. The mushroom cloud rising in August 1945 from the last of the major Tillamook reburns was, apart from its colossal size, indistinguishable from those that towered over the 1871 Peshtigo fire, the 1903 Adirondacks fires, or the Big Blowup of 1910. But the mushroom cloud that rose over Hiroshima that same month was to give the expression "blowup fire" a new meaning, fire research new topics, and fire protection new responsibilities.

FIRE AND WATER:
A FIRE HISTORY OF THE NORTHWEST

We came to a section further up the slopes towards the mountains that has no trees more than fifty years old, or even fifteen or twenty years old. These last show plainly enough that they have been devastated by fire, as the black melancholy monuments rising here and there above the young growth bear witness. Then, with this fiery suggestive testimony, on examining those sections whose trees are a hundred years old or two hundred, we find the same fire records, though heavily veiled with mosses and lichens, showing that a century or two ago the forests that stood there had been swept away in some tremendous fire at a time when rare conditions of drought made their burning possible.
—John Muir, at Puget Sound, ca. 1912[9]

The distribution, growth, and economic management of timber, the greatest natural resource of the Northwest, are largely controlled by fire. . . . —William G. Morris, 1934[10]

I

It is frequently joked that the Pacific Northwest was first described in terms of rain. Had the timing been slightly different, that description might equally have been of fire. Fire occupied the droughty breaks in the region's maritime climate, and its effects could be enormous, from the Olympic Peninsula to the coastal redwoods and from the Coast Range across the rain shadow of the Cascades. The immemorial competition between fire and water focused here on the phenomenal forests of the region, some of the finest timber in the world and some of the heaviest fuel loads in North America. Those magnificent forests, too, have determined the peculiar character of fire protection in the Northwest. The struggle between fire and water was to be mediated by the imperatives of industrial logging. Large-scale logging arrived in the early twentieth century, in the era of the conservation movement, amid the fears of a timber famine and the growing acceptance of fire control as a necessity for a successful timber industry. By the 1920s the Northwest took the lead in national logging statistics and, not coincidentally, in national programs for cooperative fire control.

The Northwest's celebrated moisture has helped to mask somewhat the prevalence of fire in its history, and its contributions to national fire programs have been similarly camouflaged by its very drive for cooperation. Time and again fire in the Northwest has been overshadowed by fire in the two more famous provinces that border it to the south and east, California and the Northern Rockies. The east wind that propels the worst conflagrations has been upstaged by its more widely advertised cousin, the Santa Ana. Its great lightning busts are less awesome than those in the Rockies and less persistent than those in the Southwest. Its incendiary fire problem is not residual, as in the South, nor psychopathic, as in Southern California. The Northwest has frequently introduced innovative fire protection programs—from prevention to manpower to equipment—only to see the products exported to adjacent regions or adopted, with a loss of regional identity, as a national endeavor.

The very concept of cooperative fire protection may be the Northwest's most enduring legacy. For this program the almost simultaneous arrival of industrial logging, the forest reservation system, and a timely series of holocausts may take the credit: private interests, the states, and the federal government came to control about equal portions of the Northwest forest and came to perceive a mutual adversary in fire. Unlike control efforts in the South and Lake States, fire protection in the Northwest did not confront cutover lands; unlike the forest lands of the South, timber wealth did not pass into private hands. Nor did tax-delinquent lands revert to the states as they had in the Great Lakes region, and the federal government did not possess a virtual monopoly on land use, as it did in the Northern Rockies. From the beginning the ownership pattern was mixed, fire control was central, and cooperative forestry became a hallmark. Out of the Northwest came the origins of the fire weather service, the Keep Green movement, the "closest forces" doctrine of initial attack, and the Western Forestry and Conservation Association, with its interstate fire research councils. More important, the region established a model in cooperative fire protection that in advance of the Weeks Act became an exemplar in the extension of systematic protection across the country. It is no coincidence that when William Greeley, the great spokesman for cooperative fire protection, resigned from the U.S. Forest Service, he became the executive secretary of the West Coast Lumbermen's Association, headquartered in Portland. Logging came to the Northwest after the agricultural reclamation largely abandoned the attempt to occupy the northern forests but before the industrial counterreclamation could be firmly established. It was the peculiar role of the Northwest to oversee the transition and to provide a national model for the exchange of fire practices it entailed.[11]

The Pacific Northwest, moreover, has had at least one fire in modern times so typical and yet so extraordinary as to command both regional and national attention. The history of the celebrated Tillamook Burn of August 1933 is archetypal: it began in association with logging operations, was controlled, then was blown into mass fire by east winds and left in its wake a tinderbox of snags and heavy downed fuels to stoke decades of violent reburns. What catapulted the Tillamook Burn into national prominence was, as a historical event, its timely appearance on the arena of national politics and, as a fire, its unbelievable intensity. The Tillamook Burn was one of the series of holocausts that encouraged the adoption of the 10 A.M. Policy. It committed the CCC to the fireline for the first time. Forest Service researchers, notably Richard McArdle and William Morris, reported on the fire in depth. It may have affected McArdle as the 1910 burn did Greeley and Silcox; McArdle's tenure as Chief Forester saw fire reforms and developments equaled only by those two illustrious predecessors. Tillamook was an ordeal by fire that brutally tested the region's fire programs and with them the experiment in cooperative fire protection that the region conducted for the nation at large.[12]

The fire began under curious circumstances. For two months the Northwest had endured drought. On August 14, 1933, the relative humidity plummeted before an east wind, and runners were sent out to close down logging operations in the Coast Range. All but one shut down without incident. On the Crosset and Western lands in Gales Creek Canyon, an area logged and laden with slash, a gyppo operator continued to cut timber. One final log, so the story goes, would be dragged by steam donkey before abiding with the closure order. Somehow the friction of rubbing cable and timber or of fresh timber on cedar slashings resulted in ignition. Within minutes the fire was hopelessly out of control. However bizarre the reported cause for the fire, its symbolism was appropriate: it was as if the sheer volume of fuel, catalyzed by logging, spontaneously created its own fire. Though the original fire may well have ignited in this manner, before the day ended incendiary fires had started a few miles away. Job hunting fires had reached epidemic proportions in some areas of the Northern Rockies and the Northwest where the Depression hit the logging industry especially hard. Whether the original Tillamook fire was of this sort is not known, but the fires that surrounded it, at times assisted it, and always complicated its suppression were incendiary.[13]

The fire burned on lands under the fire protection jurisdiction of the Oregon Department of Forestry. Control was aggravated by lack of access; Tillamook County contained the last great stand of virgin timber in the state, roads were few, and the logging companies gnawing into the edges of the forest relied on narrow gauge railroads. Eager laborers poured into fire camp at Forest Grove, hoping to be hired, but many were turned away because there were not enough tools to support them. Loggers rolled into camp, too, bringing their own tools and foremen and their professional experience of the business at hand. But this fire was extraordinary. Stoked by the heaviest fuels in the country, blistered by the rushing east wind, the fire became a conflagration. The Northwest had not seen a fire with such potential since 1902, some said since 1868. Larger numbers of organized and equipped crews were needed. In a move that was at once desperate, brilliant, and yet almost reflexive, forestry officials turned to the Civilian Conservation Corps. A nearby camp of 100 men was requested. A second was soon mobilized. The camps not only came outfitted with tools but also brought, indirectly, the logistical resources of the U.S. Army. Although their normal work routine was directed by civilian agency officials, the camps were run and supplied by the Army. Within four days 10 CCC camps had been relocated to the fire, and military truck convoys connected them to Vancouver and Portland. In some cases Army officers led the crews to the line. The Forest Service, meanwhile, rushed in men and special overhead, and in what was perhaps a first, researchers from its Pacific Northwest Experiment Station were assigned duties as advisors—an early experiment in the role of the fire behavior officer. The Southern Pacific Railroad patrolled its right-of-ways with a special engine and tank car.

For the first 10 days the fire rose and fell with the east wind like a furnace fanned by an enormous bellows. More fires, accidental and incendiary, appeared. The main fire was alternately contained and lost. Fog rolled in on the twenty-third, leaving the burn at 40,000 smoldering acres—the largest since the 1902 cycle. On the twenty-fourth the east wind returned, and fire officials grimly ordered an immediate evacuation of firefighters and civilians from the west end of the fire as the burn smoked like a giant fuse creeping toward a mountain of explosives. At last the inevitable occurred: the Tillamook Burn blew up. Within 20 hours it swelled from 40,000 to 240,000 acres. The mountains roared with a deep, chilling boom like the magnified hiss and blast of a furnace with its door flung open. Winds reached the hurricane velocity typical for firestorm perimeters, mowing down stands of enormous Douglas firs like a scythe to feed the flames further. Flames as high as 1,600 feet were reported. Pastures in Tillamook County were buried under three feet of ash. The tremendous timber that had fueled the Northwest's industrial growth now stoked a conflagration. But it was the smoke column that was most often remembered. The smoke had risen to 9,000 feet and leveled out in a thick blanket. With the blowup, however, the superheated air punched through the inversion in a great mushroom-shaped dome and shot upward to the ceiling of the troposphere itself. Not until September 5, when heavy rains signaled the reestablishment of the normal flow of marine air, did the fire subside into steam, and then into history.

Big fires are rarely singular events. The Tillamook Burn was not only a complex of fires but a cycle of fires, large in acres and intensity and long in time. There is a legend in the Northwest of an almost 3-million-acre fire that burned for 4 years. In a sense, Tillamook was a 300,000-acre fire that burned intermittently for 18 years. In August 1939 the scenario of 1933 was

15. CCC crews en route to the Tillamook fire, 1933.

repeated: 200,000 acres of the old burn was fired along with an additional 20,000 acres of green timber. The reburn began in almost the same location as the original. In the initial fire some 3,000 men were organized for the line; in the reburn, 4,000. Rains on August 28 finally extinguished it. In August 1945 another reburn swept the site. Again it was ignited by logging activity in Gales Creek Canyon. Two fires merged, crossed almost 200,000 acres of the original burn, and swept into 30,000 acres of green timber. Though some 3,500 firefighters were rushed to the scene, including Army troops, the manpower was ineffective until rain extinguished the fire after eight weeks. But the six-year curse was wavering. In July 1951 fire once more broke out on the burn, but heavy equipment, presuppression preparations, and almost two decades of timber salvage and fuel management caught the fire at 20,000 acres after only 4 days.[14]

In the 18-year fight between man and fire over the Tillamook Burn, the focus of contention was, appropriately enough, the forest itself. Whether the charred forest and its green perimeter would be timber or tinder depended on whether man or fire became temporarily ascendant. Salvage on the burn continued as long as the fires. To professional foresters, the burn was a shocking economic catastrophe. Over 10 billion board feet of timber had been destroyed and many operators burned out—this in the midst of a national depression that eroded deeply into the logging industry. Experts from the Forest Service voiced collective wisdom when they glumly predicted that only a fraction of the charred timber was marketable or could be culled before insects, disease, and further fire would invade the rest.

Stumpage prices were low, even for prime timber. Tillamook County faced impoverishment. Two months after the burn, Lynn Cronemiller, state forester and Tillamook fire boss, informed a national audience that "the only hope lies in salvage, but the enormous amount of timber involved makes it impossible to log it all before decay and insects take a large toll. Isolation and expense of development under present conditions place an impossible barrier before the private interests. It is a stupendous problem and necessitates concerted private and public action to remove all possible values."[15]

"Concerted private and public action" was exactly what the Pacific Northwest specialized in. The problem was perhaps unique: it was one thing to make capital improvements for reproduction; it was something else to invest heavily in burned timber. Over 125,000 acres were abandoned, reverting to the county in lieu of taxes and ending in a legal limbo. But for the independent companies without other lands, there was no alternative but to log the burn. A number of firms formed the Consolidated Logging Company, the county revised its tax structure to accommodate the situation, and, by operating its own mills, the company could salvage what it wanted. With $4 million in new equipment, the greatest salvage operation in American history began. In culling out only the most select timber, the company harvested over 1 million board feet in six years. Business in the burn boomed: probably more timber was removed and more loggers hired under the impress of the burn than could have been achieved without it during the Depression. Though some small operators had been literally burned out, many others survived because of the gigantic labor of salvage.[16]

Reburns swept away capital equipment as well as salvagable timber, but the strange harvest continued. Wartime demands increased the value of snags missed in the first selection. Nor had disease and insects corrupted the burn as experts had forecast. Demand increased, and so did the number of companies willing to log. In 1939 fewer than 50 outfits worked in the burn; by the 1950s, there were 200. Successive harvestings took virtually all the sound timber; of the 10 billion board feet originally dismissed as lost, about 7 billion was shipped to mills. And all this activity was in addition to specific fire control improvements made largely with CCC labor: over 700 miles of roads were punched through the burn, and long corridors of snags were felled as fuelbreaks. Only the emergency salvage of debris left by the 1938 New England hurricane can be compared with the Tillamook story.

Reburns, however, obliterated refugia left by the original fire, devastated reproduction, and pushed back the green borders from which reseeding might occur. The Tillamook had been burned by man and logged by man; it would also have to be restocked by man. Again, this would be a cooperative endeavor. In 1948, after spontaneous efforts by citizens' groups to plant on the burn, Oregon voters approved a constitutional amendment to authorize bond issues to pay for reseeding on state forest lands. A year later

16. Site of the Tillamook Burn, 1953. Note extensive salvage, roads, and corridors of felled snags.

reforestation on the Tillamook Burn began under the direction of the Oregon Department of Forestry. With painstaking labor and at a cost of $12 million, 255,000 acres of the burn again supported young timber. Trees were reaching 50 feet in height when in 1973 Oregon's governor dedicated the Tillamook State Forest—obliterating in name, as the reclamation had veneered in appearance, the Tillamook Burn.[17]

Fire protection arrived in the Northwest after the double blows dealt by the 1902 and 1910 conflagrations. For the next sixty years the story of large fires in the Northwest is basically the story of the Tillamook cycle. The reclamation of the Tillamook Burn was perfectly in keeping with those dimensions that had made the Northwest unique: its commitment to industrial logging, its extraordinary fuels, and its willingness to engage in cooperative fire protection. In the South, the Lake States, and the Northeast lands were burned *after* they had been logged, and reforestation met with public hostility; cutover lands were to be maintained as pasturage, converted to farmland, or abandoned to old field pine and coppicing. The Northwest reversed this scenario: the land was logged after it had burned, and it was deliberately protected and stocked within the confines of industrial forestry. As much as the use of the CCC, this unhesitating, collective commitment to the reclamation of cutover and burned-over land propelled the Tillamook Burn into national prominence. It was during the 1930s, after all, that the question of cutover lands in the Lake States and the South was finally resolved in favor of reforestation. The salvage and replanting of the Tillamook Burn marked a turning point not only in the methodology of fire control but also in its purposes.

II

As a fire regime, the Pacific Northwest is bounded on the west by the Pacific Ocean, on the east by the Cascade Range, and north and south by the realm of the foehn winds, known locally as the east or north winds. Its northern border extends to the Sitka spruce, a species sensitive to fire in a climate where fire is rare; its southern border includes the coastal redwoods, whose survival is intimately connected with periodic fire and whose habitat merges into the endemic brushfields of the California Coast Range. The fire season commonly extends from June through September. Thanks to strong maritime influences, temperatures and humidity are moderate in the summer and precipitation light. It is during droughts that major fires come, often with terrific intensity.[18]

The pattern of historic fires conforms closely to that of the Tillamook cycle. Even under natural conditions fuel loads can exceed several hundred tons an acre. Blowdowns have leveled large swaths. A hurricane that struck the Olympic Peninsula in 1921 rivaled in damages the New England hurricane of 1938. An Oregon windstorm in 1962 felled 7 billion board feet, an amount roughly equivalent to the timber salvaged off the Tillamook Burn. Insects and diseases have killed even more timber, especially where exotics were involved. These die-offs are in addition to the prodigious natural deposits that slowly carpet the forest floor. Blowdowns and bugkills are prime fire traps; indeed, until the advent of industrial logging no other mechanism could dispose of the organic debris that collected on such a scale. The fires that liberated these deposits came in cycles of reburns that could extend over decades; each reburn culled out more of the original debris and swept aside any reproduction that had sprouted in the interim. Reproduction, in fact, provided a carpet of fine fuels to assist the periodic fiery flushing of the old burn. Reburns became kindling from which fires could spread into green timber, thus assuring the perpetuation of a fire regime. Once initiated, the pattern of fire radiating slowly from reburns became self-reinforcing. With the advent of industrial logging, potential sites for fire proliferated, and so did ignition sources. The historical result was a mosaic of vast, even-aged stands of forest growing on the sites of cleaned-out reburns—a prime target for fire, wind, and insects.[19]

Though all the fires of historic notoriety in the Northwest were man-caused, lightning fires are common in some portions of the region. If one imagines the Olympic Peninsula as an epicenter, the frequency of thunderstorms and lightning fires increases slowly to the south and to the east, with the heaviest concentrations occurring in the great pine belts along the borders of the Cascades. Even on the Olympic Peninsula, as studies of 60 years of fire records for Olympic National Park show, some 747 fires have started from lightning, though researchers confess that the burns have had "negligible impact" and that "at the present rate, nearly ten thousand years would be required for the whole Park to burn over once." But they also observe that there is "ample evidence of much more extensive burning."[20] That is, the fire regime was shaped—as John Muir recognized when he visited the region—by anthropogenic fire. The same may be said generally of the entire Northwest. There is little reason to believe, for example, that the even-aged tracts of Douglas fir for which the region is renowned are due to natural fire. All existing stands are almost certainly man-caused. And once initiated, such a pattern of fire becomes self-reinforcing.

The evidence clearly shows that the frequency of large fires increased with the appearance of American settlers in the 1840s. Furthermore, plenty of records confirm that big fires existed in presettlement days, the product of deliberate burning by Indians. The reasons for burning were the common ones. David Douglas (after whom the Douglas fir was named) observed extensive burned-over lands on the west side of the Willamette Valley, from Fort Vancouver to the Umpqua River, in the autumn of 1826. The fires were so widespread as to cause considerable hardship for his party: burned stubble made for difficult walking, game was absent, and forage for the horses was lacking. "Some of the natives," Douglas wrote, "tell me it is done for the purpose of urging the deer to frequent certain parts to feed, which they leave unburned, and of course they are easily killed. Others say that it is done in order that they might the better find wild honey and grasshoppers, which both serve as articles of winter food." Elsewhere Douglas described open savannahs of pine and oak not unlike those in the South and the Ohio Valley. After more than a week of steady travel south, he wrote: "Camped on the side of a low woody stream in the center of a small plain—which, like the whole of the country I have passed through is burned."[21]

Here as elsewhere fire had replaced forest with grassland, and the effect of agricultural settlement was to reverse this process. Speaking of the Champoeg area, a treeless prairie of 150 square miles, Harvey W. Scott (for four decades the editor of the Portland *Morning Oregonian*) observed that "the prairie had been an Indian grazing area, and was evidently keep clear of trees by fires set by the Indians for that purpose. At present [ca. 1900] it contains groves of trees in many places, obscuring the view and changing the old-time condition."[22] Throughout the nineteenth century smoke was so thick along the Columbia River that in the fall it often brought navigation to a standstill. In 1868 it was suggested that lighthouses be established along the Willamette to guide steamers, so opaque was the smoke from clearing fires.[23]

The Oregon grasslands, notably the Willamette and Tualatin valleys, were fired annually. One purpose was to harvest wild wheat. The technique was not one to which the first white settlers in the 1840s were accustomed. "We did not know that the Indians were wont to baptize the whole country with fire at the close of every summer; but very soon we were to learn our first lesson," one recalled. "The Indians continued to burn the grass every season until the country was somewhat settled up and the whites prevented them; but every fall, for a number of years, we were treated to the same grand display of fireworks."[24] Another purpose was to keep down the undergrowth—in the words of a settler from Grande Ronde, "so that no hostile war party could approach unseen." This function was combined with another, "the occasion for a grand hunt to secure an ample meat supply for the winter."[25] S. A. Clarke recorded the character of such a hunt at great length in his *Pioneer Days of Oregon History*. "The bands in the eastern part of the Willamette Valley from the Molalla to the Santiam united in this annual roundup. . . . At a given signal, made by a fire kindled at some point as agreed, they commenced burning off the whole face of the country and driving wild game toward a common center. . . . There was considerable skill required to do this correctly and effectively. . . . When the circle of fire became small enough . . . the best hunters went inside and shot the game they thought should be killed. . . ." Within a few years, of settlement by whites, however, "the hills and prairies had already commenced to grow up with a young growth of firs and oaks. . . ."[26]

The earliest of the great historic fires burned sometime between 1845 and 1849 and probably fired 500,000 acres between the Siuslaw and Siletz rivers of Oregon. Accounts of the burn conflict as to date, time, and place. It probably occurred in 1849, and the most plausible version is that pieced together by William G. Morris.[27] What is reported as one burn may in fact have been several, occurring a year or so apart and consuming over a million acres. The reported origins of the original burn are all man-caused ignitions: escaped slashing fires, embers from an open fireplace, traditional Indian fall burning. There may, in fact, have been a complex of fires, all fanned by the dreaded east wind into a hurricane of flame. One record of the fire comes from an Alsea Indian, William Smith, as recorded by an ethnologist from the Smithsonian Institution. "We were coming back from Siuslaw," Smith recalled, "when long ago, the world was in flames. Then it seemed to be getting dark all over. . . . Although the sun stood high, nevertheless it threatened to get dark . . . 'what on earth was nature going to do?' The fire was falling all around us. Wherever it would drop, another fire would start there. Everybody was staying near the ocean on the beach. The fire was flying around just like the birds. All the hills were on fire. Even the hills that were near the sea were burning as soon as the fire arrived at the sea. . . . For probably ten days it was dark all over. . . . Even the trees [that] lay in the water caught fire." At Yaquina Bay another informant related that the flames leapt across the river and that much game and many natives perished in the fire.[28]

Another version of probably the same complex of fires comes from Lieutenant Theodore Talbot, who explored the Coast Range in 1849. Searching for a land route between the Willamette and the coast, Talbot recorded a large fire burning south of the Yaquina River in late August. The smoke obscured his view, and later his troops passed through miles of tangled burned forest, which made travel almost impossible. "These fires," the lieutenant noted, "are of frequent occurrence in the forests of Oregon, raging with violence for months, until quelled by the continued rains of the winter season."[29]

Chronicles report another large fire of about equal acreage along the Yaquina in 1853. It may represent a confusion of dates or a reburn; the record is unclear. Either way, continued Indian burning and landclearing fires by American settlers annually turned the major river valleys of Oregon into a pall of smoke during August and September. Large wildfires raged throughout the southern region in 1857 and again in 1864. Fires in 1864 burned to the suburbs of Victoria on Vancouver Island. But public attention was most fixed on smoke, not fire. The *Oregonian* reprinted an article that claimed that "much of the sickness which prevails among us at present is attributed to the heated state of the atmosphere and the immense volumes of smoke created by the vast fires."[30] A few years later, the paper ran another, more comical blurb from the south, which was again rimmed by fire. "We read about Egyptian darkness, but it is smoke, Josephine smoke—smoke in the morning, at noon and at night. Meet a neighbor, it is smoke; parting from one, it is smoke. Hogs running around are smoked through and through—live, running bacon . . . so you see we live in the days of smoke. It is smoke, smoke, smoke!"[31] Miraculous cures were even attributed to the smoke. One sufferer declared that his throat became tanned like buckskin from the hot, charged air and that he had never since been able to distinguish whiskey from spring water. The *Oregonian* wryly editorialized that "it is not certain, however, that he has tried the latter within the last 20 years. . . ."[32] Oregonians became acclimated to fog in the winter and smoke in the summer.

The sources of fire changed as agriculturalists replaced hunting and gathering societies. Fire moved steadily out of the grasslands and into the surrounding timber; landclearing and logging added intensity as well as frequency. It even became something of a sport among certain elements to fire the woods and then to boast over the size of the various fires. Critics of the practice were more concerned with the smoke that suffocated the valleys than with the forests that burned on the distant hills. Only in exceptional years did such fires, however discomforting their smoke, reach holocaust proportions.

Such a year was 1868. In perhaps the worst fire season on record, fires rampaged throughout the entire Pacific Northwest. Probably over a million acres burned, from the Olympic Peninsula to the California redwoods, from the middle of August to the beginning of October. Hundreds of small fires— landclearing, camp, and incendiary—swelled into conflagrations under the impress of the east wind. Fires burned around Victoria, Seattle, Olympia, Yaquina Bay, and Coos Bay, where a single fire raged through 125,000 to 300,000 acres. Near the mouth of the Columbia a slashing fire escaped control, a backfire was set by a panicky neighbor, the two fires merged under the breath of the east wind, and despite efforts by considerable numbers of volunteers "an advancing line of fire extended from the very edge of the bay to the mountain tops." "Dark days" settled on the principal cities. Navigation periodically came to a halt. Despite their immensity, the fires burned largely in country not yet settled or logged, and the loss of life and property was not so great as the fires' size suggested.[33]

For the remainder of the nineteenth century the Northwest was spared such holocausts. Fire and smoke continued, of course, localized along slashings, landclearing, highways, and railroads and, especially along the Cascades, where sheepmen appeared with herd and torch, firing the mountain landscape for pasturage. John Muir scathingly satirized such practices as a primary threat to the Sierra forests, but his remarks are equally applicable to the Cascades. The process was described by W. J. Lord of Tuolomne County, California.

In the 70s those who came later had to go to less accessible places for feed. They encountered considerable difficulty in herding their flocks and where brush and timber interfered it was common practice to burn. Most of the burning was done as the sheep were taken from the mountains in the fall to the plains to be put on stubble. This was usually in September.

Burning at that time became such a practice that people knew when sheep were leaving the mountains by the number of fires set. Smoke from the fires was so thick at times that it was hard to see at midday. No attempt was made to stop the fires unless someone's place was threatened, then back fires were set and usually the fire went some other direction. These fires burned thousands of acres almost everywhere where timber and brush grew in the mountains.[34]

Smoke was notoriously thick throughout most of the 1880s. So dense was it in summer and fall that geological survey parties to the Cascades abandoned attempts at topographic mapping. Not until the late 1880s did the smoke elicit a public outcry. Suppression came, with great difficulty, only after park and forest reserves were established, often with military administration. The U.S. Geological Survey parties sent to inventory the forest reserves in the 1890s had better luck, and their maps record vast stretches of burned terrain. Pinchot at the turn of the century was both distressed and amazed to see the evidence of large fires even in the virtual rain forest of the Olympic Peninsula. Examining the Puget Sound region, John Muir concluded: "Fire, then, is the great governing agent in forest-distribution and to a great extent also in the conditions of forest growth."[35]

Industrial logging arrived in the Northwest around the turn of the century. Almost immediately it had to confront the fire menace, and, like northern forests elsewhere, the Northwest experienced some of its most devastating fires during the transition. In 1902 autumn slash and debris fires from Seattle to the Siskiyou Mountains were suddenly fanned by the treacherous east wind; at least 80 major fires resulted. A host of seemingly innocuous campfires left by hunters and berry pickers swelled to mammoth proportions. Breaking out on September 7, these fires and many landclearing fires contributed to the problem. In fact, even as horror stories from the woods filtered into metropolitan areas, farmers east of Portland continued to fire brush. Frontier fire brigades battled bravely but hopelessly. Townsfolk turned out en masse. From the eleventh to the fourteenth of September, farmers and villagers dug and ploughed firelines, formed bucket brigades, and backfired day and night. A score of communities throughout the Northwest suffered from fire; 16 settlers perished in a valley around Mount St. Helens; for 50 miles around Portland fires poured out smoke that turned noon into midnight. Around Mount Hood, 170,000 acres burned. The Yacolt fire in the Lewis Valley of southern Washington, a merger of two fires, destroyed so many thousands of acres (200,000 plus) that it lent its name to the whole complex of 1902 fires. In all, perhaps 650,000 to 700,00 acres burned.[36]

The usual relief effort was mounted by a shocked citizenry. But a more unusual and more nationally significant step was taken by alarmed timbermen. In 1904 the Booth-Kelly Lumber Company initiated a cooperative fire patrol in Lane County, Oregon. Fire protective associations were in fact appearing across the country: Idaho soon had several, and, following the 1903 fires, so did the Northeast.[37] In the 1902 fires the greatest success had been achieved by fire guards hired by the GLO to patrol the forest reserves, an example that was not lost on the thoughtful. A Bureau of Forestry agent, William Cox, observed that "commendable work has been done by government rangers in the forest reserves, and the absence of serious fires in them should serve as an object lesson to the states of Oregon and Washington."[38] The 1902 fires foreshadowed uncannily the 1947 Maine fires, where forest

17. Part of the Yacolt burn (1902) some 32 years later; Mt. Hood in the distance.

districts received more adequate protection than did rural communities. The events of 1910 confirmed the value of the pattern of cooperative fire protection that emerged after 1902, much as the fires themselves reburned and extended the fire scenes of 1902.

Mutual protection took hold in the Northwest with particular tenacity, culminating in 1909 with the Western Forestry and Conservation Association. Elsewhere protection had followed fire and ax; in the Northwest, fire control and logging evolved in symbiosis, establishing a basis for all manner of cooperative forestry. The problem of cutover land and reproduction was less an issue than the harvest of existing forests; fire control was less a bid for the future than it was insurance for present resources, and it was not in serious competition with other land uses. As late as 1940 the cost of planting one-year seedlings was twice the going price for well-stocked stands of 50-year-old Douglas fir. Not until the tremendous demands generated by World War II did the costs come into parity. Thus, for the greater part of the twentieth century it was industrial forestry, not agriculture, that became the chief means for the penetration and settlement of forests in the Northwest.

From 1910 to the Tillamook cycle, the story of big fires is basically the familiar story of reburns. The fuel reduction brought about on the Tillamook Burn by salvage relogging came about on the sites of earlier holocausts by reburning. Each new fire culled out more snags, consumed heavy downed timber, obliterated reproduction, and encroached into previously untouched forests around the perimeter. Large fires appeared in 1918 (Washington called out its national guard for fire patrol), 1922, 1924 (overshadowed by California fires, as the 1918 burns were by the holocaust at Cloquet, Minnesota), 1927, and 1929 (when, despite a nearly 200,000-acre loss in two days, still larger reburns were in progress in the Northern Rockies). But even as the cycle of the 1902 and 1910 fires continued, so did the protection machinery they had brought into being. Both Washington and Oregon enacted compulsory patrol statutes, which required landowners either to establish fire protection over their forested lands or to contribute to state support for it. The Clarke-McNary Act was promptly embraced. The early fire weather service issued forecasts from Portland and Seattle. "Showboat" prevention programs carried fire messages to rural communities. And, under the aegis of the Forest Protection Board, a regional committee of private, state, and federal protection organizations met to eliminate duplication of services and to pool resources.[39]

The Tillamook cycle dominated the fire history of the Northwest for a quarter of a century. But there were other bad years. Fires in September 1936 burned some 144,000 acres in Coos and Curry counties, Oregon and swept the seacoast town of Bandon into oblivion. In 1939 the town of Pine Ridge, Oregon, was similarly devastated. Meanwhile, the incompatibility of fire and logging in the coastal redwoods was demonstrated through studies done by Emmanuel Fritz, professor of forestry at the University of California. Accepted practice called for felling the giant redwoods, followed by *in situ* broadcast burning of slash and forest, which made extrication of the redwood boles easier. Loggers relied on the relatively impervious bark of the redwood to withstand the often severe burns, and since logging was selective, the consequence of intense fires for other species was ignored. Along with foresters in general, Fritz argued that fire damaged the redwoods as well, to say nothing of the surrounding environs. The practice was stopped, and the subject was not reopened until the question of natural fire in redwood forests appeared in the mid-1960s.[40]

Meanwhile, the CCC was flexing its muscle as a firefighting force. The Tillamook Burn was its first mass mobilization. The site of the burn—and others like it—furnished plenty of presuppression jobs as roads and fuelbreaks were prepared.[41] The Northwest remained the great proving ground for organized fire crews. The two most important successors to the CCC appeared in 1939: smokejumping underwent operational testing on the Chelan National Forest in Washington, and the 40-man crew was developed on the Siskiyou National Forest in Oregon. Their example was followed by the creation of other organized crews, including Willamette Flying 20, the Red

Hats, and the Forest Service Reserves in the 1940s, and by helicopter rappelling crews in the 1970s.

As the economics of logging steadily improved, other cooperative fire and
forestry ventures were undertaken in the 1940s. The Keep America Green
and the Tree Farm movements both emerged out of the Northwest in 1941,
only to lose their regional identity in the successful drive to expand them
into national campaigns. With the deliberate restocking of the Tillamook
Burn in 1949, fire protection achieved a new plateau: reproduction and even
snags were accorded equal status with mature timber. In the Northern
Rockies the issue had been between front- and backcountry lands. In the
Northwest, as in the Lake States and the South, it was a question of existing
versus future stands of timber. Loggers were not in competition with farmers or herders (though they would come into conflict with recreational interests), and different fire practices were not at issue. The question of fire protection was one of means, not ends.

The Northwest also made its contribution to the debate over prescribed
fire. Experiments in prescribed burning in the pine-belted Indian reservations east of the Cascades were conducted in the 1940s—the most important
trials outside the South—but they acquired national importance only after
their investigator, Harold Weaver, was transferred to the Southwest.[42] Pinchot claimed that he first heard the concept of light burning enunciated here
and the light-burning controversy made famous in California also occurred
at the southern limits of the Cascades in the region around Mount Shasta.
At this ecological, no less than historical, crossroads the debate about fire
and silviculture might have gone either way: fire protection might have
looked south instead of north, east instead of west. But by the end of the
controversy, the Northwest had become the primary timber region of the
United States, and it became a place where the useful role of fire was less
obvious and where controlled burning meant the firing of piled logging slash,
not underburning. The mixture of ownerships contributed other problems,
and the 1951 Forks fire on the Olympic National Forest led to a Supreme
Court ruling that the government could be held liable for negligent fire control practices and be sued for escape fires. Growing alarm over air pollution
further limited the use of fire even for slash disposal. The region thus has
the highest fuel loads in the country and among the lowest quantity of controlled broadcast burning.

Smoke has always been a serious matter in the Northwest. Farmlands
and metropolises collected in the valley troughs running between the Coast
Range and the Cascades. As with northern forests elsewhere, agriculture
was rarely successful in occupying the true forest belts, settling instead on
former prairies. The occupation of the woods was left to industrial forestry.
But this settlement pattern meant that the effluent of the logging industry
would be funneled into the sites of greatest human density. To burn under

conditions that encouraged smoke dispersal was to invite escape fires, but to burn under a stable atmosphere was to channel smoke into valleys capped by atmospheric inversions.

For most inhabitants of the Northwest the historic fires have been known by their smoke, and before industrial logging became the foundation of the regional economy, the chief argument for fire control was protection from smoke. Smoke was endemic in autumn. Residents blasted it as an unnecessary irritant, and physicians and folk practitioners alike roundly criticized it for a host of seasonal distempers and diseases. Agricultural burning in the Willamette Valley was thus banned early, not as a fire hazard, but as a health hazard because of its cloying, foglike smoke. Slash smoke from logging was similarly regulated. Tough standards for slash disposal and air quality were enacted in 1965 and are overseen by interagency boards.

In the late 1960s the cooperative spirit so characteristic of the Northwest was again manifest, with both regional and national consequences. In 1965 private, state, and federal organizations in Oregon signed an agreement that adjusted protection boundaries to coincide with fire environments rather than with political or administrative units. Guidelines were promulgated for slash burning and air quality. The next year saw, for the first time, a major fire controlled in the face of an east wind. The Oxbow fire, like the Tillamook Burn, was a state operation, begun in the Western Lane Forest Protective District and directed by the Oregon Department of Forestry. The largest fire since the Tillamook cycle, it burned over 46,000 acres through mature and decadent stands of Douglas fir under the propulsion of an east wind. It was stopped at a cost of $1 million and was probably a bargain at that. The Oxbow fire was followed in 1967 by a rash of fires, many of them lightning-caused, but these fires were dwarfed for the most part by an even larger bust in the Northern Rockies. None of the fires developed into a conflagration.[43]

It was otherwise with the 1970 Wenatchee complex, which burned along the fire regime boundary between the Northwest and the Northern Rockies. Drought prevailed in the rain shadow of the Cascades from April to August. In the early morning hours of August 24, dry lightning crashed into parched stands of Douglas fir and lodgepole pine. Fanned by strong downdrafts from passing thunderheads, the fires flared, merged, and eventually formed into four groups, a complex totaling nearly 120,000 acres. It was the worst fire in the Northwest since the Tillamook cycle and the worst in the Northern Rockies since the Selway debacle. Even more devastating fires followed in California. What the Selway fires were to the adoption of the 10 A.M. Policy, the Wenatchee and California complexes were to the 1971 policy reforms.[44]

Few fire incidents in the Northwest fail to stimulate cooperative programs, and the Wenatchee complex was no exception. The 1902 and 1910 fires in the Northwest had encouraged the development of private timber protective associations, culminating in the Western Forestry and Conservation Association. The 1970 fires became the immediate stimulus for the National Wildfire Coordinating Group and the total mobility concept. Regional liaisons were strengthened. In 1975 a "closest man" approach was worked out through agreements between the federal, state, and private protection associations. Regardless of political jurisdictions, the closest suppression unit to a fire would, in theory, respond to it. Two years later, outraged at high "administrative surcharges" imposed by the BIFC, California, Oregon, and Washington entered into an interstate compact to supply necessary services and equipment, especially along state borders.

From 1868 to 1970 major fire cycles had been initiated at roughly 33-year intervals. In 1868 the fires were almost wholly unconfined, running the length of the Coast Range. Protection was limited to frontier fire brigades and backfiring. Even the size of the complex is unknown; it consumed perhaps 600,000 to 1 million acres. In 1902 the fires burned about 600,000 acres but were largely confined to settled or forested areas northeast and southeast of Portland. In addition to farmer-firefighters, government rangers patrolled more or less successfully for fire on the new forest reserves. By 1933 private protective associations, a state forestry department, and a federal fire organization, bolstered by the CCC, were ready to battle a major conflagration in the backcountry. Despite weather similar to that of 1868 and incendiary fires totaling 100,000 acres, the Tillamook fires did not break out regionwide. No villages were lost. Thirty-three years later, a solitary fire of 46,000 acres—a tenth of the size of the original Tillamook Burn—was confined despite almost impossibly adverse weather and fuel. Logging and slash fires had been reduced to a mere 3 percent of total wildfire ignitions. The 1967 problem fires were ignited by lightning; none became a holocaust or reached even a tenth of the acreage of the Oxbow fire. In 1970, another bust of lightning fires burned along the border of the Northwest and Northern Rockies for a total of 120,000 acres. Logging is still the primary justification for fire control, but it is no longer the chief catalyst of fires. Most major fires are now ignited by freak or high-intensity lightning storms rather than by man.

In achieving this record, the Northwest often became the source region for important innovations in technique, though it commonly saw its experiments adopted with greater fanfare by neighboring fire provinces. The Northwest pioneered in the use of organized crews, beginning with the CCC at Tillamook. The first smokejumper trials were in Washington, but the Aerial Fire Depot was established in Missoula. The 40-man crew was developed in Oregon but became more widely known as hot shot crews in Southern California and interregional crews in the Northern Rockies. Its promotion of intermittent crews like the Red Hats provided an exemplar for the the vastly larger Southwest Forest Fire Fighters (SWFFF) program in the Southwest. Its prevention program, Keep Green, was overshadowed by the popularity of Smokey Bear. Its prescribed broadcast fire experiments became most famous only in retrospect, after other programs on its model were adopted in California and Arizona. Early fire research reached promising proportions through the Pacific Northwest Station of the Forest Service and even contributed one Forest Service Chief, Richard McArdle—but the fire labs went to Missoula and to Riverside, California and the transfer of fire research to Alaska came through the Intermountain Station. Owing in part to its ties with the logging industry, the Northwest became fertile ground for experiments in mechanized line construction equipment, and the WFCA held an annual conference on fire equipment for many years. The 1936 Spokane conference sponsored by the Forest Service proposed an equipment lab at Portland. But the radio lab went to Beltsville, Maryland, and the two equipment centers went to Missoula and to Arcadia (later, San Dimas), California. And so it has gone, leaving the Northwest in undeserved anonymity.

But the region's technical triumphs were far from the most significant advances it contributed to national fire history. Nowhere else perhaps was the transition from the fire practices of the reclamation to those of the counterreclamation made more smoothly. In its commitments to fire control and by its innovative administrative arrangements, the Northwest provided a paradigm for fire management that might be emulated and expanded on a national scale. Not the least part of that model was the cooperative programs on which it was based. These, too, were to be exported and assimilated into the texture of a national and even international fire establishment. The WFCA, for example, soon expanded to include British Columbia. The cooperative concept proved popular, and it was from the British Columbia Forest Service that the U.S. Forest Service in the Northwest adopted the idea of helicopter rappelling as a means of initial attack. It was from such seemingly humble liaisons, too, that it became possible to negotiate cooperative agreements for protection of the international border from fire and to sign international treaties with both Canada and Mexico for mutual support on large fires—a novel addition to the old doctrine of "hot pursuit." When large fire complexes appeared in Manitoba and Ontario in 1980, not a little of the suppression supplies came through the warehouses at BIFC, but the belief in cooperation emerged out of the long history of the Northwest.[45]

CROSS FIRE: A HISTORY OF
FIRE PROTECTION BY THE STATES

The immediate effect of the Weeks Law was to make the administration of fire protection a recognized field for the state forestry departments generally. It gave them something specific to do, and the same thing to do everywhere.—Copeland Report, 1933[46]

No achievement in state forestry stands out more clearly or merits greater commendation than the progress that has been made in the prevention and suppression of forest fires.
—Joseph Illick, 1950[47]

I

Shortly after Franklin D. Roosevelt assumed office, the Forest Service presented the new administration with the 1,677-page Copeland Report, *A National Plan for American Forestry*. It left no aspect of forestry unexamined. Of the lands it discussed, however, none were more at heart of the counterreclamation than those for which the states were rapidly assuming responsibility, and no topic was more vital to that concern than was fire protection. Rural lands were ravaged by environmental no less than economic ills. The report determined that "lack of protection, rather than lumbering, has created out of former timber lands most of our near hopelessly bankrupt no-man's land." But it noted, too, that "in a considerable number of States, forestry departments were first brought into existence to provide for organized systems of protection against forest fires," that "in nearly all States the maintenance of the protective system is the most outstanding form of forestry activity," and that "the problems of protection are still for most States urgent, and in many cases will probably continue for years to demand major attention." Fire protection, it concluded, "is the greatest accomplishment of State forestry. More constructive effort has been put into it and more money spent on it than on any other form of forestry activity."[48]

The aptness of these observations was highlighted by the Tillamook fire five months later. Under state direction, the massive strength of the CCC joined the fireline for the first time, a move that had national repercussions. But it was state fire protection efforts, especially the long liaison with the Forest Service, that made the CCC available to the states in the first place and, through the states, to private cooperators. The effects of the Depression on patterns of land ownership and the presence of the CCC together made the era a great one for state forestry, and through the states the domain of organized fire protection was enlarged to an extent rivaled only by the creation of the federal forest reserves.

Fire protection had long been a concern of the states, though the enforcement of fire codes and the suppression of wildfires remained largely in local hands. State forestry, too, had long antecedents, though forest administration was also largely a matter for local determination, and the objectives were often related to arboriculture and reforestation. The forest policies of the states, as the Copeland Report noted, "are conspicuously diverse." Yet state forestry did not evolve piecemeal state-by-state. The various policies had "an integral connection with the general forestry movement, which has been national in character."[49] Thus the Lake States in the 1860s appointed special boards of inquiry, the Northeast followed in the 1870s and 1880s; and in 1885 California, Colorado, Ohio, and New York created boards of forestry.

Within a few years all of these boards had withered away except New York's. The reason is instructive: alone among the group, the New York Forestry Commission was not merely an advisory board but a bureau with explicit instructions and a land base. The Adirondacks Reserve, which was established by the same Forestry Act as the commission, meant that the commission had something to manage. It did for New York state forestry what the Transfer Act did for the federal Bureau of Forestry.[50] The Forestry Act, however, allowed only a circumscribed range of management techniques. The land was to be administered more or less as a wilderness preserve; reforestation was encouraged, logging was not allowed. The primary function of the commission was to guard against timber trespass and to protect against fire. The commission chose to expand the old fire warden system—previously a local or at best a county unit—to statewide dimensions, an innovation that would be copied by many of its northeastern neighbors. The real lesson of the New York Forestry Commission was, as it would be on the federal level, that forestry could become a political force only when it had acquired a degree of authority, and almost always that came when it assumed responsibility for fire control. The reverse was equally true: namely, that by reaching for fire protection authority, state forestry came into existence and became able subsequently to practice other forms of traditional forestry. Until they achieved that authority, state forestry boards were figureheads, more tied to arboriculture than to forestry and rarely staffed with professional foresters.

By 1900, 12 states had established some form of forestry board, though 3 had already expired and none was active in industrial forestry or included professional foresters. Only in New York and Pennsylvania, where land acquisition had also begun, was forestry a political presence. In the great flush of reforms between 1900 and 1910, 17 additional states created forestry units. Most were concerned with the collection and dissemination of information—advising on ornamentals suitable for landscaping, amassing statistics on forest uses, and so forth. But by 1909, 11 states owned a total of 3 million acres. The size of each state's holdings ranged from 60 acres in New Hampshire to 1.6 million acres and 800,000 acres in New York and Pennsylvania, respectively. On the national example, such lands were referred to as forest reserves or preserves, and most again on the federal example, were tied to watershed protection. States could become involved in fire control in other ways, too. For example, Maine established its famous Forestry District in 1909. In the Far West, where federal landholdings were extensive, the Western Forestry and Conservation Association successfully organized private timber protective associations into state and interstate bodies. In New Jersey, a railroad fire line law gave the state forestry commission authority to compel the railroads, then the greatest source of fire, to construct fuelbreaks along their right-of-ways (though the act was later declared unconstitutional). In some states, such as California, statutes provided for county fire districts, and in others, such as Oregon in 1913, compulsory fire patrol laws required that a private owner either belong to a fire protective association or pay taxes toward a state organization for the same purpose.

By 1913 two principal fire control systems were in effect in the states.[51] One, on the New York model, was directed by the state through a fire warden or chief forester; the other, like that in California or in those portions of New York and Maine outside the reserves or forestry districts, promoted fire protection for townships, counties, or special fire districts. Though the latter type was often required by statute, the state had no further administrative responsibility and contributed no assistance; the program was in reality a somewhat more formal version of traditional rural fire protection, which aimed at the suppression of wildfire near fields and towns rather than the regulation of fire practices in forests. The history of state fire and forestry organizations is basically the chronicle of how, through statute, land acquisition, and federal assistance, the state-directed system expanded, and, where the local system was retained, how it tended to be absorbed within the first.[52]

Controlling fire was no easier on the state preserves than on the federal lands of the West. Even the Copeland Report lamented that "the habit of firing the woods for one reason or another has persisted in many parts of the United States, although the original purpose for doing so as a rule no longer exists."[53] The national forests could simply evict or prosecute reckless fire users. It could argue with some plausibility that it managed lands for the national interest and that, where conflicts arose, local economies were a secondary consideration. The states, on the other hand, were necessarily more sensitive to local sentiment. They could not easily sacrifice local economies to the larger imperatives of the counterreclamation. Even in the 1930s the burning of the woods for pasture improvement remained a spring ritual in Iowa. In the 1919 gubernatorial campaign in Kentucky, a candidate ran on a platform to "fire the fire putter-outers." He won the election.[54]

If the states, in short, were to participate effectively in the grander objectives of the counterreclamation, they had to establish a larger constituency. Participation in the national conservation movement, which by the 1910s was under Pinchot's direction and well focused on forestry issues, and formal contacts with the U.S. Forest Service achieved both purposes. Organized conservation and forestry looked first to fire protection. Through mutual fire control programs, the states and the federal government could join in the "continental experiment" proposed by the nation's Chief Forester. Equally, the ubiquitous concern with fire protection gave the states themselves, and among themselves, a unity of interest that no other issue could have provided. The states not only participated in the national drive to extend fire protection to new lands; they also became an object of special interest in that drive as they were caught up in a potentially deadly cross fire between federal forestry and industrial logging.

II

To the Forest Service, the state of forests nationwide called for an extension of the national forest system program to the states. This could be accomplished in two ways: through outright acquisition of land to be added to the national forest system, or by an extension of methods and objectives from the national forests to state and private lands. The state and federal governments thus became interdependent with respect to fire protection: the state foresters required the example (and often the financial assistance) of the federal foresters in order to establish adequate fire control on state lands or on lands for which the state had assumed fire protection responsibility, and the federal foresters, finding land acquisition or a nationalization of the timber industry a thorny political problem, needed to work through the states in order to enlarge the realm of protected lands. With the burst of new forestry bureaus in the early years of the century, many foresters with the national bureaus resigned to head state or private organizations. For example, W. T. Cox became state forester for Minnesota, T. S. Gaskill, forester for New Hampshire, and E. T. Allen, executive director of the WFCA. The transfer of professional forestry from the federal to the state level, that is, involved personnel as well as ideas, objectives, and techniques.

The national forest system furnished a national experiment, a model for forest administration and a graduate school for professional foresters. But it was the Weeks Act of 1911 that made possible the extension of national standards of fire protection.[55] In the West, where federal and state lands often bounded each other, cooperative agreements were adequate. The

Weeks Act brought the national forest system to the East as well. Indeed, it created a mechanism by which state fire protection might be in a sense nationalized in much the same way that the creation of state fire wardens had generalized local fire protection. The act led directly to the creation of many state organizations and eliminated some of the need for state reserves, like those amassed in Pennsylvania and New York, with their elaborate state bureaus, nurseries, and even training schools. Participation required only a fire plan, not a land base. Instead of a conglomeration of state forest systems, a miniature of the national forest system, a grand mosaic was arranged under the guidance of the Forest Service. The act thus made fire protection both more uniform and more widely applicable. It established a form of forest management acceptable to conservation and industry both. When the Forest Service in the mid-1930s, with the Copeland Report as a blueprint, prepared for its experiment on a continental scale, its program relied heavily on the states and was constructed on the foundation laid by the Weeks Act.

Western opposition in the Senate to the Weeks Act had, as Harold Steen notes, "gone up in the smoke of the 1910 holocaust in Idaho."[56] Among its 15 sections the act included three broad provisions concerned with fire protection. The act appropriated funds for "use in the examination, survey and acquirement of lands located on the headwaters of navigable streams or those which are being or which may be developed for navigable purposes." It authorized states to enter into agreements or compacts with one another for the purpose of conserving forests and water supplies. And it allotted $200,000 for Forest Service cooperation with any state or group of states in the protection from fire of any private or state forest lands on the watersheds of navigable streams. For the first objective, the act created a National Forest Reservation Commission to oversee acquisitions, which were to be managed as national forests, though in 1916 the President was authorized to establish wildlife refuges on any of the lands purchased under the act. The second provision was not used until 1949, when further enabling legislation provided for the Northeastern Interstate Forest Fire Protection Compact.[57]

For fire protection as a national enterprise and for state forestry as a medium for fire control, the act's third provision was fundamental. The acquisition of national forest lands was in a sense an alternative to state control, but a cooperative assistance program, in which federal aid was extended directly to the states, was an incentive to it. Though the initial appropriation was small, as seed money its impact was great: states almost immediately created new programs or revitalized old ones in order to qualify under the terms of the act. To be admitted, state forestry bureaus had to be politically active, not merely advisory. At the same time, the prospect worked to expand the horizons of the Forest Service as administrator of the act. To oversee the program, it had to create an independent Office of State and Private Cooperation, which became in time a separate branch of the Service, equal to the administration of the national forest system and to forest research.

Admittance to the program was contingent upon satisfying three requirements: the land under protection had to occupy the forested headwaters of a navigable stream; the federal contribution could not exceed the amount appropriated for the same purpose by the state; and the state must have provided by law for a system of forest fire protection. Forest Service inspectors approved the adequacy of the latter. An approved fire plan and legislated mechanisms for its administration generally became the basis for a statewide system of wildland and rural fire protection, and this pattern of financial assistance, which, as S. T. Dana observes, "inevitably involved a certain degree of Federal leadership, was one that could be, and presently was, extended to other activities." From now on, "the government was to participate at least indirectly in the formulation and administration of state policies in the overwhelmingly important field of forest fire control."[58] By 1913, 20 states met these provisions, and the Forest Service, appreciating the potential magnitude of the experiment, scheduled a Weeks Law Forest Fire Conference.

In a keynote address, Chief Forester Graves noted that the act "initiated a new policy," that it was "an experiment to test the efficacy of this kind of Federal aid."[59] The participants shared experiences, and not a few observers from states still outside the program rushed home to encourage legislation that would allow entry. Originally the Forest Service administered the federal funds directly, assigning them, for example, to certain "lookout watchmen" or "patrolmen." After 1921 the Service evolved a complex allotment formula by which it made awards to the various participants; the money entered a state forestry department's general fund to be spent as the state saw fit. J. Girvin Peters directed the program out of the Service's Washington office. With time, inspections for state compliance came to be made by the regular fire officers of the nearby national forests. Greeley had informed the 1913 conference that, although the Forest Service would not "insist on the adoption of [its] ideas," it "would not simply serve as a clearing house for ideas and methods, but . . . should make the best ideas and methods effective by putting them into effect as far as they are locally applicable."[60]

The tiny annual appropriation for the program was the least of the Service's problems. The restriction that only the forested headwaters of navigable streams could qualify excluded precisely those states most in need of strengthened fire protection, notably the Lake States and the South, with their vast cutover pineries. The Capper Report, submitted by the Forest Service in 1920 at the request of the Senate, emphasized the tremendous fire losses on just such nonfederal lands. Whatever its practical problems, the fire protection program for the national forests was at least an administrative reality. For 80 percent of the country no such presence existed. Another survey quickly followed the Capper Report, this one with statistics on areas in need of protection, on the projected costs of adequate fire control, and on the amounts the states would be willing to contribute. In November 1920 Greeley, now Chief Forester, arranged for a second major policy meeting for the states cooperating under the Weeks Act. The original grants-in-aid were intended to encourage states to establish protective programs. With typical foresight, Greeley anticipated increases in the total federal appropriations available for the program and wanted to work out an equitable formula that would provide maintenance funds and would allot funds differentially according both to need and to the size of the state operation. Thus those states most desperate for assistance in order to establish a program and those that had established programs and were willing to invest most heavily were the states most rewarded. A follow-up conference was held two years later. By then, 26 states were participating.[61]

The Forest Service had long aspired to regulate the conduct of the timber industry, either through acquisition of lands or by tough guidelines. It was Greeley's political genius to realize that, however acrimonious the various factions might be on the regulatory issue, on fire they were all in agreement and that with cooperative fire programs the Forest Service could achieve better forest management on state and private lands. Nor was fire simply an opportune instrument for Machiavellian politics: Greeley's own experience in 1910 testified to the magnitude of the threat it posed; he considered fire (along with poor tax laws) a chief cause for the migratory behavior of industrial logging; and he was powerfully impressed by the successful example of cooperative fire protection in the Northwest.[62] In 1920, as he prepared for the second Weeks Act conference and the Mather Field conclave, Greeley represented the Forest Service at a national conference of professional foresters, timber companies, and timber protective associations gathered together to recommend legislation for an adequate national forest policy. Eight points were agreed on, the first advocating that the "chief, although not entire emphasis for the time being, should be on fire prevention as the most important single step, and not less than a million dollars annually [should be available] for such cooperation with States."[63] Soon afterward participants formed the National Forestry Program Committee, which submitted model legislation embodying the endorsed principles to Congress.

When at last Congress acted on the proposals, the eight points were boiled down to three: fire protection, reforestation, and regulation of cutting. In 1923, as the California Forestry Committee laid light burning officially to rest, a Senate committee traveled across the country, held 24 hearings in 16 states, amassed 1,447 pages of testimony, and listened to statements from 274 of the nation's most prominent foresters, timber owners, and lumbermen. It soon became apparent that, with the exception of fire control, the issues under debate were politically insoluble. Greeley accompanied the committee on most of its travels, and he recalled unabashedly how he helped to pack the hearings with advocates of better fire protection.[64] "Master strategist" E. T. Allen of the WFCA contributed measurably to committee deliberations. The example of Weeks Act cooperation shone like a beacon as the committee navigated treacherous political shoals. In the end it agreed with R. S. Kellogg, spokesman for the National Forestry Program Committee, who reported: "The first step is effective fire prevention on both the public and privately owned forested lands of this country. It makes no difference who owns it. The first thing is fire prevention."[65] On cooperative fire protection would hinge other, future forms of cooperative forestry. In the words of the committee report, "foremost among such practicable forms of assistance is the extension of Federal aid in the protection of forested and cutover lands from fire. If the hazard of loss from this source can be reduced to an insurable risk, a large part of the forest problem of the United States will be solved."[66]

The committee's findings led to the adoption of the Clarke-McNary Act in 1924. The soul of the act was a great extension of the cooperative fire experiment begun under the Weeks program. It incorporated most of those recommendations advocated by the Weeks Act collaborators at their national conferences and increased significantly the amount of the federal largesse. It permitted cooperative fire protection for any forested watershed, thus incorporating the pineries of the South and the Lake States. A 1925 amendment extended the program to municipal watersheds as well, absorbing the brushfields of Southern California. It thus encompassed wild lands of all sorts, so long as they came under the scope of an approved state fire control program. The states were allowed to include certain private timber protective association funds within the amount upon which each state's particular allotment would be based. The Forest Service continued to administer the program through the state foresters; more than remaining simply a disbursing agent, the Service became a working partner. It established general standards, inspected state programs for conformity, and conducted research on topics of mutual interest. For many states the initial Clarke-McNary survey, undertaken by the Forest Service to determine the conditions of admittance to the program, was the first systematic study of their fire protection needs. Thanks to the Mather Field Conference, the Forest Service could furnish a reasonably standardized and systematic package of techniques and objectives.[67]

The Clarke-McNary Act was the crown jewel of Greeley's administration. As a method of extending the influence of federal forestry, it is surpassed only by the Forest Reserve Act and the Weeks Act. Harold Steen notes that "Greeley ranks second only to Pinchot in stamping his personal philosophy on American forestry, and the law was his great personal victory after four years of effort." The essence of the Clarke-McNary Act—"cooperation to inspire voluntary action—has been the essence of the Forest Service."[68] It brought under organized protection (in conformity with federal standards) vast regions of the United States that had hitherto been excluded. Much as Show's and Kotok's studies prepared the national forest system to receive the emergency money and manpower made possible by the New Deal, the Clarke-McNary Act prepared the states to absorb the new lands and responsibilities that would come to them during the upheaval of the Depression. In 1930, at the time the regional foresters conference set standards for fire protection in the national forest system based on hour control, the Service's cooperative forestry branch conducted a thorough inventory of state protection needs, which were incorporated into the Copeland Report.

It was the Clarke-McNary program, moreover, that gave the states access to the federal largesse legislated under the New Deal, especially the CCC. On state and private lands, the CCC constructed 1,314 lookout towers, 315 lookout dwellings, 39,431 miles of telephone lines, 43,782 miles of truck trails, 8,247 miles of foot trails, and 42,708 miles of fuelbreaks. Hazard reduction work covered more than a million acres. The CCC furnished primary suppression crews, particularly significant in areas where private and rural fire control had never been well developed. About 1.5 million man-days were spent on fire prevention and presuppression and about 2.3 million man-days in actual suppression. As Earl Pierce and William Stahl note, "the States owe a large, and in many cases a major, part of their fire protection improvements to the CCC."[69] Chief Forester Silcox explained to Roosevelt in 1934 that "it is in the public interest to insure adequate fire protection of all the timberlands in this country."[70] Under the arrangements established by the Clarke-McNary program, 250 CCC camps were located on privately owned timberlands. Silcox estimated that between Clarke-McNary funds and the CCC, $22 million would be appropriated for fire protection on private forest lands. When Roosevelt questioned the propriety of federal support for private lands in 1937, CCC Director Robert Fechner replied that the camps were engaged in fire protection work and that such labor was legal under the provisions of the Clarke-McNary Act. The private lands were cooperating with the states under fire plans approved by the Forest Service.[71]

Not least among its effects, the act confirmed the principle that federal influence would spread through cooperative programs and not through the nationalization of either forested land or the means of harvesting it. Greeley's brilliant compromise showed Pinchot's ritual jeremiads for what they were: the idiosyncratic ideology of an aging Progressive. Even though Forest Service data showed that more forest devastation resulted from fire than from logging practices, Pinchot contended that the fire issue, on which the

Forest Service and industry were united, was a smoke screen for tolerating a wasteful logging economy, for which the only solution was nationalization. Greeley's success was thus a victory for more than fire protection: it set a pattern for forest management at large. The program was greatly enlarged following World War II and became, in turn, the administrative mechanism and model for rural fire defense beginning in the 1950s.

The states, meanwhile, were acting on their own initiative. Prior to 1924, under the lesser stimulus of the Weeks Act and the internal needs of the states themselves, 29 states had adopted some form of state forestry department authorized to cooperate with the Forest Service. Fire protection, as Ralph Widner summarizes, "was inevitably the prime concern as soon as the office of state forester was created."[72] Some states, like those along the West Coast, had adopted compulsory fire patrol statutes, which were the state forester's responsibility to oversee. Others, like New York and Pennsylvania, had acquired significant landholdings, which demanded a minimum of fire control. In still others, like California, the state forester assumed fire responsibilities for local fire districts or counties under contract.

The greatest burst of state fire and forestry work came after Clarke-McNary. The act coincided with an enormous redistribution of land. Timber prices began to fall ruinously around 1926, leaving many private timberlands unmanaged or tax-delinquent. Agriculture, too, began to collapse as a market economy around the same time. From marginal farmlands like the semiarid High Plains or the cutover podzol soils of the north woods, there began a slow exodus of farmers, leaving yet more lands to slide into tax delinquency and idleness. In the South the boll weevil decimated the single-crop economy of cotton, aggravating the exhaustion of the logging boom. The collapse of the stock market in 1929, moreover, coincided with terrible drought across the United States, which pushed the productivity of farms still lower and drove off farmers working marginal land. Under the New Deal still other farmlands were deliberately transferred through programs like the Resettlement Administration. What had been a trickle of refugees became a flood. Indeed, where lands had not been ravaged by drought, they were often hit by floods, beginning with the awesome Mississippi deluge of 1927.

The national forest system expanded terrifically in the East and South, acquiring abandoned farmland or timberland with the idea of reforesting and regenerating it. A good portion of the land was purchased so that the CCC would have sites to work on, particularly in the South, where camps could work through the winter. The Taylor Grazing Act closed and reorganized the public domain, at least in part to bring systematic protection to it. The process was no less pronounced for the states. Some states acquired land through purchase (like Pennsylvania), from gifts (like Baxter State

Park in Maine), from federal land cessation (an exchange of school lands within national forests for public land outside the forests), or from other federal grants. With the Depression, many more states came into land inadvertently, land that they turned over to forestry. By 1933 nearly 4.5 million acres were incorporated into state forests, the bulk in Pennsylvania, Minnesota, and Michigan. State parks, like the Adirondacks and Catskills preserves in New York, totaled 2.7 million acres. Other state-controlled lands not formally incorporated into forests or parks amounted to 6.1 million acres. In the Northeast, much of the land had been acquired by purchase; in the Lake States, by tax delinquency and federal grants; in the Northwest, through federal grants resulting from expansion of national forests. Southern states took over land, but private timber concerns purchased much of the abandoned land after the discovery in 1931 that southern pines could be used for pulp. Other events were locally or regionally significant. In Oregon the replanting of the Tillamook Burn became the occasion for the Tillamook State Forest. The fire hazards left by the 1938 New England hurricane gave purpose to the fire organizations of Massachusetts and Rhode Island. Military and civil defense concerns accelerated fire protection programs in coastal states during World War II. What Raymond Clar wrote of California might be generalized: "the State fire protection agency has thrived upon adversity, whatever the cause—wildfire, flood, war or economic depression."[73]

Most states, in short, could no longer afford advisory boards for forestry or ornamental bureaus appointed to appease popular conservation enthusiasms. The state forestry departments acquired real responsibility. For some, this included substantial land management charges; for all, it meant fire protection. The Clarke-McNary program supplied the administrative mechanism for helping the states to absorb this expansion and gave them a national point of reference and shared standards. In 1944 the maximum appropriations under Clarke-McNary expanded from $2.5 million to $6.3 million, and additional increases were allowed, up to a maximum of $9 million for 1948. Another act authorized the Secretary of Agriculture to spend not more than $1 million a year to assist states with suppression and emergency presuppression costs; beyond that, state and private cooperators would have to contribute. National conferences of Clarke-McNary cooperators gave the program new horizons, and the Society of American Foresters and the Charles Lathrop Pack Foundation independently funded studies of selected state protection needs.

By 1949 the authorized appropriations under Clarke-McNary had increased to $11 million, and the theory had developed that national interests in forestry and fire control were such as to justify a contribution by the federal government of approximately half the total cost of adequate protection on state and private lands, then estimated at $20 million. The 1950

Cooperative Forestry Management Act expanded the range of cooperative endeavors beyond fire. In 1956 the Forest Service contracted with the private Batelle Institute of Ohio for a multiyear study of the entire cooperative fire program. Further increases were recommended as a result. But by then a whole range of cooperative forestry programs had been worked out, and new dimensions had been added to federal-state cooperation in the field of fire protection, centered on the question of rural fire defense.

Between 1950 and 1954, under the direction of the Office of Civil Defense, the Forest Service gradually assumed responsiblity for coordinating wildland and rural fire protection in the United States. The Forest Service naturally built its national plans out of the Clarke-McNary plans; thus state foresters, even more than previously, became responsible for coordinating rural fire protection, at least to the extent of preparing fire disaster plans. A further form of assistance also growing out of World War II came through the Federal Excess Property Program, which gave the Forest Service priority access to surplus equipment, largely military, which it could pass on to cooperators. By the 1970s over $200 million worth of equipment had been distributed to state and rural fire agencies. Locally, the effect could be great; nationally, the success of this program became the basis for the Rural Community Fire Protection Program authorized in 1972. Not only had fire protection opened the door to all forms of cooperative forestry; it had even left wildlands for the landscape of rural and suburban America. The state forestry departments were the greatest beneficiaries. By 1966 the Clarke-McNary program extended to every one of the 50 states, and thanks to it, systematic fire protection had spread on a scale that the architects of the Weeks Act would have found barely imaginable. The Cooperative Forest Management Act of 1980 recognized this fact and began a withdrawal of federal funds that it considered no longer necessary.

III

The states were never solely dependent on the federal example and on federal assistance. In 1908 the Association of Eastern State Foresters had pooled regional resources, and in 1920 the concept was generalized into the National Association of State Foresters. After World War II, the states began to take advantage of a provision tucked away in the Weeks Act that allowed them to form interstate compacts for the promotion of better forest protection. The Maine fires of 1947 furnished the immediate stimulus, and two years later the Northeastern Interstate Forest Fire Protection Compact resulted. The compact, signed by all the northeastern states and New York, promoted the development of integrated—eventually regional—fire plans and provided for mutual assistance during fire emergencies, for the maintenance of firefighting services, for uniform training, and for the establishment of a central agency to coordinate the provisions of the compact. Quebec joined in 1969 and New Brunswick in 1970.

On this example, and prodded by the rural fire protection objectives of the Forest Service and the Office of Civil Defense, the southeastern and south central states adopted similar compacts in 1954, the mid-Atlantic states in 1956, the Pacific Northwest states and California in 1978. To bring even more states at least informally into cooperation, two still larger groups were created: the Northeast Forest Fire Supervisors (1973) and the Southern Forest Fire Chiefs' Association (1974). The former included the Great Lakes and north central states, as well as the states under the northeastern and mid-Atlantic compacts; the latter, both the southeastern and the south central compact states. In keeping with its cooperative mission, the Forest Service signed as well.[74] In 1971 the northeastern states agreed to underwrite the Roscommon Equipment Center, on the site of the Michigan Forest Fire Experiment Station, for a trial period, and the arrangement was extended in 1974 for another five years. The center emphasized plans for the conversion of surplus military hardware into equipment suitable for state and rural fire protection. In 1972 the first intercompact agreement was signed between the mid-Atlantic and southeastern states.[75]

Among the western states, where the federal presence was so much greater, there was generally less urgency for direct state cooperation. Most interstate needs were satisfied by cooperative agreements with the Forest Service that allowed for the transfer of men and equipment during fire emergencies and provided for a relatively uniform training schedule. Here the trend was to emphasize mutual research interests, which gave rise to fire councils incorporated within the Western Forest Fire Committee of the Western Forestry and Conservation Association. By the early 1970s participants included the California-Nevada, Intermountain, Northwest, Rocky Mountain, Southwest, and Alaska forest fire councils. In the mid-1970s, again thanks to longstanding contracts with the Forest Service, a state representative from NASF became a working partner in the National Wildfire Coordinating Group, and the states could draw on the resources of the Boise Interagency Fire Center.

"Nearly everywhere," Dana concluded, "the states have taken over major, and in some cases practically exclusive, responsibility for the control of forest fires."[76] Most fires, and most fires involving unambiguously high values at risk, have been handled by state departments. In the late 1970s, even as federal support under Clarke-McNary began to shrivel up, the vision of the early prophets of cooperative fire protection nonetheless seemed fulfilled: cooperative fire control had extended the range of fire protection and forestry to near maximum limits, and most lands under protection were under state protection. Born out of the cross fire between federal foresters and private timber owners, state fire protection had come to exceed the scope of each. It expanded the realm of Forest Service influence, giving it a hegemony that other federal agencies conspicuously lacked. It brought confidence to private industrial forestry, buffering it from the dual threats of fire and nationalization. And in the process the state program had achieved a degree of autonomy, becoming a third estate in the world of the counterreclamation. For most states, thanks to their links to the Forest Service and to the support of industrial forestry, fire protection responsibilities had given permanency to forestry departments and allowed for a range of fire services that might not have been possible otherwise. Their absorption of rural fire protection, for example, often encouraged the bureaus to become truly statewide organizations, not dependent on the presence of state lands or on the tumultuous fortunes of either populist conservation or industrial forestry.

By the 1970s the states were often prepared to absorb the burdens borne previously by others. In Alaska tens of millions of acres were actually being ceded to the state from the public domain, and in the Great Basin the so-called Sagebrush Rebellion advocated an extension of this process. The Forest Service turned to its state cooperators in the East for assistance on large fires on the national forests of the West. Major interagency concern—as epitomized by FIRESCOPE—shifted from the federal land boundaries to the rural landscape and to the boundary between wildland and urban land. This process was well symbolized by the maturation and dedication of the Tillamook State Forest in Oregon: the burn had been fought by a state bureau, and the rehabilitated landscape was remade into a state forest. Like so many state forest agencies, Oregon's had been created in the cross fire between federal forestry and industrial logging, and like so many other state forests, this one had achieved national prominence and power in the crucible of flame and Depression. Some 40 years later, the investments were ready to pay off.

UNDER FIRE: A HISTORY OF
MANPOWER IN FIRE CONTROL

"Where's Smith and Hennessy, Edwards, Stowe—
  Where's Casey and Link and Small?"
The Ranger listened, and murmured low:
  "They're missing, Chief, that's all.

"Where the smoke rolls high, I saw them ride—
  They waved goodby to me:
Good God! they might as well have tried
  To put back the rolling sea.

"I rode for aid till my horse fell dead,
  Then waded the mountain stream:
The pools I swam were red, blood red,
  And covered with choking steam.

"There was never a comrade to shout 'hello'
  Though I flung back many a call:
The brave boys knew what it meant to go—
  They're missing, Chief—that's all."

  —Arthur Chapman, "The Fire Fighters," 1910[77]

  Some say the world will end in fire,
  Some say in ice.
  From what I've tasted of desire
  I hold with those who favor fire. . . .
  —Robert Frost, "Fire and Ice"[78]

I

When the first call went out for manpower at Tillamook, state and federal officials promptly turned to the nearby CCC camps. By the second day two camps (200 men) were on the line; by week's end, some seven camps. Presented with something of a fait accompli, the agency's director, Robert Fechner, officially authorized the use of CCC enrollees in firefighting on August 17, three days after they had been sent to firelines. The CCC had been legislated into existence barely five months before, and although enrollees had fought small fires in Arizona, California, and Wyoming, the Tillamook Burn was collectively their real baptism by fire.[79] The commitment of

18. Forest ranger on patrol, Montana, ca. 1915.

CCC camps to labor-intensive presuppression activities was foreordained by the nature of the enabling legislation, but the use of the camps as frontline firefighters was not. It came perhaps as something of an afterthought, an accident precipitated by the crisis at Tillamook. But of the two assignments, notwithstanding the CCC's sweeping presuppression programs, its suppression role was perhaps the more important.

To a peculiar degree—and to an extent without parallel in the control of other natural disasters—fire management has meant the management of firefighters. Equipment commonly supplements manpower, but it rarely substitutes for it. One finds roughly the same (or even larger) numbers of firefighters shipped out to fires in a given region in 1970 as were ordered out in, say, 1950 or 1930, despite vast improvements in the caliber of crews and equipment. Only in the South and the Lake States has mechanization replaced the need for crews. In the mountainous West, with a settler class excluded from the reserves, it was necessary to create a special cadre of firefighters. Mobilizing the CCC for the Tillamook fire introduced a revolution in the history of fire manpower. For the first time crews rather than individual firemen were routinely dispatched to fires. For the first time adequate companies of civilians were available for fireline duty. And not least of all, as the CCC swelled in size, as casualties mounted, as camps acquired better organization and training, and as medals were created for heroism under fire, the CCC strengthened the implicit claim of firefighting to be a moral equivalent of war.

Prior to creation of the CCC the problem of developing reliable crews was among the knottiest faced by fire protection. Pinchot noted in his *Primer on Forestry* that equipment and improvements without good men were "as little use against dangerous fires as forts without soldiers against invading armies."[80] In *Systematic Fire Protection* duBois stated flatly that "the success of protection work depends, first, last, and all the time on the men who compose the rank and file of the control force."[81] But it was not easy to get and keep top men. In rural areas, a warden system was generally employed. The warden would organize in advance—or legally impress during an emergency—able-bodied settlers in the region; the system was in effect an extension of the the volunteer fire brigade and a formalization of those forces already in existence to protect ranches, mills, and farms. In the classic evolution of settlement from frontier to city, one could imagine this pattern of protection gradually extending over most of the United States, with wildland coming under rural fire districts.

The question was what to do with land deliberately removed from settlement, where fuel types would not be converted from forest and range to farm and town and where manpower would not exist in sufficient quantities to staff a warden system or man a volunteer department. The national parks could rely on Army troops, but the new forest reserves in the West had to look elsewhere. For small, routine fires, fire guards could be hired at slight cost. Many were contracted on a per diem schedule, paid only for those days they were actually activated for fire control or patrol. The Forest Service also looked to its permittees, charging the loggers and ranchers with mandatory assistance on going fires as part of the price for a permit. Indeed, the need for manpower was often an argument for issuing permits in the first place. Cooperative agreements further supplemented local forest resources. Lookout-firemen were common; once a fire was sighted, the lookout would abandon his perch to make the initial attack. Where highways or railroads were under construction, road gangs were organized for fire duty as necessary. When large crews were hired for blister rust control in the Northern Rockies, they often doubled as firefighters. The proposal for a "sit-tight" crew of firefighters on a city model was commonly dismissed as too expensive. Not until after the CCC experience was it considered preferable to hire fire crews who could do project work than to have project work crews interrupt assignments to fight fire. By this time, too, the logic of adequate presuppression was recognized, and an emergency presuppression fund helped to make crews dedicated primarily to fire control a fiscal possibility.

Where large fires threatened, the most common recourse was simply to tap the abundant pool of unskilled, semiemployed laborers common to the

19. Typical crew of the 1910 fires, here on Mt. Hood, Oregon.

20. Lookout firemen, Idaho, ca. 1940.

frontier, especially from mining camps and railroad gangs or from nearby metropolises. The state of New York shipped hundreds of such laborers from New York City to the Adirondack fires of 1903; the Forest Service drafted thousands out of the labor pools at Spokane and Missoula for the 1910 fires. In the words of one exasperated official, "I want men to fight the fire—any kind of men will do. I can use a thousand of them."[82] Dissatisfaction with the emergency firefighter (EFF) was widespread among all firefighting agencies, however. He offered little more than a warm body—untrained in hand tool use; ignorant of fire behavior strategy and tactics; ill-equipped in clothing and footwear; insubordinate by nature and generally unknown to those who worked beside him; of dubious health and not infrequently seized by *delirium tremens* after a morning on the line. Often too little water went on the fire, and too much firewater went into the gullet. In the early days, when dispossessed miners were signed up, crews were polyglot assemblies where English was often, at best, a second language. Construction gangs were preferable to pure pickup labor. Success depended on the tough, seasoned foremen who ramrodded them. After World War I the IWW was rumored to have infiltrated the work pools and to have agitated EFFs to further insubordination and, even more ominously, to outright sabotage. Small wonder that most fire officers looked upon EFF labor as a last resort, though for a major conflagration it was more often than not the only resort.

Small wonder, too, that fire officers looked admiringly, even covetously, on the Army troops stationed in the national parks. The success of the Army in the 1910 fires made a profound impression on many rangers. Almost to the eve of the Depression the Forest Service periodically revived the suggestion that regular troops be stationed on the national forests during the summer, primarily for fire suppression.[83] The Army, in truth, never completely left fire control—the Forest Service, for one, never wanted it to. A pattern of cooperative agreements persisted, to the agencies' mutual benefit, though the thrust shifted from manpower to logistical support, from the commitment of troops for fireline duty to the use of specialized equipment (aircraft or troop transports, for example) and the staffing of support facilities. As the Army mechanized and specialized, its troops proved less useful on the fireline, little better in fire suppression than civilian fire crews would be in a military engagement.

II

What ended proposals to use the military in fire protection was the advent of the CCC. Despite Army management of its camps, the CCC was a profoundly civilian phenomenon. The idea for Roosevelt's Tree Army was already in the air throughout the United States and the Western world: several European countries organized similar institutions, as did California, Wisconsin, and New York.[84] About three-fourths of all the CCC camps were assigned to the Department of Agriculture, which in turn divided their work load about equally into forest protection and forest improvement. The implication of this body of men for fire protection was staggering. In a typical year—1936, for example—enrollees stretched 44,750 miles of telephone line, cleared 11,402 miles of truck trails, maintained an additional 62,920 miles of trail, constructed 611 lookout towers, labored on the Ponderosa Way and other firebreaks (among them, the corridors of felled snags at Tillamook), and built fire caches by the score. Moreover, dozens of other CCC projects affected fire protection indirectly. Half of all the trees planted in the United States were planted by the CCC, ironically contributing to fuels problems 30 to 40 years later. Swamps in the Lake States, once drained for agriculture, were reflooded. Control of forest pests reduced dead fuel concentrations. Small wonder that in his 1936 annual report as Chief Forester Silcox observed with considerable understatement that "perhaps the largest and most important contribution of the Civilian Conservation Corps has made during the three years of its existence has been in protecting the forests from fire."[85] In fact, the full consequences of the CCC over its nine-year existence are almost incalculable.

Perhaps even more important than its creation of facilities was the CCC's staffing of them. "By 1942," as John Salmond summarized, "the CCC had spent nearly 6.5 million days fighting fires, a period equivalent to the constant efforts of more than 16,000 men, working for a whole year on the basis of an eight hour day."[86] Its foundation in emergency relief programs meant that cost, one of the great difficulties in the establishment of regular fire crews, was not an issue and that the other traditional dilemma of fire manpower, its episodic and season nature, was irrelevant. Job-hunting fires—so prominent in the early years of the Depression—retreated to a more acceptable percentage as CCC camps instead of EFFs rolled out to fires. At many camps special fire companies, often known as "hot shots," were formed, and selection became a mark of distinction for enrollees. Previously, initial attack had been made on fires by individuals, not by crews, and training had been directed at individual guards and specific skills. Now large crews had to be trained and equipped. The era of the CCC inaugurates the real development of organized line construction methods, of crew production figures, and of reliable crews stationed more or less permanently at critical locales. The great experiment was conducted on a prodigious scale, and its lessons about the strength of organized manpower were imprinted deeply—a lesson reinforced by wartime experiences with massive troop movements as ex-GIs returned from the barracks to the fire camp. The presence of the CCC allowed fire control to make the critical transition from a frontier militia to a disciplined army. The transformation required considerable planning and imagination. Upon the grand experiment of the CCC in fire control would be based that "continental experiment" in fire control proposed by the 10 A.M. Policy.[87]

21. CCC crew attacking a fire in South Carolina, 1935.

22. Broadcast burning of slash for site preparation, Idaho, 1935.

23. Constructing the Ponderosa Way, a gigantic fuelbreak and road system traversing the west slope of the Sierra Nevada, 1935.

But with new promises there surfaced old problems. Beginning with the Tillamook Burn, in which 1 enrollee lost his life, a total of 47 CCC workers were killed in fireline duty. At nearly the same time that Silcox praised the CCC as a fire control force, the weekly CCC newspaper, *Happy Days,* ironically ran as its lead article a nightmarish account of a Chatsworth, New Jersey, fire that claimed the lives of 3 enrollees. The episode faded amid public praise from civilians whose property was saved and from soldiers out of Fort Dix who worked on the fireline beside the enrollees.[88] It was less easy to pass over the Blackwater fire tragedy of 1937. Lightning ignited the fire on the Shoshone National Forest, Wyoming. Crews arrived at sunset, only to find that the fire had left its original drainage basin and slopped over a ridge. The next day the slopover was attacked with an undercut line along the slope. But there was an undetected spot fire below the crews, and when afternoon conditions favored a blowup, the main burn and the spot flared at about the same time. One crew scrambled up to a rock pile and waited out the firestorm. An enrollee perished, however, when, as flames leaped about on all sides, he panicked and ran from the rocks. A second crew of 14 found itself trapped between the main fire above it and the spot fire below it. The crew elected to attack and control the rapidly flaring spot; all died in the attempt.[89]

The Blackwater tragedy shook the whole CCC fire program to its foundations. The Forest Service immediately commissioned a board of review to investigate the incident. Members included such luminaries as David Godwin, later director of the Division of Fire Control, and A. A. Brown, first director of the Division of Forest Fire Research. "Not since 1910," their report noted, "have so many lives been lost on a single national forest fire. . . ." Yet no blame was assessed. Godwin could "find no reason for criticism or organizational change." It must be recognized, he regretted, "that in man's control of forest fires some accidents will occur—just as in city fire protection—without fault or failure on the part of anyone." "It is reassuring," he concluded lamely, "to know that such circumstances are infrequent." The forest supervisor of the Shoshone cited as "one bright memory" the way in which the CCC crews subsequently turned against the fatal fire and suppressed it. "The CCC men rose to the occasion like veterans, and at daylight three crews were on the fireline carrying on—determined to whip the fire that had beaten them so terribly the afternoon before."[90] Moved by the tragic spectacle, John Guthrie, a Forest Service official who had transferred to a high post with the CCC, proposed that a medal be created for heroism in firefighting. Silcox concurred and in 1938 made detailed proposals. "We need a movement," he argued, "that may serve to articulate some of the finest traditions of the Forest Service. Our Pulaskis, Ingrams, Everetts, and Claytons have gone too long without tangible recognition." The AFA agreed to head an American Forest Fire Foundation, which would administer the program.[91] The earliest earliest recipients were Urban Post, Bert Sullivan, and Paul Tyrell—all overhead on the Blackwater fire. Tyrell's medal was posthumous.[92]

Battlefield rhetoric meant little to Robert Fechner, director of the CCC. The CCC had been created, in the words of FDR, "to relieve distress, to build men, and to build up the nation's forest resources." Fatalities achieved none of these aims, whatever they might contribute to the traditions of the Forest Service. And however much intellectuals might applaud the CCC as a moral equivalent of war, the public was less charmed by the proposition: the charge of militarism was a constant criticism, and as fascism spread war clouds over Europe, the belief grew among prospective enrollees that the CCC camp was but a steppingstone to the Army barracks. The fear of becoming cannon fodder reduced CCC enlistments in its final years, a condition that the specter of fire fatalities did nothing to ameliorate.

In 1938 Fechner released new regulations governing the use of the CCC in fire control. The CCC had in a sense inaugurated the New Deal, and as criticism mounted against the failure of the New Deal to end the Depression, the CCC suffered accordingly. The Blackwater tragedy was but one of many incidents that decayed the remarkable espirit de corps of its early years. Desertion rates ran high, mutinies broke out at several camps, there

24. The Blackwater fire, Wyoming, 1937.

were occasional riots and minor disturbances; racial segregation further increased tensions. Nor was the Blackwater fire an isolated episode. Three enrollees had died the year before in a New Jersey fire, and 8 would perish on the Pepper Run fire in Pennsylvania two years later. Death by fire advertised a spectacular breakdown in the goals of the institution, but reforms of the CCC as a firefighting force must be seen as part of a larger package of new regulations intended to revive a valuable program that, like the New Deal itself, was becoming both moribund and desperate.[93]

The reforms were overdue. In 1933 it seemed entirely suitable to hurl companies of young men—many from cities and all but recently enrolled— into the breaches of the Tillamook Burn. Fire control agencies were accustomed to semiskilled pickup labor on the line and treated the enrollees accordingly. By 1938, however, the presence of the CCC had contributed to improved standards. To commit green laborers to the fireline of a holocaust was as inexcusable as to order raw, untrained militia into a major battle. Both the Forest Service and the CCC were ready for more sophisticated procedures. The 1938 guidelines set better standards for training and restricted use of CCC enrollees in certain environments. A 1939 memorandum went even further: it prohibited the dispatch of CCC crews for initial attack in many areas. The prohibition was a response in part to fire-related fatalities and in part to growing dependence on the CCC for all fire protection needs. As John Guthrie explained, "when the CCC started, the idea was that as far as fire fighting went, the Corps was to be a last resort, an emergency agency, the reserves. . . ." Instead, however, "no other item in their list of daily activities during the first five years shows a larger expenditure of their time."[94]

The latter decree was a serious matter. Local crews had melted away before the hordes of CCC enrollees, and programmed fire funds had all but vanished in favor of ECW money; yet a strong initial attack remained the essence of fire control strategy. The CCC was still present, still a force to be reckoned with and counted on for fire control, but one more closely circumscribed. The Forest Service, aided by emergency presuppression funds, began a search for substitutes. In 1939 it experimented in the Northwest with two versions of specialized crews: the smokejumpers, primarily intended for initial attack in backcountry regions, and the 40-man crew, employed predominantly for heavy-duty use in the construction of handlines during large fires. The creation of special, organized fire crews was a CCC legacy as great as, and in many respects more enduring than, its lookout towers, firebreaks, and trails. Fire protection agencies were determined, if possible, to find an alternative to mere mobs of EFFs. If the CCC could no longer function as front-line defense, then substitutes would be invented.

III

World War II interrupted the manpower experiment. Despite pleas by the Forest Service to retain the CCC for forest protection, the demands of the military draft were stronger, and the CCC was disbanded in 1942. Throughout the war, citizen reserves were organized to chink the stupendous gaps left not only by the abolition of the CCC but also by the exodus of veteran fire guards on the local districts. The manpower shortage was met in part by new power equipment—the tractor-plow appeared in the South; the bulldozer, ground tanker, and chain saw in the West. In part it was satisfied by cooperative agreements with the military, especially in locales around military bases. In 1945 Army bases were instructed to supply men and material for fire control in the face of Japanese fire balloons. And in part new sources of manpower were tapped: the Civilian Public Service (CPS), consisting largely of conscientious objectors from various "peace churches," kept the smokejumper program aloft, and the Office of Civilian Defense organized the Forest Fire Fighters Service (FFFS) in 1942.[95]

Under the direction of the Forest Service, the FFFS became particularly active along the coasts, where smoke interfered with shipping and where saboteurs were a constant, if—as it turned out—only imaginary, threat. The FFFS organized high school and college students, women, and draft-exempt males nationwide. It was during the war and in the spirit of Rosie the Riveter that the first all-women fire crew was organized at Soledad, California. In keeping with the region's cooperative tradition, the FFFS program was particularly active in the Pacific Northwest; indeed, the idea spontaneously germinated in the Northwest well in advance of the rest of the nation. In January 1942 three citizens approached the Forest Service offering volunteer assistance. It was agreed to have the three cooperators rally interest in a corps to be known as the Forest Service Reserves. Publicity was favorable, hundreds of citizens responded, training was conducted, and the reserves helped to man lookouts, prepared to fight fires, and organized eight teams of 10 women each to take over SOS (Service of Supply) duties for 200-man fire camps.[96]

Some states turned to prison inmate labor, and indeed, nearly all states with large fire problems and inadequate manpower have turned to prison labor at one time or another. In this California excelled. Its fire season in the brushfields could be year-long, and its fire protection strategy was heavily dependent on labor-intensive improvements like fuelbreaks. These could not be economically maintained or firelines manned except by appeal to the CCC or to inmates. The California Department of Corrections had long maintained road gangs, but not until the war emergency prompted the Enabling Act of 1941, which liberated manpower for resource protection, were prisoners used on the fireline. In 1942 the first conservation camp was established in Southern California, followed by 19 others—almost all on national forest lands and usually housed in old CCC buildings. The program continued after the war and in the early 1960s underwent tremendous expansion under the joint direction of the California Department of Forestry and the Department of Corrections. Nearly 45 percent of the camps' entire work load went to fire protection, and special mobile units were placed on call. By the late 1960s, however, the program was beginning a long decline.[97]

Of more significance was the outcome of experiments begun after restrictions were placed on the use of the CCC and before the CCC vanished. The wartime mobilization of volunteers and prisoners was an effort to replace the ranks of firefighters left gaping by the military draft; it was intended to sustain the status quo. The smokejumper and 40-man crew programs were mainline, if novel, advances in the evolution of specialized fire suppression crews, exemplars for a host of other organized crews that would follow. Especially in the case of the smokejumpers, the experiments had the effect of romanticizing and publicizing fire control to the public. They not only made firefighting more efficient; they made it glamorous.

Among the reforms that appeared after the adoption of the 10 A.M. Policy was the Aerial Fire Control Project under the direction of David Godwin, a daring advocate of mechanized, technologically sophisticated fire control and a man who began his career under duBois. Godwin's mission was to discover ways to bomb fires with chemical retardants, but the trials proved disappointing, and in 1939, amid new constraints on the use of the CCC, Godwin transferred the project to the Northwest and reoriented it from the aerial delivery of chemicals to the aerial delivery of smokechasers.[98] Godwin later claimed that the idea originated with T. V. Pearson, a ranger in Utah who experimented with some jumps in 1934. But Godwin was more impressed perhaps with the spectacle of mass paratroop movements by the German and Russian armies in 1936. (The American military had investigated paratroop possibilities in the 1920s but had abandoned them as impractical.) The Soviets, moreover, had apparently experimented with "parachute firemen." Godwin established correspondence with the Russian legation and commissioned translations of some of the literature describing the trials. In the course of the exchanges he also learned that the Russians had successfully stopped some steppe fires by smothering them with dry chemicals dropped from planes.[99] Godwin immediately saw the applicability of parachuting men, not merely supplies or water, into fires. Not the least of the flaws in the aerial tanker concept was that, as in military actions, bombing must be followed up by ground control. The limiting factor of firebombing in remote areas, where the scheme could be most useful, was still the difficulty of getting firefighters on the scene. Parachuting smoke-jumpers into the backcountry was an ideal, if somewhat brash, scheme. No one had ever attempted to parachute into mountainous terrain.

With Godwin as supervisor, experimental jumps were conducted on the Chelan National Forest in Washington in October and November 1939. Under a competitive contract the Eagle Parachute Company of Lancaster, Pennsylvania, furnished consultants, equipment, riggers, and jumpers. With Frank Derry in charge of actual jumping, 58 live jumps were made, some by two local smokechasers with no prior parachute experience. The crucial experiment, however, came with jumps into the mountainous country, where rock slides, snag patches, and heavy timber magnified the dangers. The first two smokechasers to make the transition were Francis Lufkin, a local fire guard, and Glenn Smith, a professional jumper and rigger with the Eagle Company. The next year operational tests were scheduled, and two bases were established at Winthrop, Washington, and Missoula, Montana. The 1940 fire season was light, though the Missoula group added to the program's capabilities by making a successful rescue jump to a plane crash on the Bitterroot National Forest—the first in a long tradition of paramedic and search and rescue operations carried on by smokejumper teams.

Already the smokejumping program was on the brink of tremendous practical and public success: the nine jumps during the summer of 1940 were so encouraging that plans were made to expand the project to include 26 men, to station it at Missoula, and to detail jumpers for fires outside the region as availability dictated. The Army belatedly recognized the military potential for parachuting into backcountry, and in June 1940 four staff officers, led by Major William Cary Lee, visited the training site at Missoula. What they learned went into organizing the first paratroop training facility at Fort Benning, Georgia; Lee later became chief of the Airborne Command, and the military trained many of its rescue units at smokejumper bases. Despite light fire loads, the smokejumper program attracted media coverage: writers clamored for copy, newsreel producers flocked to Missoula, and a movie studio even had a film in the works. In an era before Smokey Bear, the Aerial Fire Control Project kept firefighting vividly before the public eye.

But just when it appeared that smokejumping might enjoy unprecedented success, America declared war. Anyone fit enough to parachute out of an airplane over the Northern Rockies with logger boots and a leather football helmet was destined for the draft. In 1942 only five experienced jumpers reported for work. Equipment was sparse, too, though Derry had invented a slotted chute, which vastly increased maneuverability and dampened oscillation. By 1943, when again only five experienced men returned, the manpower situation was desperate. Responding to inquiries from conscientious objectors stationed at CPS camps, the Forest Service solicited volunteers and finally selected 62 candidates. For 1944 about 110 men (60 percent) of the jumper force were drawn from the CPS program. So successful was the program that the smokejumper operation was at last financed through regular appropriations as an integral unit of fire control and not, as had been the case so far, as a special, auxiliary force supported by experimental funds. Indeed, as manpower and funds shrivelled, a number of national forests became primarily dependent on smokejumpers. In 1945, out of apprehension over Japanese fire balloons, jumper bases were supplemented by paratroopers of the 555th Battalion, who were used on several project fires with mixed results.

Despite the war, smokejumping had in fact arrived. The major training base and depot at Missoula was supplemented with three other principal bases at McCall, Idaho, Winthrop, Washington, and Cave Junction, Oregon. So confident were the architects of the jumper program, and so vividly had the war demonstrated the significance of air superiority, that an experimental "air control area" was established within the Northern Rockies. Incorporating some 2 million roadless acres across the Continental Divide, the zone became known as the Continental Unit. Within it detection and suppression were to be conducted exclusively by air, with jumpers as primary firefighters. Major expansions were planned for 1947, and the depots in Washington and Idaho developed independent training facilities. Prompted by a Canadian who had been trained as a military rescue jumper at Missoula during the war, officials from Saskatchewan visited the Missoula facility and brought smokejumping to Canada. The staggered appearance of fire seasons throughout the West made possible a system of satellite bases manned out of Missoula, and a first detachment traveled to Silver City, New Mexico, to handle the spring fires there.[100] Joint trials with the

25. Smokejumpers descending to a fire in the roadless backcountry of Washington, 1957.

military in the Rockies, meanwhile, tested the feasibility of converting aircraft into firebombers, while experiments with helicopters in Southern California culminated in the tactical use of helicopters during the Bryant fire
and eventually developed into an analogous program for the aerial delivery
of firefighters—the helitack project. Sadly, in that same year the chief
architect of these developments, Godwin, newly promoted to head the Division of Fire Control, died in an airplane crash.

By 1949 smokejumpers had achieved such fame that they staged an
exhibition on the Ellipse outside the White House to the delight of the President and members of Congress. But when smokejumpers returned to
national headlines a few months later, it was in tragedy, not triumph. On
the Mann Gulch fire near the Gates of the Wilderness in Montana a blowup
fire overran a crew of 16 men; 3 lived, but the rest died, 12 of them jumpers.

The story prompted a popular movie, *Red Skies of Montana,* and led to a serious examination of the entire program.[101] The result was Operation Smokejumper, a general overhaul that culminated in the construction of the Aerial Fire Control Depot in Missoula and its dedication by a beaming President Dwight D. Eisenhower in 1954.[102] The depot became a focus for all aspects of aerial fire control, and when the Forest Service scheduled an all-Service Air Operations Conference in 1959, it was naturally held in Missoula. Meanwhile, other permanent smokejumper bases were established at Boise, West Yellowstone, Redding, California, and, most notably, at Fairbanks, Alaska (1959), where jumpers became the core of BLM Alaska operations.

Yet, despite its high visibility, the smokejumper concept may have exhausted itself. Perhaps its greatest source of dynamism came from its own growth, its rapid expansion into new regions and even into new countries. By the 1970s there were few frontiers left, and many areas for which smokejumping was especially well designed—remote, rugged country with heavy lightning concentrations—were being withdrawn from traditional fire suppression. Much of the lands included in the old Continental Unit, for example, have been, or soon will be, redesignated as wilderness areas. The interior of Alaska is being removed from BLM jurisdiction and rezoned for national parks, wildlife refuges, native corporations, and state lands—all with different fire objectives. Attempts by the Forest Service to expand smokejumping into the Ozarks and Appalachians (1971–1973) proved unconvincing, and so did discussions by BLM about possible jumper deployment in the Great Basin and Utah.[103] In 1980 the jumper base at Boise was closed.

Compared with the alternatives possible in the late 1930s, particularly the fixed wing aircraft available to the Aerial Fire Control Project, smokejumping was a dramatic innovation. But the concept faced serious competition, technical as well as strategic, from the capabilities of the 1970s, especially the extraordinary potential of rotary wing aircraft. Smokejumpers cannot be used in winds over 15 mph; helicopters can function in winds up to 30–35 mph—conditions more likely to produce critical fires. Helicopters can operate out of a landing pad; airplanes require a landing field. Most districts can afford a heliport, and nearly all with any sort of fire hazard can justify a helicopter during periods of extreme fire danger. Helicopters can deliver goods as easily as paracargo and with greater accuracy, reducing the need for special tree climbers to recover chutes lodged in treetops. Helicopters can move masses of firefighters who do not require the rigorous training demanded of jumpers, and they are better suited to deploying the interregional fire crews that now form the backbone of campaign fire suppression. The same machine that delivers people and cargo, moreover, can also deliver fire retardant with the use of a bucket or belly tank—something that fixed wing aircraft cannot do. Jumpers, moreover, still face the ancient problem of returning from a fire. Ironically, the transfer of the helicopter program from California to the Northern Rockies in the late 1940s was dictated in part by the necessity to retrieve jumpers and their gear, and the first use of a helicopter on a fire came during the evacuation of injured jumpers from the Mann Gulch fire. As helitack increases in sophistication—allowing for night flights, retardant drops, and the delivery of crews by rappelling—the merit of smokejumping recedes. Its main value lies in long-distance hauls (over 50 miles), in which it can still compete economically with the more flexible helitack. More and more, the formidable smokejumper corps resembles an elite military unit that—mindful of its past triumphs and glorying in its esprit de corps—has continued to specialize in a complicated technique of largely historic value. It may be that smokejumping will be maintained in much the spirit that the United States Army retained a remnant cavalry unit even past World War I.

Of greater impact has been the progeny of that other 1939 experiment, the 40-man crew. The idea was prompted as much by the restrictions on CCC availability as by the likelihood of an invasion of chromium prospectors (and fires) into the Siskiyou National Forest. It was proposed to Headley in 1938 and was given clearance for trial operations the following year. Rolfe Anderson directed the experiment, with strong support from Edward Cliff, the Siskiyou supervisor. The crew would be composed of specially screened Forest Service employees, not CCC workers; it would be trained in the new, "progressive" methods of line construction developed out of the CCC experience; it would pack enough food for three days in the field and carry its own cooks for longer tours; it would specialize in handline construction; and, when not on fires, it would engage in the heavy-duty project work, such as road construction, that had occupied the CCC earlier. The 40-man crew would be trucked into large fires, wherever started. If the smokejumpers were the commandos of fire control, operating in small units on isolated burns, the 40-man crews were its shock troops, intended to stiffen resistance on tough, prolonged campaign fires.[104]

The 40-man crew was a rousing success. It so impressed Godwin that he requested a lengthy report, which was later published in *Fire Control Notes*. Among its recommendations, the report observed that "it is believed that this system cam be applied to other crews organized from picked CCC enrollees. . . ."[105] Here, perhaps, was a means to revitalize CCC involvement in fire and still stay within announced guidelines. No more raw recruits would be, or need be, shipped willy-nilly to fires. In 1940 special 40-man CCC crews of carefully screened enrollees and top foremen were organized on the Siskiyou, again under the direction of Anderson. Training was intensive, and organization was prized: the one-lick method was used exclusively in line construction; each man was assigned a number and a pack; and whenever "the boys" went to town, their gear went with them. Not least of all, each member was entitled to wear a special badge of red felt with the logo "CCC, 40," which designated him as part of an elite corps. When not on fires, the crew labored on road construction projects—much like the Roman or French foreign legions. Similar crews were established throughout most of the forests of the Northwest, and the idea spread to other Forest Service regions. The Willamette National Forest developed a smaller version, the Willamette Flying 20, composed of carefully selected fire guards and used for initial attack on the heads of threatening fires. The Forest Service used the crews' success as a major argument in its plea to Congress to retain the CCC program throughout World War II, because fire protection was considered to be an essential war-related industry.[106]

But, even more than the smokejumpers, the 40-man crew experiment wilted under the demands of the military draft. The slack was taken up in part by the emergence of cilivian crews organized along the principles of the 40-man crew, which were available on call. Not surprisingly, the first of such crews, the Red Hats, came from the Northwest, sponsored by the School of Forestry at Oregon State College, the National Youth Administration, the Oregon Department of Forestry, the U.S. Forest Service, and the Oregon Forest Fire Protection Association. Initially, the crew was composed exclusively of forestry students and was stationed at MacDonald State Forest. Their daily routine consisted of two hours of study, two hours of training, and four hours at hard labor on project work. Later, the program was broadened to include anyone interested. Over the summer 113 men were enrolled; at one time three 26-man crews were fielded, and two crews were always on call. But the presence of the Red Hats was actually larger: experienced assistant foremen were often detailed to supervise pickup crews, and the Red Hats themselves easily absorbed stray EFFs to bolster their ranks to 30 or 40 men. Unlike the CCC, the program had the additional advantage that it was not subject to national legislation, and in the postwar period many forests organized similar crews out of local labor pools or colleges—leading to white-, black-, and red-hatted crews.[107]

In that champagne year, 1947, the concept of fire crews manned by regular Forest Service employees revived. There were already the self-styled "hot shot" crews stationed in the mountainous brushfields of Southern California. Fire protection of all sorts had always assumed a magnified scale here, but in the forests, where developments were largely restricted to fire protection improvements and where roads were often scarce, hand crews were needed rather than engine companies. This the hot shot units provided. It is no accident, moreover, that the earliest helicopter experiments were conducted in Southern California: the helicopter could deliver hot shot crews by air into the brushfields, just as the transport airplanes of the Northern Rockies could bring smokejumpers into the tall timber. Within a decade, assisted by a research program into mass fire control sponsored by the Forest Service and the Army Corps of Engineers, hot shot crews were manning helitack units, and the nearby Arcadia (later, San Dimas) Equipment Development Center was outfitting choppers to lay hose, sling pumps, and drop retardant. In 1954 the Chilao Helishots were formed on the Angeles National Forest, and three years later the first full-fledged helitack program was undertaken by the Forest Service. Among the new techniques was the helijump, in which helitack crewmen, suited up like smokejumpers, would jump a short distance from a hovering helicopter to the ground, most often into brush. They would then construct a helispot, and the helicopter could transport crewmen and supplies in the orthodox way. A scant decade after its first deployment on a fire, the helicopter had suddenly blossomed into a multipurpose tool of seemingly limitless capabilities. By the 1970s, when medium-size helicopters became widely available, its range was even greater, and helirappel crews in Oregon and Washington posed an even more direct challenge to smokejumping.[108]

Meanwhile, other crews were formed in the postwar era outside of Forest Service rosters. In 1948 the Bureau of Indian Affairs organized a 25-man crew of Mescalero Apaches in New Mexico. The next year the crew assisted the Forest Service on a fire on the Lincoln National Forest. The Forest Service was impressed and decided to supervise a larger program of crews manned from local reservations. Thus was born the Southwest Forest Fire Fighters (SWFFF) program. Originally restricted to the Indian tribes of the Southwest, the program expanded in 1953 to include Hispanic crews from northern New Mexico and in turn became a model for the organization of Indian crews in the Northern Rockies by the Forest Service, for the establishment of the Snake River Valley, Alaskan Indian, and Eskimo crews by the BLM, and for the organization of Mexican nationals (braceros) in California during the 1950s. The 25-man SWFFF crews were specially trained and in strong demand throughout the West. Their presence was no less important to the natives: fire pay became a significant cash income for many individuals and tribes. Working on farms, pasturing herds, or practicing handicrafts—fire call could be easily accommodated to all of them in ways not possible for warden crews or other laborers with fixed schedules. And for some tribes fire duty perhaps recalled their warrior past as raiders: they would be gone from home for days or perhaps weeks, to return rich with booty, stories, and tales of daring. But improving conditions on the reservations and pueblos made assignments less attractive, and by the early 1970s the SWFFF program was in decline.[109]

The same was true for other bodies of men on which the Forest Service had come to rely. Large crews had traditionally been maintained for project work—brush and slash disposal, thinning, blister rust control, and road construction—and such crews were trained in and assigned to fireline duty

26. Helijumper, Southern California, 1959.

when campaign fires were in progress. Many received special training and were supervised by foremen drawn from the regular fire organizations or from smokejumper ranks. As the SWFFF crews had, some took on distinctive names and insignias, like the Redmond Raiders, a timberstand improvement crew from Oregon. Districts often organized their full-time and seasonal employees into crews, regardless of their regular assignment; such crews became known as "regulars." But by the late 1950s the blister rust control programs were being terminated, and mechanization took over such labor-intensive chores as slash disposal and road maintenance. By the 1970s most such work within the forests was contracted out, further reducing the manpower pool available for local fires.

Long before this happened, however, the Forest Service had recognized that a premium had been placed on organization and, above all perhaps, on mobility—on flexibility of skills, transportability, and interchangeability. At the 1950 fire conference in Ogden, Utah, fire officers recommended the creation of "aerial shock troops" consisting of 50 to 100 men stationed at key airports and ready for dispatch to campaign fires.[110] A Service publication in 1957 rhetorically, if rather mystically, advanced the argument: "How valuable is an atom bomb or a battleship? This is a perfectly sensible question. And the answer is in keeping with the fiscal philosophy justifying the organized crew."[111] The concept of a rapid deployment force, which the American military had debated and abandoned, the Forest Service adopted with the creation in 1961 of the Interregional Fire Suppression Crew. The Chief Forester at the time was Edward Cliff, forest supervisor of the Siskiyou National Forest during the 40-man crew experiments. Like the 40-man crew, the interregional (IR), or "hot shot," crew would exhibit the élan and dedication expected of Forest Service employees. Like the organized crews of the 1950s, the IR crews would be employed on a national arena. The essence of the crew was its combination of mobility and strength. It could go anywhere, and it arrived as a complete package, outfitted with individual packs, sleeping bags, hand tools, saws, radio, and overhead. The program quickly expanded, filling the partial vacuums left by the departure of regulars, inmates, and the SWFFF. By 1963 there were 9 such crews; by 1970, 15; by 1977, over 30. As with other fire control resources, there developed a trend toward interagency utilization under the doctrine of total mobility and the complete interchangeability of crews and overhead. By the end of the 1970s, BIFC assumed control over the dispatch of organized crews among regions and agencies. Ironically perhaps, the mobile crews were not so much beneficiaries of interagency harmony as they were a prod to it. As had happened with air tankers and helicopters; until such specialized and expensive tools became available, there was little necessity to agree on guidelines for their use or to establish coordination centers through which to dispatch them.[112]

But however novel their conception and however daring their application, the great armies of organized fire crews are designed for the big fire, and one of their primary functions is to release district fire crews, which can then restore their initial attack capabilities. The basic strategy of fire control is still to suppress fires at their origin, and this is the job of the local guards—the small tanker crews, helitack squads, and handline smokechasers. Smokejumpers, helitack crewmen, helirappellers, ground tankers—all are less new ideas in crew composition than they are innovative means to transport smokechasers faster to small fires. Only when they fail are the interregional divisions moved in. The creation of special district fire crews on call for the immediate suppression of small fires is of greater significance and of greater cost benefit than the proliferation of large crews for campaign fires.

It was this special problem that the Increased Manning Experiment, con-

27. Redding Hot Shots, Wenatchee fires, 1970.

ducted between 1955 and 1959, sought to address. Its "primary purpose was to determine whether additional expenditures to strengthen prevention, detection, and initial attack would result in a corresponding or greater reduction in suppression costs."[113] Amplified funds were made available for regular accounts in an effort to reduce heavy drafts from the emergency suppression and presuppression accounts. District organizations on the pilot forests received a whopping 80 percent increase in fire guards, from a normal force of 723 to 1,317. Both costs and burned acreages were dramatically reduced. Indeed, costs had been rising so fast that resource protection, the program conceded, was a secondary consideration; it was the growth of the emergency accounts that had to be controlled. The program concluded on the eve of the interregional fire suppression crew project, and it gave added urgency to plans to improve the skills and training of fire crews as well as to enlarge their numbers.

For all their effectiveness, the hot shot crews are like galley slaves, digging powerfully but mindlessly to the beat of orders from above. The crucial decisions usually rest with the district smokechaser, with the tanker or helitack crewman who must evaluate and outmaneuver the fire. Except in areas designated in advance for natural prescribed fire, an evaluation of alternatives takes place, according to the 1978 Forest Service fire policy, only if initial attack fails. Only some fires are visited by the large organized crews, but every fire is attacked with district guards—whether it turns into a vast panorama of fire, like the 1910 holocaust, or remains on isolated snag fire; whether it races sublimely through mountain suburbs of California or smolders miserably in the dry dung of extinct ground sloths, like the burn in Ramparts Cave that the fire guards of Grand Canyon were sent to contain.[114]

IV

The evolution of organized fire manpower did not always come as the result of organizational logic and economic efficiency. Many reforms—some of the most important—came at the price of blood. One CCC enrollee died on the Tillamook Burn, and another was seriously injured. A month later the worst tragedy since 1910 came to Griffith Park in Los Angeles. At the report of a brushfire hundreds of ECW (but no CCC) laborers were sent from road projects to the fire scene. There was little direction at the fire despite the arrival of city firemen, and soon men were strung out around a draw busily encircling the burn with a fireline. But someone down in the draw panicked when the brush momentarily flared; backfires appeared; pandemonium resulted as men up the slope thrashed helplessly through the chaparral. Some made it out; 125 were hospitalized with burns or injuries; another 26 perished in the fire; and 2 more succumbed subsequently from burns received. When finally controlled, the notorious Griffith Park fire totaled 47 acres. A coroner's jury took a dim view of the debacle, freely dispensing charges of gross negligence.[115] It was fortunate, in retrospect, that CCC boys were not involved: the reaction following the Blackwater fire might have paled beside it, and the CCC might never have developed into a significant force for fire protection.

Most firefighters scoff at their occupational hazards, considering C rations a greater threat to their health. One could also cite the case of the hot shot crewman spending yet another anniversary on a fireline and thinking about his "old lady—she ain't gonna like this one bit." And in fact, most fatalities result from snag falls, vehicle accidents, and heart attacks rather than from burns or from asphyxiation. Paralleling fires themselves, there is a low-intensity background count of fire-related death and only a few cases of multiple fatalities from burning. Between 1926 and 1976, some 145 men died from burns on 41 fires on national forests, and another 77 fatalities occurred on 26 fires elsewhere.[116] These statistics are not complete, but they include most of the larger episodes. Of the total, 14 fires accounted for 149 deaths, or nearly 70 percent. Thus fire-induced fatalities show the same logarithmic distribution as fires themselves. The correlation is not coincidental: the same, still unpredictable elements that cause otherwise unremarkable fires to explode are the same ones that take lives. Nearly every instance of multiple deaths from burning has resulted in major reforms in crew organization, firefighter and overhead training, and personal safety equipment. In the old days it was enough, in the words of one safety publication, to exhort EFFs to "stay with your foreman" if there was danger: "he has been to many fires and is still alive."[117] That was no longer good enough with organized district and interregional crews.

The worst disasters involved pickup crews. The CCC—organized but not always adept at firefighting—took losses totaling 47 men, in part attributable to the sheer number of fires they attacked. The Blackwater fire was the worst. Marines called out in 1943 to fight the Hauser Creek fire on the Cleveland National Forest in Southern California were trapped by shifting winds created when Santa Ana conditions faltered; 11 died, and 72 were injured. EFF losses were easy to explain away, a result of primitive organization. The CCC casualities, especially in the late 1930s, when the CCC as an institution suffered miscellaneous ills, resulted in new guidelines for the use and new standards for the training of enrollees; none of the new 40-man crews, for example, was involved in a disaster. The Hauser Creek disaster dispelled any lingering sentiments that military units untrained in fire suppression were automatically superior to civilian crews designed for fire control.

It was less easy to dismiss costly losses suffered on the Mann Gulch fire in 1949, the Inaja fire of 1956, or the Loop fire of 1966. The first took the lives of 13 smokejumpers; the second, an organized crew of 11 inmates supervised by Forest Service overhead; and the last, 12 members of a hot shot crew. In all three episodes the men were caught when spot fires or an arm of the main fire worked below them, blew up, and trapped them as they fled up the slopes. The Mann Gulch fire came soon after the establishment of a Division of Fire Research by the Forest Service, and the Chief Forester ordered the new division to investigate fire behavior as a means of better educating crews to fireline hazards. The 11 fatalities of the Inaja fire gave new urgency to the quest. The fire became a hot political issue in Southern California. Lawsuits were filed, and the Chief Forester, Richard McArdle, an old hand at fire research, appointed a special task force to make recommendations for improving fireline safety. The Fire Task Force reevaluated the whole state of the art, from research to training to protective clothing. The 10 Standard Fire Fighting Orders were promulgated as a guide to fireline safety; in 1958 the first national training course in fire behavior was held in Missoula; and the desirability of a national training center was openly debated.[118] The fire behavior class became a model for a host of sub-sequent suppression training courses aimed at improving overhead skills. With the creation of interregional fire crews came interregional overhead teams and the need for national training courses to establish standards. The first of these courses, fire generalship, appeared at Santa Barbara in 1961, and in that same banner year the new crews were outfitted with orange, flame-resistant shirts—the first of a series of devices that would effectively armor individual firefighters.

In 1966 a hot shot crew was mauled by the Loop fire on the Angeles National Forest. Twelve died when the fire crept below them in a draw, blew up, and trapped them in a narrow chimney. The fire flashed up a chim-ney 2,200 feet long in about a minute, overrunning the crew before the men had time to take any defensive action. It was in many ways a distressing repetition of the Inaja tragedy, and another task force was assembled. It took as its point of reference its 1957 predecessor and the recommendations made by an analysis group that investigated the specifics of the Loop fire tragedy. The task force concluded that "the greatest opportunity to prevent these tragedies lies in the management and organizing on the job to insure and require the use of what we already know and what we already have."[119] The Loop fire was incorporated as a case study into basic firefighter and fire behavior training courses. The proposals of 1957 were reexamined and reaffirmed; a checklist of progress in fire behavior knowledge was included; and new emphasis was placed on fuels evaluation rather than on further crash research programs into mass fire behavior. The concept that fuels guidelines could help to alert firefighters to blowup conditions thus merged nicely with the belief, then gathering momentum, that active fuels manage-ment was the best method for the control of conflagrations. Another line of research was also being explored: the possibility of carbon monoxide poison-ing. Tests determined that carbon monoxide could indeed accumulate in hazardous quantities.[120]

The Loop fire tragedy accelerated demands for a formal fire academy. In 1967 the Forest Service sanctioned a National Fire Training Center (later renamed the National Advanced Resources Technology Center) at Marana, Arizona. The site was a former military airfield, later abandoned, then con-verted by the Southwest Region of the Forest Service into a regional train-ing facility, zone dispatch office, and fire warehouse. National-level courses in suppression, prevention, and fire management were scheduled at the cen-ter. With the establishment of BIFC, prospects brightened for interagency training and certification. For the most part, Forest Service manuals served as the foundation for the proposed new series. In 1974 the fire training cen-ter at Marana became interagency as well as national when agencies from the Interior Department began contributing toward its operating costs. With the formal chartering of the NWCG, a training committee figured prominently, both to oversee an ambitious scheme of training for virtually every fireline position and nearly every dimension of fire management and to determine interagency standards for certification. The latter became the National Interagency Fire Qualifications System, a necessary component of the total mobility scheme.[121]

But most of this training was for the big shots, the staff and overhead, not the hot shots on the line. As with an army, the real strength of fire control organizations lay less with their commissioned officers than with their NCOs—their tanker and helitack foremen, their crew bosses, veterans of years on the fireline. And it rested, too, on the firefighters. In the South, the Lake States, and parts of the Northeast, mechanization and integration with volunteer fire departments largely eliminated the need for sizable hand crews, but not so in the West, where the ranks of fire organizations are filled with seasonal laborers. The seasonal nature of fire control remained a source of both strength and weakness—strength, because of perennial youth, low wages, absence of unions, and ability to capitalize on highly motivated collegians eager for summer employment; weakness, owing to the perennial need for training and a reluctance to invest too heavily in advanced training. For nearly all such firefighters, fire is a temporary assignment, rarely extending beyond two or three seasons; despite pretenses to the contrary, rank-and-file firefighters are not career professionals. Nor could most become so if they wished: there are few career ladders in fire protection. Even in the Forest Service one needs a degree in forestry to advance up the administrative ranks. Such requirements are, if anything, strengthened by the trend in federal agencies to subordinate the fire function to larger land management objectives. This fact accounts for the paucity of college-level courses on fire: only a tiny handful of universities offer curricula in wildland fire management, and these are often most used by established officials returning to school on company time to improve their fire credentials. The absence of a college curriculum has, in turn, reinforced the drive to create training courses from within the organizations themselves. But this certification is only meaningful on the fireline.

What, then, keeps the ranks full? Writing in 1925 Forest Service ranger H. A. Calkins offered a simple theory: "I have met a few fellows that claimed they enjoyed fighting fire, but I have always thought there was something wrong with their heads."[122] The truth is more complex. Some are attracted by a youthful sense of adventure, a desire to participate in swashbuckling feats in romantic wilds. Some are proud of their skills, welcome a tough physical challenge, and take pride in their crew traditions. All look at the money, which for a seasonal job is not bad; and in a heavy fire year with lots of overtime and hazard duty, pay may be downright lucrative. Whether or not, in a macabre way, the presence of fatalities strengthens the unvoiced belief that firefighting is a moral equivalent of war, fire control nonetheless perpetuates the belief in the value of the strenuous life. Ranger Calkins notwithstanding, whatever the reasons given for fighting fire, they are less likely

28. Lone smokechaser riding off to a fire, North Carolina, 1923.

to come from the head than from the heart. And though for some firefighters the world will end by fire, for most, frozen out of any hope of future
advancement, their careers will more likely end in ice.

> The noonday sun was coming strong,
> As toward his station passed along,
> A man who bore upon his back
> A most ill-fitting heavy pack.
>
> His is a trail where drinks are few—
> On the sunny side of the mountain too—
> His eyes are bloodshot, his back near broke,
> For he has been chasing a distant smoke.
>
> Twenty-four hours since he left his station—
> He has been eating Emergency Ration,
> But now on his face is the start of a smile
> As he gets near the end of the last, long mile.
>
> . . .
>
> But he would not barter his place in the hills
> For life in the city with all of its ills,
> Its tinsel, and noise, its dust and its smell,
> For one of his mountains, his vales or his dells.
> —Charles H. Scribner, "A Smoke Chaser," 1929[123]

7 THE COLD WAR ON FIRE

There is no reason why we can't develop fire extinguishing forces as strong and effective in their way as atomic weapons have been in their field.—Keith Arnold, 1956[1]

The second great reburn of the Tillamook cycle was frustrating fire control efforts when the atomic bomb was dropped on Japan. The event mapped a distinctive watershed in the culture of fire. Both as a tool and as a weapon, atomic power had symbolic and practical significance for fire history. In peaceful forms, as reactors, it offered the benefits of fire without flame, smoke, and combustion and without recourse to forest resources, modern or fossil, for fuel. As a military weapon, however, it restored broadcast fire to the strategic arsenal of industrial warfare. It is in fact a striking coincidence that the rediscovery of broadcast fire in the form of prescribed burning and the reestablishment of broadcast fire as an incendiary weapon appeared at very nearly the same time.

Fire was little used in World War I; it was considered both inhumane and impractical. Most of the combatants, however, did experiment with some fire weapons, such as the German *Flammenwerfer*. The exception was the American military, whose reluctance to use fire thus paralleled the attitude of American foresters, who maintained that fire had little scientific justification in the conduct of forestry. For both groups fire as either weapon or tool had a stigma of primitivism. In World War II, however, fire was applied on a scale unprecedented in conflicts between industrialized states. The images of the war are those of a world in flame and the war's worst horrors were those evoked by fire—the London blitz, the burned and asphyxiated bodies of German civilians dragged from shelters in Dresden, the ghastly ovens of Auschwitz, the atomic fireball over Hiroshima.

World War II restored fire as an active weapon of warfare and frequently forced fire control into a genuine paramilitary role. Fire protection was nationalized, first through the Office of Civil Defense, then through the Forest Service as its responsibilities were enlarged to include rural fire protection. The example of a mechanized military spurred the Forest Service to create fire equipment development centers. The wartime lessons derived from large bodies of disciplined troops gave added momentum to the drive for organized crews. The atom bomb created a new research problem, mass fire, while at the same time it symbolized the power of organized science to find an answer; new research programs and fire laboratories were a partial response. And not least of all, the wartime experience and its somber legacy revived determination to control the "red menace." The cold war on fire had begun.

The national fire scene shifted, too, leaving the remote backcountry for the city perimeter. Southern California seemed to dominate national attention in fire as in other things, and it inaugurated an era of great fires, half-wildland and half-urban. The peculiar nature of the conflagrations suggested that these brushfields and suburbs best approximated the conditions of fire warfare that a thermonuclear confrontation might bring. It was in Southern California that the research on mass fire was conducted, that the first Forest Service equipment development center was established and the air tanker created, that hot shot crews and helitack were developed, and that firefighting fatalities began to accumulate. In the 1930s and 1940s the major fatality fires, like the Blackwater and Mann Gulch, occurred deep in the forested wildernesses of the Rocky Mountains; in the 1950s and 1960s they came in Southern California brushfields—at the Rattlesnake, the Inaja, the Loop, and the Canyon fires. The Forest Service saw an astonishing 25 percent of its entire fire budget funneled into the four national forests of Southern California. The region seemed to develop an autonomous existence and peculiar patterns of interagency cooperation. The war, as Raymond Clar observed, put the California Division of Forestry's fire protection program "in business." FIRESCOPE is only the latest of a long line of innovative, cooperative programs.

The atomic bomb also introduced a new era of environmental philosophy. As a weapon, the bomb had three effects: fire, blast, and nuclear radiation. The first was ancient; the second had constantly expanded since the invention of gunpowder; but the third was qualitatively new—disturbingly new. The bomb became an archetype for the concerns of the new conservation movement. The world of the bomb was, first of all, a world of the synthetic landscape, whose catastrophes were the product of scientific technology rather than of nature. The bomb furnished metaphors for the era's chief concerns. The ambivalence of nuclear power, for example, was closely followed by concern over the "chemical fallout" of pesticides, herbicides, and industrial pollutants and over the "population bomb." It is not mere coincidence that in 1954 a major symposium was held on the subject of "Man's Role in Changing the Face of the Earth," the hydrogen bomb was tested, and Operation Firestop, an early study of mass fire and fire control technology was conducted.

The status of broadcast fire was, like that of atomic energy, ambivalent. The initial response of the Forest Service was to retain aggressive fire protection as a visible sign of conservation at a time when many of the traditional concerns of the conservation movement were fading. But gradually fire lost its premier position as a paradigm of conservation philosophy. The revived interest in it had been sustained, in a sense, by the Cold War; unlike most old conservation programs, fire control was readily identified with the holocaust of World War II and the horrors of atomic arsenals. As the new conservation movement both expanded and defined its particular concerns, especially as it absorbed and refined new concepts of wilderness, fire control was eclipsed in importance. Like many other projects of an earlier era of conservation, aggressive fire suppression even appeared as something of a nemesis to proper environmental management. The fireball that rose over Hiroshima was on the one hand the dying flames of fire seen as the Dragon Devastation and, on the other, the rising, if ominous, sun of a new era in the culture of fire. Wildland fire was destined to be regarded less as the fiery Armageddon of thermonuclear holocaust and more as the friendly flame in the wilderness.

FIREPOWER: FIRE AND WARFARE

It's a gigantic operation. Much like a military mission in wartime.
—General Donald Pierce, on the 1967 fires in
the Northern Rockies[2]

He who would make his nation strong must look to its fire
defenses as well as to his armed forces.
—Horatio Bond, National Fire Protection Board, 1946[3]

I

As soon as fire was available as a tool, it became equally accessible as a weapon. Fire hunting was readily adapted to the tactics of war, and examples abound of its use in this way by primitive societies. More significantly, fire was probably humanity's first strategic weapon. It could destroy crops, pastures, hunting grounds, and villages. It could change the environment itself, destroy the capacity of an enemy to wage war, and terrorize warriors and noncombatants alike. Fire could subdue a hostile enemy as it could a hostile environment. When used in conjunction with other weapons and strategies, the scope of fire could be magnified still further. With the exception of fire hunting, the most common use for broadcast fire cited by Indians was to modify the terrain so that foes could not easily launch suprise attacks and ambuscades. "Incendit et vastavit," wrote a chronicler of the Hundred Years War, and to burn and lay waste has been a well-nigh universal characteristic of war from its inception. Warfare must be considered as an episodic, though potent, means by which fire has been applied to the landscape and distributed around the globe, along with the diseases, exotic organisms, and foreign institutions that conquering armies have traditionally left in their wake. If its ability to manipulate fire is a measure of a culture's technological sophistication, its capacity to control fire for the ends of war is a measure of its military might.[4]

Indians applied fire to flush out enemies from heavy brush, grass, or timber; to harass bodies of troops too large for frontal assault; to deny cover to enemy patrols engaged in guerrilla raids; to obliterate tracks or cover movement with smoke; to deprive enemy cavalry of fodder; and to communicate messages over long distances. The more primitive the society, the more likely it was to resort to broadcast fire. The use of incendiary weapons, like the practice of fire hunting, was widespread. Not infrequently, the hunting grounds of hostile tribes were fired as an economic weapon, much as the

Cree and Assiniboine attempted against the Hudson's Bay Company and as the Sioux tried against Dakota cattlemen.[5] In the Rogue River wars of 1853, retreating Indians fired the woods for miles around to obliterate trails, confuse their pursuers, deprive cavalry horses of pasturage, and cover ambushes.[6] Hiding behind trees, especially at night, Indian archers fired on troops illuminated by the flames and confused by smoke. Apaches used similar tactics in the Southwest.[7] During his campaigns against the Sioux, General George Crook found it necessary to construct firebreaks at every camp in lieu of earthwork fortifications. Frustrated by their inability to lure Crook's command into a trap as they had Custer's, the Sioux finally decided simply to burn them out. Something like a half a million acres around the Big Horn Mountains were fired. The strategem failed, however, when supply trains proved adequate for stock and the troops, much to their delight, took to fishing and for weeks lived off mountain trout.[8]

In the ancient world fire was common in war. Assyrian bas-reliefs from the eight century B.C. show warriors projecting and extinguishing fires. The first treatise on warfare, *The Art of War* (500 B.C.), by Sun Trun Wu, describes the use of fire arrows. In the Bible the prophet of the Lord proclaimed that "a fire goeth before him, and burned up his enemies round about." Samson tied firebrands to the tails of three hundred foxes to burn the shocks, corn, vineyards, and olives of the Philistines—anticipating incendiary devices attached to birds and animals by the eleventh-century Chinese and experiments with bats by the U.S. Army in World War II.[9] In the Jewish Wars of the first century A.D., according to Josephus, firebrands and burning were widely used for offensive and defensive purposes, both tactical and strategic, so that "Galilee from end to end became a scene of fire and blood. . . ." Fire was also directed into caves harboring bandits and guerrillas, much as it was in World War II against the Japanese.[10]

In Europe Lucretius told how

Weapons of ancient times were hands and nails and teeth.
Then axes hewn from the trees of the forest.
Flame and fire as soon as men knew them.[11]

The Greek tactician Aeneas, who assembled the first European treatise on war (circa 350 B.C.), listed sulphur, pitch, pine, wood, incense, and tar as basic incendiary devises. Herodotus described how the Scythians protected themselves by a scorched-earth policy. The strategy has long been a successful one on the steppes: in the seventeenth century a massive cavalry invasion planned by Peter the Great abruptly halted when the steppes were burned out ahead of him and the invaders lacked the fodder to continue independently; during World War II German and Russian armies turned the plains into a vast burned wasteland.[12] In the Peloponnesian War the Spartans encircled the Platean army with an enormous fire, but a freak rain storm (perhaps induced by the fire) extinguished it before the Plateans became desperate. Elsewhere the Spartans displayed mechanical ingenuity by manufacturing a flame thrower as a siege machine.[13]

The Byzantine invention of Greek fire became the scourge of besieging armies and of wooden naval vessels. It continued as a prominent naval weapon until about 1200 and was still active at the siege of Constantinople in 1453. To the English it was known as "wild fire" and was used at least into the fourteenth century. At the siege of Malta (1565) various antipersonnel incendiary devices were in use and proved especially effective against the Turks, whose long robes readily flared. Earlier medieval Arabs had outfitted their incendiary forces with special flame-resistant armor.[14] Wildfire was eventually superseded by the tamer fire of gunpowder.

Europeans used fire to modify cover and terrain for military purposes. In some places, this meant a conversion of farm to forest. It is said that the devastation wrought by English fire during the Hundred Years War "gave France back its forests." In Bosnia, the Turks determined to eliminate the forest cover used so effectively by guerrilla insurgents and began a systematic program of felling and firing.[15] In the American Civil War General Sheridan burned out the Confederate breadbasket, the Shenandoah Valley, while General Sherman achieved infamy by burning across Georgia and the Carolinas. Bruce Catton describes how elements of Sherman's troop

> casually burned towns and looted plantations and set fire to pine forests just for the fun of seeing the big trees burn—and came up north. . . . An Indiana soldier remarked that the men set fire so much that some days the sun was almost entirely obscured by the smoke of the consuming buildings, cotton gins, etc. But when they marched through the turpentine forests, the stragglers who continued to fringe the moving army set fire to the congealed resin in notches on the trees, and for mile after mile the army moved under a pall of odorous pine smoke. An officer wrote that the flames in the forest aisles "look like a fire in a Cathedral," and one soldier remembered "the endless blue columns swaying with the long swinging step," and said that above the crackle of the flames could be heard the massed singing of "John Brown's Body."[16]

Incendiary weapons have remained remarkably constant through the ages, dividing into two general categories: the fire missile and flammable liquid. Projectiles included fire clubs, catapulted pots, and fire arrows. Fire arrows were ubiquitous and especially potent in an age of wood and grass structures. Raleigh reported them in Virginia; the Incas used them with terrifying success against the Spanish at the siege of Cuzco; the Indian sepoys who mutinied in 1857 shot flaming arrows against the British. As a projectile, the arrow was replaced by the gun. Gunpowder made fire more an agent of propulsion than a warhead, though in time incendiary shells were developed. Black powder was originally regarded primarily as an incendiary device and only secondarily as a means of propulsion. Flammable liquids have shown similar durability, especially as mechanical devices were invented for propelling them. The Spartan flame thrower was one among many. Greek fire could be poured out of vats, it could soak fire arrows, or it could be launched by artillery. Tamerlane outfitted "flame projectors" atop elephants, thus achieving maximum terror as well as efficiency and foreshadowing the flame-throwing tanks of the twentieth century, as Molotov cocktails recall the bubbling cauldrons of Greek fire poured on siege machines.

The trend in weaponry has paralleled fire practices at large. Broadcast fire disappeared as cultures evolved in economic status, as disciplined armies replaced marauding bands of raiders, and as fire was itself progressively confined. The great problem with broadcast fire was its indiscriminate nature. In a pitched battle it was likely to obscure the field as much for the attacker as for the defender. Broadcast fire became less a tactical weapon than a strategic one, applied as part of a scorched-earth policy by frantic defenders or by a determined attacker after he had seized the land or city in question. Open fire was replaced by confined fire, torches by arrows, arrows by artillery shells, and shells by bombs of phosphorus, whose heat could ignite surrounding fuels without itself flaming. Ultimately, with the industrialization of armies, the effects of fire were created without flame at all.

Open burning was most commonly found in encounters between industrialized and primitive societies. When Captain Cook landed his chief naturalist, Joseph Banks, on an Australian beach, suspicious natives promptly fired the surrounding brush, and the offshore wind drove the scientific party back to the boat. Fire was a medium of conflict between different cultures or socioeconomic groups in remote colonies. The suppression of traditional cultures often meant a suppression of traditional fire practices. When the French crowded Algerian pastoralists off their ancient grazing lands, the pastoralists took their flocks to the hills—burning the woods before them and causing the French to create an elaborate network of fuelbreaks and guard stations to save valuable cork forests. In the forests of the American South a similar pattern of economic colonialism was acted out. Northern foresters and northern capital wanted fire excluded so that the forests would be encouraged at the expense of pasturage on the open range. To protest the resulting enclosure movement that threatened traditional piney woods hunting and herding, natives often turned to fire. Broadcast fire was an especially adaptable weapon for use by dedicated guerrilla bands or a sullen population intent on resisting enclosure, relocation, or the transformation of a traditional landscape whose economic value for the natives depended on its proper firing. Fire and fire control were in a sense a measure of the ebb and flow of these larger political and economic conflicts.

In conflicts between the industrial states, open fire became increasingly rare. In World War I, for example, artillery replaced fire arrows altogether, and mustard gas substituted for flammable liquid agents. A few flame throwers were deployed experimentally, but their chief value was psychological: broadcast fire was too uncontrollable (and in a landscape of trench and mud, not very plausible), and flame propellants lacked range. Smoke screens were generated mechanically rather than through open burning. German dirigibles dropped a few incendiaries on Britain, but their purpose was as much to terrorize as to damage. There seemed, moreover, to be a general aversion to open fire within Western civilization. Americans went further, giving little value to incendiary devices at all.[17] The images of the war are those of mud, not of fire.

World War II was a fire war. Incendiary weaponry became commonplace. Flame throwers, napalm, firebomb clusters, and the atomic bomb were among the successful innovations; incendiary leaves and the batbomb—in which delayed ignition devices were tied to the legs of drugged bats—among the most curious failures. Fire was used against ground troops, against vegetation, and against civilian populations in great industrial centers. Increasingly fundamental to all these uses, however, was aircraft. Napalm, for example, revived the ancient uses of Greek fire, but it was the airplane that overcame the chief limitation of flammable liquids, their lack of range, and the primary drawback, their uncontrollability. Aircraft allowed Allied fire to become an offensive weapon, delivered far from the site of a particular battlefield, while the use of fire balloons by the Japanese showed the vulnerability of the United States to an incendiary attack by air, even across the distant Pacific. The great civilian death tolls of war resulted from saturation firebombing at Dresden, Tokyo, Kassel, and Hamburg; at Dresden alone at least 135,000 perished in the resulting conflagration. An old word was revived to describe a new weapon: firestorm.[18]

The rediscovery of broadcast fire for military purposes brought new threats to civilian populations during the war, but it also promised new support for fire research and fire control by civilian agencies during the uneasy peace that followed. A liaison between civilian fire authorities and military strategists was established that would profoundly affect the character of civilian fire protection for years to come. The firebombings of Europe had followed elaborate experiments with full-scale models of German and Japanese cities constructed at Eglin Field, Florida, at the Edgewood Arsenal, Maryland, and at Dugway Proving Ground, Utah, under the architectural direction of émigré German Jews. These trials furnished an important precedent for later mass fire experiments conducted by the Forest Service for the Office of Civil Defense. Many of the principal investigators became leaders in the field of fire research after the war ended. The question of ignition and suppression, of firebombing and fire defense, were complementary. When the time came to reverse the program—to substitute aerial retardants for aerial incendiaries—the experiments occurred at the same locations, notably Eglin Field. On the problem of mass fire, urban and wildland fire agencies found common cause. All of these matters, however, were technical concerns. Unanswered by military proponents of fire weaponry was the question that, in an analogous form, had begun to preoccupy foresters who were rediscovering the potential of broadcast fire for wildland management. The reintroduction of free-burning fire as a weapon brought with it the need for new limits or prescriptions, both military and moral, for its use. But just as the new inventions of chlorine gas and mustard gas obscured the possibility of including incendiary devices among the weapons prohibited after World War I, so the atomic bomb obscured by its novelty the perhaps greater horror of conventional fire warfare.

Consciousness of fire weapons was reflected in accelerated civil defense measures against fire. Little was done during World War I to augment fire services. A few states sought to improve rural fire protection as a means of preserving crops needed in the war effort, but nothing special was done to protect forests. When forestry journalists compared the Cloquet fire with an enemy invasion, everyone knew this was mere poetic license. By World War II, however, the possibility of enemy fire attacks was real. OCD established special fire defense zones along the East Coast, alerted fire lookouts on the West Coast, organized the Forest Fire Fighters Service, and nervously sweated out the Japanese fire ballons. It was during the war, moreover, that the two most prominent national fire prevention campaigns were launched: Keep America Green and Smokey Bear, the latter evolving directly out of wartime propaganda.[19]

The fire balloon was a Japanese equivalent to the German V-2 bomb, an attempt to tie up resources that otherwise might be directed to the war effort, to demoralize the civilian population, and, of course, to encourage outright destruction. The analogy is enhanced by the apparent randomness of the targets. The Japanese balloon, in fact, may have inaugurated the "strategy of large-scale modern warfare, which leans very heavily upon the Intercontinental Ballistic Missile."[20] The Doolittle raid on Tokyo in April 1942 was the immediate stimulus for the balloon project: to save face the Japanese were required to make a similar attack on the American mainland, but the Battle of Midway ensured that Japan would have neither island airstrips nor carriers from which to launch such a retributive attack. The concept of saturation firebombing by balloon had been first proposed in 1933 by Lieutenant General Reikichi Tada of the Japanese Military Scientific Laboratory. By early 1943 balloons capable of being launched from submarines were in place for operational tests, but all submarines were then diverted to the fierce island warfare developing in the southwest Pacific. The balloon project soon acquired new advocates, however; Japanese meteorologists discovered the jet stream; and the possibility of intercontinental free-flight balloons became very real. For two years research on the Fu-Go weapon accelerated, promoted by the Imperial Army and Navy. The final design called for a paper balloon, 33 feet in diameter, with one antipersonnel bomb and two thermite incendiary bombs.

Official attacks began on November 3, 1944, the birthday of the Emperor Meiji. The assault thus coincided with the German V-2 rocket attacks on Britain. By the end of the war nearly 9,000 balloons had been released, of which about 1,000 reached North America, scattered from the Aleutians to Mexico, from California to Michigan. Some 285 incidents were reported in all. The balloons killed six American civilians in Oregon but did little damage otherwise, though their range effectively blanketed the major fire regions of the Cordillera. Japanese propaganda made much of a "something big" that would happen to the United States, but in a perhaps unprecedented case of largely self-imposed censorship by the American media, Japan learned nothing about the success or failure of its taxing experiment and stopped at a time when it might, in fact, have been on the verge of success.

The balloon threat was taken seriously by American military authorities. Puzzled by early sightings, the Air Force speculated that the balloons might contain chemical or bacteriological agents as well as bombs and incendiaries. The long-dreaded air assault by the Japanese was apparently underway, though with balloons rather than with aircraft carriers. Countermeasures were promptly initiated. Active searches for balloons began in December 1944 and intercepted balloons were shot down by fighters wherever possible. But sightings were poor, weather unfavorable, and the balloons too high (30,000 to 37,000 feet). To improve detection the Air Force conceived the Sunset Project, in which balloons were to be located by radar and fighter planes guided by VHF ground equipment. Technical difficulties limited the effectiveness of the scheme, however, and in fact it began (April 1945) at a time when the Japanese had abandoned their launches. The Lightning Project, meanwhile, prepared for the possibility that the balloons might carry bacteriological agents.

The chief threat posed by the balloons, the military concluded, was fire. The upshot was Project Firefly, with the Forest Service and the military joined in a cooperative effort. The Army and Air Force detailed Stinson L-5 and Douglas C-47 transport planes, an engineering brigade, and nearly 2,700 troops, including 300 paratroopers, for fire suppression. Military bases were ordered to assist the Forest Service in fire control. The example of Germany, where some half a million men were tied up in fire protection against Allied fire bombing, showed what the magnitude of involvement could be. The incendiary potential of the balloons was never realized, however, and only a portion of the troops were used on fires—including the 1945 Tillamook reburn. Officials gave the troops mixed reviews: some areas considered them indispensable; others, an uncertain blessing. Nonetheless, the balloons brought home to fire officers and civil defense strategists the threat of fire warfare in ways that even photos of the atomic devastation of Hiroshima and the incendiary gutting of Dresden could not by themselves have done.[21]

Firebombing against both military and civilian targets had been applied widely to Germany and Japan, and it was the impact of a terrifying new incendiary device, the atomic bomb, that precipitated the capitulation of Japan. Urban and wildland fire protection officials felt justified in their alarm, and the war experience prompted a full reassessment of fire control technique, equipment and strategy. Civilian security from wartime fire became a major responsibility for OCD, one that was prolonged beyond the wartime emergency as the Cold War seemingly prolonged the war itself. A major study of urban fire released by the National Fire Protection Association in 1946 grimly concluded that the next war would be a fire war. Its author, Horatio Bond, had earlier advised the RAF in its firebombing campaign gainst Germany and had inspected the consequent havoc with some care.

It was the OCD and the military who revitalized the national program of fire protection after World War II. The mechanization of fire control meant, for the most part, the conversion of military hardware. OCD funded studies that led to infrared mapping and resource locators and that investigated fire physics and fire weather. Fire research became fire behavior research. Previously fire had been the province of forestry, an economic and silvicultural problem; now it became a problem for physics. The first major study concluded by the new Division of Forest Fire Research in the Forest Service was a survey of the blast and fire effects of atomic weapons in the forests of western Europe. Mass fire behavior and control was the object of Operation Firestop. After the Cuban missile crisis, OCD appropriated money for detailed studies of mass fire behavior, like Project Flambeau (1962–1967), and of mass fire control, like the National Fire Coordination Study (1964–1965). The Forest Service received both contracts. Until World War II broadcast fire had been for the most part a decreasing phenomenon in both rural and urban environments. Only on vast wildland reserves were conflagrations still relatively common. Ironically, the Forest Service was practically the only agency with considerable expertise in holocaust fires. The Forest Service thus became responsible for both rural and wildland fire protection on a national scale. And not least of all, the forces that tended to prolong the Cold War also worked to preserve the vision of enemy fire. The problems that atomic energy faced by being identified with a bomb, broad-cast fire confronted in part because of its use as a military weapon during World War II.[22]

Since World War II the most extensive application of incendiary weapons has been in Indochina. The French experimented with napalm early in 1950. The bombs, which sent sheets of flame over wide areas, were used as an antipersonnel device, as a means of cover reduction, as a ready technique to destroy crops, and as a psychological weapon. The psychological effect, it turned out, was ephemeral. Of greater value, the French concluded, was the potential of napalm for broadcast burning of forests and crops; French Air Force Commander General Chassin strongly recommended this strategy for future guerrilla wars. The Japanese had employed fire for this purpose on Bataan, and Americans returned the practice in kind during the final strug-gle for the Philippines. The experience confirmed the value of fire for the removal of tropical vegetation. When the British tried it in Malaysia, how-ever, they found that the heavy jungle cover was too moist for active burn-ing. Herbicides were used instead.[23]

It remained for the American military to combine these strategies. United States armed forces practiced the most extensive deployment of incendiary weapons in its history. Not since the Indian Wars of the American West had fire for strategic and tactical purposes been so routinely applied. Thus fire returned to the battlefield at almost the exact time that broadcast fire was being championed for American forests and wildlands. The country that was so eagerly rediscovering broadcast fire for forestry also waxed most enthusiastically over its military applications. New ground combat weapons were devised using flame as a munition; napalm was abundant.

The most spectacular and controversial use of fire power was in support of the policy of "area denial." The concept is ancient and one with which fire has been long associated. It is, moreover, strictly analogous to fire pro-tection systems dependent on fuels modifications—firebreaks, type conver-sion, fuel reduction. Rather than attacking a fire directly, one modifies the environment in which it occurs. In Vietnam bulldozers and plows stripped cover; chemical defoliants reduced unwanted vegetation, either in strips or in blocks; and fire removed debris left after clearing or defoliating. The scale was unprecedented, but the technology was that which had been worked out for decades to maintain fuelbreaks in brush, to remove slash after logging or landclearing, and to convert timber cover such as juniper to grass. Fire was no stranger to Indochina; slash-and-burn agriculture had been practiced for millennia in the highlands, and the great savannahs of the plains were largely a case of type conversion perpetuated by broadcast fire.[24] But the specific technology of broadcast fire used by the American Army in Viet-nam came directly out of the American West, most prominently from the brushfields of Southern California.

In Vietnam at least three attempts were made to generate mass fires by which to burn off vast blocks of chemically treated vegetation. The war was largely an air war for the Americans, and cover removal was necessary for adequate visibility. Two studies were commissioned by the Advanced Research Projects Agency (ARPA): one, from the Natick Army Laboratories, surveyed "Fire in Tropical Forests and Grasslands" (1966); the other, from the Forest Service, looked at "Forest Fire as a Military Weapon" (1966–1970). The mass fire experiments in Vietnam coincided with the National Fire Coordination Study and the field and theoretical work of Project Flambeau. ARPA sent Forest Service experts to Vietnam to advise on the operations. Reports on the success of these missions vary: by some accounts, the ambient humidity and lack of ground litter in tropical forests prevented mass fire ignition, despite simultaneous saturation bombing with incendiaries over broad areas; according to others, the firestorms only generated rainstorms, which suppressed their fiery source; still others insist that as much as 100,000 acres burned. The official reports are classified, and Defense Department spokesmen declare that the forest did not burn.[25]

In 1968, however, broadcast fire did prove successful. Fires of unknown origin began in the U Minh Forest (Forest of Darkness), a Viet Cong stronghold since the first Indochina War. Whatever the source, the fires burned stubbornly; obviously, conditons were right for *in situ* burning without chemical treatment or slashing. Major conflagrations have after all occurred in tropical environments throughout history, particularly when accompanied by landclearing. In the 1960s, for example, thousands of acres burned in the Dominican Republic and millions in the Brazilian highlands. Once the extensiveness of the U Minh fires was finally recognized, aerial, naval, and land-based artillery rained incendiaries onto the sites to frustrate Viet Cong fire control efforts. The fires burned for six weeks and "destroyed 20 years' worth of Communist building and hoarding, setting off secondary explosions of ammunition or fuel at a rate of three an hour, denuding the enemy's protective cover completely."[26] The episode clearly demonstrated the strategic value of fire under the proper military and environmental circumstances.

It has been speculated that "greater quantities of incendiary munitions have been used in South Viet Nam than in any other country in any other war."[27] The vast proportion was used for tactical support of ground troops, not as broadcast fire in either urban or wildland contexts. The protest against incendiary weapons, however, arose as part of a general revulsion against chemical and environmental warfare. Burns are an especially hideous form of personal injury, and the wholesale use of defoliants to prepare sites for burning came at a time of general public outcry over the application of pesticides and herbicides. The war was unpopular, and fire and the war became indelibly linked; nearly all the literature based on the war, for example, relies on fire imagery. In 1972 a United Nations resolution condemned the use of incendiary weapons.[28] The emphasis was on antipersonnel devices such as napalm, but especially disturbing to the authors of the report was the potential for civilian and environmental destruction that little related to military objectives, that is the use of fire as a weapon of terror. When the firestorm experiments in Vietnam were finally made public that same year, the U.S. Senate promptly voted to curtail all military expenditures destined for the creation of broadcast fire as a weapon.[29] This came about a year after the Forest Service—amidst great support from environmental groups—amended the 10 A.M. Policy to accommodate greater use of prescribed fire. Thus a curious spectacle resulted: on one hand broadcast fire was condemned as an especially insidious instrument of warfare and environmental destruction; on the other hand it was being vigorously promoted as a means of restoring ecological health to forests, brushfields, and grasslands.

That paradox followed from another, established 30 years earlier. In the same year that Hamburg suffered the first great firestorm from Allied bombing, Lyle Watts approved broadcast fire for the management of the southern pine, and Harold Weaver published experiments on the use of broadcast fire for the western pines. The rediscovery of fire was a broad one, and it demanded a wide-ranging search for prescriptions on its uses. The Cold War seemingly entangled fire control in a geopolitical strategy of containment through the erection of fire-free zones, just as wild and urban lands apparently polarized and mutually threatened each other. The concepts (and even the language) underwriting the strategy of modern fuels management are reflected in the larger military strategy that it is better to tolerate or even encourage "brushfire" wars in remote lands than to risk a major "holocaust."

Common to the new attitudes, as to the old, was an aversion to the indiscriminate nature of broadcast fire. As a weapon, fire per se was less damaging than other modern inventions. Few natural environments have not to some extent and with anthropogenic assistance adapted to it. What makes most modern fire weaponry especially abhorrent are the auxiliary effects—the toxic chemicals and radiation—just as it is smoke and pretreatments that often limit the usefulness of prescribed burning. The trend to replace open burning with more confined fire, broken first by the development of aircraft and then sustained by the introduction of thermonuclear arms, may be reestablished with more experience, with more precise control over targeting, and with the development of weapons, like the neutron bomb, that do not produce such an array of side effects. To outlaw fire in warfare is a bit quixotic: civilian foresters have abandoned the similar proposition to exclude fire from wildlands. Fire is power, and the question is not so much whether it will be applied, but how, to what purposes, and with what side effects—in short, within what prescriptions. To paraphrase an ancient adage, fire makes a good soldier but a bad warlord.

II

A passage in the *Iliad* compares the assault of Achilles with a forest fire:

> On went Achilles: as a devouring conflagration rages through the valleys of a parched mountain height, and the thick forest blazes, while the wind rolls the flames to all sides in riotous confusion, so he stormed over the field like a fury, driving all before him, and killing until the earth was a river of blood.[30]

This may be the first time in literature that fire and fighting were linked metaphorically. Yet the analogy remains a common one in the twentieth century. The image of firefighting as a battle springs readily from the pens of journalists, and the story of a fire is usually told in the style of that literary set piece, the battle scene. War historians from Homer to Liddell Hart, in turn, commonly describe battles with fire imagery. But in addition to its superficial appeal, the analogy has a certain historical validity and perhaps a deeper conceptual significance beyond the fact that fire is often a weapon of offensive war and that fire control is often an arm of paramilitary defense.

The strategy and tactics of fire control bear a certain resemblance to battlefield maneuvers. The fire boss on the line (or "fire general," as he often pretentiously styles himself) deploys men and equipment under emergency conditions. The problems of logistical support are so similar between battlefield and fireline that national guard units are frequently activated to organize fire camps and furnish transportation. A general staff gathers intelligence, predicts anticipated fire behavior, calculates control force requirements, and issues plans of attack. The battle might be staged with ad hoc skirmishes along a fireline or with counterattacks (backfires) from existing fortifications (for example, fuelbreaks). The day-by-day challenge of quelling a fire regime, whether set by lightning or by incendiarists, is not unlike a chronic guerrilla war fought in rural or wildland surroundings amid a backdrop of low-intensity raids broken by occasional flare-ups into pitched battles.

Of course, what is lacking in the analogy of fire control to war is the human opponent. This simple fact makes the gulf between the two both fundamental and unbridgeable. Forest fires are environmental events; wars are tragedies. Fire may occur as an entirely natural phenomenon, but wars are distinctly and exclusively human. In fire control there are no prisoners, no feints or decoys, no agents provocateurs or codes, no truces or negotiated peaces, no diplomacy. There is no declaration of hostilities and no surrender. War is a cultural phenomenon, though acted out on a natural landscape; fire is a natural phenomenon, though one that for millennia has been assimilated in fundamental ways by human societies.

Yet there may be a reality to the metaphor that saves it from mere literary convention, journalistic hyperbole, and bureaucratic pretense. In a

Jamesian sense, firefighting may be considered a moral equivalent of war. Both seem to share a common reservoir of moral energy. It is no accident that James's essay appeared in the same era that supported a heavy commitment to fire control or that the morale of fire control has waxed and waned with enthusiasm for, and a sense of purpose by, the military—a fact that accounts for much that is commendable and puzzling about the evolution of fire protection in America, especially in the hands of the federal government. Fatalities ensure that firefighting is not merely an exciting or expensive sport. Firefighting seems to offer the romance of battle without its moral ambivalence.

Like the American military in the great wars of the century, fire control rarely had either its methods or its goals questioned by the intelligentsia. Once past the era of frontier fire, it waged total war against an enemy from whom unconditional surrender was demanded. Yet at about the time that the American military entered the quagmire of Vietnam, fire control encountered the new environmentalism. Suddenly it was asked to wage a limited war—limited in geographic scope, limited in armaments, limited in objectives. It had to tolerate privileged sanctuaries for wildfire (wilderness areas), to restrict its use of major weapons (especially bulldozers and aerial retardant), to tolerate review by a skeptical public, and to conduct its operations in strict conformity to larger "political" considerations. Fire control had to subordinate its objectives to resource management goals as the military had to make its plans subservient to the national electorate. It had to abandon its hope of total fire containment; fire was deemed desirable in many cases, and a totally protected (or "friendly") landscape was often regarded as artificial and ecologically oppressive. It was sometimes difficult to discriminate between fire as friend and fire as foe. Environmental critics attacked fire protection as war protesters did the conduct of the military in Vietnam. The two wars were declared both immoral and hopeless; the enemy, both natural and inevitable; attempts at suppression, more damaging to the environment or to society than partial capitulation; success, either impossible or achievable only at the price of destroying the object in question. No forester claimed that it was necessary to destroy the forest in order to save it, but the actions of more than one fire boss seemed to come close. And still the interminable war against fire went on, ever more costly, more embarrassing, and—with the intelligentsia looking skeptically over the foresters' shoulders—ever more dubious. In 1943 Godwin could unblushingly declare war against the "red menace"; by 1970, it was felt that the woods were perhaps better red and burning than dead and unrejuvenated.[31]

That the rhetoric of fire and war is often similar is superficial; that the techniques and attitudes toward fire in peacetime and in war are shared signifies only that they belong to a common culture; but that the two seem to tap a similar moral energy is fundamental to the peculiar evolution of fire protection in the United States. Why fire control took the direction it did, why it acquired such perseverance and intensity is the product of a series of historical accidents—like the compounding of meteorological and biological coincidences that lead to a major conflagration—and one of these events has apparently been the equation of the fireline with the battlefield. For better or for worse, it has given fire protection a vitality and dedication that is otherwise enigmatic. And it has given modern man a source of metaphor distinct from his scientific theories to replace the philosophy and myth that men in more ancient times used to describe their relationship to fire. As a physical phenomenon, fire has long been a common weapon of war, and fire control, a necessity of defense. But it is in a richer and more specifically human sense that firefighting has come to mean a moral equivalent of war itself. Fire is not merely a part of mankind's technology but, to borrow another Jamesian concept, an expression of its will to believe.

A BURNED-OUT CASE: A FIRE HISTORY
OF SOUTHERN CALIFORNIA

The forest fire problem in southern California is without parallel
in the United States.
—Charles Connaughton, U.S. Forest Service, 1957[32]

I got out of the car and stamped on the cigarette. "You don't do
that in the California hills," I told her. "Not even out of season."
—Philip Marlow, in Raymond Chandler, *Playback*[33]

I

The fire began when sixteen-year-old Gilbert Paipa of the Inaja Indian
Reservation "just got a mad, crazy idea to do it. I threw a match in the
grass to see if it would burn."[34] That was on the morning of November 24,
1956. The fire was attacked within 25 minutes of report, but it burned with
Santa Ana winds and four years of drought behind it. Before it was controlled on the twenty-eighth, the fire consumed nearly 44,000 acres and took
the lives of 11 firefighters caught by a flare-up that resulted in a flash-over
of volatilized gases. Within a month the Inaja fire was followed by a fire
complex in the Santa Monica Mountains that drew even more national
attention. The three Malibu fires began from incendiarism and ended by
burning out onto the beaches of the Pacific Ocean. The fires swept over some
38,000 acres, but where the Inaja fire claimed only 5 structures, the Malibu
fires took 120. Cooperative agreements turned the suppression action into
a multiagency affair involving state, county, municipal, and federal
resources.[35]

The story of the two fires did not end when they were extinguished. The
Inaja fire was followed by a Forest Service board of review, by reforms in
national fire training programs and fire research, by lawsuits, by a review
of air tanker possibilities (the fire was among the first on which air tankers
were used), and by congressional hearings. The Malibu complex was similarly analyzed. The Forest Service studied its fire behavior, noting that the
"Malibu fires combined most known elements of violent, erratic, and
extreme fire behavior ... fire whirls, extreme rates of spread, sudden
changes in speed and direction of fire spread, flash-overs of unburned gases
complicated by intense head and impenetrable smoke held close to the
ground."[36] The Red Cross critiqued its own role in the disaster, concluding
that "to our knowledge, every formal request made of the Red Cross which

fell within the scope of our responsibilities was honored."[37] The Los Angeles County Civil Defense Authority reported on its substantial support operations during a conflagration that it termed "the first major fire disaster of national scope," and President Eisenhower formally concurred. The Inaja and Malibu fires became the subject of the Engle Committee hearings on behalf of the U.S. House of Representatives.[38] The National Board of Fire Underwriters also made a special, detailed investigation, which applauded the élan and expertise of the fire services but concluded glumly that "with the encroachment into these areas [mountain brushfields] of more homes without a corresponding increase in the amount of quickly available fire protection, even greater losses are probable."[39]

Considered simply as fires, the Inaja and Malibu burns were not particularly unusual. More fatalities had resulted from the Hauser Creek fire (1943) and the Rattlesnake fire (1953); more acres had burned in the Big Dalton fire (1953) and the Refugio fire (1955). The Maine fires of 1947 had been larger in size and damages by an order of magnitude, and brush fires that threatened developments had been commonplace on the California scene since Spanish mission days. But the timing of the two fires helped to give them national significance. Doubling its population every decade, Southern California had become a national force in American life. In part its fire problem was something the region simply carried into the national arena with it, another unique curiosity from the land of sunshine and Hollywood. In part, however, it became the emblem of a new and growing fire hazard that confronted fire services nationwide. By the mid-1950s the center of gravity of the Forest Service fire commitment was shifting to Southern California, and so was the attention of urban fire services. Coming within a month of each other, the Inaja and Malibu fires effectively symbolized the transition from backcountry to mass fire, from the problems of fire in the hinterlands to those along the newly fashioned boundary of wildland and suburb. In subsequent years the Los Angeles Fire Department became "intensely sensitive to the mountain danger" and the city adopted its first "major emergency" plan in 1958. "Thirty years ago," an official report noted, "firemen battled watershed fires on these lands under many of the trying conditions that are faced today. . . . The one significant change is in the amount of structural development that has taken place during the past three decades."[40]

A vast demographic shift had occurred in the decades following World War II, and Southern California was one of the prime benefactors. As with other settlements in the national history, this one would be accompanied by fire. The emerging fuel arrangements were a nightmare. The modern suburb can be relatively conflagration-proof. Buildings are dispersed, fuels are light; pavement covers streets, driveways, and sidewalks; mowed lawns and deciduous trees further diminish flammability. The type conversion of brush to

houses should have stopped conflagrations; instead, it all too often stoked
them. Particularly in the more exclusive developments, the conversion in
Southern California was often incomplete. The brush was considered part
of the natural aesthetics of the scene, a barrier between wealthy and merely
affluent suburbs and a stabilizing agent for watersheds. The very exclusiveness of the districts precluded further developments as undesirable. Many
residents allowed—even encouraged—brush to engulf their structures. At
the same time, construction and design practices—notably the reintroduction of wood shingle roofs—made the houses far more receptive to the firebrands than they should have been and contributed an unusually high proportion of firebrands in return. The greatest cause for urban conflagrations
has been wooden roofs. "The *largest* nonwood shingle conflagration is
smaller than the average wood shingle conflagration," one critic observed.
Another put the matter even more simply by stating that "the largest wood-
shingle fire has razed more structures than all the non-wood shingle conflagrations combined."[41]

The situation so alarmed local fire agencies that in 1958 they requested
an investigation by the National Fire Protection Association, which issued
a "conflagration warning." A year later another survey labeled the fuel complex in the Santa Monica Mountains a "design for disaster." The expansion
of recreational areas as well as residential suburbs multiplied the potential
for tragedy. In 1961 the explosive Basin fire along the chaparral foothills of
the Sierra National Forest showed the potential fire threat for the one, while
the famous Bel Air–Brentwood conflagration in Los Angeles showed the
reality of fire threat for the other. They were followed by a numbing succession of large, costly, and damaging fires on an almost annual basis. In 1970
and 1977 fire complexes reached such proportions that they threatened to
paralyze national fire resources. The 1970 complex was followed, not acci-
dently, by reform of the 10 A.M. Policy, and the 1977 complex—ignited by
freak lightning barrages—helped to catalyze the adoption of a new fire policy by the Forest Service a year later. In 1979, against fires burning in the
Hollywood Hills and Santa Monica Mountains, the Forest Service spent
$1 million a day to maintain an army of 7,000 firefighters and mountains of
material. Total suppression costs reached an astronomical $28 million. Dur-
ing drought years, fuel moisture plunged, and the chaparral burned. During
wet years, the rains of winter produced lush grasses that burned during the
long summer and fall. The dry years sent bulldozers into the heart of the
Ventana Wilderness, and the wet years saw an air tanker (albeit accidentally) drop retardant on downtown Palm Springs. Fire protection confronted
not a single, more or less homogeneous fuels environment, like that in the
Lake States or in the Northern Rockies, but a bizarre ensemble of wild and
urban landscapes, of mixed and overlapping jurisdictions, and of contradictory and uncompromising purposes.

There was much to make the Southern California fire scene unique on the national arena. Its chaparral fuels, its famous Mediterranean climate, and the presence of Santa Ana winds from September to December made the fire season a year-long possibility. In northern forests the natural reburn cycle allowed a steady reduction of heavy fuels left after a conflagration. In Southern California, as in the South, the reburn cycle referred to the time it took the chaparral to reestablish itself. That is, in the North the natural cycle of reburns was defined by fuel attrition and in the South by fuel accumulation. But if Southern California's fuel cycles resembled the South's, its types and range of ignitions differentiated the region. In the South, the fire regime was perpetuated by the persistence of a source of ignition; in Southern California, by the persistence of a fuel complex—the chaparral—one that could not be easily managed by prescribed fire. In the early years of systematic fire protection, it was northern California that had commanded national attention, and new means of fire control had been devised primarily to keep brush from encroaching on timber lands. In the postwar era Southern California came into prominence, and new programs were devised to cope with the encroachment of suburbs onto brushlands. In the Lake States after the Civil War there had been a deadly mixture of farmers and timber, catalyzed by the railroad. In Southern California after World War II there developed a similarly lethal mixture of homeowners and brush, one catalyzed by the automobile.

Prior to the emergence of this regime the history of fire control on the region's national forests most closely resembled that of the Northern Rockies. Large, episodic, and perhaps untamable conflagrations in a remote mountain backcountry plagued both regions. For each, the salient concern of fire protection was accessibility, and both regions were effectively opened up and developed in the name of fire control. Both mechanized wherever possible—the Northern Rockies relied more and more exclusively on air attack; Southern California, on a mixture of air and ground equipment. When equipment development centers appeared, one went to each of these regions. Of the three fire labs established by the Forest Service, two similarly went to these regions. When interregional crews were promoted, the first were stationed in the Northern Rockies and Southern California. The smokejumper program for the Rockies and Northwest found a counterpart in the helitack program of Southern California. Nowhere else was firepower so concentrated. And nowhere else, perhaps, was there a more pronounced need to determine the economic limits that could be placed on fire protection.

Above all, California's political history made its program of fire protection unique. Unlike the Northern Rockies, where the Forest Service often exercised sole responsibility over vast areas, Southern California was a great jumble of jurisdictions. In the Northwest, cooperative fire protection evolved among private owners of forested lands; in Southern California fire control required elaborate cooperation among all levels of government and between forest fire agencies and urban fire services. For this the political history of the forest reserves is largely responsible. By the 1880s, as urban development and irrigation agriculture began to emerge, promoters recognized that the fate of the region as an agricultural domain or as a metropolis depended on the reliability and abundance of its water supply. From the beginning, the establishment of watershed reserves in the mountains went hand in glove with fire control. The first attempt to preserve the watershed came in 1885, the same year that New York set aside lands in the Adirondacks for much the same reasons. Furthermore, the argument for federal reserves was first and foremost to protect watersheds, and the citizenry of Southern California seized on this principle to argue for the creation of reserves in the mountain ranges that framed the basins. The Southern California reserves were among the first established. Elsewhere the reserves typically came at the price of considerable bitterness among local residents; those in Southern California were met with exultation. What was true of the national scene was magnified in Southern California: the reserves had been created for watershed first .and forest second, but the presence of the reserves brought a demand for forester administrators. Those origins have been obscured in other areas by subsequent events, but they have been unblinkingly preserved in Southern California, and it is only with time that the regional reserves have appeared so anomalous a feature in the national system.[42]

Citizen pressure since then has been unending. It was thought that the creation of reserves would prevent fires. Instead, they worked to preserve one of the most conflagration-prone environments in the world, though changing fire practices have decreased the amounts burned annually. But the explosion of residential developments in the 1950s introduced another paradox into the scene. Unlike the northern forests, where settlement created anomalous but temporary conditions of fuel and fire that would end with the conversion to either farm or forest, and unlike the grasslands, where settlement resulted in conversion to a less fire-prone landscape, the residential settlements of Southern California were held in dynamic suspension, a mercurial mixture of suburb and brush. Neither brush nor houses could seemingly drive the other out. The necessities of watershed protection meant that suburbs had to stop at the boundaries of the reserves, flanked by fire and flood. The logic of exclusiveness meant that in many neighborhoods the process of conversion would be deliberately arrested, held in unstable equilibrium. The situation resembled that created during the great era of holocausts in the Lake States, and it persevered in the region's political and economic environment. Boosterism had always been a California specialty, and the state's abundant natural disasters—fire, flood, earthquake—have never been allowed to deter the flow of settlement. Just as the Lake States promoted relief programs aimed at keeping the farmer on his land, Southern California sought still more intensive fire control programs to keep the suburbs where they were. A holocaust was less a warning than an invitation to further development: new houses rose on the ashes of chaparral as farms had on the ashes of the north woods.

In the north woods the creation of national and state forests worked to disengage farms from forests. Ironically, the existence of watershed reserves in Southern California often served to encourage population growth, which then initiated new cycles of wildfire. Originally, the city and the forest were bound indirectly through the medium of water, not directly through the medium of fuel. In the older cycle the effects of fire and flood were removed from direct contact with the metropolis. The new cycle of wildland fire, however, directly threatened the metropolitan areas with destruction by fire, with air pollution by smoke, with flooding from mudslides following intense burns, and with water shortages due to watershed deterioration. Fire officials found themselves in a hopeless dilemma: they could not modify the environment without damaging the watershed or threatening housing developments; yet to leave the watershed unmodified was to preserve an inevitable cycle of holocaust fires.

Fuels built up as a result of settlement—not through logging or landclearing, but through fire protection. At the same time, restraints were placed on fuel management programs: too much reduction of brush meant a loss of valuable watershed, too little meant an increased hazard from fire. A variety of fuel modification schemes have been attempted, including the search for a plant that could provide the watershed protection afforded by the chaparral but without its fire potential. The treatment of choice, however, has consistently been fuelbreaks, and with the advent of the new fire regime a Fuelbreak Program (1958–1972) sought to break up expanses of chaparral into more manageable units, to provide conflagration barriers within the reserves, and to construct a sort of fire-free zone between the reserves and the developments that crowded to their boundaries. The Forest Service also established the San Dimas Experimental Forest to explore different techniques and to determine in particular the nature of fire and watershed deterioration, much as the Service had created the Shasta Experimental Fire Forest in an earlier era. In 1972 Congress appropriated funds for more fuel management work. The resettlement and rezoning that finally disentangled the north woods have apparently proved unacceptable in Southern California. The presence of the reserves has in a sense sustained the settlement that followed, and it remains to be seen whether the new Santa Monica National Recreation Area can work to segregate fire from hazards or to ensure a permanent exchange between them.[43]

The greatest controversies in Southern California fire control have focused on the possibilities for prescribed fire. During the Engle Committee hearings the state forester explained that "when it is safe to burn the brush, it will not burn; when it is burnable, it cannot always be controlled."[44] Fire was often used in connection with fuelbreak construction and maintenance, but only after extensive pretreatment by mechanical crushing or chemical defoliation; by the 1970s such methods had fallen into disfavor. Broadcast burning has always been difficult in mountain environments, but the Southern California situation added immeasurably to its complexity. The mobile soil left little margin for error. Rigid air pollution standards restricted open burning. The absence of an understory of litter in the brushfields made underburning difficult. Though stockmen insisted that fire did little damage and that fire protection only led to an inevitable holocaust, the geochemistry of the environment was unique—and suburbs had intervened in the traditional quarrel between foresters and herders over fire practices.[45] Fuelbreaks and greenbelts of various sorts are often the only acceptable compromise. Southern California adopted a policy of fire exclusion long before other regions did, and it continues to adhere to it long past the time that others have abandoned the goal. From the beginning, foresters were not so much the architects of the fire program as the instruments chosen to effect it.

It was inevitable in this context that fire would become a political issue. Nearly all major fires from the turn of the century to the present have been reviewed by legislative committees. The Southern California Watershed Fire Council and various citizen advisory boards work with the four national forests. In a move perhaps unprecedented in American fire history, local groups agitated actively for better fire control in the mountain wilds around them and even raised money for that purpose. Their concern with fire control as a means of watershed protection was in good measure responsible for the establishment of the original Southern California reserves. As early as 1905, citizen groups pressured the state legislature into approving special appropriations for fire trails and patrols. Of the five counties in California that maintain their own fire services, all are in brushlands and four are in the south. Fire control became a fundamental charge on all levels of government.[46]

The political arena was in fact one of the mechanisms by which the Southern California cycle became nationalized. The Fire Task Force set up by Chief Forester McArdle after the Inaja tragedy reviewed the entire state of fire control and recommended programs that reformed Forest Service fire administration. The House of Representatives Committee on Interior and Insular Affairs, chaired by Claire Engle, held wide and controversial hearings in 1957 on the subject of fire control in Southern California. Chairman Engle, a Californian who as a state senator had assisted the passage of the 1945 Brush Burning Law, growled ominously that "if the Forest Service had been half as good at preventing fires as they have been fighting fires, they would not have had to fight so many fires." Yet challenges gradually dissipated during the course of the hearings. The spirit of the eight major findings that emerged from the committee is perhaps best summed up in the recommendation that "the Forest Service intensify its present program for complete fire prevention and protection in southern California." The two reports, the Forest Service Fire Task Force and the Engle Committee hearings, thus complemented each other.[47]

Already in the 1950s Southern California was becoming a hotbed of Forest Service equipment development and fire research and a proving ground for new control devices, from crews to air tankers and helitack, from concepts in fuel management to insights into mass fire behavior. Practically all of the conflagrations of this era were followed on the 1956 example by political hearings at federal, state, county, and municipal levels. The California situation came to be recognized as symptomatic of a global syndrome common to lands that exist under a Mediterranean climate. During the big bust of 1977, California played host to an international symposium on exactly this subject.[48]

The California fire cycle existed as well in a political context that extended beyond local or even national bounds. The appearance of these seemingly new fires coincided with the emergence of the Cold War and a deadly strategic arms race in thermonuclear weaponry. The Malibu and Inaja fires coincided with the development of ICBMs; the Bel Air conflagration was preceded by the Berlin crisis and followed by the Cuban missile crisis. A post-mortem of the Bel Air debacle by the National Fire Protection Association concluded that "considerations of large scale fire fighting emanating from this conflagration are worthy of thought and careful review by every fire department." Though similar fuel assemblages were not common, "this fire does suggest some of the fire fighting problems that will confront fire departments in time of war. It is time to start 'large scale' thinking about communications, emergency water supplies, fire equipment mobilization, and movement."[49] Operation Firestop, a 1954 cooperative effort involving the Forest Service, state and county fire services, the military, and civil defense had been staged in Southern California. The project offered a daring new concept in fire research, one destined to liberate fire from the intellectual confines of academic forestry, and it proposed a dramatic new context: a film describing the experiments opened with the picture of a tremendous smoke cloud rising out of the California mountains, while a narrator assured the audience that this was not an atomic explosion but a forest fire. McArdle echoed somewhat similar sentiments when he prefaced the official report on the Inaja tragedy with a panegyric to its victims: "Surely these men gave their lives in defense of their country, for without the strength of our forests, water, and other natural resources, this Nation would not have been a leader in the free world today."[50] When in the early 1960s the Office of Civil Defense advanced money for research into mass fires, both urban and wildland, and into fire control systems capable of coping with mass fire, the Forest Service received contracts on both topics. More often than not, collective interests intersected in the peculiar landscape being fashioned in Southern California.

Research was necessary, too, to cope with the causes of fire in the new regime. To match the confusion of fuels and jurisdictions, there developed a medley of ignition sources. Lightning accounts for 5 to 10 percent of fires in the national forests, though most of these occur along the forested crests, and thunderstorms are rare under conditions that favor foehn winds. An unusual storm in 1967 started 13 fires in chaparral on the Angeles National Forest, but none became large. The big fires of 1977, however, were largely the product of lightning, though the biggest came in the Coast Ranges and the Klamath Mountains.[51] The cavalcade of people and life styles throughout the region has resulted in a similar profusion of exotic ignitions. New arrivals literally did not know how to behave in the Southern California fire environment. The almost instantaneous injection of homes into wildlands, for example, suddenly placed children amidst a tinderbox, and for a while in the recent cycle children (overwhelmingly boys) started nearly a third of all fires. Another third came from the intensive use of equipment—the bulldozers that prepared sites, the powerlines that brought them electricity, the automobiles that transported new residents. But a disturbing number— many of those that occurred under the most extreme conditions—were incendiary. The major fires of 1956 nicely illustrate the variety of ignition sources: the McKinley fire (10,000 acres) resulted from a military plane crash in Santa Ana winds; the Highland fire (15,000 acres) began when powerlines swaying rhythmically in Santa Ana torrents came into contact with each other and arced to the ground; the Malibu fires were of unknown cause, though at least some were incendiary; and the Inaja fire began with a true pyromaniac. In 1955 the Forest Service created a special prevention program at its Pacific Southwest Station, the only such project outside the South.

Perhaps not the least of the elements that contributed to this fire cycle was that Southern California had become a national media center. Fire as entertainment spectacle—as thrill, as celebrity—was an old California tradition, and like other aspects of the scene, this one, too, was nationalized. Even in the 1920s state troopers and local deputies had to be mobilized to keep spectators from interfering with actual suppression efforts on the line. An account of the San Marcos fire of August 1940 recorded that the blaze "provided a thrilling spectacle for the thousands of Santa Barbarans and outside visitors celebrating the annual Spanish Days festival."[52] The scene that pleasure seekers observed in the hills was perhaps a truer depiction of Spanish days than the one they saw in the streets. The region that had nurtured Hollywood and Disneyland now seemingly brought, for everyone's viewing enjoyment, an extravaganza of holocausts.

In the ancient world, fire and sunshine were often joined. In Southern California, wildfire and sunshine came to be linked in the national mind as part of a tumultuous tapestry, both burned and bleached, that apparently symbolized the region. Nathaniel West climaxed his 1939 novel *The Day of the Locust* by imagining a future Hollywood consumed by flames spreading across the city while the psychologically burned-out populace rioted madly in the streets, as if the Southern California experience even before the post-war boom were one vast *danse macabre* leading to self-immolation. By 1979, whatever the psychological validity of that literary vision, the prospect of a conflagration through the Hollywood Hills was no longer fiction. It had become history.

II

Though the fire history of Southern California took on special national significance in the 1950s, the presence of fires can be traced as far back as written and physical records exist. Varves in the Santa Barbara channel record charcoal from wildfires as old as 2 million years.[53] Its fire environment is no less ancient. Like Mediterranean climates the world over, California is subject to periodic heat waves from subsiding high pressure cells.[54] The Santa Ana winds in Southern California and the Mono winds in central California belong to a family of winds in North America that spill out of the Great Basin during late fall and early winter, pouring over the Rockies to the east (the chinook) and over the Cascades (the east wind) and the Sierra (the north, Mono, and Santa Ana winds) to the west. All are foehn winds, part of the family of dry, high velocity, usually warm winds that originate in continental interiors—the sirocco of North Africa, the mistral of the Mediterranean, the bora in the Balkans. All compete with moderating marine air, temporarily driving out its moisture and quickly generating conditions that favor fire. When such winds appear, as they do in California, after a long, dry Mediterranean summer, fire conditions can become extreme.[55]

For this reason, fire and climate have come to favor in California, as elsewhere in Mediterranean environments, a peculiar, pyrophytic fuel complex. The brush that is endemic to the mountains of Southern California, that clothes the Coast Ranges well north of San Fransisco Bay, that rises out of the Central Valley into the Sierra, and that invades the forested pinelands of the north is given the generic name *chaparral*. The early Spanish took the Basque expression *chabarra,* which refers to the scrub oak common in the Pyrenees, corrupted it to *chaparro,* and applied it to the brushfields of California, one component of which is scrub oak. Americans in turn corrupted the word to *chaparral.* Similar vegetative communities are known wherever there exist similar Mediterranean climates—in France (garrigue, maquis), Spain (tomarillares), the Balkans (phrygana), Chile (matarral), South Africa (macchin fynbosch), and South Australia (mallee-scrub,

29. Fire burning across a chaparral-covered mountain in Southern California, 1948.

mulga scrub, brigalow). Actually a composite of various plants, including chamise, oak, and manzanita among others, chaparral is often found in three distinct brushlands: in those that exist with coniferous forests, in those that form woodland-grass ecotones, and in those pervasive brushfields that give rise to uncontrollable conflagrations and by themselves result in no long-term deterioration of the community. The first type is found in northern California and in middle slopes of the Sierra; the second, in the lower slopes of the Sierra and in much of the northern and central Coast Ranges; the last, in the mountains of Southern California.[56]

The three types respond differently to fire. In the case of the first type, frequent burning can often replace forest and brush with grass; infrequent burning can expand the range of brush at the expense of forest; and fire exclusion can result in the establishment of reasonably well-stocked forests. For the second type, the elimination of fire either by direct suppression or by overgrazing tends to allow the woodland and brush to expand at the expense of grasslands. For the third type, the brush endures—with fire it drives off competitors, and without fire nothing succeeds it until fuel loads eventually stoke a conflagration. It is important to remember that light burning and systematic fire protection fought over the first type of brush, the one arguing for more fire and the other for less. If the fire cycle could be broken, foresters realized, the forest could more or less permanently reclaim the brushlands.[57] Foresters, too, had the hard lessons of historical geography before them—the permanently deteriorated lands of Greece and Palestine, for example, where the fire-brush cycle had exhausted the frail soils of the region. Stockmen, by contrast, wanted prescribed fire for the second brush type, where oak savannahs might result from proper firing.[58] But for the third type, where fire practices came under intense political scrutiny, there was considerable confusion over proper means and ends. Herders and hunters hoped to maintain pasturage, browse, and habitat through periodic fire. Foresters hoped to break the fire cycle, thereby allowing reclamation of the land by trees, or at least its preservation from ruinous soil losses. To a professional class inculcated from Pinchot's day onward to the horrors of soil destruction, the latter remained a compelling reason even in the new regime. "A sobering truth that should not be taken lightly," wrote a forest supervisor, "is that the structures lost can, if necessary, be quickly replaced while the soil, once gone, cannot be replaced in a lifetime."[59]

The citizens of the metropolises wanted the reserves for watershed: they wanted chaparral without fire. The chaparral could not be removed or replaced. Moreover, fire and flood were recognized as tandem catastrophes. Soil erosion accelerates on the steep slopes and poorly consolidated soils after a fire, silting in water storage reservoirs and flooding the fields and suburbs below with debris. Large debris floods followed the Matilija fire (1932) and the Arrowhead Springs fire (1953).[60] The problem became even more complex as suburbs crowded into the brushfields themselves, creating a suburb-chaparral ecotone to match the woodland-grass ecotone and forest-brush ecotone further to the north. For the forest-brush ecotone, fire suppression is the preferred treatment; for the woodland-grass ecotone, prescribed fire and grazing control. But for the suburb-chaparral ecotone so recently assembled, there is as yet no generally accepted treatment. Whereas it was often remarked that chaparral, particularly that composed largely of chamise, is a fire-climax community, it is now joked that the same is true of the Southern California mountain suburb.

How this modern ecotone was created is a study in cultural history. The use of fire by California Indians is documented and remarkably uniform, considering the medley of peoples that even then crowded into the region. The Spanish explorer Cabrillo found so many fires around San Diego that he named it the Bay of Smokes.[61] Santa Barbara may owe its name less to the accident that its discovery coincided with the saint's feast day and more to the fact that Barbara is the patron saint of thunder, lightning, and fire. Intentionally or not, the association of name and place has proved devastatingly apt. In October 1774 Captain Fernardo Riveray Moncada, stationed at Monterey, described

> a fire from the west that was burning the forage of the countryside and as it neared the Presidio, the soldiers, servants and even I went out to fight it, not because of the danger to the houses but to preserve the grass for our animals. We managed to extinguish it. The heathens are wont to cause these fires because they have the bad habit, once having harvested their seeds, and not having any other animals to look after except their stomachs, they set fire to the brush so that new weeds may grow to produce more seeds, also to catch the rabbits that get confused and overcome by the smoke.[62]

In 1792 the naturalist José Longinos Martinez noted the persistence of the practice: "in all of New California from Fronteras northward the gentiles have the custom of burning the brush, for two purposes: one, for hunting rabbits and hares (because they burn the brush for hunting); second, so that with the first light rain or dew the shoots will come up which they call *pelillo* (little hair) and upon which they feed like cattle when the weather does not permit them to seek other food."[63] A year later Vancouver witnessed a spectacular fire south of Los Angeles in December. "These clouds of smoke containing ashes and dust soon enveloped the whole coast. The easterly wind prevailing brought with it from the shore vast volumes of this noxious matter. . . . Under these circumstances it cannot be a matter of surprise that the country should present a desolate and melancholy appearance."[64]

Apparently this was not the first fire of that season. On May 31, 1793, Governor Arrillaga had written from Santa Barbara to Father Lasuen, president of the missions, a proclamation apparently born of exasperation and zeal both. "With attention to widespread damage which results to the public from the burning of the fields, customary up to now among both Christian and Gentile Indians in this country, whose childishness has been unduly tolerated, and as a consequence of various complaints that I have had to such abuse," the governor wrote, he found it necessary "to prohibit for the future . . . all kinds of burning, not only in the vicinity of the towns but even at the most remote distances, which might cause some detriment. . . ." The commandantes of the presidios were charged with fire suppression and the mission padres with fire prevention.[65]

Evidently the fires originated from fire hunting (rabbits), from the harvesting of grass products, and for the management of mesquite and oak; many were the responsibility of women. Not only around the missions themselves but also for some distance away, annual firing was a serious matter. During sudden weather changes, as with the onset of Santa Ana winds, such controlled fires could quickly become conflagrations. In his reply, Father

Lasuen spoke of the "horribly destructive fires which are experienced every year in this country," and he applauded the governor's stern measures.[66] It was the goal of the missions to convert the traditional Indian societies from hunting and gathering to sedentary agriculture. This required a similar conversion of fire practices. It has in fact been speculated that the use of fire for harvesting natural products was so successful that it prevented the establishment of crop agriculture among the California tribes. By using fields of grass for pasture, the Spanish destroyed the "cereal crops" of the natives. Those Indians not assimilated into mission agriculture were driven further into the mountains to continue their traditional harvesting of grasses, acorns, berries, and small rodents, and they took fire into the mountains with them—just as similarly displaced peoples have done the world over.

It would take acculturation or destruction to stop the fires, not merely proclamations. Spanish pastoralism was not hostile to broadcast fire—Basque shepherds a century later literally burned their way across the Cordillera—but the cycle of Indian fires did not coincide with the cycle of fire for pasturage. Broadcast fire for range is rarely practiced except under seminomadic conditions, and until the missions were secularized and dissolved in 1833, most Spanish livestock occupied pastures near the missions. Until the Spanish introduced stock, too, the Indians had only wildlife to deal with, and fire hunting for rodents was not exactly the same as range management for domesticated animals. Where European or American settlement was light, Indian fire practices persisted; records from San Diego indicate that the tribes of that area were still burning widely as late as the 1870s.[67]

Fires were common around settlements, too. Richard Henry Dana visited Santa Barbara in 1835 and recorded that a great fire had swept the area about 12 years before, "a very terrible and magnificent sight. The air of the whole valley was so heated that the people were obliged to leave town and take up their quarters for several days on the beach."[68] In 1841, when the Duflot de Mofras passed from Santa Cruz to Santa Clara, he observed that "occasionally the traveler is amazed to observe the sky covered with black and copper colored clouds, to experience a stifling heat, and to see a fine cloud of ashes fall." Such burns "seriously handicap travelers," he continued, but since the Mexican Revolution and the secularization of the missions there had been little effort to control them. When fire threatened Monterey, "not the slightest attempt was made to extinguish the flames."[69]

By the 1840s, too, the hide and tallow industry was in full flush. Stockmen burned to improve pasturage for cattle, as shepherds would burn later in the century. The livestock interests concentrated on the grasslands of the valleys and on the vegetative ensembles of the Coast Ranges. Sheepmen came several decades later and flocked mostly to the Sierra, where, in true Mediterranean fashion, they could use the high mountains for summer pasture and the lower valleys for winter feed. John Muir, for one, raged about the "running fires [that] are set everywhere, with a view of clearing the ground of prostrate trunks to facilitate the movement of the flocks and to improve the pastures. The entire forest belt is thus swept and devastated from one extremity of the range to the other."[70] Fire and hoof were controlled by the expulsion of shepherds from parks and forests, but the problem of cattle was more intractable. Stockmen, no less than loggers, were advocates of light burning, and after the lumbermen were squelched in 1923, the torch in the timber passed to herders. Portions of the Coast Range (for example, the Mendocino National Forest) that were traditionally used for herding seethed with resentment against forestry, which had become a political arm of the state, and incendiary fires were widespread for many of the same reasons they persisted in the open ranges of the South. In 1945 California compromised by passing a brush-burning law that allowed clearing by broadcast fire under a permit system administered by the state Division of Forestry.[71]

Fires for pasturage and browse were set in the Southern California mountains as well, and even in the nineteenth century they were sometimes fought in order to save stock and pasture. In 1877, for example, the residents of Santa Barbara County turned out en masse to suppress a fire that destroyed a large number of cattle, imperiled winter feed, and even threatened the suburbs of the town. Nor was this action unusual.[72] Newspaper accounts from the period show a similar turnout whenever property was threatened, with perhaps hundreds or even thousands of residents committed to the fire-lines. In Southern California, from San Diego to Santa Cruz, such spectacles became almost annual events. Using a prescribed cycle, the stockmen burned even in chaparral. For two years after a clean burn the sites would be rich in annual flowers and grasses. Then, as the brush recovered, the range would deteriorate. For another two years it was usable, then abandoned. By the seventh to tenth year the brush had accumulated in sufficient volume to carry fire. The pattern, however, required free access to a large public domain and a seminomadic pattern of herding, neither of which were available after the mountains were placed in watershed reserves. Combined with overgrazing, for which there was no mechanism for regulation, even stockmen freely admitted that prescribed burning contributed to range degeneration and widespread soil erosion.[73] The practice, of course, was condemned by foresters.

Although such fires were satisfactory for maintaining pasturage and browse, they resulted in accelerated soil erosion, occasional floods, and frequent conflagrations as weather conditions changed. Fall burning coincided with the onset of the Santa Ana season. During the 1880s irrigation agriculture spread into the valleys, especially for fruit and citrus growing, thus corresponding to the national interest in irrigation generated by Powell and the Geological Survey. Large dams were constructed for reservoirs and, by the 1890s, for hydroelectric power. As urban communities developed in the basins, the necessity for watershed protection led to a demand for forest reserves, valuable both to ensure that important reservoir sites did not become private monopolies and to protect the watersheds themselves. The San Gabriel Reserve was authorized in 1892, second only to the Yellowstone Reserve in order of priority. Within a year three immense reserves were established, which eventually became four national forests. California's sensitivity to water resources was sharpened, too, by the spectacle of hydraulic mining in the Sierra. Until shut down by the Caminetti Act of 1884, hydraulicking had dislocated a quantity of earth equal to eight times the amount moved in excavating the Panama Canal. Disturbances on such a vast scale threatened both the agricultural interests of the Sacramento Valley and the commercial interests of San Francisco Bay.[74] Residents of Southern California were aware that fire and flood could induce similar consequences in the loosely consolidated soils of the nearby mountains. The creation of reserves was intended to forestall all such possibilities. The reserves could eliminate some fires, such as those associated with herders and Indians, but it was not recognized that, by sustaining the chaparral, they would work to perpetuate the general fire environment.

The Geological Survey report (1900) on what became the Los Padres National Forest noted that most of the lands had been burned over during the preceding two or three decades, mainly by herders.[75] Logging and land-clearing had become locally important on other reserves, though markets were strictly local, and the process did not create a fire regime so much as it extended one.[76] Citizen pressure was strong for fire control, another Southern California anomaly. Typically, officials of the reserves had to battle traditional fire practices of forest residents. In Southern California, however, there were few residents on the reserves, and the desires of visionary urbanites forced officials into a tough fire control posture, even as awareness grew among fire officers that fire exclusion might be hopeless. Following passage of the Forest Management Act (1897), the Los Angeles Chamber of Commerce, for example, organized a committee for the express purpose of improving fire control on the reserves.[77] And the GLO, indeed, had some successes in reducing fires. A June 1903 report from the Los Padres noted exultantly that "during the past five years during which there has been almost total exemption from forest fires, cover is being established in many places where there was practically no covering on account of previous periodical fires."[78] Stockmen were not particularly keen on the program, but they were rapidly becoming a minority. A 1904 report by the Committee on Forest and Water for the San Bernardino Board of Trade noted that "with the mountains denuded of their covering of trees and shrubs this mighty

region, known to all the world as Southern California, would become a veritable desert waste." The committee insisted that "no location in the Nation
more imperatively demands governmental action."[79]

Groups proliferated to ensure that such action was forthcoming. In 1906
the Tri-Counties Reforestation Committee enlarged the scope of the older
Los Angeles organization and affiliated with the California Water and Forest Association. A year before, the state had passed a new Forest Protection
Law largely to promote better fire control. In 1922 a joint committee in Los
Angeles County furthered the domain of its predecessors, and after the fires
of the 1924 season, the joint committee was reorganized into the Conservation Association of Los Angeles County. The 1924 fires set a standard for
havoc—environmental and administrative—that endured for decades.
Chief Forester Greeley himself traveled to the regional board of review;
within a year the Clarke-McNary Act was amended specifically to include
chaparral watersheds, and congressional hearings soon followed on the subject of fire protection for the four national forests of Southern California.
Local reforms did not lag far behind: a statewide reorganization of chambers of commerce resulted in the California Development Association,
which included a department of natural resources. Like self-exciting dyna-
mos, such groups not only complemented but also reinforced one another.[80]

Citizen reviews extended even to the conduct of actual fire suppression.
The Waterman Canyon fire (1911), which led to a special study of large
fire organization by Roy Headley for duBois, also witnessed the establishment of a citizens' board. The blaze began when a boy set fire to a rat's
nest. Santa Ana winds fanned the brush into a conflagration. The sheriff
was called upon to collect men, as specified by state law, but no one would
go; finally, under guard he impressed a group of reluctant Mexicans, who
continually tried to evade the assignment. State law still allowed for backfires by individual property owners, and many where set; one man alone
strung out three miles of backfire. The backfires enlarged the total perimeter, endangered firefighters, and reduced several houses to ash. The Forest
Service was strongly criticized for its lackluster performance, and it lamely,
if correctly, replied that its backcountry was inaccessible and its ranks
understaffed. There were even some voices of dissent. One participant
remarked, "I think you will always have fires in the mountains. You will
have big fires." Another rather breathlessly wanted "to give just one idea.
The pathos, the sublimity, and the grandeur of the great force going up the
mountain side was nothing like anything that had ever fallen under my
observation before. It made one think of Dante's inferno. It comes right
up like a blast of a furnace feeding as it comes and to talk of fighting it is
out of the question. You have got to get there before it gets into that
condition."[81]

The region that had already reached out to the Owens Valley for water and was on the verge of completing an aqueduct considered to be an engineering marvel and a companion to the Panama Canal, was more than prepared to "get there before." There was little point in reaching across half the state for water supplies if the city could not even protect the watersheds in its backyard. Southern California was on the verge of creating one of the most intensive wildland fire control programs in the world. At the beginning of each new outbreak of fires, the fire establishment was strengthened. In the early 1920s Riverside, San Bernardino, and Los Angeles counties established fire programs and entered into cooperative agreements with state and federal fire agencies. The bulldozer and the mountain fire truck were invented. In northern California and along the forest-brush ecotones of the Sierra, the hour control concept was developed. In the 1930s the CCC worked with tremendous effect throughout the state, particularly—under the eye of S. B. Show—in the brushfields. Some 17 camps were working on the Los Padres National Forest in 1933, and their overwhelming concern was fire. CCC labor opened up the backcountry for fire control, increased fuelbreaks, and improved detection systems.[82] A national conference on fuelbreaks was held in 1936 in Los Angeles, and the Ponderosa Way etched its path through the Sierra foothills. During the 1940s military alliances were formed, the Smokey Bear program was launched, and the California Division of Forestry—founded on fire control—went into the big leagues. Nearby Army, Navy, and Marine Corps bases furnished large consignments of troops for firefighting, and the Forest Service, in turn, was given fire control responsibilities over some of the bases. In the 1920s, more than 900,000 acres burned in Southern California; by the 1930s, less than 400,000; by the 1940s, 320,000.

As a new fire regime appeared in the 1950s the old alliances were strengthened and new ones created. The Los Angeles County Watershed Commission was organized for the express purpose of bolstering fire control in the mountains. The Southern California Watershed Council followed, a citizen advisory group to the national forests in the region. Beginning in 1955 the California-Nevada Fire Council provided a forum for cooperative research. The Forest Service concentrated its national fire resources heavily in the area and eventually saw some 25 percent of its total fire budget siphoned, like the waters of the Owens Valley, into Southern California. FIRESCOPE is only the latest, if highly sophisticated, episode in the long evolution of an intense, cooperative enterprise. During the mammoth fire complex of 1970, some 773 fires burned under Santa Ana conditions; 741 (96 percent) were contained at less than 300 acres. About 4 percent of the fires accounted for 93 percent of the damages. Similar figures hold for the 1977 season.[83] But the advent of the new regime in the 1950s had changed something besides quantity: the character of the fires had metamorphosed equally with the landscape.

The largest single fire of regional historic record burned on the Los Padres National Forest in 1932.[84] The origin of the Matilija fire is uncertain. It was sighted on the morning of September 7, and a crew was dispatched long miles by trail into the backcountry to control it. Containment was expected at about 600 acres when winds caused the fire to blow up. Within an hour after flare-up the fire traveled 15 miles; by mid-afternoon it had cleaned out 20,000 acres. For 11 days the flames swept 10,000 to 20,000 acres daily. On the September 10, gale force Santa Ana winds drove the fire in a wild rush to the sea on a front 12 miles long and 5 miles wide. Time and again control lines were emplaced, only to have the wind throw fire across them and through camps with a deafening roar. Flames licked the sky like biblical pillars of flame. "The rapid runs made by the fire," noted the official report, "were of such intensity that fire-fighters on a greater part of the fire line were in constant danger."[85] One of the rangers on the line put it more colorfully when he exclaimed, "The smoke was blinding, the roar of the flames terrific, the very mountains shook underfoot. Man, it sure put the fear of God into you!"[86]

The Forest Service mobilized 2,500 firefighters and brought in overhead from throughout the West. State, county, and city firemen worked alongside Forest Service crews. Seventeen major fire camps were organized, 12 of which would be serviced only by pack mules. Most of the mountains were inaccessible except by trail, and airplanes were used for scouting. Four hundred fifty miles of line were cut for perimeter control. Horses, boats, automobiles, gasoline speeders, tractors, and railroad trains were used for transportation, though the inaccessibility of the site left the fire more or less secure in its mountain fastness. Tales of daring and heroism filtered out of the mountains, but only when marine air reclaimed the slopes after the cessation of Santa Ana winds did the fire subside sufficiently for control. Reseeding by airplane followed, and highway construction into the interior was expedited. When CCC workers arrived a year later, they had ample presuppression projects waiting for them.

For all its size and the high drama of its fireline exploits, the saga of the Matilija fire has an old-fashioned flavor to it. Substitute lodgepole pine for chaparral, and the events surrounding the Matilija are interchangeable with those of the 1910 or 1934 fires in the Northern Rockies. Writing for *American Forests* immediately after the event, Wallace Hutchison inadvertently made the same point. "The Matilija fire," he declared, "will go down in the annals of California forestry as a counterpart of the Winter of the Blue Snow in the mythical days of Paul Bunyon."[87] The Matilija holocaust was a big fire, but a typical one. It was a regional phenomenon, an event that could be easily placed in an evolutionary chain from the north woods to the Northern Rockies; it was not yet a new kind of fire or a fire with broad implications for national policy. That would come after World War II, when fires much smaller in size and less grand in setting would sweep Southern California into national prominence. In creating a synthetic landscape of unprecedented intensity, Southern California has fashioned a synthetic fire regime without parallel in the United States. The Matilija fire cost $400,000 to suppress; the September 1979 fires, some $28 million.

The intensification has been more than one of cost alone. Southern California has nurtured a cultural watershed in which fire is as endemic as it is in the natural watersheds that frame and sustain the region. Nor is the transition simply a bizarre transformation from an extreme natural fire environment to an extreme synthetic one, with wood shingles substituting for chamise, for example. Especially disturbing is the pattern of incendiarism that has evolved as part of the new regime. It is not an incendiarism like that in the South, where old fire practices persisted maliciously beyond their time. It seems rather to be tied, like that of the Lake States, to the peculiar pattern of settlement. It has more in common with the psychology of mass murders. And it is not a source about which fire control is knowledgeable.

As much as the encroachment of suburbs and the revival of wood shingle roofs and the intensification of watershed values, this pattern of random, violent incendiarism—typically fired under Santa Ana conditions—fashioned a qualitatively new fire regime. The Matilija fire epitomized a past regime; it belonged to a tradition of big smokes in the wilderness, one in which foresters could unblushingly allude to past folk legends of the North Woods and Old West. The Malibu fires inaugurated a new fire cycle in which fires were media events, a suburban invention, started for reasons that had nothing to do with traditional fire practices and unhesitatingly compared with the energy dynamics of Hiroshima-type atomic bombs.

"There was a desert wind blowing that night," recalled Philip Marlow in Raymond Chandler's short story "Red Wind." "It was one of those hot dry Santa Anas that come down through the mountain passes and curl your hair and make your nerves jump and and your skin itch. On nights like that every booze party ends in a fight. Meek little wives feel the edge of the carving knife and study their husbands' necks. Anything can happen. You can even get a full glass of beer at a cocktail lounge."[88] In Chandler's day the Devil Wind might indeed cause parties to end in fights and wives to look for knives; in more contemporary times, the parties seemingly end in fires and find the meek reaching for a torch.

THE BEST TOOLS:
A HISTORY OF FIRE EQUIPMENT

The best tools with which to fight a fire are the rake, hoe, shovel, and axe.—Fred Besley, Maryland State Forester, 1911[89]

The situation left one choice: Mechanize or burn.
—A. A. Hartman, U.S. Forest Service, 1949[90]

In the California fires of 1956 an extraordinary arsenal of mechanized equipment was brought to bear on the fireline. Stake trucks carried men and supplies, rugged pumpers transported water where needed, and bulldozers cut fireline over ridge and ravine. Above all (literally) were helicopters, used for scouting, and air tankers, making their debut as a means of delivering water and chemicals to remote sections of line. If the fires publicized a new fire problem, they revealed equally a new strategy of fire control, a dedication to heavy equipment.

Agencies could no longer rely on the labor-intensive strategies afforded by the CCC; the CCC's legacy, the organized crew, flourished, but its numbers were small. A crew numbered 20 to 25 firefighters; a CCC camp, 100. Instead, agencies looked to heavy equipment to supplement the loss of muscle. The mechanization of the military in World War II was an important model, especially the doctrine of air superiority. Equally, military hardware became an important source of supply. Soon after the war, the Forest Service redoubled its equipment development program, established an equipment center, and investigated the ways and means of converting jeeps, trucks, aircraft, and even halftracks to the fireline. After 1955, as part of its general responsibilities in rural fire defense, the Forest Service directed the distribution of excess federal property, largely military, to its cooperators, especially the state foresters.[91] The upshot was to make the deployment of concentrated power in the form of mechanized equipment the central strategy of fire control.

Equipment development, of course, had been around since the Transfer Act of 1905, when Ranger Malcolm McLeod of the Sierra National Forest designed the rake and hoe hand tool that bears his name. Equipment has subsequently passed through three general stages: a period of tools powered by human or animal muscle; a period characterized by the elaboration of the internal combustion engine into virtually every aspect of fire control equipment; and a period in which electronic information technology dominates development programs. The first phase helped to give fire control a

30. The mechanization of fire control. A staged photo on the Shasta Experimental Forest, 1955.

distinctiveness. Originally tools, like the firefighters who wielded them, had come from the farm, logging camp, ranch, or railroad. The development of unique tools for fire control emphasized the distinctiveness of the Service's policy: it was not merely a transfer of frontier skills but a systematic adaptation and reorganization of them. The special tools symbolized in a simple and popular way the special policy they were designed to implement. The internal combustion engine of the second phase supplied added muscle, deferred for a while by the collective strength of the CCC, but in the postwar era mighty enough to constitute in itself a new and glamorous strategy for fire control. The information technology of the last phase corresponds to the problem of fire management rather than that of mere fire control. None of its "tools" were applied directly to the fireline; instead they attempted to manage better what was already available and to subordinate the use of heavy equipment to the larger goals of fire policy. Equipment development began with tools designed to control fire on the frontier; it is maintained by the sophisticated electronics with which man put fire on the moon.

I

The Transfer Act put the Forest Service in the business of forest management, but the 1910 fires made it a fire protection agency. Both missions, as the early rangers recognized, were a new idea, and they were eager to show how professional forestry diverged from frontier practices. Rangers and forest guards were different from ranchers, farmers, and loggers, and they wanted their tools to reflect that distinctiveness. The pine bough and wet blanket of the Indian, the long-handled shovel and pick of the miner, the double bit ax and misery whip of the logger, the mattock and plow of the farmer—all were worked over by zealous foresters with the same intensity with which they sought to rework and regulate frontier habits of fire and forestry. Previously fire control had been undertaken by institutions primarily established to do something else—to log, lay track, or survey land. After 1910, the Forest Service set out to create hand tools specifically for fire control in much the same spirit as it invented institutions and techniques to achieve its conservation mission.

The Service took pride, moreover, in its local Edisons, its continuation of the American tradition of "ingenious mechanics" and tinkerers. And the development of equipment was no less popular with field men, who realized that they could immortalize themselves by contriving a novel fire tool. In 1905 the *Use Book* declared that "generally the best tools for fighting a fire are a shovel, mattock, and axe."[92] Those were the tools still in use during the 1910 fires in the Northern Rockies. In 1911 Maryland had added a rake to that list and a 16-quart "moonshiners bucket"; Michigan got by with a shovel, bucket, and matches; ranchers on the Plains relied on the drag, a bucket, and a swatting device; Wisconsin favored a horsedrawn plow and collapsible canvas pail; and North Carolina preferred the ax, shovel, and rake.[93] Travel was by foot, horseback, canoe, or velocipede. Detection was made from a tree or barren peak with at best a plane table for sighting.

By 1910 things were already changing. W. B. Osborne invented a fire finder—the first of a long, distinguished series of devices contributed by nearly every region in the country. The brush hook evolved out of hay-cutting tools. John Clack invented a smokechaser pack frame. H. J. Eberly of Texas produced the modern fire flapper. Elers Koch devised the Koch tool, a combination shovel and hoe. Ranger Kortick revised the McLeod tool so that it could be disassembled. In fact, the development of combination tools, first concocted by McLeod, became something of a rage, culminating in the Perkins tool, whose interchangeable heads rather improbably combined the ax, hazel hoe, pick, shovel, and rake. Less extravagantly, the Forest Service came up with a shorter handled, more maneuverable shovel. The backpack pump was in use by 1914, though from its inception it was notorious (as is still is today) for leaking. A variety of torches was produced. And, of course, there was an abundance of special rakes—the Barron tool from California;

31. Fire guard on a "speeder," North Carolina, 1923.

32. Fire shield and special extinguisher outfit, Arkansas, 1912. When not in use, the extinguisher unit was kept suspended in readiness for quick attachment to mules.

the Rich or Council tool; the wire broom; and the Pennsylvania fire rake. The "Modoc double rock drag" replaced the beef drag. Daniel W. Adams of the Arkansas National Forest devised a portable plow that could be carried by pack mule, a new metal horse pack, and a chemical fire extinguishing system also for horse transport. The Beatty grader, drawn by a mule, greatly expedited the construction of trails. Fire shields that enabled firefighters to approach close to flames were available by 1912.[94]

For communication single wire telephone lines were used with an earth ground return wire. During the 1910 fires, particularly in California and the Northwest, phones were used extensively. In 1915 the Forest Service published a handbook on phones, and the Northern Rockies had its own telephone engineer, R. B. Adams, who later invented a portable phone for field use. Other devices were borrowed from the Army—heliographs, semaphores, signal lights, and even carrier pigeons. In 1919 the Forest Service was experimenting with radio sets borrowed from the Army Signal Corps. Until the 1930s, however, the hand phone was to communication what hand tools were to the fireline.[95]

Mechanization was limited, but occasionally spectacular. Perhaps the most outstanding illustration was the railroad. Even in the 1903 Adirondack fires trains had been used for the transportation of men and equipment and were outfitted with special pumps and tank cars. Considering that trains were a frequently prolific source of fire, that they were often under obligation to remove debris along their right-of-ways, and that they were a common mode of transport for logging, it should not be surprising that considerable mechanical ingenuity was expended toward making trains an effective tool for fire control. By 1910 the Canadian Pacific Railroad had developed special fire trains. Among his other inventions, Daniel Adams published detailed plans in 1912 for a "gasoline-propelled fire fighting engine."[96] Nonetheless, the steam engine remained the prime mover of these devices.

The Southern Pacific Railroad undertook an exemplary program of fire control based on fire trains. Lookouts were established, tank cars positioned on spur tracks at key locations, and special fire locomotives kept under steam. By 1931 a typical fire train had a locomotive outfitted with duplex pumps, two tank cars with 12,500 gallons combined reserve, hand tools, fire ladders, portable telephones, chemical fire extinguishers, and 1,000 feet of 2 1/2-inch hose. Though the fire trains were primarily designed to protect wooden ties, trestles, and snow sheds along the right-of-way, more than once they saved small towns along their routes. Once dispatched to major fires, the fire trains were given priority routing. More typical during periods of high fire danger was the velocipede, or tracked motor car, carrying a patrolman who followed behind the puffing engines swatting out minor blazes along the right-of-way.[97]

33. Southern Pacific fire train, California, 1924.

The most celebrated of all fire tools is the pulaski. Its story nicely sum-marizes the evolution of hand tools, both as a practical and as an adminis-trative invention typical of this era. Folklore has it that Ed Pulaski designed the popular combination tool (ax and mattock) after his trials in the 1910 fire. The mountain fires were thus the forge and Pulaski's experiences the hammer that gave fire control its first distinctive tool. Firefighters through-out the West seized on the tool as they did on the daring tale of its inventor. Like American forestry, the pulaski tool was born of fire.

There is a symbolic truth to the story. But, as with the 1910 fire, it was not the event in itself but its context that was most significant. The pulaski tool is at best a case of independent rediscovery. An exhibit by the Collins Tool Company at the 1876 Centennial included a combination ax and hoe tool that more closely resembles the modern pulaski than the one Big Ed originally fashioned. The logic of the tool was born of the frontier and, like many fire practices, had to be rediscovered by foresters. When he recovered from his fire-induced ailments, Pulaski, something of an amateur black-smith, may have attempted to work on some sort of new tool. If he did, he put it aside. Meanwhile, his supervisor, W. G. Weigle, was interested in a new combination tool suitable for both fire patrol and replanting—after the summer's conflagrations there was plenty of opportunity for both. Whether

34. Pulaski tool, Idaho, 1967.

Pulaski and Weigle got together and compared notes is uncertain; probably they did. However, Joe Halm claimed that in the winter of 1910 Weigle came to him and Ed Holcomb with instructions to build a shovel, ax, and hoe combination tool. This device was exhibited at the 1911 supervisors' meeting in Butte but was not well received. Pulaski reportedly went back to his forge and played further with the particulars. By 1913 Pulaski's tool was ready for field trials. It was this model that was introduced at an annual meeting of the Western Forestry and Conservation Association. By 1920 the Forest Service contracted for commercial production of the tool.[98]

Its annual meetings, in fact, made the association a major forum for introducing new ideas, techniques, and tools for fire control. The Forest Service as yet had no institutional procedure for handling equipment concepts outside of conferences like those at Mather Field or Spokane, which were relatively rare, and no vehicle for publicizing such ideas and tools outside of special bulletins. Consequently, such important devices as the Osborne fire finder, the portable back pump, and the pulaski were announced at association meetings. Beginning in 1931, the association established an equipment research committee in cooperation with the Forest Service, half a dozen state and private fire agencies, and several equipment manufacturers. W. B. Osborne was both an important catalyst for the association effort and a liaison with the Forest Service.[99]

The Forest Service was interested in equipment, but in a peculiar way. At its first concourse of fire officers, the Mather Field Conference of 1921, several topics related to tool management. Three types of tools were identified: regular tools available through commercial manufacturers; specialized tools, probably limited to fire control operations; and experimental tools, largely proposed from the field. When Greeley approved the tool committee's recommendations, he wished to "mention particularly the importance of the parts of the report dealing with inspections, physical inventories and distribution." He further directed the Ogden Supply Depot to "do all it can to make centralized purchase of equipment a success."[100] The problem was seen as one of administration, not invention.

A decade later a triumvirate of meetings went a long way toward firming the institutional base of fire protection. The regional foresters conference addressed policy; the Shasta Fire Control Meeting, research; and the Spokane Equipment Standardization Conference, management of the proliferating number of tools. The 1936 Spokane conference assigned dozens of topics to different regions for reports. It was at this conference that the pulaski tool was proposed for national distribution, and the conference instructed Region One (Northern Rockies) to develop and test further a prototype suitable for Service-wide use.

The 1931 conference had sought less to introduce new tools—new devices were being proposed in profusion—than to select the best models from among the existing stocks and to standardize them. The fundamental hand tools were more or less already out in the field; the problem was not one of further invention but of pruning and purchasing for Forest Service warehouses at Ogden, Portland, and Spokane, for example. Several years later Headley was still vocal in assuring participants that standardization did not mean conformity, that new equipment and improvements in old models were solicited. At the 1936 conference hand tools were thought to be a "fertile field for investigation," and there was agreement that the "whole subject needs systematic investigation."[101]

But in fact, the great outburst had spent itself. Creative energy was directed more toward the administrative debates leading to the 10 A.M. Policy, to the management of CCC labor, and, where equipment was concerned, to mechanization. The 1936 Spokane conference mostly confirmed the equipment program of 1931. With respect to the pulaski, fire officers from the Northwest wanted a heavier grub end than the existing prototype allowed; this proposal was accepted, and subsequent modifications produced the pulaski in its modern form. Further development has been in the realm of materials specification and standardization rather than design.

At the 1936 conference, moreover, it was proposed that the Forest Service sponsor a national equipment laboratory. The idea had gathered momentum since the 1931 meetings, but in 1936 it strangely floundered. There was simply not enough left to do that a national lab could do well and too much that it would not be able to do. The 10 A.M. Policy had been founded on CCC manpower, not on equipment. Foresters from California and the Northern Rockies successfully argued that the money for the proposed laboratory should go instead into an aerial fire control project—something that could perhaps better satisfy the 10 A.M. Policy's directive to attack fires in the backcountry.[102]

The genius of early fire protection lay in its capacity to organize, in the simple establishment of administrative units and basic policies. In the realm of equipment this meant, on one level, the creation of rudimentary tools; on a deeper level, it demanded the standardization and coordination of those inventions. The elaboration of a detection system based on fixed lookouts, for example, required a fire locating device like the Osborne model, but even more it demanded a means of coordinating individual lookouts into a suitable network and of standardizing the procedures for reporting smokes. The pulaski tool was a useful invention for the smokechaser, but elapsed-time studies for getting the firefighter rapidly into the backcountry were more significant. The answer to light burning was not a hand tool like the McLeod, no matter how clever or popular, but Coert duBois's *Systematic Fire Protection in the California Forests,* which demonstrated how the coordinated use of these tools made adequate fire protection possible. Perhaps the most prolific of all the ingenious mechanics, W. B. Osborne, best represented this paradox: it was less his fire finder and sundry tools that revolutionized the fireline than it was his fireline manual, the *Western Fire Fighter,* published by the WFCA. The newly invented hand tools were just that—tools. It was their systematic deployment that most occupied the talent of the age. Of all the inventions of this enterprising era, the greatest was organization.

II

The mechanization of fire control was effectively announced around 1928. The Forest Service established a radio laboratory at Beltsville, Maryland, and proceeded to conduct fundamental, pioneering research into the transmission of radio waves over mountainous terrain. The mountain fire truck appeared for the first time on the remote roads of Southern California.[103] The Forest Service acquired its first aircraft, severing ties with the Army Air Service. *American Forests* ran an article on tractor-drawn plows, and the Forest Service successfully tested a "trail builder." After the 1924 fires in California, Show in particular had urged better equipment for road construction and patrol. Successful experiments with tractor-graders took place on the Shasta Experimental Forest in 1926, but although the machines moved brush effectively on level surfaces, they failed in later trials on the steeper slopes of the Los Padres. A special budget was established to finance experiments in making an adjustable blade for the tractor unit, and Earl Hall, an engineer with the Service, was assigned the project. He completed a working, full-scale model in 1929, and field trials were scheduled for the Northern Rockies. The "trail builder" was the first modern bulldozer. The success of this "invention to order," as Show labeled it, led to recommendations for a national fire equipment laboratory—the first in a long succession of such suggestions.[104] The Forest Service "disapproved" the idea. Instead, the state of Michigan created its Forest Fire Experiment Station in 1929 at Roscommon, soon to become a major center for heavy equipment research. The Forest Service elected for regional equipment centers.

The new technology promised to revolutionize every dimension of fire control, and so it did—generating power hand tools; mechanizing air and ground transportation and fire reconnaissance; developing heavy equipment for line construction; refining pumper units, both portable and integral; converting military hardware for assorted fire protection duties; and, of course, extending to an extraordinary commitment to aerial fire control. Many of these innovations were forced on the Forest Service by its desperate need to open up the backcountry for fire control. The studies on elapsed time and hour control done by Show and Kotok in the early 1920s, for example, furnished guidelines for a road network into undeveloped areas. The existence of such road systems required, first, an efficient means of constructing roads and, second, a satisfactory means of patrolling them for fire. The first led to Hall's invention of the bulldozer; the second, to the mountain fire truck. As Hall politely put it after the 1930 trials in the Rockies, "They surely need roads up there."[105] The role of aircraft for observation and supply was self-evident. This commitment to mechanization, in turn, demanded expertise and facilities for developing, converting, testing, and establishing specifications for the purchase of mechanical equipment.

By 1931 the whole problem of equipment warranted its own conference. The sessions devoted to the standardization of hand tools and commonly purchased gear were in many respects a postmortem on the heroic age of hand tool invention. Everyone recognized that the future lay in the internal combustion engine. But although the engine was promising, it was also complicated. It required testing facilities, adaptations that could no longer be accommodated at a nearby forge under the inspiration of ranger-blacksmiths, and funding on an order that demanded better technicians and more centralized procedures for evaluating and ranking projects.

The Forest Service debated how best to institutionalize such a development program. In part, proposals for an equipment laboratory reflected a genuine desire to ensure that the Forest Service got the equipment it needed, that it did not live in ignorance of industry developments. But in part, too, they merely reinforced the Service's old predilection for ideas that had come through its own ranks and for control over developments that might in some way touch on its management mission. The creation of regional fire caches with only part-time fire duties suggested the possibility of combining a cache and a laboratory, where the mechanized hardware could be improved as well as repaired and stored. A number of regions had moved in this direction already. The Northwest eventually created an equipment lab in Portland that addressed a wide range of subjects, from heavy tractor-plows to duff hygrometers and radio parts. The region confidently expected that it would become the foundation for a national laboratory. After its success with the "trail builder," the California region had also established an equipment center.

A 1931 conference in Spokane proposed that the Forest Service circulate an equipment information bulletin, prepare an equipment handbook, and establish a national laboratory at Spokane. Some 93 topics on equipment themes were farmed out to specialists across the country, and these items constituted the opening agenda of the 1936 Spokane conference. Animated by the 10 A.M. Policy, the 1936 meeting approved the various recommendations of its predecessor: the bulletin became the quarterly publication *Fire Control Notes,* with its strong bias toward equipment; a purchasing procedures handbook was approved; and a committee was appointed to oversee a Service-wide equipment handbook, which appeared a year later.[106]

The big question, however, was power equipment. And it was "during the discussion of the proposed gas engine," the conference chronicler recorded, that debate "led definitely to the question of the proposed equipment laboratory." Opinion was divided. Frank Jefferson protested that "the laboratory would inevitably lead to duplication of what manufacturers all over the country are doing, and perhaps have already done." Jay Price countered by observing that if there were a sufficient volume of "engine work," the laboratory would be justified. Headley, as moderator, noted that the "establishment of Equip. Lab. [had been] publicized more or less." The purpose of a central equipment laboratory was "to provide special personnel that Regions [are] not in position to employ; also specialized equipment to enable operators to function, experiment, develop models, carry on tests, etc." The director of the Division of Fire Control concluded, "I hope this conference will ponder well its thought in the matter before recording its decision as to whether the Service does not need or cannot afford a laboratory, so far as the future can be foreseen." Kotok reminded the participants that most of the Shasta Experimental Forest had been dedicated to fire research and could be used as a proving ground.[107]

But the proposal once again fell through. The reasons were many. Gasoline power technology was still young and its specific value to the Forest Service uncertain. The 2-cycle engine, for example, had a dubious future, and 4-cycle engines were cumbersome and demanded a considerable investment in shop machinery. Most of the immediate equipment needs could be handled through the existing regional facilities for hardware and fire research. Nor was the field one for which foresters were especially well prepared. With the CCC, moreover, labor-intensive projects were a more worrisome issue, and the ECW money could not be diverted into a laboratory.

The main cause for rejection, however, was a brilliant presentation by David Godwin on the promise of aircraft and chemicals for aerial firebombing. Godwin's performance electrified his audience. The dream, of course, was an old one, and the assembled multitude of veteran firefighters wanted to believe it. Price declared that "airplane experimentation in general as outlined by Godwin [is the] most important next step in fire control." Headley believed it was "highly desirable to push chemicals and explosives." Even Kotok was "anxious to see the whole field explored." The reasons for undertaking an experimental program in aerial fire control were legion, and, in contrast to other power equipment projects, it would be a process of conversion and adaptation, not of basic research in mechanical engineering. Besides, as the conference chronicler summarized, there was wide "recognition" of this subject "as a field of fire control effort into which [the] Forest Service must enter to maintain public confidence, if for no other reason." Aerial research and development was not only eminently practical but also potentially glamorous. So long as the CCC flourished, ground suppression strategy would necessarily emphasize hand crews. Money for capitalized heavy equipment purchases was lean during the Depression, and equipment was scarce during the war years. Aircraft adaptation, moreover, recalled the glory days of the Army forest fire patrols begun in 1919. In a frank trade-off the lab was scrapped in favor of the Aerial Fire Control Project.[108]

Changing circumstances after World War II revived the old hope for a national equipment laboratory. The Aerial Fire Control Project failed to match its dazzling promise; the CCC had permanently disbanded. The tractor-plow was not the least of the developments that had at last broken the awful cycle of fire in the Lake States, and its successful use in the South was more than a little influential in leading to a revised position by the Forest Service on prescribed burning.[109] And finally, of course, there was World War II itself, which gave not only the example of a mechanized military but also the prospect of tremendous quantities of surplus hardware at its conclusion. In 1938 Ted Flynn of the Portland regional lab had undertaken basic research and development with air-cooled 2-cycle engines because no industrial designer bid on the proposed project. The end of the war released plenty of hardware that needed only adaptation or conversion for use in fire control. In 1944 Chief Forester Lyle Watts circulated questionnaires designed to beef up Forest Service research. One set led to a plan for a Division of Forest Fire Research; the other, to an important equipment development meeting in Washington, D.C., in 1945.

The 1945 conference opened with remarks by Watts that "equipment is one of the really important features of the Forest Service." The Chief Forester reported that he was "disturbed" that "the Service has lacked facilities for carrying knowledge of developments of equipment throughout the organization." To the conferees, chaired by Godwin, Watts offered assurances that "there was never a greater need or opportunity for the work that you men are undertaking to do." With this invocation, the conference introduced a broad spectrum of reforms: it confirmed the need for a national fire control equipment committee; it broadened the scope of equipment concerns from the national forest system to include Forest Service cooperators in state and private forestry; it approved a Washington office equipment fund, to be allocated by region; it revised the equipment development handbook and the master specification file; it resuscitated *Fire Control Notes,* left dormant during the war; it endorsed a proposed joint committe on forest protection to be sponsored by the SAF and the American Society of Mechanical Engineers; it urged the creation of a fire control equipment committee in each region; and it established liaisons with the military, both to test equipment of possible mutual value and to ensure access to surplus equipment after the war. A report on Army research at Eglin Field, where military equipment had been adapted to fire control in the South, was widely circulated, and Arthur Hartman informed the conference of ongoing cooperation between the southern region and the Army. In fact, the prospect of surplus military hardware seemed to promise for the mechanization of fire protection what CCC labor had accomplished for its physical plant and labor pool: an inexpensive source that could be readily adapted to the needs of fire control and that could perhaps put it decades ahead of normal evolution. Industry, too, had made rapid strides in mechanization during the war years, and items such as the chain saw had been developed in response to the needs of commercial forestry. "To take full advantage of new equipment," Kotok shrewdly reminded the conferees, "is likely to require much change in current methods." The 1945 meeting set those changes in motion.[110]

Perhaps its outstanding recommendation was for the creation of four all-Service equipment development centers: one at Portland, where a regional lab had once pioneered in the fields of fire, road, and radio equipment and where there were close ties with the timber industry; one at Tuscaloosa, Alabama, to be a cooperative program with the University of Alabama School of Engineering; another as a joint facility with the state of Michigan at its Roscommon Station; and one at Arcadia, California, to be known as the Arcadia Fire Control Equipment Development Center. It was hoped that the Arcadia facility would tap the expertise of UCLA, Cal Tech, and nearby heavy industry; the site was also located dead center in a highly visible fire regime. At the same time, many states continued their own backyard forge approach, though two, Michigan and Wisconsin, maintained special centers. The Roscommon facility had the advantage of proximity to Detroit, and in time it became the locus for state and rural equipment research. Meanwhile, the Forest Service retained its regional equipment programs; inspectors on special assignment wrote detailed reports on regional needs and inventions; and the Forest Products Laboratory at Madison, Wisconsin, undertook occasional equipment and design problems at the request of fire control.

A largely pro forma follow-up conference in 1948 once again linked research into equipment with research on fire as an economic, physical, and biological phenomenon. With Godwin's untimely death, the subject lost a charismatic champion, but the major ideas were already in the air. Subsequent meetings focused more and more exclusively on aircraft, and of the proposed national centers, only those connected with aircraft really flourished. The Portland and Alabama labs never materialized; the Roscommon facility remained with the state; but the Arcadia Center survived, thanks to the ancient liaison between California and military aircraft. The multifaceted aircraft needs of the Northern Rockies led eventually to an all-Service equipment center at Missoula. In the final analysis the promised postwar efflorescence of equipment meant an affluence of aircraft.

The Arcadia Center promptly launched cooperative projects on the two topics of longstanding interest to the region: the mountain fire truck (or ground tanker) and aircraft (aerial tanker). A cooperative program with the Army tested the adaptability of helicopters to the fireline, while a joint project with UCLA explored hoses, pumps, and chemicals. The 1954 Operation Firestop in Southern California, which combined fire physics with tests on fire suppression equipment, including air tankers, was succeeded by the helitack program undertaken in cooperation with the Army Corps of Engineers. With the 1956 breakthrough in air tanker use and helitack, aerial equipment development entered a golden era. When an all-Service conference on aerial techniques was arranged for 1959, it was appropriately scheduled for Arcadia.

Missoula based its claim to a center on old habits of tinkering with paracargo equipment, on 1947 Army tests using a B-29 and a P-47 to drop chemical bombs on fires, on the smokejumper program revived with the Aerial Fire Depot, and on Project Skyfire, the lightning research program that depended heavily on aircraft and special mechanical devices. In 1960 the Missoula Equipment Development Center (MEDC) was formally dedicated as an all-Service lab. Concomittant with aircraft development came wide-ranging research into chemical retardants.[111]

Time and again fire control had broken trail for technical developments that afterward spread to other areas of forest management. It may not be too much to say that its relative success with fire control legitimated professional forestry in the first place. This is what happened with equipment development. In 1950 the Arcadia Center assumed not only a multifunctional role but also an interagency one when the BLM contracted with it to develop range management equipment. It dropped the sobriquet "fire control" from its title, and in 1963 national control over the equipment development centers passed from fire control to engineering. In 1965 the Arcadia Center moved to San Dimas. MEDC had undergone a similar evolution by 1964, becoming a multifunctional facility. Both centers retained their traditional biases nonetheless: San Dimas emphasized heavy equipment, ground tanker and pump testing, and helicopter accessories; Missoula, with its smokejumper heritage, specialized in parachute accessories, portable hand and power tools, disposable fireline items, protective clothing, and physiological testing.

Despite the expressed hopes of fire control chief Merle Lowden in 1958 that an equipment center would be established east of the Mississippi, none was. Instead, the Forest Service contributed some funds to the Roscommon Station. In 1971 the northeastern states agreed to fund a Roscommon Equipment Center at the Michigan station over a three-year trial period. During this time the center experimented mostly with machinery for fire control, the bulk involving the conversion of surplus military hardware into ground tankers and plows suitable for rural fire protection—in a way, the proverbial beating of swords into plowshares. The arrangement was extended for another five years in 1974. For further interagency coordination, the Forest Service maintains an equipment specialist at BIFC and the NWCG sponsors an equipment committee.[112]

The evolution of—and meaning of—equipment development is perhaps best summarized in the history of aircraft. Aerial fire control was the most dazzling chimera of early protection hopes, its most spectacular form of public advertisement, and, in terms of mechanical developments, the most influential in revamping fire control strategy and tactics. Of all the events of 1928 that announced the advent of mechanization, perhaps the most significant was the Forest Service's decision to abandon its shaky dependence on military planes and to contract out commercially for its aerial fire patrols. The airplane thus moved out of the category of novelty and became an integral part of fire detection and fiscal programs. The experience with Army patrol planes had been fundamental, however, and in a peculiar way "equipment development" for the Forest Service came to mean the adaption of aircraft to forestry. Those programs survived best that were in some sense dependent on aircraft—research on communications, chemical retardants, accessory hardware for adapting planes to fire control and for adapting firefighters to aerial attack and deployment.

Aspirations soared early. In 1909 William Cox, then state forester for Minnesota but formerly with the U.S. Bureau of Forestry, witnessed a demonstration flight by the Wright brothers and forecast a glowing future for aircraft and fire control.[113] That same year, when the Southwest regional supervisors of the Forest Service met in El Paso, they approved a resolution that the use of aircraft for fire patrol was something that should be anticipated and explored. In 1909 the Army acquired its first plane. Two years later the WFCA went so far as to draw up contracts with Curtis Aircraft for flying boat tests to assess their possible use in detection. In 1915 aerial patrols were conducted with a flying boat in Wisconsin, though by a sort of honorary warden working without pay. From then on, prospects multiplied and ambitions swelled.[114]

What is especially significant in this early outburst of official interest is that almost from their invention aircraft were seen as a fundamental component of forest protection. Canada, too, showed an immediate interest, and several provinces contracted to have special patrol planes constructed. American foresters contemplated a similar program. Graves even wrote a long letter to *Aviation and Aeronautical Engineering* outlining how planes could be used in fire control. The Forest Service had successfully spared large expanses of western land from settlement, only to discover that it, too, was often effectively excluded from its own backcountry. Roads and trails were slow to build and expensive to maintain; aircraft seemed to offer an ideal solution.

The big break came in 1919. In March the Secretary of War received a letter from the Secretary of Agriculture noting that "one of the most difficult problems in the administration of the National Forests is the prevention of forest fire," that both airplanes and balloons could be used to good advantage, and that a system of fire patrols on the national forests could be combined with Army aviator training. Apparently the letter had been prepared by the Forest Service, but the Army Air Service, eager to discover peacetime duty, was favorable to the idea. According to Coert duBois, a chance meeting between himself and Henry H. ("Hap") Arnold of the Army Air Service early in the year set up the chain of events. Arnold, later commanding general of the U.S. Air Force in World War II, recalled that "we were trying to find ourselves during this period—trying to find out how the airplane could be used and what value it might be to the public." A cooperative agreement resulted. The patrols began that summer in California and Oregon with an Army pilot and a Forest Service observer in each plane.[115]

The results were sensational. "No event in American conservation since the Ballinger-Pinchot controversy," Henry Clepper has written, "stimulated as much public interest as the aerial forest patrols of 1919."[116] Requests for patrols poured in from Montana, Idaho, Washington, Wyoming, and Arizona; landing strips were constructed by citizens eager to have a patrol nearby; and, not incidentally, the flights proved useful in locating fires deep

35. The famous Army fire patrols, Washington, 1921. The Army supplied the pilots and the Forest Service the observers.

in the backcountry. The limiting factor was adequate communications: without radios, observers had to drop messages in special cans or by small parachutes to lookouts outfitted with a telephone. They even used carrier pigeons. A more formal Forest Service–Army agreement was signed in 1920, radio sets were installed, and public pressure mounted to make the patrols permanent. In 1921 arrangements were made to test dirigibles, though, as the reports admitted, their explosive gases among other things made their utility in fire control "somewhat limited." Correspondence was also initiated with the Army Chemical Warfare Service about prospective foams, froths, and gaseous retardants.[117]

But an unfavorable Army report at the end of the 1920 season and a Congress balky at appropriating funds to a peacetime Army ended the program in 1922. Special requests released planes for some flights in Oregon, but there was no more funding until a little was temporarily restored in 1926. A year later Maine became the first state to authorize public money for aerial fire patrol. The War Department finally withdrew altogether from the cooperative program, and in 1928 the Forest Service began contracting for commercial aircraft.

The love affair between the airplane and the Forest Service had gone beyond infatuation. Aircraft had demonstrated both its utility and its inestimable value for public relations; nor was it accidental that the first great use of aircraft in forestry was for fire control. It is perhaps not too harsh to observe that the equipment development programs tied to aircraft were, on the whole, those that succeeded and that no aspect of equipment development shows greater continuity than that related to aircraft. In a fundamental way fire control on the national forests became wedded to aviation. There were practical reasons, of course, but the collective memory of Army patrols was perhaps the decisive argument. It may be that the 1919 episode came too early and that its public success was perhaps too complete. The Forest Service came to expect a succession of similar coups, or, what was worse, expected that its public anticipated that it would produce further spectaculars. Even where personnel were involved, the elite corps like the helitack crews and the smokejumpers were distinguished by their reliance on aircraft. It was not the national equipment laboratory that produced aircraft; rather, aircraft made the long-dreamed-of lab possible. The first Service-wide meeting of fire officers took place within the shadow of Army patrol planes at Mather Field in California. And at a 1950 conference intended to consolidate the gains after World War II as the Mather Field Conference had done after World War I, conferees were warned again that "it is essential to control enthusiasm that outruns justification in the use of aircraft."[118] But that enthusiasm was already too ingrained to be warded off with bureaucratic incantation. The concept of total mobility is an administrative one founded on the geographic mobility made possible by aircraft. When the Forest Service elected to change the title of its Division of Fire Control to something more in keeping with its true organizational objectives, it was redesignated the Division of Aviation and Fire Management.

The Army patrols stirred official and public imaginations to propose that fires could be bombed with chemical extinguishing agents or that, at the least, supplies could be dropped by plane into remote fire camps. By 1925 fire officers in Washington were dropping mail and food free fall out of a plane, and in California experimenters dropped buckets of free fall water directly on fires. In 1929 more free fall deliveries were made in California and Washington; in the case of one fire they supplied the entire suppression effort. Within a decade free fall deliveries became commonplace. In 1932 an autogiro rather than an airplane was employed, and suggestions were subsequently entertained on how to adapt the autogiro to backcountry work. The Army's selection of the helicopter over the autogiro in 1939, however, ended such experiments by foresters. Throughout the 1930s, particularly in the stubborn backcountry of the Northern Rockies, experiments continued with parachutes. In 1935 Howard Flint of the Forest Service dropped equipment with homemade chutes. A year later Flint and Robert Johnson invented the static line for automatically opening cargo chutes, a device later adapted for chutes carrying people. The two foresters also acquired some condemned Army parachutes and adapted them for paracargo delivery. Meanwhile, further aircraft trials were conducted on national forests in Washington, Oregon, and Arizona. In innumerable experiments bags, buckets, and kegs of water were dropped from aircraft. The Forest Service itself was innundated with public suggestions on aerial efforts to quench fires. In 1933, for example, Master Appliances Company described to the Washington office its two years of trials with a "bomb fire extinguisher" suitable for forest fire control. Here, too, the Army patrols had set the example. The cover of the August 1919 issue of *Scientific American* had displayed a full-color picture of an Army biplane dropping chemicals on a fire.[119]

The 1936 Spokane conference again confronted the choices to be made among the directions of mechanization and its elected aircraft. This left the development of power pumps, saws, vehicles, and line construction equipment in industry hands or at Roscommon. The Aerial Fire Control Project under Godwin conducted experiments with both aircraft and retardant, and as part of these investigations the Forest Service purchased its first plane. Despite promising results reported by the Russians in the early 1930s, the outcome of the American trials was disappointing. The retardant experiments were also equivocal. Still committed to the concept of air attack, Godwin directed the remaining funds into another project with Russian precedents: the smokejumper program. The smokejumpers did for public relations immediately after World War II what the Army patrols had done after World War I.[120]

The 1945 conference revolutionized the field of equipment development, but the essence of that reform was a reaffirmation of the commitment to aircraft. The doctrine of air superiority, demonstrated so decisively during the war, seemed equally applicable to fire protection. In 1945 the Forest Service set up its Continental Unit in the Northern Rockies, where detection and control were conducted entirely by air. The Forest Service also entered cooperative agreements with the military to adapt World War II aircraft to fire control. Experiments were conducted at Arcadia with helicopters (1945–1947) and at Eglin Field, Florida, and at Missoula with fixed wing craft (1947). A B-29 and a P-47 tested various fire extinguishing bombs. The bombing trials were favorable in Florida's light fuels; in the Northern Rockies, they left much to be desired, though results were labeled as "definitely promising." Once again the questions of deployment and control were complementary. The experience of Allied pilots with jettisoning spent fuel tanks on enemy positions had encouraged the development of a specific firebomb, outfitted with jellied gasoline, and the success in delivering this new substance, napalm, suggested that the same techniques could be applied to the delivery of a fire retardant. The guiding genius behind these arrangements was Dave Godwin.[121]

The 1950 Ogden fire conference recognized that the commitment to aircraft had gone so far that "resultant radical changes in fire control plans" were apparent, and it admitted that "retreat from the airplane is not possible." The conferees in fact had no wish to retreat. Instead, they recommended experimentation with helicopters as a "parallel operation to the smokejumper project," the possible establishment of an "aerial shock troop system" modeled on local hot shot crews, and fuller participation with the Army and Air Force. A year later the first all-Service *Air Operations Handbook* was published. The imperatives of rural fire defense made the possibility of retreat even more remote and revived the old alliance between fire control and the military. Air tanker experiments, helicopter adaptions, and chemical retardants were a major concern of Operation Firestop. Again the trials were promising without being decisive.[122]

The breakthroughs came in 1956. Joe Ely, a veteran of the Firestop program and fire officer on the Mendocino National Forest, succeeded in making "tactical air support for ground fire fighters," in his words, "a reality." After Firestop, Ely requested the Willows Flying Service, a company concerned primarily with agricultural flying, to develop a method of delivering water on fires. Trials in 1955 relied on Stearman and N3N biplanes, recalling with a powerful sense of déjà vu the earliest proposals for air attack. Successful conversion owed much to pilot experience in crop dusting and the use of a free fall cascade of liquid rather than a containerized bomb. A chemical fire retardant (sodium calcium borate) was deemed suitable for use. The first fireline operations came in 1956, including drops on the Inaja fire.[123]

The 1956 commitment to air attack went even deeper. The Navy gave the Forest Service a surplus fleet of TBMs, which soon replaced the Stearmans as air tankers. Rotary wing aircraft were equally boosted, thanks again to military liaisons. A Helicopter Fire Fighting Program was launched under a cooperative agreement between the Forest Service and the Army Corps of Engineers. The Corps had undertaken a national defense subproject on "Atomic Fire Fighting Techniques and Equipment," and it in turn naturally contacted the Forest Service as the rural fire defense coordinator and as a former cooperator on similar test programs. Since the tests would be in Southern California, the Forest Service brought the California Division of Forestry into the organization, though the Services supplied the project leader, Carl Wilson. The result was the helitack program—a consolidation of existing uses, an exploration into new techniques, such as helijumping, and a tremendous stimulus to helicopter equipment development for the Arcadia Center.

Fire officers and researchers immediately recognized that the successes of summer 1956 had introduced a new order of complexity into fire control. In February 1957 a select conference was held at the Pacific Southwest Station of the Forest Service on the subject of "Handling Aircraft on Forest Fires."

36. B17 dropping slurry, Wenatchee fires, 1970.

A more widely publicized meeting in April, the Western Air Attack Review, released its reasoned conclusions to the public and to the fire community at large. The popular enthusiasm shown for earlier innovations paled beside the prospect of a forestry air force: aircraft again showed both its eminent practicality in tactical suppression and its equally inestimable value as an agent of publicity. Equipment development over the next decade was largely directed toward improving helitack capabilities and air tanker attack.

It was a golden age, the culmination of the drive for mechanization based on the internal combustion engine. Not accidentally, the outstanding example of technology transfer—the extension of fire control into interior Alaska—occurred in the late 1950s; the strategic possibility was almost wholly predicated on aerial transport, support, and direct attack. Nor is it surprising that the debate over prescribed fire and fuel management did not appear until the terrific expansion and elaboration of air attack technology had exhausted itself in the late 1960s. A new strategy of fire protection had been promoted during this era. It amalgamated the 10 A.M. Policy, with its imperative to stop fires in the backcountry, and rural fire defense, with its requirement that fires be kept out of the suburbs. But it was founded on the copious abundance of surplus military hardware and the very real success of mechanization in general and of aircraft in particular.[124] The successful mechanization of fire control had meant, too, the control of wildland fires ignited by internal combustion engines, and it demanded considerably greater attention to standardization of parts, specifications, and performance.[125]

Perhaps the climactic expression of mechanization came when the commitment to hardware extended even to the firefighter himself. Since the smokejumper program, much research had gone into protective clothing, and with the advent of interregional crews in 1961, the process was greatly generalized. In a little over a decade the basic firefighter was completely refitted with flame-resistant shirts and pants, with special fire shelters, with hardhats, faceshields, nomex hoods, and gloves. But also reaching back to the smokejumper program—this time to its motorability tests—MEDC undertook a programmatic study of "man the machine," measuring the firefighter's caloric intake against his fireline production, as if his work cycle imitated the adiabatic cycle of a heat engine, and developing exercise techniques and accessories to improve his performance, as if he were a model of helicopter in need of adaptation.[126]

III

The great flaw in the 1919 aerial patrol scheme, its communication system, was corrected a year later by the installation of wireless radios in the planes. Similarly, the 1956 breakthrough in aircraft was followed a year later by a major overhaul of Forest Service communications to transistorized circuits. Electronics made possible the information technology that Norbert Wiener described as a second industrial revolution, a revolution destined to outmode the internal combustion engine as the prime mover of Forest Service fire equipment. Remote sensing, data processing, simulation, telecommunications, and decision making were the first fire tools not created for physical use on the fireline. Hand tools had been designed to enhance a fixed source of energy, muscle; power tools, from trenchers to air tankers, were designed to develop a new source of power; but information technology was exploited to organize the tools and prime movers already in existence. Hand tools helped to implement and announce the early innovation in fire policy. Mechanization intensified and extended the dictates of the 10 A.M. Policy and rural fire defense. The introduction of electronic devices for making decisions coincided, appropriately enough, with major decisions by the Forest Service leading to its 1978 fire policy and with the adoption of a new strategy for fire management, the modification of the fire environment (largely fuels), rather than a further increase in the strength of suppression forces.

As a means by which to collect, process, and interpret information, the new technology gradually penetrated more and more provinces of fire control. It was often tied to aircraft of some sort, though with a critical twist:

the aircraft merely assisted information flow (for example, by transporting a remote sensing device); the technological innovations were not used as a means of justifying further aircraft development. Satellites could detect smokes and act as communications relays; infrared mapping became commonplace; remote weather stations automatically radioed data for fire danger forecasts; and computers entered nearly all dimensions of planning, presuppression, and even suppression.[127] FIRECAST and FIRESCOPE are prime examples of how information technology could underwrite both the current strategy of fuel management and the cooperative integration of resources characteristic of contemporary fire suppression. In the early days of fire protection, decisions had to be made on the basis of past statistics; they can now be based on simulation of future scenarios.[128] It is not the application of force or the origin of force that is at stake, but its management. In the processing of information, the decision itself is mechanized. The major exemplars of fire policy revisions during the 1970s, like DES-CON and wilderness fire management plans, require an almost constant sequence of decisions about how to manage a particular fire. These decisions commonly follow a flow chart. It is through and around the library of computer programs, like FIREMODS at the Northern Forest Fire Lab, that the entire spectrum of fire management is being remodeled.

Plenty of heavy equipment remained on the nation's firelines, and there were considerable attempts to integrate new machinery with fuel management programs. But the internal combustion engine was no longer, as it had been in the postwar era, a prime mover of fire strategy. The equipment centers were not inventing new technologies so much as they were completing the ambitions of an older one and overseeing its radiation into further niches, such as rural fire protection. A great demand continues for converted military surplus for use in the rural fire protection program, but the range for heavy equipment on federal lands may actually be shrinking. The contraction is manifest on several dimensions. Geographically, environmental considerations restrict certain types of chemicals and machinery, especially heavy tracked vehicles; in most wilderness or primitive areas, they are banned. Whereas mass fire control relied on mechanized suppression forces, wilderness fire management often requires the removal of such devices. Technically, the problem of converting military hardware has become one of production rather than design. Special fire equipment—for example, the BLM's Dragon Wagon—is often too expensive to manufacture in quantity and too limited in use. Most air tankers are still World War II vintage planes. Economically, too, a point of diminishing returns has seemingly been reached. The 10 Acre Policy showed that, even with sumptous increases in funding, added equipment could not overcome the irreducible quantum of large fires. Strategically, though heavy equipment can (and has been) adapted for fuel management, in many critical areas it is excluded, and current thinking prefers prescribed fire, whose chemistry produces effects that heavy machinery cannot duplicate.

When amendments to the 10 A.M. Policy appeared in 1971, fire equipment had reached a limit of gigantism and a point of consolidation. The Modular Airborne Fire Fighting System (MAFFS) outfitted the military C130 Hercules as an air tanker to carry 3,200 gallons of retardant, but the sheer size of the plane required to heft such a load limited its effectiveness in rugged terrain. The Dragon Wagon promoted by the BLM led to a similar paradox for ground tankers: its sheer size and complexity limited its use to the sagebrush of the Great Basin, and its cost restricted its mass production. Only a year before the new policy, a Fire Equipment Seminar was held at Marana, Arizona. It was the first since 1956, and it seemed to bracket an era. A year later the National Academy of Sciences Committee on Fire Research, set up in 1956 to coordinate and evaluate fire research sponsored by the Office of Civil Defense and the Department of Defense, held a symposium on air operations in the fire services. The proceedings were more a summary of past accomplishments than a preview of future activities.[129]

The chief innovations in technology are in a sense not coming from the equipment centers but from the fire labs next to them at Riverside and Missoula. Previously the equipment centers (often regional) had developed the instruments on which fire research relied; now fire research is creating the tools that have displaced heavy equipment from the avant-garde of fire management. The air tanker performance program at the Northern Fire Laboratory, for example, is a product of ambitious fire research, development, and application. But it has nothing to do with drop tank design, equipment adaptation, or engines. Rather it has explored different strategies for dropping in the context of existing hardware capabilities and particular fuel types. Its drop prints are an aid to the management of tankers and to decisions about their use, not a means to further the conversion of new tankers or to extend their range into new environments.[130]

The policy revisions of the 1970s were founded on an abundance of research information, much as earlier strategies arose from copious reserves of manpower and a surplus of machinery. The thrust of this new era was to consolidate and redirect information. Such information was less likely to come through an equipment center developing further mechanical fire power than, for example, from a computer bibliography like FIREBASE— also a product of the fire research, development, and application program— which could assimilate the exponentially increasing data on fire behavior and ecology. The contemporary equivalent to the 1945 equipment conference may be the 1978 National Fire Effects Workshop, which aimed to collect, organize, and publish the known effects of fire, wild and controlled.

Fire management had changed its need for evaluation in fundamental ways. Decisions about how to handle fires had typically been decided in national conferences and reported in the *National Forest Manual*. It was the duty of fire officers simply to implement those policies. Performance was subject to the judgment of a board of review. Current trends, however, have multiplied and localized the need for decisions. Whereas the 10 A.M. Policy emphasized the application of technique and tools to a fire, and not strategic decisions about whether to attack the fire at all, the 1978 policy required better, ongoing methods for making decisions according to wilderness fire plans, DESCON, escaped wildfires, and prescribed burning schedules. The question is often not how to apply power and tools, but when and whether to apply them. For immediate needs there are flow charts and logic trees for decisions and computer programs for data processing; on a larger scale, there is a dedication to systems analysis, evident in management by objective approach. Even on the fireline the pressing need is often not for more equipment or more crews but for better management of existing resources. Too often massive mobilizations have actually threatened to swamp suppression efforts by the sheer volume of equipment, crews, and emergency money. The internal management of suppression forces threatens to be more problematic than the control of the fire. Total mobility has made scarcity even more unlikely; the real problem is one of organization, accountability, and adherence to complex objectives requiring informed strategic decisions on a local level. Information technology promises to advise those decisions.

The transformations in equipment development from mechanical to electronic technology reflect an even more fundamental change. The information on which the new policies and strategies are based represents not only an increase in quantity but a change in type. The systems analysis and cybernetics that underwrite many management practices also underscore a revolution in theories of nature; they have mapped an intellectual watershed in how people think about the relationship of fire and nature. The two great ideas of nineteenth-century science were evolution and entropy. Both shared a belief in historicism, or time's arrow, in which something was progressively added or progressively lost. When the energy balance of nature was analyzed, it was compared with the great machine of the age, the heat engine. The life cycle characterized individuals and the successional cycle typified communities, but the energy dynamics of both was modeled on the closed thermodynamic cycle of an engine. Fire was seen as an interruption of this evolutionary progress, as a source of intolerable biological waste and of physical entropy. Now, however, in keeping with the information technology of cybernetics and systems theory open rather than closed systems typify nature. Rather than manifesting evolutionary epicycles, biological systems exist in a kind of steady state. Rather than eroding toward a state of entropy, the system is organized by information, a kind of negative entropy. Nature is an information processing machine. Fire is an agent of recycling, not an occasion for degeneration. It was this view of nature's energy budget that helped to force a reevaluation of fire management's escalating economic budget.

The Fire Task Force that followed the 1956 season addressed not only the new fire regime being created in Southern California and the old agony about fatalities but also the new mechanical possibilities for fire control so magnificently publicized that summer on the fireline. Ten years later, when the tragedy at Inaja was reenacted at the Loop fire, a second task force reexamined the program proposed by its predecessor. The hot shot crew on the Angeles was in fact an outstanding product of that new technology manifest a decade earlier, with its commitment to fire protection through equipment application and with its vision of man the machine. The sad conclusion of the 1967 task force, however, was both to reaffirm and to question the old program: the solution, it suggested, did not lie in the production of more machinery or better techniques so much as in better management of what already existed. In that phrase is summarized the changes of the decade that followed. The best fire tools for the immediate future consist of information technologies that can assist management, that rely on software rather than hardware and on *homo sapiens* rather than *homo machina.*

THE RED MENACE: A HISTORY
OF RURAL FIRE DEFENSE

That the Federal Government shows more interest in protecting
its trees than its citizens from fire merely reflects the long-stand-
ing indifference of Americans to the problem of fire loss.

Firefighters are, by temperament and training, people-
rescuers. . . .
—National Commission on Fire Prevention and Control, 1973[131]

I

The new fire regime advertised in Southern California had properties of
both wildland and urban fires, and the fires were attacked by agencies having primary responsibility for one or the other. In a way, this dramatic juxtaposition of wildland and urban environments only accentuated the mur-
kier and more widespread problem of fire in rural environments. On the eve
of the Inaja and Malibu fires, two events occurred that more systematically
directed the attention of both wildland and urban fire agencies toward the
question of rural fire. The Forest Service established a National Rural Fire
Defense Committee and made preparations for Operation Alert, an elaborate civil defense exercise to test the firefighting capabilities of the country
if it were subjected to a thermonuclear attack. At the same time, the
National Fire Protection Association (NFPA) released a revised edition of
its publication 295, "Community Organization and Equipment for Fighting
Forest, Grass, and Brush Fires." Prepared by its Forest Committee, the
pamphlet became the NFPA's major document on rural fire protection.
Together, the two events showed a determination both to invigorate the
ancient art of rural fire control and to contain distinctly modern threats
posed by the "red menace."

That rural fires attracted both urban and wildland fire organizations is
another way of saying that rural fire protection had no real organization of
its own. This was ironic for both wildland and urban fire services had originally evolved out of rural fire protection. Rural, wildland, and urban fires
behaved alike. They relied on the same fuels: for example, colonial towns
contructed of wood and thatch were merely a rearranged complex of forest
fuels. Fires in each environment were attacked with the same techniques:
the direct application of an extinguishing agent if the fire was small; firelines

and backfires if it was a conflagration. Fire control relied on the same body of knowledge, often depended on the same people, and was organized along similar lines and in a similar voluntaristic way. The initial settlement of the country in the name of agricultural reclamation meant that the landscape was rural and that fire practices and fire protection were rural. But with the advent of industrialization, this rural landscape began to melt away. Urbanization drained population from the farms into the cities and suburbanized adjacent lands, while the counterreclamation reserved public lands from reclamation and encouraged the reversion of marginal farmland to forest, marsh, and wilds. After World War II, moreover, the creation of rural *qua* rural communities became increasingly popular. "Rural" often designated a residential community, not necessarily an agricultural one.

The traditional scenario assumed that with time wildland fire would be transformed into rural fire, and rural fire into urban fire. Rural conflagrations were a transitory phenomenon of settlement. No systematic program of rural fire protection really evolved, nor was one considered necessary. Inherent in the new patterns of rural residency and land use, however, was a positive desire not to convert the fuel complex to a less intense load, and in many instances there was a determination to increase it in the name of forestry or aesthetics. Rural land had been the great beneficiary of the American reclamation and the chief victim of industrial counterreclamation. In defiance of traditional folk wisdom it was converted into residential sites and wildlands, both of which required more intense fire protection and neither of which would evolve into a less fire-prone state. Thus, when rural fire protection became a national goal as a means of improving the quality of rural life, as a necessity for the protection of recreational suburbs, and as a defense against the prospect of thermonuclear war, no organization really existed to cope with it specifically. Instead, fire protection moved in from the wilds and out from the cities.

The Forest Service might have remained an indirect observer of rural fire, much like the NFPA, contributing its knowledge through bulletins and fire codes. But the atomic bomb and the Cold War forced it into direct participation. Its contacts with the state foresters made it an ideal medium for nationalizing fire control into the rural environments for which state foresters were largely responsible. Clarke-McNary plans could form the basis for a national fire plan. The Service's long-time liaisons with the military on matters of fire weaponry and fire control expanded to include civil defense and were extended beyond World War II for both equipment development and mass fire research. Whether it wanted it or not, the Transfer Act had compelled the Forest Service to become something of a national encyclopedia on fire research and techniques, a national institute for conflagration control, and a national oracle on fire. More than a mere repository of knowledge or a regulatory agency intent on reducing hazards to human life, the

Forest Service had to manage fire actively. The agency had little choice but to become a national fire service as well as a national forest service.

The worst civilian disasters in World War II had resulted from firebombing. Reports by the Strategic Air Command and surveys like *Fire and the Air War* by Horatio Bond for the NFPA agreed that the next war would be a fire war and that civilians would be prominent targets. Nor was it lost to observers that the Battle of Britain had been won on the ground by aggressive fire brigades as well as in the air by RAF pilots. Retaliatory raids against German cities attempted to overwhelm fire services by crippling water supplies and communication systems as well as by sheer volume of fire starts. The combination of blockbuster bombs, which shattered buildings, followed by incendiary devices ensured that the urban firestorms would behave much like forest fires in heavy slash. This similarity of fire behavior linked Forest Service research interests with those of urban fire services.

The Hopley Report of 1948 for the Office of Civil Defense outlined in detail a plan for an agency that could be activated either in case of enemy attack or in the event of a natural disaster. Fire was likely to be common to both. The Soviet Union exploded an atomic bomb in September 1949, and six months later congressional hearings began on an appropriate civil defense organization. Before the year ended, the Korean War broke out, the Atomic Energy Commission released its "Effects of Atomic Weapons," which included sections on fire, Executive Order 10186 created a Federal Civil Defense Administration (FCDA) within the Office of Emergency Management, and the Federal Civil Defense Act of 1950 was readied. The FCDA submitted model interstate compacts for mutual defense based on the Northeast States Forest Fire Compact and released a publication on "Fire Effects of Bombing Attacks." Its administrator promptly directed letters to the state governors and to the Secretaries of the Interior and Agriculture with instructions to develop plans to cope with enemy fire attacks on the wildlands of the Lower 48 and Alaska. The example of the 1947 Maine fires showed the potential for fire disaster in rural lands as, within the decade, did the fire regime in Southern California. A National Wild Land Planning Committee administered the program under the direction of the Forest Service. Plans were completed in 1951.[132]

In September 1954 the President and the FCDA administrator formally delegated to the Forest Service responsibility for "a national program for the prevention and control of fires caused by enemy fire attacks in rural areas of the U.S."[133] The existing wildland plans were revised to include all nonurban land. A National Rural Fire Defense Committee oversaw the project from Washington, while the regional foresters were instructed to serve as general coordinators for planning a rural fire defense program in the states within their respective areas. Guidelines for the preparation of plans were issued; the wartime authorization for the Forest Service to serve as a distributor of excess federal property to state fire agencies was reinstated and strengthened; tests of the new arrangements, such as Operation Alert, were scheduled. By 1958 the fire plans were incorporated into a "National Plan for Civil Defense and Defense Mobilization."[134]

For the Forest Service there were privileges as well as responsibilities in its new assignments. It enjoyed a virtual hegemony over nonurban fire. Civil defense and military funding revitalized wildland fire research, and the Forest Service became a charter member in the Committee on Fire Research set up by the National Academy of Sciences–National Research Council. Its interests and institutions were integrally involved in the "National Plan for Fire Research," which the committee sponsored. The infusion of surplus military equipment made possible a new commitment to fire suppression and resulted in novel strategies. The reevaluation of the 10 A.M. Policy that might have been expected after the demise of the CCC was postponed, first by World War II and then by the imperatives of the Cold War. Sustained by its added directives and its newly found largesse, the Forest Service established two equipment centers and three fire labs, and it expanded into new geographic territory and jurisdictional terrain. On the issue of mass fire wildland, urban, and rural fire protection interests could intersect.

Following international crises in the early 1960s, FCDA (hereafter abbreviated OCD, Office of Civil Defense, as the agency continued to be called) appropriations for research increased dramatically, especially for fire research. The Forest Service participated in contracts to investigate the physics of mass fire behavior and to explore the adequacies of national fire defense. Additional contracts with the Defense Atomic Support Agency (DASA) and ARPA inverted the question of fire defense into one of the feasibility of fire attack, that is, the possibility of broadcast fire as a military weapon. The National Fire Coordination Study (1964–1965) functioned for OCD on a national scale as the 1957 Fire Task Force had for the Forest Service on an agency basis. The new task force outlined 19 subject areas for analysis. "Obviously fire services throughout the United States have great fire prevention and suppression strengths," the prospectus for the program announced. "But given all these strengths and others, how effectively can our fire services join together and cope with the fires that might result from a serious nuclear attack?"[135] That inquiry led the task force into a national inventory of fire suppression resources, to the agreements that made them available to cooperators, and to their actual functioning in situations as varied as the Malibu fires of 1956 and the Watts fires of 1964. The investigation, moreover, generalized the question into one of civilian disasters at large, including within its ambitious survey of emergency preparedness such events as the Alaska earthquake and Hurricane Carla. It was in conjunction with this survey that several experiments for improving rural fire protection were undertaken, among them an instructor training school and a Pilot

Rural Fire Defense Project in Florida, Colorado, Oregon, Kentucky, and Missouri. In this regard OCD sponsored research toward the development of a fire simulator for training at the electronics lab maintained by the Forest Service at Beltsville, Maryland.[136]

By the time Forest Service research into the behavior and control of mass fires ended in 1967, the Office of Emergency Planning (OEP) prepared a report to Congress on "Forest and Grass Fires." The immediate issue was whether to expand the air operations arm of fire control, but the report also urged that money be allowed for the construction of the Boise Interagency Fire Center, that fire research programs of the Forest Service, the BLM, and the National Bureau of Standards be better coordinated, and that the Department of Agriculture "take necessary action to obtain legislative authority and budgetary support to place in operation the rural fire defense program."[137] But although rural fire defense contributed to the larger national fire establishment, it did not yet benefit from it. Rural fire protection had greater strength on paper than in the field, and its rationales continued to be those of defense rather than of civilian disaster relief.

By the late 1960s, the prospects for detente with the Soviet Union improved, the afterglow of the Great Society sought programs to improve rural life, and there appeared a renewed interest in urban programs, including urban fire. Even as Congress entertained the OEP report on forest and grass fires, it initiated hearings that led to the Fire Research and Safety Act (1968). Among other things, the act provided for an augmented program of fire research under the auspices of the National Bureau of Standards. A year later the Committee on Fire Research proposed a National Fire Research Program, which sought to distribute and coordinate the array of federally sponsored research on the fundamentals of fire behavior. In particular, the congressional hearings noted the disparity in emphasis between forest and urban fire. Some $3.5 million went to Forest Service fire research. About $1.5 million went to the Department of Defense and about $1 million to Civil Defense—though much of these sums also ended up in contracts to the Forest Service. Half a million dollars went to the National Bureau of Standards, which maintained a research facility, somewhat on the model of the Underwriters Laboratory, dedicated largely to flammability studies. The Committee on Fire Research had provided a medium for the exchange of information among these various agencies, but the congressional committee was startled to discover the apparent disparity between urban and wildfire commitments.[138] What it failed to recognize, first, was that, insofar as the research emphasis was on mass fire behavior, the Forest Service was the only organization really equipped to undertake it. Urban conflagrations were largely a thing of the past except in Southern California, but wildland conflagrations were an annual occurrence.

Even more to the point, the federal government had direct responsibilities for the management of federal lands but not for the management of national metropolises. The National Commission on Fire Prevention and Control allowed by the act failed to emphasize this point in its 1973 report, *America Burning*. Although the report addressed the problem of rural fire and somewhat cursorily, the issue of wildland fire, it was overwhelmingly directed toward urban fire. Out of the report came further hearings and the establishment in 1975 of the National Fire Prevention and Control Agency, renamed the U.S. Fire Administration in 1979.[139] The Fire Administration proposed to do for urban fire what the Forest Service had achieved for wildland fire, but it proceeded more on the example of the Department of Housing and Urban Development than on that of the Forest Service. That is, it thought in terms of municipal services rather than in terms of land management. It imagined the federal role in fire protection as analogous to that which had developed for law enforcement. It was as if the federal government were a vast city hall.

Rural fire protection, however, tended to fall between administrative slats. As a technical question, rural fire suppression was a hybrid, requiring many of the skills and much of the equipment of both urban and wildland fire. It could not rely entirely on seasonals or on a paid cadre of professionals. It could use the specialized equipment and manpower of each system only sparingly. It required the ability to handle extensive fires in natural fuels, but also the capacity to rescue lives. More fundamentally, rural fire posed a jurisdictional and philosophical question, and here the relationships and programs developed in the name of rural fire defense promised the best answers. Even as the National Commission on Fire Prevention and Control began preparations for its report, Congress enacted the Rural Community Development Act of 1972, Title IV of which authorized assistance for a Rural Community Fire Protection Program aimed at organizing, training, and equipping local forces. The program would be administered by the Forest Service through the state foresters—in a sense yet another extension of the Weeks and Clarke-McNary programs. In 1975 Title IV was funded for a three-year trial period, and in 1978 the Cooperative Forestry Assistance Act included appropriations for the program.[140] The Forest Service and Fire Administration together contracted with the Research Triangle Institute for a basic survey on the rural fire problem, the first documented study of its kind.

The national commission voiced dismay in its report that the federal government seemingly cared more for its trees than for its citizens. The problem was not one of trees and people, however, but one of land. Both wildlands and municipalities were competing for a rural landscape at once decaying and transforming. For fire protection the rural scene was one of mixed jurisdiction and mingled techniques. The two poles of land use—wilderness and metropolis—are fixed; the rural lands in between are in a state of flux, and it is the transitional environment that is generally most susceptible to fire. The core metropolises are relatively free of conflagrations; the old pattern of urban rejuvenation through conflagrations has been replaced by a deliberate program of urban renewal. In wildlands, too, though conflagrations continue, their number is small and seemingly at an irreducible quantum. The intervening lands—not yet fully occupied by either the counterreclamation or by suburbanization—hold the greatest promise for a potential holocaust.

Thanks to industrialization, the modern farm is out of the woods and onto the grasslands, and the total acreage under production is contracting. Urbanization encroaches on these abandoned lands from one side, hurling suburbs and recreational residences (such as summer homes and retirement communities) into the rural landscape around it. The counterreclamation encroaches from another direction, promoting arable land into tree farms, creating recreational wildlands and wildlife preserves, and enlarging state, federal, and corporate forests. Unlike the reclamation, which initially increased fuel loads but ultimately reduced them, the processes of industrialization have often increased fuel loads with the idea of arresting them in a new status quo. Many reforested lands, for example, remain as woodlots, never to evolve into state or federal forests. Many recreational communities will likewise never evolve into incorporated municipalities. And, of course, to these developments there must be added the unmanaged fields and decaying hamlets of a rural landscape in decline. In each case an absence of jurisdictional control is reflected in a partial vacuum of organized fire protection.

The tendency to date has been to expand the apparatus developed by the Forest Service and state foresters, a system founded on principles of land management, rather than to elaborate urban fire machinery, a program founded on a concept of municipal services. The appearance of private companies, such as Rural/Metro in Arizona, that contract for both suburban and wildland fire protection may fill local voids.[141] But the national machinery will probably remain that developed over most of the century through cooperative programs under the aegis of the U.S. Forest Service. As an institution, these programs were immeasurably strengthened by the perceived wartime necessities of rural fire defense. They have become permanent through the recognition that the rural fire threat is less from an enemy without than from demographic and economic processes within.

To the National Commission on Fire Prevention and Control this evolution seemed both odd and vaguely degrading to the traditions of urban fire services. It wanted to unify national fire services in an organization based on urban experience. It thus echoed the common political sentiment of the day that America's cities were the nation's great problem and promise, that people services should prevail over all other concerns. But, as Abraham Lincoln observed, a nation consists of a people, its laws, and its land, and of these, the greatest is its land. Urban fire has always emphasized the protection of people; wildland fire, the protection of land. America's rural lands are the great mortar of national history and geography. The problem of rural lands is a problem of land management first, and of municipal services second. That fire protection came to the rural environment from the forests did not reflect American indifference to the problem of fire loss, as suggested by the national commission, but the American preference to reach people through the land rather than through the city.

II

Despite their often fundamental differences, wildland and urban fire protection have advanced stride for stride.[142] Before the industrial revolution, fire in both environments showed remarkable similarities. As a physical phenomenon, fire frequency exhibited a similar logarithmic distribution, largely as a function of weather. Fire behavior in each was indistinguishable, given that most urban fuels consists of wildland fuels only rearranged in form and still sensitive to meteorological processes. Fire functioned in the natural history of cities as it did in the natural history of forests. It was, on the one hand, an interruption in the evolutionary succession of urban development and, on the other, a mechanism for recycling and rejuvenating. Fire most often attacked the very young, closely stocked portions of a city and its very old, decadent slums—the same forest stages most susceptible to fire. But it did more than interrupt the life cycle of a city: it often invigorated it. Urban conflagrations frequently converted slums into posh residential sites or factory zones. Fire most likely struck abandoned areas or sections decimated by plague, much as in natural contexts it prefers dead patches left by former fires, plagues, or insect invasions. In both cases fire attracts vandals and predators. Yet in both it is an extravagant and devastating tool for the management of an environment. Prescribed fire or surrogate techniques now substitute in wildland settings as do urban renewal programs in municipal contexts.

The strategies of fire control for both environments bear strong resemblances. Both rely on prevention, strong initial attack, and modification of the fire environment. It is the question of fuels management, however, that most readily distinguishes the two approaches. Urban fire protection deals with a synthetic environment, and nearly all phases of the urban environment are—in theory at least—subject to fuels management. Flammability tests and the combustion of individual fuel units, rather than the behavior of fire in large fuel complexes, dominates urban fire research. Urban fire strategists rely on building and fire codes to regulate the kind and arrangement of construction materials. Zoning laws can create built-in conflagration controls. Particularly hazardous substances may be banned. Inspectors

may use the force of law to remove unwarranted fuel accumulations. And
for various fuel loads or complexes, a certain level and type of fire protection
is required. In theory one could construct an environment almost entirely
out of noncombustible materials and design it for maximum fire security.
Buildings are no longer designated as fire-proof, but they do have careful
standards for compartmentalizing and retarding fire spread. A wildland
environment too, must by definition remain flammable, though the fuel load
and its properties may be altered. In urban environments one can specify
what materials may be used and excluded, but in wildland environments,
one can only modify what already exists. In both cases, however, there is a
problem of exotics—new flora and fauna for the one; new plastics and syn-
thetics for the other.

Urban conflagrations in America were common up to and through the
early decades of the twentieth century. Especially devastating was a series
of holocausts from the Baltimore (1904) and San Francisco (1906) fires to
the Berkeley fire (1923). That urban and wildland fire protection share a
common origin in industrialization is shown by the close, parallel steps taken
in each arena. In 1896 the NFPA was incorporated as a nonprofit private
organization to give a common focus for urban fire problems; that same year
the National Acadamy of Sciences began its investigation into the management of the forest reserves.[143] In 1905, prodded by the Baltimore fire, the
National Board of Fire Underwriters (later the American Insurance Company) sponsored the first model building code. This was the same year of
the Transfer Act and the first edition of the *Use Book*. Reforms intended to
eliminate urban conflagrations were promoted vigorously after the 1923
Berkeley fire, much as the 1924 forest fire season in California helped to
pass the Clarke-McNary Act. For both environments conflagrations diminished in frequency and extensiveness, though concern about both revived as
a product of the firebombing in World War II and the massive deployment
of incendiary weapons. The chairman of the NAS Committee on Fire
Research, Hoyt Hottel, had been acting chief on fire warfare for the
National Defense Research Committee during the war, and he described to
congressional hearings in the late 1960s how "three years in charge of intensive wartime research effort—and I mean lying awake nights trying to figure out how to start more fires—an effort costing millions of dollars, this
effort on how to start fires convinced me of the inadequacy of our knowledge
of building-fire hazards."[144] In recent years the worst conflagration problems
have occurred where the wildland and urban environments intermingle. The
incomplete suburbanization of Southern California, for example, is not
unlike the mixture of barns and buildings that resulted during the rapid
urbanization of Chicago and contributed to the 1871 inferno.

The fire problem in America has evolved out of a cultural environment as
well as a natural one. The coming of American fire was a process charac-

terized by great haste, an almost constitutional instability, and a lingering heritage of frontier violence. When Washington Irving ventured onto the Great Plains, he hoped to witness a prairie fire; tourists expected fires as they did buffalo and wild Indians. Irving's guide was ready to oblige, promising to start one himself if none was encountered naturally. The same attitude could make urban fire a similar species of entertainment. A Swedish visitor to New York in the 1840s wrote that "when the fire breaks out tonight . . . we'll go out and take a look at it. It was like deciding to go to the theatre to see a play that had been announced and that could be counted on with certainty to come off. And sure enough, we did not have long to wait for the spectacle."[145] American fire statistics parallel its crime statistics. The United States has far and away the worst fire record of any industrialized nation. Its fire death rate is 200 times greater than Canada's, 400 times greater than England's, and 650 times greater than Japan's. Its fire losses are exorbitant by a similar order of magnitude.[146]

Nonetheless wildland and urban fire protection systems show fundamental, even irreconcilable differences. Urban fire protection has tended to treat fire as an economic problem; wildland fire protection regards it as a political problem. Urban fire history tends to read like a series of case studies, and as a political phenomenon, urban fire protection represents a loose confederation of city-states. Union and private organizations, such as the NFPA, have given these city-states a national focus, but until the U.S. Fire Administration there was no real political mechanism for unifying them. Effective integration has come through fire insurance companies.

The insurance movement arrived early in colonial history, precipitated by the great London fire of 1666, which followed on the heels of the plague. The fire led to a host of firefighting reforms, of which insurance was the most important. Insurance companies promoted early mutual fire brigades, sought model uniform building and fire codes, promoted research establishments like the Underwriters Laboratory, and encouraged modernization of equipment. Through their grading schedule, which evaluates fire hazards and protection strength in order to establish rate premiums, the companies have been the great force for the elaboration of urban fire protection. Likewise, the driving mechanism for extending municipal-type fire services into rural environments has been the grading schedule. Arson for profit, the chief source of urban incendiary fires, presupposes an insurance structure, which wildlands lack. Fire insurance companies predated the creation of most American cities, and they may also have engendered the attitude that fire (or arson) is economically or legally wrong, rather than morally wrong. In Japan, for example, as in European countries where cities of antiquity still stand, it is shameful to allow one's house to burn, because the entire community is also threatened. In America a fire is disgraceful only if the property was not properly insured. The metropolises of the New World were so rapidly constructed that custom could not readily be brought to bear on the fire problem, and fires were so common amid the constant turmoil of expansion that they became as much a part of urban life as of rural life.

A system of insurance could never be applied to American wildlands. The fire threat alone was too great, the migratory impulse too compelling, and the transitions of growth too sudden. Instead, a political solution developed. The creation of forest and park reserves made the federal government a permanent landholder. Though it sought to apply economic criteria by which to assess the appropriate investment it should make in fire control, its administrative programs were necessarily dictated by politics. Wildland fire agencies had direct political control over protection funds, whereas urban fire services had to rely on a tax base set indirectly through fire insurance companies. Thus urban fire protection found itself with a national economic unit based on the insurance grading schedule but with little political organization beyond the municipal level. Wildland fire protection, conversely, fashioned a marvelous national political framework but found itself searching for better economic criteria. The urban organizations consequently campaigned in the 1970s for a national fire administration, while wildland agencies sought to revamp policies to reflect economics and ecology more accurately.

Nowhere are the peculiar qualities of each fire service better highlighted than in the differences between the firefighters themselves. They constitute two different personality types, and there is little interchange between them on the line or in temperament. Urban firefighters see themselves as lifesavers. Urban fires are by definition fires in a human environment; wildland fires are a natural process, though one often directed toward human ends. Urban fire services see their fire problem in terms of human life: they measure fires by their cost in lives, those of firefighters and citizens both. The great challenges to urban fire control are not conflagrations sweeping block after block of cityscape but fires in buildings densely populated with people. The insurance companies estimate the property loss; the fire department, the loss of life. The core of the uniform building and fire codes are the life safety codes; they are designed not to improve access for firefighters or to reduce fire intensity so much as to improve egress for occupants and to lessen hazards that might compromise their exit. Urban fire strategies conceived fire protection as a basic service for human life support, along with water, power, and hospitals.

Its people, too, make wildland fire control distinct. Its ranks tend to be made up of seasonals—young, non-unionized, and with a high turnover. Urban firemen pursue their profession for life; wildland firefighters for a tour of duty en route to other careers, much as they might do military service. Wildland fire success is measured in resource protection, not in lives. Its crews are not lifesavers or even, on most cases, savers of private property.

They see themselves as an army waging a moral equivalent of war on nature. Wildland fire control does not promote a life-support service so much as it defends land from attack, and it is appropriate that Civil Defense turned to wildland fire agencies rather than to urban fire services to protect rural lands from the red menace. Urban fire is in many ways an act of interpersonal behavior between firefighter and victim, with the fire as the medium of communication. Wildland fire pits man against nature. The cherished image of the urban firefighter is the picture of a daring rescuer with a helpless victim nestled in his arms. The ruling images of wildland fire control are those of retardant dropping from the belly tanks of a B-17, of long rows of tough, begrimed crews marching along a burned-out fireline, of a ranger at the entrance to an abandoned mineshaft magnificently defying the tidal wave of flames hurled against him.

8 FIELDS OF FIRE

Firefighting is a matter of scientific management, just as much as silviculture or range improvement.—William Greeley, 1911[1]

In the annals of Alaskan forest fire history and in the memory of the old-timers, there has never been a fire season to match the one of 1957. It is estimated that between 5 and 6 million acres have already burned this year, and the season is not yet over. —Charles Hardy, U.S. Forest Service, 1957[2]

In the history of wildland fire protection in the United States there are two years within and around which events cluster naturally, and one around which events cluster as a result of administrative fiat. The first two are 1910 and 1956; the third, 1936. The significance of these years is both practical and symbolic. On each occasion virtually every dimension of fire protection underwent a fundamental transformation. The events surrounding 1910 are organized, if not catalyzed, by the great fires of that summer; those around 1936, by the 10 A.M. Policy and its promise of fire exclusion; and those around 1956, by the fires of Southern California and the prospect of a global holocaust.

The complex in 1956 opened up new fields of fire. Old techniques and new equipment extended fire protection into new geophysical realms like the counterreclaimed rural landscape and the unclaimed interior of Alaska. The development of equipment centers, moreover, was paralleled by the establishment of fire research laboratories. Equipment dominated fire strategy because it relied primarily on the conversion of existing hardware; scientific knowledge exerted a controlling influence only later, because it demanded the creation of new data and in some cases the invention of new sciences. If the events of 1956 succeeded an important Service-wide meeting on fire research, they were in turn followed by the first Service-wide meeting on air attack. Both the technology for fire control and the understanding of fire effects were poised to begin the exploration, penetration, and assimilation of the fire geography of the planet. While the products of equipment development led to the annexation of new domains for fire protection, the products of fire research revealed new fields for fire management.

The Forest Service came to exert intellectual dominance over the subject of wildland fire in the same way that it exercised institutional control. From the beginning the Service recognized that fire research was essential to forest management. The first study by the Bureau of Forestry under Pinchot had been an investigation into fire damages and a study of fire in the Pine Barrens of New Jersey. In 1910 Graves admonished the nation that "the first measure necessary for the successful practice of forestry is protection from forest fires."[3] In 1940 H. T. Gisborne reiterated this belief: "Fire research is therefore intended to serve as directly as possible the fire-control men who must first be successful before any of the other arts or artistry of forestry can function with safety."[4] So little was known about systematic fire protection that fire research had to investigate virtually every dimension of the subject, from the proper placement of lookout towers to the calculus of fire damages, from the distribution of roads and tool caches to the mechanisms of fire spread.

The rejection of folk wisdom by professional forestry meant that knowledge of fire use had to be rediscovered, translated, and codified within the apparatus of modern science. Knowledge, as Francis Bacon put it, is power, and despite its political jurisdiction over the forest reserves, the Forest Service found itself strangely powerless against the chief threat to its management of the public lands. Research was expected to supply whatever knowledge was needed to implement systematic fire protection successfully, and in fact it invented the concept in the first place. By 1955 nearly three-fourths of all the scientific research in the world came from the United States, and this was as true of fire research as of other topics. By the mid-1950s the Service was prepared for a major transformation in wildland fire research, though it was not always so well prepared for some of the consequences. During the previous era, researchers were acutely aware that fire was a part of forestry, and they looked to forestry for objectives and methodology. During the coming era of Big Science, however, researchers saw fire as a problem worthy of investigation in its own right, a phenomenon that required techniques drawn from engineering and the physical sciences. Forestry was merely one field among many suitable for its application.

New geographic fields opened as well as intellectual ones. In 1955, while the BLM attempted for the first time to control a large fire in the Alaskan interior, it invited Forest Service researchers to survey the fire protection needs of the interior. Two years later perhaps as much as 5 million acres burned in one season. With the Forest Service example and advice, the BLM prepared for a massive buildup of control forces to guard the public lands of Alaska. The task was enormous; half of all the public lands in the United States were located in Alaska, then still a territory. The attempt was a spectacular example of technology transfer—one made possible by the development of aircraft for fire control in the Lower 48, one made desirable by the prospect of statehood, and one made especially meaningful by its compressing within a 20-year span the 100-year history of fire protection in the United States. It was as if in its regional ontogeny Alaska recapitulated the national phylogeny. Alaska, moreover, showed both the capacity for and the limitations upon the export of fire expertise into new lands. The experience put the BLM into fire protection in a big way and led the agency a decade later to push for the Boise Interagency Fire Center.

Its directives for rural fire defense also took the Forest Service into the rural landscape of the continental United States, and the Rural Community Fire Protection program even brought it to the island trust territories of Samoa, Guam, Puerto Rico, and the Virgin Islands. Wildland fire also became a subject of international concern. For many rapidly developing countries, landclearing for increased agricultural production created unprecedented fire hazards, and overgrazing aggravated the ancient alliance of fire and hoof. For industrialized countries, new fire hazards appeared with the abandonment of marginal agricultural land, with deliberate reforestation, and with the purposeful increase in wildlands. The prospect of fire warfare through either conventional or thermonuclear weaponry made prescriptions on fire use and knowledge of its control of widespread interest.

The United States tended to become an international point of reference in fire management as in other things. The original export of industrial forestry in the late nineteenth century had largely followed the British flag around the globe; the dissemination of fire control knowledge by the mid-twentieth tended to emanate from the United States. In establishing its forest reserves, America had looked to India and Canada for possible models. World War II, however, left the United States as a global economic, political, and scientific power; fire control was among its many experiments in technology transfer. Fire officers with the Forest Service wrote the bulletins on fire control that the Food and Agriculture Organization (FAO) published for the United Nations.[5] It participated in elaborate field experiments on mass fire under a joint program with Canada, Australia, and Britain. It proposed that the North American Forestry Commission (NAFC), a subsidiary of the FAO that included the United States, Canada, and Mexico, establish a Forest Fire Control Committee and publish an international newsletter.[6] Under its own and FAO auspices the Forest Service sponsored study tours to Australia, the Soviet Union, Canada, and Mexico. Forest Service fire specialists have been sent under FAO and Peace Corps auspices to South American and Europe. Advisers went to Brazil during the fire crises of 1964 and to the Dominican Republic in 1965.[7] (In the Dominican situation the Forest Service found itself almost elbow to elbow with U.S. Marines in what might seem to be a case of firetruck diplomacy, a mutual effort to suppress the fire and the sword of insurrection.) The United States in particular hosted an astonishing array of symposia on the subject of fire in various landscapes.

Not everyone of course was pleased with this deployment of American firepower in what might seem to be a global strategy of containment. Until the 1970s American foresters emphasized fire suppression. Indeed, Show himself assisted in the transition, exporting the lessons of systematic fire protection developed in the United States to new lands. In the American experience, controlled fire had seemed a barrier to the introduction of professional forestry, but to many developing nations it promised to be a useful means of transition from traditional forest practices (usually dependent on fire) to the techniques of industrial forestry. When the FAO held a study tour of the United States and Mexico in 1975, for example, the topic that most interested the participants was precribed burning. By then the United States had converted to the doctrine of prescribed fire and was able to direct the tour to its showcases, the corporate and federal forests of the South.

In this process the United States got as much as it gave. The experiences of Canada and Australia, for example, were useful models for extending fire protection to the taiga of Alaska and for adopting prescribed broadcast fire for fuel reduction. Perhaps even more important have been the global surveys of fire practices and fire environments, a process begun with the Tall Timbers Fire Ecology Conferences (1962–1974). Much as early man brought fire to all the world's terrestrial environments, modern man has radiated across the globe in an effort to assess the effects of fire. Fire in Africa, fire in Southeast Asia, fire in North America, fire in northern environments, fire in tropical and arid environments, fire in Mediterranean environments—the surveys constitute a global inventory, both biological and anthropological. They are part of a new endeavor to define fire, its meanings and limits.

It can hardly be doubted that the sheer panorama of fire practices revealed by these symposia affected the revision of Forest Service fire policy. The effect was not unlike the consequences that resulted from global geographic explorations conducted by earlier ages or from contemporary explorations of the ocean basins and the planets, which have revolutionized the earth sciences. Before the onslaught of new information, beneath the sheer multiplicity of peoples, places, and practices, the logic of the old systems crumbled. The surveys compelled recognition that fire and fire policies, like other cultural practices, are relative, that they are not founded on absolute environmental standards. In the early days, Western Europe provided the example of what a properly managed, fire-free forest should look like. By the mid-1970s there were dozens of counterexamples in which controlled fire advanced the interests of industrial forestry and in which, even within the United States itself, aggressive fire control had resulted in peculiar problems of its own. The United States could no more aspire to be a global fireman than a global policeman. The revised fire policies were in a sense a recognition of the pluralism of fire. In extending itself to Alaska and to the lands of its allies, America learned the limits to its fire policy and firepower.

In looking out, it became necessary to look inward, too. To those pivotal years in which nearly all aspects of fire protection were transformed it is possible to add the period 1970–1972. Within three years, not only in the United States but around the world, the problems and possibilities of free-burning fire worked their way to a more or less common definition in a grand ensemble of politics, science, economics, and environmental concern. Wilderness fire replaced mass fire as an informing topic of policy. Fire ecology replaced fire behavior as a defining concern of fire research. The programs of one era came to a triumphant conclusion while those of the next era rapidly took shape. By the end of the 1970s it was time to assimilate rather than to expand, to reevaluate policy rather than to export it. The geographic emphasis would shift from the new fields of fire posed by Southern California and Alaska to that unique blend of old and new regimes offered by the Southwest. If Alaska was a recapitulation of national experience, the Southwest was a palimpsest of it, a mosaic of American wildland fire problems and practices, at once unique and representative. Like the United States in the global arena, the Southwest both expanded and consolidated American fields of fire.

The current fire policy of the Forest Service, with its recognition of a pluralism of fire regimes and of fire management objectives, has long existed in the Southwest by virtue of its complex historical geography. The region not only adapted and absorbed but also assimilated and preserved. In 1956, when striking breakthroughs in equipment development were ready to launch fire control into the Alaskan interior, when the prospect of an atomic war led to the extension of fire protection into the rural landscape, and when fire research was poised to open new realms to the intellectual understanding of wildland fire, the Southwest experienced a record-setting bust of lightning fires. The most recent of fire cycles in American history was a tribute to the most ancient.

THE THERMAL PULSE: A HISTORY OF
WILDLAND FIRE RESEARCH

All things are an exchange for fire and fire for all things. . . .
—Heracleitus[8]

The new Philosophy calls all in doubt.
The Element of fire is quite put out. . . .
—John Donne, "An Anatomy of the World"[9]

I

To the ancients fire was both a substance and a process. It was a universal phenomenon and a universal principle of explanation. It inspired analysis and philosophical reverie. Heracleitus made fire the essence of the natural world, a principle of endless change and a phenomenon of universal extent. Empedocles transformed this oracular vision by making fire one of four elements, and Aristotle accepted this primacy of earth, air, water, and fire. Fire was thus a fundamental substance, a cosmogenic principle, and a procedure by which to induce change. *Philosophus per ignem,* as it was known to medieval alchemists: knowledge through fire. Even in the age of the Enlightenment fire was seemingly broadcast throughout the cosmos. The central fires accounted for geologic events; the ethereal fire, for celestial phenomena; electrical fire, for lightning and electricity; dephlogistinated fire, for chemical events. Fire powered the new engines of the industrial revolution, suggesting an equivalence of mechanics and heat. Fire was involved in some form with virtually every scientific revolution, especially chemical, from the Middle Ages on, and its failure to cope satisfactorily with fire was a primary reason for the relative retardation of chemistry. Even in the eighteenth century Boerhaave wrote, "If you make a mistake in your exposition of the Nature of Fire, your error will spread to all the branches of physics, and this is because, in all natural production, Fire . . . is always the chief agent."[10]

Gradually, though, fire lost its primacy. It became an object of investigation rather than a principle of explanation. Chemistry separated fire from the elements; mechanics separated it from heat; optics, from light. With the development of thermodynamics, the concept of energy assumed the role previously held by fire. Once fire was explained in terms of mechanical philosophy, the possibilities of fire as an explanation itself vanished. The industrial revolution, moreover, transformed the physical realm of fire: confined fire belonged in the engines of the new heat machinery, and free-burning fire was loosed onto the fields of the counterreclamation. Industrial fire preoccupied most investigators, and to it they applied the techniques of mechanical and chemical engineering.

Fire had once been an enigma of universal significance, but the last vestiges of its ancient intellectual primacy were dissolved in 1905 with two publications by Albert Einstein. His paper on the photoelectric effect gave great impetus to the development of quantum theory, a chemistry not based on combustion and one for which transmutation came through radioactive decay, not *per ignem*. His special theory of relativity, meanwhile, destroyed the last trace of the ether, a concept traceable at least back to Anaxagoras, who had identified it with elemental fire. Fire had nothing to do with the revolutionary discoveries of the microcosm, the atom and gene, or with the macrocosm of relativity. The intellectual reconstruction of the physical world on the basis of the atom completely replaced the pre-Socratic construction based on fire. The atom promised, too, an alternate source of energy to replace those based on combustion.

Yet it was in that *annus mirabilis*, 1905, that the Transfer Act presented forestry with its great mandate to manage fire. Just as free-burning fire appeared to vanish from serious scientific inquiry, it reappeared as a fundamental, practical problem for forestry. The outcome was the birth of a new and largely American science, though one that eclectically proceeded by analogues to other sciences. "Forest fire research apparently originated in the United States," Earle Clapp wrote, "undoubtedly as the direct result of a forest-fire situation which is more serious than in almost any other country."[11] It opened new intellectual as well as administrative terrain. "Here is a challenge to the brains, the imagination, and the investigative ability of the Forest Service," Roy Headley exulted during a debate about whether the Service should establish an independent fire research program. "We have a concrete and definite problem of far-reaching importance which has us all baffled. It can not be studied by the conventional methods employed in physics, chemistry, or biology. Only trail blazers need apply for the job."[12]

In blazing that trail, the Forest Service came to assume an intellectual hegemony over the science of wildland fire as it had over its practical administration. As an administrative problem, forestry studied fire protection within the methodology of operations research. As a scientific question, it applied the techniques developed for heat transfer and confined fire to the peculiar properties of free-burning fire. The incentive for the first research was the Transfer Act, but the catalyst for the second was the challenge of mass fire. Here, in a strangely appropriate way, fire and atom were reunited. The great transformation in fire research did not look for inspiration to the cosmogenic fire of Heracleitus but to the anthropogenic firestorms over Germany and Hiroshima.

During the period in which systematic fire protection was developed for frontcountry lands and then extended into the backcountry, fire research was envisioned as an administrative process. Research was intended to preserve policy from political attack, to put protection systems on a firm administrative base, and to assess the economic investment appropriate to adequate fire control. Researchers and administrators exchanged positions freely. The forester, insisted Coert duBois, "must combine the functions of the economist and engineer. He must state the problem and then solve it."[13] The methods used were, by and large, those of silviculture and forestry economics, adapted to new topics. By committing folk knowledge for the most part to the rubbish heap of outworn superstitions, forestry found it necessary to relearn everything in a form it considered suitably scientific. The preoccupation of fire research, insofar as it enjoyed an independent identity, was to supply the information with which the Forest Service could establish administrative control over its new domains. Fire research was conceived of as research into fire protection, an operational problem whose management required systematization more than experimental discovery. The great products of this era were manuals of occupation, working plans for the establishment of a fire protection system. In fact, the most striking thing about early fire research may be the extent to which it does not deal with fire at all.

Fire research was often indistinguishable from administration. In the absence of folk knowledge and in a general vacuum of European precedents, research supplied the basic information upon which working plans could be based. Equally, it furnished standards upon which performance in executing those plans could be evaluated. The breakdown of the fire organization during 1910 showed the necessity for fundamental analysis, and the period that followed was no less a time of experimentation for research than it was for administration. In both cases the most important trials came from California, where they were largely under the direction of one man, Coert duBois. As regional forester, duBois recognized the extent to which new administrative plans and methods had to be founded on precise information. He initiated a host of investigations: Roy Headley analyzed the 1911 Arrowhead fire, the origin of individual fire case studies; E. I. Kotok examined fire reports in 1914, the origin of statistical methods in fire research; and duBois and S. B. Show investigated fire behavior and damages through field trials at the Feather River Station, the origin of fire behavior and fire effects research.[14] The search for a scientific methodology that underlay these efforts was thus complementary to the search for administrative methods that motivated concomittant experiments in light burning, let burning, and so forth. Both were in a sense amalgamated and solved with the publication of duBois's brilliant monograph, *Systematic Fire Protection in the California Forests* (1914), which was as much a triumph of early research as it was of early administration. Not incidentally, most of the great names in fire protection and fire research for the next 30 years came under the personal influence of duBois or his successor, Show.

DuBois's publication was well timed. The year after it appeared, the Service established a Divison of Fire Control within its Operations unit and created a new agency, the Branch of Research. The new chief of research, Earle Clapp, appealed to the experiment stations to initiate investigations on fire. The prospectus envisioned studies on fire and climate, rates of spread, fuels, topographic influences, and the prediction of dangerous fire conditions. Individual stations differed widely in their response, though the nearly universal question of light burning tended to provide a common focus, and duBois proceeded immediately to direct the California research in that direction. In December 1920 Clapp transmitted a memorandum to all regions concerning "Research Projects in Fire Protection." Admitting that earlier classifications of topics had been unsatisfactory, he divided the subject into a hierarchy of economics, origin and behavior, and technics of fire control.[15] The memorandum coincided with the succession of William Greeley as Chief Forester and did for fire research what the Mather Field Conference, which Greeley convened the next year, would do for fire administration. As early as 1911 Greeley had endorsed the value of fire research, and throughout his tenure as Chief he reaffirmed that commitment. At Mather Field he oversaw the standardization of the individual fire report, the fundamental source of fire research data; confirmed the value of duBois's methodology by supporting elapsed time studies; and promoted the work of Stuart Brevier Show, duBois's heir apparent. What Greeley was to do for the administration of fire control over the next decade, Show was, by his example more than by his administrative position, to do for research. Already Show had demonstrated himself to be a consistent winner. His research had exposed equally the fallacies of T. B. Walker's light-burning program and the failures of Roy Headley's let-burning schemes. His collaboration with E. I. Kotok in 1920 resulted in the influential "Forest Fires in California, 1911–1920," which, along with the California Forestry Commission's report—much of it based on field experiments by Show—gave the coup de grâce to light burning. Together Show and Kotok defined the range, methodology, and themes of fire research, culminating in the hour control concept, which underwrote the extension of duBois's concept of systematic fire protection into the backcountry. When in 1926 Kotok left administration to assume the directorship of the California Forest Experiment Station and Show left research to become regional forester, the transfers made not a whit of difference to their collaboration—a fine example of the degree to which administration and research interests were intermingled.

It was the Shovian conception of fire research that came to dominate. In contrast to Clapp, who eventually proposed that fire research begin with fire, Show insisted that it begin with fire control. Fire behavior meant the calculation of rates of spread for the determination of hour control standards, not the physics and meteorology of free-burning fire. Fire effects meant the study of fire damages, not fire ecology. The primary source of data was to be the individual fire report, not laboratory or field experiments, and the chief methodology was the statistical analysis of hundreds of such reports. Fire research was promoted, too, as a "tool for perfecting performance." It did more than merely standardize; it furnished standards by which fire problems in different forests could be evaluated and individual competency judged. Both administration and research agreed that their shared concern was to establish, direct, and evaluate fire protection systems. "We accepted that the route of research was from the general laws and relations to the particular, seeking to refine and measure the arithmetical values," Show explained. "In no sense did we accept the route of the particular to the general." The "general" was the program established by administration; fire research was meant to answer questions, not generate them. And, over the next decades, whenever fire research was asked to supply answers, it was Show who generally spoke, the Shovian conception of research that generally prevailed, and the Shovian solution that was largely adopted.[16]

After Mather Field, Greeley sent the team of Show and Kotok center stage to the light-burning controversy—in Show's words, to "tackle the application of the general to the particular on our old training ground, the Shasta National Forest."[17] Greeley, meanwhile, sought out a more truly national framework for fire research. Two programs were proposed. Early in 1924 a meeting of experiment station directors at the Forest Products Laboratory created a Committee on Forest Fire Research Program, chaired by S. T. Dana, which issued a lengthy summary of recommendations. The committee included Show, and its conclusions were fundamentally his.[18] In 1926 Clapp published a rejoinder of sorts in his *National Program of Forest Research,* the report of a special committee set up by the Society of American Foresters. Clapp lamented that "we have fewest and least valuable precedents from European and other foreign practice in fire protection, and this is particularly true in research." He then outlined the basis for an adequate research program. At its nucleus, he concluded, was the determination of the "laws of combustion of forest fuels in the open." On these laws all else rested, and "closely connected with these should be the necessary supplemental research to carry the results through to the point of final practical application in fire suppression." The "final expression" of research should be the "determination of protection standards themselves." The total package envisioned research on combustion, fire prevention, fire suppression, fire damage, fire as an agent, protection standards, and organization.[19]

In September 1926 Clapp wrote memorandums to Greeley and Headley on the possibility of enlarging fire research. Greeley and Headley in turn exchanged correspondence over the next year. With the establishment of the Forest Protection Board, a special division of fire research had the opportunity to integrate research on a national scale. In 1928, moreover, Congress passed the McSweeney-McNary Act, which authorized the Forest Service to engage in extensive forest research, including fire research. All federal research on forest fire, in fact, was to come through the Forest Service. Greeley favored the bold stroke. In March 1928 he circulated a proposal for an all-Service fire research program. Headley, however, felt that such a move was premature, unnecessary, and probably expensive.

Greeley resigned before the results could be assembled, but as Headley communicated them to his successor as Chief, Robert Stuart, in 1929, they strongly favored a regional and practical approach. In striking contrast to Clapp's belief that fire research should begin with fire, fire officials almost unanimously insisted that it should begin with fire protection. Show and Kotok argued for administrative analyses rather than "an academic conception of relationships which no doubt exist but the demonstration of which would have little practical or usable value." Raphael Zon confessed that "eventually fundamental studies of combustion may yield some results of practical value," but he wanted to put the money into problems "of direct and concrete character." Evan Kelley stated the case simply: "research into such questions as the physical behavior of fire, or to seek refinement of knowledge in respect to weather factors, is not believed to be worthwhile at this time from an administrative standpoint." E. T. Allen wanted to assemble "information largely from the administrative forces" instead of as "a wholly independent research project." Another forester opted for an even more direct approach: "the spirit of 'go and get 'em' will probably continue to be more important than abstract findings about what fires might do." Headley himself of course concurred. He cited in particular the publications of Show and Kotok as an example of what research ought to do and how researchers ought to think. He approved, too, of the proposal for a demonstration forest—not a place for experimentation so much as a model administrative unit, "a test" of what "could be accomplished under conditions deemed by the Service adequate for protection." Various types of management would be evaluated, not assorted fires or fire effects.[20]

Headley's selection of Show and Kotok for special approbation was to be expected. In this matter, as in all others he entered, the remarkable Show came out on top. The research program would emphasize fire protection, not fire. It would be founded on the statistics of fire behavior, not on simplistic economic theories of damages or theoretical investigations into the laws of combustion. In 1928 the California Experiment Station set up a separate fire research division and a year later created the Shasta Experimental Fire Forest as part of the larger Shasta Fire Control Project. Thus the all-Service research program that Greeley had proposed came to California under the administrative jurisdiction of Show and the research influence of Kotok; the demonstration forest that Headley had recommended appeared on the site of Show's earliest research on light burning; and the program of laboratory and field research that Clapp had considered fundamental was scrapped in favor of the statistical methodology so brilliantly employed by Show and Kotok. Show's and Kotok's work underwrote the programs adopted by the 1930 regional foresters conference, animated the Foresters' Advisory Committee on Forest Fire Control in its meetings from 1929 to 1931 (most of which occurred at Shasta) and contributed to the fire section of the Copeland Report. One of the few men who appreciated the full significance of the CCC, Show put the legions of Roosevelt's Tree Army to their most intensive and programmatic use. The extension of systematic fire protection into the backcountry proceeded in accordance with the hour control program, and it was Show's grasp of the big fire problem and the inescapable importance of initial attack that was instrumental in deciding the 1932 debate about fire protection in the low-value backcountry and in persuading the regional foresters to adopt the 10 A.M. Policy in 1935. What had occurred in California during the 1910s was thus repeated on a national scale, with S. B. Show as the nexus. In *The Determination of Hour Control for Adequate Fire Protection in the Major Cover Types of the California Pine Region,* published in 1930, Show and Kotok noted approvingly that "after 16 years" their research and the hour control concept had "supplied the rational base which duBois postulated for essential fire control."[21]

The direction of fire research followed that of fire administration. At a 1930 conference at the Forest Products Laboratory, the five "major needs of immediate importance" listed by the Committee on Fire Research were those demanded by the concept of hour control.[22] "The fire problem can no longer be regarded as a simple proposition for which a ready solution in terms of things to be done can be set forth," Show and Kotok cautioned. "It is, on the contrary, an exceedingly complex group of overlapping and intertwined problems, and each of these must be isolated and a separate solution worked out before fire-control organizations can guarantee reasonable and systematic success under every set of circumstances."[23] The various regions divided up the research effort, both to eliminate duplication and to find just what accommodations were needed in order to apply the hour control program to their particular needs. Meanwhile, in an attempt to drum up support for more fire research, the California group under Show sponsored a national meeting in San Francisco. A further refinement of research tasks resulted, with California taking the lead in the development of hour control plans and "fire fundamentals" and with the Northwest and Northern Rockies emphasizing the development of a fire danger rating method. A 1936 national fire research conference reaffirmed this direction.[24]

While Show and Company extended hour control studies from the forests of northern California into the chaparral of Southern California, Lloyd Hornby and H. T. Gisborne brought the method to the Northern Rockies. In the process, and with the organized crews of the CCC as a new presence, Hornby introduced several practical and theoretical refinements into the design of hour control. Rather than consider "cover type" as a single property defined by a certain rate of spread, Hornby described it in two dimensions: by its rate of spread and by its resistance to control (that is, the difficulty of cutting line through it). Preliminary results were ready in 1932, and final publication came in 1936.[25] Gisborne's work on fire danger rating followed the same scenario. In a sense what Hornby did for fuels, Gisborne did for weather. Rate of spread was not a constant but a dynamic condition, and difficulty of control depended on the character of the fire as much as on the arrangement of fuels. Using statistical methods, Gisborne devised a method of forecasting fire behavior. The resulting meter was seemingly a philosopher's stone for forest administrators: it offered a quantitative measurement by which to compare fire seasons and to contrast fire problems among different regions. With it one could better allocate funds and evaluate individual performances. The meter perfectly complemented the purposes and methodology of the hour control program, and for the next two decades fire research meant largely the refinement of these two concepts and their adaptation for new regions.[26]

The alliance was not accidental. As Charles Hardy notes, a 1922 meeting with Show "helped Gisborne find his bearings and set his course."[27] A year before, Show had published his "Physical Controls on Forest Fires," which related fire behavior to fire weather. Using careful climate data assembled over a decade for silvicultural purposes, Gisborne immediately determined to work out the physical factors that could lead to the prediction of high fire danger. He wrote a friend that "the field is so new that we have nothing to help us except our own imagination and what little ingenuity we possess." A 1926 description of his approach to fire behavior showed the power of enthusiasm and the limitations of imagination. "It consists of getting inside the fire and as close to the main front as possible, or else on a promontory from which I can see all fronts, and then measuring everything I can measure. . . ." "Actually," he confessed later, "I haven't any method in this work. . . ."[28] But the statistical methodology developed by Show and Kotok for analyzing fire reports was applicable to Gisborne's interests, and it was their example that helped to cool his Baconian fever for measurements into a disciplined system. The division of fire topics among the regions had left the invention of a prediction system to the Northern Rockies and the Pacific Northwest. But whereas the Northwest looked to relative humidity alone, a single measurement that could be easily understood by loggers and used as a standard for forest closures, the Rockies sought, on Show's example, a multivariate analysis.

The meter was first demonstrated in the disastrous 1934 season. So impressed were administrators with its possibilities that it was incorporated into the 10 A.M. Policy. Emergency presuppression funds would be made available on the basis of the computed fire danger. Both Hornby and Gisborne expressed skepticism over the 10 A.M. Policy, but it was their work—amplifications of Show, Kotok, and duBois—that formed the basis for the national fire planning program set up to promote the policy in 1937. Responding to the new policy as did all other branches concerned with fire protection, fire research convened a national conference at Shasta in 1936 to put its own house in order. The Washington planning conference that followed coordinated the new fuel and fire danger classifications and ultimately nationalized the data into a single system of universal application.[29] So successful were these endeavors that the next national fire research meeting was held at the Priest River Station in the Northern Rockies.

The Show and Kotok team, meanwhile, intensified their old interests and added new ones. One reason for their continued importance was that, in Show's words, "the California Region went much farther than others" in exploiting the CCC program. The allowance for "facilitating personnel" to be used in conjunction with camp projects enabled the Forest Service to hire some researchers, including a chemist. The volume of ECW money encouraged the construction of research facilities and allowed for some large-scale field research. The sheer magnitude of the CCC largesse forced research into detailed planning based on existing knowledge rather than into the exploration of new subjects. The program was more a test of research, an application of it, than a vehicle for further investigations. Kotok had shown the magnitude of the challenge when he authored the section of the Copeland Report on fire, and Show soon demonstrated how the hour control concept and CCC labor could combine to attack it.[30]

Nor did Show leave refinements of the hour control concept to Hornby's work on fuel and Gisborne's on weather. Throughout the 1930s he oversaw a range of studies that sought to better define fire behavior. It was recognized that experimental verification of the rate of spread figures required by hour control planning was desirable. One approach, used intermittently through the decade, was to expand the field trials that Show had initiated under duBois's direction in 1915. Up to 1936 most of this work was done by Charles Buck and John Curry at Shasta; afterward, C. A. Abell continued it in the fuel types of Southern California. Closer to Show's own heart perhaps was a large statistical study he began in 1937 in the belief "that the sine qua non for effective action in control was a full understanding of fire behavior and fire pattern from past occurrence."[31] The result, based on case studies and fire reports, was a typology of fire behavior patterns and a matrix of average rates of spread. But in the search to discover still better data and to improve the credibility of fire research as a species of scientific inquiry, the California Station by the end of the decade undertook even more fundamental investigations. Under the guidance of Wallace Fons, a mechanical engineer, it established laboratories and instruments for the study of test fires under controlled conditions.[32] At the 1940 Ogden Fire Conference the big fire was the topic of most interest, and it was agreed that the Service needed "badly to know the fundamentals of combustion (fire behavior) and its meteorological effects on forest materials in the open air."[33] The committees did not specify anyone or any method to accomplish this but subordinated such research to the demands of hour control and fire danger rating. Despite Fons's example, there was no reason to think that knowing the "fundamentals of combustion" meant adopting new methods any more than it meant adopting new objectives.

The Priest River Fire Meeting in 1941 was less a foreword to a new chapter of fire research than it was the opening of a long obituary to the great era of pioneer researchers. It concluded literally on the eve of Pearl Harbor and in the deepening twilight of the CCC. The conference heard reports on a host of traditional topics, but two presentations in particular stood out. In a sense, they offered two possible directions for fire research, given that the hour control program and the fire danger rating system had more or less achieved administrative success. Gisborne, the host of the meeting, introduced the session with a review of "Milestones of Progress" over the past 35 years. Alarmed by the 10 A.M. Policy and fearful that under its grandiloquent ambitions fire control would leave the intellectual and institutional confines of forestry, Gisborne implied that the future of research depended on a return to past objectives and methods. Consequently, he strenuously argued that economics, including the calculation of damages, had to be restored to preeminence among the priorities of research and administration as a means of recapturing the equilibrium of means and ends upset by the New Deal programs.[34]

To this Charles Buck, in many ways a successor to Show, replied eloquently for a "fundamental method of approach" based on "quantitative estimates of the behavior characteristics of fires." All aspects of "the technical phases of fire control," Buck asserted, will "in their ultimate state" require such information. "We here do not have to, nor can we, define the exact limit of desired knowledge. To me, the fundamental question is not how important is fire behavior information to fire control, but rather, *when* will further fire control development on any unit be critically hampered by lack of the necessary kind and quantity of behavior information."[35]

It was a ringing valedictory for the past decade of research in California, and in its way it was as daring a presentation as that with which Godwin had galvanized the 1936 Spokane conference behind the Aerial Fire Control Project. The Ogden conference had observed that "administrative practices have caught up with and surpassed the original objectives of any and all existing danger measuring systems," and Gisborne privately confessed that further developments in fire danger rating and planning "are believed to be jobs for administrative solution."[36] Buck admitted that it was not yet possible to write a good program for the study of large fires, and his committee recommended that "fire danger rating research be temporarily suspended as soon as current work reaches a reasonable stopping point."[37] Instead of further consolidation, it proposed a program of exploration into fundamentals. But the conferees concluded that a subject like fire was perhaps too premature, probably too visionary, and certainly too theoretical. Any such program must first return fire to its roots in silvicultural effects and forest economics. Fire behavior was an *ignis fatuus* that would lead research into a miasma of pure and laboratory science.

Though the meeting recommended that the Service establish a Division of Forest Fire Research, its purpose was essentially conservative: the institution would eliminate duplication among the regions and return fire research—and fire administration—to its original goals. It seemed in truth that with the success of the hour control concept as a planning program, with the creation of regional fire danger rating meters, and with the 10 A.M. Policy dictating decisions, fire research had reached a state of exhaustion, perhaps lost its vision, or even experienced a failure of nerve.

In July 1944 Chief Forester Watts put his office behind the drive for an independent Division of Forest Fire Research. Suggestions were solicited, as they had been in 1928; but this time the response was favorable. A fire conference convened at Asheville to review postwar prospects. Charles Buck was assigned to collate the assembled needs and objectives into a single package, and his report was ready by mid-1945. "The present status of fire protection research," Buck underscored, "is wholly unsatisfactory."[38] For the entire country, a scant nine researchers were working full-time on fire problems, three of them in California. The figure was somewhat deceptive, since much of the research—such as that of Show, Kotok, duBois, and Hornby—had been done by people ostensibly in administration, not research. Other work, including the fire damage studies in the South, had been done under the guise of silviculture or range improvement. A good deal of prescribed burning, moreover, had been done *sub rosa* in the name of "administrative experimentation." But for a subject of such fundamental concern to the Forest Service, the dearth of professional researchers in fire was surprising. Buck also included a proposal for a place to house an augmented research staff, a "centralized fire laboratory." Fire research had often had occasion to use the Forest Products Laboratory, which furnished one sort of model for the fire laboratory, and the national equipment centers being promoted at the same time provided another.

The goals proposed for the new division were those recommended at Priest River. The contrast with what was occurring simultaneously in equipment development is striking. In retrospect, it is apparent that a "fundamental method of approach" to fire behavior had the capacity to redefine all aspects of fire protection profoundly by creating a physical standard with which fires and fire effects could be compared and against which the effectiveness of fire control programs could be measured. But whereas the equipment program looked to the future, fire research looked to the past, and it was equipment that consequently revolutionized fire protection strategy in the immediate postwar era. Fire research evidenced a strong desire to return to its origins in forestry rather than to explore an area into which physics, chemistry, biology, and meteorology had hardly penetrated.

All of these tendencies were apparent at the Washington Fire Conference of 1948. Chief Forester Watts asserted his belief that fire control was *"the foundation of forestry"* and that it had two tenets: to catch fires while small, and to involve research more fully as a guide to expenditures. The conference, like fire research, was profoundly ambivalent: it wanted new data for old problems, and it sought to intensify with new techniques the old objectives. The Chief set the tone by noting fondly that the meeting recalled the Mather Field Conference, for him the beginning of organized fire protection. Among the present conferees, he informed the audience in a folksy way, he himself was one of four who had also been at Mather Field. Another of the holdovers, Kotok, listed three fields for basic study: fire behavior, indicators of fire behavior usable by practitioners, and tests on protection results by "periodic analysis of records." He stressed furthermore the desirability of "establishing a career ladder in fire research." But despite jeremiads against complacency, most of the attendees, like Watts, looked to past achievements. Changes would take the form of restoration and intensification. Even Kotok wanted "closer ties between fire research and fire control men," and he advocated more fireline duty for the scientists. Show, then a forester with the Foreign Agricultural Organization, pointed out "that systematic fire control has been pioneered for the whole world by our own Forest Service," that the "answers we get are useful in all fire countries," and that "you can't get on top of the forestry job anywhere until you're on top of the fire job." It was thus fully in the spirit of the meeting that the Fire Research Committee urged simultaneously that the old Fire Control Advisory Committee be reestablished and that "fire research be raised to divisional status in the Washington Office." The goals of fire research did not arise from new techniques, new voices, or a new problem fire; they came from the desire to restore the old values more intensively and economically.[39]

In late 1948 the Division of Forest Fire Research at last became a reality, with A. A. Brown as director. Brown immediately toured the various regional facilities—much as his counterparts in administration surveyed regional equipment developments—and then sought to consolidate and coordinate the various programs already in existence.[40] However, two events in 1949 dramatically redirected the embryonic program. In August the Mann Gulch fire blew up, and in September the Soviet Union exploded its first atomic device. The Mann Gulch tragedy was the worst since the Blackwater fire, and the prospect of Cold War antagonists armed with atomic weapons raised the specter of a global holocaust. The effect of both was to underscore the need for better knowledge of fire behavior. The "fundamental method of approach" was no longer a quixotic vision with exotic methodologies but a necessity for both fire control and national defense. Mass fire became the ruling question of fire research, and the traditional methodology of forestry was simply not equipped to handle it. Fire research had to invent new techniques, incorporate new personalities, search for new informing concepts. The new policies would pervade forestry in the 1950s as much as the hour control program had in the 1930s.

The changing of the guard is poignantly illustrated in the story of H. T. Gisborne. Gisborne was in many ways typical of that early generation—first and last a forester, a man inspired by Show, and a man who managed to equate fire research with the doctrine of the strenuous life. When a tall pine needed to be topped for weather instruments, for example, Gisborne gleefully topped it himself. He approached the mental challenges of fire research with the same physical gusto. Early researchers in fact freely exchanged roles with administrators and fire officers. Shortly after completing his masterpiece on fire control planning, Lloyd Hornby—like many a forester warhorse—found a fire call irresistable, only to perish tragically in the Toboggan Creek fire. No less interchangeable between research and management were the ideas and objectives of this pioneer group, and researchers even insisted that their ideas be tested on the fireline. Looking back in 1936 John Curry observed that "treatises on military strategy are not developed during the heat of battle but only after the war is over. Our war to control fire has no let-up. The emergency character of fire control activities has worked against the scientific method of approach." Research, indeed, had been in the form of battle plans rather than of philosophies.[41]

Fire control had made fire research, not vice versa. The Service's ordeal by fire had been practical and political, not theoretical and scientific. Fire research had invented the concept of systematic fire protection. When the time came to extend the idea into the backcountry, fire researchers wanted to be on hand to blaze the trees, and when the time came for that first generation to retire from the scene, they did so with their boots on. Despite a bad heart, Gisborne had insisted on inspecting the site of the Mann Gulch disaster in person before contributing a report on the fire's explosive behavior. His knowledge of fire behavior was, like the Service's, intensely personal and empirical, the product of on-site experience. But an afternoon of hiking proved too much. Gisborne collapsed, then died near the burn. Later, in honor of his contributions, a mountain in the Northern Rockies was named after him. For that first era of fire research, it was on all counts an especially fitting end.

II

As the events of 1949 dramatized, the pressures for a reformation of wildland fire research came from both inside and outside the Forest Service. At the Ogden Fire Conference of 1950, Brown recalled, participants "gave more emphasis to research than had been given at any previous meeting." The Washington office insisted that "a thorough understanding of fire behavior by every responsible fire man must come close to being the number one prerequisite to competent fire fighting by the Forest Service."[42] The Division of Forest Fire Research, then just over a year old, was directed to strengthen its facilities, put its research results into working manuals, and "get at the blow-up fire."[43] Using traditional methodology, Jack Barrows assembled his comprehensive *Fire Behavior in the Northern Rocky Mountain Forests* (1951), and relying on longstanding precedents, fire research proposed to put its knowledge on the line in the form of a fire behavior officer, a new staff position on active fires. Both, however, were summaries of past data, processed to support hallowed concepts of planning and suppression. Simultaneously, the realization grew that fire control was a three-dimensional problem, measured against fire intensity, not a two-dimensional problem as envisioned by the tenets of hour control and as analyzed by the statistical methodology of Show and his successors. But more than intellectual insight was required, and among the institutional reforms that sustained a reformation in fire research were the ascendancy of Richard McArdle to Chief Forester in 1952 and the charge to direct rural fire defense. A man who had begun his Forest Service career in fire research and who had received a rare doctorate in forestry with a dissertation on the history of fire protection, McArdle was temperamentally disposed to oversee a fundamental expansion in fire research. The charge to oversee rural fire defense, meanwhile, gave the Forest Service a new landscape in which to direct fire research, and this new intellectual terrain brought with it new liaisons and new methodologies. For two decades the specter of atomic warfare would be the ghost in the machinery of fire research. The consequences would profoundly affect Forest Service perception of the entire fire problem.

Results were apparent by the 1952 fire meeting in San Francisco. Brown reported excitedly that since the last meeting "we have been given a classified military project of broad scope." Its significance, he informed an audience well aware of what the military connection had meant for equipment development, "is that it enables us to do a lot of highly technical work we need for our own programs, though nothing can be published on it until declassified. One of its primary objectives is to measure the thermal and blast effects of A bombs in a natural environment. This project has already gained a great deal of prestige for the Forest Service and is regarded as highly successful by the Military." It also helped "to gain recognition by both civilian defense and military authorities of the importance of forest, brush, and grass fires in any air war."[44] With the DASA as the principal contractor, the Forest Service had actually conducted several projects. Buck evaluated the blast and fire effects of atomic weapons in the forests of Western Europe; Forest Service observers attended the Nevada and Pacific thermonuclear bomb tests with the objective of determining their urban and wildland conflagration potential; and, with the assistance of the Forest Products Laboratory, a detailed series of experiments measured the thermal properties of forest fuels and the dynamics of the ignition process as initiated by the thermal pulse of a nuclear explosion. With OCD and Army funding, moreover, the Forest Service began Project Firescan, a program that sought to use infrared sensing equipment to map fires over extensive areas.[45]

The thermal pulse of the atom bomb not only presented fire research with new problems, it also ignited its imagination and soon led to new methodologies. Outfitted with its additional missions and monies, fire research made the transition from field to laboratory, from little science to Big Science, from the study of forestry economics and the damages of light burning to the thermal mechanics of mass fire and the postattack environment of thermonuclear war. Around the country studies appeared suggesting the physical causes for mass fire behavior, including the jet stream, subsidence, fuels, and the effects of upper-level winds on convective column formation.[46] In 1954 two bold research programs were devised to demonstrate the new approaches. Operation Firestop moved against the newest of fire challenges, mass fire, while Project Skyfire studied the oldest, lightning.

Under Keith Arnold of the Pacific Southwest Station (formerly the California Station), Operation Firestop proposed a broadside attack on the question of mass fire. Researchers and fire officers from all levels of California fire protection convened for a grand, year-long investigation, much in the spirit of the International Geophysical Year. Its organization plan presented Firestop as a "spontaneous action on the part of the fire control agencies of California to attempt to find some new tools to combat the mutual problem of mass fire." Specifically, Operation Firestop was to explore two research topics: "civil defense against fire and reduction of loss from large wildland fires through the development of new or unconventional fire control measures." The work was used to argue for a national fire research laboratory, such as Britain and Canada had established. Independently of national approval the station prepared for a "complete fire physics experimental facility" at the mechanical engineering department of the University of California, under the direction of Wallace Fons.[47]

Project Skyfire was an equally ambitious program to explain and control lightning fire. Directed in its first phase by Jack Barrows, Skyfire, like Firestop, demanded technological as well as scientific support. Firestop had relied on the Arcadia Equipment Development Center for the first and led to the Western (Riverside) Forest Fire Laboratory for the second. Skyfire was probably the deciding factor in locating an equipment development center (MEDC) and a fire lab (Northern) at Missoula. When a national fire research meeting was scheduled for 1956, it was held at Missoula. Like mass fire research, Skyfire would find itself supported during the period of conflagration control and challenged by the era of wilderness fire.[48]

Though the Forest Service in one sense strove to maintain its traditional themes, sustaining the format of objectives approved at Priest River and at the 1948 Washington conference, it saw them rapidly reconstituted. The basis of planning became conflagration control, not hour control. Fire danger rating became nationalized, founded on fundamental physical measurements, not on administrative statistics. Fire weather meant the dynamics of atmospheric elements interacting with mass fire, not statistical correlations between fire size and climate. In technics, too, mass fire compelled new strategies and tools. Firestop, for example, proposed a new direction in techniques and objectives for fire research for the first time since duBois. From afar Show recognized this transformation, and it disturbed him. He and Kotok were "struck by the fact that during this period no attempt at comprehensive analysis of the individual fire reports and collateral material was undertaken."[49]

Mass fire was destined to take Forest Service research even further from its traditional base than Firestop and Skyfire suggested. Free-burning fire had attracted widespread interest, which took such diverse forms as the annual meetings of the Combustion Institute and a directory of fire research by the NFPA. After the Engle hearings, which followed the Malibu and Inaja fires, the Forest Service was authorized and encouraged to "conduct a comprehensive program of forest fire research and to establish laboratories."[50] Meanwhile, OCD contracted with the National Academy of Sciences to establish an advisory board to oversee its investment in national fire research. The concept grew out of an exploratory conference between the Forest Service and OCD on mass fires and firestorm research, an aftermath of Firestop. The Forest Service immediately became a "supporting participant" in the developments. The contracts between OCD and NAS were renegotiated in 1956 and 1958. The Department of Defense and the National Bureau of Standards joined. The National Science Foundation provided some direct support and approved funding for a number of projects subsequently proposed by the NAS committee. Under the contracts a Committee on Fire Research consisting of 9 members and a Fire Research Conference composed of 34 members who would advise the committee were established within the Division of Engineering and Industrial Research of the National Research Council. Among its charges, the committee was to develop and maintain a directory of fire research; to organize research correlation conferences on specific factors in fire research; to provide the opportunity for the voluntary coordination of research among the various agencies in fire research; to assist in planning specific research projects; to provide "guidance for the placement of research programs or projects"; and to advise "on the adequacy and significance of progress and reports." The chairman of both the committee and the conference was H. C. Hottel, then of MIT, but formerly chief of wartime fire research. Thus, both urban and wildland fire research grew out of the response to a common problem: mass fire.[51]

A year after its first charter, the committee sponsored an initial correlation conference that focused even more specifically on the topic of commanding interest: "Methods of Studying Mass Fires." That was followed in 1958 by the publication under committee auspices of *Fire Abstracts and Reviews* and "A Proposed Fire Research Program." The proposal stated unequivocally "that our only hope for being able ultimately to cope with large fires lies in a major extension of our fundamental research on fires."[52] It recommended funding from the new government sources and solicited proposals from universities, government laboratories, and private research labs. A third correlation conference, an international symposium on the "Use of Models in Fire Research," followed in 1959. For less broad subjects the committee organized a series of technical meetings, one of which was jointly staged by the Forest Service and the Naval Radiological Defense Laboratory (NRDL) on ignition and thermal degradation of wood. By 1961 the program of the committee was virtually completed with the publication of the *Directory of Fire Research* (revised annually until 1977) and with *A Study of Fire Problems,* the proceedings of a summer institute that reviewed some of the most promising lines of research. In the words of its chairman, it was the committee's belief that, although a "considerable effort of some 20 government laboratories and research-support offices presently goes into fire research," what was demanded was "research of a more fundamental character—research that makes more use of modern developments in heat transfer, combustion, fluid mechanics, and computational methods." The committee outlined seven problem areas and assigned them to the contributing members.[53]

The Forest Service was designated as the "working agency" in three areas: modeling homogeneous complexes for the study of fire front movements, collection and analysis of field data on fire spread, and operations

research on firefighting. It was an "interested agency" in two others: interaction of fire with the atmosphere and development of instrumentation for
field and laboratory research. A. A. Brown represented the Forest Service
on the committee proper. Arnold, Barrows, Buck, and Fons sat on the conference. OCD envisioned that the committee would advise it on the best
allocation of its research money and on the merit of the products. The Forest
Service projects funded by OCD were by and large those recommended by
the committee. What the Clarke-McNary program was to many state organizations, the Committee on Fire Research and the OCD were to fire
research in the Forest Service. Many of the programs founded with OCD
seed money continued subsequently with support by other agencies or by
the Forest Service proper.

The Forest Service had plenty of reasons of its own to pursue an intensive
and much renovated program of fire research. It had, after all, established
a Division of Forest Fire Research prior to any military or OCD connections. The big fire had been a steady topic of debate among researchers since
at least the 1940 Ogden conference. Fire problems such as the Inaja tragedy
and the emerging fire regime in Southern California might have led to a
national fire lab on its own merit. The Fire Task Force of 1957 had recognized the need for better fire behavior knowledge a year before the Committee on Fire Research released its "Proposed Fire Research Program."
Independently of its rural fire defense mission, fire research was moving into
the Northeast, the Lake States, Alaska, and other lands where it had not
previously existed.

Yet the interests of rural fire protection and fire research converged won-
derfully. The fire laboratories illustrated this nicely. Between 1959 and 1963
the Forest Service established three laboratories for forest fire research: the
Southern, at Macon, Georgia; the Northern, at Missoula; and the Western,
at Riverside, California.[54] Project Fire Model, for example, was recommended by the Committee on Fire Research and funded by the OCD.
Research began in the engineering lab at Berkeley under Fons. With the
completion of the Southern Lab, the project was transferred there and combined with related research on firewhirls and scaling laws. After 1960, on
committee recommendations, funding for the project came through the
National Bureau of Standards.

By 1962 the basic themes, programs, and facilities for fire research by the
Forest Service were in place. But as Project Fire Model suggested, even
where old subjects were retained, they were fundamentally reconstituted.
The physics of fire and its environment dominated virtually all aspects of
research, replacing, it seemed, the forestry of fire, which had animated the
previous era. Fire behavior, for example, meant fire physics, and the Committee on Fire Research recommended a series of projects to investigate the
thermal mechanics of free-burning fire. Foresters had to search out engi-

neers, physicists, and meteorologists to conduct such a program, and the new researchers brought with them the techniques, concepts, and experimental methodology of the physical sciences. Some programs, like Project Fire Model at the Southern Lab (1958–1966), started under OCD funding. Some, like the Mechanisms of Fire Spread project at the Northern Lab (1964–1969), came under military funding by ARPA. Project Fire Model created mathematical and physical models for the "uncontrolled aero-thermodynamic fire" burning in steady-state conditions through fuel beds made of wooden cribs. These models were accompanied by others dealing with assorted propagation mechanisms associated with mass fire, such as fire whirlwinds and aerial firebrands. The Northern Lab added a mathematical model for fire spread under steady-state conditions but with wildland fuels. From such modeling it was hoped that, with appropriate scaling laws, it would be possible to generalize from laboratory steady-state fires to free-burning mass fires. But growth models—as distinct from steady-rate models—proved elusive, and so did scaling laws. George Byram likened the problem to that of "making miniature thunderstorms in the lab."[55]

After the international crises in the early 1960s, money became available from OCD, ARPA, and NRDL for direct, experimental studies of mass fires in the field. From 1962 to 1964 OCD contracted with the Forest Service for heavily instrumented quantitative research on the interaction of mass fire with its environment. The program was reapproved as Project Flambeau (1964–1967) and was complemented on the international scene through The Technical Cooperation Program (TTCP) with Project Euroka in Australia. For the experiments wildland fuels were arranged into fuel complexes that resembled the suburban housing developments that were burning almost annually in the Southern California fire regime. Thus the experiments again echoed their World War II antecedents, for which whole mock cities had been constructed and burned. Under contracts Forest Service researchers calculated the probabilities of ignition and spread on fires in both urban and wildland environments as a result of nuclear attack—a dramatic reconstruction of the old fire effects studies and one that would have astonished the old guard at Priest River who wished to restore economic and damage studies to primacy in fire research.[56]

Much as fire behavior moved beyond statistical correlations and into physical and mathematical modeling, so virtually all dimensions of the fire problem were similarly reworked. Fire weather research ceased to deal with climate in the way that silviculturalists were accustomed and began technical investigations into the dynamics of foehn winds, the microclimates of particular sites, the mesophenomenon of mountain wind fields, and so on. Its meteorological component became so large, in fact, that in 1970 the division title was changed to Division of Forest Fire and Atmospheric Sciences. Modern principles of operations research and computer models

entirely revamped the processes of planning.[57] Though appropriately stationed in California, where duBois and Show had done their pathbreaking
work, programs like FOCUS and FIRESCOPE showed that fire spread was
something that need not be derived from past statistics but could be projected by simulation. Systematic fire planning did not follow from a methodology of statistical analysis directed toward the Shovian concept of hour
control; rather, the futuristic gaming made possible by computer analysis
guided toward the objective of conflagration control. So fundamental were
these changes that when a National Fire Research Workshop was held in
1964, the Committee on Large Fire Behavior Research went so far as to
insist that "very little of the fire behavior data as recorded on Form 5100–
29, Individual Fire Report, is useful for fire behavior research purposes.
. . . Fire research is developing to a stage where large quantities of subjective
or unreliable data are becoming of lesser value."[58]

The extent of the retooling is perhaps best illustrated in the program of
greatest administrative interest, the development of a national fire danger
rating system (NFDRS). Not only did such a project integrate a variety of
traditional and contemporary needs, but it also gave a practical focus for
the variety of research projects the Forest Service had assumed. McArdle
initiated the program in 1957 with a circular letter, and a year later a committee began serious planning. By 1961 a four-phase rating system based
on ignition, risk, fuel energy, and spread was ready for testing. Reevaluation
began almost immediately after its publication in 1964 as the first flush of
research poured out of the labs. In 1968 a special NFDRS work unit was
established to reconstitute the project. Using the systems engineering format
that had become typical of planning, and relying on the physical modeling
of fire behavior and fuels that had emerged from the labs, the new NFDRS
was published in 1972 and had important amendments published in 1978.
As explained in the accompanying statement on project philosophy, it was
determined that "ultimately, the system would be purely analytical, being
based on the physics of moisture exchange, heat transfer, and other known
aspects of the problem." In stark contrast to Gisborne's conceptualization,
"the system would not be empirically or statistically based."[59] In place of
cover types, the system relied on fuel models—defined not by their biological characteristics but by their physical and chemical properties. The
research into fire fundamentals was thus integrally bound up with the needs
of the NFDRS.

Nor was the old prospect of fire as an agent lost; rather, it, like everything
else in this era, was transformed. Prescribed fire, for example, was applied
to the construction of fuelbreaks and to the disposal of slash for conflagration control. The research on mass fire had as its objective "to write a prescription for mass fire." In 1962 ARPA established liaisons with the Forest
Service on the possible uses of broadcast fire as a tactical and strategic

weapon, and on the eve of the massive American buildup in Vietnam those liaisons ripened into contracts. Forest Service research undertook a general survey of incendiary weaponry in wildland environments: what had been considered the "worst case" in its studies for OCD now became the "best" case. Conflagration control through fuelbreaks was in a sense the obverse of area denial through defoliation and mass fire. Forest Service advisors traveled to Vietnam to advise on the application of its newly discovered prescriptions for fire behavior. The success of the venture was apparently mixed; the official reports and the ARPA manual remain classified; and the Senate banned such research in 1972.[60] Interestingly at the same time that this era of fire research had come into final definition, satisfactory mathematical models for fire behavior appeared from the Forest Service fire labs, fire weather research was consolidated, and TTCP published its final summary on "The Effects of Nuclear Weapons in a Forest Environment—Thermal and Fire."

Its involvement with the military and OCD had come as something of a surprise to the Forest Service. But its own internal needs had apparently converged with important national needs. The intensification of the Cold War, the beginning of a space and arms race with the Soviets, the establishment of a national scientific commitment—all coincided with the appearance of a particularly disturbing fire regime in Southern California, with an increase in fire-related fatalities, and with the entrance of a new problem fire, mass fire. Forest Service interest in fire research joined with a broad cultural renaissance of concern about fire, symbolized by the Committee on Fire Research of the NAS and by annual fire research contractors conferences under the aegis of OCD. By 1962 the Forest Service was committed to a radical reconstitution of its fire program in both conceptual understanding and methodology. It participated in a host of research projects under the guidance of the Committee on Fire Research and with the sponsorship of OCD, ARPA, the National Science Foundation (NSF), and the National Bureau of Standards. The Forest Service saw itself integrated into a national fire research program and itself became a major instrument for the institutional integration of national fire services. By 1965 three-fourths of its fire research program, according to its director, was located in the labs. The contrast to the sentiments of the Priest River Fire Meeting and even to the charter of the Division of Forest Fire Research is striking.

But as the Forest Service slowly learned, it had perhaps made another Faustian bargain. The CCC bonanza had created a nearly fatal schism between policy and programs; the new infusion of funds from outside the organization had a similar effect on fire research. As had the 10 A.M. Policy, mass fire took the Forest Service into new geographic arenas and into new intellectual territory from which it could extricate itself only with difficulty. The OCD projects were proposed by fire researchers, initiated by an NAS committee, and reviewed, on practical grounds, by military or civil defense authorities and, as science, by physicists, engineers, and meteorologists. Free-burning fire had become a subject independent of forestry, and Forest Service research had to decide whether it was primarily interested in fire or in fire protection. The shift in funding and viewpoint helped to break up the alliance of research and administration that had compromised Forest Service handling of prescribed burning in the South. It helped to divorce fire research from the methodology and purposes of forest research. The transformation in fire research made possible by the new funding offered a badly needed opportunity to see old problems in new ways, but it also forced fire research to be evaluated by new standards and to be applied to new areas.[61]

Conspicuous by its absence on the research agenda was any reference to fire as a biological phenomenon. The old argument about effects had embraced two concepts: fire as an economic question (damages) and fire as an ecological event (agency). But under the direction of the Committee on Fire Research, the Service had retooled its research program for mass fire, not controlled burning, and for the uses of fire as a military weapon, not its uses as a silvicultural tool. The Service had broadened its fire damage assessments to include the rural landscape, but it assumed that the fire would be what the Committee on Fire Research had termed "hostile fire."[62] Its great expansion of facilities, responsibilities, and subjects had come in the name of national defense, and this reorientation had reworked not only the physical study of fire behavior but also its metaphysical foundation, as it were. During the 1963 review of Project Firescan, for example, the Northern Lab opened its presentation with the image of a "blinding flash of a 50-megaton weapon" ripping "across the sky at Seattle, Washington. Two minutes later the terrible brilliance of a second thermonuclear explosion seared the stratosphere above Portland, Oregon. This was the beginning of the greatest fire on earth. This was the spawning of massive disaster—of thousands of fires—of families of fire storms—of hundreds of miles of sweeping flames—or more than 200,000 square miles of near zero visibility in the gigantic smoke pall. It presented the supreme test of Civil Defense preparedness."[63]

It was just such supreme tests that were accepted by Forest Service fire research. The author reminded reviewers that the 1910 fire and the 1957 Alaskan conflagrations had come close to approximating this vision of thermonuclear holocaust. But the Firescan scenario was sketched in the same year in which the Leopold Report proposed a research program on the biology of fire in the national parks, especially as it related to wildlife. As the nation prepared for the Wilderness Act of 1964, the Forest Service still looked toward the prospect of urban conflagrations, investigating, for example, the Watts fires. As the new conservation movement rallied around the sanctity of wilderness, the Forest Service was poised to enter the conflicts in

Vietnam. Just when the emergence of fire ecology suggested the value of prescribed burning for habitat maintenance and the validity of natural fires in wilderness areas, the Forest Service was bringing to a climactic test its search for a "prescription for mass fire" both in the United States and in Southeast Asia.

In 1966 Forest Service researchers presented a state of the art summary on mass fire behavior research to a NATO Scientific Working Party. "I am disturbed," wrote Craig Chandler, "by the fact that we have paid more attention to the possibility that forest fires will enable the world to be taken over by the insects than we have to determining the degree of impact of forest fires on the postattack communications, transportation, and power distribution networks. Here again, I believe reasonable extrapolations can be made from peace-time forest fire experience. Someone should make them."[64] As often as not, that "someone" was the Forest Service. The 1967 fires, for example, were analyzed as natural phenomena by a team from the Northern Lab outfitted with the latest analytical tools and conceptual models and by fire officers under contract to OCD to study the relevance of fire control problems for possible wartime fire conditions.

Its acceptance of new funds, missions, and methodologies revitalized a moribund Forest Service fire research program. Without the subsequent research on the physics of fire, which provided a means of comparing fires, the biology of fire might have remained the abode of naturalists. But acceptance gave fire research an identity outside forestry, an autonomy beyond the Forest Service, and a momentum beyond administrative control. The Forest Service discovered that its fire program was immeasurably strengthened and strangely isolated. Responding to a cultural environment with a favorable climate of opinion, the growth in the Division of Forest Fire Research had in a sense mirrored the behavior of its informing topic. It made the transition to Big Science in the way a fire, responding to the proper complex of fuel and weather, might make the transition to a mass fire. There it stood, its rising pillar of intense, sophisticated research resembling the convective column of a mass fire, oblivious to the ambient winds that swirled around it. But the popular winds of environmental concern were more likely to focus on the insects that Chandler satirized than on postattack effects, and by the end of the 1960s the Forest Service discovered that it knew little more about fire and insects than it had in the days of Stewart Edward White.

The policy reform of 1971 was ambivalent, looking to two problem fires simultaneously. It was at once a culmination of the conflagration control concept—a final bid to manage mass fire—and a harbinger of the fire of greatest importance to the 1970s, wilderness fire. Conflagration control had demanded knowledge of fire behavior, but wilderness fire required knowledge about fire ecology. Prescribed burning in particular had to take on new definitions and new functions. It could be either natural or anthropogenic, but it had to be a fire somewhere between the discredited light burning of the frontier and the mass fire of thermonuclear holocaust.

III

In 1962, the same year in which OCD convened the first of its annual conferences of fire research contractors, the Tall Timbers Research Station in Tallahassee, Florida, inaugurated the first of its annual fire ecology conferences. The object of the former was the physics of fire and the effects of its military use; the purpose of the latter was the biology of fire and the effects of its application in land management. The conference chairman for Tall Timbers, E. V. Komarek, expressed a general quandary. "I have spent about half of my life influenced, taught, and educated against fire in nature," he wrote, "and then I have spent the other half of it using fire and trying to understand it." While the Forest Service groped toward a definition of mass fire, Komarek proposed a definition of fire ecology. While the Forest Service was completing the last of the fire labs, Komarek extended acknowledgments to "natural laboratories," "where most of the studies of fire ecology must be made."[65] While the Southern Lab worked painstakingly to make explicit the physical and meteorological circumstances under which prescribed fire would be ignited, Herbert Stoddard, a research associate at Tall Timbers, recommended that "we've got to take calculated risks in this world, and I don't think we can possibly ever use fire on a large scale and get the results we ought to unless we do take those calculated risks."[66] Within a year the Leopold Report was published; within two years, the Wilderness Act became law; and for two decades the defining problem of land use remained wilderness. The program of the Tall Timbers conferences was far better suited to answer these questions than was that developed by OCD. Once again administrative policy and research programs were to be retooled. When the Forest Service changed the title of Fire Control to the Division of Fire Management, and when it outlined its plans for reconstituting its objectives in conformity with land management and ecosystem concepts, it did so at a joint meeting with the Tall Timbers conference held at Missoula in 1974.

As a field of study, forest ecology was not new. It was considered well established in 1916, when R. H. Boerker summarized its 50-year history. From the beginning, investigation into fire effects had been deemed basic to the Forest Service research effort. At the Priest River Fire Meeting, W. G. Morris declared that "fire effects are the most important part of the mud in which we are floundering in an attempt to specify objectives of fire control. We can't appraise damage until we know the qualitative and quantitative effects."[67] It was a common sentiment. But as nice as Morris's etiology was in the ideal, in practice the Forest Service usually had its fire protection objectives set for it from the outside. With or without adequate knowledge of fire effects, the Forest Service had to administer the reserves. Fire protection created the study of fire ecology, not vice versa; not until a problem fire was encountered that could not be managed without knowledge of fire ecology did that relationship reverse.

When change did come, it was less the result of new data than it was the product of new perceptions. The conversion reflected a change in philosophy rather than additional information. The major programs in fire ecology came after the conversion, not before. That fire was "natural" was taken by early investigators as a sign of nature's lamentable drive toward entropy, one that could and ought to be corrected; later investigators regarded fire as a commendable and necessary mechanism of nutrient recycling. The thick bark on a sequoia was interpreted by early students as a defensive measure, a necessary adaptation to protect against fire; by later students, as an opportunistic adaptation by which to thrive in a fire environment. It was originally felt that forests existed in spite of fire; at a later date, because of it. The heavy growths of reproduction to which early foresters pointed with pride, later foresters considered stagnant and hazardous; the drive toward a climax state that the former considered desirable because it was natural and that fire protection could encourage, the latter lamented as unproductive and unnatural. To argue their case researchers could point to the same data and often the same photo transects. But each group understood its information against a different conceptual, cultural, and metaphysical frame of reference.

In 1910 Frederick Clements published his *Life History of Lodgepole Pine Forests.* It presented a wealth of data, and its conclusions for management sound remarkably modern. But the real influence of Clements's study came in other ways and persisted for a long time. It was Clements who adapted European ideas of ecology to the American landscape. It was the Clementian concept of ecology that entered American forestry and land management. In its essence, the Clementian theory examined the phenomenon of biological succession, according to which a series of biological communities (or types) would occupy a particular site in conformity to a particular historical pattern (or cycle). Beginning with a bare field, it was possible to predict this evolutionary epicycle as it proceeded from an early "pioneer" stage to a final "climax" stage. The climax stage was self-replicating and would continue indefinitely unless disturbed by some catastrophe, such as clearing or firing. Then the cycle would be initiated again. It was possible, say through light burning, to arrest this progress at a subclimax stage. The concept of succession became fundamental to American ecology, the driving force in ecosystem dynamics, and much of the literature of ecology concerned the adaption of the concept to a great variety of environments, each of which conformed to the general pattern but differed in its particulars.

Clements himself modeled his theory after the geomorphic theories of William Morris Davis, and, like Davis, he even invented a taxonomic declension to describe the sequence of events. The idea saturated intellectual thought of the day: nearly every theory of significance in the biological, geological, and social sciences relied on an inevitable life cycle of events that progressed under natural conditions but could be arrested or reversed by catastrophic intervention or trauma. As a system of biological classification, the Clementian theory seems itself a logical climax to a succession of ever more complex concepts. The systematist Linnaeus had multiplied the great Chain of Being into several parallel chains; Humboldt, the geographer, had added another axis to the chain, making them into a cartesian map, no longer united by a hierarchy of characteristics but by geographic suites or associations; Darwin added a historical axis: species and communities tended to evolve through time. Clements broke this great historical chain of a single grand evolution into a kaleidoscope of evolutionary epicycles in which the ontogeny of individual communities recapitulated the phylogeny of organic evolution. The concept of a deterministic life cycle—be it of an individual, community, stream, planet, mountain, or civilization—entered nearly all areas of thought. The concept of the Nebraskan Clements, in fact, bears an uncanny resemblence to the frontier hypothesis of Frederick Jackson Turner, according to which an area advances from wilderness to a pioneer stage to civilization.

The Clementian concept of "nature's economy" underwrote much of conservation through the mid-twentieth century, and it was a concept especially gratifying to foresters. If the changing amount of sunlight and shade controlled the direction of succession as more shade-tolerant vegetation replaced the less tolerant, then the climax state tended to consist of large trees. Forestry saw as its greatest challenge the reclamation of cutover land and the conservation of existing forests. The Clementian theory vindicated fire protection in the frontcountry as a means of assisting the succession of deforested land to forest climax and in the backcountry as a means of promoting the innate and "natural" drive for successional climax. That lightning set many fires was really irrelevant: it was well known that nature suffered from waste and entropy, which human engineers could, and ought to, eliminate. Ecosystems adapted defensively against this challenge in their drive for a climax state. The great questions about fire protection were not whether it was scientifically vindicated but whether it was technically feasible and economically justifiable. The study of fire effects meant the calculation of fire damages. Prescribed fire might be used to advance human objectives, but it was not nature's way.

The transformations of the early 1970s came only after the Clementian theory was replaced with a view of fire that legitimated the use of prescribed fire as a "natural" tool and presented a new conception of the natural state.

The second Tall Timbers conference illustrated well this dual conversion: it opened with excerpts from the Leopold Report and concluded with remarks by Eugene Odum. Odum's research and his textbook on ecology were destined to make him Clements's successor as the dean of American ecologists. The change was part of a broad reconstruction of intellectual thought. The revision of the Clementian theory, for example, paralleled a similar attack on Davisian geomorphology. Both models simply broke down amid the multiplicity of landforms and biotas. They lost the ability to forecast that had made them so attractive. Rather than the entropy clock of closed systems, the newer theories preferred the steady state of open systems. Rather than an invariant succession of forms, there were stochastic probabilities and an endless cycling of nutrients. Instead of a management program that merely advanced or retarded successional tendencies, managers saw communities that fluctuated about a mean, that formed complex assemblages, and that beat to biological rhythms often powered by fire energetics.[68]

Odum's remarks at the conference showed the potential lines of research for the subject of fire ecology. The "great unknown in the field," he concluded, was the "question of *the effect of fire on mineral regeneration.* In many terrestrial ecosystems the cycling of the scarce elements necessary for life is the bottleneck which controls the rate of production." He observed that most of what was known about fire in the environment was "empirical," based on "correlations that may or may not represent cause and effect, and observations on situations not well controlled from the standpoint of science"; that fire management was a test of man's ability to manage the ecosystem; that "fire is and has been in the past an important factor in many environments and that it can be used as a tool in management on a much wider scale than is generally realized"; and that he was distressed by the "lack of projects on basic fire ecology" submitted to NSF and other granting agencies.[69] Perhaps more important than his particular observations, Odum demonstrated that the redefinition of fire ecology had come from a redefinition of fire and a redefinition of ecology and that both were part of a general transformation of conservation thought and of the intellectual understanding of nature.

Komarek added some significant observations of his own. Before fire could be explained as an ecological process or used as a technology it had to be understood. Even Aristotle in his day complained that "what we call fire conventionally is not fire."[70] Western logic has held that like causes produce like effects; but until fire was adequately defined, its biological effects could not be scientifically determined. Much of the confusion about fire under the Clementian theory resulted from the simple failure to recognize that the effects were not comparable because the causes were not comparable. Lamenting, like Odum, the absence of fire ecology among academic subjects, Komarek wanted "to point out also, that fundamental ideas and information can be gathered by anyone with good powers of observation—from the man on the land to the scientist in the laboratory." Citing the example of Darwin, he recommended the "close observation of many natural experiments" and the use of "natural areas."[71]

It was a shrewd commentary: the revolution in fire ecology and fire management came from a reexamination of basic facts in natural areas. The Forest Service commanded the lion's share of fire research, but it had nothing programmed in fire ecology. Komarek, conversely, related his surprise at discovering the physics research being conducted at the Northern Lab. The revisions in Forest Service fire policy resulted from a complex of political incentives and philosophical convictions, but the transformation of Forest Service fire research at that same time came about through the reconciliation of the biological and physical manifestations of fire, especially as presented in wilderness systems.

In 1972 military research into broadcast fire was terminated, the National Commission on Fire Prevention and Control promised to transfer responsibility for urban fire to a new agency, and the National Science Board of NSF recommended the quantitative modeling of the ecological effects of fire as a high-priority research activity. Within a year, under the aegis of the Coniferous Forest Biome program, NSF sponsored the Fire Ecology Project for the primary purpose of studying the natural role of fire in the functioning of western coniferous forest ecosystems. The Forest Service, too, applied for grants, and in short order its research machine was once again redirected. With the elimination of the emergency presuppression account and with the repeal of the 10 Acre Policy, the NFDRS lost much of its historical raison d'être as a guide to presuppression, but it was readily adapted to the decision models guiding the new fire plans. Fuel models became dynamic rather than static. Fire weather data and models were applied to questions of prescribed fire: the Southern Lab, for example, began a work unit on fuel treatment, prescribed fire, and air pollution, resulting in a smoke management handbook. Fire behavior models, with their emphasis on steady-state fires, proved ideal for prescribed burning. Modeling was extended to biological associations, partly for better dynamic fuel models, partly as a means of coping with fire in problem ecosystems, such as wilderness areas and the Southern California chaparral. Planning guidelines and appraisal studies calculated the beneficial aspects of fire as well as the traditional damages. A "resource advisor" joined the fireline staff, just as in an earlier age the fire behavior officer had and as in a still earlier age of research into economics a comptroller had. The uses and effects of prescribed fire—initiated by natural or anthropogenic causes—became the informing theme of virtually all fire research.

In the early days of industrial forestry, prescribed fire was used because there seemed to be nothing better at the time; by the 1970s, to the minds of most land managers, there was nothing that could be better. Previously, when a fire was shown to be valuable for silviculture or rangeland (as with brownspot disease in longleaf pine), the discovery was considered a freak, a sport of nature. But the identification of fire with wilderness had made fire "natural." Fire, it was assumed, was good and necessary; it was fire exclusion that was peculiar and in need of justification. With wilderness fire as a new exemplar, prescribed fire was extended to every landscape of the counterreclamation. The distribution of prescribed fire came to be synonymous with fire management, and the study of prescribed burning virtually came to be equated with fire research. The perception of wilderness fire as natural encouraged its use, much as the Clementian vision of fire as an interruption of succession had discouraged it. The outcome of the mass fire experiments added their weight to the general conviction that prescribed fire was valuable. The conclusion of Project Flambeau, for example, stated that "fuel characteristics, including those associated with both fuel elements and fuel beds, are the major controlling factors in fire behavior." Conflagration control required fuel treatment, and the natural properties of prescribed fire made it the treatment of choice.

By 1974 the main strands of the new era were coming together. At a joint convention in Missoula the Tall Timbers Fire Ecology Conference, the Intermountain Fire Research Council, and the Forest Service's Fire and Land Management Symposium consolidated their new and shared interests.[72] The Tall Timbers Station had been chartered in 1958, making it contemporaneous with the NAS Committee on Fire Research and the great buildup of Forest Service fire research facilities and programs. By the mid-1970s, however, NSF and the reformed Forest Service had absorbed the station's original fire research goals, and its influential conferences ceased. In 1974, too, the Forest Service established a Fire in Multiple-Use Management Research, Development, and Application Program (RD&A) at the Northern Lab. The goals of the program were to define the role of fire in forest and range ecosystems and to develop means by which fire management could be better assimilated into land management. Among a wide array of projects, it sponsored FIRELAMP, a complex computer model that could predict fire effects on various components of an ecosystem, and FIRE-BASE, a computerized retrieval system for information on fire. In 1976 the Service promoted a National Fire Effects Workshop "as a first step in responding to the most recent changes in policies, laws, regulations, and initiatives." Six working groups began preparation of state-of-knowledge reports for the effects of fire on soil, air, water, flora, fauna, and fuels. The reports had the dual purpose of consolidating the published literature and defining priorities for future research.[73]

On the surface it seemed that the old vision of forest management following from forest knowledge had triumphed, that the repudiation of folk knowledge would bring an abundance of scientific knowledge. But there was an ironic note. By the mid-1970s the information industry had become the largest industry in the United States. Fire management, like other fields, was both a beneficiary and a victim of that growth. As had the overabundance of manpower in the 1930s and the surplus of equipment in the 1950s, a surfeit of information threatened to distort the equilibrium of policy and programs. Studies on fire behavior and ecology had grown to conflagration proportions, and information control, it seemed, had replaced conflagration control and hour control as principles of planning. Just as ever more burdensome invesments in control technology confronted a nonetheless irreducible quantum of big fires, the further production of information did not seem likely to improve control over big fires, to assist the administration of problem fires, or to determine the proper range of policy objectives.

Science is infinite, management finite. Policy has to be based on broad cultural perceptions and political paradigms, not solely on ecological or economic investigations; scientific research is only one component among many that contribute to it. Early research had not lowered the cost of fire suppression, but it had extended its range. Modern research, likewise, does not seem destined to lower costs but will extend the range of prescribed fire. Like administrators surveying the legacy of successful fuel buildup, researchers might well contemplate the proliferating products of their investigation with a sense of ambivalence. The technology most typical of the era is that which collects and processes information; the research programs most pressing are those which assimilate, store, and distribute information. The policy of prescribed natural fire threatens to create a crew of research technicians to "monitor" fires as large as the crews that the old policies had dispatched to extinguish them. The liberation of wildland fire early in the century has led, with remarkable success, to the invention of a new science. But it has also confirmed an old adage: fire research, like fire itself, makes a good servant, but a bad master.

FIRE AND FROST:
A FIRE HISTORY OF ALASKA

The forests of interior Alaska, as is true of all forests of the far
north, are very susceptible to destruction by fire. . . . Once started,
fires may burn for weeks or even months, spreading over
hundreds of thousands or even millions of acres.
—H. J. Lutz, Alaska Forest Research Center, 1956[74]

You don't fight Alaska. You learn to live with it.
—R. R. Robinson, Fire Control Officer, BLM, Alaska[75]

After he toured Alaska briefly in 1915, an outraged Henry Graves observed
that "the interior forests of Alaska are being destroyed at an apalling rate
by forest fires. Conditions existing in the western United States 25 years ago
are repeating themselves in Alaska."[76] What Graves could not foresee, of
course, was that 50 years later Alaska would recapitulate the national history of fire control, much as it repeated a history of wildland fire practices.
Effective fire protection has to wait for certain conditions—technological,
for the availability of efficient air attack; military, for the recognition of
Alaska's strategic importance and the military value of smoke control; political, for the imminent admission of Alaska as a state; environmental,
for renewed interest in the "Last Frontier's" bounteous wildlands and
resources. And, of course, there was the mandatory fire complex, which
burned 5 million acres or so in the summer of 1957. Within the next 20
years, fire control in Alaska recapitulated the entire twentieth-century
experience of fire control in the Lower 48.

The process came with a few special variations. It was dependent on aircraft from the start. It began in the era of mass fire, bypassing the debates
and concepts of earlier eras. Alaska's remoteness, climate, peculiar political
status, and reliance on the "northern economy" of fur, fish, forests, and minerals meant that the region would never really experience wholesale reclamation by agriculture. The fire history of Alaska passed almost directly
from frontier to counterreclamation. But in overstriding the full evolution
of fire control in the Lower 48, fire protection in Alaska never acquired some
of its traditional justifications. Within two decades Alaska compressed the
policy debates that had informed the past seven in the Lower 48. Alaska
began the era as America's low-value backcountry; it concluded it as the
nation's most spectacular wilderness.

The physical geography of the region shows three broad provinces. To the south are great mountain ranges, arcing from the Aleutians to the Alaska Range to the Rockies. The climate is maritime, not unlike that in the Pacific Northwest. To the north rises a somewhat parallel band of mountains, the Brooks Range, whose north slope glides into the Arctic Ocean. Between these two ranges lies the interior, with its relatively arid, continental climate. The rivers that drain it, like the Yukon, have their headwaters outside the interior. To a remarkable degree, Alaska's interior topography of basins and ranges, its continental climate and summer storm pattern, its relative inaccessibility, its fuels of moss, grass, and "stringers" of timber along watercourses, its value as rangeland, and its hydrology of permafrost muskegs (an Arctic playa) all resemble the Great Basin of the Far West. Thanks to the BLM, which has administrative responsibilities for both interiors, their political histories are similarly analogous. Fire control for Alaska and the BLM became self-reinforcing: the BLM absorbed the charge to protect Alaska from fire, and it was Alaska fire control that made the BLM a serious fire agency in the Lower 48.

The two great forces shaping the vegetative mosaic of interior Alaska are fire and permafrost, an incessant antagonism of heat and cold, aridity and moisture. Permafrost—that is, permanently frozen ground—underlies much of the taiga. In the southern ranges the permafrost is discontinuous, but it becomes more continuous in latitudes approaching the Arctic Circle. The permafrost forms an impermeable boundary between surface and ground waters: it holds ground water near the surface; its active layer (the zone of thaw) allows for shallow soil, the establishment of rooted vegetation, and can create a thermokarst landscape of thaw ponds and muskegs. Vegetation and permafrost are thus in a delicate equilibrium: the distribution of one determines to an extent the distribution of the other. Typically the biota divides into wet and dry sites. What makes this equilibrium a dynamic one is the presence of fire.[77]

Fire induces changes like those resulting from a change of climate. The fires comes naturally with the lightning of summer thunderstorms and burn fiercely in the long summer days of the midnight sun. The period for growth is short, and so is the time for decomposition. Fire is perhaps the sole mechanism capable of massive decomposition and nutrient recycling under near-Arctic conditions. Fires are episodic and large. A complete continuum of fuels makes for frequent, large fires. Grasses, which range extensively from muskeg to ridgetops, and moss, which is ubiquitous in layers as deep as three feet, provide flashy ground fuel. Scrub reproduction, often from past fires, functions as brush. White and black spruce, assisted by curtains of lichens, burn with crackling explosiveness that makes crown fires common. The black spruce functions in Alaska like other fire weeds in the Lower 48—the lodgepole, pitch, and jack pines, for example. It occupies the poorer habitats, seizes sites frequently burned, and encroaches on the range of the more valuable timber, such as white spruce.[78]

In removing vegetation fire can profoundly, if only temporarily, alter the active zone of the permafrost. After a fire, the zone can increase; with the accumulation of insulating vegetation, it contracts again. As with successional models, the relationship between fire and frost is stochastic, not deterministic. Wildfire nonetheless helps to control the moisture available for the biota, not merely its typology of successional stages. Water is cycled as well as nutrients. The mechanics of a "pyrogenic tundra" have even been described, whereby fire initiates a frost cycle that retards tree seedlings and maintains tundra grasses. The complex mosaic of biotas in the taiga is a product of microclimates, but fire and frost are its two broad determinants. Every observer of interior Alaska has been struck forcefully by the extent to which fire has shaped the texture of the taiga.

Fire and frost encouraged a kind of natural nomadism among Alaska flora and fauna—a shifting agriculture of biotas and a pastoralism among migratory herds. Nomadism and fire, here as elsewhere, were mutual causes and effects. It was a pattern imitated by aboriginal man. Of Alaska's three broad native groups, the Eskimos and Aleuts had a maritime economy, and the Athabascans of the interior were hunters and gatherers who used fire for typical reasons. Campfires, signal fires, and fires for the preparation of gum for canoe repair were common; many became wildfires. Writing of a military reconnaissance in 1885, Henry Allen described the upper Tanana River, where "heavy smoke, caused by extensive timber fires, obscured the sun the entire day, so that an observation was impossible. This smoke had originated from signal fires which were intended to give warning of our presence in the country. When we first arrived at Nandell's there was only an occasional smoke around, but as his guests departed for their different habitations each marked his trail by a signal fire. The prevailing wind was from the east and carried the smoke along with us. In answer to the fires on the south bank new ones started on the north, so that for nearly two days we barely caught a glimpse of the sun except through the heavy spruce smoke." Rarely were such fires carefully extinguished, and one observer of the boreal forest voiced surprise that "conflagrations should not be more general and frequent," "considering every Indian and traveller usually lights his fires against the trunk of some prostrate tree, and leaves it burning."[79]

Accounts vary as to the deliberate carelessness of the Indian. Robert Bell, at the conclusion of some 30 years spent among the boreal tribes of Canada, probably guessed the correct explanation in 1888. "One of the reasons for the growing frequency of forest fires," he decided, "is that the Indians travel more than they did formerly (and thus make more fires than when they were accustomed to stay longer in one place), along with the fact that they are less careful to extinguish than when they are not on or near their own hunting ground."[80] The differences in carelessness noted by observers simply varied with distance from the homeland. Signal fires, campfires, canoe repair fires are all products of travel; resident white observers would, on the contrary, see practices around more settled camps. But in both cases escape fires were common, whether deliberate or accidental, and often immense tracts of land burned.

Fire was deliberately used in the service of hunting. Few reports describe game being actively driven by fire, but many note the use of fire to shape habitats. Broadcast fire removed underbrush, improving mobility and vision; it created clearings especially suitable for moose hunting; it was often used for hunting or trapping muskrats; it deadened large stands of spruce and birch reproduction, which could then be uprooted from the shallow soil and aligned into "fences" to manipulate migratory caribou. By banging loud beaters and torching trees, hunters could drive caribou along the fences into places of slaughter. The lichens on which caribou fed could be ruined by fire and, unlike grasses, required as long to reestablish themselves as a mature forest. The odor of grass smoke could mask the scent of a human hunter approaching a herd upwind. The great expanses of grass in the interior—the corridors for migration—may well have been maintained by broadcast fire just as they were elsewhere in North America.[81]

The clearings improved the military position of a tribe by making ambush less easy, and smoke was welcomed as an insect repellant. Smudge fires were ubiquitous during the summer months, and many, set during travels, were left like campfires to become conflagrations on occasion. To a seminomadic people this was not a catastrophe. Where broadcast fire did not result accidentally, it was often induced deliberately. It was a common saying that mosquitoes were the greatest cause for the firing of Alaskan forests. Describing a military reconnaissance to the interior in 1884, Frederick Schwatka wrote that "evidences of conflagration in the dense coniferous forests were everywhere frequent" and that the cause was incendiary fires set by the Indians "with the idea of clearing the district of mosquitoes."[82]

Controlled fires were used for felling and bucking timber. The careful application of fire could fashion a raft, for example, without stone or steel tools. Broadcast fire was employed for the procurement of firewood. Firing a spruce forest in one season promised a supply of dead, cured wood for the next. With their shallow roots, moreover, the spruce would commonly fall to the ground soon after a fire, sparing the native forester the labor of felling. It was a fire practice—one among many—readily adopted by the whites who followed. Fire in Alaska was also a source of entertainment. An Inglik informant from the lower Yukon told how land might be fired "sometimes just for fun."[83]

In summarizing years of careful observation for the U.S. Geological Survey, Alfred Brooks wrote in 1906 that "large quantities [of forest] are annually destroyed by fire, for which the natives must largely be held responsible. The writer has remarked again and again that the Alaska Indians are utterly careless about forest fires. It seems probable that they deliberately burn over large tracts in order to somewhat reduce the insect pest. That this indifference to forest fires was not learned of the white man is shown by the fact that many tracts are found which must have been burned over long before the appearance of any foreigner." Brooks failed to consider lightning, a prolific fire source in the interior, but his reasons for aboriginal firing were accurate.[84]

They were reasons equally applicable to the white man, many of whose fire practices must have been learned from the natives. Europe discovered Alaska in 1728 on Bering's first voyage, but it was not visited until the second voyage in 1741. Small settlements and vast fur trading ventures followed, stretching along the coastlines from the Aleutians to California. In 1867 the United States purchased Alaska from Russia. Until the land received territorial status in 1912, it was indifferently administered by the Army and the Treasury Department. Most of the white population remained on the coasts. Military reconnaissances, an aborted telegraph line to Siberia, and prospectors were the only source of information about the interior. A Russian prospector who attempted to ascend the Kenai River in 1850 was blocked by forest fires.[85]

Then came the gold rush in 1896—and with it, an epidemic of forest fires. An observer of the Copper River region remarked that "during the late gold fever flames were to be seen in the summer months on all the mountain sides, where they looked at night like the outpost lamps of a great city."[86] In the Canadian taiga the list of "fire-setting travelers," as Robert Bell termed them, was impressive even before the rush: "fur traders, missionaries, surveyors, explorers, prospectors, etc. and, nearer to civilization, railway builders, common-road makers, lumbermen, bush-rangers, and settlers."[87] Gold seekers were by definition transients and added their campfires to those of the Athabascans. They extended the range of firing for fuelwood, using much of the charred timber to melt through the permafrost for mining. Like the Athabascans of the interior, the Laplanders of Scandinavia, and the tribes of Siberia, the newcomers to the taiga carelessly distributed smudge fires and fired vast expanses to drive off tormenting insects. They used fire, too, for moose drives. The principle was the same as the surround or fire hunting in grasslands, but instead of creating a moving fire front, the hunters, with the assistance of (and probably on the advice of) the Athabascans, fired a ring of spruce trees. The trees were individually torched, and by advancing inward with successive rings of fire, the hunters could corral a goodly population of moose. "It became obvious," wrote one witness, "that the plan was absolutely perfect for the extermination of moose."[88]

The newcomers also introduced fire practices unknown to the natives. To provide forage for their pack stock, they commonly fired the regions around habitations for pasture improvements. Modest clearing for agriculture led to slash fires. And like their predecessors from Roman Spain to pre-Columbian South America, the miners fired the vegetative cover to expose outcrops and mineral deposits. The construction of the Alaska Railroad (1915–1923) from Anchorage to Fairbanks wasted the land to all sides and became the focus for early fire protection programs. The newcomers, in turn, learned about the unbelievable flammability of Alaska's fuels. Throwing a lighted match down into the moss of the Kobuk River, an 1885 explorer was alarmed at the subsequent rate of spread. "Nowhere in the world," he exclaimed, "probably will forest fires spread so quickly as here, and I felt considerable anxiety to know where this conflagration would end."[89] Like the Athabascans around them and their frontiersman counterparts to the south, the new Alaskans also burned for fun.

Territorial status brought Alaska under the administrative control of the GLO in the Interior Department. Information from the early resource exploration conducted by the Geological Survey was consolidated into Brooks's *Geography and Geology of Alaska*, published in 1906. Foresters soon took an interest. The forest wealth of Alaska promised to alleviate the forecast national timber famine, and national forests were created south of the Alaska Range and along the southeastern panhandle. Furthermore, on Pinchot's example, foresters had come to see themselves as spokesmen for conservation. It was Pinchot's clash with Ballinger and Taft over the leasing of Alaska coal mines for which he was eventually dismissed. R. S. Kellogg inspected the Alaska forests in 1909, and Graves himself followed in 1915. Both were appalled by the amount of destruction, especially by fire. Brooks reiterated his earlier observations with the publication of a study of Mount McKinley in 1911. "In the inland of the province," he concluded, "the supply of timber is at best rather scanty but would probably be sufficient for local use were it not subjected to rages by forest fires nearly every year."[90] One million acres is probably a good estimate for the size of the annual burn in the interior from 1898 to 1939. It is estimated that 80 percent of Alaska burned at least once during that period.

Nationwide the Transfer Act effectively removed the GLO as a fire protection agency. Without a fire control mission in the forest reserves, it had little incentive to create one in the unreserved public lands of Alaska. The GLO's role in Alaska remained its traditional one: it sought to manage legal title and to prevent trespass. But that was not good enough for the Forest Service, which insisted that protection meant protection of resources and not just of title. The construction of the Alaska Railroad was the catalyst. Though the railroad promised to open up the resource wealth of the interior, it also threatened to become a fire hazard, and some of the tracks passed

37. Tanana District, Alaska, 1905. A burned landscape, with the timber in the foreground crudely cut for firewood.

through national forest land. In 1916, a year after Graves's tour, the Forest Service produced a fire protection plan that would extend into the interior. Little of it was accomplished, however. Some fire patrols were conducted by the GLO's Division of Field Investigations near towns and along the tracks beginning in 1921, and a feeble effort was made to educate a skeptical population to the value of prevention. In 1929, the railroad from Seward to Fairbanks was patrolled, but it was not considered advisable to "attempt other interior patrols this season." Officially, fire protection for the interior had been "a matter of grave concern for years," GLO staffers assured an inquisitive Forest Service.[91]

With the completion of the Alaska Railroad in 1923, the Forest Service again demanded action. It preferred that the Department of the Interior take the initiative, but insisted that some agency must do something. The countryside along the tracks was becoming a burned wasteland. John Guthrie, a Forest Service official, described the dangerous condition for the *Journal of Forestry* in 1922. He urged that the Forest Service, a "logical choice by reason of its 17 years' experience in handling government forest lands," take responsibility for protection of the interior forests. He did not recommend that the interior be incorporated into the national forest system, only "that its protection for fire should be delegated to the Government service whose special function is the protection and administration of Federal forest lands."[92]

The GLO continued its patrols through 1933. To the Forest Protection Board it revealed that, out of 71 patrolmen hired in 1929, only 2 were in Alaska; most were on the O&C lands in Oregon and California. When fires outside Fairbanks threatened both the town and the Alaska Agricultural College, it was the chief of the Biological Survey who extended funds for their suppression and for a "local fire patrol" under the guise of protecting the biological station at the college. The Forest Protection Board included a fire plan for Alaska among its general package. With the collapse of the board and the coming of the CCC, it appears that no patrols were conducted from 1934 to 1938. Most of the New Deal's interest in Alaska centered on the Matanuska Colony, an attempt to reclaim a valley north of Anchorage into arable land. The colony wilted away, but not before the old problem of slash fires was added to the Alaska landscape.

By the summer of 1939, the Interior Department was prepared for something more. On July 1 it established the Alaska Fire Control Service (AFCS) which was charged with fire control statewide, with forestry and recreation programs, and, in 1940, with the administration of the CCC program. It had an operating budget of $37,500. Despite frantic messages, approval of the state fire plan did not come until September. For the interim, commanders of the CCC camps around Anchorage and Fairbanks offered to provide "a first line of defense" in case of fire. Under the AFCS, however, the camps served primarily on construction in native villages and for military airfields. A field handbook prepared in 1940 boldly proclaimed that the purpose of the AFCS was "to protect the forests and vegetated areas of the Alaskan public domain lands from the depredations and ravages of uncontrolled fires. It is pledged to use its every resource, and those at its command, toward the prevention of fire, and, within its scope of authority and means of practical accomplishment, toward the suppression of all fires regardless of origin."[94] The handbook was written by foresters.

The official fire plan was somewhat more cautious. "The policy includes prevention of the forests, woodlands and tundra regions of Alaska and the detection and suppression of fires occurring in and adjacent to the more populated districts." The GLO did not yet have an emergency fire account, and "in lieu of limited suppression funds, fires occurring in remote regions will not be touched. Strick adherence to the policy of not acting on a fire unless it can be corralled and put out will be maintained."[95] The AFCS would emphasize patrol along rivers and roadways and the protection of villages. Not until the late 1950s did the BLM attempt blanket coverage over the interior.

What it lost with the termination of the CCC, the AFCS gained by affiliation with the military. The Japanese had attacked the Aleutian as well as the Hawaiian Islands, and Alaska suddenly acquired strategic importance. The interior was again opened up, this time by military occupation. The

Alaska Highway was constructed; new bases and airfields were laid out; and air travel to the continental United States was established. According to Alaska historian William Hunt, "Probably no other part of America was affected more by World War II than Alaska."[96] The enlarged military presence was especially favorable to fire control. The AFCS made contact initially by supervising CCC construction of airfields; later, with the help of the Alaskan Defense Command, it fought every fire within range. Under the impress of the war, it entered into a wide range of cooperative agreements among state and federal agencies and undertook a prevention and education program. Smoke was considered a serious military obstacle, and with military aircraft and personnel to assist, fire losses dropped from an estimated 4.5 million acres in 1940 to 117,000 in 1945. The Cold War strengthened the military presence in Alaska, and cooperative agreements between the military and the BLM assisted the transition to civilian fire protection. The two evolved reinforcing strategies of protection through airpower. Even today the primary BLM base in the interior is stationed next to the runways of Fort Wainwright in Fairbanks.

In 1946 the AFCS was absorbed into the newly formed BLM. The next year it was disbanded as a semiautonomous agency and incorporated within the BLM's Division of Forestry. The immediate consequences were not encouraging. Fire losses again swelled to over a million acres per year, and it seemed that fire control would be lost in a labyrinthine bureaucracy dedicated to other concerns. With time, however, the BLM's management objectives officially broadened, and it became apparent that the AFCS's incorporation into a much larger organization (and one with a Washington, D.C., office) had strengthened the capabilities of the fire control program. In 1949 the BLM was granted fire protection funding similar to that enjoyed by the Forest Service, and it contributed money, along with the Forest Service, toward H. J. Lutz's investigations into the effects of fire in Alaska. Lutz, a professor of silviculture at Yale, was surpervised by the Alaska Forest Research Center of the U.S. Forest Service. His study, *The Ecological Effects of Forest Fires in the Interior of Alaska,* was published in 1956 as a technical bulletin by the Department of Agriculture. It concluded that "uncontrolled wildfires have no place in either forest or wildlife management," that "the ultimate place of prescribed burning in Alaska cannot now be stated," and that "uncontrolled fires, sweeping over vast areas of the interior nearly every summer, place in jeopardy the future economic development of that portion of Alaska." "It seems apparent" that the widespread burning that "has occurred during the past half century or more has not resulted in a land teeming with furbearers and big game. It may be doubted that more burning will produce such a result in the future."[97] Starker Leopold and Fraser Darling, who inspected the causes of wildlife depopulation in the interior, agreed. In 1953 they concluded that in central

Alaska "it appears to us that range destruction by fire is principally responsible." One fire, they reminded readers, "could undo the work of decades
protecting a local caribou population from men and wolves."[98]

The reports were timely. The early 1950s were big fire years. Slowly the
BLM developed a protection force based on civilian aircraft, although the
Army and the Air Force contributed heavily at times under cooperative
agreements. In 1954 the BLM acquired three Grumman Goose model aircraft from the Coast Guard and several small planes for detection. When
an amendment extended the McSweeney-McNary Act to the territories, the
Forest Service acquired a research interest in Alaska. In 1955 the BLM
published its "Comprehensive Forestry Program for Interior Alaska," in
which, naturally, fire protection loomed large; the Western Forestry and
Conservation Association, reflecting the opinion of informed foresters,
passed a resolution calling for adequate protection of Alaskan forests; the
BLM requested assistance from the Forest Service in preparing a fire danger rating system for Alaska; and, with the help of the Grummans, for the
first time a major interior conflagration (the Galena fire) was attacked. In
1956 a formal cooperative agreement between the Forest Service and the
BLM gave the Forest Service a presence in the interior overwhelmingly
focused on fire. All that was required for a major BLM investment in fire
control throughout the interior was a catalyst.

By any standard the fires of 1957 were enormous. Most were lightning
caused. At least 5 million acres burned, but the full extent remains
unknown. Suppression forces were quickly overwhelmed. BLM overhead
from the Lower 48 were shipped north with minimal success. Hundreds of
thousands of cubic miles of smoke saturated the Alaskan skies. Smoke shut
down air traffic in the interior, isolated villages, and alarmed the military.
For two weeks the smoke was so intense it forced the airfields at Anchorage
and Fairbanks to close. Fires could reach tens or hundreds of thousands of
acres before they were even detected. A Forest Service observer sent to test
a fire danger rating meter noted that the smoke had a "pyramiding effect."
"New fires could not be located; the extent of current fires could not be
determined; fire crews could not be dispatched; crews or fires could not be
supplied; fire bosses could not see or get around enough to determine the
best methods to use or find the most critical locations for control work."
Forestry magazines publicized the fires as a matter of shameful neglect.
Politicians recognized that the territory, soon to be a state, demanded more
attention. The military worried that such conflagrations might compromise
the stragetic value of its Alaska outposts. And the Forest Service, building
on its earlier cooperative agreements and the amended McSweeney-
McNary Act, established the Institute of Northern Forestry at Fairbanks.
The fires, in short, had a regional effect not unlike that of the 1956 fires in
Southern Califormia.[99]

By 1958 modernization of fire research and suppression programs came with almost breathtaking speed. Research recapitulated its Forest Service history, moving from statistics to fire behavior to effects. The Interior Alaska Forest Survey Program let to broad reconnaissances and to the 1963 publication of a massive statistical compendium by Charles Hardy (Forest Service) and James Franks (BLM), *Forest Fires in Alaska*. Protection acquired a P-51 Mustang for high-speed detection, adopted a 10 A.M. Policy, organized crews of Indians and Eskimos, and, in 1959, opened a smokejumper base at Fairbanks. Beginning in 1961, a jumper contingent from Missoula supplemented the BLM force during early summer. Aerial water bombing was attempted during the 1957 conflagrations, and as early as the 1940s, thanks to military equipment, helicopters had been used. By making its rite of passage during the era of mass fire, Alaska fire control bypassed the hour control concept. Like the Lower 48, it looked for conflagration control through heavy equipment, especially aircraft. A typical dispatch sequence began with aerial detection, a call for an aerial retardant or water drop, followed by smokejumpers, followed, at last, by helitack or ground crews. Supplies came by helicopter or paracargo (the BLM runs the largest civilian paracargo operation in the world).[100]

Following Alaska's admission to the Union in 1959, the American Forestry Association, expressing the concerns of professional foresters, published a fire plan for the new state.[101] But almost in spite of itself the BLM was already well on the way. The next decade would witness a costly display of technology transfer and development as impressive in its way as the concurrent space race from Sputnik to Apollo. The program had a significant impact on the interior—both on its natural biota and resources and on its native populations. As EFF money poured in, native economies became at least partly dependent on the seasonal cash flow. It was an era of adventurism and expansiveness. Fire control in Alaska was but one example of a restless American affluence that would put men on the moon and bring Great Society social programs to the natives of the Marshall Islands.

By 1963 the BLM's Alaska fire operation was greater than its counterpart in the Lower 48. Alaska foresters participated in the BLM Forestry Conference at Coos Bay, Oregon, and Alaska fire overhead traveled for the first time to troublespots on BLM land in the Great Basin. In 1965 Alaska developed an ambitious five-year plan for fire control over its interior, while on the Alaska example and with Alaska advice after the Elko fire debacle, the BLM created the Great Basin Fire Center. Thereafter the two regions tended to reinforce each other: the Great Basin Fire Center became the Boise Interagency Fire Center, and the five-year plan culminated in a flurry of suppression activities in 1969—the Swanson River fires on the Kenai Peninsula, enormous wildfires in the interior (over 1 million acres), the cloud-seeding experiments of Project MOD and Project Skywater, and a major reevaluation of Alaska fire control organization. The Swanson River fires coincided with the opening of BIFC and were coordinated through the center. Their suppression costs were the greatest of any fire up to that time, and the spectacle of heavy tracked vehicles punching firelines the size of football fields through a moose refuge raised environmental questions. The Swanson River fire, in short, anticipated nearly all the problems experienced at the Seney Wildlife Refuge fire in the upper peninsula of Michigan in 1976. The understanding of wildland fire in Alaska also made important strides in 1969: the Alaska Forest Fire Council was created, the Interior Alaska Forest Survey Program published the findings it had accumulated over the past decade, and the Interior Department library released a comprehensive bibliography of fire in the far northern regions.[102]

Alaska's economic history is a series of booms and busts. The rush of fire protection into the backcountry repeated that tendency. With the exception of the national forests of the south and southeast, the BLM protected, either directly or through contract, nearly all the lands of the state—about 19 percent of Alaska. Its commitment was enormous. Most fires came along lines of transportation or near population centers. By 1970 perhaps 70 percent occurred on state and private lands covered by reimbursable contract. The BLM, its state fire officer explained, served as "the Department of Interior's fire department" for Alaska. Even during its most expansive period, the BLM recognized that a priority system was necessary. During periods of low activity, action was taken on all fires. During critical periods, suppression resources were allocated on the basis of values at risk, fire danger rating, and prior commitment to going fires. For the fires it selected to suppress, a 10 A.M. Policy was the working norm. In a perceptive analysis from the era of conflagration control, J. H. Richardson of the BLM exclaimed that "we have not been able to identify any area where fires can safely be left to burn without serious consequences and high costs. The only effective means of control we have is to take all-out action on these fires while they are small and controllable. Wildfire is not a precision tool that can be allowed to burn to certain predetermined limits, then stopped. It is not like a gas stove that can be turned up, then turned down when the pot starts to boil. It is more like an atomic reaction that we either stop at the beginning or don't stop at all." Prescribed fire, he noted, could "be a highly useful management tool when used within proper limits."[103] But the defining problem was the remote lightning strike.

Both the BLM and Alaskan fire were anamalous. Alaska had become a territory when the forty-eighth state, Arizona, achieved statehood. Its peculiar political situation and the Ballinger-Pinchot controversy suspended it from the impact of the early conservation reforms. Conservation efforts stayed along the maritime perimeter, with the national forests of the south, the Pribiloff Island preserves for fur seals, and the parks and refuges in the

Alaska Range and the Aleutian Islands. The Forest Service—and with it the programs of the first era of conservation reform—never entered the interior of Alaska. When fire control exploded into the interior in the mid-1950s, it came through the BLM and made the BLM the nouveau riche of national fire control. But by 1970 the boom was faltering, and by the end of the decade the BLM program was in jeopardy.

The very success that helped to propel BLM fire control into the front ranks and to integrate it institutionally with fire protection in the Lower 48 also subjected BLM fire policy to the same reevaluation and redefinition experienced by the Forest Service. Effective fire control began with events surrounding one of the pivotal fire years nationwide, 1956–1957, and underwent major reforms during 1970–1972. Put simply, BLM suppression strategy was challenged, its objectives redefined, and its land base eroded. In 1970 it adopted a new state fire plan. Studies based on the suppression of a fire that had occurred a few years previously warned that firelines in permafrost might be more damaging than the fire, and the experiences on the Kenai National Moose Range reinforced that alarm. Restrictions on tactics, especially those involving heavy equipment, were written into the new plan. New techniques, such as aerial backfiring, were also developed to minimize potential suppression damages and to contain escalating costs.

Fire research was reorganized. Originally, it had emphasized statistical compilation and the adaptation of a fire danger rating system. Much of the first objective was accomplished with the 1963 report by Hardy and Franks, and the latter culminated in the NFDRS guidleline published in 1972. Researchers then explored fire behavior in Alaska fuel types. The next stage, as it was nationwide, was research into fire ecology. In 1971 the Alaska Forest Fire Council organized a symposium on "Fire in the Northern Environment," and the BLM contracted with the Institute of Northern Forestry for a four-year "Fire Effects Study." Exploiting a fire about 50 kilometers from Fairbanks, the institute established a fire ecology work unit. The information accumulated since Lutz's masterpiece was summarized at another symposium on the fire ecology of western and northern forests of North America, a part of the NSF program in forest and fire ecology. Another part was the Taiga Ecosystem Project, conducted in 1974 through the University of Alaska. A year later the Fire Management Study Group of the North American Forestry Commission sponsored a seminar on research needs in the far north.

The Department of the Interior adopted a new fire policy in 1974. For Alaska this required a team evaluation on all fires that escaped initial attack, and it led to a new dispatching system. A classification of interior lands by value was incorporated into a decision flow chart to guide dispatchers, a system not unlike the charts being developed in the South and California at the same time. In 1976 congressional legislation at last eliminated the vestiges of the BLM's origin in interdepartmental restructuring and made it more of an integrated department. The agency undertook environmental impact statements for its national fire policies and began installation of an automatic electronic lightning detection system in Alaska and the Great Basin.[104] A year later major fires broke out in the interior. Nearly 3 million acres burned, the largest burn since 1957 and 1969. Doubts were naturally expressed about the effectiveness of the BLM's 20-year investment in fire control.[105]

Similar doubts after the 1957 and 1969 fires had led to reorganizations and a redirection of policy. But the 1977 fires came under more troubling circumstances. The BLM's land base was rapidly eroding. It was a custodian, a trustee: it was never really charged with land management in perpetuam as were the Forest Service and the Park Service. Instead, like the British raj in India, the BLM in Alaska was destined to oversee the partition of its once vast and homogeneous empire. The Native Claims Settlement Act of 1971 granted 44 million acres to native corporations. Fire policy for the land is still uncertain, and for the present the natives continue to contract with the BLM. This will probably change. Immense portions of the interior are being rezoned into national parks and wildlife refuges. The Park Service wants to establish its own research and protection system, turning to the BLM only for assistance under extreme circumstances. The Fish and Wildlife Service continues to debate the nature of an appropriate policy. Both it and the Park Service want prescribed fire, natural or anthropogenic—a technique in which the BLM never acquired much expertise.

Nor were the reforms limited to federal land ownership. The legislation admitting Alaska to statehood granted it 25 years to decide which lands among the public domain it wanted to retain as state lands. Discussions in 1970 let to a five-year plan. The first land concession came in 1973, and the process has accelerated since then. Its constitution imposes on the state all responsibility for fire protection on state lands. The high cost of BLM protection makes it unlikely that the state will continue to contract with it for services. The state cannot hope to match the emergency presuppression account the BLM still holds. Nor will it require the most expensive of BLM services: the aerial attack on remote fires. Most state fires will be around developed sites and will be handled much as rural fires elsewhere. The state operations are in line with the Clarke-McNary and rural fire protection programs, particularly as formulated by the Cooperative Forestry Act of 1978. In 1974 a joint federal-state land use planning commission issued preliminary conclusions about future fire protection problems, and in 1977, while the BLM struggled with its backcountry fires, the state sponsored a study of California as a possible model of a state-run protection system with both wildland and rural jurisdictions.[106] Unlike most Forest Service agreements, moreover, the relationship of the state and BLM was not really reciprocal but contractual. The BLM had enjoyed a de facto monopoly.

By the late 1970s the BLM was returning to its historic role as the dispenser of public lands, and it promised to be left, as in the past, with the lands no one else wanted—lands for which it may be difficult to justify a heavy commitment to fire protection. The morale and effectiveness of a fire control organization, like wildfire itself, are difficult to modulate. Fire organizations typically do well with problems of growth but poorly with problems of decay. Only heavy fire activity can sustain a quality organization: fire services tend to grow exponentially in quality as their fire load increases. How much reduction in activity the BLM can endure and still field a tough suppression organization is difficult to determine. Meanwhile, though it was on the whole excluded from the land breakup in the interior, the Forest Service promises to augment its own presence through research contracts and cooperative programs with the state. In 1978 an Interagency Land Managers Task Force was formed to coordinate activities over the balkanized interior. The fire subcommittee worked on a pilot fire management plan for the 40 Mile region that was subsequently presented at national training courses in Marana.[107]

Alaska has been described as a land of edge. Far from showing a dense and frowning uniformity, its tundra and taiga form a complex kaleidoscope. Its political composition is coming to show a similar heterogeneity. Its abundance of edge is a major reason for its almost mythical abundance of wildlife—a natural nomadism of migratory fauna and shifting cover types—and the management of that edge may well be the focus of land and fire policy. Like the Great Basin, the economy may depend on a government presence (largely military) and on grazing animals. In the Great Basin this means domesticated livestock; the Alaska interior, the natural wildlife. The seminomadism of hunting tribes may be replaced by the seminomadism of recreational tourism and sport hunting. No real agricultural settlement ever took place on the tundra and taiga, nor have domesticated stock been introduced successfully. Despite a few famous attempts, reindeer and musk oxen have never seriously challenged the native fauna. (In the Great Basin, conversely, domesticated livestock, such as burros and horses, have gone wild and now enjoy special legislative protection for maintenance of their newly acquired "natural" status.)

The relationship of fire to wildlife is complex. Clearly, in Alaska, as elsewhere, the two have come into mutual accommodation through millennia of interaction. But the mechanics of that detente are not at all certain. In a broad way, it is known that moose habitat can be improved by fire and caribou range ruined. But the relationship is stochastic, not determinant. The etiology of fire and flora is complicated enough; the etiology of fire and fauna superimposes at least another order of indeterminacy. The historical record shows, for example, that at any given time moose may be attracted to burned areas, indifferent to them, or repelled by them. Part of the problem no doubt is that the fires under consideration are not comparable. But the confusion leaves the exact role of anthropogenic prescribed fire uncertain. There is little doubt, nonetheless, that wildlife will be fundamental to the economy and land use of Alaska and that fire management, taken at large, will be integral to the habitat management of the interior.[108]

In the Lower 48 most of the federal lands reserved from the public domain were set aside during the first era of conservation. They were removed from agricultural reclamation or pastoralism in an attempt to conserve natural resources important to industrial devleopment. The biggest gainer in terms of land acquisition was the Forest Service. Alaska came later. The debate over its development did not begin until the second era of conservation. Agriculture and livestock were not serious problems, but the presence of the synthetic landscape was. The issue was preservation, not conservation. Wilderness was pitted against industry, native cultures against industrial society. Its cause célèbre was not the Dragon Devastation, not the reckless consumption of natural resources aptly symbolized by fire, but the threat to wilderness posed by the Trans-Alaska pipeline. In the balkanization of the interior, the biggest federal land gainers were the National Park Service and the Fish and Wildlife Service. The fire protection problems, technically speaking, were different only in scale from those of the early days in the Northern Rockies, the Great Basin, and Southern California. But the political and cultural contexts were entirely different. The Forest Service presence was important only as an example and a source of advice. Fire protection in the Lower 48 began against a backdrop of frontier fires; in Alaska, wilderness fire dominated. The BLM imitated the Forest Service example, but it copied the tradition of the first era of conservation reform. The 1957 fires were in a sense an Alaska equivalent of those of 1910. But the broader context had changed, and almost immediately the BLM was forced to realize that it was not fighting the frontier fire, the backcountry fire, or even the mass fire. It had to confront wilderness fire.

Alaska was the last major fire regime of the United States to receive organized fire protection. The BLM experience with all-out suppression was another epicycle in the boom and bust economy so typical of Alaska's history. But it left permanent features, too. It put the BLM into big-league fire protection, inspired the BIFC, and created a working protection force that the new landholders of Alaska could adapt, modify, or contract for. It was an essential transition from a condition of laissez-faire fire practices to systematic fire management. It was a means of discovering the interior and its resources. If wildlife and wilderness indeed become a foundation for the land management of the interior and of its economy, fire management will exercise a crucial, perhaps unique role. The BLM effort was, in the words of its most prominent fire officer, R. R. Robinson, a means of learning to live with Alaska. As it integrated with other fire agencies and national objectives, BLM fire policy in Alaska came under sharp criticism from advocates of wilderness values.

The charge was not without irony. Since the gold rush, Alaska had been subjected to an unprecedented wave of anthropogenic fire. Between the rush and 1940 it is estimated that 80 percent of Alaska burned at least once. The respite offered by organized fire suppression probably gave many regions an opportunity to restore some of the features that existed before the interior was opened up so violently to industrial civilization. But even these prior conditions were far from purely "natural," and because fire protection was imported just prior to the ascendancy of wilderness as an informing concern, the debate over fire policy in Alaska perhaps shows better than in most regions the paradoxes of contemporary expectations. It was through the interior of Alaska that anthropogenic fire entered North America during the Pleistocene, and the region served as a training ground for the use of fire by early man in the New World. Far from being the last pure wilderness, Alaska was the first on the continent to feel the impact of anthropogenic changes and has borne for the longest time the impress of anthropogenic fire.

FIRE ON THE MOUNTAIN:
A FIRE HISTORY OF THE SOUTHWEST

Beyond the hot springs [Ojo Caliente] they made a fifty-five mile
jornada and, on the far side, got themselves into a prairie fire.
One of Gilpin's campfires spread through the mountains, where it
burned beside them throughout a day's march. Lieutenant Gibson
remembered an old song, "Fire on the mountains! run, boys, run!"
and that night they had to run, when a gale drove the flames
down to their camp. There was a wild half hour when the army
set backfires, galloped the horses and wagons about, and swore at
one another in pyrotechnic light till the show was over.
—Bernard DeVoto, *The Year of Decision: 1846,* describing
an American column near the Sierra Madre during the
Mexican War.[109]

From every peak now curled the omnious smoke signal of the
enemy, and no further surprises could be possible. Not all of the
smokes were to be taken as signals; many of them might be signs
of death, as the Apaches at that time adhered to the old custom
of abandoning a village and setting it on fire the moment one of
their number died, and as soon as this smoke was seen the adjacent villages would send up answers of sympathy.
—Captain John G. Bourke, *On the Border with Crook*, 1891[110]

In 1972 the Tall Timbers Research Station sent a special Task Force on
Ponderosa Pine Management to the Apache reservations of Arizona. It
found in the Malay Gap region of the San Carlos Reservation a natural
preserve of ponderosa forest, inviolate from fire protection, and it discovered
in the prescribed broadcast fire practices of the Fort Apache Reservation an
examplar for ponderosa management throughout the Cordillera. It noted
approvingly that "more controlled or prescribed burning has been done on
these three Reservations (mainly the Fort Apache) and over a longer period
of time than in any other forested area of the western United States."[111]

   Its ideas, and even its selection of Malay Gap, had not originated with
the task force. But in advertising an idea whose time had come, it also dramatized a region, the Southwest, whose time on the national stage had also
come. The Southwest was not the site for the creation of a new regime, like
Southern California, nor the scene for the recreation of an old one, like

Alaska. Its value lay in the vegetative and cultural mosaic it exhibited, and it had quietly come into national prominence through the sheer pluralism of its fire regimes. Fire management involved far more than prescribed fire or wilderness fire: it demanded a variety of fire practices, a mixture of management responses. This the Southwest, by virtue of its unique historical and geographical situation, had long known and was especially prepared to demonstrate. The national forests of the region average more fires per year than any other region; they have the second highest rate of burned acreage, from both wild and prescribed fire; and critical fire weather occurs there with greater frequency and persistence than anywhere else in the United States.

A comparison to Alaska is striking. The fire history of Alaska, like that of most regions, tended to emphasize a particular cycle of fire or a distinctive type of fire. The Southwest by contrast was an ensemble of fire types, an assimilation of fire practices preserved over cycles of conflagrations. Its fuel complexes range from desert to grassland to chaparral to pine and finally to taigalike tangles of spruce and fir. It harbors one the the heaviest concentrations of lightning fire in the world. Whereas the short Alaskan summer demands rapid decomposition by large fires, the southwestern lightning complex results in frequent low-intensity fires over the course of a long season. In Alaska, fire management is dictated by considerations of lightning and wildlife; in the Southwest, by lightning and livestock. The pre-European fire history of both regions was controlled by Athabascan Indians and tribes related to them linguistically. In fact, the fire history of the Southwest was in good measure shaped by the migration of fire practices from the Alaskan interior and their transformation over centuries on the Great Plains.

Interior Alaska was the last region of the United States to be explored by Western civilization; the Southwest was among the first. Whereas Alaska consequently suggests perpetual rediscovery, an endlessly new and final frontier, a *de novo* wilderness, the Southwest has always recognized the long span of its natural and cultural past. More than an *omnium gatherum* of history, however, the Southwest, unlike Alaska, has tended to assimilate and preserve. The rocks exposed in the inner gorge of the Grand Canyon are among the oldest in North America. The continent's most ancient forests are those petrified in shale deposits of the Colorado Plateau. In Alaska nothing seemed to predate the Athabascan tribes who were there when the Europeans and Americans arrived. In the Southwest even the Indians recognized the relics of those who went before. When the desert and rivertine tribes, like the Pima and Papago, discovered the ruins at Casa Grande, they gave the Pima name *Hohokam,* "the ancient ones" or "those who have gone," to their builders. The Athabascans used the word *Anasazi,* "ancient ones," to refer to the creators of the cliff dwellings and ruins they found across the Colorado Plateau. This characteristic overplay of old and new is reflected in the region's fire history. More than a mosaic of fuel types, the Southwest shows a palimpsest of fire regimes, a preserved historical geography of fire in America. The first Spaniard to enter the region, Cabeza de Vaca, recorded fire practices of Indians in Texas. Appropriately, as Bernard DeVoto records, one of the first American columns into the Southwest during the Mexican War found that fire on the mountain was a Southwest tradition. As their successors learned, it was a fire regime controlled equally by natural and cultural history.

The geography of the Southwest breaks into three large provinces, bordered on the east by the Rio Grande and on the west by the Colorado River. To the north, there is the Colorado Plateau, a high tableland of mesas, canyons, and volcanic features. To the southeast, there is the Chihuahua Desert, and to the southwest, the Sonora Desert, both in the geologic province known as the Basin Range. In between, a broad mountain belt rises from the Sierra Madre of Mexico, extends northward along the border of Arizona and New Mexico to the Colorado Plateau, and then follows the rim of the plateau northwesterly through central Arizona. The topography is that of a series of mountains and valleys, defined by the fringe of the Colorado Plateau, the famed Mogollon (or Tonto) Rim. The valley floors are elevated over 4,000 feet. The rim varies from 5,000 to 7,000 feet. Evidence of vulcanism is rich along its border. To the interior, volcanic peaks rise to over 12,000 feet. The concentration of lightning fires in this middle providence of the Southwest is the heaviest in the United States and among the heaviest in the world.

The climate of the Southwest shows strongly seasonal patterns both within and between years. Drought cycles are common, and most annual precipitation comes in the course of a summer rainy season. The climate is continental—arid in lower elevations, semiarid along the mountain flanks and rim, and humid only atop the higher mountains. Rainfall is rare from March until the summer rains. Lightning comes with the summer storms. Favorable synoptic patterns call for a subtropical high pressure cell to form north of the region, bringing a moist, unstable southeasterly flow out of the Gulf of Mexico, which interacts with the mountain topography and the violently heated desert surfaces. Dry lightning is possible in May and early June. In July and August the "monsoon" arrives; lightning fires are abundant but typically small, owing to generous precipitation. The fire load is commonly one of great numbers of fires, not fires of great size.

The basic fine fuel is grass. In the valleys, the grass forms a true grassland. It continues up the mountain flanks, where it underlays the forest in a quasi savannah. At lower elevations, the grass gives way to desert succulents and woody species; at higher elevations, to the needles and branchwood of dense, mixed conifer forests. The line of demarcation between desert and grassland and between savannah and forest is largely determined, as Robert

Humphrey remarks, "by the ability of the vegetation to carry fire."[112] The vegetation in question is grass. Where drought retards it, the desert encroaches; where rains replenish it, the grasslands expand, often through the agency of their old ally, fire. Grass is intimately related, moreover, to the ponderosa pine, the primary forest cover on the Mogollon Rim and mountains. Together they are coextensive with the great zone of lightning fire that arcs across the region. "In many respects," the Tall Timbers group wrote, "the ponderosa pine forest is the western counterpart of the longleaf pine forest of the South, and their fire relationships are strikingly similar."[113] The introduction of prescribed broadcast burning in the West came from the study of fire and ponderosa, much as in the South it came from fire and longleaf. In both environments, too, the association of pine with grass—and thereby with livestock—was fundamental to the regional fire history and to the conduct of prescribed burning.

The sources of fire were both natural and cultural. The ponderosa forest along the Mogollon Rim is famous for the density of its annual lightning fires, but lightning has also ignited widespread conflagrations in the inter-mountain grasslands. In the summer of 1979 lightning fires swept over 90,000 acres of grassland in southern Arizona, and the scene was repeated the next year for the grasslands of the north. But to this source must be added extensive anthropogenic burning over thousands of years.

The record of Indian occupation is plentiful. A paleo-Indian culture of big game hunters, dating as far back as 30,000 years, is evidenced by artifacts at the Folsom site in New Mexico and in the lower deposits of Ventana Cave, Arizona. By 10,000 B.C. the hunters had disappeared. Between 5,000 and 9,000 years ago, hunting and gathering societies known as the Desert Culture (locally, Cochise Culture) occupied the same sites. By 200 B.C. they, too, had vanished. Other civilizations flourished between A.D. 600 and 1400 and left spectacular remains in the form of well-preserved ruins. Three distinct groups are recognized, corresponding to the three geographic provinces of the region: the Mogollon, along the rim; the Hohokam, a desert and rivertine group; and the Anasazi, on the Colorado Plateau. The first European explorers discovered only their abandoned dwellings and a more primitive linguistic group of Ute-Aztecan peoples on the plateau and a Yuman-speaking group along the rivers and deserts. Into this complex, perhaps simultaneously, two further invaders arrived: from the south came the Spanish, and from the east came Athabascan-speaking tribes.[114]

The Athabascans were apparently the last peoples, until the Eskimos, to migrate across the Bering Strait. From the Alaskan interior they moved eastward toward Lake Athabasca; then groups began a long migration south, through the Great Plains. When the Spanish arrived in northern New Mexico, they met the vanguards of a linguistic people stretching from the Southwest to the central Plains. Already the newcomers had earned the

emnity of the tribes settled in the region. *Apache* is probably derived from
the Zuni word *apachu,* meaning "enemy." The invaders were detested
wherever they went, and their fiercest foes were other Indians, including the
Pueblos, Pimas, Papagos, and Opatas, many of whom formed military units
in cooperation with the Spanish.[115]

In time, two broad groups of Athabascans were recognized: the Navajos
occupied the Colorado Plateau regions; the Apaches, a conglomeration of
clans often at feud with one another, eventually made their range coextensive with the lightning fire regime ranging from the Sierra Madre to the
Mogollon Rim. How long it took to move from the Great Plains of eastern
New Mexico to the southwestern mountains and Mogollon Rim is unknown.
Probably some groups were already in place when the Spanish arrived. But
whether the movement was caused by Spanish intrusion or only catalyzed
by it, within 300 years the Apaches took up an autonomous residence in the
region without connections to their Plains relatives.

The key was the introduction of Spanish livestock. The Apaches learned
about horses from Pueblo herders in the 1630s and 1640s. By the 1670s they
had adopted careers as predators on the herds of Spanish sheep, cattle,
horses, and mules. The southward movement of the Comanches, who
acquired horses around 1700, cut off the southwestern Apaches from their
linguistic kin on the Plains, from the buffalo, and from the Plains cultures
based on the horse. Instead, the southwestern Apaches preserved a kind of
grassland economy of hunting, gathering, and seminomadism and began to
stalk the domesticated herds introduced by the Spanish. They were as likely
to eat a horse as to ride it. More often than not, the Apache traveled on
foot, and his running endurance was legendary.

The Apaches, in short, never really left the grasslands or really abandoned a prehorse grassland economy. They simply left the Great Plains for
the high grasslands of the Sulphur Springs, San Pedro, and Santa Cruz
valleys and for the forest savannahs of the Chiricahuas, Mogollon Rim, and
White Mountains. Cut off on the one hand by the developing Plains cultures, they were attracted on the other by rich herds of domesticated livestock to the south and west. Spanish accounts report Apaches in the San
Pedro Valley in 1698, in the Tonto Basin in 1750, and in the White Mountains in 1808. From their mountain fastnesses they began intense predation
on the livestock introduced by the Spanish and maintained by missionized
tribes. So widespread were Apache raids that they virtually depopulated
portions of the American Southwest, Sonora, and Chihuahua, effectively
retarding Spanish settlement and keeping livestock herds in check. By the
time of the Mexican Revolution (1810–1821) the Apaches were in control
of the mountains and grasslands of the region. By the time of the Mexican
War, Mexican authority over the area in the face of the Apache menace
was negligible. Most military protection came from surrounding tribes, such

as the Opatas and Pimas, who detested the Athabascan intruders. The American occupation of the region would likewise depend on control over the Apaches, and that was not seriously possible until after the Civil War.

The Apaches worked to create and perpetuate a grassland environment. Their heavy predation on Spanish livestock prevented overgrazing, and they even practiced a cynical theory of conservation. By leaving enough stock behind in a raid to allow the herds to replenish, they could maintain a perpetual source of livestock. Not until the Apaches were finally subdued in the late 1870s did overgrazing become a serious problem in the southern grasslands. By contrast, the more northerly Athabascan branch, the Navajos, settled down as permanent herders. Charles Best estimated that the Navajos possessed 30,000 cattle, 500,000 sheep, and 10,000 horses, mules, and burros at the time of the Mexican War.[116] Emulating the herding economy brought by Spain, the Navajos adopted goats, too, with the result that their traditional lands came to resemble the devastated terrains of the Mediterranean. Shifting red sands and towering cliffs transformed the landscape that Father Escalante had declared in the 1770s to be the finest grazing land he had ever seen into a vision of windswept Mars.

The uses of broadcast fire by the Apaches were those typical of grassland tribes. They used smoke signals extensively, burned to cover trails made during a retreat, and broadcast burned as an inducement for rain. They baited traps with smoke to attract deer plagued by flies. W. A. Bell noted in 1870 that "the Apaches also have a very destructive habit amongst their long catalogue of vices of firing the forests of their enemies." S. J. Holsinger, describing the state of forest reserves in Arizona, explained in 1902 that "the most potent and powerful weapon in the hands of these aborigines was the firebrand. It was used alike to capture the deer, the elk and the antelope, and to vanquish the enemy. It cleared the mountain trail and destroyed the cover in which their quarry took refuge." The "high pine forests," he elaborated, "were their hunting grounds, and the vast areas of foothills and plateaus, covered with nut-bearing pines [pinyon,] their harvest fields. . . ."[117] For all these purposes fire was the ideal instrument, though, especially in the mountain fastnesses of the Apache—where goats and sheep could not prevent regeneration—the open forests nonetheless showed a mixture of age classes. It is surely no coincidence that after the tribe's centuries of migration the range of the Apache came to be identical to the range of that great lightning fire regime that nature had shaped over millennia.

It was in this same zone of the Southwest that Aldo Leopold began his career as a forester. Leopold turned his knowledge of the Southwest to a successful campaign for the creation of the first Forest Service wilderness area, the Gila, in 1924. On his experiences in the old range of the Apacheria, Leopold based one of his most celebrated essays, "Thinking Like a Mountain." The focus of the essay was the wolf, whose howl was hated and feared by man but not by mountains. Leopold described the "green fire" in the eyes of a dying she-wolf, the victim of his bullet. The meaning of that strange fire on the mountain was a lesson in wilderness ecology. With the extirpation of wolves, the deer population swelled; overbrowsing followed, leading to erosion, habitat degeneration, and starvation. Thus, Leopold argued, the mountain feared its deer, not its wolves.[118] In the southwestern history the virtual extinction of the Apache had an effect similar to the eradication of the wolf: domestic livestock reached unprecedented numbers and the grasslands deteriorated.

As early as the 1920s Leopold himself realized that overgrazing had profoundly altered the southwestern landscape: forest stocking was increasing; juniper and pinyon occupied sites that had formerly been grasslands; brush was encroaching on other grasslands; and marginal grasslands were reverting to deserts. Surely, he reasoned, "one is forced to the conclusion that there have been no widespread fires during the past 40 years."[119] The absence of fires was due to intensive grazing, but the solution was not simply to think in terms of the natural ecology of the mountain. What Leopold failed to realize in his essay was that the problem involved an interaction among peoples, not merely between people and wildlife. The reason for the sudden success of livestock after nearly 300 years was the elimination of the predatory Apache. It would be a cruel misreading of history to say that the relationship of the Apache to the Southwest was harmonious. Like the wolf, he was hated and feared by all who had contact with him. It was rather a case of one exotic, the Athabascans, controlling another exotic, Spanish livestock. Where Indians adopted domestic stock, overgrazing became endemic. Even in the case of burros and horses left wild, overgrazing and competition with native wildlife have become serious. The fire regime of the Southwest has been shaped by lightning and livestock, but the Apache was for centuries the intervening variable.

The United States acquired most of the Southwest from the Treaty of Guadalupe Hidalgo (1848), which concluded the Mexican War. The Gadsden Purchase (1853) added the lands of present-day Arizona south of the Gila River. Army engineers conducted some explorations through the region, and fur traders were active, but it was not until after the Civil War and the failure of Grant's peace policy that the military seriously addressed the Apache problem. By the early 1870s General George Crook had settled the western Apache menace in the Tonto Basin. Crook was then transferred to fight the Sioux, and the southern Apaches (Mimbrenos and Chiricahuas) continued spasmodic outbursts or raiding. Not until 1886, and after a short return by Crook, were the Apaches killed, exported, or settled on reservations. It was at this time that the great mining and livestock interests spread over what had been Apacheria. The Tombstone strike, for example, came in 1888.

The Apaches were eventually settled on three reservations: the Mescalero in New Mexico and the San Carlos and Fort Apache in the White Mountains of Arizona. The locations were center stage in the lightning and ponderosa pine zone of the Southwest, and their significance to the fire practices of the region continues. That significance has a symbolic as well as a practical dimension. It is possible to draw a loose analogy between control over the Apaches and control over the fire regime they occupied. "In those days," Leopold recalled, "we never heard of passing up a chance to kill a wolf."[120] As long as the Spanish and Mexicans had been in the Southwest, a very similar philosophy had guided their relationship to the Apaches. The Apaches were considered intractably hostile and unregenerative. For centuries the Southwest has witnessed a war of extermination, often waged through scalp bounties or the deportation of captives as slaves to Mexico or Cuba. The policy was not without its successes in terms of Apaches killed, but as a policy of frontier management it failed: despite substantial losses, the Apaches had virtually depopulated the region by the time of the Mexican War. The warfare was one of relentless, chronic, low-intensity raids and ambuscades, never of pitched battles.

The great insight by Crook into the control of renegades was to use Apaches as scouts, and at times the scouts on an expedition far outnumbered the regular troops. The idea of using Apaches on search and destroy missions against other Apaches was reportedly first suggested to Crook by Albert Banta with the argument that he had to "fight fire with fire."[121] And indeed, the frequent, low-intensity, lightning raids of the Apache were not unlike the lightning fires of the landscape they occupied. The control of wildfire in this environment could not be conducted through a war of attrition or extermination. As Crook learned, one had to use fire against fire, to use controlled burning against wildfire. The successful management of fire would come to resemble the management of the Apache.

When Arizona achieved statehood in 1912, most of its forests were already included within the national forests system. Indeed, about 75 percent of its total land was managed by federal agencies. The volatile competition between loggers and farmers so disastrous elsewhere never developed. Logging remained tied to local and regional markets, not national ones, and slash never reached the proportions found in many northern forests. Agricultural reclamation, moreover, meant irrigation, and, like the forests of Southern California, many of the early reserves in the Southwest, especially around the Tonto Basin, were established primarily for watershed protection. Most of the white communities along the Mogollon Rim were an extension of Mormon settlement, and they brought with them an irrigation culture. As they moved into the Salt River Valley in particular, the modern agriculturalists were able to reoccupy the fields and reactivate the canals once used by the Hohokam. The first major dam constructed by the

Bureau of Reclamation, Roosevelt Dam on the Salt River, required a secure watershed if it was to succeed. Citizens' organizations were formed, and in 1910 the officials of the Salt River Valley Water Supply Protective Association surveyed the Tonto Basin in the company of Gifford Pinchot.[122] Agricultural reclamation meant the irrigation of desert lands, not the clearing of forested lands.

The southern connection goes beyond analogy. Except for the Mormons working down from Utah, most of the new settlers came from the South. Many of the mine discoveries, such as that in Prescott, were made by southerners dislocated by the Civil War. The developing agriculture relied on adapting cotton cultivation to irrigation technology. The lumber industry along the Mogollon Rim began when a Louisiana logging company left the cutover southern yellow pine for its western cousin and even took the company town (McNary) with it. The livestock industry was bolstered by émigré Texans driving herds of cattle before them. The cattle migration had several causes: reformed land laws in Texas (1879, 1883), which helped to close the open range; a market created by military forts, Indian reservations, and mining camps; drought on the southern plains during the 1870s and 1880s; and, above all, the elimination of Apache predation. The consequences were widespread. In 1876, particularly in the northern areas populated by Navajos and Mormons, sheep outnumbered cattle 10 to 1, and to these flocks were added others spilling out from the grand California circuit around the Sierra Nevada, much as cattle had poured out of the southern Great Plains. Combined with drought in the early 1890s, the effects of intensive browsing became severe. To the south, cattle began decades of acute overgrazing, and even the Apaches took up herding. Settlement concentrated on the grasslands, using the mountain savannahs in the summer and extending, through the medium of the watersheds, into the irrigated deserts.[123]

The frontier fire practices of the territory were essentially those of herders, and they operated over vast public commons. Open range laws still continue. Not surprisingly, light burning came to be advocated for the pine belt of the Southwest. Without corporate timber ownership, however, the arguments never reached the intensity of those in California, and without a large population in the woods, burning did not continue surreptitiously as in the South. Nevertheless, in 1920, it became necessary for the Forest Service to detail the fallacies of light burning. It noted that "conditions of reproduction of yellow pine in the Southwest are generally so unfavorable" that nothing that might aggravate the problem should be permitted. Especially "under a system of periodic burning every two or three years," the Service concluded, "it is safe to predict that yellow pine will not restock at all."[124] Nonetheless, the remoteness of the national forests led to experiments in the 1920s with let burning, "take a chance" management, and "loose herding" rather than all-out suppression. The experiment was unsuccessful and contributed to the national drive for better initial attack in the backcountry.[125] Having little state land to manage, Arizona did not appoint a state forester until 1974, the last state to participate in the Clarke-McNary program. Meanwhile, through default and calculation both, the Apache reservations, like the piney woods of the South, effectively maintained a quasi frontier economy of hunting, herding, and fire.

Though its typical fire problem is lightning—chronic, low-intensity, unpreventable—the Southwest has had its big fire years. Large fires in 1956 supplemented those in Southern California and the South. Exceptionally heavy burns in 1970 and 1971 coincided with the influential complexes in Southern California and Washington. Big fires raged in June 1974. In 1977 every national forest in Arizona had a project (class I) fire, anticipating the even larger complexes in California and Alaska. To meet these challenges the Southwest has known an unusually wide array of suppression technologies. The region showed an early interest in aerial attack, experimenting with paracargo and autogiros in the 1920s. From 1947 onward a smokejumper contingent has been stationed at Silver City, New Mexico, for use primarily on the Gila National Forest. Helitack units and aerial tankers operate on nearly all national forests; the first use of the MAFFS system was in the Southwest in 1974. In a semiarid land often lacking surface streams, ground tankers are a prerequisite; model 60 and 70 tankers provide some structural as well as wildland protection. In the SWFFF program the Southwest contributed a model for the development of organized crews. In the private fire protection firm, Rural/Metro, which contracts for both wildland and urban fire services, it may have furnished an important precedent for future suppression along the boundary between city and forest. Even one of the great brainstorms in fire prevention, Little Smokey, came out of the Southwest.[126]

But success here, as in other regions, brought problems. Even before fire control became effective on the national forests, the explosion of livestock was working to change dramatically the vegetative cover of the region. As early as 1902 an examiner with the GLO observed that "when first invaded by white man the forests were open, devoid of undergrowth, and consisted in the main of mature trees, with practically no forest cover." It was "not an uncommon thing for the early setttlers to cut native hay in the pine forests and fill large government contracts at the different military posts."[127] But where intensive grazing occurred, thereby eliminating broadcast fire, woody vegetation sprang up or the grassland dissolved into desert. The Navajos achieved these effects without the help of American settlers. Elsewhere in Apacheria, the process did not begin until the 1880s. Drought decimated forest and flock alike during the 1890s, but the return of a wet cycle and the elimination of predation spread the effects by the time of statehood. The first consequence of the livestock explosion, particularly since so much of it consisted of sheep, had been to reduce woody reproductions. But as cattle replaced sheep and goats, as fire control became effective on the public domain, as a wet cycle returned in the early years of the century, the outcome was a dramatic expansion of woody vegetation and a concomitant decay of the grasslands.[128] Mesquite and brush encroached on the grasslands of the south. Desert grasslands gave way to woody succulents. Chaparral took over mountain slopes and much of the Tonto Basin. Pinyon and juniper reclaimed the grassy plateaus of the north. Ponderosa reproduction reached unprecedented densities. The cycle of reproduction was unpredictable, responding to climatic cycles that had little to do with grazing or fire. But the elimination of fire removed the natural mechanism for thinning the stands.

These changes did not go unnoticed. Leopold remarked on them in 1924. Holsinger went so far as to claim that few forests antedated the advent of white settlement. More accurately, Leiberg reported for the U.S. Geological Survey examination of the San Francisco Mountain Forest Reserve that, owing to "the numerous fires" that had swept over the region for centuries, the ponderosa stands, even where "untouched by the ax, do not carry an average crop of more than 40% of the timber they are capable of producing."[129] The restocking or reproduction by fire control was noted with satisfaction by early foresters. The Forest Service conducted experiments in broadcast fire along the Mogollon Rim between 1950 and 1958 but concluded, as with light burning, that the value of fire was marginal in comparison with intensive silvicultural practices. Not everyone was pleased with the outcome, however, and as the long-term effects of overgrazing and fire control became apparent, agitation mounted from many sides for some form of prescribed fire.[130]

The Arizona Watershed Program, an interagency group organized in 1956, took a keen interest in prescribed fire in the Tonto Basin. The basin had once been prized for its fabulous pasturage and protected for its value as watershed; the streams that drained it were essential for the reclamation by irrigation of the Salt River Valley. But the range had been lost to chaparral; runoff was decreasing; destructive fires were frequent. The objectives of the watershed program aimed at multiple use management, but it wanted "an intelligent application of fire" in addition to mechanical and chemical treatments. "Our first need in Arizona," its director informed the second Tall Timbers conference, "is to get men to develop experience and confidence in the art of applying fire for useful purposes." Yet the only source of expertise was that on the Fort Apache Reservation. Under prodding by the watershed program, the Southwest Interagency Fire Council (SWIFCO) was incorporated "to create an interagency committee to encourage an exchange of information and ultimately develop more men capable of using

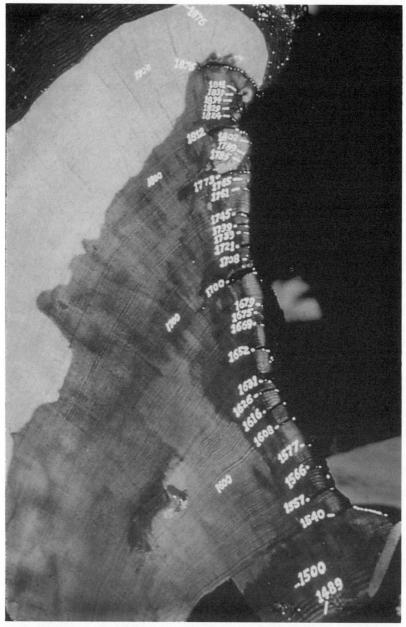

38. Cross section of fire scars from a ponderosa pine, Chimney Springs area, northern Arizona. The first scar, in 1540, corresponds with European entry into the region (the Coronado Expedition), but the scars cease with permanent American settlement in the late 19th century.

fire as a tool." Among its field excursions was a trip to Fort Apache to witness the fall burnings. The council noted approvingly that fire had been successfully applied in the ponderosa at Fort Apache with a reported reduction in the number of wildfires by 82 percent, of burned area by 44 percent, and in the average size of wildfires by 65 percent.[131] Researchers like Charles Cooper and Robert Humphrey were also making the the pilgrimage to Fort Apache to document the undesirable changes that had occurred since settlement and to argue for prescribed fire in pine forest and desert grassland. Broadcast fire was brought to the Hualapai Reservation to reclaim grassland from an invasion of pinyon and juniper, and as a result of a cooperative experimental program in the Tonto Basin, broadcast fire was applied for type conversion of chaparral to grass.[132]

The reemergence of the Apache in the fire history of the Southwest was the product of a number of historical peculiarities. Not until 1910 did the Indian Service (later renamed the Bureau of Indian Affairs) organize a Forestry Branch, which was largely dependent on cooperative agreements with the U.S. Forest Service; not until 1930 were the responsibilities for administering grazing activities transferred to the Forestry Branch; and not until the 1934 Indian Reorganization Act was the practice of conservation in forest and range management made mandatory.[133] During the terrible drought and fires of 1903 and 1904, the superintendent of the Fort Apache Reservation explained that "the practicable and effective remedy is to employ six or eight Indians to ride the mountain ranges at appointed places and be on the lookout for fires."[134] Instructions issued nationally in 1918 required the control of all fires, but costs prevented the BIA from extending this policy into the backcountry as ruthlessly as had the Forest Service. The Indian Service was the poor stepchild of the Forest Protection Board, strapped for funds and demoralized by confused legal status and uncertain directives. Many areas, including the famous Malay Gap on the San Carlos, escaped fire protection entirely. Even though the CCC constructed numerous improvements to assist in fire control, protection came haltingly, when it came at all. The reservation forests changed their character, but rarely with the intensity of change found in the surrounding national forests.

The reservation forests, moreover, are unique in that they are owned by the tribes but managed in trust by the BIA. The BIA charges the tribes for its forestry services and can be held legally liable for improper management. In this context, fire protection has had a legal accountability in the BIA not felt by other federal agencies. Fire suppression was often just too costly, whereas prescribed fire promised more economical protection through fuel reduction and encouragement of pasturage for livestock herds (which became the foundation of many tribal economies), and it continued fire practices existing prior to reservation life. Prescribed burning developed on reservation lands in the Lake States, the Seminole Reservation in Florida, the Colville and Warm Springs reservations on the east side of the Cascades, and, perhaps most spectacularly, on the Apache and Hualapai reservations of Arizona. The core of the program came from the western reservations, whose primary forest was of ponderosa pine. The fire regime there was nearly identical to the one that inflamed the old light-burning controversy, but the revived program was advocated by a forester with the BIA, Harold Weaver, and applied to special preserves offered by reservation forests.

Weaver experimented with controlled burning on the Colville Reservation in Washington. The program began in 1942 with the idea that broadcast fire could be used to reduce fire hazards and to serve such silvicultural objectives as thinning. Initial conclusions were published in 1943—the same year the Forest Service approved prescribed fire on its southern forests. But the BIA, perhaps worried that unorthodoxy might make it vulnerable to lawsuits, tacked a disclaimer on to the end of the article.[135] By 1948, however, Weaver was transferred to the Southwest as area forester with the understanding that he would initiate a program of broadcast burning. Especially at Fort Apache, fire hazards had developed locally as a result of protection, and pasturage, on which the Apache economy had come to depend, was degenerating. It was a big year for BIA fire management: in addition to official broadcast burning under Weaver's direction, destructive wildfires raged across several thousand acres on the Fort Apache Reservation, and the Mescalero Apaches organized the Red Hats, the beginning of the SWFFF program. Weaver left in 1951, and a year later the BIA approved controlled burning as a presuppression measure. Fort Apache was well on its way to becoming a showcase of prescribed broadcast fire for the Southwest.[136]

But there were qualifications, especially from foresters. Much of the burning was advocated by stockmen looking for better pasturage. Industrial forestry arrived late on the reservation—in the early 1950s—and then it came into competition with tourism and recreational uses. The foundation for Apache economies devolved largely from livestock; like the Navajos, they found it difficult to abandon their dependency on stock and so adopted a new relationship as herders rather than predators. Prescribed fire was welcomed by reservation herders, as by stockmen on the national forests, as a means to increase grass at the expense of woody vegetation. The motives were suspect to foresters. The close cropping of grass had induced a type conversion often more effective than that which followed from fire suppression organizations thrown together by the federal government. The loudest frontier critics of early forestry were herders, not farmers or loggers. Much of the erosion and arroyo cutting, the destruction of range, and the encroachment of brush had come from ruthless overgrazing; for this, the suppression of the Apaches rather than the suppression of fire was the larger cause. The identification of forest with watershed, moreover, was at the foundation of the national forest system. To suggest that woody vegetation was less effective than grass for watershed management was an anathema greater than the questioning of organized fire protection. Much of the experimental burning in the Tonto Basin was, on the California example, an exercise in the construction of conflagration control barriers.

Nevertheless, these hesitations were more a revival of old doubts than a creative modern protest. By the late 1960s prescribed burning had acquired a new scientific and cultural meaning. Fire exclusion seemed as serious a menace as overgrazing. Prescribed fire advocates from the Southwest were among the earliest to appear at the Tall Timbers conferences, and it was the fire program at Fort Apache that was seen as an exemplar for the management of ponderosa pine forests throughout the West by the Tall Timbers task force. In 1969 the Arizona Watershed Program sponsored a symposium on "Fire Ecology and the Control and Use of Fire in Wild Land Management." SWIFCO consolidated its experiences into publications on fire planning and a handbook of prescriptions for fire in ponderosa pine. Prescribed fire came to be widely distributed throughout the old range of Apacheria. In fact, so widespread did broadcast fire become in late autumn that smoke at least twice seriously threatened the program, once in October 1975 and again in November 1979. The Tonto Basin, Mogollon Rim, and White Mountains are watersheds for the Salt River. When high pressure cells stagnate into a regional inversion, smoke rather than water flows down the Salt to the metropolitan area around Phoenix, causing air pollution alerts. Smoke, it is felt, may be the great threat to broadcast fire on Fort Apache. The most productive uses of the land are recreational tourism, not forestry, and smoke is an ugly public relations message to sent to the source of that revenue.

The national policy reforms of 1970–1972 were prompted by the puzzle of wilderness fire, probably symbolized on the grand scale by interior Alaska. But among the earliest sites to adopt natural prescribed fire plans were those in the Southwest—the Gila Wilderness, Saguaro National Park, and, de facto, the Malay Gap region of the San Carlos Reservation. Yet by the end of the decade the guiding concern of fire management was with no particular problem fire but with the sheer pluralism of fire regimes, the range of possible responses, and the coordination of agencies having different land management objectives. No other region could compete with the historical geography of the Southwest. Its superimposition of land uses on an intense natural regime of lightning fire is without equivalent elsewhere. It contains the oldest and the newest—not only a geographic variety of fuel types and fire agencies but also a historical variety that makes neighbors of culturally different fire practices and fire regimes. The first experimental range in the United States was at Santa Rita (1903) on the desert grasslands of southern Arizona, and the first experimental forest was at Fort Valley (1908) on the ponderosa savannahs of northern Arizona. Appropriately, the national wildland fire training center is also located in the Southwest. The region that began the grand experiment, it seems, is ready to report on the outcome.

EPILOGUE:
THE FORBIDDEN FLAME

These four bodies are fire, air, water, earth.
—Aristotle, *Meteorologica*

Fire is the most tolerable third party.
—Henry David Thoreau, *Journal*

Fire is an event, not an element. It exists within a fire environment, without which it would perish. To modify that environment is to change the nature of fire. Equally, to change the nature of fire is to modify the fire environment. That environment is as much cultural as natural, and fires are only truly comparable when their physical, biological, and cultural environments are comparable. Ecologically speaking, the counterreclaimed landscape may or may not be new; culturally speaking, it is unprecedented. The fire regime fashioned by the industrial revolution is a novel one, and the assimilation of fire required new definitions and new prescriptions for fire's use and suppression.

Fire is plural, not singular. One can accurately speak about fire only in conjunction with something else—fire and flora, fire and fauna, fire and earth, fire and water. As an agency, fire can be as effective by its absence as by its application. For this reason one is inevitably led back to that most fundamental of all associations with fire, the relationship between fire and man. Mankind's use of fire has likewise been plural—fire and ax, fire and sword, fire and hoof. If man alone can create fire, he alone can extinguish it deliberately. He can alter a landscape as much by excluding fire as by introducing it.

O Fire, you are the countenance of all the Gods and of all learned men. Yours is the power to penetrate the innermost recesses of the human heart and discover the truth.—Hindu mantra before the Sacred Fire

Every man's work shall be made manifest: for the day shall declare it, because it shall be revealed by fire; and the fire shall try every man's work of what sort it is.—St. Paul, I Corinthians

Like matter and energy, fire and fuel are neither created nor destroyed, only transformed. The history of fire is the study of these transformations, but for millennia this history has often been indistinguishable from the history of mankind. Well before the appearance of *homo sapiens,* mankind's evolutionary ancestors made a pact with fire, establishing a symbiosis between fire and culture. It was an irrevocable commitment. The prominence of mankind is largely attributable to his ability to manipulate fire, but the prevalence of fire is in good measure dependent on the success of mankind. Man is the foremost source of fire, the primary vector for its distribution throughout the world, the greatest modifier of the fire environment, and the arbiter of those values and perceptions that select fire regimes. Fire cannot exist independently of its environment, and for most of the earth that environment is one arranged, consciously or accidentally, by man. Notwithstanding the longevity of lightning fire and the adaptations it has selected for, it is probable that most terrestrial environments are shaped by anthropogenic fire practices, not by natural fire. Equally, it was fire that presented man with his greatest agency for the management of the world around him and, until recent times, with the great instrument for the comprehension of that cosmos.

The Kachins of Burma say that in the beginning men had no fire; they ate their food raw and were cold and lean. . . . So they sent Kumthan Kunthoi Kamam to Wun Lawa Makam to borrow some of his fire. The messenger crossed the river on a raft and soon came to Wun Lawa Makam, and said, "Great Father, we are cold, we eat our food raw, and we are very lean. Give us then your fire." The spirit answered, "You men cannot possess the Fire-spirit; he would cause you too many misfortunes."
—James Frazer, *Myths of the Origin of Fire*

We only live, we only suspire
Consumed by either fire or fire.
—T. S. Eliot, *Four Quartets*

Fire is power. Whether or not fire can be considered as an essential human trait, it has surely been the foremost medium through which mankind has shaped its environment. But fire was not invented, it was discovered. Fire power was consequently not merely something that extended outward from human cultures but something that profoundly challenged that culture. Fire had to be assimilated, explained, regulated. Time and again fire origin myths depict man as human only after he has acquired fire. Yet, as often as not, fire was refused early man because its possession would prove too dangerous. In the Judeo-Christian myth of genesis, mankind acquires the knowledge of good and evil by tasting a forbidden fruit. But in

the myths of most cultures, that knowledge—and the power it brings—
comes by seizing the forbidden flame.

Behold, how great a matter a little fire kindleth!—St. James, III

His special firepower has created for man a special obligation, a unique
responsibility to manage fire and its effects. Our ancient pact with fire has
built fire into the technics of civilization, from the hearth to the dynamo,
from the campfire to the afterburners of interplanetary space probes. It is
an obligation that cannot be ignored or abdicated: to remove fire abruptly
may be as serious a cultural and ecological event as to introduce it suddenly.
When early man grasped fire, he irreversibly altered the course of his evolution. It is a relationship that is long from ended. Even after tens of millennia man is far from exerting the control over fire, or acquiring the understanding of fire, that he would like to have. By constantly changing fire's
environment, he has repeatedly changed the nature of the fire he wishes to
examine.

The problem of fire for early man was probably not unlike the discovery
of atomic energy by modern man. Both are natural phenomena of great
mystery and power. Both in their day underwrote the intellectual comprehension of the cosmos. The effort to define and prescribe uses and limits for
each is remarkably similar and promises to be similarly insoluble. But if
never ultimately resolved, our relationship to fire suggests that accommodations can be made, that it is possible to live with great natural processes.
If our ancestors, with little more than river cobbles and chipped flint, could
seize fire, surely modern mankind can cope with its own versions of the forbidden flame.

Then the seven angels that held the seven trumpets prepared to blow
them.

The first blew his trumpet; and there came hail and fire mingled with
blood, and this was hurled upon the earth. A third of the earth was burnt,
a third of the trees were burnt, all the green grass was burnt.
—John, Revelations

We live in a natural and social world sustained by fire. It is a world—as
so many legends, apochryphal religions, and astrophysical sciences
profess—that may well end in fire. The assimilation of fire has not always
been easy, and its use has often proved ambivalent. In his allegory of the
cave, Plato imagines humanity in the bonds of dark ignorance, guided by
the treacherous illumination of fire. "They see only their own shadows or
the shadows of one another, which the fire throws on the opposite wall of

the cave." Yet, however uncertain the ultimate effects, his pact with fire has guided man as perhaps nothing else could through the inevitable darkness that surrounds him. Whether or not the world will end in fire, for humanity it surely began with fire. For both fire and mankind the history of one will largely decide the history of the other.

Fire and People do in this agree,
They both good servants, both ill masters be.
—Fulke Greville, *Inquisition Upon Fame*

BIBLIOGRAPHIC ABBREVIATIONS

AF	*American Forests, American Forestry, American Forestry and Forest Life*
AFA	American Forestry Association
Clar, *California Government and Forestry*	C. Raymond Clar, *California Government and Forestry,* Vol. 1: *From Spanish Days until the Creation of the Department of Natural Resources in 1927,* Vol. 2: *During the Young and Rolph Administrations* (Sacramento: California Division of Forestry, 1959–1969)
FCN, FM, FMN	*Fire Control Notes, Fire Management, Fire Management Notes*
GPO	Government Printing Office
JF	*Journal of Forestry*
NA	National Archives, Washington, D.C.
RFS	Records of the Forest Service
RG	Record Group
SAF	Society of American Foresters
TTFECP	*Tall Timbers Fire Ecology Conference Proceedings,* no. 1–15 (Tallahassee: TTRS, 1962–1976)
TTRS	Tall Timbers Research Station
WNRC	Washington National Records Center, Suitland, Maryland

NOTES

PROLOGUE: THE SMOKE OF TIME

1. Archibald MacLeish, *The Collected Poems of Archibald MacLeish* (Boston: Houghton Mifflin, 1963), p. 171.

1: NATURE'S FIRE

1. Robert Burns, "First Epistle to J. Lapraik," st. 13, in John Bartlett, *Bartlett's Familiar Quotations,* 14th ed. (Boston: Little, Brown, 1968), p. 493a.
2. William Wood, *New England Prospects* (1634), quoted in Daniel C. Thompson and Ralph H. Smith, "The Forest Primeval in the Northeast—A Great Myth?" in *TTFECP,* no. 10, p. 259.
3. Lucretius, *On the Nature of the Universe (De Rerum Natura),* trans. R. E. Latham (London: Penguin, 1977), p. 204.
4. Henry Wadsworth Longfellow, *Evangeline* (Boston: William D. Ticknor and Co., 1847).
5. *The Travels of William Bartram* (1791), ed. Mark Van Doren (New York: Dover, 1955), p. 59.
6. J. Bronowski and Bruce Mazlish, *The Western Intellectual Tradition* (New York: Harper and Brothers, 1960), p. 367. For a good summary of scientific ideas about electricity at the time, see Bernard Cohen, *Franklin and Newton* (Philadelphia: American Philosophical Society, 1956).
7. E. V. Komarek, "Lightning and Lightning Fires as Ecological Forces," in *TTFECP,* no. 8, p. 175.
8. Peter Viemeister, *The Lightning Book* (Cambridge, Mass: MIT Press, 1972), p. 123.
9. See Viemeister, *Lightning Book,* pp. 122–123, 177–187; Alan R. Taylor, "Lightning Effects on the Forest Complex," in *TTFECP,* no. 9, pp. 127–150, and Taylor, "Ecological Aspects of Lightning in Forests," *TTFECP,* no. 13, pp. 455–482; Martin A. Uman, "The Physical Parameters of Lightning and the Techniques by Which They Are Measured," in *TTFECP,* no. 13, pp. 429–454; Donald Fuquay et al., "Lightning Discharges That Caused Forest Fires," *Journal of Geophysical Research,* 77 (1972) 2156–2158, and Fuquay, "Predicting Ignition of Forest Fuels by Lightning," in *Fifth Joint Conference on Fire and Forest Meteorology* (Boston: American Meterorological Society, 1978), pp. 32–37. See also various reports for Project Skyfire, U.S. Forest Service, listed in the bibliographies in note 19, below.
10. See Taylor, "Ecological Aspects of Lightning in Forests," pp. 455–462; E. P. DuCharme, "Lightning—A Predator of Citrus Trees in Florida," *TTFECP,* no.

13, pp. 483–496. The best overviews of lightning and lightning fire as ecological events are those in articles by E. V. Komarek: "Introduction of Lightning Ecology," in *TTFECP*, no. 13, pp. 421–427; "The Natural History of Lightning," in *TTFECP*, no. 3, pp. 139–183; "Lightning and Lightning Fires as Ecological Forces," in *TTFECP*, no. 8, pp. 168–198; "The Meterological Basis of Fire Ecology," in *TTFECP*, no. 5, pp. 85–125; "The Nature of Lightning Fires," in *TTFECP*, no. 7, pp. 5–42; "Ancient Fires," in *TTFECP*, no. 12, pp. 219–240; and "Lightning and Fire Ecology in Africa," in *TTFECP*, no. 11, pp. 475–478.

11. See Komarek, "The Natural History of Lightning," pp. 139–184, "Ancient Fires," pp. 219–240, and "Lightning and Fire Ecology in Africa," pp. 475–478; Viemeister, *Lightning Book,* pp. 138–141, pl. XLII–XLIV.

12. Komarek, "Meteorological Basis of Fire Ecology," pp. 96–98.

13. Jack S. Barrows, "Lightning Fires in Southwestern Forests" (U.S. Forest Service, 1978); Jack S. Barrows, David V. Sandberg, and Joel D. Hart, "Lightning Fires in Northern Rocky Mountain Forests" (U.S. Forest Service, 1977); Earl Kauffman, "Lightning Batters Western Forests," *AF,* 46 (April 1940), 391–392. For explanations, see Komarek, "Meteorological Basis of Fire Ecology"; and Charles Buck and Mark J. Schroeder, *Fire Weather: A Guide for Application of Meteorological Information to Forest Fire Control,* U.S. Department of Agriculture, Handbook 360 (Washington: GPO, 1970).

14. Komarek, "Natural History of Lightning," pp. 157–158.

15. Viemeister, *Lightning Book,* pp. 21–30, summarizes much folklore about lightning, and my references derive for the most part from him. See also James Frazer, *The Golden Bough: A Study in Magic and Religion,* abr. ed. (New York: Macmillan, 1922), pp. 708–710.

16. Lucretius, *Nature of the Universe,* p. 222.

17. Quoted in Fred G. Plummer, *Lightning in Relation to Forest Fires,* U.S. Forest Service, Bulletin 111 (Washington: GPO, 1912), p. 12.

18. Viemeister, *Lightning Book,* p. 29.

19. For Skyfire research, see the synopsis by J. S. Barrows, "Weather Modification and the Prevention of Lightning-Caused Forest Fires," in W. R. Derrick Sewell, ed., *Human Dimensions of Weather Modification* (Chicago: University of Chicago Department of Geography, 1966), pp. 169–182. In addition to the works listed in Sewell's fairly complete bibliography of Skyfire reports, see Donald Fuquay and Robert G. Baughman, "Porject Skyfire Lightning Research: Final Report" (U.S. Forest Service, 1969). Syrus is quoted in Viemeister, *Lightning Book,* p. 17.

20. Books on Americans and nature are legion. Among the most useful are Roderick Nash, *Wilderness and the American Mind* (New Haven: Yale University Press, 1967); Hans Huth, *Nature and the Americans* (Berkeley: University of California Press, 1957); Roderick Nash, ed., *The American Environment: Readings in the History of Conservation* (Menlo Park, Calif.: Addison-Wesley, 1968). For literary interpretations, see Henry Nash Smith, *Virgin Land: The American West as Symbol and Myth* (New York: Vintage, 1950), and Leo Marx, *The Machine in the Garden: Technology and the Pastoral Idea in America* (New York: Oxford University Press, 1964).

21. For a survey of contemporary concepts about wilderness, see John C. Hendee

et al., *Wilderness Management* U.S. Department of Agriculture, Forest Service Miscellaneous Publication no. 1365 (Washington: GPO, 1978). The book also includes a synopsis of the scientific rationale for wilderness fire management by Miron Heinselman, "Fire in Wilderness Ecosystems," pp. 249–278.

22. A good summary of wilderness legislation is contained in Roderick Nash, "Historical Roots of Wilderness Management," in Hendee et al., *Wilderness Management*, pp. 27–40; and Nash, "Wilderness Acts: Legal Mandate for Wilderness Classification and Management," in ibid., pp. 61–91.

23. Aldo Leopold, *A Sand County Almanac* (1949; rpt. New York: Ballantine, 1970), p. 265. For an overview of Leopold's philosophy, see Nash, *Wilderness and the American Mind*, and Susan L. Flader, *Thinking Like a Mountain: Aldo Leopold and the Evolution of an Ecological Attitude Toward Deer, Wolves, and Forests* (Columbia; University of Missouri Press, 1974). For the larger cultural significance of wilderness sites, see again Nash, *Wilderness and the American Mind;* for the national parks in particular, see Alfred Runte, *National Parks: The American Experience* (Lincoln: University of Nebraska Press, 1979).

24. General descriptions of fire behavior can be found in A. A. Brown and Kenneth Davis, *Forest Fire: Control and Use,* 2d ed. (New York: McGraw-Hill, 1973), pp. 155–213; R. H. Luke and A. G. McArthur, *Bushfires in Australia* (Canberra: Australian Government Printing Service, 1977), pp. 23–118; George M. Byram, "Combustion of Forest Fuels" and "Forest Fire Behavior," in Kenneth Davis, *Forest Fire: Control and Use* (New York: McGraw-Hill, 1959), pp. 61–123; Buck and Schroeder, *Fire Weather*; and assorted fire training manuals, especially "Intermediate Fire Behavior" (U.S. Forest Service, TT–80–5100).

25. Robert Bell, "Forest Fires in Northern Canada," in *Report of the American Forestry Congress* (Washington, D.C., 1889), p. 53.

26. See the following case studies: Craig C. Chandler, "Fire Behavior of the Basin Fire, Sierra National Forest, July 13–22, 1961" (U.S. Forest Service, 1961); E. V. Komarek, "Meteorological Basis of Fire Ecology," pp. 97–99; Hal E. Anderson, "Sundance Fire: An Analysis of Fire Phenomena," Research Paper INT–56 (U.S. Forest Service, 1968); Rodney W. Sando and Donald A. Haines, "Fire Weather and Behavior of the Little Sioux Fire," Research Paper NC–76 (U.S. Forest Service, 1972); Dale D. Wade and Darold E. Ward, "An Analysis of the Air Force Bomb Range Fire," Research Paper SE–105 (U.S. Forest Service, 1973).

27. The Miramichi and Maine fires of 1825, the North Carolina fire of 1898, the 1910 fires in Idaho and Montana, and the Alaska fires of 1957 exceeded 3 million acres. The Wisconsin fires of 1894, the Michigan fires of 1871, and the Washington and Oregon fires of 1910 exceeded 2 million acres. The largest grass fire on record is the 1906 burn in eastern New Mexico. A partial list of large fires can be found in Brown and Davis, *Forest Fire,* pp. 24–25; National Fire Coordination Study, "Large Fires in the World Since 1825" (U.S. Forest Service, 1965); Fred G. Plummer, *Forest Fires: Their Causes, Extent and Effects, with a Summary of Recorded Destruction and Loss,* U.S. Forest Service, Bulletin 117 (Washington: GPO, 1912); and John D. Guthrie, "Great Forest Fires of America" (U.S. Forest Service, 1936).

28. Barrows, "Lightning Fires in Southwestern Forests," p. 110.

29. Glossaries followed the Mather Field Conference (1921), which had as one of its objectives the standardization of technique and language; the reforms of 1936–1938, which followed the adoption of the 10 A.M. Policy; the complex of events around 1956, which culminated in the reform of line operations and fire research; and the reformation that has occurred in the 1970s and is still to some extent unresolved.

30. James Frazer, *Myths of the Origin of Fire* (1930; rpt. New York: Hacker Art Books, 1974), pp. 605–106.

31. See, for example, the simple descriptions in Marie Morisawa, *Streams: Their Dynamics and Morphology* (New York: McGraw-Hill, 1968).

32. C. B. Belt, "1973 Flood and Man's Constriction of the Mississippi River," *Science,* 189 (August 29, 1975), pp. 681–684.

33. Joel 2:30–31. Joel came from a rural background, and his fire references are nearly always to rural and wildland fires. Many of the other prophets flourished in urban settings, and their appeal to fire imagery, though frequent, is usually to the fires of furnace and crucible.

34. E. V. Komarek, "Fire Ecology," in *TTFECP,* no. 1, p. 97.

35. Negro spiritual, in *Bartlett's Familiar Quotations,* p. 1101a.

36. Stephen H. Spurr and Burton V. Barnes, *Forest Ecology,* 2d ed. (New York: Ronald Press, 1973), p. 347.

37. E. V. Komarek, "Fire Ecology—Grasslands and Man," in *TTFECP,* no. 4, p. 204.

38. The literature on fire ecology is vast and still multiplying. Useful studies include Spurr and Barnes, *Forest Ecology;* U.S. Forest Service, "National Fire Effects Workshop," General Technical Reports WO–6, 7, 9, 10, 13 (U.S. Forest Service, 1978–1980); T. T. Kozlowski and C. E. Ahlgren, eds., *Fire and Ecosystems* (New York: Academic Press, 1974); Heinselman, "Fire in Wilderness Ecosystems," pp. 249–278; and innumerable entries in FIREBASE and the multitudinous conferences and symposia held throughout the 1970s. The best single source remains the *Tall Timbers Fire Ecology Conference Proceedings* (no. 1–15) and assorted works in them by E. V. Komarek, including his "Fire Ecology" (no. 1, pp. 95–108) and "Fire Ecology—Grasslands and Man" (no. 4, pp. 169–220).

39. Spurr and Barnes, *Forest Ecology,* p. 347.

40. Komarek, "Ancient Fires," pp. 219–240.

41. Spurr and Barnes, *Forest Ecology,* p. 348.

42. Paul D. Brohn, "Mine Fires Still Burn," *FCN,* 33 (Winter 1971–1972), 10, 14.

43. Thurston Clarke, *The Last Caravan* (New York: G. P. Putnam's Sons, 1977), p. 203.

44. Robert B. Batchelder, "Spatial and Temporal Patterns of Fire in the Tropical World," in *TTFECP,* no. 6, p. 174, and Komarek, "Fire Ecology—Grasslands and Man," pp. 169–220.

45. E. V. Komarek, "Principles of Fire Ecology and Fire Management in Relation to the Alaskan Environment," in C. W. Slaughter et al., eds., "Fire in the Northern Environment: A Symposium" (U.S. Forest Service, 1971), p. 15.

46. Timothy Dwight, *Travels in New England and New York,* ed. Barbara Miller

Solomon, 4 vols. (Cambridge, Mass.: Harvard University Press, 1969), vol. 1, Letter 7, p. 72.

47. Quoted in Whitney R. Cross, *The Burned-Over District: The Social and Intellectual History of Enthusiastic Religion in Western New York, 1800–1850* (Ithaca: Cornell University Press, 1950), p. 210.

48. For a general survey of the fire environment in the Northeast, see Silas Little, "Effects of Fire on Temperate Forests: Northeastern United States," in Kozlowski and Ahlgren, eds., *Fire and Ecosystems*, pp. 225–250. For a synopsis of forest history in the region, see Spurr and Barnes, *Forest Ecology*, pp. 475–493; for a summary of wildfire records, see Donald Haines et al., "Wildlife Atlas of the Northeastern and North Central States," General Technical Report NC–16 (U.S. Forest Service, 1975).

49. For a description of the hurricane itself, see Bernard L. Gordon, ed., *Hurricane in Southern New England* (Watch Hill, R.I.: Book and Tackle Shop, 1976), though its concern is primarily with shorelines, not forests. For the cleanup, see Samuel T. Dana, *Forest and Range Policy: Its Development in the United States* (New York: McGraw-Hill, 1956), pp. 251–252; Ralph Widner, ed., *Forests and Forestry in the American States: A Reference Anthology* (Missoula, Mont.: National Association of State Foresters, 1968), pp. 206–231 passim; and E. G. Amos, "Final Report on the New England Forest Emergency Fire Hazard Reduction Project," *JF*, 39 (1941), 749–752.

50. Quoted by Gordon M. Day, "The Indian as an Ecological Factor in the Northeastern Forest," *Ecology*, 34, no. 2 (1953), 329. Day summarizes a host of similar romantic effusions.

51. The best summary of a contrasting view is Day, "The Indian as an Ecological Factor," pp. 329–346. See also Thompson and Smith, "The Forest Primeval in the Northeast—A Great Myth?" pp. 255–265; and Spurr and Barnes, *Forest Ecology*, pp. 475–477. A useful edition of early exploration accounts is H. S. Burrage, ed., *Early English and French Voyages (Chiefly from Hakluyt), 1534–1608* (New York: Barnes and Noble, 1934).

52. Day, "The Indian as an Ecological Factor," p. 330. As have all researchers on this topic, I have followed Day's analysis closely.

53. Spurr and Barnes, *Forest Ecology*, p. 487.

54. Day, "The Indian as an Ecological Factor," pp. 335, 337.

55. Thomas Morton, *New English Canaan: or, New Canaan* (1637; rpt. New York: Arno Press, 1972), pp. 52–54. For the extensiveness of Indian firing for agriculture, see Carl O. Sauer, *Seventeenth Century North America* (Berkeley: Turtle Island, 1980), pp. 228–229.

56. Peter Kalm, *Travels into North America . . .* , trans. John Reinhold Forster (1770–1771; rpt. Barre, Mass.: Imprint Society, 1972), pp. 210, 361. Kalm offers other observations, equally mixed, on the extensiveness of Indian alteration of the landscape, but as an emissary of Linnaeus and a forerunner of the romantics, Kalm's evaluations should be read with care.

57. Dwight, *Travels*, vol. 4, Letter 3, pp. 37–38.

58. Ibid., p. 39.

59. Ibid., p. 40.

60. Ibid.

61. Thaddeus Harris, *The Journal of a Tour into the Territory Northwest of the Alleghany Mountains; Made in the Spring of the Year 1803* (1805), reprinted in vol. 3 of Reuben Gold Thwaites, ed., *Early Western Travels, 1748–1846* (Cleveland: A. H. Clark, 1904), p. 327.

62. Dwight, *Travels,* vol. 2, Letter 13, pp. 321–322.

63. Ibid., Letter 14, pp. 325–326.

64. Quoted in Clarence J. Glacken, *Traces on the Rhodian Shore* (Berkeley: University of California Press, 1967), p. 697. See, moreover, Glacken's summary of John Lorain's *Nature and Reason Harmonized in the Practice of Husbandry* (1825) for further descriptions of swidden farming in the American reclamation, especially for the contrast between Yankee and Pennsylvanian practices.

65. Dwight, *Travels,* vol. 2, Letter 14, p. 327, and Letter 13, p. 322.

66. Quoted in Spurr and Barnes, *Forest Ecology,* p. 478.

67. Dwight, *Travels,* vol. 2, Letter 2, p. 206.

68. Ibid., vol. 1, Letter 8, pp. 74–75.

69. See Widner, ed., *Forests and Forestry,* pp. 24–25.

70. See Philip T. Coolidge, *History of the Maine Woods* (Bangor: Furbush-Roberts Printing, 1963), pp. 127–135, for a summary of historic Maine fires and their consequences; and Charles B. Fobes, "Historic Forest Fires in Maine," *Economic Geography,* 24 (October 1948), 269–273.

71. See Widner, ed., *Forests and Forestry,* and Henry Clepper, *Professional Forestry in the United States* (Baltimore: Johns Hopkins University Press, 1971). For some statistical data on production by regions, see Eliot Zimmerman, "A Historical Summary of State and Private Forestry in the U.S. Forest Service" (U.S. Forest Service, 1976); and Bureau of the Census, *Historical Statistics of the United States,* 2 vols, House Document no. 93–78, 93d Congress, 1st sess. (Washington: GPO, 1975), pp. 526–564.

72. Dwight, *Travels,* vol. 3, Letter 7, pp. 350–351. For a list of Dark Days, see Plummer, *Forest Fires,* p. 18.

73. Stewart Holbrook, *Burning an Empire: The Story of American Forest Fires* (New York: Macmillan, 1943), p. 57.

74. For summary accounts of the Miramichi fires, see ibid., pp. 54–60, and Franklin B. Hough, *Report on Forestry,* vol. 3 (Washington: GPO, 1882), pp. 228–231.

75. See Hough, *Report on Forestry,* pp. 155–165. For comparison, see William A. Main and Donald A. Haines, "The Causes of Fires on Northeastern National Forests," Research Paper NC–102 (U.S. Forest Service, 1974).

76. Dana, *Forest and Range Policy,* p. 6; Hough, *Report on Forestry,* pp. 135–165 passim.

77. James Fenimore Cooper, *The Pioneers* (New York: Holt, Rinehart, and Winston, 1967), p. 431; Hough, *Report on Forestry,* pp. 156, 158–161.

78. Archibald MacLeish, "Where the Hayfields Were," in *The Collected Poems of Archibald MacLeish* (Boston: Houghton Mifflin, 1963), p. 163.

79. Widner, ed., *Forests and Forestry,* pp. 227–228; Clepper, *Professional Forestry,* pp. 216–218. Clepper shrewdly notes the alliance of tree planting with fire control: "The point to be made is that American forestry, which began in the eastern states, stemmed from public interest in two aspects of forest conservation: protection from fire and tree planting."

80. See Dana, *Forest and Range Policy*, pp. 92–93; Widner, ed., *Forests and Forestry*, pp. 8–12, 92–99. For general background, see Frank Graham, *The Adirondack Park: A Political History* (New York: Alfred A. Knopf, 1978); and the classic Alfred Donaldson, *A History of the Adirondacks*, 2 vols. (New York: Century, 1921).

81. There are several accounts of the fires. In addition to the general histories of the region already cited, see especially H. M. Suter, *Forest Fires in the Adirondacks in 1903*, U.S. Bureau of Forestry, Circular no. 26 (Washington: GPO, 1904); and A. Knechtel, "Forest Fires in the Adirondacks," *Forestry Quarterly*, 2 (November 1903), 2–13.

82. For a good account of the regional response to the 1908 fires, see Austin H. Wilkins, *Ten Million Acres of Timber: The Remarkable Story of Forest Protection in the Maine Forestry District (1909–1972)* (Woolwich, Me.: TBW Books, 1978); Widner, ed., *Forests and Forestry*, pp. 92–100, 120–123; A. S. Hopkins, "Forest Fire Hazard in the Adirondack and Catskill Regions," *JF*, 20 (1922), 629–632; and William G. Howard, *Forest Fires*, State of New York, Conservation Commission, Bulletin 10 (Albany: J. B. Lyon, 1914).

83. There are many accounts of the 1947 fires, published and unpublished. See Austin H. Wilkins, "The Story of the Maine Forest Fire Disaster," *JF*, 46 (1948), 568–573; H. H. Chapman, "Local Autonomy Versus Forest Fire Damage in New England," *JF*, 47 (1949), 101–105; Daniel B. Tierney, "New England's Indian Summer Disaster" (National Agricultural Library, n.d.).

84. See in particular Chapman, "Local Autonomy." For a description of reforms, see A. D. Nutting and Austin H. Wilkins, "Maine's New Forest Fire Protection Program, *FCN*, 14 (January 1951), 14–19.

85. See Dana, *Forest and Range Policy*, pp. 309–310; W. J. Stahl and J. N. Diehl, "Interstate Forest Fire Protection Compacts," *Quarterly of the National Fire Protection Association*, 55 (October 1961), 152–160; Fire Working Team, Society of American Foresters, "Interstate Wildfire Organization Directory," includes the text of the compact; R. M. Evans, "Northeastern Interstate Forest Fire Protection Compact," *FCN*, 11 (October 1950), 44–46; "25th Anniversary of Holocausts Remembered," *FCN*, 33 (Fall 1972), 15; A. E. Eckes, "The Quebec Joinder," *FCN*, 31 (Summer 1970), 9–10; Arthur S. Hopkins, "The Fire Control Training Program of the Northeastern Forest Fire Protection Commission," *FCN*, 20 (April 1959), 26–28.

86. William G. Herbolsheimer, "Roscommon Equipment Center: A 20-State Approach to ED&T," *FM*, 36 (Fall 1975), 6–7.

87. For an overview of fire in southern New Jersey, see Little, "Effects of Fire on Temperate Forests;" Silas Little, "Fire Ecology and Forest Management in the New Jersey Pine Region," in *TTFECP*, no. 3, pp. 35–60; James A. Cumming, "Prescribed Burning on Recreation Areas in New Jersey: History, Objectives, Influence, and Techniques," in *TTFECP*, no. 9, pp. 251–269; Hough, *Report on Forestry*, pp. 158–161; Widner, ed., *Forests and Forestry*, pp. 140–144, 520–522; and Gifford Pinchot, "Study of Forest Fires and Wood Protection in Southern New Jersey," in *Annual Report of Geological Survey of New Jersey* (Trenton, 1898), appendix.

88. John McPhee, *The Pine Barrens* (New York: Farrar, Straus and Giroux, 1967), p. 111.

89. Wayne G. Banks and Silas Little, "The Forest Fires of April 1963 in New Jersey Point the Way to Better Protection and Management," *FCN,* 25 (July 1964), 3–6.
90. Banks and Little, "Forest Fires of April 1963," pp. 4–6; Little, "Fire Ecology and Forest Management"; Cumming, "Prescribed Burning on Recreation Areas."
91. The controversy is summarized in Aime Gauvin, "As Baxter Park Burns, So Burns Maine," *Audubon,* 79 (January 1977), 146–153.

2. THE FIRE FROM ASIA

1. Walter Hough, *Fire as an Agent in Human Culture,* United States National Museum, Bulletin 139 (Washington: GPO, 1926), p. xii.
2. Loren Eiseley, "Man the Fire-Maker," *Scientific American,* 191 (1954), 57.
3. E. V. Komarek, "Fire Ecology—Grasslands and Man," in *TTFECP,* no. 4, pp. 169–220, and "Fire—and the Ecology of Man," in *TTFECP,* no. 6, pp. 143–170. See also Hough, *Fire as an Agent.* Except for Hough, arid especially Komarek, there are few treatments of early man and fire apart from mentions of charcoal at cooking sites and the like. See, however, Carl O. Sauer, "The Agency of Man on the Earth," in William L. Thomas, ed., *Man's Role in Changing the Face of the Earth* (Chicago: University of Chicago Press, 1956), 1:54–56.
4. Komarek, "Fire Ecology—Grasslands and Man," p. 209.
5. Sauer, "Agency of Man," p. 55.
6. Quoted in Louis Barrett, "A Record of Forest and Field Fires in California From the Days of the Early Explorers to the Creation of the Forest Reserves" (U.S. Forest Service, 1935), p. 51.
7. Hiram Chittenden and Alfred Richardson, eds., *Life, Letters, and Travels of Father DeSmet Among the North American Indians* (New York: F. P. Harper, 1905), p. 1047.
8. "Fire," in Maria Leach, ed., *Funk and Wagnall's Standard Dictionary of Folklore, Mythology, and Legend* (New York: Funk and Wagnall's, 1972).
9. No one has yet compiled a summary of Indian uses of fire, though some regional studies have been made. My references for the most part derive from these works and from the accounts of early explorers and ethnographers and will be documented in detail for the fire histories of the various regions. Among general studies, see Omer C. Stewart, "Fire as the First Great Force Employed by Man," in Thomas, ed., *Man's Role,* pp. 115–133, Stewart, "Burning and Natural Vegetation in the United States," *Geographical Reveiw,* 41 (April 1951), 317–320, and Stewart, "Barriers to Understanding the Influence of Use of Fire by Aborigines on Vegetation," in *TTFECP,* no. 2, pp. 117–126; Hough, *Fire as an Agent;* Komarek, "Fire—And the Ecology of Man," pp. 143–170, and "Fire Ecology—Grasslands and Man," pp. 169–220; Henry T. Lewis, *Patterns of Indian Burning in California: Ecology and Ethnohistory,* Ballena Press Anthropological Papers, no. 1 (Ramona, Calif.: Ballena Press, 1974); Harold J. Lutz, *Aboriginal Man and White Man as Historical Causes of Fires in the Boreal Forest, with Particular Reference to Alaska,* Yale School of Forestry, Bulletin no.

65 (New Haven: Yale University, 1959); Gordon Day, "The Indian as an Ecological Factor in the Northeastern Forest," *Ecology*, 34, no. 2 (1953), 329–346. For related overviews on fire practices of aboriginal peoples, see H. H. Bartlett, "Fire, Primitive Agriculture, and Grazing in the Tropics," in Thomas, ed., *Man's Role*, pp. 692–720; Homer Aschmann, "Aboriginal Use of Fire," in "Proceedings of the Symposium on the Environmental Consequences of Fire and Fuel Management in Mediterranean Ecosystems," General Technical Report WO–3 (U.S. Forest Service, 1977), pp. 132–141; Robert B. Batchelder and Howard F. Hirt, "Fire in Tropical Forests and Grasslands," U.S. Army Natick Laboratories, Technical Report 67–41–ES (Natick, Mass.: U.S. Army Materiel Command, 1966); and Sylvia J. Hallam, *Fire and Hearth: A Study of Aboriginal Usage and European Usurpation in South-western Australia* (Canberra: Australian Institute of Aboriginal Studies, 1975). See also the *TTFECP* volumes for many, many other references to the use of fire by primitive peoples.

10. *The Journals of Lewis and Clark*, ed. Bernard DeVoto (Boston: Houghton Mifflin, 1953), p. 419.

11. Hough, *Fire as an Agent*, pp. 152–153; Josiah Gregg, *Commerce of the Prairies* (1844), ed. Max L. Moorhead (Norman: University of Oklahoma Press, 1954), p. 240.

12. Another curious fire device used to collect insects is given by Hough, *Fire as an Agent*, p. 66.

13. Lewis, *Patterns of Indian Burning*, p. viii.

14. See Gregory Thomas, "Fire and the Fur Trade," *The Beaver*, Autumn 1977, pp. 34–35.

15. Edmund Morris, *The Rise of Theodore Roosevelt* (New York: Ballantine, 1979), p. 310.

16. Edwin James, *Account of an Expedition From Pittsburgh to the Rocky Mountains ...* (1822; rpt. Ann Arbor: University Microfilm, 1966), pp. 254–255; Lutz, *Aboriginal Man and White Man*, p. 17.

17. Quoted in Herbert E. Bolton, *Pageant in the Wilderness; The Story of the Escalante Expedition to the Interior Basin, 1776* (Salt Lake City: Utah State Historical Society, 1972), p. 179.

18. David A. Clary, "'I Am Already Quite a Texan': Albert J. Myer's Letters from Texas, 1854–1856," *Southwestern Historical Quarterly*, 82 (July 1978), 41, 46–47.

19. Col. Richard Dodge, *The Plains of the Great West and Their Inhabitants* (1877; rpt. New York: G. P. Putnam's Sons, 1959), p. 79.

20. Stewart, "Fire as the First Great Agent," p. 119; Barrett, "Record of Forest and Field Fires," passim.

21. H. J. Spinden, quoted in Lewis, *Patterns of Indian Burning*, p. viii.

22. Hough, *Fire as an Agent*, pp. 152–153.

23. John Lankford, ed., *Captain John Smith's America* (New York: Harper and Row, 1967), p. 22.

24. John Lawson, *A New Voyage to Carolina ...* (1709), ed. Hugh T. Lefler (Chapel Hill: University of North Carolina Press, 1967), p. 215.

25. Adolf Bandelier, ed., *The Journey of Alvar Nuñez Cabeza de Vaca ... 1528–1536*, trans. Fanny Bandelier (1905; rpt. New York: AMS Press, 1973), pp. 92–93.

26. Carl O. Sauer, *Seventeenth Century North America* (Berkeley: Turtle Island, 1980), pp. 211, 237, passim. References to fire are so abundant that it is difficult not to find them, and a complete listing is hopeless. See, for example, Reuben Thwaites, ed., *The Jesuit Relations and Allied Documents . . . ,* vol. 69 (1896–1901; rpt. New York: Pagant, 1959), "Letter by Vivier," p. 209.

27. Quoted in Hough, *Fire as an Agent,* pp. 62–63.

28. Jefferson to Adams, May 27, 1813, quoted in "Thomas Jefferson on Forest Fires," *FCN,* 13 (April 1952), 31.

29. See Douglas Branch, *The Hunting of the Buffalo* (Lincoln: University of Nebraska Press, 1962), esp. pp. 52–63; Erhard Rostlund, "The Myth of a Natural Prairie Belt in Alabama," *Annals of the Association of American Geographers,* 47 (1957), 392–401, 407, 409, and "The Geographic Range of the Historic Bison in the Southeast," ibid., 50 (1960), 407; Gilbert Roe, *The North American Bison* (Toronto: University of Toronto Press, 1970).

30. The nature of the prairie peninsulas has been the subject of considerable discussion, though the arguments generally divide into two camps: those who feel fire was sufficient to create them and those who look for larger climatic effects associated with the Pleistocene. For an exposition of the latter view, see H. E. Wright, "Late Quaternary Vegetational History of North America," in Karl K. Turekian, ed., *The Late Cenozoic Glacial Ages* (New Haven: Yale University Press, 1971), pp. 425–464, Wright, "History of the Prairie Peninsula," in Robert Bergstrom, ed., *The Quaternary of Illinois* (Urbana: University of Illinois Press, 1968), pp. 78–88, and Wright, "Vegetational History of the Central Plains," in Wakefield Dort and J. Knox Jones, eds., *Pleistocene and Recent Environments of the Central Great Plains,* Department of Geology, University of Kansas, Special Publication 3 (Lawrence: University of Kansas Press, n.d.), pp. 157–172. Such analyses rely heavily on palynology. Whatever climatic changes were taking place, however, once Indian firing was removed, the grasslands quickly gave way to trees—showing the sufficiency of fire as a causative agent, though perhaps not its necessity.

31. James, *Account of an Expedition,* p. 405.

32. Washington Irving, *A Tour of the Prairies,* ed. John Francis McDermott (Norman: University of Oklahoma Press, 1956), p. 128.

33. Thomas, "Fire and the Fur Trade,", p. 34.

34. See *The West of Alfred Jacob Miller* (Norman: University of Oklahoma Press, 1968), p. 197. Miller's interesting description of the incident is also reproduced.

35. James Fenimore Cooper, *The Prairie* (New York: New American Library, 1964), p. 267. For an interesting depiction of a prairie fire, see pp. 252–266, which, despite Cooper's embellishment, is drawn from original accounts.

36. Robert Wells, *Fire at Peshtigo* (Englewood Cliffs, N.J.: Prentice-Hall, 1967), p. 91.

37. Lewis, *Patterns of Indian Burning,* p. viii.

38. Conrad T. Moore, "Man and Fire in the Central North American Grassland 1535–1890: A Documentary Historical Geography" (Ph.D. diss., University of California, Los Angeles, 1972), p. 115. Along somewhat similar lines see J. G. Nelson and R. E. England, "Some Comments on the Causes and Effects of Fire in the Northern Grasslands Area of Canada and the Nearby United States,

1750–1900," in Connie M. Bourassa and Arthur P. Brackebusch, eds., *Proceedings of the 1977 Rangeland Management and Fire Symposium* (Missoula, Mont.: University of Montana School of Forestry, 1978), pp. 39–47.

39. Chittenden and Richardson, eds., *Life, Letters, and Travels*, p. 1047.

40. John Wesley Powell, *Report on the Lands of the Arid Region of the United States* (2d ed., 1879; rpt. Cambridge, Mass.: Harvard University Press, 1962), pp. 24–29, 113.

41. Ibid., p. 28.

42. Barrett, "Record of Forest and Field Fires," p. 9.

43. Clar, *California Government and Forestry*, 1:7.

44. Irving, *A Tour on the Prairies*, p. 15.

45. Cordia Sloan Duke and Joe B. Frantz, *6,000 Miles of Fence: Life on the XIT Ranch of Texas* (Austin: University of Texas Press, 1961), p. 58.

46. *Journals of Lewis and Clark*, ed. DeVoto, p. 19.

47. John Charles Frémont, *Report of the Exploring Expedition to the Rocky Mountains . . .* (1845; rpt. Ann Arbor: University Microfilm, 1966), p. 10.

48. James, *Account of an Expedition*, p. 178.

49. The literature on the ecology of grassland fires is very large. For a good summary, consult T. T. Kozlowski and C. E. Ahlgren, eds., *Fire and Ecosystems* (New York: Academic Press, 1974), especially Richard J. Vogl, "The Effects of Fire on Grasslands," pp. 139–194, and Robert R. Humphreys, "Fire in the Deserts and Desert Grasslands of North America," pp. 366–400. Many of the proceedings of the Tall Timber Fire Ecology Conferences deal with grasslands; among the most useful for both their science and history are V. W. Lehmann, "Fire in the Range of Atwater's Prairie Chicken," pp. 127–144, E. V. Komarek, "Fire Ecology—Grasslands and Man," pp. 169–220, and A. S. Jackson, "Wildfires in the Great Plains Grasslands," pp. 241–260, in *TTFECP*, no. 4; Thadis W. Box, "Brush, Fire, and West Texas Rangeland," in *TTFECP*, no. 6, pp. 7–20; John Blydenstein, "Burning and Tropical American Savannas," in *TTFECP*, no. 8, pp. 1–14; H. J. Van Rensburg, "Fire: Its Effects on Grasslands, Including Swamps—Southern, Central, and Eastern Africa," in *TTFECP*, no. 11, pp. 175–200; Grasslands Session, *TTFECP*, no. 12, pp. 5–240. In addition, there are Henry A. Wright and Arthur Bailey, "Fire Ecology and Prescribed Burning in the Great Plains—A Research Review," General Technical Report INT–77 (U.S. Forest Service, 1980); Henry A. Wright et al., "The Role and Use of Fire in Sagebrush-Grass and Pinyon-Juniper Plant Communities: A State-of-the-Art Review," General Technical Report INT–58 (U.S. Forest Service, 1980); and Robert Daubenmire, "Ecology of Fire in Grasslands," *Advanced Ecology Research*, 5 (1968), 209–266. For an interesting overview on the history of the central grasslands, see James C. Malin, *The Grassland of North America: Prolegomena to Its History* (Lawrence, Kan., 1947).

50. Quoted in Alan Gussow, *A Sense of Place: The Artist and the American Land* (New York: Seabury Press, 1971), pp. 71–73. Most of the major nineteenth-century landscape artists painted grassland fires, and most, like Catlin, gave verbal accounts of what they saw.

51. See the entries in C. T. Onions, ed., *The Oxford Dictionary of English Etymology* (Oxford: Oxford University Press, 1966).

52. Robert Wells, "On the Origin of Prairies," *American Journal of Science and the Arts,* 1, no. 4 (1819), 335.

53. Francois André Michaux, *Travels to the West of the Allegheny Mountains* (1805), reprinted in vol. 3 of Reuben Gold Thwaites, ed., *Early Western Travels, 1748–1846* (Cleveland: A. H. Clark, 1904), pp. 221–222, 268. See also N. S. Shaler, *Kentucky: A Pioneer Commonwealth* (Boston: Houghton Mifflin, 1884).

54. James, *Account of an Expedition,* pp. 99, 404–405.

55. Gregg, *Commerce of the Prairies,* pp. 360–362.

56. John T. Hughes, *Doniphan's Expedition . . .* (1848; rpt. Chicago: Rio Grande Press, 1962), pp. 298–299.

57. Wells, "On the Origin of Prairies," pp. 336–337.

58. Laura Ingalls Wilder, *Little House on the Prairie* (New York: Harper and Row, 1953), pp. 276–282.

59. For accounts of important grass fires, see Ralph Widner, ed., *Forests and Forestry in the American States: A Reference Anthology* (Missoula, Mont.: National Association of State Foresters, 1968), esp. pp. 338–356; Jackson, "Wildfires in the Great Plains Grasslands," pp. 241–260. In addition, there are numerous locally produced publications on grass fires, some of which contain a history; see, for example, J. H. Foster, "Grass and Woodland Fires in Texas," *Bulletin of the Agricultural and Mechanical College of Texas,* 3d ser., 3 (May 1916).

60. See Duke and Frantz, *6,000 Miles of Fence,* pp. 44–59; J. Evetts Haley, *The XIT Ranch of Texas . . .* (1929; rpt. Norman: University of Oklahoma Press, 1967), pp. 170–180. The techniques were common throughout the Plains; see, for example, Theodore Roosevelt, *An Autobiography* (New York: Charles Scribner's Sons, 1925), pp. 108–109.

61. Haley, *XIT Ranch,* pp. 179–180.

62. Duke and Frantz, *6,000 Miles of Fence,* pp. 47–48.

63. Ibid., pp. 58–59.

64. Vogl, "The Effects of Fire on Grasslands," pp. 177–178.

65. Like the evidence for Indian firing, the story of the reclamation of grassland by forest is found in fragments. Most of the articles already mentioned under fire and grassland ecology refer to these changes, and detailed documentation will be contained in regional fire histories. An excellent photographic treatment for the Black Hills is Donald R. Progulske, *Yellow Ore, Yellow Hair, Yellow Pine,* Agricultural Experiment Station, Bulletin 616 (Brookings, S.D., 1974).

66. Barrett, "Record of Forest and Field Fires," p. 34.

67. Gregg, *Commerce of the Prairies,* pp. 360–362.

68. Quoted in Humphreys, "Fire in the Deserts and Desert Grasslands," pp. 393–394.

69. Aldo Leopold, *A Sand County Almanac* (1949; rpt. New York: Ballantine, 1970), pp. 31–32.

70. John Muir, *The Story of My Boyhood and Youth* (1913; rpt. Madison: University of Wisconsin Press, 1965), p. 183.

71. Wilmon H. Droze, *Trees, Prairies, and People* (Denton: Texas Women's University, 1977), relates the history of the Shelterbelt program.

72. See Andrew H. Clark, "The Impact of Exotic Invasion on the Remaining New World Mid-lattitude Grasslands," in Thomas, ed., *Man's Role,* pp. 737–762;

John T. Curtis, "The Modification of Mid-lattitude Grasslands and Forests by Man," in ibid., pp. 730–731; A. C. Hull and Joseph Pechanec, "Cheatgrass—A Challenge to Range Research," *JF,* 45 (1947), 555–564.

73. Hu Maxwell, "The Use and Abuse of Forests by the Virginia Indians," *William and Mary Quarterly,* 19 (October 1910), 73–103.

74. "Lara" in *The Complete Poetical Works of Lord Byron,* ed. Paul Elmer More (Boston: Houghton Mifflin, 1905).

75. Quoted in Thurston Clarke, *The Last Caravan* (New York: G. P. Putnam's Sons, 1977), p. 185.

76. See, for example, B. Ribbentrop, *Forestry in British India* (Calcutta: Superintendent of Government Printing, 1900), pp. 148–166. For an overview of forestry and politics in British India, see Richard Tucker, "Forest Management and Imperial Politics: Thana District, Bombay, 1823–1887," *Indian Economic and Social History Review,* 16, no. 3 (1980), 273–300. For a survey of nineteenth-century colonial fire problems, see Franklin Hough, *Report on Forestry,* vol. 3 (Washington: GPO, 1882), pp. 252–255.

77. See Charles R. Young, *The Royal Forests of Medieval England* (Philadelphia: University of Pennsylvania Press, 1979). A somewhat different view of the forester and his status can be found in William Greeley, *Forest Policy* (New York: McGraw-Hill, 1953). Gifford Pinchot, too, was keenly aware of the popular view of foresters and went to great lengths to demonstrate how, according to his political formulas, the forest reserves could be used by local residents.

78. Hough, *Report on Forestry,* p. 147. As a sample of British lecturing on controlled fire, though at a later date, see R. Maclagan Gorrie, "Protective Burning in Himalayan Pine," *JF,* 33 (1935), 807–811. Many similar articles, much earlier, appear in the *Indian Forester.*

79. Gifford Pinchot, *Breaking New Ground* (1947; rpt. Seattle: University of Washington Press, 1972), p. 46.

80. A good survey of light-burning experiments is contained in NA, RG 95, RFS, Research Compilation File, though most of the studies on record were made in the late 1910s and 1920s.

81. The best summary of the controversy is contained in Clar, *California Government and Forestry,* 1:209–212, 298, 320, 323–324, 339, 343, 488–494; 2:48–49. Clar includes a good bibliography of the major publications involved in the affair, though the citations are not complete and do not include archival materials (see NA, RG 95, RFS, Division of Fire Control, especially Box 98, and Research Compilation File). My interpretation differs from Clar's on a number of points. For a good summary of fire practices in northern California at the time, see Barrett, "Record of Forest and Field Fires." For a sample of conservation sentiment at the time, see Marsden Manson, "The Effect of Partial Suppression of Annual Forest Fires in the Sierra Nevada Mountains," *Sierra Club Bulletin,* 6 (January 1906), 22–24.

82. From the Plumas Boundary Report of 1904, quoted in Barrett, "Record of Forest and Field Fires," p. 48.

83. Quoted in Andrew Denny Rodgers III, *Bernhard Eduard Fernow: A Story of North American Forestry* (New York: Hafner, 1968), p. 154. The same incident is recorded in Pinchot, *Breaking New Ground,* p. 24.

84. See Clar, *California Government and Forestry,* 1:21n; John Muir, "The American Forests," *Atlantic Monthly,* 80 (August 1897), 145-157; National Academy of Sciences, *Report of the Committee Appointed by the National Academy of Sciences Upon the Inauguration of a Forest Policy for the Forested Lands of the United States to the Secretary of the Interior* (Washington: GPO, 1897).

85. Ernest Sterling, "Attitude of Lumbermen Toward Forest Fires," in *Yearbook of the United States Department of Agriculture, 1904* (Washington: GPO, 1905), pp. 133-140. See also Clar, *California Government and Forestry,* 1:202-213, for the general context and for comments on Sterling. The kinds of plans that Sterling and the other Bureau foresters proposed are exemplified in "Manual of Technical Methods. Report of Subcommittee on Fire Protection Plans. May, 1906," in NA RG 95, RFS, Research Compilation File, and *Control of Forest Fires at McCloud, California,* U.S. Forest Service, Circular no. 79 (Washington: GPO, 1907).

86. T. B. Walker in Clyde Leavitt, "Forest Fires," in *Report of the National Conservation Commission,* vol. 2 (New York: Arno Press, 1972), pp. 424-425. Other descriptions of light-burning preparations can be found in Stewart Edward White, "Woodsmen, Spare Those Trees!" *Sunset,* 44 (March 1920), 115, which relates how White and Joseph Kitts conducted their burning.

87. G. L. Hoxie, "How Fire Helps Forestry," *Sunset,* 34 (August 1910), 145-151. See also Clar, *California Government and Forestry,* 1:343, for Hoxie's opinions as stated to a special forestry commission set up to review the fire problem in California.

88. For Walker's problem fire, see Coert duBois, *Systematic Fire Protection in the California Forests* (Washington: GPO, 1914), p. 18; for Ballinger's comments, see *New York Times,* August 26, 1910, p. 4. See Clar, *California Government and Forestry,* 1:342-345, for additional insight. The imbroglio appeared just before the Weeks Act and the adoption of a compulsory patrol law by California; both issues heightened the intensity of the debate.

89. Henry S. Graves, *Protection of Forests from Fire,* U.S. Forest Service, Bulletin 82 (Washington: GPO, 1910), pp. 7, 26-27. Among the others who wrote about fire control after 1910 are Ferdinand Silcox, "The Forest Fire Problem in District 1," *Journal of Agriculture,* 4, no. 3 (1916), 100-101, and "How the Fires Were Fought," *AF,* 16, (November 1910), 631-639; Earle Clapp, "Fire Protection in the National Forests," *AF,* 17 (October 1911), 583-618; William Greeley, "Better Methods of Fire Control," in SAF, *Proceedings,* 6 (1911), 153-165. In addition, see the November 1910 issue of *American Forests,* which is devoted exclusively to the 1910 fires. In his annual report for 1910, Graves went so far as to criticize "Indian" and "settler" fires as "enormously destructive. Many of them were much more destructive than the fires of the season of 1910 or of any other recent year."

90. Harold Steen, *The U.S. Forest Service: A History* (Seattle: University of Washington Press, 1976), pp. 135-136. The episode is a curious one, showing how foresters were willing to meet light-burning proponents in debate until the controversy became more political and more public—as it did by entering the pages of *Sunset.* At this point, stunned by the magnitude of the 1910 fires, foresters closed ranks.

91. See Coert duBois, *National Forest Fire-Protection Plans* (Washington: GPO, 1911), which constructed a model fire plan for a California forest, "Organization of Forest Fire Control Forces," in SAF, *Proceedings,* 9 (October 1914), 512–521, and *Systematic Fire Protection.* For duBois' own version of how he came to write the latter work, see his autobiographical *Trailblazers* (Stonington, Conn., 1957). Olmstead's contribution can be found in "Light Burning in California Forests" (U.S. Forest Service, 1911), and "Fire and the Forest: 'The Theory of 'Light Burning,' " *Sierra Club Bulletin,* 8 (January 1911), 43–47. In the latter, Olmsted stated that "the theory of 'light burning' is sound. The Forest Service uses it in practice, has done so for years, and will continue to do so." What Olmstead meant was that the Service had used fire to dispose of logging debris. His confusion over light burning is understandable; only after another decade would Service opinion irreversibly stiffen toward such practices.

92. See Inman Eldredge, "Fire Problem of the Florida National Forest," in SAF, *Proceedings,* 6 (1911), 166–170. For the larger setting of the southern fire problem, see Ashley Schiff, *Fire and Water: Scientific Heresy in the U.S. Forest Service* (Cambridge, Mass.: Harvard University Press, 1962); and R. J. Riebold, "Early History of Fires and Prescribed Burning," in "Prescribed Burning Symposium Proceedings" (Asheville, N.C.: U.S. Forest Service, 1971), pp. 11–20.

93. For the nature of this early research, see duBois, *Trailblazers,* pp. 75–81; Clar, *California Government and Forestry,* 1:343, 461, 488–489; Stuart Brevier Show, "Light Burning at Castle Rock," *JF,* 10 (1915), 426–433, which offers a sample; and Show and E. I. Kotok, History of Fire Research in California, unpublished report, title page missing, U.S. Forest Service, ca. 1956, WNRC, RFS, pp. 8–11. See Show and Kotok, *The Rôle of Fire in the California Pine Forests,* U.S. Department of Agriculture, Bulletin no. 1294 (Washington: GPO, 1924), for a summary of research findings, and consult Steen, *Forest Service,* pp. 135–137, for the larger administrative context of these moves and some of the attendant problems. The general discussion about what sort of research was appropriate can be followed through the medium of the Forest Service Minutes of the Service Committee, History Office, U.S. Forest Service. Among the early returns, consider R. H. Boerker, "Light Burning vs. Forest Management in Northern California," *Forestry Quarterly,* 10 (June 1912) 184–194; S. B. Show and R. F. Hammatt, "Will Fire Prevent Fire? A Discussion of 'Light-Burning' " (U.S. Forest Service, ca. 1919); M. B. Pratt, "Results of 'Light Burning' Near Nevada City, California," *Forestry Quarterly,* 11 (December 1911), 420–422.

94. For the course of the controversy, see Clar, *California Government and Forestry,* 1:488–489. A curious sidelight of the controversy is the number of civil engineers who became involved, notably Kitts, Hoxie, and Hall. Though they upheld what were basically adaptations of frontier fire practices, the exponents of light burning were not semiliterate frontiersmen.

95. See Clar, *California Government and Forestry,* 1:488–489. Steen, *Forest Service,* p. 136, describes the threat of lawsuits. The *Sunset* articles include Stewart Edward White, "Woodsmen, Spare Those Trees!" 44 (March 1920), 23–26, 108, 110, 112, 114–117; Henry S. Graves, "The Torch in the Timber," 44 (April 1920), 37–40, 80, 82, 84, 86, 88, 90; Paul G. Redington, "What Is the Truth?" 44 (June 1920), 57–58.

96. "Graves Terms Light Burning Piute Forestry" *The Timberman*, January 1920; William Greeley, "Piute Forestry or the Fallacy of Light Burning," *The Timberman*, March 1920.

97. Donald Bruce, "Light Burning: Report of the California Forestry Committee," *JF*, 21 (February 1923), 129–130. For the context, see Clar, *California Government and Forestry*, 1:489–492; Show and Kotok, History of Fire Research in California, pp. 12–14; Show and Kotok, *Rôle of Fire in the California Pine Forests*, summarizes the actual experiments involved.

98. Clar, *California Government and Forestry*, 1:490.

99. Bruce, "Light Burning," p. 132.

100. Ibid., pp. 130–131.

101. S. B. Show and E. I. Kotok, *Fire and the Forest (California Pine Region)*, U.S. Department of Agriculture, Department Circular 358 (Washington: GPO, 1925), pp. 11–12, 19. The circular is a distillation of the longer *Rôle of Fire in the California Pine Forests*.

102. Graves, "Torch in the Timber," p. 82.

103. Show to John Coffman, November 12, 1927, NA, RG 95, RFS, Division of Fire Control, Region 5, Correspondence.

104. Clar, *California Government and Forestry*, 2:47–50.

105. Ibid., p. 48n.

106. The best synopsis of the southern version of the light-burning controversy is contained in Ashley Schiff, *Fire and Water*, pp. 15–115, which also contains a substantial bibliography of the major sources. See also Riebold, "Early History of Fires and Prescribed Burning"; and Elwood R. Maunder, ed., *Voices from the South: Recollections of Four Foresters* (Santa Cruz, Calif.: Forest History Society, 1977), especially the interviews with Inman Eldredge and Elwood Demmon. A summary of Forest Service research for the period under consideration can be found in David Bruce and Ralph M. Nelson, "Use and Effects of Fire in Southern Forests: Abstracts of Publications by the Southern and Southeastern Forest Experiment Stations, 1921–55," *FCN*, 18 (April 1957), 67–96.

107. Eldredge, "Fire Problems on the Florida National Forest," p. 166. See also Eldredge, "Administrative Problems in Fire Control in the Longleaf-Slash Pine Region of the South," *JF*, 33 (March 1935), 342–345.

108. See "Memorandum for Acting Forester," January 21, 1919, NA, RG 95, Division of Fire Control, Region 7. The memo was a covering letter for "attached correspondence" that "fully covers the fire protection situation on the Florida Forest."

109. H. O. Stabler, "Memo. for Acting District Forester Reed, January 3, 1919," p. 8, NA, RG 95, RFS, Division of Fire Control, Region 7. See also citation in note 108.

110. H. H. Chapman, *Factors Determining Natural Reproduction of Longleaf Pine on Cut-over Lands in LaSalle Parish, Louisiana*, Yale School of Forestry, Bulletin no. 16 (New Haven: Yale University, 1926). This study was but an opening salvo.

111. William B. Greeley, "Light Burning Policy for the South" (1927), NA, RG 95, RFS, Research Compilation File. The statement is followed by an equally interesting "Memorandum for PR" that criticizes the essay.

112. S. W. Greene, "The Forest That Fire Made," *AF*, 37 (October 1931), 583–584, 618; Herbert L. Stoddard, *The Bobwhite Quail: Its Habits, Preservation and Increase* (New York: Charles Scribner's Sons, 1931). For Greene's work on ranges, see Schiff, *Fire and Water*, pp. 42–43. One should also mention Austin Cary, "Some Relations of Fire to Longleaf Pine," *JF*, 30 (May 1932), 594–601. Like Chapman, Cary helped to unite northern and southern forestry, despite different technologies of fire. That such a widespread interest in prescribed fire should become manifest in nearly the same year and that the Forest Service should, however tentatively, revise the Clarke-McNary program to accommodate new fire practices show with what intensity and suddenness the problem was presented. For the Forest Service revisions of Clarke–McNary, see Memorandum from R. Y. Stuart, June 16, 1932, "Federal Policy Relating to Controlled Burning in Cooperative Fire Protection in the Longleaf Pine Region," NA, RG 95, RFS, Division of Fire Control, General Correspondence.

113. Eldredge, "Administrative Problems in Fire Control," p. 344. The context of the meeting is well explained in Schiff, *Fire and Water*, pp. 64–67. The proceedings of the convention were printed in *JF*, 33 (March 1935).

114. A. A. Bickford and L. S. Newcomb, "Report on 1943–1944 Prescribed Burning on the Osceola National Forest" (U.S. Forest Service, 1944), describes the aftermath of the Lake City conference. For the larger setting, again see Schiff, *Fire and Water*, pp. 98–101. An informative personal view of the fires and Forest Service decisions is given in John W. Squires, "Burning on Private Lands in Mississippi," in *TTFECP*, no. 3, pp. 1–5.

115. Arthur W. Hartman, "Fire as a Tool in Southern Pine," in *Yearbook of Agriculture, 1949* (Washington: GPO, 1949), pp. 517–527.

116. Schiff, *Fire and Water*, pp. 77–78, 115.

117. Charles Buck, "A National Program of Fire Research," [June 27, 1945], draft, WNRS, RFS, pp. 15–16. Buck insisted that "this is a first-priority problem in all regions." See also "Fire Effects and Use," in "Proceedings of the Priest River Fire Meeting" (U.S. Forest Service, 1941), pp. 46–65.

118. H. T. Gisborne, "Forest Protection," in Robert K. Winters, ed., *Fifty Years of Forestry in the U.S.A.* (Washington: SAF, 1950), p. 37. For examples of prescribed burning, see Kenneth Davis and Karl A. Klehm, "Controlled Burning in the Western White Pine Type," *JF*, 37 (1939), 399–407; Joseph F. Pechanec and George Stewart, *Sagebrush Burning—Good and Bad*, U.S. Department of Agriculture, Farmer's Bulletin no. 1948 (Washington: GPO, 1944); Homer Shantz, *The Use of Fire as a Tool in the Management of the Brush Ranges of California* (Sacramento: California Division of Forestry, 1957). It should be noted that the purpose of pine forest burning was slash disposal and that brush burning was range improvement, not broadcast underburning for silviculture.

119. Quoted in Schiff, *Fire and Water*, p. 96. For a similar view, see the two articles by Arthur Hartman in the *Yearbook of Agriculture, 1949:* "Fire as a Toll in Southern Pine" and, immediately following, "Machines and Fires in the South."

120. Earl Loveridge to A. A. Brown, Memorandum, "F-Control-Prescribed Burning, 01–01–53," WNRC, RFS, Division of Fire Control.

121. See Arthur Hartmen, "Letter to a Professor in Forest Protection," *FCN,* 11

(July 1950), 34–38. A sample: "People are like pendulums—they tend to swing from one extreme to the other. One finds the pure fire exclusionist; after learning just enough to know fire can produce some marked benefits (but not enough to appreciate its complexities), this type tends to swing too far over, get 'match happy' and overdo it." Elsewhere Hartmen emphasized that "a fairly widespread concept has been that prescribed burning is a form of 'light' burning, done to keep an area from being burned by wildfire. Such a primary use of fire is rare. . . ."

122. See H. H. Biswell, "Prescribed Burning in Georgia and California Compared," *Journal of Range Management,* 11 (1958), 293–297, and "Forest Fires in Perspective," in *TTFECP,* no. 7, pp. 43–63. Brown's comments can be found in A. A. Brown to Arthur Hartman, October 4, 1956, WNRC, RFS, Division of Forest Fire Research. Useful surveys of prescribed fire in the mid-1950s are found in Kenneth Davis, *Forest Fire: Control and Use* (New York: McGraw-Hill, 1959), pp. 494–533; A. A. Brown and A. W. Folweiler, *Fire in the Forests of the United States* (St Louis: John S. Swift Co., 1953), pp. 28–34; and Silas Little et al., "Choosing Suitable Times for Prescribed Burning in Southern New Jersey," *FCN,* 14 (January 1953), 21–25.

123. See Rayonier Incorporated v. United States, 352 U.S. 315 (1957); and Richard E. McArdle to Regional Foresters and Directors, Memorandum, April 15, 1957, WNRC, RFS, Division of Fire Control. McArdle noted that "planned burning is a technical job requiring knowledge of fire behavior and experience in fire control"—which is to say that only foresters and experts should really engage in it.

124. For the Ponderosa Way, see J. H. Price, "The Ponderosa Way," *AF,* 40 (September 1934), 387–390; Show and Kotok, History of Fire Research, p. 37; Clar, *California Government and Forestry,* 2:252–258. The literature of fuelbreaks is vast. Much of the early history in California is included in Clar, passim. For the period under consideration, see Verdie E. White and Lisle R. Green, "Fuel-Breaks in Southern California, 1958–1965" (U.S. Forest Service, 1967); Lisle R. Green and Harry E. Schimke, "Guides for Fuel-Breaks in the Sierra Nevada Mixed-Conifer Type" (U.S. Forest Service, 1971); Lawrence S. Davis, *The Economics of Wildfire Protection with Emphasis on Fuel Break Systems* (Sacramento: California Division of Forestry, 1965); and Lisle R. Green, *Fuelbreaks and Other Fuel Modification for Wildland Fire Control,* Agriculture Handbook no. 499 (Washington: GPO, 1977), which contains a large bibliography. The popularity of the concept in this period is reflected by the large number of articles in *FCN* that describe the adaptation of the fuelbreak concept to local fuels and topography.

125. Francis Marion National Forest, "DESCON Plan" (U.S. Forest Service, 1974); David Jay, "DESCON: Utilizing Wildfires to Achieve Land Management Objectives," in "Proceedings of the Fire by Prescription Symposium" (U.S. Forest Service, 1976), pp. 67–68; David D. Devet, "DESCON: Utilizing Benign Wildfires to Achieve Land Management Objectives," in *TTFECP,* no. 14, pp. 33–45. The program was adopted after the policy reforms of 1971 and received formal approval in 1974.

126. Robert E. Martin et al., "Report of Task Force on Prescribed Burning," *JF*, 75 (May 1977), 297–301.
127. The literature on smoke management has swelled rapidly. The best compendium is Committee on Fire Research, *Air Quality and Smoke from Urban and Forest Fires: Proceedings of International Symposium* (Washington: National Academy of Sciences, 1976), esp. pp. 296–373. Also useful is Southern Forest Fire Laboratory Personnel and Southeastern Forest Experiment Station, "Southern Forestry Smoke Management Guidebook," General Technical Report SE–10 (U.S. Forest Service, 1976).

3: THE FIRE FROM EUROPE

1. Quoted in Clarence Glacken, *Traces on the Rhodian Shore* (Berkeley: University of California Press, 1967), p. 312.
2. Quoted in ibid., p. 692.
3. This episode is related in Homer Aschmann, "Aboriginal Use of Fire," in "Proceedings of the Symposium on the Environmental Consequences of Fire and Fuel Management in Mediterranean Ecosystems," General Technical Report WO–3 (U.S. Forest Service, 1977), p. 134; George Perkins Marsh, *Man and Nature* (1864), ed. David Lowenthal (Cambridge, Mass.: Harvard University Press, 1965), p. 114
4. Bernard DeVoto, *The Course of Empire* (Boston: Houghton Mifflin, 1952), p. 7.
5. See Julius Klein, *The Mesta: A Study in Spanish Economic History, 1273–1836* (Cambridge, Mass.: Harvard University Press, 1920).
6. For an overview of the reclamation, see Glacken, *Traces*, pp. 318–347; H. C. Darby, "The Clearing of the Woodland in Europe," in William L. Thomas, ed., *Man's Role in Changing the Face of the Earth* (Chicago: University of Chicago Press, 1956), pp. 183–216; and N.J.G. Pounds, *An Historical Geography of Europe, 450 BC–AD 1330* (Cambridge: Cambridge University Press, 1973), esp. pp. 170–433 passim.
7. Charles E. Randall, "Crazy Blazes," *AF*, 69 (July 1963), 8.
8. For a summary of Mediterranean fire and forests, see Darby, "Clearing of the Woodland"; Pounds, *Historical Geography*, esp. pp. 126–169; Stephen H. Spurr and Burton V. Barnes, *Forest Ecology*, 2d ed. (New York: Ronald Press, 1973), pp. 372–373; Glacken, *Traces*, pp. 122–128; Z. Naveh, "Effects of Fire in the Mediterranean Region," in T. T. Kozlowski and C. E. Ahlgren, eds., *Fire and Ecosystems* (New York: Academic Press, 1974), pp. 401–434; L. G. Liacos, "Present Studies and History of Burning in Greece," in *TTFECP*, no. 13, pp. 65–96; Henry LeHouerou, "Fire and Vegetation in the Mediterranean Basin," in *TTFECP*, no. 13, pp. 237–278. See also numerous articles, some by the above authors, in "Proceedings of the Symposium on the Environmental Consequences of Fire and Fuel Management in Mediterranean Ecosystems," in particular: Morris H. McCutchan, "Climatic Features as a Fire Determinant," pp. 1–11; Lucio Susmel, "Ecology of Systems and Fire Management in the Italian Mediterranean Region," pp. 307–317; H. N. LeHouerou, "Fire and Vegetation in

North Africa," pp. 334–341; and Vasilios P. Papanastasis, "Fire Ecology and Management of Phrygana Communities in Greece," pp. 476–482. Also very revealing is Ellen Churchill Semple, *The Geography of the Mediterranean Region: Its Relation to Ancient History* (New York: Henry Holt, 1931).

9. Quoted in Darby, "Clearing of the Woodland," p. 185

10. Virgil, *Aeneid,* 10.405–411.

11. See Klein, *The Mesta.* Paul Sears has observed that "in the valley of Mexico it was not simply that the Spaniards had found a use for timber and needed the space it occupied. Their action, according to contemporary testimony, was influenced by their values. They were homesick for the treeless plains of Castile. They preferred the type of landscape in which they had grown up and which had been produced by the activities of the Mesta. . . ." From Thomas, *Man's Role,* p. 406.

12. See Carl C. Wilson, "Protecting Conifer Plantations Against Fire in the Mediterranean Region," FO: FFM/77/3–06 (New York: Food and Agriculture Organization, 1977), and "Comparing Forest Fire Problems in the Mediterranean Region and in California," FO: FFM/77/1–03 (New York: Food and Agriculture Organization, 1977).

13. Pounds, *Historical Geography,* p. 145

14. Lucretius, *On the Nature of the Universe (De Rerum Natura),* trans. R. E. Latham (London: Penguin, 1977), pp. 221–222.

15. Matthew 13:40.

16. Lucretius, *Nature of the Universe,* p. 209.

17. See Bernhard Fernow, *A Brief History of Forestry in Europe, the United States, and Other Countries,* 3d ed. (Toronto: Toronto University Press, 1913).

18. See Darby, "Clearing of the Woodland," pp. 189–190; P. J. Vior, "Effects of Forest Fire on Soil," in Kozlowski and Ahlgren, eds., *Fire and Ecosystems,* pp. 7–8; Evald Uggla, "Fire Ecology in Swedish Forests," in *TTFECP,* no. 13, pp. 171–190; Gustav Siren, "Some Remarks on Fire Ecology in Finnish Forestry," in *TTFECP,* no. 13, pp. 191–210; J.G.D. Clark, *Prehistoric Europe: The Economic Basis* (Stanford: Stanford University Press, 1966), pp. v (plate), 91–100.

19. Quoted in Darby, "Clearing of the Woodland," p. 193; Glacken, *Traces,* p. 334.

20. Glacken, *Traces,* p. 309.

21. Ibid., pp. 326–330, 336–340. For reservations as preserves, see also Charles R. Young, *The Royal Forests of Medieval England* (Philadelphia: University of Pennsylvania Press, 1979).

22. See Young, *Royal Forests,* passim; Glacken, *Traces,* pp. 322–330.

23. See William Greeley, "Foreword," in Stewart Holbrook, *Burning an Empire: The Story of American Forest Fires* (New York: Macmillan, 1943); Samuel T. Dana, *Forest and Range Policy: Its Development in the United States* (New York: McGraw-Hill, 1956), esp. pp. 2–4, 11–16, a description of how folk practices clashed with colonial politics.

24. See Glacken, *Traces,* p. 345. William Greeley, *Forest Policy* (New York: McGraw-Hill, 1953), offers a global survey of forest fire legislation and attitudes.

25. Glacken, *Traces,* pp. 320–321, 335, 342–345; Darby, "Clearing of the Woodland," p. 191. For heath and moor in Britain, see A. J. Kayll, "Moor Burning in

Scotland," in *TTFECP,* no. 6, pp. 29–39; and C. H. Gimingham, "British Heathland Ecosystems: The Outcome of Many Years of Management by Fire," in *TTFECP,* no. 10, pp. 293–322.

26. Darby, "Clearing of the Woodland"; Young, *Royal Forests,* is especially good on the Forest Charter. See also Greeley, *Forest Policy,* pp. 107–114.

27. See Glacken, *Traces,* pp. 484–491 for Evelyn, and pp. 491–494 for the Code Colbert.

28. National Academy of Sciences, *Report of the Committee Appointed by the National Academy of Sciences Upon the Inauguration of a Forest Policy for the Forested Lands of the United States to the Secretary of the Interior* (Washington: GPO, 1897), pp. 11–13. Britain, France, Germany, and Canada were also examined. An earlier, if similar, exploration for models can be found in Franklin Hough, *Report on Forestry,* vol. 3 (Washington: GPO, 1881), pp. 245–256.

29. Such allusions are endless and almost constitute a ritual incantation. As a sample, see Robert Riley, "Ancient Mayas Burned Their Forests, A Practice That May Have Been the Source of Their Decadence," *AF,* 38 (August 1932), 442–443. Pinchot was particularly instrumental in propagandizing the China story and succeeded in passing on its lessons to such political luminaries as Franklin D. Roosevelt.

30. For the revival of fire protection problems in Europe, see, for example, Ricardo Velez, "Environmental Difficulties in the Fight Against Forest Fires in Spain," in "Symposium on the Environmental Consequences," pp. 326–333; Jane M. S. Robertson, "Land Use Planning of the French Mediterranean Region," in ibid., pp. 283–288; Johannes George Goldammer, "Forest Fire Problems in Germany," *FMN,* 40 (Fall 1979), 7–10; Wilson, "Protecting Conifer Plantations" and "Comparing Forest Fire Problems."

31. The best compendium of fire myths is James Frazer, *Myths of the Origin of Fire* (1930; rpt. New York: Hacker Art Books, 1974). Unless otherwise specified, references to myths derive from this work.

32. Frazer's ages come from *Myths of the Origin of Fire,* pp. 201–203. See also Walter Hough, *Fire as an Agent in Human Culture,* U.S. National Museum, Bulletin 139 (Washington: GPO, 1926), pp. 124–283; John Burnet, *Early Greek Philosophy,* 4th ed. (Cleveland: World Publishing Co., 1964); *Encyclopedia Britannica,* s.v. "Fire"; Winifred Blackman, "The Magical and Ceremonial Uses of Fire," *Folklore,* 27 (1916), 352–377. Among the stranger expositions on fire as an object of fascination and reverie, there is Gaston Bachelard, *The Psychoanalysis of Fire* (Boston: Beacon Press, 1964). Sigmund Freud's analysis is both irrelevant and silly: "The Acquisition and Control of Fire," in *The Standard Edition of the Complete Psychological Works of Sigmund Freud* (London: Hogarth Press, 1953–), 22:187–193.

33. See James Frazer, *The Golden Bough: A Study in Magic and Religion,* abr. ed. (New York: Macmillan, 1922), pp. 609–657.

34. Ibid., pp. 641–642.

35. John Shea, "Our Pappies Burned the Woods," *AF,* 46 (April 1940), 159–162, 174. The article was based on Shea's much longer field studies for the Forest Service, notably "Man-Caused Forest Fires: The Psychologist Makes a Diagnosis" (U.S. Forest Service, 1939) and "Getting at the Roots of Man-Caused

Forest Fires: A Case Study of a Southern Forest Area" (U.S. Forest Service, ca. 1939).

36. Hamilton Pyles, "Fire Prevention Strategy in the East," *Forest History,* 16, no. 3 (1972), 23.

37. Shea, "Our Pappies," p. 159.

38. Ibid., p. 162.

39. Quoted in Gordon M. Day, "The Indian as an Ecological Factor in the Northeastern Forest," *Ecology,* 34, no. 2 (1953), 334.

40. *The Travels of William Bartram* (1791), ed. Mark Van Doren (New York: Dover, 1955), p. 149.

41. Quoted in Elwood R. Maunder, ed., *Voices from the South: Recollections of Four Foresters* (Santa Cruz, Calif.: Forest History Society, 1977), p. 35.

42. Among the many studies done on woodsburning in the South, see Lucy W. Cole and Harold F. Kaufman, "Socio-Economic Factors and Forest Fires in Mississippi Counties," Mississippi State University, Social Science Research Center, Preliminary Report no. 14 (December 1966), for a good summary with a useful bibliography; Thomas Hansbrough, "A Sociological Analysis of Man-Caused Forest Fires in Louisiana" (Ph.D. diss., Louisiana State University, 1961); John B. Morris, "Preliminary Investigation of Human Factors in Forest Fires" (U.S. Forest Service, 1958); J. E. Dunkelberger and A. T. Altobellis, "Profiling the Woods-burner: An Analysis of Fire Trespass Violations in the South's National Forests," Auburn University, Agricultural Experiment Station, Bulletin 469 (May 1975); Arthur R. Jones et al., "Some Human Factors in Woods Burning," Louisiana State University and Agricultural Experiment Station, Bulletin no. 601 (August 1965); Alvin L. Bertrand, "Attitudinal Patterns Prevalent in a Forest Area with High Incendiarism," Louisiana State University and Agricultural Experiment Station, Bulletin no. 648 (Novemeber 1970); George R. Fahnestock, "Fire Prevention in the South: A Study in Diversity," in SAF, *Proceedings* (1964); Hilliard Henson, "Why Incendiary Fires in the Southern Appalachians," *AF,* 48 (September 1942), 419, 430. See also note 35 above for the voluminous studies made under the direction of John Shea for the Forest Service. Good summaries are also given in George R. Fahnestock, "Southern Forest Fires: A Social Challenge," *FCN,* 26 (April 1965), 10–12, 16. Alvin L. Bertrand and Andrew W. Baird, "Incendiarism in Southern Forests: A Decade of Sociological Research," Mississippi State University, Social Science Research Center, Bulletin 838 (1975), contains an excellent summary of the research, with full bibliography up to the mid-1970s.

43. H. H. Chapman, "Forest Fires and Forestry in the United States," *AF,* 18 (July 1912), 516.

44. Ronald Harper, *Forests of Alabama,* Geological Survey of Alabama, Monograph 10 (Tuscaloosa: University of Alabama, 1943); Erhard Rostlund, "The Myth of a Natural Prairie Belt in Alabama: An Interpretation of Historical Records," in *Annals of the Association of American Geographers,* 47 (1957), 392–411; Merle C. Prunty, "Some Geographic Views of the Role of Fire in Settlement Process in the South," in *TTFECP,* no. 4, pp. 161–168.

45. Mark Twain, *Life on the Mississippi* (1883; rpt. New York: Airmont, 1965), pp. 65–66.

46. Ray Allen Billington, *Westward Expansion,* 4th ed. (New York: Macmillan, 1974), p. 313. For a somewhat different view, see David M. Potter, *The Impending Crisis, 1848–1861* (New York: Harper and Row, 1976). In the chapter entitled "On the Nature of Southern Separatism" (pp. 478–484), Potter argues that the fear of slave insurrections may have been the motive perpetuating the plantation system. Either way, the plantation persisted, with the geographical results described.

47. Quoted in Billington, *Westward Expansion,* p. 60.

48. Prunty, "Some Geographic Views," pp. 165–166.

49. Clarence King discovered one such family in the Sierra Nevada in the late 1860s. See King, "The Newtys of Pike," in *Mountaineering in the Sierra Nevada* (1872; rpt. Lincoln: University of Nebraska Press, 1970), pp. 94–111.

50. H. L. Stoddard, "Use of Fire in Pine Forests and Game Lands of the Southwest," in *TTFECP,* no. 1, pp. 32–35. See also Herbert L. Stoddard, *Memoirs of a Naturalist* (Norman: University of Oklahoma Press, 1969), for an interesting autobiography of an important figure in American fire history.

51. For information on naval stores forests, see Henry Clepper, *Professional Forestry in the United States* (Baltimore: Johns Hopkins University Press, 1971), pp. 231–234; Norman R. Hawley, "Burning in Naval Stores Forest," in *TTFECP,* no. 3, pp. 81–88, for a modern description; Charles Mohr, *The Timber Pines of the Southern United States,* Division of Forestry, Bulletin no. 13, rev. ed. (Washington: GPO, 1897), pp. 62, 67–72, for a depiction of turpentining at the turn of the century; and Ralph Widner, ed., *Forests and Forestry in the American States: A Reference Anthology* (Missoula, Mont.: National Association of State Foresters, 1968), pp. 412–491 passim.

52. For an overview of southern logging, see Clepper, *Professional Forestry,* pp. 231–254; Widner, ed., *Forests and Forestry,* pp. 412–491; Stoddard, "Use of Fire," pp. 32–37.

53. Inman Eldredge, "Fire Problems on the Florida National Forest," in SAF, *Proceedings,* 6 (1911), pp. 166–168. For a similar treatment, see Eldredge in Maunder, *Voices from the South,* pp. 49–50: "The big thing about it was that nobody cared. The people in the city didn't care; the people in the small towns didn't care. They were accustomed to having the air full of smoke at certain times of the year and of riding along on the train and seeing mile after mile of fire along both sides. Railroads didn't care. Politicians not only didn't care, but pooh-poohed fire protection."

54. Eldredge, "Fire Problems," p. 170. For a different analysis from elsewhere in the South at this time, see J. S. Holmes, "Forest Fires and Their Prevention," North Carolina Geological and Economic Survey, Economic Paper no. 22 (Raleigh, 1911).

55. Quoted in Clepper, *Professional Forestry,* p. 243.

56. Ibid.

57. The story of the hunting plantations can be found in E. V. Komarek, *A Quest for Ecological Understanding,* TTRS Miscellaneous Publication no. 5 (Tallahassee: TTRS, 1977), pp. 15–21. For the British equivalents, see Gimingham, "British Heathlands Ecosystems," pp. 298–300.

58. For the history of the quail study program, see Komarek, *Quest for Ecological*

Understanding, pp. 15–21. For Stoddard's report and its controversies, see Stoddard, "Use of Fire," pp. 38–39. The general context is conveyed in Ashley Schiff, *Fire and Water: Scientific Heresy in the Forest Service* (Cambridge, Mass.: Harvard University Press, 1962), pp. 64–78 passim.

59. Komarek, *Quest for Ecological Understanding,* pp. 15–21.

60. The fire was investigated by W. W. Ashe of the state geological survey and reported in the *Raleigh News and Observer,* February 27, 1898. Ashe reports that 14 lives were lost and perhaps 250,000 turpentine trees destroyed; damages totaled about $1 million. Moreover, "a repetition of this fire" could be expected. But the blowup in the backwoods of North Carolina could not compete with the blowing up of the battleship *Maine* in Havana Harbor, which was reported in the headlines that same day.

61. Quoted in Widner, ed., *Forests and Forestry,* p. 468. Widner's is the best compendium of large fires in the South, though the material is somewhat scattered. The main fire years are also recorded in Schiff, *Fire and Water,* passim. Of further use are the Clarke-McNary studies after 1924, which surveyed fire protection needs and some of which were subsequently published.

62. The best accounts of the droughts and fires are in Schiff, *Fire and Water,* pp. 51–100 passim, and Widner, ed., *Forests and Forestry,* pp. 412–491 passim.

63. Widner, ed., *Forests and Forestry,* p. 490; records on file, Kentucky Division of Forestry, Frankfort. An interesting repetition, though on about half the scale, occurred during late autum of 1980.

64. Widner, ed., *Forests and Forestry,* pp. 204–205; files, North Carolina Forest Service, Raleigh.

65. G. K. Schaeffer, "Report on Buckhead Fire—March 24, 1956. Osceola National Forest," WNRC, RFS, Division of Fire Control.

66. Dana, *Forest and Range Policy,* pp. 310, 423; "First Intercompact Agreement Signed," *FCN,* 33, no. 3 (1972), 16. Fire Working Team, Society of American Foresters, "Interstate Wildfire Organization Directory," includes the texts of the agreements.

67. George Byram, "Atmospheric Conditions Related to Blowup Fires," Research Paper SE–35 (U.S. Forest Service, 1954). The files and annual reports at the Southern Forest Fire Lab are the best source for its history, but see also K. W. McNasser, "Forest Fire Research," *FCN,* 24, no. 1 (1963), 11.

68. The mid-1950s and early 1960s fires are described by state in Widner, ed., *Forests and Forestry.* Many of the larger fires were described in *FCN*; for example, Ernest A. Rodney, "Forest Fires and Fire Weather Conditions in the Asheville, N.C., Fire Weather District—Spring Season, 1963,"*FCN,* 25, no. 3 (1964), 7–9, 15. Some of the fires were reported on in depth. See, for example, Schaeffer, "Report on Buckhead Fire"; Keith Argow, "The Carolina Blowup," *AF,* 72 (July 1966), 14–17, 51; Dale D. Wade and Darold E. Wade, "An Analysis of the Air Force Bomb Range Fire," Research Paper SE–105 (U.S. Forest Service, 6973); Eugene Cypert, "Plant Succession on Burned Areas in Okefenokee Swamp Following the Fires of 1954 and 1955," in *TTFECP,* no. 12, pp. 199–218; Charles F. Evans, "Flaming Florida," *AF,* 38 (June 1932), 344, 376; "Florida—The Uncivilized" (editorial), *AF,* 38 (June 1932), 351.

69. Associated Press, November 1978.

70. John D. Guthrie, ed., *Forest Fire and Other Verse* (Portland: Dunham Printing Co., 1929), p. 34–35.
71. Mark Twain, *Roughing It* (1872; rpt. New York: New American Library, 1962), pp. 139–140.
72. Col. Richard Dodge, *The Plains of the Great West and Their Inhabitants . . .* (1877; rpt. New York: Archer House, 1959), pp. 79–80.
73. H. Duane Hampton, *How the U.S. Cavalry Saved Our National Parks* (Bloomington: Indiana University Press, 1971), p. 100.
74. Quoted in Harold F. Kaufman, "A Psycho-Social Study of the Cause and Prevention of Forest Fires in the Clark National Forest" (U.S. Forest Service, 1938), pp. 12–16. See also Widner, ed., *Forests and Forestry*, p. 485.
75. Paul D. Brohn, "Mine Fires Still Burn," *FCN,* 33 (Winter 1971–1972), 10, 14; Clar, *California Government and Forestry,* 1:275–307 passim; "The Capitan Tragedy" (U.S. Forest Service, History Office, n.d.).
76. National Academy of Sciences, *Report of the Committee,* p. 19. For similar acts of vandalism, though on a vaster scale, see William G. Morris, "Forest Fires in Western Oregon and Western Washington," *Oregon Historical Quarterly,* 35 (December 1934), 325–326.
77. Dana, *Forest and Range Policy,* p. 6. Hough, *Report on Forestry,* pp. 130–155, lists existing fire codes for the states and territories.
78. Quoted in J. Evetts Haley, *The XIT Ranch of Texas . . .* (1929; rpt. Norman: University of Oklahoma Press, 1967), p. 171.
79. Louis Barrett, "A Record of Forest and Field Fires in California From the Days of the Early Explorers to the Creation of the Forest Reserves" (U.S. Forest Service, 1935), pp. 71, 120.
80. Greeley, *Forest Policy,* passim; George Vitas, "Unholy Fire," *AF,* 53 (September 1947), 395.
81. Quoted in Andrew Denny Rodgers III, *Bernhard Eduard Fernow: A Story of North American Forestry* (New York: Hafner, 1968), p. 167.
82. Gifford Pinchot, "Study of Forest Fires and Wood Protection in Southern New Jersey," *Annual Report of Geological Survey of New Jersey* (Trenton, 1898), appendix, p. 21.
83. Charles S. Cowan, *The Enemy Is Fire* (Seattle: Superior Publishing Co., 1962), p. 67. For the Forest Service response to the "rising wave of incendiarism," see Roy Headley to Regional Foresters, April 15, 1932, "Incendiarism," NA, RG 95, RFS, Division of Fire Control, General Correspondence.
84. Charles J. Gleeson, "The Man Who Played with Fire," *AF,* 59 (August 1953), 21–22, 40, 48.
85. For an interesting description of the persistence of even old industrial fire causes, see Michigan Department of Conservation, "Railroad Fire Control Conference" (1960, 1965). Railroads were first held legally liable for fire starts in 1837 by a Maryland court.
86. At least once, firefighters walking into a fire stumbled onto a still, whereupon they surrendered the onerous burdens of smokechasing for the pleasures of sipping—and were discovered hours later in a deep, contented stupor. See Cowan, *The Enemy Is Fire,* p. 71; also E. John Long, "Menace of the Stills," *AF,* 62 (May 1956), 36–39.

87. Emma H. Morton, "Cupid Goes Incendiary," *Service Bulletin,* 37 (1935), 6.
88. Eloise Hamilton, *Forty Years of Western Forestry* (Portland: Western Forestry
    and Conservation Association, 1949), pp. 15, 21. For a sample of how fire pre-
    vention was taught to children, see *A Forest Fire Prevention Handbook for the
    Schools of Washington,* Department of Agriculture, Miscellaneous Publication
    no. 40 (Washington: GPO, 1928). For a sample of posters, see J. Girvin Peters,
    "Development of Fire Protection by the States," *AF,* 19 (November 1913), 721–
    737.
89. J. A. Mitchell and D. Robson, "Forest Fires and Forest Fire Control in Mich-
    igan" (Michigan Department of Conservation, 1950), pp. 30–35.
90. Widner, ed., *Forests and Forestry,* p. 369. The Showboat concept was not lim-
    ited to Missouri or even to the South; it became common practice wherever rural
    communities and fire problems coexisted.
91. Uralsky won first place with the painting in the art competition at the Loui-
    siana Purchase Exposition in St Louis, 1903. The painting enjoyed wide popu-
    larity as well as critical acclaim, and prints of it—sans the fire prevention mes-
    sage that foresters tacked onto it—were commonly framed and exhibited.
    Grandma Moses made several copies; the most famous and the closest to the
    original can be found in Otto Kallir, *Grandma Moses* (New York: Harry N.
    Abrams, 1973), p. 46. For a partial reproduction of the original, see *AF,* 45
    (April 1939), 151.
92. Guthrie, *Forest Fire and Other Verse.* Guthrie had published an earlier anthol-
    ogy of Forest Service poetry, *The Forest Ranger and Other Verse* (Boston: Gor-
    ham Press, 1919), which also included a few fire poems.
93. For Wheeler's peripatetic crusade, see Schiff, *Fire and Water,* pp. 67–71. See
    also Roy Headley, Memorandum, January 21, 1937, NA, RG 95, RFS, Division
    of Fire Control. Wheeler's own version appears in his unpublished autobiography
    and papers, in the H. N. Wheeler Collection, Conservation Library, Denver Pub-
    lic Library; for a sample of his message, see "Fire," the text for a speech, within
    the collection. "Forest and Flame in the Bible" was written by George Vitas of
    the Forest Service's Information and Education Division and was promoted
    under the Cooperative Forest Fire Prevention Campaign, which was also respon-
    sible for the Smokey Bear program. William Greeley also published objections
    to the "circus publicity education" and wanted a more factual approach through
    bulletins and other printed means; see Greeley, "Forest Fire—The Red Paradox
    of Conservation," *AF,* 45 (April 1939), 153–157.
94. A. A. Brown and A. D. Folweiler, *Fire in the Forests of the United States* (St.
    Louis: John S. Swift, 1953), pp. 66–67; Massachusetts Forestry Association,
    "The Cape Cod Forest Fire Prevention Experimant" (1928). The first edition of
    Fire in the Forests was written by Folweiler in 1936 and "published" in loose-
    leaf form. It constitutes the first textbook on fire in the U.S. and perhaps in the
    world. The 1953 edition remains a valuable source of historical information.
95. W. C. McCormick, "The Three Million," *AF,* 37 (August 1931), 479–480;
    "Through 1930 with the Dixie Crusaders," *AF,* 36 (March 1930); "Association
    Launches Fire Prevention Campaign," *AF,* 45 (May 1939), 266–267; Henry
    Clepper, *Crusade for Conservation: The Centennial History of the American For-*

estry Association (Washington: American Forestry Association, 1957), pp. 46–48; Schiff, *Fire and Water*, pp 36–37; "Two Years with Dixie Crusaders, "*AF*, 36 (September 1930), 582; "Southern Forestry Educational Project," *AF*, 35 (September 1929), 569; Erle Kauffman, "The Southland Revisited,"*AF*, 61 (August 1955), 33–42.

96. Quoted in "Two Years With the Dixie Crusaders," p. 582.

97. Quoted in Schiff, *Fire and Water*, p. 44.

98. Ibid., p. 71.

99. "Fire Control Meeting, Spokane, Washington, February 10–21, 1936. Brief of Minutes and Papers" (U.S. Forest Service, 1936), pt. 2, p. 16.

100. Quoted in John Shea, "Man-Caused Forest Fires," p. 7. One might contrast this analysis with those done for true pyromaniacs; see, for example, Donald Scott, *Fire and Fire-Raisers* (London: Duckworth, 1974) and *The Psychology of Fire* (New York: Charles Scribner's Sons, 1975).

101. Shea, "Man-Caused Forest Fires," pp. 7, 25–26. A list of the anthropologists, sociologists, and psychologists involved in the project is given in Shea's acknowledgments to his "Getting at the Roots of Man-Caused Forest Fires" (U.S. Forest Service, 1938).

102. Synopses of the reports were compiled in Shea, "Man-Caused Forest Fires." Other reports in the series include Shea, "Getting at the Roots of Man-Caused Forest Fires"; Harold F. Kaufman, "A Psycho-Social Study of the Cause and Prevention of Forest Fires in the Clark National Forest" (U.S. Forest Service, 1938); and George H. Weltner to J. F. Brooks, "Fire Prevention Study in Chickasawhay and Bienville Districts," March 20, 1943, WNRC, RFS, Division of Fire Control. The general records can be found in WNRC, RFS, Division of Fire Control, 60–A–931, Box 73.

103. For a summary of this research, see the citations in note 42 above, especially the bibliography given in Cole and Kaufman, "Socio-Economic Factors and Forest Fires," p. 1n.

104. See *AF*, 45 (April 1939), esp. 173, 200. For more on Flagg and his fire paintings, see Susan E. Meyer, *James Montgomery Flagg* (New York: Watson-Guptill, 1974); and "Association Launches Fire Prevention Campaign," pp. 266–267, 286.

105. Ovid Butler, "For a Public Will," *AF*, 45 (April 1939), 173.

106. For a sample of wartime fire protection concerns, see Secretary of Agriculture Claude R. Wickard, "All Forest Fires Are Enemy Fires," address broadcast July 25, 1942, text in History Office, U.S. Forest Service; and "Fires Delay Victory!" *AF*, 49 (August 1943), 14. A summary of the abortive involvement of the Red Cross in fire prevention can be found in "American National Red Cross Forest and Range Fire Prevention, Summary—1946 Program," WNRC, RFS, Cooperative Fire Program.

107. See Arthur W. Prilaux, "The Story of Keep Oregon Green," *JF*, 48 (February 1950), 87–91.

108. Norma Ryland Graves, "The Little Green Guards," *AF*, 48 (November 1942), 506–508; Prilaux, "The Story of Keep Oregon Green," pp. 89–90.

109. See Wickard, "All Forest Fires Are Enemy Fires." For the story of the

CFFP, see Mal Hardy, "Smokey Bear—A Biography," Cooperative Forest Fire Prevention Program Files; and Ellen Earnhardt Marrison, *Guardian of the Forest: A History of the Smokey Bear Program* (New York: Vantage Press, 1976). The CFFP files include a complete collection of visuals and graphics developed since the program's beginning. Perhaps the best published source for the program is Clint Davis, "Smokey Is Convincing a Nation: Only You Can Prevent Forest Fires," *AF,* 57 (April 1951), 6–10, 40–41.

110. Hardy, "Smokey Bear," p. 7.

111. "Truman Calls Fire Prevention Conference," *AF,* 52 (May 1947). For the SCFFP, see "Southern CFFP Program," U.S. Forest Service, Southern Region Files, Atlanta; W. T. Ahearn, "Progress and Objectives of the Southern Cooperative Forest Fire Prevention Project," U.S. Forest Service, Southern Region Files, Atlanta.

112. For Skyfire information, see Chapter 1, note 19. The scheme was transported to Alaska as a part of Project MOD, which was aimed at both lightning suppression and rain induction; see BLM, Division of Fire Control, "Annual Reports, 1969–1973" (Anchorage, 1969–1973).

113. William W. Herman et al., "Man-Caused Forest Fire Prevention Research" (University of Southern California, 1961); U.S. Forest Service et al., "Wildfire Prevention Analysis: Problems and Progress" (U.S. Forest Service, 1975). In 1955 the Pacific Southwest Station of the U.S. Forest Service established a fire prevention unit much like that at the Southern Station, and reports developed along similar lines; see, for example, John R. Christiansen and William S. Folkman, "Characteristics of People Who Start Fires . . . Some Preliminary Findings," Research Note PSW–251 (U.S. Forest Service, 1971) and William S. Folkman, "Fire Prevention in Butte County, California . . . Evaluation of an Experimental Program," Research Paper PSW–98 (U.S. Forest Service, 1973). Much of the latter coincided with a similar prevention research program by the California Division of Forestry. For overviews, see William W. Hermann, "A Psychologist Looks at Forest Fire Prevention," pp. 1–3, and Carl C. Wilson, "Our Fire Research Program and the Western Forest Fire Laboratory," pp. 3–6, in "Western Forest Fire Research, 1961," *Proceedings of the Annual Meeting of the Western Forest Fire Research Committee* (Portland: Western Forestry and Conservation Association, 1961).

114. Fire Prevention Working Team, National Wildfire Coordinating Group, "Multi-Agency Fire Prevention Cooperation. Report from Advanced Fire Prevention, February 1976–February 1977" (1977). The lesson plan for the course is filed at the National Advanced Resources Technology Center, Marana, Ariz. For an excellent overview, see James L. Murphy, "The National Interagency Wildfire Prevention Task Force," in "Global Forestry and the Western Role," *Proceedings of Permanent Association Committees* (Portland: Western Forestry and Conservation Association, 1975), pp. 30–33. The story of Smokey in Zaire is in "Smokey Bear Travels Far," *FCN,* 15 (July 1954), 29. By the mid-1960s Smokey enjoyed incredible popularity; see Haug Associates, Inc., "Public Image of and Attitudes Toward Smokey the Bear and Forest Fires" (Los Angeles, 1968).

4: THE GREAT BARBECUE

1. Quoted in Louis Barrett, "A Record of Forest and Field Fires in California From the Days of the Early Explorers to the Creation of the Forest Reserves" (U.S. Forest Service, 1935), p. 37.
2. Vernon Louis Parrington, *Main Currents in American Thought,* vol. 3 (New York: Harcourt, Brace, 1958), p. 23.
3. The literature on conservation reform is large. Good overviews can be found in Samuel P. Hays, *Conservation and the Gospel of Efficiency: The Progressive Conservation Movement, 1890–1920* (New York: Atheneum, 1969); Donald Worster, ed., *American Environmentalism: The Formative Period, 1860–1915* (New York: John Wiley, 1973); Roderick Nash, ed., *The American Environment: Readings in the History of Conservation* (Menlo Park, Calif.: Addison-Wesley, 1968); A. Hunter Dupree, *Science in the Federal Government* (New York: Harper and Row, 1957); Henry Clepper, *Origins of American Conservation* (New York: Ronald Press, 1966). For works dealing more specifically with forested lands, see Samuel T. Dana, *Forest and Range Policy: Its Development in the United States* (New York: McGraw-Hill, 1956); Henry Clepper, *Professional Forestry in the United States* (Baltimore: Johns Hopkins University Press, 1971) and *Crusade for Conservation: The Centennial History of the American Forestry Association* (Washington: American Forestry Association, 1975); John Ise, *The United States Forest Policy* (New Haven: Yale University Press, 1920); Gifford Pinchot, *Breaking New Ground* (1947; rpt. Seattle: University of Washington Press, 1972); Ralph Widner, ed, *Forests and Forestry in the American States: A Reference Anthology* (Missoula, Mont.: National Association of State Foresters, 1968).
4. Pinchot, *Breaking New Ground,* p. 29.
5. Ibid., p. 36.
6. "The Forest Service History Line," vol. 1, no. 1 (March 1971), p. 3.
7. Gifford Pinchot, *The Fight for Conservation* (1910; rpt. Seattle: University of Washington Press, 1967), p. 40.
8. This fact is so apparent to every historian of the conservation movement that it would not need restatement here except for the widespread ignorance of it among foresters in general and the avoidance of the precedent by forester-historians, who tend to accept Pinchot's analysis—as they do in nearly all other matters. To those citations given in note 3, add William H. Goetzmann, *Exploration and Empire: The Explorer and the Scientist in the Winning of the American West* (New York: Alfred A. Knopf, 1966), pp. 572–601; Wallace Stegner, *Beyond the Hundredth Meridian: John Wesley Powell and the Second Opening of the West* (Boston: Houghton Mifflin, 1954); Thomas G. Manning, *Government in Science: The U.S. Geological Survey, 1867–1894* (Lexington: University Press of Kentucky, 1967). Even in the new edition of Pinchot's autobiography, *Breaking New Ground,* an introduction by James Penick recognizes that Pinchot was a politician and publicist, not an originator.
9. A good chronology of the major events can be found in Dana, *Forest and Range Policy,* pp. 372–425.

10. See Manning, *Government in Science;* Goetzmann, *Exploration and Empire;* Mary Rabbitt, *Minerals, Lands, and Geology for the Common Defense and General Welfare,* vol. 1: *Before 1879* (Washington: GPO, 1978); John Wesley Powell, *Report on the Lands of the Arid Region of the United States* (2d ed., 1879; rpt. Cambridge, Mass.: Harvard University Press, 1962). For the significance of role models, see Fernow's relationship to Rossiter Raymond, in Andrew Denny Rodgers III, *Bernhard Eduard Fernow: A Story of North American Forestry* (New York: Hafner, 1968). For an interesting comparison of objectives and professional mores, see Clark Spence, *Mining Engineers and the American West* (New Haven: Yale University Press, 1970), and Henry S. Graves, *The Profession of Forestry,* U.S. Forest Service, Circular 207 (Washington: GPO, 1912).

11. Pinchot, *Breaking New Ground,* p. 325.

12. See Stewart Holbrook, *Burning an Empire: The Story of American Forest Fires* (New York: Macmillan, 1943), pp. 147–155, and Holbrook, "The Tragedy of Bandon," *AF,* 42 (November 1936), 494–497.

13. Clar, *California Government and Forestry,* 1:301–305.

14. Ibid., pp. 304–305; National Board of Fire Underwriters, "Report on the Berkeley, California Conflagration of September 17, 1923" (New York, 1923); Donald Matthews, "The North Oakland Hills Fire of September 22, 1970," report, Oakland Fire Department; U.S. Congress, House, Subcommittee on Forests, *Predisaster Assistance for Eucalyptus Tree Fire Hazard, Hearing on H. R. 7545, H. R. 7669, H. R. 7664,* 93rd Cong., 1st sess., May 29, 1973 (Washington: GPO, 1973), esp. pp. 19, 50–52.

15. Robert Marshall, "Mountain Ablaze," *Nature,* June–July 1953, Forest Service reprint, p. 4.

16. For some data, see Bureau of the Census, *Historical Statistics of the United States* (Washington: GPO, 1975). See also John C. Fisher, *Energy Crises in Perspective* (New York: John Wiley, 1974), pp. 11–17. To these assessments should be added the charcoal industry, which by the late 19th century was extensive in many regions and which offered a product in some respects intermediary between wood and coal.

17. Clepper, *Professional Forestry,* p. 270.

18. William Greeley, "Foreword," in Holbrook, *Burning an Empire,* pp. vii–ix.

19. Ibid., p. ix.

20. Quoted in Elwood R. Maunder, ed., *Voices from the South: Recollections of Four Foresters* (Santa Cruz, Calif.: Forest History Society, 1977), pp. 17–18. Eldredge also observed that "it was the rarest thing in the world at that time for anybody to leave the Service for private employment. There was a real spirit, a deep-down-in-you feeling."

21. Elers Koch, "Region One in the Pre-Regional Office Days," in *Early Days in the Forest Service,* 4 vols. (Missoula, Mont.: U.S. Forest Service, 1944-1976), 1:102.

22. Rodgers, *Fernow,* p. 29. Fernow did his part to keep the issue foremost, however; see, for example, Fernow, *Is Protection Against Forest Fires Practicable?,* Department of Agriculture, Division of Forestry, Bulletin no. 14 (Washington: GPO, 1897). For antecedents to Fernow's sentiment, see Charles Sargent, "Report on the Forests of North America," in *10th Census of the United States*

(Washington: GPO, 1884); Franklin Hough, *Report on Forestry*, vol. 3 (Washington: GPO, 1882).

23. John Muir, "The American Forests," *Atlantic Monthly*, 80 (August 1897), 154. The 10 to 1 ratio was not original with Muir; most accounts of forest destruction in the 19th century also use it.

24. Powell, *Report on the Lands of the Arid Region of the United States*, pp. 25–29; William Culp Darrah, *Powell of the Colorado* (Princeton: Princeton University Press, 1951), pp. 132–133, 231–232.

25. Darrah, *Powell*, pp. 312–313.

26. National Academy of Sciences, *Report of the Committee Appointed by the National Academy of Sciences Upon the Inauguration of a Forest Policy for the Forested Lands of the United States to the Secretary of the Interior* (Washington: GPO, 1897), pp. 5, 17–18.

27. Ibid., pp. 12–13, 23. Pinchot dismissed the committee with some contempt as being practically worthless, but few would agree with that judgment. The report of the committee, moreover, crystallized many of the ideas then prevalent about forestry and fire and is invaluable as a historical document. But then, Pinchot was skeptical about written documents, too.

28. Henry Gannett, ed., *Twentieth Annual Report of the United States Geological Survey*, part 5: *Forest Reserves* (Washington: GPO, 1900), p. 50. See also Henry S. Graves, "Black Hills Forest Reserve," in ibid., esp. pp. 83–86.

29. F. E. Town, "Bighorn Forest Reserve," in Gannett, ed., *Twentieth Annual Report*, p. 179. Elsewhere Town observed that "fire has been and is the greatest enemy of this forest. There are large areas burned over and destitute of timber."

30. John B. Leiberg, "Priest River Forest Reserve," in Gannett, ed., *Twentieth Annual Report*, p. 233.

31. Ibid., p. 235.

32. Pinchot, *Breaking New Ground*, p. 144.

33. Gifford Pinchot, *A Primer of Forestry, Part I: The Forest* (Washington: GPO, 1899).

34. A synopsis of the fire entries of the *Use Book* is contained in Pinchot, *Breaking New Ground*, pp. 276–279. Quoted instructions are from p. 266.

35. Pinchot, *Fight for Conservation*, pp. 44–46.

36. William James, *The Varieties of Religious Experience* (New York: Modern Library, 1902), p. 3.

37. John R. Curry and Wallace Fons, "Forest-Fire Behavior Studies," *Mechanical Engineering*, 62 (March 1940), 219. See also H. T. Gisborne, "Forest Pyrology," *Scientific Monthly*, 49 (July 1939), 21–30.

38. H. H. Chapman, "Forest Management," in Robert K. Winters, ed., *Fifty Years of Forestry in the U.S.A.* (Washington: Society of American Foresters, 1950), p. 80.

39. Clepper, *Professional Forestry*, p. 270.

40. Quoted in Ralph Hidy et al., *Timber and Men: The Weyerhaeuser Story* (New York: Macmillan, 1963), p. 382.

41. See Harold K. Steen, *The U.S. Forest Service: A History* (Seattle: University of Washington Press, 1976), pp. 187–188; Earl Pierce and William Stahl, "Cooperative Forest Fire Control: A History of Its Origin and Development

Under the Weeks and Clarke-McNary Acts" (U.S. Forest Service, 1964), pp. 42–47. Greeley is quoted in Clar, *California Government and Forestry,* 1:559.

42. Pinchot, *Breaking New Ground,* p. 285.

43. Quoted in Robert W. Wells, *Fire at Peshtigo* (Englewood Cliffs, N.J.: Prentice-Hall, 1968), p. 42.

44. For early fire histories of the region, see "Fire! Fire Control and Forest Protection in Wisconsin" (Madison: Department of Natural Resources, 1970); J. A. Mitchell, "Forest Fires in Minnesota" (Forest Service, State of Minnesota, 1927); J. A. Mitchell and D. Robson, "Forest Fires and Forest Fire Control in Michigan" (Michigan Department of Conservation, 1950); J. A. Mitchell and Neil Lemay, "Forest Fires and Fire Control in Wisconsin" (U.S. Forest Service, n.d.); Holbrook, *Burning an Empire,* which includes histories of many of the worst fires; and Widner, ed., *Forest and Forestry,* pp. 492–515.

45. Among the original accounts, most were published locally or under subscription, and there was often considerable repetition and plagiarism. The most useful volumes include C. D. Robinson, "Account of the Great Peshtigo Fire of 1871," in Hough, *Report on Forestry,* pp. 231–242; Frank Tilton, *Sketch of the Great Fires in Wisconsin* . . . (Green Bay: Robinson and Kustermann, 1871); Frank Luzerne, *The Lost City! Drama of the Fire-Fiend!* . . . (New York: Wells and Co., 1872); Rev. William Wilkinson, *Memorials of the Minnesota Forest Fires in the Year 1894* . . . (Minneapolis: Norman E. Wilkinson, 1895); Lucy Kelsey, *The September Holocaust, A Record of the Great Forest Fire of 1894, by One of the Survivors* (Minneapolis: A. Roper, 1894); Detroit *Post* reports, 1881 fires in Michigan, in vol. 4, no. 266, Files of Michigan Department of Natural Resources; Rev. Peter Pernin, "The Great Peshtigo Fire," *Wisconsin Magazine of History,* 54 (Summer 1971), 246–272; Sgt. William O. Bailey, "The Michigan Fires of 1881," in *Signal Service Notes No. 1* (Washington:. GPO, 1882); Elton T. Brown, *The History of the Great Minnesota Forest Fires* (St. Paul: Brown Brothers Publishers, 1894). Many later histories rely on primary accounts other than the above; see Holbrook, *Burning an Empire,* pp. 13–30, 46–53, 61–107; Wells, *Fire at Peshtigo;* Antone Anderson, *The Hinckley Fire* (New York: Comet Press Books, 1954); Gerald Schultz, *Walls of Flames* (privately published, n.d.). After the turn of the century, accounts of the large fires reside in official reports and in the copy of reporters, both of which are after the fact and of less use in reconstructing the behavior of the fire as distinct from its damages. See, for example, E. G. Cheyney, "A Holocaust in Minnesota," *AF,* 24 (November 1918), 643–647; J. F. Hayden, "The Great Minnesota Fire," *AF,* 24 (November 1918); and C. C. Andrews, *Sixteenth Annual Report of the Forestry Commissioner of Minnesota for the Year 1910* (St Paul: Pioneer Co., 1911). See also citations in note 44. For a modern analysis of the meteorological conditions surrounding the fires, see Donald Haines and Earl Kuehnast, "When the Midwest Burned," *Weatherwise,* 2 (June 1970), 112–119; Donald A. Haines and Rodney W. Sando, "Climatic Conditions Preceding Historically Great Fires in the North Central Region" (U.S. Forest Service, 1971), which includes a map of the major burns; and Donald A. Haines et al., "Wildfire Atlas of the Northeastern and North Central States," General Technical Report NC–16 (U.S. Forest Service, 1975).

46. Stephen H. Spurr, "The Forests of Itasca in the Nineteenth Century as

Related to Fire," *Ecology,* 35 (January 1954), 21–25. Something of a resurvey was published by Sidney S. Frissell, "The Importance of Fire as a Natural Ecological Factor in Itasca State Park, Minnesota," in H. E. Wright and Miron L. Heinselman, eds., "The Ecological Role of Fire in Natural Conifer Forests of Western and Northern America," *Quaternary Research,* 3 (October 1973), 397–407.

47. Holbrook, *Burning an Empire,* p. 31. For general information about fire in the region prior to the great holocausts, see C. E. Ahlgren, "Effects of Fire on Temperate Forests: North Central United States," in T. T. Kozlowski and C. E. Ahlgren, *Fire and Ecosystems* (New York: Academic Press, 1974), pp. 195–224; Stephen H. Spurr and Burton V. Barnes, *Forest Ecology,* 2d ed. (New York: Ronald Press, 1973), pp. 467–469; Miron L. Heinselman, "Fire in the Virgin Forests of the Boundary Waters Canoe Area, Minnesota," pp. 329–382, Albert M. Swain, "A History of Fire and Vegetation in Northeastern Minnesota as Recorded in Lake Sediments," pp. 383–396, and Sidney S. Frissell, "Importance of Fire as a Natural Ecological Factor in Itasca State Park, Minnesota," pp. 397–407, all in Wright and Heinselman, eds., "Ecological Role of Fire in Natural Conifer Forests"; and Richard J. Vogl, "Fire and the Northern Wisconsin Pine Barrens," pp. 175–209, C. E. Van Wagner, "Fire and Red Pine," pp. 211–219, and J. H. Cayford, "The Force of Fire in the Ecology and Silviculture of Jack Pine," pp. 221–244, all in *TTFECP,* no. 10.

48. Quoted in Spurr, "The Forests of Itasca," p. 21.

49. Schultz, *Walls of Flames,* p. 3.

50. Detroit *Post,* vol. 4, no. 266, p. 32, Files of Michigan Department of Natural Resources.

51. Quoted in ibid., p. 56.

52. Quoted in Hough, *Report on Forestry,* p. 191.

53. Rev. E. J. Goodspeed, *History of the Great Fires in Chicago and the West* (New York: H. S. Goodspeed and Co., 1871), p. 554.

54. C. D. Robinson summarized the October 1871 conflagration for the state legislative manual, and his report is reprinted in Hough, *Report on Forestry,* p. 231. This synopsis is probably the most accessible and concise of the many versions, and my description will follow Robinson's account where it can be generalized as typical of the era.

55. Report by Robinson, reprinted in Hough, *Report on Forestry,* pp. 231–236.

56. Ibid., p. 233.

57. Ibid., p. 234.

58. Ibid., p. 236.

59. Ibid., p. 232. Tilton, *The Great Fires,* p. 38 describes the showering of fire flakes by analogy to a snowstorm.

60. Report by Robinson, reprinted in Hough, *Report on Forestry,* p. 233.

61. There are several versions of Root's ride, though the fullest is the compendium of personal accounts assembled by Wilkinson, *Memorials of the Minnesota Forest Fires,* pp. 127–164. See also the simpler account in Holbrook, *Burning an Empire,* pp. 20–22, which also includes short versions of other railroad rescues from Hinckley, pp. 16–17. Other accounts of the railroad rescues can be found in Brown, *A History,* and Anderson et al., *The Hinckley Fire.* "Stony ground" quote from Brown, p. 43. As a curious sidelight, it seems that Root had been an

engineer for William Tecumseh Sherman during the traverse through Georgia, and one may surmise that he was not unfamiliar with fire and trains.

62. Hough, *Report on Forestry,* p. 235. Descriptions of similar programs for relief can be found in all the major accounts of the fires. An especially full version, albeit for a later fire, is contained in *Final Report of Minnesota Forest Fires Relief Commission* (Duluth, 1921).

63. Oliver P. Newman, "A Forest Fire That Cost Uncle Sam Fifteen Million Dollars," *AF,* 31 (June 1925), 323–326.

64. Detroit *Post,* vol. 4, no. 266, September 13, 1881, Files of Michigan Department of Natural Resources.

65. *Lake States Forest Fire Conference* (American Lumberman, 1911), p. 19.

66. See *Lake States Forest Fire Conference;* Widner, ed., *Forests and Forestry,* pp. 492–515; and other citations in note 44.

67. "Minnesota's Forest Fire Disaster" (editorial), *AF,* 24 (November 1918), 651. H. W. Richardson, "The Northeastern Minnesota Forest Fires of October 12, 1918," *Geographic Review,* 7, no. 1 (1919), 220–232, gives fuller treatment of the 1918 fires.

68. For an account of Cloquet's aftermath, see *Final Report of Minnesota Forest Fires Relief Commission,* and Holbrook, *Burning an Empire,* pp. 31–46.

69. For examples of this melancholy process, see John H. Dietrich, "Fire and Sand," *Wisconsin Conservation Bulletin,* 28 (July–August 1963), 18–19; Aldo Leopold, *A Sand County Almanac* (1949; rpt. New York: Ballantine, 1970), p. 160. For control over the cause, see D. J. Parker et al., *Peat-Bog Fires—Their Origin and Control,* U.S. Bureau of Mines, Information Circular (Washington: GPO, 1947).

70. Holbrook, *Burning an Empire,* pp. 40–41.

71. Gilbert I. Stewart, "Of Fires and Machines," *Michigan Conservation,* 24 (January–February 1955), 15–18; "The Michigan Forest Fire Experiment Station" (Michigan Department of Conservation and U.S. Forest Service, 1932) provides the best summary of the formative years; William G. Herbolsheimer, "Roscommon Equipment Center: A 20-State Approach to ED&T," *FM,* 36 (Fall 1975), 6–7. The Roscommon facility maintains a good record in its files, and many of its reports are held in the files of the Department of Natural Resources in Lansing.

72. Sgt. William Bailey, "The Michigan Fires of 1881." Schultz, *Walls of Flame,* pp. 65–73, describes the involvement by the Red Cross.

73. Michigan Department of Natural Resources, "Seney Fire" (Lansing, 1977); Roswell K. Miller, "The Keetch-Byram Drought Index and Three Fires in Upper Michigan, 1976," in *Fifth Joint Conference on Fire and Forest Meteorology* (Boston: American Meteorological Society, 1978), pp. 63–67; Von J. Johnson, "The Lake States 1976 Fire Season," in *Fire Management in the Northern Environment* (Anchorage: BLM, 1979), pp. 79–88.

74. For the rationale behind the fire, see Les Line, "The Bird Worth a Forest Fire," *Audubon,* November–December 1964, reprinted by the Michigan Audubon Society as "The Jack-Pine Warbler Story." A summary of more or less successful prescribed fires is given in Linda R. Donoghue and Von J. Johnson, "Prescribed Burning in the North Central States," Research Paper NC–111 (U.S. Forest Service, 1975).

75. Quoted in H. Duane Hampton, *How the U.S. Cavalry Saved Our National Parks* (Bloomington: University of Indiana Press, 1971), p. 72.

76. Quoted in Hough, *Report on Forestry,* p. 189.

77. Ibid., pp. 172, 171, 176, 179, 194, 144, 147.

78. Ibid., p. 179.

79. Ibid., pp. 184–185, 158.

80. Barrett "Record of Forest and Field Fires," passim.

81. Hough, *Report on Forestry,* pp. 197, 221, 157.

82. Barrett, "Record of Forest and Field Fires," p. 137.

83. Hough, *Report on Forestry,* p. 160.

84. Ibid., pp. 129–130.

85. Ibid., p. 183.

86. See *Early Days in the Forest Service,* 3:213–216, 4:60–63.

87. National Academy of Sciences, *Report of the Committee,* p. 19.

88. Clar, *California Government and Forestry,* 1:203n.

89. The significance of the Canadian example can be found in National Academy of Sciences, *Report of the Committee,* pp. 13–16; Hough, *Report on Forestry,* pp. 215–219; and Fernow, "Is Protection Against Forest Fires Practicable?"

90. See Fernow, "Is Protection Against Forest Fires Practicable?" pp. 1–3.

91. Dana, *Forest and Range Policy,* p. 6; Hough, *Report on Forestry,* pp. 147–149.

92. See Chapter 1, note 80, for appropriate references. See also Dana, *Forest and Range Policy,* pp. 92–94.

93. H. M. Suter, *Forest Fires in the Adirondacks in 1903,* U.S. Bureau of Forestry, Circular no. 26 (Washington: GPO, 1904), pp. 10, 12. For other accounts of the fires, see Chapter 1, notes 80–81.

94. Hampton, *The Cavalry Saved Our Parks,* deals almost exclusively with Army administration. For other views, see Aubrey L. Haines, *The Yellowstone Story,* 2 vols. (Colorado Associated University Press, 1977); and National Academy of Sciences, *Report of the Committee,* pp. 18–19, 21.

95. Hampton, *The Cavalry Saved Our Parks,* p. 82.

96. Ibid., p. 83.

97. Haines, *Yellowstone Story,* 2:27.

98. Hampton, *The Cavalry Saved Our Parks,* p. 107.

99. Dana, *Forest and Range Policy,* p. 155. The memos are more fully reproduced in William S. Brown, "History of Los Padres National Forest" (U.S. Forest Service, 1945), pp. 81–82.

100. Quoted in Dana, *Forest and Range Policy,* p. 111.

101. Ibid., p. 113. For a good analysis of the situation, see Edward T. Allen, "The Application and Possibilities of the Federal Forest Reserve Policy," SAF, *Proceedings,* 1 (November 1905), 41–52.

102. Memorandum of J. H. Fimple, acting commissioner, General Land Office, June 2, 1903, Archives, Rogue River National Forest, Oregon.

103. William T. Cox, "Recent Forest Fires in Oregon and Washington," *Forestry and Irrigation,* November 1902, p. 466.

104. Edwin A. Tucker and George Fitzpatrick, *Men Who Matched the Mountains: The Forest Service in the Southwest* (Washington: GPO, 1972), pp. 49–50.

105. Dana, *Forest and Range Policy,* p. 113.

106. Ibid., p. 112.

107. Tucker and Fitzpatrick, *Men Who Matched the Mountains,* p. 52.

108. Roy A. Phillips, "Recollections," in *Early Days in the Forest Service,* 2:21.

109. For an excellent summary of the protective organizations in the Northwest, see George T. Morgan, "The Fight Against Fire: The Development of Cooperative Forestry in the Pacific Northwest, 1900–1950" (Ph.D. diss., University of Oregon, 1964), pp. 70–75; see also, Eloise Hamilton, *Forty Years of Western Forestry* (Portland: Western Forestry and Conservation Association, 1949); Widner, ed., *Forests and Forestry,* pp. 253–270, 441–446; "Timberland Protective Associations," *AF,* 17 (1911), 660–670; E. T. Allen, "What Protective Cooperation Did," *AF,* 16 (November 1910), 641–643. Among the earliest of the organizations was the Pocono Protective Association in Pennslyvania, 1902.

110. Hamilton, *Forty Years,* p. 3.

111. Contra Costa Hills Fire Protective Association, records in Bancroft Library, Berkeley. See Clar, *California Government and Forestry,* vol. 1, passim, for many similar organizations.

112. Pinchot, *Breaking New Ground,* p. 23.

113. See Steen, *Forest Service,* pp. 54–55; and citations in Chapter 2, note 81.

114. *The Use of the National Forests,* rev. ed. (Washington: GPO, 1907), p. 31. The *Manual* had stipulated that "Your principle duty will be regular patrol service, which consists of riding through the reserve to protect it from fire and trespass . . ."; quoted in Tucker and Fitzpatrick, *Men Who Matched the Mountains,* p. 9.

115. Clarence B. Swim, in *Early Days in the Forest Service,* 1:198.

116. Robert L. Hess, in ibid., p. 97.

117. R. L. Woesner, "U.S. Forest Service, 1909–1920," in ibid., pp. 209–210.

118. Ibid., p. 210.

119. Pinchot, *Fight for Conservation,* p. 15.

120. William James, "The Moral Equivalent of War," in *The Writings of William James,* ed. John J. McDermott (New York: Modern Library, 1967), pp. 663, 668–669. For similar treatments of the strenuous life during this era, see Edward White, *The Eastern Establishment and the Western Experience: The West of Frederic Remington, Theodore Roosevelt, and Owen Wister* (New Haven: Yale University Press, 1968), and George Frederickson, *The Inner Civil War* (New York: Harper and Row, 1965).

5: THE HEROIC AGE

1. Theodore Roosevelt, "The Strenuous Life," in *The Works of Theodore Roosevelt* (New York: Charles Scribner's Sons, 1923–1926), 13:20–21.

2. William James, "On the Moral Equivalent of War," in *The Writings of William James,* ed. John J. McDermott (New York: Modern Library, 1967), pp. 669–670.

3. Elers Koch, "History of the 1910 Forest Fires in Idaho and and Western Montana," in Koch, ed., "When the Mountains Roared" (Coeur d'Alene National Forest, 1942), p. 1.

4. For interesting insights into this obsession, see Roderick Nash, *Wilderness and the American Mind* (New Haven: Yale University Press, 1967), chap. 9. Also, consult citations in Chapter 4, note 119, above.

5. Elers Koch, "The Lochsa River Fire," in *Early Days in the Forest Service,* 4 vols. (Missoula, Mont.: U.S. Forest Service, 1944–1976), 1:114.

6. "A Monument to Bravery," *AF,* 29 (August 1923), 486.

7. Pinchot is quoted in Roy Headley, "Beating Fire," *AF,* 36 (July 1930), 449. Clyde Leavitt, "Forest Fires," in *Report of the National Commission* (New York: Arno Press, 1972), 2:438, 446.

8. There are many accounts of the 1910 fires. Archival material exists in "Claims for Damages, Fire Records, Region 1, 1910," NA, RG 95, RFS, Division of Fire Control, Claims, General, and in Files, Region One, U.S. Forest Service, Missoula, Mont. See also the report on *Interment of Bodies of Employees Killed in Fighting Fires,* House Document 1271, 61st Cong. 2d sess. (Washington: GPO, 1912). Among published accounts, see Koch, ed., "When the Mountains Roared"; Betty G. Spencer, *The Big Blowup* (Caldwell, Idaho: Caxton Printers, 1956); Stan Cohen and Don Miller, *The Big Burn* (Missoula; Mont.: Pictorial Histories Publishing Co., 1978); Stewart Holbrook, *Burning an Empire: The Story of American Forest Fires* (New York: Macmillan, 1943), chap. 11; Ruby El Hult, *Northwest Disaster* (Portland: Binsfords and More, 1960); *Early Days in the Forest Service,* passim; *American Forestry,* 16 (November 1910), entire issue, especially Ferdinand Silcox, "How the Fires Were Fought"; "Excerpts Concerning the 1910 Fires as Taken From a Congressional Report of July 15, 1911," Files, Region One, U.S. Forest Service. The Koch edition incorporates material on the fires gathered as early as 1926 by Fred Morrell and adds other material. The chief participants wrote accounts of their actions; apart from those included in compendiums already mentioned, see William Greeley, *Forests and Men* (New York: Arno Press, 1972), chap. 1; Edward C. Pulaski, "Surrounded by Forest Fires," *AF,* 29 (August 1923), 485–486; and Joe Halm (who published a seemingly endless number of versions), "The Great Fire of 1910," *AF,* 36 (July 1930), 424–428, 479–480.

9. Sharlot M. Hall, *Sharlot Hall on the Arizona Strip,* ed. C. Gregory Crampton (Flagstaff: Northland Press, 1978), p. 69.

10. Quoted in Spencer, *Big Blowup,* p. 156.

11. James Wilson, "Protecting Our Forests From Fire," *National Geographic,* 22 (January 1911), 99.

12. Lt. Mapes to Maj. Logan, n.d., and Logan to C. S. Ucker, Chief Clerk, Interior Department, September 4, 1910, NA, RG 79, Records of the National Park Service, Glacier National Park.

13. Hal E. Anderson, "Sundance Fire: An Analysis of Fire Phenomena," Research Paper INT–56 (U.S. Forest Service, 1968), and Arnold I. Finklin, "Meteorological Factors in the Sundance Fire Run," General Technical Report INT–6 (U.S. Forest Service, 1973). Popularized accounts are in Stuart E. Jones and Jay Johnston, "Forest Fire: The Devil's Picnic," *National Geographic,* 146 (July 1968), 100–127; Clifford Lylie, "Black Friday—Idaho Style (Part I)," *AF,* 74 (June 1968), 14–16, and "Black Friday—Part II," *AF,* 74 (July 1968), 20–22; George S. Gorsuch, "A Firefighter Views the Sundance Fire," *AF,* 75 (March 1969), 4–5, 52.

14. The Pulaski story is commonly recounted in nearly all the histories of the 1910 fire (see note 8). For Pulaski's own version, see "Surrounded by Forest Fires." The quotations are from Koch, "History of the 1910 Forest Fires," p. 6, and Pulaski, "Surrounded by Forest Fires," p. 485.
15. Koch, "History of the 1910 Forest Fires," p. 27; P. D. Hanson, foreword to Spencer, *Big Blowup,* p. 7. See also John James Little, "The 1910 Forest Fires in Montana and Idaho: The Impact on Federal and State Legislation" (M.A. thesis, University of Montana, 1968), for an assessment of the political impact of the fires.
16. "Claims for Damages, Fire Records, Region 1, 1910," NA, RG 95, RFS, Division of Fire Control, Claims, General. Appropriately enough, these records begin the official file in fire protection.
17. See assorted letters in Files, Region One, U.S. Forest Service, for the sad story of the graves. In particular, consult L. C. Stockdale, "SPECIAL. St. Maries Burial Plot. January 27, 1933"; Edward C. Pulaski, "SUPERVISION—Coeur d'Alene. 1910 Graves. April 30, 1917"; Fred Morrell, "FINANCE—Coeur d'Alene. January 3, 1921."
18. Quoted in Koch, "History of the 1910 Forest Fires," p. 39.
19. Pulaski, "Surrounded by Forest Fires," p. 486.
20. Greeley, *Forests and Men,* p. 24.
21. Joseph Halm, "The Big Fire," in Koch, ed., "When the Mountains Roared," p. 31.
22. Spencer, *Big Blowup,* p. 155.
23. For lightning and early fire history, see Jack S. Barrows et al., "Lightning Fires in Northern Rocky Mountain Forests" (U.S. Forest Service, 1977); Stephen F. Arno, "Forest Fire History in the Northern Rockies," *JF,* 78 (August 1980), 460–465, and "The Historical Role of Fire in the Bitterroot National Forest," Research Paper INT–87 (U.S. Forest Service, 1976); James R. Habeck and Robert Mutch, "Fire-Dependent Forests in the Northern Rocky Mountains," in H. E. Wright and Miron L. Heinselman, "The Ecological Role of Fire in Natural Conifer Forests of Western and Northern America," *Quaternary Research,* 3 (October 1973), 408–424; James R. Habeck, *Fire Ecology Investigations in Selway-Bitterroot Wilderness: Historical Considerations and Current Observation,* Publication No. R1–72–001 (Missoula, Mont.: U.S. Forest Service, 1972); R. Daubenmire and J. B. Daubenmire, "Forest Vegetation of Eastern Washington and Northern Idaho," Washington Agricultural Experiment Station, Technical Bulletin 60 (Spokane, 1968); Charles Wellner, "Fire History in the Northern Rocky Mountains," in *Role of Fire in the Intermountain West* (Missoula Mont.: Intermountain Fire Research Council, 1970), pp. 42–64; David F. Aldrich and Robert W. Mutch, "Ecological Interpretions of the White Cap Drainage: A Basis for Wilderness Fire Management," Progress Reports and Summary Statement (U.S. Forest Service, 1971–1972). Some useful material is contained in Harold Weaver, "Effects of Fire on Temperate Forests: Western United States," in T. T. Kozlowski and C. E. Ahlgren, *Fire and Ecosystems* (New York: Academic Press, 1974), pp. 279–302; Martin E. Alexander and David V. Sandberg, *Fire Ecology and Historical Fire Occurrence in the Forest and Range Ecosystems of Colorado: A Bibliography* (Ft. Collins: Department of

Forest and Wood Sciences, Colorado State University, 1976); George E. How, "The Evolutionary Role of Wildfire in the Northern Rockies and Implications for Resource Managers," in *TTFECP*, no. 14, pp. 257–266; and Donald Despain, "Forest Fires in Yellowstone National Park," *JFH*, 10 (July 1974), 68–79.

24. F. V. Hayden, *Preliminary Report of the United States Geological Survey of Montana and Portions of Adjacent Territories* (Washington: GPO, 1872), p. 99.

25. F. E. Town, "Bighorn Forest Reserve," in Henry Gannett, ed., *Twentieth Annual Report of the United States Geological Survey, pt. 5: Forest Reserves* (Washington: GPO, 1900), p. 178.

26. John B. Leiberg, "Bitterroot Forest Reserve," in Gannett, ed., *Twentieth Annual Report*, p. 275, and "Bitterroot Forest Reserve," in *Nineteenth Annual Report of the United States Geological Survey, pt. 5: Forest Reserves* (Washington: GPO, 1899), p. 388. Consider, also: "Forest fires, ancient and modern, have everywhere devastated the basin. None of the zones have escaped."

27. H. B. Ayres, "The Flathead Forest Reserve," in *Nineteenth Annual Report*, p. 283.

28. Henry Graves, "Black Hills Forest Reserve," in Gannett, ed., *Twentieth Annual Report*, p. 81.

29. From the Missoula *Daily Gazette*, August 16, 1889: "From the northernmost boundary of British Columbia to the Mexican line, the broken mountain range is all ablaze . . . the latter part of the dry season upon the Pacific Slope is always marked by forest fires more or less extensive. The entire absence of rain over long periods presents a predisposing condition. This year the destruction is unparalleled."

30. See, for example, Graves, "Blacks Hills Forest Reserve," p. 85; George B. Sudworth, "White River Plateau Timberland Reserve," in *Nineteenth Annual Report*, p. 150. The whole panorama of forest reserve reports by the USGS is informative in this regard.

31. Theodore Shoemaker, "Memories," in *Early Days in the Forest Service*, 2:42–43. A good survey of evolving techniques is contained in Theodore Shoemaker, "Fighting Forest Fires—Then and Now," in ibid., pp. 32–41. The four volumes of the *Early Days* collection provide a useful, if anecdotal, source.

32. See "Report of Fire Conditions. District No. 1. Seasons of 1919," Files, Region One, U.S. Forest Service, Missoula, Mont., and NA, RG 95, RFS, Division of Fire Control, for reports and findings of the board of review. Another useful assessment from this period comes from "Forest Fires in Idaho: A Brief Survey of the Problems and Remedies" (Boise: Idaho State Cooperative Board of Forestry, 1926), a result of the Clarke-McNary survey. Also relevant are Wellner, "Fire History in the Northern Rocky Mountains," and Region One, *Annual Narrative Fire Report, Calendar Year 1967* (Missoula, Mont.: U.S. Forest Service, 1967), pp. 14, 16, 19, which contains convenient graphs of fire starts and burned acreages.

33. See "Board of Review Report. Selway Fires—1934," "A History of the Pete King, McLendon Butte and Eighteen Other Selway Forest Fires. August and September, 1934," "Fire Weather," "Elapsed Time Record," and other items relevant to the fires, in Files, Region One, U.S. Forest Service. See also C. B.

Sutliff, "Selway Forest Fires of 1934," in *Early Days in the Forest Service,* 4:73–87. The fires enjoyed prominence in the debate about low-value backcountry lands, which began in the Northern Rockies in the early 1930s, and the board of review, in turn, led to another round of policy criticism. See Chapter 5, "The Forester's Policy."

34. A good chronology of fire seasons can be found in *History of Smokejumping* (Missoula Mont.: U.S. Forest Service, 1976). The 1961 fire season is described in Ernest R. DeSilvia, "Forest Fires in the Northern Rocky Mountains in 1961," in SAF, *Proceedings* (1962), pp. 54–55. The other developments will be analyzed in the sections of this book dealing with particular themes, such as equipment development.

35. Accounts of the season are available in Northern Region, "Fire Emergency 67" (U.S. Forest Service, 1968); Region One, *Annual Narrative Fire Report, Calendar Year 1967;* John W. Chaffin, James W. Jay, and William R. Moore, "Fire in the Northern Rockies, 1967: Some Lessons for Wartime Fire Defense" (U.S. Forest Service, 1967); Jones and Johnston, "Forest Fire: The Devil's Picnic." For the Sundance fire in particular, see Anderson, "Sundance Fire," and Finklin, "Meteorological Factors." One might also note that both the Sundance and Trapper Peak fires began on lands under the jurisdiction of the Priest Lake Timber Protective Association. After the fire blew up for 2,000 acres, the association found its funds exhausted, but only on the morning of September 1 did the association request its cooperator, the Forest Service, to assume responsibility for the fire. Four hours later the fire made its major run. Had the transfer been effected earlier, even the Sundance burn might have been vastly reduced.

36. Arthur L. Roe et al., "Fire and Forestry in the Northern Rocky Mountains: A Task Force Report," *JF,* 69 (1971); Robert W. Mutch, "'I Thought Forest Fires Were Black!'" *Western Wildlands,* 4 (Summer 1977), 68–74; and Orville L. Daniels, "Test of a New Land Management Concept: Fritz Creek 1973," *Western Wildlands,* 4 (Summer 1977). See also John F. Chapman, "The Teton Wilderness Fire Plan," pp. 11–19, and Don G. Despain and Robert E. Sellers, "Natural Fire in Yellowstone National Park," pp. 20–24, in the same issue of *Western Wildlands.*

37. Gifford Pinchot, "Study of Forest Fires and Wood Protection in Southern New Jersey," *Annual Report of Geological Survey of New Jersey* (Trenton, 1898), appendix, p. 11.

38. Henry Graves, *Report of the Forester for 1913* (Washington: GPO, 1914), p. 16.

39. R. M. Conarro, "Fire Effects and Use," in "Proceedings of the Priest River Fire Meeting" (U.S. Forest Service, 1941), p. 57. The general history of the Forest Service can be found in Harold K. Steen, *The U.S. Forest Service: A History* (Seattle: University of Washington Press, 1976); Michael Frome, *The Forest Service* (New York: Praeger, 1971) and *Whose Woods These Are: The Story of the National Forests* (Garden City, N.Y.: Doubleday, 1962); Samuel T. Dana, *Forest and Range Policy: Its Development in the United States* (New York: McGraw-Hill, 1956); Henry Clepper, *Professional Forestry in the United States* (Baltimore: Johns Hopkins University Press, 1971); and A. A. Brown and Kenneth Davis, *Forest Fire: Control and Use,* 2d ed. (New York: McGraw-Hill,

1973), which contains a wealth of historical information. See also annual reports of the Chief Forester, the proceedings of innumerable fire conferences, and the archival sources mentioned in the bibliographic essay. A pictorial history is available in *100 Years of Federal Forestry*, Agriculture Information Bulletin no. 402 (Washington: GPO, 1976).

40. B. M. Huey, "The First U.S. Forest Ranger," *JF*, 45 (October 1947), 765.
41. Edwin A. Tucker and George Fitzpatrick, *Men Who Matched the Mountains: The Forest Service in the Southwest* (Washington: GPO, 1972), p. 43. For similar stories, consult *Early Days in the Forest Service.*
42. *New York Times*, August 26, 1910, p. 4.
43. E. Deckert, "Forest Fires in North America: A German View," *AF*, 17 (May 1911), 275, 279.
44. Coert duBois, *Systematic Fire Protection in the California Forests* (Washington: GPO, 1914), p. 3.
45. See *The Principal Laws Relating to Forest Service Activities*, Agriculture Handbook no. 453 (Washington: GPO, 1974), p. 72; James Wilson, "Protecting Our Forests from Fire," p. 99; Henry Graves, *Report of the Forester, 1912*, p. 46.
46. Roy Headley, "Plans, General, 1941," p. 51, WNRC, RFS, Division of Fire Control.
47. I. F. Eldredge, "Fire Problem on the Florida National Forest," SAF, *Proceedings*, 6, no. 2 (1911), 166–170. See also Graves, *Annual Report of the Forester*, for the years 1911–1914 for a description of the breadth of planning; Barrington Moore, "The Essentials in Working Plans for National Forests," pp. 117–123, and William Greeley, "Better Methods of Fire Control," pp. 153–165, SAF, *Proceedings*, 6, no. 2 (1911); John D. Guthrie, "Standardization of Fire Plans, Organization, Equipment, and Methods in District 3," *Forestry Quarterly*, 12 (1914), 381–389, describes the reforms in the Southwest; H. E. Woolley, "What Has Been Accomplished on the National Forests," *AF*, 19 (November 1913), 760–768, gives a national overview.
48. Daniel Adams, *Methods and Apparatus for the Prevention and Control of Forest Fires, as Exemplified on the Arkansas National Forest*, U.S. Forest Service, Bulletin 113 (Washington: GPO, 1912). A forgotten masterpiece, the work has been lost in favor of the solutions proposed for the more vigorous and politically sensitive fire regimes in the South and in California.
49. Coert duBois, *Trailblazers* (Stonington, Conn., 1957), pp. x, 61, 76–77, 79. See duBois, *National Forest Fire-Protection Plans* (Washington: GPO, 1911) and "Organization of Forest Fire Control Forces," SAF, *Proceedings*, 9 (October 1914), 512–521, for early versions of *Systematic Fire Protection.* Compare this course of development with S. B. Show and E. I. Kotok, History of Fire Research in California (unpublished report, title page missing, U.S. Forest Service, ca. 1956), WNRC, RFS, for a different perspective on the history of planning, and with Roy Headley, "Rethinking Forest Fire Control" (U.S. Forest Service, 1943), for a slightly different analysis by a participant.
50. See Steen, *Forest Service;* Graves, *Annual Reports of the Forester*, 1915–1916; Greeley and Headley, letters from 1915, NA, RG 95, RFS, Division of Fire Control, "Supv. (1913–1922)."

51. Graves, "Standard Instructions for the Determination of Fire Damages," February 6, 1914 (U.S. Forest Service, 1914).

52. Graves, Memorandum, "Standard Instructions for Determination of Fire Damages," May 8, 1916, NA, RG 95, RFS, Division of Fire Control; Roy Headley, "Fire Suppression Manual," Region 6 (U.S. Forest Service, 1916); Ferdinand Silcox, "The Forest Fire Problem in District 1," *Journal of Agriculture*, 4, no. 3 (1916), 100–101; H. T. Gisborne, "Review of Problems and Accomplishments in Fire Control and Fire Research," in "Proceedings of the Priest River Fire Meeting," p. 10.

53. A good bibliography of the successive attempts to modify the theory can be found in Julie K. Gorte and Ross W. Gorte, "Application of Economic Techniques to Fire Management—A Status Review and Evaluation," General Technical Report INT–53 (U.S. Forest Service, 1979). I believe that this particular analysis shares the flaws of its precedessors in that it fails to recognize the forests as a political institution first and an economic device second and makes unreal assumptions about the ways in which fire behaves.

54. S. B. Show and E. I. Kotok, *Forest Fires in California, 1911–1920*, U.S. Department of Agriculture, Department Circular 243 (Washington: GPO, 1923), p. 7. The first 10 pages contain an excellent summary of developments in California during this period of experimentation. See also Show and Kotok, History of Forest Fire Research, pp. 11–12, 34–35. See also Paul Reddington to M. Jotter, Memorandum "Fire-Trinity," April 21, 1921, NA, RG 95, RFS, Division of Fire Control, Region 5, which describes how "common sense judgment" must be used by the local fire officer in deciding how to attack a fire.

55. "Committee Reports. Approved by the Fire Conference and Comments by the Forester. Conference Held at Mather Field, California, From November 14–27, 1921" (Mather Field Conference), NA, RG 95, RFS, Division of Fire Control, Fire Conferences, General.

56. Greeley, *Forests and Men*, pp. 24, 18. For additional information about Greeley and his administration, see Steen, *Forest Service*, pp. 173–195; and George T. Morgan, *William B. Greeley: A Practical Forester* (St. Paul: Forest History Society, 1961).

57. Mather Field Conference, p. 39. See also comments in NA, RG 95, RFS, Division of Fire Control, "Supv. Misc."

58. See *Principal Laws*, and Steen, *Forest Service*.

59. See Greeley, Memorandum, NA, RG 95, RFS, Division of Fire Control, "Supv. Meetings" and "Financial Conference, 1927."

60. H. T. Gisborne, "Forest Protection," in Robert K. Winters, ed., *Fifty Years of Forestry in the U.S.A.* (Washington: SAF, 1950), p. 35. Actually a de facto policy of control by 10 A.M. existed in the California and Pacific Northwest regions as well. See, for example, Charles Buck, memorandum, April 10, 1935, NA, RG 95, RFS, Division of Fire Control.

61. Earl W. Loveridge, "The Fire Suppression Policy of the U.S. Forest Service," *JF*, 42 (1944), 550–551. "Loose herding" was also tried in the Northern Rockies in the early years, but without conclusive success, and it was formally dismissed after hearings in 1932.

62. See Greeley, Memorandum, "General Plan for Protection of the National For-

ests from Fire," March 10, 1927, NA, RG 95, RFS, Division of Fire Control. This memo became the basis for the revised policy in the *Manual.*

63. Loveridge, "Fire Suppression Policy," p. 551.

64. "Committee Reports. Approved by the Washington Conference of District Foresters and Comments by the Forester" (Regional Foresters Conference, 1930) (U.S. Forest Service, 1930), p. 7.

65. Quoted in Regional Foresters Conference, 1930, p. 16. See also T. W. Norcross and R. F. Grefe, "Transportation Planning to Meet Hour Control Requirements," *JF,* 29 (1931), 1019–1033.

66. Regional Foresters Conference, 1930, p. 14.

67. Ibid., p. 50. For the development of the hour control program, see Show and Kotok, History of Forest Fire Research, pp. 26–27; Show and Kotok, *The Determination of Hour Control for Adequate Fire Protection in the Major Cover Types of the California Pine Region,* U.S. Department of Agriculture, Technical Bulletin no. 209 (Washington: GPO, 1930), which defined the practical meaning of the concept and served as a model for its application; and Show, "Recent Technical Advances in Forest Fire Control," *JF,* 29 (February 1931), 207–213, which describes the objectives of hour control in more popular form.

68. Regional Foresters Conference, 1930, p. 11.

69. References to the Forester's Advisory Committee on Fire Control can be found in NA, RG 95, RFS, Division of Fire Control, "Operations File," Accession 799, Drawer 259, Report of Fire Control Committee, and Shasta Meeting; "Fire Control Meeting. Spokane, Washington, February 10–21, 1936" (Spokane Fire Control Meeting, 1936) (U.S. Forest Service, 1936), especially topics 1 and 3; and Regional Foresters Conference, 1930, pp. 50–65. See also Show and Kotok, History of Forest Fire Research, passim.

70. For the backcountry debate, see NA, RG 95, RFS, Division of Fire Control, "Supv. Misc."; Loveridge, "Fire Suppression Policy"; Show and Kotok, History of Forest Fire Research, pp. 34–35. For the 1935 conference, see NA, RG 95, RFS, Division of Fire Control, Box 85. For the backcountry discussions, see "Backcountry Fire Policy," and E. W. Loveridge, "Low Value Area Policy—A Record from Discussions Held During the Low Value Area Expedition" (September 3–20, 1932), NA, RG 95, RFS, Division of Fire Control, Box 30.

71. Show and Kotok, History of Forest Fire Research, p. 35.

72. For the range of conferences held during this period, consult in particular Roy Headley, "Developments in Forest Fire Control," Report of Forest Committee, National Fire Protection Association, Annual Meeting, (rpt., U.S. Forest Service, 1937), and "New Devices and Ideas in Forest Fire Control," *JF,* 36 (February 1938), 136–138.

73. John A. Salmond, *The Civilian Conservation Corps, 1933–1942* (Durham: Duke University Press, 1967). For a fuller analysis of the CCC and fire control, see Chapter 6, "Under Fire: A History of Manpower in Fire Control."

74. Contrast Regional Foresters Conference, 1930, p. 50 (and pp. 50–65 passim), with Ferdinand Silcox, Circular to All Regions, May 25, 1935, NA, RG 95, RFS, Division of Fire Control, Box 110.

75. For the emergency presuppression account, see Silcox, Circulars to All Regions, May 7 and May 25, 1935, NA, RG 95, RFS, Division of Fire Control,

Box 110. For the quotation, see Spokane Fire Control Meeting, 1936, pt. 2, topic 4, p. 9.

76. Elers Koch, "The Passing of the Lolo Trail," manuscript in NA, RG 95, RFS, Division of Fire Control, Box 93, pp. 1–2. The entire debate about the article and its issues is contained in this archival unit. See, in particular, Earl W. Loveridge, "Is Back Country Fire Protection a 'Practical Impossibility?' " and Roy Headley, "Memorandum for Mr. Silcox, September 17, 1934," which summarize the debates that followed the board of review and include a list of the participants.

77. Koch, "Passing of the Lolo Trail," p. 3.

78. Koch to Silcox, November 20, 1934, NA, RG 95, RFS, Division of Fire Control, Box 93.

79. Loveridge, "Back Country Fire Protection," p. 5.

80. Headley, "Memorandum for Mr. Silcox," pp. 4, 7.

81. Ibid., pp. 7–10.

82. See Chapter 2, "Paiute Forestry: A History of the Light-Burning Controversy," and particularly citations in notes 93–103. See also *JF,* 33 (March 1935), for most of the proceedings of the meeting.

83. Loveridge, "Fire Suppression Policy," pp. 551–552. An earlier version, which preceded the 1935 conference, is the reply to Koch, "Back Country Fire Protection."

84. Silcox, Circulars of May 7 and May 25, 1935. For criticisms, see E. W. Loveridge to P. A. Thompson, Memorandum on Fire Control Policy, March 17, 1943, WNRC, RFS, Division of Fire Control. The policy was modified on March 10, 1943, to accommodate wartime realities, though it was to be "applicable for the war period only." For other summaries of criticism, see Spokane Fire Control Meeting, 1936, pt. 2, topics 4, 6; and "Proceedings of the Priest River Fire Meeting," especially papers by H. T. Gisborne, "Review of Problems and Accomplishments in Fire Control and Fire Research," Kenneth P. Davis, "An Economic Approach to Fire Control," and Gisborne, "Economics of Fire Control."

85. See the elaborate minutes and papers of the Spokane Fire Control Meeting, 1936.

86. John Curry, "Notes on a Theory for Forest Fire Protection Expenditures" (unpublished report, U.S. Forest Service, 1938), WNRC, RFS, Division of Fire Control.

87. Headley, "Rethinking Fire Control," p. 11.

88. Davis, "An Economic Approach to Fire Control," p. 27.

89. Gisborne, "Review of Problems and Accomplishments," pp. 16–17.

90. H. T. Gisborne, "Comparison of Intensive Versus Limited Forest Fire Control Action," Northern Rocky Mountain Forest and Range Experiment Station, Research Note no. 10 (September 1940).

91. See Jack Wilson, "History of NWCG," *FMN,* 39 (Summer 1978), 13–16. Temporary adjustments were also made in the 10 A.M. Policy; see Loveridge to Thompson, Memorandum on Fire Control Policy. A survey of fire protection as it existed at this time can be found in P. A. Thompson to Chiefs, Division of Fire Control, Regions 1–6, Memorandum on Fire Control Plans, January 8, 1944, WNRC, RFS, Division of Fire Control.

92. Watts to All Directors of Forest Experiment Stations, Memorandum, July 12, 1944, WNRC, RFS, Division of Fire Control. See also E. I. Kotok, to All Directors of Forest Experiment Stations, Memorandum, October 6, 1944, WNRC, RFS, Division of Fire Control. Some of the replies are included with the requests.

93. See "Act of September 21, 1944," in *Principal Laws*, p. 81.

94. For fuller treatments see Chapter 7, "The Red Menace: A History of Rural Fire Defense," and Chapter 8, "The Thermal Pulse: A History of Wildland Fire Research." Much of the material on rural fire defense is contained in WNRC, RFS, 63–A–4027, Box 48, and 70–A–6533, Box 49. A thumbnail sketch of Firestop is contained in Alva Neuns, "Operation Firestop," *AF*, 61 (January 1955), 8–12.

95. The feeling that forestry was being subordinated to fire control was a common one from the beginning, but it seems to have been given extra emphasis during this period. As an illustration, consider this passage by George Stewart, a professor of forestry at Berkeley, in his novel *Fire* (1949; rpt. New York: Ballantine, 1972):"'Here I am,' he thought, 'managing an estate for Uncle nearly as big as—well, bigger than Rhode Island anyway—with a hundred ideas of things to do to make it a better estate. Yet for four months a year I, and everybody else, drop silviculture, grazing improvement, erosion control, roads, recreation projects, fish and game, and everything else, just because we have to live, eat, think, and dream *nothing but fire!'*"

96. "Fire Policy and Procedure Review Committee. February 10, 1967" (U.S. Forest Service, 1967). A brief history of the committee is contained in a covering memo attached to the report.

97. "Fire Policy Meeting. Denver, Colorado. May 12–14, 1971" (U.S. Forest Service, 1971). Many interpretations and popularizations of the 10 Acre Policy followed, among them: William R. Moore, "From Fire Control to Fire Management," *Western Wildlands*, 1 (Summer 1974); John R. McGuire, "Fire as a Force in Land Use Planning," in *TTFECP*, no. 14, pp. 439–444. The entire volume 14 of *TTFECP* is in a sense a testament to the policy reforms evidenced not only in the Forest Service but also in all major fire agencies.

98. Policy Analysis Staff Report, "Evaluation of Fire Management Activities on the National Forests" (U.S. Forest Service, 1977).

99. See Henry W. DeBruin, "From Fire Control to Fire Management, A Major Policy Change in the Forest Service," pp. 11–17, and John McGuire, "Fire as a Force in Land Use Planning," pp. 439–443, in *TTFECP*, no. 14.

100. Richard Barney and David F. Aldrich, "Land Management—Fire Management Policies, Directives, and Guides in the National Forest System: A Review and Commentary," General Technical Report INT–76 (U.S. Forest Service, 1980). Perhaps the most important piece of legislation was the Forest and Rangeland Renewable Resources Planning Act of 1974. Forest Service response can be seen in Policy Analysis Staff Report, "Evaluation of Fire Management," and in various reports made as required by the act.

101. Policy Analysis Staff Report, "Evaluation of Fire Management," pp. 75–80.

102. As measures of the confusion, see Headley, "Rethinking Forest Fire Control," and Brown and Davis, *Forest Fire*, pp. 613–635.

103. See Policy Analysis Staff Report, "Evaluation of Fire Management," p. 96, for a summary of the various types of fire plans in use. For the new policy, see "Revised Fire Management Policy Fact Sheet" (U.S. Forest Service, 1979), and Lynn R. Biddison, "Legislative and Economic Realities: Impacts on Federal Agencies," in Richard G. Barney, ed., *Fire Control for the 80's* (Missoula, Mont.: Intermountain Fire Council, 1980), pp. 34–47.

104. "Coordination of Forest Fire Control Policy," April 18, 1929, NA, RG 95, RFS, Division of Fire Control, Records of the Forest Protection Board, p. 6.

105. H. Duane Hampton, *How the U.S. Cavalry Saved Our National Parks* (Bloomington: University of Indiana Press, 1971), pp. 82–83.

106. Leavitt, "Forest Fires," p. 450; National Academy of Sciences, *Report of the Committee Appointed by the National Academy of Sciences Upon the Inauguration of a Forest Policy for the Forested Lands of the United States to the Secretary of the Interior* (Washington: GPO, 1897), pp. 17, 23–28.

107. See correspondence relating to the fires in NA, RG 79, Records of the National Park Service, Glacier National Park.

108. For information about Mather, see Robert Shankland, *Steve Mather of the National Parks,* 3d ed. rev. (New York: Alfred A. Knopf, 1970). For general information about the Park Service, see William C. Everhart, *The National Park Service* (New York: Praeger, 1972); John Ise, *Our National Park Policy: A Critical History* (Baltimore: Johns Hopkins University Press, 1962); and Alfred Runte, *National Parks: The American Experience* (Lincoln: University of Nebraska Press, 1979).

109. Arno Cammerer to J. R. Eaken, November 23, 1921, NA, RG 79, Records of the National Park Service, Glacier National Park.

110. See correspondence, "Fires to 1925," NA, RG 79, Records of the National Park Service, Glacier National Park.

111. Clepper, *Professional Forestry,* pp. 120–121. A summary of authorizing legislation is in "Fire Management Guideline, NPS–18" (National Park Service, 1979), chap. 2, exhibit 2, p. 1. See also the plans and descriptions of protection resources in the records of the Forest Protection Board, NA, RG 95, RFS, Division of Fire Control, Box 83, and Ise, *Our National Park Policy,* p. 206.

112. Clar, *California Government and Forestry,* 2:47. The Forest Service had instituted and institutionalized annual boards of review for the region after the disastrous 1924 season.

113. A five-year plan was developed by the board in 1927, and the Park Service was an integral part of it. Park Service reports were included in the general series of reports issued by the Board. Particularly useful is the "General Plan for Protection of National Parks and National Monuments From Fire" (n.d.), which gives a synoptic history of fire legislation and activity, a park-by-park breakdown of protection resources, and a description of the debris clearing in Sequoia; NA, RG 95, RFS, Division of Fire Control, Forest Protection Board reports, Box 83. See also a more popular version by Curtis K. Skinner, "Fire, the Enemy of Our National Parks," *AF,* 38 (August 1939), 519–520, which describes fire control efforts at Rocky Mountain Park.

114. Clepper, *Professional Forestry,* p. 121. A similar distillation of events is in Edward E. Hill, comp., *Records of the National Park Service,* Preliminary Inventory no. 166 (Washington: National Archives, 1966), p. 12.

115. Everhart, *National Park Service,* p. 32. See also correspondence in "Central Classified File, Gen.," NA, RG 79, Records of the National Park Service, for information about the emergency accounts in effect.

116. "Coordination of Forest Fire Control Policy," NA, RG 95, RFS, Division of Fire Control, Records of the Forest Protection Board, Box 83, p. 5.

117. Arno B. Cammerer, "Outdoor Recreation—Gone With the Flames," *AF,* 45 (April 1939), 185.

118. Skinner, "Fire," p. 519.

119. Quoted in Ralph Smith to Demaray, NA, RG 79, Records of the National Park Service, no. 558. Similar sentiments were common on the Forest Protection Board. See, for example, "The areas of government lands withdrawn for National Park purposes are the choicest gems of mountain scenery on the continent. As such they render an economic service in the form of national education and recreation of a value probably already even greater than an equivalent area of the choicest commercial forest. Accessibility and grade of material have nothing to do with the valuation of such a forest." "Coordination of Forest Fire Control Policy," p. 5.

120. A case in point, from Justice William O. Douglas: "Protection of our forests from fire is basic conservation. Planless burning is one of the great destroyers of our wealth. When the woods burn, the wilderness economy becomes bankrupt. Wildlife disappears. Water once held on spongy, plant-carpeted land rushes off. Erosion starts. Nature's balance is upset; and all living creatures suffer." "Man's Inhumanity to Land," *AF,* 62 (May 1956), 9.

121. See Superintendent Reports and other records in the archives of Grand Canyon National Park, at Grand Canyon. A special unit deals with CCC activities at the canyon and includes an excellent array of photographs.

122. For the Forest Service and a number of state organizations, the war brought some special emergency funding, thus helping to compensate for the loss of ECW funds. See Robert Thompson, acting chief of forestry, to Director Newton Drury, May 21, 1943, and Memorandum from Acting Director, May 6, 1943, NA, RG 79, Records of the National Park Service, "Central Classified File, General."

123. A. S. Leopold et al., "Wildlife Management in the National Parks," in *Compilation of the Administrative Policies for the National Parks and National Monuments of Scientific Significance,* rev. ed. (Washington: GPO, 1970), p. 101.

124. Ibid., p. 106.

125. Ibid., p. 105. The NAS-NRC committee made its recommendation in 1964.

126. For summaries of the transformation, see Bruce M. Kilgore, "Fire Management in the National Parks: An Overview," in *TTFECP,* no. 14, pp. 45–58; and the numerous papers delivered in the "National Parks Session," *TTFECP,* no. 12. For the Everglades story, see William Robertson, "A Survey of the Effects of Fire in Everglades National Park" (National Park Service, 1953), p. 86.

127. H. H. Biswell, "Forest Fire in Perspective," pp. 43–64, and R. J. Hartesveldt and H. T. Harvey, "The Fire Ecology of Sequoia Regeneration," pp. 65–78, in *TTFECP,* no. 7.

128. Pinchot, *Breaking New Ground,* p. 44.

129. Quotation in Don G. Despain and Robert E. Sellers, "Natural Fire in Yellowstone National Park," *Western Wildlands,* 4 (Summer 1977), 21. The policy

restatement is contained in *Compilation of the Administrative Policies* and is explained in assorted popular articles, among them: Gary Everhart, "Fire Management Perspectives: National Park Service," pp. 2–3, and James K. Agee, "Fire Management in the National Parks," pp. 79–85, *Western Wildlands,* 4 (Summer 1977). A good summary of the policy transition is also available in David B. Butts, "Fire for Management in National Parks," in *Occasional Papers in Park and Recreation Administration,* Department of Recreation and Watershed Resources (Ft. Collins: Colorado State University, 1968), esp. pp. 51–65.

130. The evolution of these plans is documented in the various revisions of NPS–18, "Fire Management Guideline," and assorted staff directives (e.g., 76–12) on the subject of fire management—all on file, Fire Management, Resource Management, National Park Service, Washington.

131. John E. Cook to Directorate, Staff Directive 76–12, December 1, 1976, "Fire Management," p. 1.

132. For general histories and descriptions of the GLO and BLM, see Marion Clawson, *The Bureau of Land Management* (New York: Praeger, 1971); and Office of the Federal Register, *United States Government Manual* (Washington: GPO, annual publication), which includes important legislation, such as the 1976 organic act, that the other work does not. The closest thing to a history of BLM fire operations can be found in Jack F. Wilson, "Fire Management Perspectives: Bureau of Land Management," *Western Wildlands,* 4 (Summer 1977), 4–6. Clepper, *Professional Forestry,* pp. 109–116, summarizes the history of forestry in the agency. The authorizing and governing legislation for BLM fire control is given in *BLM Manual,* "9210–Fire Management."

133. See references in Chapter 4, "From Fire and Ax: Early Private and Government Fire Protection," especially notes 98–106. Also relevant are the monthly fire reports made by superintendents of the reserves, NA, RG 79, Records of the General Land Office, no. 795, and the analysis by Clepper, *Professional Forestry,* pp. 102–105.

134. See reports by the GLO in NA, RG 95, RFS, Division of Fire Control, Forest Protection Board, Boxes 81–83, and "Fire Control Activity in the Bureau of Land Management, Alaska" (BLM, 1959), Files, BLM Alaska State Office, Anchorage, which includes a short chronology. Accounts of the O&C lands are in Clepper, *Professional Forestry,* pp. 109–112; Clawson, *Bureau of Land Management,* pp. 89–108; and W. H. Horning, "The O&C Lands: Their Role in Forest Conservation," *JF,* 38 (1940), 379–383.

135. "Coal Fires on the Public Domain," in "Report of Forest Protection Board, Part II," April 1, 1929, pp. 15–16, NA, RG 95, RFS, Division of Fire Control, Box 83.

136. Office of the Chief Coordinator, Circular Letter No. 6, April 18, 1929, in "Report of Forest Protection Board," p. 4, NA, RG 95, RFS, Division of Fire Control, Box 83.

137. Clepper, *Professional Forestry,* pp. 109–113; Clawson, *Bureau of Land Management,* pp. 89–108; R. R. Robinson, "Forest and Range Fire Control in Alaska," *JF,* 58 (1960), 448–453.

138. Robinson, "Forest and Range," pp. 448–453, is the best synopsis of the his-

tory of early fire protection in Alaska. For a full treatment see Chapter 8, "Fire and Frost: A Fire History of Alaska."

139. Clawson, *Bureau of Land Management,* pp. 38–39.

140. Personal communication, Fred McBride, BLM, Washington office, 1978.

141. J. W. Jay, "Fire Study Report: Nevada Fires," National Fire Coordination Study (U.S. Forest Service, 1964).

142. See "Bureau of Land Management Progress Report in Fire Control Planning" (unpublished report, 1965), Files, Boise Interagency Fire Center.

143. Director, Great Basin Fire Center, to Director, BLM, "Progress Report—Fire Control Planning," September 3, 1965, pp. 12–14, Files, Boise Interagency Fire Center.

144. Gordon J. Stevens, "A Chronology of BIFC Development" (unpublished report, 1976), Files, Boise Interagency Fire Center (my copy was annotated by Fred McBride, BLM); "Dedication Ceremony Officially Opens Fire Center," *FCN,* 31 (Fall 1970), 3–6.

145. Stevens, "A Chronology," and "Dedication Ceremony" have details. Also useful, though it emphasizes a somewhat different aspect of interagency coordination, is Jack Wilson, "History of NWCG," *FMN,* 39 (Spring 1978). See also R. L. Bjornsen, "Fire Management . . . Toward an Expanded Dimension," *FMN,* 35 (Winter 1974), 14–16.

146. A synopsis of fire weather services is provided in Donald R. Whitnah, *A History of the United States Weather Bureau* (Urbana: University of Illinois Press, 1965), pp. 151–153. See also Harry L. Swift, "A Short History of the Fire Weather Service and the 'Federal Plan for a National Fire Weather Service,'" in C. W. Slaughter et al., eds., "Fire in the Northern Environment: A Symposium" (U.S. Forest Service, 1971), pp. 117–120, and Alice J. Svorcek, "50 Years of Fire Weather Service," *FCN,* 26 (April 1965), 8–9. For an assessment of the earliest warning program, see three articles by Edward A. Beals, "Value of Weather Forecasts in the Problem of Protecting Forests from Fire," *Monthly Weather Review,* 42 (February 1914), 11–119, "Fire Wind Forecasts," *The Timberman,* December 1914, pp. 33–36, and "Forecasting Fire Winds," *The Timberman,* December 1913, pp. 22–23.

147. Svorcek, "50 Years"; "Minutes of Conference of Weather Bureau Officials and Representatives of Cooperating Agencies on the Organization of a Fire-Weather Warning Service" (U.S. Weather Bureau, 1926); L. N. Gray, "Fire-Weather Service," *AF,* 36 (August 1930), 512–514, 524; E. B. Calvert, "Weather Forecasting as an Aid in Preventing and Controlling Forest Fires," *Monthly Weather Review,* 53 (May 1925), 187–190.

148. Svorcek, "50 Years," p. 9.

149. "Federal Plan for a National Fire-Weather Service" (Office of the Federal Coordinator for Meteorological Services and Supporting Research, 1967); Dee F. Taylor, "Current Status National Fire-Weather Plans" (unpublished report, Southern Fire Chiefs' Meeting, 1965); Warren B. Price, "The Role of the Corporate Meteorologist in Fire Control," *FM,* 36 (Fall 1975), 12–13, 23; Charles Buck and Mark Schroeder, *Fire Weather: A Guide for Application of Meteorological Information to Forest Fire Control,* U.S. Department of Agriculture, Handbook 360 (Washington: GPO, 1970).

150. Gerald A. Petersen, "A Review of National Weather Service Policy on Services," in "Proceedings of the Fourth National Conference on Fire and Forest Meteorology," General Technical Report RM–32 (U.S. Forest Service, 1976), p. 26.

151. A summary report of the board's activities was prepared and sent by R. Y. Stuart to the Chief Coordinator in 1932; see Stuart to Craven, "Fire, Cooperation, Forest Protection Board," April 16, 1932, NA, RG 95, RFS, Division of Fire Control, General Correspondence.

152. "Forest Protection Board (Summary)," p. 2. The idea for the board was evidently proposed by a "committee composed of timber land owners" who visited the President and requested better cooperation in fire control—an extension of the Clarke-McNary spirit. The committee may well have been the National Forestry Program Committee with which Greeley had been so involved and which had helped to promote the Clarke-McNary Act.

153. Office of the Chief Coordinator, Circular Letter No. 6, "Forest Protection Board," p. 2.

154. Wilson, "History of the NWCG," pp. 13–16.

155. Ibid., p. 13.

156. See, for example, Al Duhndrack, "Wildfire Coordination in Colorado," *FMN,* 39 (Spring 1978), 8–9.

157. Richard A. Chase, "Firescope: A Regional Solution to Multi-Agency Coordination Problems," *International Fire Chief,* 43 (November 1977), 18–21, and "Firescope: A New Concept in Multiagency Fire Suppression Coordination," General Technical Report PSW–40 (U.S. Forest Service, 1980); Randall J. Van Gelder, "A Fire Potential Assessment Model for Brush and Grass Fuels," *FMN,* 37 (Summer 1976), 14–16; and the assorted publications of FIRESCOPE (unnumbered), such as "Standards for Establishing Incident Command Posts."

6: A CONTINENTAL EXPERIMENT

1. Edgar B. Nixon, ed., *Franklin D. Roosevelt and Conservation, 1911–1945* (Hyde Park: Franklin D. Roosevelt Library, 1957), 1:143–144.

2. William E. Leuchtenburg, *Franklin D. Roosevelt and the New Deal* (New York: Harper and Row, 1963), p. 35.

3. Nixon, ed., *Roosevelt and Conservation,* 1:342.

4. Quoted in Leuchtenburg, *Roosevelt,* p. 136.

5. Nixon, ed., *Roosevelt and Conservation,* 1:430–431, 299.

6. Joseph P. Lash, *Eleanor and Franklin* (New York: Norton, 1971), pp. 267–268.

7. Nixon, ed., *Roosevelt and Conservation,* 1:140. See also "Reports of the Chief Forester" (U.S. Forest Service) for the years 1933–1939.

8. E. I. Kotok et al., "Protection Against Fire," in *A National Plan for American Forestry,* 2 vols., Senate Document no. 12, 73d Cong., 1st sess. (Washington: GPO, 1933), 2:1395, 1397.

9. John Muir, *Steep Trails* (Boston: Houghton Mifflin, 1913), p. 238.

10. William G. Morris, "Forest Fires in Oregon and Washington," *Oregon Historical Quarterly,* 35 (December 1934), 313.

11. General histories of fire and forestry in the Northwest include: Harold K. Steen, "Forestry in Washington to 1925" (Ph.D. diss., University of Washington, 1969); George T. Morgan, "The Fight Against Fire: Development of Cooperative Forestry in the Pacific Northwest, 1900–1950" (Ph.D. diss., University of Oregon, 1964); Harold Weaver, "Effects of Fire on Temperate Forests: Western United States," in T. T. Kozlowski and C. E. Ahlgren, eds., *Fire and Ecosystems* (New York: Academic Press, 1974), pp. 279–320; William G. Morris, "Forest Fires in Oregon and Washington"; Henry Clepper, *Professional Forestry in the United States* (Baltimore: Johns Hopkins University Press, 1971), esp. pp. 269–286; Ralph Widner, ed., *Forests and Forestry in the American States: A Reference Anthology* (Missoula, Mont.: National Association of State Foresters, 1968), esp. pp. 271–292; Robert E. Martin et al., "Fire in the Pacific Northwest—Perspectives and Problems," in *TTFECP,* no. 15, pp. 1–24; Stewart Holbrook, *Burning an Empire: The Story of American Forest Fires* (New York: Macmillan, 1943), pp. 134–155; Charles S. Cowan, *The Enemy Is Fire!* (Seattle: Superior Publishing Co., 1962); Oregon Department of Forestry, "Forest, People, and Oregon" (Salem, 1977), which includes a master chronology.

12. Among many histories of the Tillamook Burn, see Lynn Cronemiller, "The Tillamook Burn," *AF,* 67 (May 1961), 29–31, 51–52, and "Oregon's Forest Fire Tragedy," *AF,* 40 (January 1934), 487–490, 531; J. Larry Kemp, *Epitaph for the Giants* (Portland: Touchstone Press, 1967); Holbrook, *Burning an Empire,* pp. 134–141. Additional records reside in Files, Oregon Department of Forestry, Salem, but there was evidently no official report on the fires, and whatever original records may have existed were lost in a 1935 fire in the state building. Cronemiller, however, was the fire boss and state forester at the time, and his description for *AF* can take the place of other reports. Also useful are Richard E. McArdle, "The Tillamook Fire," U.S. Forest Service, *Service Bulletin,* 17, no. 21 (October 1933), and assorted articles in *Forest Log,* the newspaper of the Oregon Department of Forestry; see in particular, "Fire Loss Runs to Staggering Totals," October 1, 1933, and "Huge Fire Spreads Ruin and Disaster," September 1, 1933.

13. See Cronemiller, "Oregon's Forest Fire Tragedy," and Kemp, *Epitaph.* For a description of one of the attendant fires, see Richard E. McArdle, "Wolf Creek Fire in August Hits 1½ Billion Feet of Timber," *Four L Lumber News,* October 15, 1933.

14. Harold Olson, "Tillamook Burns Again!" *AF,* 51 (September 1945), 431–433; H. V. Simpson, "Holy Old Mackinaw in the Smoke," *AF,* 57 (November 1951), 10–13, 53, 55; Widner, ed., *Forests and Forestry,* pp. 286–288; records in Files, Oregon Department of Forestry, including "Northwest Oregon Scorched by Blaze" and "Tillamook Blaze Traps Fire Crew," *Forest Log,* 15 (August 1945), and "Drenching Rains End Fire Hazard," *Forest Log,* 15 (September 1945).

15. Cronemiller, "Oregon's Forest Fire Tragedy," p. 523.

16. Accounts of the salvage are given in Holbrook, *Burning an Empire,* pp. 142–146; Widner, ed., *Forests and Forestry,* pp. 288–289; Kemp, *Epitaph,* pp. 79–83, 91–93; "Summary of the Tillamook Burn Study, 1942–1943" (Portland: U.S. Forest Service, 1944); Cronemiller, "The Tillamook Burn."

17. Oregon Department of Forestry, "From Burn to Tillamook Forest" (Salem, 1971) and "Forest, People, and Oregon," pp. 32–33; "Tillamook Planting Starts December 1," *Forest Log,* 15 (September 1945).

18. Colloquial usage of "Pacific Northwest" includes the east slopes of the Cascades and portions of the Northern Rockies; my use is more in keeping with fire weather and historical geography.

19. For an overview of debris, see Owen P. Cramer, ed., "Environmental Effects of Forest Residues Management in the Pacific Northwest," General Technical Report PNW–24 (U.S. Forest Service, 1974), especially those papers relating to fire, pp. F–1 to F–46 and G–1 to G–27.

20. S. G. Pickford et al., "Weather, Fuel, and Lightning Fires in Olympic National Park," *Northwest Science,* 54, no. 2 (1980), 92–105; William G. Morris, "Lightning Storms and Fires in the National Forests of Oregon and Washington" (Pacific Northwest Forest Experiment Station, U.S. Forest Service, 1934).

21. Quoted in Morris, "Forest Fires," pp. 313–316.

22. Ibid., p. 316.

23. Oregon Department of Forestry, "Forest, People, and Oregon," p. 5.

24. Quoted in Morris, "Forest Fires," pp. 316–317.

25. Ibid., p. 317.

26. Ibid.

27. Ibid., pp. 318–323. For the aftermath, see Thornton T. Munder, "Out of the Ashes of Nestucca," *AF,* 50 (July 1944), 342–345, 366, 368.

28. Morris, "Forest Fires," pp. 319–320.

29. Ibid., p. 321

30. Ibid., p. 329.

31. Ibid., p. 324.

32. Ibid.

33. Ibid., p. 330.

34. Quoted in Louis Barrett, "A Record of Forest and Field Fires in California From the Days of the Early Explorers to the Creation of the Forest Reserves" (U.S. Forest Service, 1935), p. 22. For more on the Northwest in particular, see National Academy of Sciences, *Report of the Committee Appointed by the National Academy of Sciences Upon the Inauguration of a Forest Policy for the Forested Lands of the United States to the Secretary of the Interior* (Washington: GPO, 1897), pp. 18–19, which shows the continuity of practices by herders in California and Oregon.

35. Muir, *Steep Trails,* p. 239. For the general fire scene in the 1880s, see Franklin Hough, *Report on Forestry,* vol. 3 (Washington: GPO, 1882), pp. 202–205. Reports on the reserves include H. B. Ayres, "Washington Forest Reserve," in Henry Gannett, ed., *Twentieth Annual Report of the United States Geological Survey,* pt. 5: *Forest Reserves* (Washington: GPO, 1900); see also, National Academy of Sciences, *Report of the Committee,* pp. 17–19, which singled out the Cascade Forest Reserve as especially devastated by fire and hoof.

36. William T. Cox, "Recent Forest Fires in Oregon and Washington," *Forestry and Irrigation,* 8 (November 1902), 462–469; F. H. Brundage, "Yacolt Fires, 1902-1927," *Forest Patrolman,* 8 (September 15, 1927), 3–4; Morris, "Forest Fires," pp. 333–337.

37. See Cowan, *The Enemy Is Fire,* p. 38; Eloise Hamilton, *Forty Years of West-
    ern Forestry* (Portland: Western Forestry and Conservation Association, 1949),
    pp. 30–32; Oregon Department of Forestry, "Forest, People, and Oregon," pp.
    8–9; Morgan, "The Fight Against Fire"; Steen, "Forestry in Washington."
38. Cox, "Recent Forest Fires," p. 466.
39. For the continuation of fires apart from the Tillamook cycle, see C. S. Chap-
    man, "Forest Fires in Washington and Oregon," *AF,* 16 (November 1910), 644–
    647; Morris, "Forest Fires"; Morgan, "The Fight Against Fire"; "The North-
    west's Worst Forest Fires," *AF,* 25 (August 1919), 1259; Holbrook, "The Trag-
    edy of Bandon," *AF,* 42 (November 1936), 494–497; George E. Griffith, "Red
    Warfare on the Western Front," *AF,* 44 (September 1938), 391–392; John D.
    Guthrie, "Ninety-Nine Red Days, The Grim Spectacle of the Pacific North-
    west," *AF,* 35 (November 1929), 675–677. For fire protection activities, see
    Morgan, "The Fight Against Fire," which is particularly good in describing the
    political context; Oregon Department of Forestry, "Forests, People, and Ore-
    gon," which outlines the main events; Cowan, *The Enemy Is Fire,* which
    recounts in some detail the progress in Washington; Widner, ed., *Forests
    and Forestry,* pp. 271–291. Early records of fire plans and inspections are also
    available for some of the national forests in Oregon, stored with the files of the
    forest.
40. Emanuel Fritz, "The Role of Fire in the Redwood Region," Agricultural
    Experiment Station Circular 323 (Berkeley, 1932). Fritz was also instrumental
    in establishing the Contra Costa Fire Protective Association, which tried to
    remove debris in advance of the 1923 Berkeley fire.
41. CCC fuel management on the burn was on an enormous scale and became the
    basis for the wholesale fuel management plans now developed for the region; see
    John D. Dell, "R–6 Preattack Guide" (Portland: U.S. Forest Service, 1972).
    The treatment of areas once burned was not different in strategy from treatment
    of slash areas not yet burned.
42. Prescribed burning means two things in the region. East of the Cascades it
    means broadcast underburning, and this is what Weaver experimented with;
    west of the Cascades it means primarily slash disposal. Summaries of these tech-
    niques can be found in *TTFECP,* no. 15, and Cramer, ed., "Environmental
    Effects of Forest Residues Management," respectively.
43. Oregon Department of Forestry, "Forest, People, and Oregon," pp. 47, 66;
    Harold M. Patterson, "State Actions for Smoke Control. Oregon," pp. 183–197,
    and Owen P. Cramer and Stewart G. Pickford, "Factors Influencing Smoke
    Management Decision in Forest Areas," pp. 231–240, in *Air Quality and Smoke
    from Urban and Forest Fires* (Washington: National Academy of Sciences,
    1976); Cramer, ed., "Environmental Effects of Forest Residues Management,"
    esp. F–1 to F–46; Oxbow fire report, Files, Oregon Department of Forestry,
    Salem.
44. Wenatchee fire reports, Files, Region Six, U.S. Forest Service, Portland. A
    documentary film of the fires was made by MGM; see "Wildfire!" *FCN,* 32
    (Spring 1971), 3.
45. For the general history of the North American Forestry Commission, under
    whose auspices most of the international agreements have been organized, see
    James Sorensen, "Seventeen Years of Progress Through International Cooper-

ation, 1962–1978" (U.S. Forest Service, 1979). International agreements include mutual assistance pacts with Mexico (1968) and Canada; Quebec and New Brunswick as partners in the Northeastern States Fire Protection Compact; mutual assistance agreements between Ontario and Minnesota and between the Yukon Territory and Alaska (BLM).

46. *National Plan for American Forestry,* 1:776.

47. Joseph Illick, "State Forestry," in Robert K. Winters, ed., *Fifty Years of Forestry in the U.S.A.* (Washington: SAF, 1950), p. 232.

48. *National Plan for American Forestry,* 1:803–805.

49. Ibid., p. 733. The best treatments of state forestry and fire protection, in addition to the 1933 summary provided by the Copeland Report, are Ralph Widner, ed., *Forests and Forestry;* Illick, "State Forestry"; Earl Pierce and William Stahl, "Cooperative Forest Fire Control: A History of Its Origin and Development Under the Weeks and Clarke-McNary Acts" (U.S. Forest Service, 1964); Eliot Zimmerman, "A Historical Summary of State and Private Forestry in the U.S. Forest Service" (U.S. Forest Service, 1976); Samuel T. Dana, *Forest and Range Policy: Its Development in the United States* (New York: McGraw-Hill, 1956). Most states have produced a history of their forestry and fire operations, and a few (such as California) have published major books on the subject. References to these works are included in discussions of regional fire histories. Also useful are the original Clarke-McNary assessments, some of which were reworked and published by the states. Another fundamental reference is "Digest of State Forest Fire Laws" (U.S. Forest Service, Northeastern Area, State and Private Forestry, 1973).

50. For a treatment of New York, see Chapter 1, note 80.

51. See J. Girvin Peters, ed., "Forest Fire Protection by the States" (Weeks Law Forest Fire Conference) (U.S. Forest Service, 1914), and "Development of Fire Protection in the States," *AF,* 10 (November 1913), 721–736.

52. Peters, "Development of Fire Protection," pp. 723–724.

53. *National Plan for American Forestry,* 2:1403.

54. Widner, ed., *Forests and Forestry,* pp. 380, 328. Widner's state-by-state accounts are extremely informative on topics like this, but his lack of references and citations makes the final value much less than it should be.

55. For information on the Weeks Act, see Dana, *Forest and Range Policy,* pp. 180–187, 222–223; Harold K. Steen, *The U.S. Forest Service: A History* (Seattle: University of Washington Press, 1976), pp. 122–131; Widner, ed., *Forests and Forestry,* pp. 211–218 and passim; Pierce and Stahl "Cooperative Forest Fire Control," pp. 1–27; J. G. Peters, "Cooperation with States in Fire Patrol," *AF,* 17 (July 1911), 383–384; and Weeks Law Forest Fire Conference.

56. Steen, *Forest Service,* p. 128. For state and national responses, see also Morgan, "The Fight Against Fire," pp. 120–160, and John Little, "The 1910 Forest Fires in Montana and Idaho: Their Impact on Federal and State Legislation" (M.A. thesis, University of Montana, 1968).

57. Dana, *Forest and Range Policy,* pp. 183–184.

58. Ibid., pp. 186–187.

59. Weeks Law Forest Fire Conference, p. 2.

60. Ibid., pp. 74–75.

61. The best account of the Clarke-McNary Act and of the events, like the Capper Report, that led up to it is in Pierce and Stahl, "Cooperative Forest Fire Control," pp. 28–86, and my version closely follows theirs. See also Steen, *Forest Service*, pp. 173–189; Dana, *Forest and Range Policy*, pp. 221–222; Widner, ed., *Forests and Forestry*, pp. 277–279 and passim, for the effects of the legislation on various states.

62. The contrast with Pinchot is effectively brought out in Steen, *Forest Service*, pp. 179–182, and Widner, ed., *Forests and Forestry*, pp. 247–249.

63. Pierce and Stahl, "Cooperative Forest Fire Control," p. 33.

64. Greeley, *Forests and Men* (New York: Arno Press, 1972), p. 110.

65. Quoted in Pierce and Stahl, "Cooperative Forest Fire Control," p. 54.

66. Ibid., p. 46.

67. A quick summary of the act and its allotment formulas is in *Forest and Grass Fires*, Senate Document No. 30, 90th cong., 1st sess. (Washington: GPO, 1967), pp. 73–77. Otherwise see citations given in note 61.

68. Steen, *Forest Service*, p. 173.

69. Pierce and Stahl, "Cooperative Forest Fire Control," p. 88.

70. Nixon ed., *Roosevelt and Conservation*, 1:278.

71. Ibid., 2:107–108.

72. Widner, ed., *Forests and Forestry*, p. 428.

73. Clar, *California Government and Forestry*, 2:310.

74. See Pierce and Stahl, "Cooperative Forest Fire Control," pp. 89–90; Dana, *Forest and Range Policy*, pp. 309–310; Fire Working Team, Society of American Foresters, "Interstate Wildfire Organization Directory," includes the texts for all the compacts; W. J. Stahl and J. N. Diehl, "Interstate Forest Fire Protection Compacts," *Quarterly of the NFPA*, 54 (October 1961), 153–160. The National Association of State Foresters has also been working to create fire disaster plans to supplement the compacts.

75. William G. Herbolsheimer, "Roscommon Equipment Center: A 20-State Approach to ED&T," *FM*, 36 (Fall 1975), 6–7. Information on the individual compacts is included in the various regional fire histories.

76. Dana, *Forest and Range Policy*, p. 340.

77. Arthur Chapman, "The Fire Fighters," *AF*, 16 (November 1910), 639. The poem was reprinted in John D. Guthrie, *Forest Fire and Other Verse* (Portland: Dunham Printing Co., 1929), p. 47.

78. *The Poetry of Robert Frost*, ed. Edward Connery Lathem (New York: Holt, 1969).

79. See "Conservation Workers Going After Forest Fires," U.S. Department of Agriculture, Office of Information, Press Release, July 31, 1933. Although "hundreds of the camps" had been organized to fight fires, few had done much fire suppression; it was major news, for example, that CCC laborers handled eight lightning snag fires on the national forests of Arizona. For Fechner's orders, see Clar, *California Government and Forestry*, 2:242.

80. Gifford Pinchot, *A Primer on Forestry, Part I: The Forest* (Washington: GPO, 1899), p. 47.

81. Coert duBois, *Systematic Fire Protection in the California Forests* (Washington: GPO, 1914), p. 35.

82. Quoted in Betty Spencer, *The Big Blowup* (Caldwell, Idaho: Caxton Printers, 1956), p. 64. For a survey of manpower resources in the East, see Austin Wilkens, *Ten Million Acres of Timber: The Remarkable Story of Forest Protection in the Maine Forestry District (1909–1972)* (Woolwich, Me.: TBW Books, 1978), pp. 71–72.

83. Greeley seemed especially taken with the idea; see "Better Methods of Fire Control," SAF, *Proceedings* (1911), pp. 159–160. The proposal was echoed by the National Conservation Congress in 1912 (see Clar, *California Government and Forestry,* 1:364), and is periodically revived even now.

84. The best history of the CCC is John A. Salmond, *The Civilian Conservation Corps, 1933–1942* (Durham: Duke University Press, 1967), though its focus is primarily political and administrative. Also invaluable is the reference data file on the CCC compiled by John D. Guthrie and C. H. Trace, 18 vols., in NA, RG 35, General Records of the Civilian Conservation Corps. For sample state programs, see Salmond, pp. 7–8; Clar, *California Government and Forestry,* 2:189–221; Widner, ed., *Forests and Forestry,* under the appropriate states.

85. *Report of the Chief of the Forest Service, 1936* (Washington: GPO, 1936), p. 14.

86. Salmond, *Civilian Conservation Corps,* p. 121.

87. John D. Guthrie, "The CCC as a Fire Fighting Unit," *AF,* 45 (April 1939), 210–211, 238. A special manual was prepared by H. R. Kylie et al., *CCC Forestry* (Washington: GPO, 1937), which includes a large section on fire control methods, along with a good dose of propaganda. As an example of how CCC manpower reformed fireline methods, see Roy Headley, "Mileposts of Progress in Fire Control and Fire Research, Comments," *JF,* 40 (August 1942), 608. Headley describes how the one-lick method of line construction was invented by Godwin in 1915 but forgotten until revived for CCC crews by Kenneth McReynolds in 1936. An example of special CCC fire crews was the 48-man unit, "the stampede fire fighters," used at Glacier National Park; see Jack S. Barrows, "Historical Perspectives in Fire Control," in Richard G. Barney, ed., *Fire Control for the 80's* (Missoula Mont.: Intermountain Fire Council, 1980), pp. 5–27.

88. "Three Flame-Trapped Men Die. Six Save Selves Thru Swamps," *Happy Days* (Washington), May 30, 1936.

89. Fred Morrell to Robert Fechner, "Report on Blackwater Fire, August 20–21, 1937. Shoshone National Forest, Wyoming," September 23, 1937, WNRC, RFS, Division of Fire Control, includes the basic documents. Distilled versions are also abundant: Division of Fire Control, "Blackwater Fire on the Shoshone," pp. 305–307, and "Statement by Ranger Urban Post," pp. 308–315, *FCN,* September 20, 1937; David P. Godwin, "The Handling of the Blackwater Fire," pp. 373–382, and A. A. Brown, "The Factors and Circumstances That Led to the Blackwater Fire Tragedy," pp. 384–387, *FCN,* December 6, 1937. A still more popular version is Earle Kauffman, "Death in Blackwater Canyon," *AF,* 42 (1937), 534–540, 558–559.

90. Godwin, "Handling of the Blackwater Fire," pp. 381, 380, 381; John Sieker, "C.C.C. Determination," *FCN,* December 6, 1937, p. 388.

91. "The American Forest Fire Medal for Heroism," *AF,* 45 (April 1939), 201,

240. Silcox is quoted in John Case, "The North American Forest Fire Medal" (U.S. Forest Service History Office, 1977), p. 1.

92. See Case, "North American Forest Fire Medal," for biographical sketches of all the recipients.

93. See Salmond, *Civilian Conservation Corps,* pp. 172–191; L. L. Colvill, "Lessons from Larger Fires on the Siskiyou," *FCN,* 3 (July 1939), 20–21, which describes cases of CCC crews striking and walking off the job, "principally because of poor leadership." Fatalities also continued; see "Forest Fires Claim Five Lives," *AF,* 45 (September 1939), 464, and Carl Wilson, "Fatal and Near-Fatal Forest Fires: The Common Denominators," *International Fire Chief,* 43 (November 1977).

94. For assessments after the Blackwater tragedy, see John Guthrie, "Memorandum for Mr. Stockdale," December 7, 1937, and "Memorandum," October 7, 1937, in "CCC Chronological Reference Materials," NA, RG 35, Records of the CCC. See also Robert Fechner, "Memorandum," June 10, 1937, NA, RG 35, Records of the CCC, "Gen. Correspondence, 1933–1942," "Forest Fires— General Information," for attempts to improve fire safety prior to the Blackwater debacle. For alarm over the replacement of the regular fire organizations by CCC camps, see Clar, *California Government and Forestry,* 2:264. The Guthrie quotation is from his "The CCC as a Fire Fighting Unit," p. 210.

95. Salmond, *Civilian Conservation Corps,* pp. 212–215; Earle Kauffman, "You Can't Win Against Fire," *AF,* 49 (March 1943), 106–108; David P. Godwin and Alan Macdonald, "The New Fire Crusade," *AF,* 49 (April 1943), 180, 182; News release, Office of Civilian Defense, September 21, 1942, announcing creation of the FFFS; Widner, ed., *Forests and Forestry,* describes many of the programs of the FFFS within his general state histories. For a sample of the lengths to which organizations could go, see "Camp Fire Girls Fight Forest Fires," *AF,* 48 (May 1942), 224–225. Wilkins, *Ten Million Acres of Timber,* pp. 177–188, describes the use of POWs to replace CCCs.

96. "Forest Service Reserves in Action," annual reports for 1942, 1943, and 1944, copies in personal possession.

97. C. Raymond Clar, *Evolution of California's Wildland Fire Protection System* (Sacramento: California Division of Forestry, 1969), pp. 32–34. Lloyd Thorpe, *Men to Match the Mountains* (Seattle: Craftsman and Met Press, 1972), deals exclusively with the California Conservation Camp Program.

98. The literature, especially the popular literature, about the smokejumper program is large. The best summaries are Henry Clepper, *Professional Forestry,* pp. 177–186, and "History of Smokejumping" (U.S. Forest Service, 1976). The records of the Aerial Fire Control Project are in NA, RG 95, RFS, Division of Fire Control. Other publications of historical significance include Division of Fire Control, "History of Smokejumping, 1939–1949," *FCN,* 11 (July 1950), 1–11; David P. Godwin, "Smoke Jumping," *AF,* 45 (December 1939), 590–592, and "The Parachute Method of Fire Suppression," *JF,* 39 (February 1941), 169–171.

99. See NA, RG 95, RFS, Division of Fire Control, General Correspondence. A blurb on the findings was published as "Aerial Control in Russia," *FCN,* September 20, 1937, p. 293.

100. A fire-by-fire account of a season's jumping out of Silver City is given in Randle Hurst, *The Smokejumpers* (Caldwell, Idaho: Caxton Printers, 1966).

101. Board of Review, Mann Gulch Fire, WNRC, RFS, Division of Fire Control; additional records are available in Files, Region One, U.S. Forest Service.

102. "Operation Smokejumper" (U.S. Forest Service, 1950) introduced the plan by observing that "we are entering a period marked by military expansion, preparation for a war economy, and an intensive national civilian defense program" and claiming that smokejumping was integral to those developments. See also "History of Smokejumping," pp. 20–21; U.S. Congress, House, *Operation Smokejumper,* House Document no. 7257, 79th Cong., 2d sess. (Washington: GPO, 1950).

103. Duane W. Myler, "Final Report on 3-Year Appalachian Parainitial Attack Test Project" (U.S. Forest Service, Southern Region, 1974); BLM, "Statewide Smokejumper Project Responsibility Increases," March 1, 1978, Files, Fairbanks District, p. 3. William D. Moody, "Smoke Jumping . . . An Expanding, Varied Role," *FMN,* 35 (Spring 1974), 13–14, sees events in a different light.

104. Edward Cliff and Rolfe Anderson, "The Forty-Man Crew" (Siskiyou National Forest, December 19, 1939), WNRC, RFS, Division of Fire Control; Edward P. Cliff and Rolfe E. Anderson, "The 40-Man Crew—A Report on the Activities of the Experimental 40-Man Fire Suppression Crew," *FCN,* 4 (April 1940), 47–62; Stewart Holbrook, "Forty Men and a Fire," *AF,* 46 (June 1940), 251–253; Rolfe E. Anderson et al., "Adapting Advanced Principles of Organization and Fire-Line Construction to CCC Suppression Crews," *FCN,* 5 (June 1941), 123–128. An overview of organized fire crews derived from the 40-man example is given in Martin E. Alexander, "High Mobility: The Interregional Fire Suppression Crew," *FM,* 35 (Summer 1974), 14–17, 19.

105. Anderson et al., "Adapting Advanced Principles," p. 123.

106. See Ralph L. Cunningham and Wesley W. Spinney, "A Sierra Ration and Equipment Outfit," *FCN,* 5 (January 1941), 22–24, for a crew on the Sierra National Forest; Anderson et al., "Adapting Advanced Principles," pp. 123–128, George H. Schroeder, "Oregon's 'Red Hats,'" pp. 129–130, and P. D. Hanson and C. A. Abell, "Determining the Desirable Size of Suppression Crews for the National Forests of Northern California," pp. 156–160, *FCN,* 5 (July 1941); Roy Elliott, "The Willamette Flying 20," *FCN,* 5 (October 1941), 179–182; Mark McMillin, "Hot Shot Fire Fighters," *AF,* 62 (August 1956), 28–29, 51.

107. Schroeder, "Oregon's 'Red Hats,'" pp. 129–130.

108. See Stanley Stevenson, "'Hot Shot' Crews," *FCN,* 12 (April 1952), 29–31; James L. Murphy, "California Helitack Report—1958," *FCN,* 21 (January 1960), 1–5; Ralph G. Johnston, "Helicopter Use in Forest Fire Suppression: 3 Decades," *FMN,* 39 (Fall 1978), 14–18; Corps of Engineers, U.S. Army Engineer Research and Development Laboratories, and U.S. Forest Service, "Helicopter Fire Fighting Program" (U.S. Forest Service, 1956), WNRC, RFS, Division of Fire Control; Ian D. McAndie, "Rappelling, An Alternative," *FM,* 34 (Summer 1973), 5–7; and Reports on Rappelling Trials, Fire Management, Region Six, U.S. Forest Service, Portland. Stevenson reports helijumping as early as 1951.

109. "Southwestern Firefighters" (U.S. Forest Service, Southwestern Region,

n.d.); K. O. Wilson, "Operation Redskin," *FCN*, 13 (July 1952), 1–3; J. L. Ball, "Angels in Hardhats," *AF*, 78 (January 1972), 12–15, 60; Clarence K. Collins, "Indian Firefighters of the Southwest," *JF*, 60 (January 1962), 87–91. For an interesting depiction of how firefighting affects the economy of tribes engaged in it, see Roger Sylvester, in C. W. Slaughter et al., eds., "Fire in the Northern Environment" (U.S. Forest Service, 1971), pp. 249–251.

110. "Topic 3. Aerial Fire Control," in "Servicewide Meeting. Fire Control, Fire Research, and Safety. Ogden, Utah, 1950" (U.S. Forest Service, 1950), p. 8.

111. "The Story Behind the Organized Fire Suppression Crew" (U.S. Forest Service, 1957), WNRC, RFS, Division of Fire Control.

112. Division of Fire Control, "The Interregional Suppression Crew," *FCN*, 24 (October 1963), 93; Alexander, "High Mobility," pp. 14–17; Jerry Ewart, "Hot Shot Crews Pay Big Dividends," *FM*, 37 (Winter 1976), 14–16.

113. "Increased Manning Experiment. Summary." WNRC, RFS, Division of Fire Control.

114. Ramparts Cave Fire reports, 1978, Grand Canyon National Park.

115. "Griffith Park Fire Report. November 20, 1933," "Report of Coroner's Committee, October 10, 1933," and "Verdict of Coroner's Jury in Griffith Park Fire," materials collected by Carl C. Wilson; George Cecil, "The Griffith Park Disaster," *AF*, 40 (June 1934), 15, 46.

116. Carl Wilson, "Fatalities," pp. 9–10, 12–15.

117. "Hot Facts of Life on the Fire Line" (U.S. Forest Service, ca. 1942), Files, Region One, U.S. Forest Service. There is considerable information about early training methods, but it is scattered among boards of review and conferences. More concentrated is S. B. Show and E. I. Kotok, History of Fire Research in California, unpublished report, title page missing, U.S. Forest Service, ca. 1956, WNRC, RFS, pp. 101–106.

118. Inaja Fire reports, Region Five, U.S. Forest Service, San Francisco; "Fire Task Force" (U.S. Forest Service, 1957), WNRC, RFS, Division of Fire Control. Copies of most of the national fire training courses are stored at the National Advanced Resources Technology Center, Marana, Arizona. Overviews of fire training since the Inaja fire are found in G. E. Cargill, "National Fire Training Is 15 Years Old and Still Growing," *FCN*, 33 (Spring 1972), 3–5; the entire issue of *FM*, 35 (Fall 1974), and of *FCN*, 20 (April 1959).

119. "Report of Fire Safety Review Team. February 3, 1967," WNRC, RFS, Division of Fire Control, p. 2; Loop Fire Analysis Group, "The Loop Fire Disaster" (U.S. Forest Service, 1966).

120. Clive M. Countryman, "Carbon Monoxide: A Firefighting Hazard" (U.S. Forest Service, 1971).

121. Jim Abbott and Mike Bowman, "Wildland Fire Goal . . . Coordination of Agencies' Courses," pp. 3–5, and Robert L. Bjornsen, "New Vistas for Federal Fire Training," pp. 10–11, *FM*, 35 (Fall 1974). See James B. David, "Building Professionalism into Forest Fire Suppression," *AF*, 77 (July 1979), 423–426, for a description of the National Interagency Fire Qualification System.

122. Quoted in *Early Days in the Forest Service*, 4 vols. (Missoula, Mont.: U.S. Forest Service, 1944–1976), 1:38.

123. Charles H. Scribner, "A Smoke Chaser," in Guthrie, *Forest Fire and Other Verse*, p. 228.

7: THE COLD WAR ON FIRE

1. Keith Arnold, "The Malibu Fires. Los Angeles and Ventura Counties, California. Fire Behavior—December 26 to 30, 1956" (California Forest and Range Experiment Station, U.S. Forest Service, 1957), p. 13.
2. Quoted in John W. Chaffin et al., "Fire in the Northern Rockies, 1967: Some Lessons for Wartime Fire Defense" (U.S. Forest Service, 1967), p. 5.
3. Horatio Bond, ed., *Fire and the Air War* (Boston: National Fire Protection Association, 1946), p. 252.
4. General surveys of fire and warfare include Stockholm International Peace Research Institute (SIPRI), *Incendiary Weapons* (Cambridge, Mass.: MIT Press, 1975); Michael McClintock et al., *Air, Water, Earth, Fire: The Impact of the Military on World Environmental Order*, Sierra Club Special Publication, International Series no. 2 (San Francisco, 1974); George J. B. Fisher, *Incendiary Warfare* (New York: McGraw-Hill, 1946); J. P. Partington, *History of Greek Fire and Gunpowder* (Cambridge: W. Heffer and Sons, Ltd., 1961).
5. See Gregory Thomas, "Fire and the Fur Trade," *Beaver*, Autumn 1977, pp. 34–35; Edmund Morris, *The Rise of Theodore Roosevelt* (New York: Ballantine, 1979), p. 310.
6. William G. Morris, "Forest Fires in Oregon and Washington," *Oregon Historical Quarterly*, 35 (December 1934), 323.
7. Franklin Hough, *Report on Forestry*, vol. 3 (Washington: GPO, 1882), p. 199.
8. Capt. John G. Bourke, *On the Border with Crook* (1891; rpt. Lincoln: University of Nebraska Press, 1971), pp. 354–355.
9. See SIRPI, *Incendiary Weapons*, pp. 15–17. Samson's feat is in Judges 15:5. For the bat bomb, see Leo P. Brophy et al., *The Chemical Warfare Service: From Laboratory to Field*, The United States Army in World War II, The Technical Services (Washington: Department of the Army, 1959), pp. 187–188.
10. Josephus, *The Jewish Wars*, trans. H. J. Thackery, Loeb Classical Library (New York: G. P. Putnam's Sons, 1927), bk. 3, p. 595.
11. SIRPI, *Incendiary Weapons*, p. 15.
12. Ibid.; Omer C. Stewart, "Fire as the First Great Force Employed by Man," in William L. Thomas, ed., *Man's Role in Changing the Face of the Earth* (Chicago: University of Chicago Press, 1956), pp. 120–121.
13. SIRPI, *Incendiary Weapons*, p. 16; L. G. Liacos, "Present Studies and History of Burning in Greece," in *TTFECP*, no. 13, pp. 68–70.
14. Partington, *Greek Fire;* SIRPI, *Incendiary Weapons*, p. 16.
15. H. C. Darby, "The Clearing of the Woodland in Europe," in Thomas, ed., *Man's Role*, pp. 199, 187–188.
16. Bruce Catton, *This Hallowed Ground* (New York: Pocket Books, 1971), p. 460.
17. See Brooks E. Kleber and Dale Birdsell, *The Chemical Warfare Service: Chemicals in Combat*, The United States Army in World War II, The Technical Services (Washington: United States Army, 1966), pp. 614–616. Some incendiaries were developed and used earlier, though for limited objectives. The chief American proponent of incendiary weaponry, Col. Zanetti, argued that between gas and fire, fire was the more deadly. "Gas *dissipates* while fire *propagates,*"

he noted shrewdly. For an internal history of the Chemical Warfare Service and fire weaponry, see John Mountcastle, "Trial by Fire: U.S. Incendiary Weapons, 1918–1945" (Ph.D. diss., Duke University, 1979); for the experimental cities, see especially pp. 147–148.

18. For accounts of the destruction, see Bond, *Fire and the Air War;* David Irving, *The Destruction of Dresden* (New York: Ballantine, 1963). Graphic, but less useful, are Martin Caidin, *A Torch to the Enemy* (New York: Ballantine, 1960) and *The Night Hamburg Died* (New York: Ballantine, 1961). Many studies were conducted on the firebombing, some based on German sources; see, for example, the listings under "Fire Storm Exploratory Analysis," in Robert G. Hahl, *DCPA Fire Research Bibliography* (Washington: Defense Civil Preparedness Agency, 1976).

19. The impact of the war on the Forest Service is described in Chapter 5 in "The Forester's Policy." The impact on manpower is given in Chapter 6 in "Under Fire," especially notes 95–96. For the impact on states, see Ralph Widner, ed., *Forests and Forestry in the American States: A Reference Anthology* (Missoula, Mont.: National Association of State Foresters, 1968), pp. 306, 466, and passim; C. Raymond Clar, *Evolution of California's Wildland Fire Protection System* (Sacramento: California Division of Forestry, 1969), pp. 30–32. Austin H. Wilkens, *Ten Million Acres of Timber: The Remarkable Story of Forest Protection in the Maine Forestry District (1909–1972)* (Woolwich, Me.: TBW Books, 1978), pp. 174–188, describes the operation of POW camps, which replaced in part the lost CCC camps.

20. See Robert C. Mikesh, *Japan's World War II Balloon Bomb Attacks on North America,* Smithsonian Annals of Flight no. 9 (Washington: Smithsonian Institution Press, 1977), p. 1; Bert Webber, *Retaliation: Japanese Attacks and Allied Countermeasures on the Pacific Coast in World War II* (Corvallis: Oregon State University Press, 1976).

21. See Mikesh, *Japan's World War II Balloon Bomb Attacks*; Webber, *Retaliation*; Neal N. Rahm, "The Fire Fly Project," *JF,* 44 (1946), 561–564; C. R. Clar materials (copies of letter on Fire Fly Project and interviews concerning Japanese incendiary balloons, 1945), History Office, U.S. Forest Service, Washington. The Clar interviews are intensive with respect to California. Rahm claims that the "total military effort for the four western regions amounted to 147,462 man-days of firefighting. They suppressed or assisted in the suppression of 282 fires." A gloomier picture was painted by P. D. Hanson to Chief, Forest Service, "Cooperation—Military," November 9, 1946, WNRC, RFS, Division of Fire Control. Hanson was "far from enthusiastic over Army assistance in fire control activities," and he opposed the extension of wartime arrangements into the postwar period.

22. A useful summary of OCD research into fire is provided in Hahl, *DCPA Fire Research Bibliography*. For the incendiary potential of thermonuclear armaments, see Samuel Glasstone and Philip J. Dolan, *The Effects of Nuclear Weapons,* 3d ed. (Washington: Department of Defense and Department of Energy, 1977), pp. 276–323. The National Fire Coordination Study released a host of reports. For a summary of its activities, see William R. Moore et al., "Defending the United States from Nuclear Fire: The Final Report of the National Fire

Coordination Study" (U.S. Forest Service, 1966). A summary of civil defense considerations can be found in *DCPA Attack Environment Manual* (Washington: Defense Civil Preparedness Agency, 1973).

23. SIRPI, *Incendiary Weapons,* p. 43.

24. For descriptions of these tactics in World War II, see Mountcastle, "Trial by Fire," pp. 166–167, 200–201. See Robert B. Batchelder and Howard F. Hirt, "Fire in Tropical Forests and Grasslands," United States Army Natick Laboratories, Technical Report 67–41–ES (1966); Charles H. Warton, "Man, Fire, and Wild Cattle in North Cambodia," in *TTFECP,* no. 5, pp. 23–66, and "Man, Fire and Wild Cattle in Southeast Asia," in *TTFECP,* no. 8, pp. 107–167; Robert B. Batchelder, "Spatial and Temporal Patterns of Fire in the Tropical World," in *TTFECP,* no. 6, pp. 171–208. Peter Kunstadter et al., eds., *Farmers in the Forest* (Honolulu: University Press of Hawaii, 1978), includes a fine series of photos showing slash-and-burn operations.

25. See SIRPI, *Ecological Consequences of the Second Indochina War* (Stockholm: Almqvist and Wiksell, 1976), pp. 58–62; Deborah Shapley, "Technology in Vietnam: Fire Storm Project Fizzled Out," *Science,* 177 (July 21, 1972), 239–241.

26. Quoted in SIRPI, *Incendiary Weapons,* p. 59.

27. Ibid.

28. Report of the Secretary-General, *Napalm and Other Incendiary Weapons and All Aspects of Their Possible Use* (New York: United Nations, 1973).

29. SIRPI, *Incendiary Weapons.* p. 70.

30. Homer, *The Iliad,* trans. W.H.D. Rouse (New York: Mentor Books, 1962), p. 244.

31. David P. Godwin and Alan Macdonald, "The New Fire Crusade," *AF,* 49 (April 1943), 180.

32. Quoted in U.S. Congress, House, *Forest Fire Control in Southern California* (Engle Committee Hearings), H. R. Serial No. 14, 85th Cong., 1st sess. (Washington: GPO, 1958), p. 16.

33. Raymond Chandler, *Playback* (New York: Ballantine, 1977), p. 90. The novel was first published in 1958, thus making it contemporaneous with the great transformation in the Southern California fire regime.

34. Quoted in "Inaja Forest Forest Fire Disaster" (U.S. Forest Service, 1957), p.1.

35. For the Inaja fire, see "Inaja Forest Fire Disaster"; Clive Countryman et al., "Fire Behavior 1930 to 2015 Hours, Inaja Fire Disaster" (California Forest and Range Experiment Station, U.S. Forest Service, 1956); and assorted reports, Files, Region. Five, U.S. Forest Service, San Francisco. The official reports stored in the WNRC have disappeared. For the Malibu fires, see National Board of Fire Underwriters and Board of Fire Underwriters of the Pacific, "The Malibu Fires" (1957); and Keith Arnold, "The Malibu Fires." See also Files, Los Angeles County Fire Department.

36. Arnold, "Malibu Fires," p. 1.

37. H. Edward Russell to Howard Earl, January 8, 1957, Files, Los Angeles County Fire Department.

38. Howard Earl, "Civil Defense: Report of Malibu Fire Operations," p. 1, Files, Los Angeles County Fire Department.

39. National Board of Fire Underwriters and Board of Fire Underwriters, "Malibu Fires," p. 13.

40. Capt. Harold W. Greenwood, "Bel-Air–Brentwood and Santa Ynez Fires: Worst Fire in the History of Los Angeles," Official Report of the Los Angeles Fire Department (1962), p. 19.

41. Rexford Wilson, "Los Angeles Conflagration of 1961: The Devil Wind and Wood Shingles," *NFPA Quarterly,* 55 (January 1962), 279. See also, Greenwood, "Bel-Air," p. 16.

42. For brief fire histories of the region, see Don R. Bauer, "A History of Forest-fire Control in Southern California," in Murray Rosenthal, ed., *Proceedings of the Symposium on Living with the Chaparral* (Sierra Club, 1974); S. B. Show et al., "Planning Basis for Adequate Fire Control on the Southern California National Forests," *FCN,* 5 (January 1941), 2–5, 27–29; Louis Barrett, "A Record of Forest and Field Fires in California From the Early Explorers to the Creation of the Forest Reserves" (U.S. Forest Service, 1935); and Clar, *California Government and Forestry.* Two histories of the national forests in the region are available: William S. Brown, "History of Los Padres National Forest" (San Francisco: U.S. Forest Service, 1945); and W. W. Robinson, *The Forest and the People* (Los Angeles: Title Insurance and Trust Co., 1946). Despite its title, Richard Lilliard's *Eden in Jeopardy* (New York: Alfred A. Knopf, 1966) helps to bring the forest histories up to date, after a fashion. Also valuable is Charles W. Philpot, "The Changing Role of Fire on Chaparral Lands," in Rosenthal, ed., *Symposium,* pp. 131–150, which provides a kind of historical geography of large fires based on fuel histories.

43. See citations in Chapter 2, note 124, for a summary of relevant fuelbreak references. Also pertinent are Philip Omi, "Long-Term Planning for Wildland Fuel Management Programs" (Berkeley: University of California, 1977); J. Michael Harrison et al., "Decision Analysis of Wildland Fire Protection: A Pilot Study" and "Decision Analysis of Fire Protection Strategy for the Santa Monica Mountains: An Initial Assessment" (Stanford Research Institute, 1973).

44. F. H. Raymond, in Engle Committee Hearings, p. 26.

45. See Harold J. Biswell, "The Effects of Fire on Chaparral," in T. T. Kozlowski and C. E. Ahlgren, eds., *Fire and Ecosystems* (New York: Academic Press, 1974), pp. 321–364; H. L. Shantz, *The Use of Fire as a Tool in the Brush Ranges of California* (Sacramento: California State Board of Forestry, 1947), though most of the brush considered is that typical of the northern part of the state; and Curtis Tunnell et al., "The Wildfire Threat" (Santa Barbara: County Board of Supervisors, 1964).

46. By far the most comprehensive summary of fire protection programs by governmental bodies is that contained in Clar, *California Government and Forestry.* A state fire protection statute adopted in 1905 made possible the county organizations. A useful distillation is available in Clar, *Evolution of California's Wildland Fire Protection System.*

47. Engle Committee Hearings, p. 207, 4; A. W. Greeley et al., "Report of Task Force to Recommend Action to Reduce the Chances of Men Being Killed By Burning While Fighting Fire" (U.S. Forest Service, 1957).

48. "Proceedings of the Symposium on the Environmental Consequences of Fire and Fuel Management in Mediterranean Ecosystems," General Technical

Report WO–3 (U.S. Forest Service, 1977). See also Carl C. Wilson, "Comparing Forest Fire Problems in the Mediterranean Region and in California," Special Paper, FAO/UNESCO Technical Consultation on Forest Fires in the Mediterranean Region (New York: United Nations, 1977).

49. Wilson, "Los Angeles Conflagration of 1961," p. 286.

50. Quotation on the title page of "Inaja Fire Disaster."

51. Douglas R. Leisz and W. A. Powers, "Fire and Drought: Bad Mix for a Dry State," *FMN,* 38 (Fall 1977), 3–7.

52. Brown, "History of the Los Padres National Forest," p. 138.

53. Roger Byrne et al., "Fossil Charcoal as a Measure of Wildfire Frequency in Southern California: A Preliminary Analysis," in "Proceedings of the Symposium on the Environmental Consequences," pp. 361–367.

54. Morris H. McCutchan, "Climatic Features as a Fire Determinant," in "Proceedings of the Symposium on the Environmental Consequences," pp. 1–11. For comparison, see Z. Naveh, "The Ecology of Fire in Israel," in *TTFECP,* no. 13, pp. 131–170.

55. A description of foehn winds in the United States is given in Charles Buck and Mark Schroeder, *Fire Weather: A Guide for Application of Meteorological Information to Forest Fire Control,* U.S. Department of Agriculture, Handbook 360 (Washington: GPO, 1970). See also William T. Sommers, "On Forecasting Strong Mountain Downslope Winds," in *Fifth Joint Conference on Fire and Forest Meteorology* (Boston: American Meteorological Society, 1978), pp. 38–43; and Michael A. Fosberg et al., "Some Characteristics of the Three-Dimensional Structure of Santa Ana Winds," Research Paper PSW–30 (U.S. Forest Service, 1966).

56. See Biswell, "Effects of Fire on Chaparral," and Philpot, "Changing Role of Fire on Chaparral Lands." Somewhat dated is Shantz, *Use of Fire.* Useful for comparative purposes is Henry N. LeHouerou, "Fire and Vegetation in the Mediterranean Basin," in *TTFECP,* no. 13, pp. 237–278.

57. See S. B. Show and E. I. Kotok, *Fire and the Forest (California Pine Region),* U.S. Department of Agriculture, Department Circular 358 (Washington: GPO, 1925), pp. 8–10; Coert duBois, *Trailblazers* (Stonington, Conn. 1957), p. 61; Don R. Bauer to Stanley N. Hirsch, comments on Stanford Research Institute Project MSU–2275, September 19, 1973, pp. 3–4, personal copy.

58. See Shantz, *Use of Fire;* Clar, *California Government and Forestry,* 2:275–307; Biswell, "Effects of Fire on Chaparral."

59. Don R. Bauer to Dennis Roth, "Re: History of Fire Mgmt. in Southern California," [ca. May 1979], p. 2, personal copy. I am indebted to Mr. Bauer for a thorough and thoughtful critique of an early draft of this subchapter. Though we did not always agree, his review measurably improved the subsequent revisions.

60. Charles J. Kraebel, "The La Crescenta Flood," *AF,* 40 (June 1934), 251–254, 286–287. See also Biswell, "Effects of Fire on Chaparral"; and L. F. DeBano et al., "Fire's Effects on Physical and Chemical Properties of Chaparral Soils," in "Proceedings of the Symposium on the Environmental Consequences," pp. 65–74. For a critical summary of Forest Service research on watershed of all varieties, including chaparral, see Ashley Schiff, *Fire and Water: Scientific Heresy in the Forest Service* (Cambridge, Mass.: Harvard University Press, 1962).

61. Lowell John Bean and Harry W. Lawton, introductory article, in Henry T. Lewis, *Patterns of Indian Burning in California: Ecology and Ethnohistory*, Ballena Press Anthropological Papers, no. 1 (Ramona, Calif.: Ballena Press, 1974), pp. xix–xx.

62. Quoted in Clar, *California Government and Forestry*, 1:7.

63. Quoted by Bean and Lawton, in Lewis, *Patterns of Indian Burning*, p. xix.

64. Quoted in Barrett, "Record of Forest and Field Fires," pp. 52–53.

65. The full proclamation is quoted in Clar, *California Government and Forestry*, 1:8–9.

66. Quoted in ibid., pp. 9–10. See also Lewis, *Patterns of Indian Burning*, for a fuller, if convoluted, presentation.

67. Bean and Lawton, in Lewis, *Patterns of Indian Burning*, p. xxii.

68. Richard Henry Dana, *Two Years Before the Mast* (New York: New American Library, 1964), p. 58.

69. Quoted in Clar, *California Government and Forestry*, 1:11.

70. John Muir, *The Mountains of California* (Garden City, N.Y.: Doubleday and Co., 1961), p. 154.

71. See Clar, *California Government and Forestry*, 2:275–307.

72. Brown, "History of the Los Padres National Forest," p. 132. See also the many entries in Barrett, "Record of Forest and Field Fires," for similar accounts of fire control.

73. Descriptions of herder fires are given in Barrett, "Record of Forest and Field Fires," esp. pp. 161–162; Brown, "History of the Los Padres National Forest." See also the reports made during the survey of the forest reserves in the late 1890s; for example, John Leiberg, "The San Gabriel Forest Reserve," pp. 411–428, and "The San Bernardino Forest Reserve," pp. 429–454, in *Nineteenth Annual Report of the United States Geological Survey*, pt. 5: *Forest Reserves* (Washington: GPO, 1899.)

74. See Robert Kelley, *Gold vs. Grain: The Hydraulic Mining Controversy in California's Sacramento Valley* (Glendale, Calif.: Arthur Clarke Publishing Co., 1959); Stephen J. Pyne, *Grove Karl Gilbert: A Great Engine of Research* (Austin: University of Texas Press, 1980), pp. 207–211, 244–254.

75. Brown, "History of the Los Padres National Forest," p. 33.

76. W. L. Vestal et al., "Report of the Committee on Forest and Water" (San Bernardino Board of Trade, 1904), pp. 2–4; Leiberg, "San Gabriel Forest Reserve" and "San Bernardino Forest Reserve," passim.

77. See Bauer, "History of Fire Control in Southern California"; Show et al., "A Planning Basis for Adequate Fire Control," pp. 3–4; Clar, *California Government and Forestry*, 1:268–291 passim.

78. See Brown, "History of the Los Padres National Forest," pp. 81–82, 134.

79. Vestal et al., "Report of the Committee on Forest and Water," pp. 1, 4.

80. See Bauer, "History of Fire Control"; Show et al., "A Planning Basis for Adequate Fire Control," pp. 3–4. Clar, *California Government and Forestry*, 1:544–545, provides an excellent summary.

81. Tri-Counties Reforestation Committee, "Forest Fires on San Bernardino Range: Review of Conflagration of 1911" (Los Angeles, 1911), pp. 9–10.

82. See Brown, "A History of the Los Padres National Forest," pp. 121–123; Show et al., "A Planning Basis for Adequate Fire Control," pp. 4–5; S. B. Show

and E. I. Kotok, History of Fire Research in California, unpublished report, title page missing, U.S. Forest Service, ca. 1956, WNRC, RFS, p. 33. For the 1924 season see George C. Henderson, "The Forest Fire War in California," *AF,* 30 (September 1924), 531–534.

83. The 1970 fire season brought major investigations and numerous reforms. Among the most important: Clinton Phillips, "California Aflame! September 22–October 4, 1970" (Sacramento: California Division of Forestry, 1971); Task Force on California's Wildland Fire Problem, "Recommendations to Solve California's Wildland Fire Problem" (Sacramento: California Department of Conservation, 1972). These may be compared with similar programs launched after the 1956 season and after the 1964 season (which featured the Coyote fire outside Santa Barbara): Senate Fact Finding Committee on Natural Resources, "Fire Problems of California" (California State Senate, 1964), and Senate Committee on Natural Resources, "Attacking California's Fire Problems" (California State Senate, 1964). Also growing out of the 1970 disaster was a follow-up study: "Alternative Systems of Providing Fire Protection to Life and Property on California's Privately Owned Wildlands" (Sacramento: California Division of Forestry, 1974). The FIRESCOPE program is well described in Richard Chase, "Firescope: A Regional Solution to Multi-Agency Coordination Problems," *International Fire Chief,* 43 (November 1977), 18–21. For the 1977 season, see Arnold Hartigan, "A Wrapup of the 1977 Forest Fire Season," pp. 4–5, and Myron K. Lee, "Marble-Cone/Big Sur Fire: From the Command Point of View," pp. 6–8, in *International Fire Chief,* 43 (November 1977); Leisz and Powers, "Fire and Drought." Also valuable is M. M. Nelson, "The Place of Southern California in the Nation's Fire Problem," *FCN,* 25 (January 1964), 4–5.

84. Accounts of the Matilija fire are in Brown, "History of the Los Padres National Forest," pp. 141–144; and Wallace Hutchinson, "The Battle of Matilija Canyon," *AF,* 39 (January 1933), 3–5, 26.

85. Brown, "History of the Los Padres National Forest," p. 142.

86. Quoted in Hutchinson, "Battle of Matilija Canyon," p. 4.

87. Ibid., p. 3.

88. Raymond Chandler, "Red Wind," in *Trouble Is My Business* (New York: Ballantine, 1964), p. 187.

89. Quoted in Widner, ed., *Forests and Forestry,* p. 236.

90. Arthur Hartman, "Machines and Fires in the South," *Trees. Yearbook for Agriculture, 1949* (Washington: GPO, 1949), p. 528.

91. The original statute is given in *The Principal Laws Relating to Forest Service Activities,* Agriculture Handbook no. 453 (Washington: GPO, 1974), p. 81.

92. Gifford Pinchot, *Breaking New Ground* (1947; rpt. Seattle: University of Washington Press, 1972), p. 277.

93. Widner, ed., *Forests and Forestry,* is a wonderful compendium of such information, but the material is widely scattered and poorly indexed.

94. For an interesting survey of early tool development, see Coert duBois, *Systematic Fire Protection in the California Forests* (Washington: GPO, 1914), and Daniel Adams, *Methods and Apparatus for the Prevention and Control of Forest Fires, as Exemplified on the Arkansas National Forest,* U.S. Forest Service, Bul-

letin 113 (Washington: GPO, 1912). Useful overviews are available in H. T. Gisborne, "Forest Fire—A Mother of Invention," *AF*, 32 (May 1926), 265–268, 300; W. B. Osborne, "Fire Protection Equipment Accomplishments and Needs," *JF*, 26 (December 1931), 1195–1201; and the survey of equipment presented in "Fire Control Meeting. Spokane, Washington. February 10–21, 1936" (Spokane Fire Control Meeting) (U.S. Forest Service, 1936), pt. 1, "Equipment Section"; David P. Godwin, "The Evolution of Fire-Fighting Equipment," *AF*, 45 (April 1939), 205–207, 235–237.

95. See J. H. Coats, "Communications in the National Forests of Region One," unpublished report, Files, Region One, U.S. Forest Service; C. J. Buck, "How Telephones Saved Lives," *AF*, 16 (November 1910), 648–651; "Carrier Pigeons Aid Forests," *AF*, 25 (November 1919); Adams, *Methods and Apparatus*, pp. 13–14; E. Clayton McCarty, "The Forest Service Heliograph," in *Early Days in the Forest Service*, 4 vols. (Missoula, Mont.: U.S. Forest Service, 1944–1976), 4:148–151.

96. Adams, *Methods and Apparatus*, pp. 18–21.

97. Harry Pollard, "A Railroad Wars on Fire," *AF*, 37 (August 1931), 486–488.

98. For information about the invention of the pulaski, see letter of Eric P. Dudle, August 24, 1955, and Ruby E. Hult, "How the Pulaski Tool Became a Popular Tool" (from the Spokane *Review*), Files, History Office, U.S. Forest Service, Washington. Also useful is Joe Halm, "Some of the Highlights of My Career in the Forest Service," in *Early Days in the Forest Service*, 1:76.

99. Eloise Hamilton, *Forty Years of Western Forestry* (Portland: Western Forestry and Conservation Association, 1949), pp. 33, 50.

100. "Committee Reports. Approved by the Fire Conference and Comments by the Forester. Conference Held at Mather Field, California, from November 14 to 27, 1921" (U.S. Forest Service, 1921), pp. 28–30.

101. Spokane Fire Control Meeting, pt. 1, "Equipment Section," pp. 3, 22.

102. See discussion given in ibid., pp. 38–39.

103. See Joseph J. Davis, "Tank Truck Fire Apparatus," *JF*, 30 (1935), 673; Clar, *California Government and Forestry*, 1:418, 2:28–29.

104. Show and Kotok, History of Forest Fire Research, p. 38; Howard R. Jones, "The Modern Bulldozer. A Forest Service Project," in *Early Days in the Forest Service*, 4:136–139; "The Bulldozer" (Missoula, Mont.: U.S. Forest Service, 1955); Arthur Victor, "The Ubiquitous Bulldozer," *Pacific Northwesterner*, 19 (Winter 1975), 1–8.

105. Quoted in "The Bulldozer," p. 2. For a description of logistics by pack train in the Rockies, see Jane Reed Benson, "Thirty-Two Years in the Mule Business: The USDA/Forest Service Remount Depot and Winter Range" (Missoula, Mont.: Lolo National Forest, 1980).

106. See Spokane Fire Control Meeting, pt. 1, "Equipment Section." The conference also led to a national survey by equipment engineer G. W. Duncan, "Power Tools for Fire Line Construction." Headley to Regional Foresters, "Fire," July 27, 1937, NA, RG 95, Division of Fire Control, General Correspondence, includes the full report.

107. Spokane Fire Control Meeting, pt. 1, "Equipment Section," pp. 38–39.

108. Ibid., p. 46.

109 For the development of the tractor-plow, see Hartman, "Machines and Fire in the South." Widner, ed., *Forests and Forestry,* p. 497, credits L. W. Lembcke with the feat in 1931, thus making his experiment contemporaneous with the bulldozer trials in the Rockies. But see, too, "Michigan Forest Fire Experiment Station" (Michigan Department of Conservation, 1932), p. 24, which describes five types of plows then under review, and J. A. Mitchell and D. Robson, "Forest Fires and Forest Fire Control in Michigan" (Michigan Department of Conservation, 1950), pp. 43–45.

110. "Fire Control Equipment Meeting. Washington, D.C. January 25 to February 3, 1945" (U.S. Forest Service, 1945), pp. 1, 10, 4.

111. For wartime developments, see David P. Godwin, "Development of Fire-fighting Equipment Since Pearl Harbor," *JF,* 40 (1945), 104–108. For postwar developments, see "Equipment Development Reports," 2 vol. (Washington: U.S. Forest Service, 1944–1957); "Report of Aerial Bombing Evaluation Board" (Missoula, Mont.: U.S. Forest Service, August 1947), WNRC, RFS, Division of Fire Control; Ira C. Funk and Fred W. Milam, "Report on U.S. Army Air Force–Forest Service Cooperative Test on Helicopters" (U.S. Forest Service, 1947), WNRC, RFS, Division of Fire Control; Corps of Engineers, U.S. Army Engineer Research and Development Laboratories, and U.S. Forest Service, "Helicopter Fire Fighting Program" (U.S. Forest Service, 1956); "Western Air Attack Review: Redding, California, April 10–12, 1957" (U.S. Forest Service, 1957); Ira C. Funk, "A Development Center for Forest-Fire-Control Equipment," WNRC, RFS, Division of Fire Control; "Missoula Equipment Development Center. Fact Sheet. December 1, 1965" (Missoula, 1965); "Fire Control Equipment Meeting. Washington, D.C., 1945"; "Fire Control Equipment Meeting, Washington, D.C., February 16 through February 19, 1948" (U.S. Forest Service, 1948); Chief to Regional Foresters and Directors, Memorandum of December 17, 1947, with Circular Letters of November 12 and 26, 1947, attached, "Inter-regional Fire Control and Equipment Development, Meeting," WNRC, RFS, Division of Fire Control; "Air Operations Meeting. Arcadia, California. March 1959" (U.S. Forest Service, 1959). Also valuable, of course, are the annual reports, photo archives, and other records held at the equipment centers themselves.

112. See citations in Chapter 4, note 71. The Roscommon Center also has some historic records, reports, and photos on file. For a popular description of its activities at mid-career, see Gilbert I. Stewart, "Of Fires and Machines," *Michigan Conservation,* 24 (January–February 1955), 15–18. In lieu of an equipment center, the Southern Region of the Forest Service maintains a regional equipment specialist.

113. Malcolm Hardy, "The Use of Aircraft in Forest Fire Control" (M.A. thesis, University of Washington, 1946), p. 124.

114. The best summaries of early aircraft use are Hardy, "Use of Aircraft in Forest Fire Control"; and Henry Clepper, *Professional Forestry in the United States* (Baltimore: Johns Hopkins University Press, 1971), pp. 166–186. Also valuable for the Army program is Clar, *California Government and Forestry,* 1:459. A history of Canadian developments is in Bruce West, *The Firebirds* (Ontario: Ministry of Natural Resources, 1974).

115. Clepper, *Professional Forestry,* pp. 168–175; duBois, *Trailblazers,* pp. 81–

82; Hardy, "Use of Aircraft in Forest Fire Control," pp. 7–37. A considerable body of correspondence from the public about the patrols is stored in NA, RG 95, RFS, Division of Fire Control.

116. Clepper, *Professional Forestry,* p. 171.

117. NA, RG 95, RFS, Division of Fire Control, General Correspondence. Some of the dirigible developments are described by Hardy, "Use of Aircraft in Forest Fire Control," p. 24.

118. "Servicewide Meeting. Fire Control, Fire Research, and Safety. Ogden, Utah. January 16 through January 20, 1950" (U.S. Forest Service, 1950), Appendix, p. 14.

119. In addition to Clepper, *Professional Forestry,* and Hardy, "Use of Aircraft in Forest Fire Control," Forest Service aircraft history is summarized in Monte K. Pierce, "Chronological History of Forest Service Air Operations," in "National Fire Chiefs Meeting Proceedings. November 30, 1972" (U.S. Forest Service, 1972).

120. Reports, Aerial Fire Control Project, WNRC, RFS, Division of Fire Control. The original plan is given in the "Comprehensive Report for the Period July 1, 1936 to July 1, 1937." Clepper, *Professional Forestry,* p. 179; "History of Smokejumping" (Missoula, Mont.: U.S. Forest Service, 1976).

121. "Report of Aerial Bombing Evaluation Board" and "Aerial Bombing of Forest Fires" (U.S. Forest Service, 1947), WNRC, RFS, Division of Fire Control.

122. "Servicewide Meeting. Fire Control, Fire Research, and Safety. Ogden, 1950," p. 9. The Firestop research was issued as a series of progress reports; see "Aerial Firefighting," Progress Report no. 9 (June 15, 1955), for a description of the air tanker experiments. The California Division of Forestry and the Los Angeles County Fire Department successfully completed bulk water drops in 1953 and hence furnished a working precedent.

123. Joseph B. Ely et al., "Air Tankers—A New Tool for Forest Fire Fighting," Western Air Attack Review (U.S. Forest Service, 1957), p. 1.

124. A useful summary of air tanker operations is in Arnold F. Hartigan, "Chronology of Air Tanker Use on Fires by the U.S. Forest Service" (BIFC, 1977). A mixture of history and technique is contained in Alexander Linkewich's very interesting *Air Attack on Forest Fires* (Calgary: Friesen and Sons, 1972), which includes Canadian as well as American developments. Nor should helicopter use be ignored; in the postwar era, it has matched developments in air tankers stride for stride. Of particular interest for the early years is Funk and Milam, "Report on U.S. Army Air Force–Forest Service Cooperative Test on Helicopters"; Corps of Engineers and U.S. Forest Service, "Helicopter Fire Fighting Program"; David P. Godwin, "Helicopter Hopes for Fire Control," *FCN,* 7 (April 1946), 16–21; Frank J. Jefferson, "The Helicopter—A New Factor in Fire Control," *FCN,* 9 (January 1948), 1–9; "Fire Control Experiments—A Story of Helitack Operation in Northwest California" (California Division of Forestry, 1963); "Aerial Firefighting—The Helitanker," Firestop Progress Report no. 10 (June 15, 1955); Ralph G. Johnston, "Helicopter Use in Forest Fire Suppression," which includes a nice bibliography.

125. See Carl Burgtorf, "The Forest Service Specification Program for Fire Equipment," *FCN,* 25 (January 1964), 9–10.

126. See "Fire Management Safety and Health" (MEDC, 1977), which summa-

rizes "Equip Tips" notices on fire protective gear, and Richard G. Ramberg, "Firefighters Physiological Study: Firefighting Efficiency of Man—The Machine" (MEDC, 1974). The study was begun in 1969, and a preliminary progress report, "Firefighting Efficiency of Man—The Machine" (MEDC, 1971), was issued two years later. A summary of the program is given in Richard G. Ramberg and Arthur H. Jukkala, "Firefighters Work Environment and Physical Demands Studied," *FMN*, 36 (Summer 1975), 16–18. The step tests, as a basic submaximal test of aerobic fitness, was made mandatory for fireline workers partly as a result of these investigations; see Jim Abbott, "Physical Fitness for Firefighters: Can You Measure Up?" *FMN*, 36 (Summer 1975), 3–5.

127. The infrared program was developed with OCD help in the early 1950s as part of the Firescan project and was transferred to ARPA in 1962. For a summary, see Ralph A. Wilson et al., "Airborne Infrared Forest Fire Detection System: Final Report," Research Paper INT–93 (U.S. Forest Service, 1971); Robert F. Kruckeberg, "No Smoke Needed," *FCN*, 32 (Spring 1971), 9–11. For a sample of remote sensing equipment extending the range of earlier equipment, see Herbert J. Shields, "Night Helicopter Operations Steering Committee: A Chronology of the Project" (U.S. Forest Service, 1975).

128. See Robert L. Bjornsen and Richard A. Chase, "Computer Simulates Fire Planning Problem," *FCN*, 32 (Fall 1971), 12–13; Theodore Storey, "FOCUS: A Computer Simulation Model for Fire Control Planning," *Fire Technology*, 8, no. 2 (1973), 91–103.

129. "Fire Equipment Seminar. Forty Years of Equipment Progress. Marana, Arizona. March 23–27, 1970" (U.S. Forest Service, 1970); National Academy of Sciences–National Research Council, *Proceedings: Symposium on Employment of Air Operations in the Fire Services* (Washington: National Academy of Sciences, 1971). For a glowing report on the Dragon Wagon, see "1976 Test Report: Twister Dragon Wagon" (Washington: BLM, 1976).

130. D. H. Swanson et al., "Air Tanker Performance Guides," General Technical Report INT–27 (U.S. Forest Service, 1976).

131. National Commission on Fire Prevention and Control, *America Burning* (Washington: GPO, 1973), pp. 111, 30.

132. Useful as a general chronology is Mary U. Harris, "Significant Events in United States Civil Defense History" (Defense Civil Preparedness Agency, 1975). Most of the documents relating to Forest Service involvement are stored at WNRC, RFS, 63–A–4027, Box 48, and 70–A–6533, Box 49.

133. See National Rural Fire Defense Committee to Chief, U.S. Forest Service, "DM-PLANS Civil Defense, Rural Fire Defense Guidelines," July 25, 1955, WNRC, RFS, Division of Fire Control. The precedent for this—like so many other liaisons with the military—grew out of World War II, in this case when the Secretary of Agriculture in 1942 assigned wartime rural fire responsibilities to state and federal agencies, primarily under the auspices of the Forest Service. See George W. Gustafson, "Montana Rural Fire Fighters Service," *FCN*, 12 (October 1951), 22.

134. A useful summary through the mid-1950s is contained in Forester, Southern Region, U.S. Forest Service, to State Foresters, "DM-RURAL FIRE DEFENSE. Planning Guidelines," October 1, 1956, WNRC, RFS, Division of

Fire Control, and "Changing Aspects of Fire Research," Report of Region Nine Fire Committee Meeting, Milwaukee, 1958 (U.S. Forest Service, 1958), pp. 15–19. Added responsibilities were assigned following the Cuban missile crisis; the results are outlined in E. Ritter, "Rural Fire Defense Project 107," in "National Forest Fire Control Workshop Report. St. Louis, Missouri. March 7–12, 1965" (U.S. Forest Service, 1965), pp. 62–63.

135. National Fire Coordination Study, "The Study of Large Fires" (U.S. Forest Service, 1964), p. 2. For a summary of the extensive publications and findings, see William R. Moore et al., "Defending the United States from Nuclear Fire: The Final Report of the National Fire Coordination Study" (U.S. Forest Service, 1966).

136. William R. Moore and James W. Jay, "Rural Fire Defense Training. Project OCD–PS–64–107: A Summary of History and Progress Through 9/30/64," National Fire Coordination Study (U.S. Forest Service, 1964); James W. Franks, "The Civil Defense Staff and Command School—A First Step in Co-ordination of Wildland and Structural Fire Fighting," SAF, *Proceedings* (1962), pp. 63–64. The appearance of the simulator was announced in Nolan C. O'Neal and Bert E. Holtby, "The Fire Control Simulator," *FCN*, 24 (April 1963), 25–31.

137. Office of Emergency Planning, *Forest and Grass Fires*, Senate Document no. 30, 90th Cong., 1st sess. (Washington: GPO, 1967), p. 19.

138. *Fire Research and Safety Act of 1967, Hearings . . . on S.1124*, Serial No. 90–5, 90th Cong., 1st sess. (Washington: GPO, 1967).

139. National Commission on Fire Prevention and Control, *America Burning; Fire Prevention and Control, Hearings . . . U.S. House of Representatives*, 93rd Cong., 1st sess., no. 15 (Washington: GPO, 1973); Public Law 93–498, 93rd Cong., 2d sess., S. 1769, October 29, 1974. The 1973 hearings opened with a statement by Robert Steele, Representative from Connecticut, in which he confessed that his active interest in the problems of fire services stemmed in large measure from reading Dennis Smith's *Report from Engine Company 82*, an account of a fire company in the South Bronx. Smith's book, that is, may actually have been more helpful in promoting the act than the report of the presidential commission.

140. For descriptions of the program, see "Rural Community Fire Protection" (U.S. Forest Service, 1977) and "How the Rural Community Fire Protection Program Can Help You" (U.S. Forest Service, 1975). For examples of the program in operation, see R. Michael Bowman, "Rural Fire Prevention and Con-trol—A Time of Awakening," *FMN*, 40 (Spring 1979), 16–18; James C. Sor-enson, "Federal Excess Property in the Rural Fire Department," *FMN*, 39 (Summer 1978), 13–14, 18; H. Ames Harrison, "The Rural Community Fire Protection Program After Two Years of Operation in the Northeast," *FMN*, 38 (Winter 1977), 12–13; Thomas R. Fontaine, "Rural Fire Defense Program Initiated in Georgia," *FM*, 34 (Summer 1973), 8–9.

141. Louis G. Jekel, "Rural/Metro—A Commercial Approach," in Richard J. Barney, ed., *Fire Control for the 80's* (Missoula, Mont.: Intermountain Fire Council, 1980), pp. 227–234.

142. For comparative history, see Paul C. Ditzel, *Fire Engines, Firefighters* (New York: Crown Publishers, 1976); Paul Robert Lyons, *Fire in America!* (Boston:

National Fire Protection Association, 1976); and, of less value, *Dennis Smith's History of Firefighting in America* (New York: Dial Press, 1978).

143. For a synopsis of NFPA history and programs, see National Fire Protection Association, "The Fire Problem: A Statement by the National Fire Protective Association," reproduced in *Fire Prevention and Control, Hearings.*

144. Statement by Dr. Hoyt Hottel, in *Fire Research and Safety Act of 1967, Hearings,* p. 60.

145. Quoted in *Dennis Smith's History,* p. 13.

146. See Ditzel, *Fire Engines, Firefighters,* p. 244.

8: FIELDS OF FIRE

1. William Greeley, "Better Methods of Fire Control," SAF, *Proceedings,* 6 (1911), p. 165.

2. C. E. Hardy, "Conflagration in Alaskan Forests, 1957" (Ogden, Utah: Intermountain Forest and Range Experiment Station, U.S. Forest Service, 1957), p. 1.

3. Henry Graves, *Protection of Forests from Fire,* U.S. Forest Service, Bulletin 82 (Washington: GPO, 1910), p. 7.

4. H. T. Gisborne, "Review of Problems and Accomplishments in Fire Control and Fire Research," in "Proceedings of the Priest River Fire Meeting" (U.S. Forest Service, 1941), p. 5.

5. A. A. Brown wrote a piece for FAO in 1947, but the major publication came from Carl Wilson, *Elements of Fire Control* (New York: FAO, 1953). For a synopsis of Forest Service involvements, see Robert K. Winters, "International Forestry in the U.S. Department of Agriculture" (Washington: Department of Agriculture, 1980). For a good survey of fire protection services globally, see *Proceedings of the United Nations Scientific Conference on the Conservation and Utilization of Resources,* vol. 5: *Forest Resources* (New York: United Nations, 1951). See in particular A. A. Brown, "Forest Fire Control," pp. 34–40. Compare this state of affairs with that described in Carl Wilson, "Developing a Global Programme in Integrated Fire Management," in "Global Forestry and the Western Role," in *Proceedings of the Permanent Association Committees* (Portland: Western Forestry and Conservation Association, 1975), pp. 4–8.

6. For an overview, see James Sorenson, "Seventeen Years of Progress Through International Cooperation, 1962–1978" (Atlanta: U.S. Forest Service, 1979), which describes the workings of the Fire Management Study Group (name changed in 1971). Its publication, *Forest Fire News,* is issued periodically. See also Henry W. DeBruin, "International Fire Control," *Annual Meeting of the Western Forest Fire Committee* (Portland: Western Forestry and Conservation Association, 1972), pp. 4–6.

7. Merle S. Lowden, "From Firefighting to Revolution in Three Days," *AF,* 71

(August 1965), 16–19, 50, and "Fire Crisis in Brazil," *AF,* 71 (January 1965), 42–44, 46.

8. G. S. Kirk, *Heracleitus: The Cosmic Fragments* (Cambridge: Cambridge University Press, 1970), pp. 345–348. Heracleitus also promised that fire would provide a final judgment, an idea picked up by the Stoics and others. It should be remembered, however, that even in his own day Heracleitus was known as Heracleitus "the Dark" or "the Obscure" for his delight in enigmatic pronouncements. His "fire," moreover, has a cosmic role not unlike the "water" proposed earlier by Thales, another Ionian pre-Socratic.

9. John Donne, "An Anatomy of the World," in *Bartlett's Familiar Quotations,* 14th ed. (Boston: Little, Brown, 1968), p. 307.

10. Quoted in Gaston Bachelard, *The Psychoanalysis of Fire* (Boston: Beacon Press, 1968), p. 60. Among general histories of fire research and metaphysics, see Joshua C. Gregory, *Combustion from Heracleitus to Lavoisier* (London: Edward Arnold, 1934); J. H. White, *The History of the Phlogiston Theory* (London: Edward Arnold, 1932); W. A. Bone and D.T.A. Townend, *Flame and Combustion in Gases* (New York: Longmans, Green, 1927), chap. 2.

11. Earle H. Clapp, "Research in the United States Forest Service, A Study in Objectives," in *A National Plan for American Forestry,* 2 vols., Senate Document No. 12, 73rd Cong., 1st sess. (Washington: GPO, 1933), 1:672.

12. Roy Headley, Memorandum, February 7, 1929, NA, RG 95, RFS, Division of Fire Control, p. 12.

13. Coert duBois, *Systematic Fire Protection in the California Forests* (Washington: GPO, 1914), p. 6.

14. See Coert duBois, *Trailblazers* (Stonington, Conn., 1957), pp. 75–77; S. B. Show and E. I. Kotok, History of Forest Fire Research, unpublished report, title page missing, U.S. Forest Service, ca. 1956, WNRC, RFS, pp. 8–14.

15. For the organization of a Branch of Research, see Harold K. Steen, *The U.S. Forest Service: A History* (Seattle: University of Washington Press, 1976), pp. 131–141; also, Clapp, "Research in the United States Forest Service," pp. 651–662. For Clapp's memo, see H. T. Gisborne, "Economics of Fire Control," in "Proceedings of the Priest River Fire Meeting," p. 36.

16. Show and Kotok, History of Forest Fire Research, pp. 19, 7.

17. Ibid., p. 14.

18. "Report of Committee on Forest Fire Research Program," in "Public Requirements and Experiment Station Directors' Conference. Madison, Wisconsin. March 10–22, 1924," NA, RG 95, RFS, Division of Fire Control, Box 72.

19. Earle H. Clapp, *A National Program of Forest Research* (Washington: SAF, 1926), pp. 60, 67–68.

20. Headley, Memorandum, February 7, 1929, pp. 4, 11–12, 16–19.

21. Show and Kotok, History of Forest Fire Research, p. 26. This invaluable document is a description of, and justification for, the Shovian program more or less in its entirety. A quick bibliography of the major Show and Kotok publications is given in Clar, *California Government and Forestry,* 1:462n.

22. References to the conference and its conclusions are given in C. C. Buck, "Fire Behavior in the Forest Fire Research Program," in "Proceedings of the Priest River Fire Meeting," pp. 66–67.

23. S. B. Show and E. I. Kotok, *The Determination of Hour Control for Adequate Fire Protection in the Major Cover Types of the California Pine Region,* U.S. Department of Agriculture, Technical Bulletin no. 209 (Washington: GPO, 1930), p. 40.

24. For the San Francisco meeting, see Show and Kotok, History of Forest Fire Research, pp. 31–32; for the 1936 meeting, see "Committee Reports of the 1936 Fire Research Conference," August 1936, NA, RG 95, Research Compilation File.

25. L. G. Hornby, "Fire Control Planning in the Northern Rocky Mountain Region: Progress Report No. 1" (Missoula, Mont.: U.S. Forest Service, 1936). See also L. G. Hornby, "Fuel Type Mapping in Region One," *JF,* 33 (January 1935), 67–71.

26. H. T. Gisborne, *Measuring Fire Weather and Forest Inflammability,* U.S. Department of Agriculture, Circular no. 398 (Washington: GPO, 1936). See also H. T. Gisborne, *Measuring Forest-Fire Danger in Northern Idaho,* U.S. Department of Agriculture, Miscellaneous Publication 29 (Washington: GPO, 1928), and "The Objectives of Forest Fire-Weather Research," *JF,* 25 (1927), 452–457; S. B. Show, "Climate and Forest Fires in Northern California," *JF,* 33 (1919), 965–979, which inaugurated the field; Charles E. Hardy, "The Gisborne Era of Forest Fire Research" (U.S. Forest Service, History Office, 1977), for useful biographical information; and Charles Wellner, "Frontiers of Forestry Research—Priest River Experimental Forest, 1911–1976" (U.S. Forest Service, 1976), for the institutional setting of much of the early fire research programs.

27. Hardy, "Gisborne Era," p. 3.

28. Ibid., p. 6.

29. For a sample of how systematic fire protection in its refurbished forms underwrote this endeavor, see P. D. Hanson et al., "A Study of the Volume and Location of the Fire Load and the Determination of an Effective Presuppression Organization to Handle It," *FCN,* 4 (October 1940), 161–172.

30. Show and Kotok, History of Fire Research, pp. 33–36; E. I. Kotok et al., "Protection Against Fire," in *A National Plan,* pp. 1395–1414.

31. Show and Kotok, History of Fire Research, p. 42.

32. See J. R. Curry and W. L. Fons, "Rate of Spread of Surface Fires in the Ponderosa Pine Type of California," *Journal of Agricultural Research,* 57 (1938), 239–267, and "Forest Fire Behavior Studies," *Mechanical Engineering,* 62 (1939), 219–225; W. L. Fons, "An Eiffel Type Wind Tunnel for Forest Research," *JF,* 38 (1940), 881–884, and "Analysis of Fire Spread in Light Forest Fuels," *Journal of Agricultural Research,* 72 (February 1946), 93–121.

33. "Minutes of Ogden Fire Conference, February, 1940" (U.S. Forest Service, 1940), p. 4.

34. Gisborne, "Review of Problems and Accomplishments in Fire Control and Fire Research," pp. 4–20, and "Economics of Fire Control," pp. 36–43, in "Proceedings of the Priest River Fire Meeting."

35. Charles Buck, "Fire Behavior in the Forest Fire Research Program," in ibid., p. 66.

36. "Minutes of the Ogden Conference," p. 44; Hardy, "Gisborne Era," p. 77.

37. C. C. Buck, "Recommendations of the Committee on Fire Behavior," in "Proceedings of the Priest River Fire Meeting," p. 144.

38. Information on the Chief's memo is in E. I. Kotok, Memorandum to Directors, All Forest Experiment Stations, October 6, 1944, WNRC, RFS, Division of Fire Control; C. C. Buck, "A National Program of Fire Research: A Draft," WNRC, RFS, Division of Forest Fire Control.

39. "Fire Control Meeting. Washington, D.C. February 9 through February 13, 1948," WNRC, RFS, Division of Fire Control, Appendix 1, pp. 29, 32, 29.

40. Brown's reports from his tour, as well as all records of the resulting fire research effort, are stored in WNRC, RFS, Division of Forest Fire Research, though some records are cross-stored with the Division of Fire Control.

41. John Curry, in "Fire Control Meeting. Spokane, Washington. February 10–21, 1936" (U.S. Forest Service, 1936), topic 26 c, p. 1.

42. "Western Regions Fire Meeting. Fire Control–Fire Research. San Francisco, California. January 28–February 1, 1952" (U.S. Forest Service, 1952), p. 47.

43. "Servicewide Meeting. Fire Control, Fire Research, and Safety. Ogden, Utah. January 16 through January 20, 1950" (U.S. Forest Service, 1950), p. 23.

44. "Western Regions Fire Meeting." p. 48.

45. Much of this research remains classified. A good sample of the early programs, however—and a roll call of those researchers who would reform Forest Service fire programs—can be found in R. K. Arnold and W. L. Fons, *Thermal and Blast Effects on Idealized Forest Fuels,* Forest Service, 20 March 1952, Operation BUSTER, Project 2.2, WT–309; Keith Arnold, *Effects of Atomic Explosions on Forest Fuels,* Forest Service, April–June 1952, Operation SNAPPER, Project 8.1, WT–506; G. M. Byram et al., *Thermal Properties of Forest Fuels,* Forest Service, October 1952, Armed Forces Special Weapons Program Interim Technical Report 404; and Fred M. Sauer, Keith Arnold, W. L. Fons, and Craig C. Chandler, *Ignition and Persistent Fires Resulting from Atomic Explosions—Exterior Kindling Fuels,* Forest Service, December 1953, Operation UPSHOT-KNOTHOLE, Project 8.11b, WT–775. All these reports have been declassified and are available through the Defense Technical Information Center.

46. See, for example, George Byram, "Atmospheric Conditions Related to Blowup Fires," Southeastern Forest Experiment Station, Station Paper no. 35 (U.S. Forest Service, 1954); W. R. Krumm, "Meteorological Conditions Which Encourage Explosive Fire Spread" (Missoula, Mont.: U.S. Weather Bureau, 1955); R. Keith Arnold and Charles Buck, "Blow-up Fires—Silviculture or Weather Problems?" *JF,* 52 (April 1954), 408–411; Vincent Schaefer, "The Relationship of Jet Stream to Forest Wildfires," *JF,* 55 (June 1957), 419–425.

47. "Firestop: Organization Plan and Work Schedule," Progress Report no. 1 (U.S. Forest Service 1954), p. 2; "Firestop: Field Review," Progress Report no. 3 (U.S. Forest Service, 1954), p. 23. In addition to the progress reports, there are general summaries of Firestop activities in R. Keith Arnold, "Operation Firestop," *FCN,* 16 (April 1955), 1–5; Alva Neuns, "Operation Firestop," *AF,* 61 (January 1955), 8–12, 54–55.

48. The Skyfire research has produced a large bibliography. See citations in Chapter 1, note 19. For a view of things at the time of operation, see J. S. Barrows,

"Lightning Fire Research in the Rocky Mountains," *JF,* 52 (November 1954), 485–587.

49. Show and Kotok, History of Forest Fire Research, p. 58.

50. The legislation followed from the Engle Committee hearings, *Forest Fire Control in Southern California,* H. R. Serial no. 14 (Washington: GPO, 1958).

51. "The Committee on Fire Research. Final Report. July 1, 1959–June 30, 1962" (National Academy of Sciences, 1962), pp. 1–2, gives the early history of the guiding contract. A complete file of reports is kept at the committee's office, National Academy of Sciences. If Firestop was comparable to the cooperative spirit and methodological emphasis of the International Geophysical Year, then the Committee on Fire Research was an effort to extend the program much as the Scientific Committee on Antarctic Research sought to continue the benefits of IGY. For consideration of the committee within the general context of the academy, see Rexford Cochrane, *The National Academy of Sciences: The First Hundred Years, 1863–1963* (Washington: National Academy of Sciences, 1978).

52. "Committee on Fire Research. Final Report," p. 3. Committee on Fire Research and Fire Research Conference, *Methods of Studying Mass Fires,* National Academy of Sciences, Publication 569 (Washington: National Academy of Sciences, 1958).

53. "Committee on Fire Research. Final Report," Appendix, p. 2. Major publications from this period include Committee on Fire Research, *A Study of Fire Problems,* National Academy of Sciences Publication 949 (Washington: National Academy of Sciences, 1961), and W. G. Berl, ed., *International Symposium on the Use of Models in Fire Research,* National Academy of Sciences Publication 786 (Washington: National Academy of Sciences, 1961).

54. Like so many reforms, the labs had their immediate genesis in Firestop. The Engle Committee recommended funding for such a facility; a "Wildland Research Plan for California" (1958), which followed upon the committee hearings, proposed a lab for Southern California; and the Firestop Executive Committee furthered the proposal at the Governors' Fire Conference (1959). Plans were made in 1961, and dedication came in 1963. The labs were originally almost identical in design—three regional clones—but they eventually divided up the national fire research program along functional lines. For information, see Carl C. Wilson, "Our Fire Research Program and the Western Forest Fire Laboratory," pp. 3–6, Karl W. McNasser, "Forest Fire Research, Southern Forest Fire Laboratory," pp. 7–8, and J. S. Barrows, "The Northern Forest Fire Laboratory Program," pp. 9–13, *Western Forest Fire Research, 1961* (Portland: Western Forestry and Conservation Association, 1961).

55. George Byram, "Forest Fire Behavior," in Kenneth Davis, *Forest Fire: Control and Use* (New York: McGraw-Hill, 1959), pp. 92, 122. For the major studies, see George Byram et al., "Final Report. Project Fire Model: An Experimental Study of Model Fires" (U.S. Forest Service, 1966); George Byram, "Scaling Laws for Modeling Mass Fires," *Pyrodynamics,* vol. 4 (1966), 271–284; Richard C. Rothermel, "A Mathematical Model for Predicting Fire Spread in Wildland Fuels," Research Paper INT–115 (U.S. Forest Service, 1972).

56. OCD was involved in one form or another with nearly all the mass fire exper-

iments. A summary of the research is available in Robert G. Hahl, *DCPA Fire Research Bibliography* (Washington: Defense Civil Preparedness Agency, 1976). The major Forest Service involvements are described in Clive M. Countryman, "Mass Fires and Fire Behavior," Research Paper PSW–19 (U.S. Forest Service, 1964); Clive M. Countryman, "Project Flambeau . . . An Investigation of Mass Fire (1964–1967): Final Report," vol. 1 (U.S. Forest Service, 1969); and J. W. Kerr et al., "Nuclear Weapons Effects in a Forest Environment—Thermal and Fire," The Technical Cooperation Program, Defense Nuclear Agency (Washington: Department of Defense Nuclear Information and Analysis Center, 1971).

57. The Operations Research program, which was housed at the Pacific Southwest Station, Berkeley, and depended on professors at the University of California, began formally in 1961. Earlier contacts had been made through OCD and the Committee on Fire Research with groups at Johns Hopkins. The emphasis was in a sense even older than that: it stems from the origins of forest fire research itself, as epitomized by duBois.

58. V. L. Harper to Directors, "Meetings (National Fire Research Workshop)," May 7, 1964, WNRC, RFS, Division of Forest Fire Research, p. 10.

59. A synopsis of NFDRS history is given in John E. Deeming et al., "The National Fire-Danger Rating System," Research Paper RM–84 (U.S. Forest Service, 1972), pp. 1–2. See also John J. Keetch, "Forest Fire Research: Chronology on the Development of a National Fire Danger Rating System," Appendix A in Keetch, "Development of the National Fire-Danger Rating System: Basic Structure and Spread Phase" (Asheville: U.S. Forest Service, 1965).

60. See citations in Chapter 7, note 24. The first formal contract between ARPA and the Forest Service for fire research came in 1965; earlier there had been contracts under Project Firescan.

61. Compare this interpretation with the prospects imagined as the era unveiled: A. A. Brown, "Forest Fire Research As It Looks in 1955," *FCN,* 16 (April 1955), 6–7. In 1945 Charles Buck complained that fire research was understaffed and overlooked. A decade later Brown could observe with pride that, owing to alarms over fire warfare, "we find that our small group of researchers are very much in demand. They are in demand by all fire agencies because they know most about the behavior of big fires that respond to the free play of weather factors and atmospheric conditions."

62. *Study of Fire Problems,* p. 97.

63. "Project Fire Scan," review draft, November 13, 1963, WNRC, RFS, Division of Forest Fire Research, p. 1.

64. Craig C. Chandler, "Wildland Fires: A 'State of the Art' Review," WNRC, RFS, Division of Forest Fire Research, p. 6.

65. E. V. Komarek, "Fire Ecology," in *TTFECP,* no. 1, pp. 95, 106.

66. Herbert Stoddard, "Discussion," in *TTFECP,* no. 1, p. 176.

67. W. G. Morris, "Discussion of Problems of Fire Effects and Use in R-6," in "Proceedings of the Priest River Fire Meeting," p. 63; R. H. Boerker, "A Historical Study of Forest Ecology: Its Development in the Fields of Botany and Forestry," *Forestry Quarterly,* 14 (September 1916), 380–442.

68. For an analysis of Davisian geomorphology and its contemporary critics, see

Ronald Flemel, "The Attack on the Davisian System," *Journal of Geologic Education*, 19, no. 1 (1969), 3–13; Stephen J. Pyne, *Grove Karl Gilbert: A Great Engine of Research* (Austin: University of Texas Press, 1980), pp. 254–261.

69. Eugene P. Odum, "Concluding Remarks of the Co-Chairman," in *TTFECP*, no. 2, pp. 178, 180.

70. Aristotle, *Meterologica*, I, 3.

71. E. V. Komarek, "Fire, Research, and Education," in *TTFECP*, no. 2, p. 187. A description of the Tall Timbers Research Station is given in E. V. Komarek, *Tall Timbers Research Station: A Quest for Ecological Understanding*, Miscellaneous Publication No. 5 (Tallahassee: TTRS, 1977).

72. The composite proceedings were published as *TTFECP*, no. 14.

73. See James Lotan, "Integrating Fire Management into Land-Use Planning: A Multiple-Use Mangement Research, Development, and Applications Program," *Environmental Management*, 3, no. 1 (1979), 7–14; National Fire Effects Workshop, "Effects of Fire on . . . ," General Technical Reports, WO–6, 7, 9, 10, 13 (U.S. Forest Service, 1979–1980).

74. Harold J. Lutz, *Ecological Effects of Forest Fires in the Interior of Alaska*, Department of Agriculture, Technical Bulletin no. 1133 (Washington: GPO, 1956), p. 1.

75. Personal communication, April 26, 1978. Robinson gave helpful commentary on the history of Alaska fire control and suggested a number of sources to investigate.

76. Henry S. Graves, "The Forests of Alaska," *AF*, 22 (1916), 33.

77. For the ecology of fire and frost, see Lutz, *Ecological Effects of Forest Fires;* Leslie A. Viereck, "Ecological Effects of River Flooding and Forest Fires on Permafrost in the Taiga of Alaska," in *Permafrost: The North American Contribution to the Second International Conference* (Washington: National Academy of Sciences, 1973); Viereck, "Wildfire in the Taiga of Alaska," in H. E. Wright and M. L. Heinselman, eds., "The Ecological Role of Fire in Natural Conifer Forests of Western and Northern America," *Quaternary Research*, 3 (October 1973), 465–495.; and Viereck, with C.T. Dryness, technical editor, "Ecological Effects of the Wickersham Dome Fire Near Fairbanks, Alaska," General Technical Report PNW–90 (U.S. Forest Service, 1979). In addition to the above symposium, three others have examined the role of fire in Alaska: Charles W. Slaughter et al., eds., "Fire in the Northern Environment: A Symposium" (U.S. Forest Service, 1971); Bureau of Land Management, "Fire Management in the Northern Environment" (Anchorage: BLM, 1976); Fire Management Study Group, *Canadian-Alaskan Seminar on Research Needs in Fire Ecology and Fire Management in the North* (North American Forestry Commission, 1975). See also Signe M. Larson, "Fire in Far Northern Regions: A Bibliography," U.S. Department of the Interior, Departmental Library Bibliography Series no. 14 (1969).

78. A good summary of fuel conditions is in R. R. Robinson, "Forest and Range Fire Control in Alaska," *JF*, 58 (1960), 448–453.

79. Lutz, *Ecological Effects of Forest Fires*, p. 12. For a companion history, see Harold J. Lutz, *Aboriginal Man and White Man as Historical Causes of Fires in the Boreal Forest, with Particular Reference to Alaska*, Yale School of Forestry, Bulletin no. 65 (New Haven: Yale University, 1959). The role of lightning

in the interior was not really recognized until the latter 1950s, but appreciation has grown since then.

80. Robert Bell, "Forest Fires in Northern Canada," in *Report of the American Forestry Congress*, Atlanta Meeting, 1888 (1889), p. 52.

81. Lutz, *Aboriginal Man and White Man*, pp. 13–16; Bell, "Forest Fires in Northern Canada," pp. 52–55. Bell also describes the hunter and forager tribes of the north as browsing through heavy burns looking for cooked wildlife. A somewhat folksy discussion of northern tribes and fires is given in Henry T. Lewis, "Indian Fires of Spring," *Natural History*, 89 (January 1980), 76–78, 82–83.

82. Quoted in Lutz, *Aboriginal Man and White Man*, pp. 18–20.

83. Ibid., p. 22.

84. Alfred H. Brooks, *The Geography and Geology of Alaska: A Summary of Existing Knowledge*, U.S. Geological Survey, Professional Paper 45 (Washington: GPO, 1906), p. 42.

85. Lutz, *Aboriginal Man and White Man*, p. 23.

86. Ibid.

87. Bell, "Forest Fires in Northern Canada, p. 52.

88. Quoted in Lutz, *Aboriginal Man and White Man*, p. 32.

89. Ibid., p. 24.

90. Quoted in Lutz, *Ecological Effects of Forest Fires*, p. 14. For similar condemnations, see Graves, "The Forests of Interior Alaska," and R. S. Kellogg, "The Forests of Alaska," in *Forest Service Bulletin 81* (Washington: GPO, 1910).

91. C. H. Flory to W. J. McDonald, telegram, May 17, 1929, NA, RG 95, RFS, Division of Fire Control, Box 104; R. R. Robinson, "Fire Control Activity in the Bureau of Land Management. Alaska, 1959," manuscript, BLM, Files, State Office, Anchorage.

92. John D. Guthrie, "Alaska's Interior Forests," *JF*, 20 (April 1922), 373.

93. See the assorted reports by the GLO to the Forest Protection Board, NA, RG 95, RFS, Division of Fire Control.

94. GLO, Division of Forestry, *Field Handbook* (1940), title page. For a view of the situation by the man who would lead the AFCS, see W. J. McDonald, "Fire Under the Midnight Sun," *AF*, 45 (April 1939), 168–169, 231.

95. GLO, Division of Forestry, *Field Handbook*, title page. See also W. J. McDonald to Commissioner, General Land Office, "The Policy of Fire Control on the Public Domain of Alaska and Its Objectives," September 1, 1939, and Commissioner to McDonald, "Approval," January 10, 1940, Records of the AFCS, Federal Records Center, Seattle.

96. William R. Hunt, *Alaska* (New York: Norton, 1976), p. 115. A considerable amount of landclearing was necessary for military roads, posts, and airstrips, and fires frequently escaped; close liaisons between the AFCS and the military resulted. As AFCS budgets decreased during the war years, this alliance became essential, though it worked poorly; for an example, see Hal Gates, "Death of the Kenai," *AF*, 54 (January 1948), 13, 45.

97. Lutz, *Ecological Effects of Forest Fires*, pp. 87, 89, 94.

98. A. Starker Leopold and F. Fraser Darling, *Wildlife in Alaska: An Ecological Reconnaissance* (New York: Ronald Press, 1953).

99. C. E. Hardy, "Conflagration in Alaskan Forests, 1957," p. 2. See also James

B. Craig, "Alaska Burns," *AF,* 75 (October 1969), 1–3, 62–63; John Clark Hunt, "Burning Alaska," *AF,* 64 (August 1958), 12–15, 40–42. The critical amendment to the McSweeney-McNary Act allowed for research in the territories, not just the states. Jack Barrows visited Alaska in 1955, and plans were made accordingly; see Intermountain Forest and Range Experiment Station, "Annual Report, 1955" (U.S. Forest Service, 1955).

100. Good histories of the buildup can be found in Robinson, "Forest and Range Fire Control in Alaska"; Dennis E. Hess et al., "Wildfires in Alaska," *AF,* 70 (February 1964); Dennis E. Hess, "Fire Problems in Alaska," *NFPA Fireman,* March 1963; Robinson, "Fire Control Activity in the Bureau of Land Management, Alaska, 1959," the Alaska Fire Control History, included in "Alaska Fire Control Organizational Study" (BLM, 1969), Files, BLM, State Office, Anchorage; Eugene Zumwalt, "Fire Control on Public Domain Lands in Alaska," SAF, *Proceedings* (1955), pp. 162–164. Also valuable for their statistics are Richard J. Barney, "Wildfires in Alaska—Some Historical and Projected Effects and Aspects," in Slaughter et al., eds. "Fire in the Northern Environment," pp. 51–60, and Nonan V. Noste, "Analysis and Summary of Forest Fires in Coastal Alaska" (Institute of Northern Forestry, U.S. Forest Service, 1969).

101. Kenneth Pomeroy, "An AFA Fire Plan for Alaska," *AF,* 65 (September 1959), 12–13, 55.

102. For the impact of Alaska on BLM fire operations in the Lower 48, see Chapter 5, "A Legal and Moral Charge." For the weather modification experiments, see Meteorology Research, Inc., "Cloud Seeding Operations on Wildfires: Alaska, 1970" (Altadena, Calif., 1970), and Lynn W. Cooper, "Alaska Weather Modification to Aid the Control of Wildfires" (Norman: Weather Science, Inc., 1971). For the Swanson River fire, see BLM, Annual Report, 1969, BLM Alaska State Office, Anchorage; and John B. Hakala et al., "Fire Effects and Rehabilitation Methods—Swanson-Russian Rivers Fires," in Slaughter et al., eds., "Fire in the Northern Environment," pp. 87–100.

103. J. H. Richardson, "Values Protected in Interior Alaska," in Slaughter et al., eds., "Fire in the Northern Environment," pp. 173–178.

104. Dale L. Vance and E. Philip Krider, "Lightning Detection Systems for Fire Mangement," in *Fifth Joint Conference on Fire and Forest Meteorology* (Boston: American Meteorological Society, 1978), pp. 68–70.

105. See Fred E. McBride, "Alaska Fire Season—1977," *FMN,* 39 (Winter 1977–1978), pp. 3–7.

106. Alaska Department of Education, "Applicability of Wildlands Fire-fighting Techniques for Structural Fires" (Anchorage: Alaska Department of Education, 1977).

107. Edo Nyland and John D. Dell, *Forest Fire News* (Washington: North American Forestry Commission, July 1977), pp. 1–7.

108. See, for example, H. J. Lutz, "History of the Early Occurrence of Moose on the Kenai Peninsula and in Other Sections of Alaska," Alaska Forest Research Center, Miscellaneous Publication no. 1 (U.S. Forest Service, 1960); and David L. Spencer and John B. Hakala, "Moose and Fire on the Keani," in *TTFECP,* no. 3, pp. 11–13.

109. Bernard DeVoto, *The Year of Decision: 1846* (Boston: Houghton Mifflin, 1961), pp. 408–409.

110. Capt. John G. Bourke, *On the Border with Crook* (1891; rpt. Lincoln: University of Nebraska Press, 1971), p. 45.

111. Harold H. Biswell et al., *Ponderosa Pine Management,* TTRS Miscellaneous Publication no. 2 (Tallahassee: TTRS, 1973), pp. 1–2.

112. Robert Humphreys, "The Role of Fire in the Desert and Desert Grassland Areas of Arizona," in *TTFECP,* no. 2, pp. 49–51.

113. Biswell et al., *Ponderosa Pine Management,* foreword. For general statements on fire ecology of the region, see Humphreys, "Role of Fire," and *The Desert Grasslands* (Tucson: Universty of Arizona Press, 1968); Dwight R. Cable, "Fire Effects in Southwestern Semidesert Crass-Shrub Communities," in *TTFECP,* no. 12, pp. 109–128. Harold Weaver, "Fire and Management Problems in Ponderosa Pine," in *TTFECP,* no. 3, pp. 61–79, summarizes the legion of studies relating to fire and ponderosa pine. For fire in the history of the Southwest, Charles F. Cooper, "Changes in Vegetation, Structure, and Growth of Southwestern Pine Forests Since White Settlement," *Ecology Monographs,* 30, (1960), 129–164, is invaluable. E. V. Komarek, "Fire and Man in the Southwest," in *Proceedings of the Symposium on Fire Ecology and the Control and Use of Fire in Wild Land Management* (Tucson: Arizona Academy of Sciences, 1969), pp. 3–22, is so general as to be of limited use. Finally, see Michael A. Fosberg and R. William Furman, "Fire Climates in the Southwest," *Agricultural Meteorology,* 12 (1973), 27–34.

114. See H. M. Wormington, *Prehistoric Indians of the Southwest,* 10th ed. (Denver: Denver Museum of Natural History, 1970). For a quick, graphic summary of the larger migrations of peoples into the region, see Donald Meinig, *Southwest: Three Peoples in Geographic Change* (New York: Oxford University Press, 1971).

115. Detailed Apache histories are given in Donald E. Worcester, *The Apaches, Eagles of the Southwest* (Norman: University of Oklahoma Press, 1979), and Jack D. Forbes, *Apache, Navaho, and Spaniard* (Norman: University of Oklahoma Press, 1960).

116. Quoted in Cooper, "Changes in Vegetation," p. 134.

117. Bell, quoted in ibid., p. 138; S. J. Holsinger, "The Boundary Line Between Desert and Forest," *Forestry and Irrigation,* 8 (1902), 23–25; Walter Hough, *Fire as an Agent in Human Culture,* United States National Museum, Bulletin 139 (Washington: GPO, 1926), pp. 59–60, 152–153. Bourke, *On the Border,* gives assorted references in passing.

118. Aldo Leopold, *A Sand County Almanac* (1949; rpt. New York: Ballantine, 1970), pp. 137–140.

119. Aldo Leopold, "Grass, Brush, Timber, and Fire in Southern Arizona," *JF,* 22 (October 1924), 2.

120. Leopold, *Sand County Almanac,* p. 138.

121. Worcester, *The Apaches,* p. 124.

122. See Cooper, "Changes in Vegetation," p. 135.

123. See ibid., pp. 134–136; Humphreys, "Role of Fire," pp. 48, 51–60. Pinchot inspected the forests in 1900 and recalled the episode in *Breaking New Ground* (1947; rpt. Seattle: University of Washington Press, 1972), pp. 177–182. The chief villain, to Pinchot's eyes, was sheep, but herders and irrigation farmers were at loggerheads over watershed use. Pinchot also described fire hunting by

an Apache, though he saw the smoke from afar. For the magnitude of the sheep invasion, see Paul Roberts, *Hoof Prints on Forest Ranges* (San Antonio: Naylor Co., 1963), and Edwin A. Tucker and George Fitzpatrick, *Men Who Matched the Mountains: The Forest Service in the Southwest* (Washington: GPO, 1972).

124. G. A. Pearson, "'Light Burning' in the Southwest," January 24, 1920, NA, RG 95, RFS, Research Compilation File, p. 3.

125. Earl W. Loveridge, "The Fire Suppression Policy of the U.S. Forest Service," *JF,* 42 (July 1944), 550–551.

126. For an overview of Southwest fire protection problems, see Lynn Biddison, "The Changing Role of Fire Management," *FMN,* 39 (Winter 1977–1978), 19–21; for the early years, see Tucker and Fitzpatrick, *Men Who Matched the Mountains*, pp. 49–69. For a description of the particular pattern of lightning and drought, see Charles Buck and Mark Schroeder, *Fire Weather: A Guide for Application of Meteorological Information to Forest Fire Control*, U.S. Department of Agriculture, Handbook 360 (Washington: GPO, 1970). For a description of big fire years, see "10 Days in June" (U.S. Forest Service, 1974) and Southwestern Region, "Annual Fire Report, 1977" (U.S. Forest Service, 1977).

127. Holsinger, "The Boundary Line," p. 22.

128. Cooper, "Changes in Vegetation," pp. 133–134; Leopold, "Grass, Brush, Timber, and Fire," pp. 1–10; T. N. Johnson, "One Seed Juniper Invasion of Northern Arizona Grasslands," *Ecology Monographs,* 32 (1962), 187–207; Robert R. Humphreys, "Fire in the Deserts and Desert Grasslands of North America," in T. T. Kozlowski and C. E. Ahlgren, eds., *Fire and Ecosystems* (New York: Academic Press, 1974), pp. 390–399; Cable, "Fire Effects"; and J. R. Hastings and R. M. Turner, *The Changing Mile: An Ecological Study of Vegetation Change with Time in the Lower Mile of an Arid and Semiarid Region* (Tucson: University of Arizona Press, 1965).

129. Quoted by Cooper, "Change in Vegetation," p. 137.

130. For an assessment of Forest Service opinions toward the burning at Fort Apache, see A. W. Lindenmuth, Jr., "A Survey of Effects of Intentional Burning on Fuels and Timber Stands of Ponderosa Pine in Arizona," Rocky Mountain Forest and Range Experiment Station, Station Paper no. 54 (Ft. Collins, Colo.: U.S. Forest Service, 1960).

131. Joseph F. Arnold, "Uses of Fire in the Management of Arizona Watersheds," in *TTFECP,* no. 2, pp. 99–111. See also Malcolm J. Zwolinski and John H. Ehrenreich, "Prescribed Burning on Arizona Watersheds," in *TTFECP,* no. 7, pp. 195–205; Charles P. Pase and Carl Eric Granfelt, "The Use of Fire on Arizona Rangelands," Arizona Interagency Range Committee, Publication no. 4 (1977).

132. See Arnold, "Uses of Fire"; Zwolinski and Ehrenreich, "Prescribed Burning on Arizona Watersheds"; J. J. Baldwin, "Chaparral Conversion Provides Multiple Benefits on the Tonto National Forest," *FCN,* 29 (Fall 1968), 8–10, and "Chaparral Conversion on the Tonto National Forest," in *TTFECP,* no. 8, pp. 203–208; John H. Dieterich, "Chaparral Management: Its Potential—Its Problems—Its Future," in *Nineteenth Annual Watershed Symposium* (Phoenix, Ariz., 1975), pp. 47–51.

133. For an overview of BIA history in forestry, range, and fire, see J. P. Kinney,

A Continent Lost, A Civilization Won (Baltimore: Johns Hopkins University Press, 1937), esp. pp. 249–321, and *Indian Forest and Range* (Washington: Forestry Enterprises, 1951); Division of Forestry and Grazing, *Forestry on Indian Lands* (Washington: BIA, 1940), describes the history and impact of the CCC (part of a congressional committee report). The BIA's *Forestry Manual* gives the statutes authorizing fire protection.

134. Quoted in Harry Kallander, "Controlled Burning on the Fort Apache Indian Reservation, Arizona," in *TTFECP*, no. 9, p. 243.

135. Biswell et al., *Ponderosa Pine Management,* pp. 12–13. For a summary of Weaver's experience and conclusions, see Harold Weaver, "Effects of Fire on Temperate Forests: Western United States," in Kozlowski and Ahlgren, eds., *Fire and Ecosystems,* pp. 281–302. For the original paper, see Weaver, "Fire as an Ecological and Silvicultural Factor in the Ponderosa Pine Region of the Pacific Slope," *JF,* 41 (1943), 7–14, with comments by A. A. Brown.

136. See Kallander, "Controlled Burning"; Pau! S. Truesdell, "Postulates of the Prescribed Burning Program of the Bureau of Indian Affairs," in *TTFECP,* no. 9, pp. 235–240; Arnold, "Uses of Fire," pp. 100–101.

BIBLIOGRAPHIC ESSAY

I have made no attempt to list within the notes all the references that might be brought to bear. Neither will I undertake an exhaustive catalogue of published sources. Such a bibliography would easily swell to over 10,000 items and would unnecessarily repeat lists given in other, published bibliographies. Instead I propose, first, to provide a register of important bibliographies; second, to describe the archival and unpublished material and important repositories, including research libraries; and, third, to list some important works, published and unpublished, that pertain to fire history. Despite considerable fire research in other countries (especially Canada and Australia), the list must remain almost exclusively American.

BIBLIOGRAPHIES AND REFERENCES

General bibliographies for literature include Ronald J. Fahl, *North American Forest and Conservation History: A Bibliography* (Santa Barbara, Calif.: American Bibliographical Center–Clio Press, 1977), and Gerald Ogden, *The United States Forest Service: A Historical Bibliography, 1876–1972* (Davis, Calif.: Agricultural History Center, 1976). For general illustrative materials, see U.S. Forest Service, Historic Photo Collection, which has a computer bibliography organized by both accession number and subject for more than half a million photographs, and the "Bicentennial Inventory of American Paintings Executed Before 1914, National Collection of Fine Arts" (Smithsonian Institution, 1976), which also has a computer-based bibliography capable of retrieval by subject.

For bibliographies on fire, see FIREBASE, a computerized storage and retrieval system operated by the U.S. Forest Service, which can furnish searches, abstracts, and copies of documents. See also Alan R. Taylor, "Transferring Fire-Related Information to Resource Managers and the Public: FIREBASE," in Harold A. Mooney et al., "Proceedings of the Symposium on the Environmental Consequences of Fire and Fuel Management in Mediterranean Ecosystems," General Technical Report WO–3 (U.S. Forest Service, 1977). Most of the more than 5,000 items included (over 75 percent) have been published since 1960, but the data base is expanding in all directions. FIREBASE absorbed the earlier bibliography by A. R. Taylor et al., "Fire Ecology Bibliography" (U.S. Forest Service, 1974), which was originally the "Fire Ecology Citations File," Coniferous Forest Biome–Fire Ecology Project (University of Montana, 1973–1974). Similar search and retrieval services are available from the National Agricultural Library (for example, "Forest Fire Control Management, 1968–1976," Quick Bibliography Series,

76–06, generated off the Agricola data base system) and from the National Technical Information Service (NTIS), which has become a general clearinghouse for government reports. Among published NTIS searches are Jack Weiner et al., "Forest Fires I: Equipment," Institute of Paper Chemistry, Bibliographic Series no. 261 (1975); Jack Weiner et al., "Forest Fires II: Methods, Meteorological Aspects, and Statistics," Institute of Paper Chemistry, Bibliographic Series no. 262 (1975); and "Fire Fighting (Oct 79)," whose two versions include building, shipboard, aircraft, and forest firefighting techniques and equipment. Most entries were indexed from *Forestry Abstracts* (1938–present) and *Abstract Bulletin of the Institute of Paper Chemistry.* Other searches can be generated on demand at cost. Also useful is *Fire Research Abstracts and Reviews* (1958–1978), published through the Committee on Fire Research, National Academy of Sciences–National Research Council, and Robert G. Hahl, *DCPA Fire Research Bibliography* (Washington: Defense Civil Preparedness Agency, 1976), which lists the fire research and programs sponsored by that agency.

Among published bibliographies—some of whose entries have been absorbed into later bibliographies and computer systems like FIREBASE—the following are national in scope: "Division of Forestry," in R. B. Handy and Minna A. Cannon, *List of Publications of the United States Department of Agriculture from 1840 to June, 1901, Inclusive* (Washington: GPO, 1902); Mildred Williams, "Effects of Fire on Forests. A Bibliography. Annotated" (U.S. Forest Service, 1938); "Forest Protection. Forest Fires," in E. M. Munns, *A Selected Bibliography of North American Forestry,* 2 vols. (Washington: GPO, 1940); Charles T. Cushwa, "Fire: A Summary of Literature in the United States from the Mid-1920s to 1966" (Washington: U.S. Forest Service, 1968); I. F. Ahlgren and C. E. Ahlgren, "Ecological Effects of Forest Fires," *Botanical Review,* 26 (1960), 483–533. A modern, select bibliography on fire ecology is contained in the series of six publications that resulted from the U.S. Forest Service National Fire Effects Workshop (1978). Each document supplies a "state-of-knowledge" review: "Effects of Fire on Fauna" (General Technical Report WO–6); "Effects of Fire on Soil" (WO–7); "Effects of Fire on Air" (WO–9); "Effects of Fire on Water" (WO–10); "Effects of Fire on Fuels" (WO–13); "Effects of Fire on Flora" (WO–16). Also useful is the annual summary, "Publications of Forest Fire and Atmospheric Sciences Research," released by the U.S. Forest Service (1949–present).

Fire research bibliographies that have a more regional, thematic, or institutional scope include Junius O. Baker, Jr., "A Selected and Annotated Bibliography for Wilderness Fire Managers" (U.S. Forest Service, 1975); "Forest Fire Protection (Pennsylvania)," Pennsylvania Bureau of Forestry, Library (Harrisburg, 1978); Martin E. Alexander and David V. Sandberg, "Fire Ecology and Historical Fire Occurrence in the Forest and Range Ecosystems of Colorado: A Bibliography," Department of Forest and Wood Sciences, Colorado State University, Occasional Report (Ft. Collins, Colo., 1976); J. H. Dietrich, "A Selected and Annotated Bibliography on Fire Behavior, Fire Danger, Fire Effects, and Fire Weather, 1940–1951" (M.A. thesis, University of Wisconsin, 1952); Signe M. Larson, "Fire in Far Northern Regions. A Bibliography," U.S. Department of the Interior Departmental Library, Bibliography Series no. 14 (Washington, 1969); David Bruce and R. M. Nelson, "Use and Effects of Fire in Southern Forests: Abstracts of Publi-

cations by the Southern and Southeastern Forest Experiment Stations, 1921–
1955," *Fire Control Notes,* 18 (April 1957), 67–96; Vincent Aitro, "Fifty Years
of Forestry Research: Annotated Bibliography of the Pacific Southwest Forest and
Range Experiment Station, 1926–1975," General Technical Report PSW–23
(U.S. Forest Service, 1977); Anita Hostetter, "Annotated Bibliography of Publications by U.S. Forest Service Forest Fire Research Staff, Their Colleagues, and
Cooperators in California, 1962–1965" (U.S. Forest Service, 1966); Constance
Brannon, "Annotated Bibliography of Publications in Fire and Atmospheric Sci-
ences Research, 1966–1972, by U.S. Forest Service Fire Research Staff, Their
Colleagues, and Cooperators in California" (U.S. Forest Service, 1973). Most of
the experiment stations of the Forest Service have bibliographies, which they
periodically update. In addition to the those listed above, see the Pacific Northwest,
Intermountain, North Central, and Rocky Mountain Forest and Range Experiment Stations. (For addresses, see "Forest Service Organizational Directory"
[U.S. Forest Service, updated annually].) Many of the documents released by the
stations, it should be noted, were not widely distributed, and the station libraries
themselves are often the only source for their examination. Among published works
with useful bibliographies, mention should be given to Robert B. Batchelder and
Howard F. Hirt, "Fire in Tropical Forests and Grasslands," Technical Report 67–
41–ES, Army Natick Laboratories (U.S. Army, 1966); Charles Hardy, "The Gisborne Era of Forest Fire Research" (U.S. Forest Service, 1977); Harold K. Steen,
The U.S. Forest Service: A History (Seattle: University of Washington Press,
1976), good for conservation, forestry, and administrative references; and the often
extensive references that accompany the voluminous published annual proceedings
of the Tall Timbers Fire Ecology Conferences (1962–1976). For an index to the
proceedings, coded to the FIREBASE key word list, see William C. Fischer,
"Index to the Proceedings of the Tall Timbers Fire Ecology Conferences: Numbers
1–15, 1962–1976," General Technical Report INT–87 (U.S. Forest Service,
1980).

LIBRARIES AND DEPOSITORIES

An invaluable, virtually definitive survey of primary materials is provided in
Richard C. Davis, *North American Forest History: A Guide to Archives and Manuscripts in the United States and Canada* (Santa Barbara, Calif.: American Bibliographical Center–Clio Press, 1977). Since not all fire history is subsumed under
forestry history, the historian will want to supplement this work with other, more
traditional references, including Philip M. Hames, *A Guide to Archives and Manuscripts in the United States* (New Haven: Yale University Press, 1961); *National
Union Catalog of Manuscript Collections,* 12 vols. to date (Washington: GPO,
1962–); and National Historical Publications and Records Commission, *Directory
of Archives and Manuscript Repositories* (Washington: GPO, 1978). There is no
analogous guide to materials relating specifically to fire, but very useful in identi-
fying institutions and programs is the *Directory of Fire Research in the United
States,* 8 editions (Washington: National Academy of Sciences–National Research
Council, 1958–1978).

U.S. Forest Service. Much of the material relevant to fire history not only relates to the Forest Service but is also held within it. Many documents and reports are either directly deposited with or indirectly duplicated by the various institutions within the Service. The History Office of the Forest Service maintains a small but useful collection of basic materials; see David R. Kepley, "Users Guide to the History Section Collection. Forest Service" (U.S. Forest Service, 1977). The Forest Service has transferred the negatives of its famous Historic Photo Collection to the National Archives, Still Photographic Collection, but it maintains a large browse file and comprehensive bibliography; see Leland Prater, "Historical Forest Service Photo Collection," *Journal of Forest History,* 18 (April 1974), 28–31. A small collection of important published documents not often available elsewhere is maintained in the Washington (Rosslyn) offices of the Division of Forest Fire and Atmospheric Sciences Research; there is no guide. Though the experiment stations are in the process of reducing their separate libraries, they still maintain collections of their own publications and are often the only source for them, especially those done in-house. For fire history, the most important stations are the Intermountain (Ogden, Utah); the Pacific Southwest (Berkeley, California); the Pacific Northwest (Portland, Oregon); Southeast (Asheville, North Carolina); Rocky Mountain (Ft. Collins, Colorado). Because these stations direct the operations of the fire labs, they not only hold lab-related publications but also deal with lab and research histories in their annual reports, which are maintained on file. Useful records of historic materials are retained at the two equipment centers: the Missoula Equipment Development Center (Missoula, Montana) and the San Dimas Equipment Development Center (San Dimas, California). The latter includes an interesting photo collection in addition to annual reports, equipment publications, and so on. Regional and even forest office files sometimes contain valuable documents not filed elsewhere; among the best kept and most extensive are those under Beverly Ayers for Region One, Missoula, Montana. The National Advanced Resources Technology Center at Marana, Arizona (formerly National Fire Training Center), has a collection of instructor guides from the national-level training courses dating back to 1958.

National Archives and Records Centers. Material stored in the National or Regional Archives belongs to the Archives, but material stored in the Federal Records Centers remains the property of the agency that generated it. The National Archives, moreover, contains records that for the most part come from or have passed through the Washington offices of the various agencies. Its holdings, that is, give the view from the top. But the records of particular regions or forests have usually been accessioned, when kept at all, in regional Federal Records Centers. The Records Centers are simply holding tanks for these documents, retaining them for a specified period of time until they are either destroyed or transferred into the Archives system. Agency permission is necessary to examine them, but the real difficulty is that the Records Centers merely file the boxes of documents as they are received, in strict accession. General guides exist for the centers, and a description is contained in Davis, *North American Forest History;* but the filing system strongly discourages browsing.

The records in the National Archives, by contrast, have been organized, and preliminary inventories (PI) and guides exist for most collections. Among the record groups (RG) valuable to fire management are:

RG 35 Civilian Conservation Corps (PI 11)
    Particularly helpful here is an 18-volume digest of principal records assembled by John D. Guthrie and C. H. Tracy of the Forest Service.
RG 49 Bureau of Land Management (General Land Office) (PI 22)
RG 57 U.S. Geological Survey
RG 75 Bureau of Indian Affairs (PI 163)
RG 79 National Park Service (PI 166)
    Includes central files, records of the Branch of Forestry, and individual park records and correspondence.
RG 95 U.S. Forest Service (PI 18, revised)
    In addition to the Preliminary Inventory, see Harold Pinkett, "Forest Service Records as Research Materials," *Forest History,* 13 (1970), 18–29, and "Records of Research Units of the United States Forest Service in the National Archives," *Journal of Forestry,* 45 (1947), 272–275. Within the group, the key holdings are the records of Division of Fire Control, 1909–1943; Division of Cooperative Forest Protection, 1915–1949; Division of Forest Management Research, 1901–1954; Office of the Chief and Other General Records, 1882–1958; and records relating to Civilian Conservation Corps Work, 1933–1942. In actuality, the records of the Division of Fire Control date from 1911 to about 1939, beginning with the claims from the 1910 fires and concluding with the ebb tide of the CCC.

The vast bulk of Forest Service records relating to fire management and fire research are stored in the Washington National Records Center (WNRC) (Suitland, Maryland). Permission to examine the materials and a general guide to the accessions are available through the Records Management Office of the Forest Service (Washington, D.C.). These materials are critical for fire history. All of the materials pertaining to fire management from World War II to the 1970s are stored here—that is, 40 years' worth as compared with 30 years' worth in the National Archives. The same holds for Cooperative Fire Protection, including rural fire defense. The complete records of the Division of Forest Fire Research reside here.

Even those official reports, scientific and administrative both, that have been published are rarely found in academic research libraries. The following government libraries, however, are invaluable for their particular areas of specialty: National Agricultural Library (Beltsville, Maryland), for Forest Service and Department of Agriculture publications; National Bureau of Standards (Gaithersburg, Maryland), for fire-related materials of all origins; Alaska Resources Library (Anchorage, Alaska), for regional scientific and historical references; and the small library maintained by the U.S. Fire Administration (Washington, D.C.), for urban fire services. One should also mention the National Fire Weather Data Library, which has information relative to the reconstruction of the synoptic conditions surrounding many historic fires; see R. W. Furman and G. E. Brink, "The National

Fire Weather Data Library: What It Is and How to Use It," General Technical Report RM–19 (U.S. Forest Service, 1977), and Donald A. Haines, "Where to Find Weather and Climatic Data for Forest Research Studies and Management Planning," General Technical Report NC–27 (U.S. Forest Service, 1977).

Other repositories of information relative to fire include the National Technical Information Service (NTIS) (Springfield, Virginia). NTIS is rapidly becoming the central clearinghouse for government-related reports on all subjects, and the Forest Service research stations are transferring their holdings for future reproduction and distribution to NTIS. Some bibliographies for particular fields exist as a result of requested computer searches, but individual searches can be made on request for cost. A list of existing searches (bibliographies) is available from NTIS. Military-related research is stored in a separate repository, the Defense Technical Information Center (DTIC), formerly, Defense Documents Center (Alexandria, Virginia). Access is controlled, but unless the items in question are classified, entry can be gained by affiliation with an appropriate government agency, such as the Forest Service or Civil Defense. Many of the reports on fire research done by the Forest Service for DOD or OCD, and those studies done by outside contractors on similar topics, are stored at the DOD Nuclear Information and Analysis Center (DASIAC) (Santa Barbara, California). Literature searches can be conducted and documents purchased.

Other repositories with particular sources of information useful to fire history include the records of the Committee on Fire Research, National Academy of Sciences–National Research Council (Washington, D.C.). A complete record of minutes and publications relative to the Committee on Fire was in its office; but the committee was abolished in 1978, and its records are now most probably in the archives of the NAS. The Forest History Society (Santa Cruz, California) has a number of relevant items in addition to its small but select library: microfilm copies of dissertations on forest history; an Oral History Collection; archives of various forest organizations; and the Harold Weaver Papers. See, in particular, Barbara Holman, *Oral History Collection of the Forest History Society: An Annotated Guide* (Santa Cruz, Calif.: Forest History Society, 1976). The Oregon Historical Society (Portland, Oregon) retains the records of the Western Forestry and Conservation Society; the E. T. Allen Papers; and a mélange of materials pertaining to the Tillamook fire and the 1910 fires. The Conservation Library of the Denver Public Library (Denver, Colorado) includes, among a wide array of manuscript materials, the H. N. Wheeler Papers. The Bancroft Library (Berkeley, California) offers oral histories pertinent to forestry (many identical to those in the Forest History Society collection) and records on the Berkeley fire of 1923 and the Contra Costa County Fire Protection Association. The Boise Interagency Fire Center files hold material pertinent to the development of the center and to the history of the BLM.

For state materials, consult the local depositories listed in Davis, *North American Forest History,* and the central files of the particular responsible agency. I examined those in Oregon, California, Michigan, Pennsylvania, and North Carolina in some depth and with good results and cooperation. Many of the forestry agencies maintain unpublished histories of their organizations and often have outstanding collections of photographs, kept both as a historical record and for training. For a guide to these agencies, see "State Agencies Cooperating with the U.S. Department of Agriculture Forest Service in Administration of Various Forestry Programs" (U.S. Forest Service, State and Private Forestry, updated periodically). Because of the particular significance of the Southern California region, I attempted to examine historical records of the Los Angeles County Fire Department but was openly discouraged; everything had to be processed through a request system that acted, in effect, as a censor. Finally, the records of the Alaska Fire Control Service (1939–1945) and of early BLM operations in Alaska are, as of this writing, in question. They had been requested from the Seattle Federal Records Center by the BLM Alaska State Office prior to my visit to Anchorage. Only by great perseverance was I allowed to examine (and photocopy) some. Their final disposition is uncertain.

SELECTED PUBLISHED REFERENCES

I found a number of publications particularly helpful and had occasion to use others many times. Among scientific and technical works, these include T. T. Kozlowski and C. E. Ahlgren, eds., *Fire and Ecosystems* (New York: Academic Press, 1974), an anthology that gives a regional and thematic summary of known fire effects; Stephen H. Spurr and Burton V. Barnes, *Forest Ecology*, 2d ed. (New York: Ronald Press, 1973), which includes some important observations about the role of fire on a macroscale and a synoptic history of the American forest in New England; and A. A. Brown and Kenneth Davis, *Forest Fire: Control and Use*, 2d ed. (New York: McGraw-Hill, 1973), which, though flawed as a text, contains a good deal of miscellaneous historical information.

There are a number of fine histories of forestry, though all are authored by foresters. Among the classics, Samuel T. Dana, *Forest and Range Policy: Its Development in the United States* (New York: McGraw-Hill, 1956), is an excellent summary of legislation and institutions up to the mid-1950s and includes a chronology of major events in the appendix. Robert K. Winters, ed., *Fifty Years of Forestry in the U.S.A.* (Washington: Society of American Foresters, 1950), presents in anthologized form a very good cross-section of activities at mid-century and includes a useful essay by H. T. Gisborne on forest protection and fire. Henry Clepper, *Professional Forestry in the United States* (Baltimore: Johns Hopkins University Press, 1971), is a thorough summary of forestry as a historical movement and of foresters in private and government positions. Ralph Widner, ed., *Forests and Forestry in the American States: A Reference Anthology* (Missoula, Mont.: National Association of State Foresters, 1968), has some anthologized portions but is largely the work of the editor. A state-by-state historical mosaic, the work is invaluable for the density and scope of its information; unfortunately, it offers no references or citations at all, and for the most part this wealth of data cannot be verified. Most evidently came from state forestry department files. A few state forestry organizations have published histories on their own. By far the most outstanding example is C. Raymond Clar, *California Government and Forestry . . .*, 2 vols. (Sacramento: California Division of Forestry, 1959–1969), which at times is exhausting as well as exhaustive, but its abundance of information makes it a microcosm of national history.

The role of the U.S. Forest Service in fire policy and programs has been incalculable. For a good popular history of the agency, see Harold K. Steen, *The U.S. Forest Service: A History* (Seattle: University of Washington Press, 1976), which in turn offers a good bibliographic essay and insight into the major archival resources available. The Forest Service has consolidated the major legislation guiding its programs in *The Principal Laws Relating to Forest Service Activities,* Agriculture Handbook no. 453 (Washington: GPO, 1974), though in the light of massive environmental legislation enacted during the 1970s both it and Dana's work need revision. Supplement them with Richard J. Barney and David F. Aldrich, "Land Management–Fire Management Policies, Directives, and Guides in the National Forest System: A Review and Commentary," General Technical Report INT–76 (U.S. Forest Service, 1980). Ashley Schiff, *Fire and Water: Scientific Heresy in the U.S. Forest Service* (Cambridge, Mass.: Harvard University Press, 1962), a study in public administration, emphasizes the structural dilemma of Forest Service research and policy. Though rhetorical at times, the study is an excellent summary of both watershed research and programs and of the prescribed burning controversy in the South. For the cooperative fire program of the Service, see Earl Pierce and William Stahl, "Cooperative Forest Fire Control: A History of Its Origin and Development Under the Weeks and Clarke-McNary Acts" (U.S. Forest Service, 1964). For the other federal agencies, a helpful start can often be made by consulting the *United States Government Manual* (Washington: GPO, annually), which includes organization charts and the principal informing legislation.

Among periodicals, *Fire Management Notes* (formerly *Fire Control Notes*), published quarterly by the Forest Service, is fundamental for the period from 1936 to the present. Indexes were published periodically at first, now annually. The quality of the issues and articles varies widely, but no other source of information rivals it for historical value. On an international scale it is now complemented by the *North American Forestry News,* a newsletter published through the North American Forestry Commission, an organ of the United Nations Food and Agriculture Organization. There are many technical journals dealing with fire physics and chemistry, but many have little to do with wildland fire; *Fire Research Abstracts and Reviews* provides an adequate survey. I frequently consulted several forestry periodicals of long standing: *American Forests* (formerly *Forests and Irrigation, American Forestry, American Forests and Forest Life*) and *Journal of Forestry* (formerly *Forestry Quarterly*) contain numerous articles of value. Special issues are devoted wholly or in large part to fire; see the May 1935 issue of *JF,* and the November 1910, November 1913, April 1939, May 1956, and June 1969 issues of *AF.* A good index to these periodicals and others, insofar as they contain material pertaining to the Forest Service (and nearly all fire topics do), is Ogden, *The United States Forest Service: A Historical Bibliography, 1876–1972.* (Agricultural History Center, University of California. Davis, 1976). Also valuable, though its fire material is leaner, is the *Journal of Forest History.*

Some works deal more specifically with fire. Walter Hough, *Fire as an Agent in Human Culture,* United States National Museum, Bulletin 139 (Washington:

GPO, 1926), tends to be overwhelmed by fire-making devices and appliances, but it does contain important references to wildland fire practices and ceremonies. Invaluable records of nineteenth-century American fire practices can be found in Louis Barrett, "A Record of Forest and Field Fires in California From the Earliest Explorers to the Creation of the Forest Reserves" (U.S. Forest Service, 1935)—a thorough gleaning from explorer accounts and newspapers, and a compendium whose value is not distorted by the author's determination to disprove that Indians burned and that light burning was therefore acceptable. Franklin Hough, *Report on Forestry,* vols. 2, 3 (Washington: GPO, 1879, 1882), actually consisted of multiple volumes. A small portion of Volume II deals with fire, but fire preoccupies most of Volume III. Hough reprints fire legislation from the states and territories, correspondence received from settlers in all parts of the country, and some comparative and historical fire material, such as on the Miramichi and Peshtigo fires. The surveys of the forest reserves conducted under the U.S. Geological Survey is equally revealing as a statement of nineteenth-century fire habits; see Part V of the *Annual Reports of the U.S. Geological Survey* for 1898 and 1899 (Washington: GPO, 1899, 1900); the remaining surveys were issued in the Professional Paper series (see Papers 8, 22, 23, 27). Stewart Holbrook, *Burning an Empire: The Story of American Forest Fires* (New York: Macmillan, 1943), was written to popularize the "Keep America Green" movement; though it often lapses into blood-and-thunder journalism, the book amalgamates a good bit of local history into the accounts of the Lake States fires in particular.

Finally, one must extend a grateful and heartfelt acknowledgment to the published proceedings of the annual Tall Timbers Fire Ecology Conferences (1962–1976). The proceedings brought together, often for the first time, a tremendous variety of knowledge about fire, not only in the United States but also throughout the world. Though hardly a synthesis, the 15 volumes represent a unique encyclopedia on the subject of wildland fire and prescribed burning.

INDEX

(Historic fires are listed by name and date under the heading "fires, historic.")

Weyerhaeuser Environmental Books
William Cronon, Editor

Weyerhaeuser Environmental Books explore human relationships with natural environments in all their variety and complexity. They seek to cast new light on the ways that natural systems affect human communities, the ways that people affect the environments of which they are a part, and the ways that different cultural conceptions of nature profoundly shape our sense of the world around us.

Faith in Nature: Environmentalism as Religious Quest,
by Thomas R. Dunlap

Landscapes of Conflict: The Oregon Story, 1940–2000,
by William G. Robbins

Weyerhaeuser Environmental Classics

The Great Columbia Plain: A Historical Geography,
1805–1910, by D. W. Meinig

Mountain Gloom and Mountain Glory: The Development
of the Aesthetics of the Infinite, by Marjorie Hope Nicolson

Tutira: The Story of a New Zealand Sheep Station,
by Herbert Guthrie-Smith

A Symbol of Wilderness: Echo Park and the American Conservation
Movement, by Mark W. T. Harvey

Man and Nature: Or, Physical Geography as Modified by Human
Action, by George Perkins Marsh

Conservation in the Progressive Era: Classic Texts,
edited by David Stradling

Cycle of Fire by Stephen J. Pyne

World Fire: The Culture of Fire on Earth

Vestal Fire: An Environmental History, Told through Fire,
of Europe and Europe's Encounter with the World

Fire in America: A Cultural History of Wildland and Rural Fire

Burning Bush: A Fire History of Australia

The Ice: A Journey to Antarctica

Fire: A Brief History